# Advances in Multi-Scale Geographic Environmental Monitoring: Theory, Methodology and Applications—Volume I

# Advances in Multi-Scale Geographic Environmental Monitoring: Theory, Methodology and Applications—Volume I


Topic Editors

**Jingzhe Wang**
**Yangyi Wu**
**Yinghui Zhang**
**Ivan Lizaga**
**Zipeng Zhang**



MDPI

Basel • Beijing • Wuhan • Barcelona • Belgrade • Novi Sad • Cluj • Manchester

*Topic Editors*

Jingzhe Wang
School of Artificial Intelligence
Shenzhen Polytechnic University
Shenzhen
China

Yangyi Wu
School of Urban Design
Wuhan University
Wuhan
China

Yinghui Zhang
MNR Key Laboratory for Geo-Environmental Monitoring of Great Bay Area
Shenzhen University
Shenzhen
China

Ivan Lizaga
Environmental Hydrology and Interactions with Climate and Human Activities
Instituto Pirenaico de Ecología (IPE-CSIC), Spanish National Research Council
Zaragoza
Spain

Zipeng Zhang
College of Geography and Remote Sensing Sciences
Xinjiang University
Urumqi
China

*Editorial Office*
MDPI AG
Grosspeteranlage 5
4052 Basel, Switzerland

This is a reprint of the Topic, published open access by the journals *Climate* (ISSN 2225-1154), *Drones* (ISSN 2504-446X), *Forests* (ISSN 1999-4907), *Land* (ISSN 2073-445X), *Remote Sensing* (ISSN 2072-4292) and *ISPRS International Journal of Geo-Information* (ISSN 2220-9964), freely accessible at: https://www.mdpi.com/topics/O6HOQ639Y3.

For citation purposes, cite each article independently as indicated on the article page online and as indicated below:

Lastname, A.A.; Lastname, B.B. Article Title. *Journal Name* **Year**, *Volume Number*, Page Range.

**Volume I**
**ISBN 978-3-7258-4709-9 (Hbk)**
**ISBN 978-3-7258-4710-5 (PDF)**
**https://doi.org/10.3390/books978-3-7258-4710-5**

**Set**
**ISBN 978-3-7258-4707-5 (Hbk)**
**ISBN 978-3-7258-4708-2 (PDF)**

# Contents

# About the Editors

**Jingzhe Wang**

Jingzhe Wang is currently a lecturer at the School of Artificial Intelligence, Shenzhen Polytechnic University, Shenzhen, China. He received a B.Sc. degree from Anqing Normal University, Anqing, China, in 2014 and a Ph.D. degree from Xinjiang University, Urumqi, China, in 2019. His research interests focus on Earth observation and remote sensing, spectral modeling, the quantitative estimation of soil properties, digital soil mapping, GIS, spatial analysis, and environmental sustainability.

**Yangyi Wu**

Yangyi Wu is currently an Associate Researcher (Assistant Professor) at the School of Urban Design, Wuhan University, Wuhan, Hubei, China. He received a B.E. degree from Wuhan University, Wuhan, China, in 2013; an M.S. degree from the University of Texas at Dallas, Dallas, U.S.A.; and a Ph.D. degree from the University of Utah, Salt Lake City, U.S.A. His research interests are concentrated on analyzing uneven urban development in the United States and China, relying on advanced spatial analysis and statistics.

**Yinghui Zhang**

Yinghui Zhang is currently an Assistant Professor at the School of Architecture and Urban Planning, Shenzhen University, Shenzhen, China. He received his B.S. degree from the Shandong Agricultural University, Taian, China, in 2014, and M.S. and Ph.D. degrees from the Institute of Geographic Sciences and Natural Resources Research, CAS, China, in 2017 and 2021, respectively. His research focuses on surface monitoring at fine spatiotemporal resolutions, including the measurement of vegetation parameters, urban environments, etc.

**Ivan Lizaga**

Ivan Lizaga is a permanent researcher at the Pyrenean Institute of Ecology (IPE-CSIC) in Spain. His work focuses on understanding human-induced land and soil degradation, with particular attention to the interaction between geomorphological processes and environmental risks such as landslides and flash floods. His research adopts a comparative approach, examining catchments around the world, with a special focus on contrasts between European and sub-Saharan African basins—particularly in the Great Lakes region—to explore how natural systems and human populations respond to environmental stressors. Ivan's scientific contributions advance knowledge on soil erosion and fine sediment transport, two major challenges in sustainable agriculture and ecosystem resilience. His areas of expertise include soil science, geomorphology, sediment fingerprinting, remote sensing, land degradation assessment, and water and risk management. Through interdisciplinary projects and international collaboration, his work supports the development of science-based strategies to address climate change's impact on vulnerable landscapes.

**Zipeng Zhang**

Zhang Zipeng is currently an Associate Professor at the College of Geography and Remote Sensing Sciences, Xinjiang University. He received his B.S. degree from Suzhou University of Science and Technology in 2017, and earned his Ph.D. degree (through a combined master's and Ph.D. program) from Xinjiang University in 2022. His research focuses on analyzing the responses of soil organic carbon to climate change and human activities using advanced spatial analysis and statistical methods.

# Preface

Geographic environments are essential for a stable Earth system, usually characterized by complex interactions among biophysical processes, ecological patterns, and human activities. However, the Anthropocene has presented a series of challenges to the geographic environment, particularly due to intensified urbanization pressures, underscoring the critical importance of effectively monitoring the interactions of environmental elements across multiple spatial and temporal scales.

Recent advancements in Earth observation technologies, sophisticated sensor networks, and the rapidly evolving field of geospatial data science have enabled researchers to observe environmental changes at unprecedented spatial and temporal resolutions. To stay at the forefront of research, this reprint compiles forty-four innovative studies, addressing a wide range of critical topics in geographical environment monitoring, including land-use change, ecosystem services, vegetation dynamics, water resource management, atmospheric processes, and the significant impacts of human activities on environmental systems. The investigations presented here span diverse geographic contexts—from densely populated urban centers and vital agricultural zones to sensitive border regions, desert–oasis ecosystems, and the environmentally significant Qinghai–Tibet Plateau. The methodologies illustrate the integration of multi-source data—including satellite imagery, UAV observations, ground-based sensors, and social media—alongside sophisticated analytical methods such as machine learning, spatial modeling, and innovative scale transformation techniques.

Rather than assembling isolated studies, this reprint offers a cohesive, forward-looking narrative on the methodological, technological, and conceptual progression of multiscale environmental monitoring. It further underscores the importance of scalable, integrative solutions for sustainable development and effective environmental governance by intentionally bridging disciplines, including remote sensing, ecology, urban studies, and data science.

Building upon these studies, we sincerely hope that this reprint will serve as a valuable reference for researchers, planners, and policymakers. We firmly believe that deepening our collective understanding of multiscale environmental monitoring is essential for fostering interdisciplinary collaboration and achieving integrated environmental governance. This reprint was jointly supported by the National Natural Science Foundation of China (42401065 and 42201347), Guangdong Basic and Applied Basic Research Foundation (2023A1515011273), the Specific Innovation Program of the Department of Education of Guangdong Province (2023KTSCX315), and the Shenzhen Polytechnic University Research Fund (6025310064K).

**Jingzhe Wang, Yangyi Wu, Yinghui Zhang, Ivan Lizaga, and Zipeng Zhang**
*Topic Editors*

land

MDPI

Editorial

# Advancing Multi-Scale Geographic Environmental Monitoring: A Synthesis of Cutting-Edge Research and Scalable Solutions

Jingzhe Wang [1,2], Yangyi Wu [3,4], Yinghui Zhang [5], Ivan Lizaga [6] and Zipeng Zhang [7,*]

1 School of Artificial Intelligence, Shenzhen Polytechnic University, Shenzhen 518055, China; jzwang@szpu.edu.cn
2 Institute of Applied Artificial Intelligence of the Guangdong-Hong Kong-Macao Greater Bay Area, Shenzhen Polytechnic University, Shenzhen 518055, China
3 School of Urban Design, Wuhan University, Wuhan 430072, China; yangyi.wu@whu.edu.cn
4 Hubei Habitat Environment Research Centre of Engineering and Technology, Wuhan 430072, China
5 MNR Key Laboratory for Geo-Environmental Monitoring of Great Bay Area & Guangdong Key Laboratory of Urban Informatics & Guangdong–Hong Kong-Macau Joint Laboratory for Smart Cities & Shenzhen Key Laboratory of Spatial Smart Sensing and Services, Shenzhen University, Shenzhen 518060, China; zyhui@szu.edu.cn
6 Instituto Pirenaico de Ecología (IPE-CSIC), Spanish National Research Council, Avenida-Montañana 1005, 50059 Zaragoza, Spain; ivan.lizaga@ipe.csic.es
7 College of Geographical and Remote Sciences, Xinjiang University, Urumqi 830017, China
* Correspondence: zp_zhang@xju.edu.cn

## 1. Introduction

The geographic environment is a complex concept that encompasses various natural elements of the Earth's surface and human activities [1,2]. Natural conditions such as climate, land, and rivers are fundamental to the emergence, survival, and development of humanity [3,4]. Conversely, human activities also significantly impact the geographic environment [5,6]. The interaction and mutual influence between natural conditions and human activities constitute the geographic environment of the Earth [7,8]. However, previous research on geographic environmental monitoring has often focused on specific objects and spatial and/or temporal scales, making it difficult to comprehensively understand the distinctive characteristics, shifts, and interconnections within the geographic environment.

Currently, field surveys, monitoring stations, sensor networks, multi-source remote sensing (satellite, airborne, and ground-based), geospatial big data, and especially the development of remote sensing technology and geographic environmental monitoring networks enable multidimensional and multi-scale observation of geographic environmental conditions over extended periods and high frequencies, obtaining data from global to local and from macro- to micro-scales [9–12]. Integrated data analysis from different scales can lead to a better understanding of the geographic environment's overall characteristics and changing patterns [13–16].

The diversification of methods for geographic environmental monitoring has expanded the depth, breadth, and accuracy of geographic process simulation and analyses [17,18]. Geographic environmental monitoring has a wide range of objectives and scientific application scenarios, such as ecosystem services [19], natural resource distribution [20], water resource management [21], climate change research [22], disaster monitoring [23], and environmental protection [24]. In summary, studies on multi-scale geographic environmental elements are of great significance for deepening our understanding of the complexity of the Earth's systems, predicting environmental changes, supporting sustainable development, and promoting interdisciplinary communication and collaboration [11,25–28].

Drawing on these research contexts, several studies were collected and compiled. The purpose of this Topic is to present the latest research related to multi-scale geographic environmental monitoring with a focus on sustainable development. The contributions incorporate diverse and creative applications of theories and methodologies across multiple fields, including geography, hydrology, remote sensing, ecology, modeling, and management.

## 2. Overviews of Topic Papers

After continuous invitation and promotion, a total of 44 papers were published and underwent a thorough peer review process, with at least two rounds of reviews. Following the guidelines and requirements of the journals (*Climate*, *Drones*, *Forests*, *Land*, *Remote Sensing*, and *ISPRS International Journal of Geo-Information* (*IJGI*)), all submitted papers were reviewed by at least two experts. These papers contributed to the Topic, entitled "Advances in Multi-Scale Geographic Environmental Monitoring: Theory, Methodology and Applications", in various ways, providing deep insights into relevant issues and presenting novel approaches, techniques, and applications. The articles cover a wide range of issues, such as water resource management, land use and land cover, vegetation change, snow cover analysis, water body extraction, and other fields. Generally, these papers can be categorized into several groups, covering studies on spatiotemporal analysis of geographic environmental elements, assessment of ecosystem services, fusion and scale conversion of multi-source heterogeneous data products, scale effects of spatial heterogeneity of geographic ecological elements, and other related themes.

With regard to paper distribution by journal, "*Land*" has the largest number, with 16 papers, followed by "*Remote Sensing*", with 14 papers. Both "*Climate*" and "*Drones*" received the lowest number, with two papers each (Figure 1).

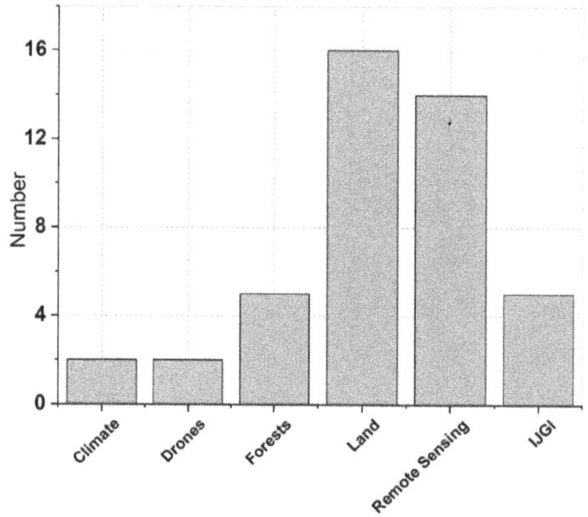


**Figure 1.** Cross-journal distribution of published papers in this Topic.

In terms of research objectives and scopes, most of the papers focus on natural phenomena, hydrogeology, the atmosphere, land use and land cover, and vegetation. These studies further establish connections with human activities, such as carbon emissions, agriculture, ecosystem services, the three life spaces (production, living, and ecological spaces), the urban heat island effect, and even the European Land Account System. Moreover, the scope of the studies covers a diverse range of geographical regions, including river basins, border regions, desert–oasis regions, the Qinghai–Tibet Plateau, some provincial regions, and even national-scale applications.

The studies included in this Topic employ a variety of approaches and data origins. These investigations involve the construction and application of remote sensing-based integrated learning conceptual models [29], soil column experiments conducted in laboratories [30], spatiotemporal variation and decoupling analyses [31], and quality comparisons of remote sensing-derived products [32].

As for data sources, platforms such as satellites and unmanned aerial vehicles were utilized [33,34]. Many studies used diverse and multi-source data. For instance, Chen et al. [35] compared the performance of Landsat-8 and Landsat-9 satellites in water body extraction. Other studies rely on multi-source data to achieve a comprehensive understanding [36].

In general, this Topic encompasses multidisciplinary research findings. These studies present a series of cutting-edge data processing techniques, innovative models, and effective strategies, which have significantly propelled the development of geographical environmental monitoring. Despite notable progress, current methodologies demonstrate persistent constraints and unexplored potential. Subsequent research should prioritize resolving these research gaps to catalyze innovative applications of geospatial technologies in geographic environmental monitoring.

## 3. Future Perspectives: Scalable Solutions

### *3.1. Advancing Multi-Scale Theoretical Frameworks and Dynamic Modeling*

The cross-scale interaction mechanisms of geographic environmental elements (e.g., climate–ecosystem–human activity cascading effects) require systematic elucidation [37,38]. Future research must transcend single-scale limitations by developing dynamic coupled models that integrate global macro-scale patterns with local micro-scale processes. For instance, satellite remote sensing-derived forest cover changes can be dynamically linked to soil moisture variations captured by ground-based sensors to reveal nonlinear threshold effects (e.g., ecosystem collapse tipping points). Concurrently, establishing a unified multi-scale classification framework is critical to resolving scale compatibility challenges between natural elements (e.g., watershed hydrology) and anthropogenic factors (e.g., urban sprawl), thereby providing theoretical support for multi-source data integration [39,40].

### *3.2. Intelligent Collaborative Monitoring Technology Systems*

To address the fusion bottlenecks of multi-source heterogeneous data (remote sensing, sensor networks, and social media), an integrated "space–air–ground" sensing network should be prioritized [41,42]. This system leverages satellites for broad coverage, drones for flexible gap-filling, and ground sensors for high-precision calibration, collectively enhancing spatiotemporal resolution (e.g., hourly disaster monitoring). Adaptive scale conversion algorithms, such as deep learning-based super-resolution reconstruction techniques, can downscale 1 km resolution precipitation data to 100 m for agricultural irrigation needs while incorporating uncertainty quantification methods (e.g., Bayesian neural networks) to mitigate error propagation risks. Furthermore, edge computing combined with cloud platforms enables real-time data processing, shifting monitoring paradigms from post-event analysis to early warning [43].

### *3.3. Precision Decision-Making and Sustainable Governance*

Translating multi-scale monitoring into actionable policies necessitates cross-scale evaluation indicator systems. For example, coupling regional carbon sequestration potential (macro-scale) with community green space distribution (micro-scale) can inform spatial planning and carbon neutrality strategies [44]. Interactive decision-support platforms visualizing multi-tiered data (e.g., urban heat island simulations under global climate change scenarios) will foster collaborative governance among governments, industries, and the public. Enhancing coupled modeling of extreme events and human activities (e.g., flood-inundation models integrated with population-mobility big data) strengthens urban resilience planning, forming a closed-loop "monitoring–simulation–decision" framework [45,46].

### 3.4. Mitigating Uncertainties and Enhancing Data Credibility

Scale dependency and data heterogeneity in environmental monitoring introduce uncertainties [47]. Addressing these uncertainties requires error source attribution frameworks to quantify contributions from sensor noise, model parameter sensitivity, and scale conversion biases. Hybrid uncertainty analysis approaches (e.g., integrating fuzzy logic with Monte Carlo simulations) can improve data product reliability [48]. Promoting global standardized validation networks—such as shared ground-truth datasets for vegetation indices—and establishing data quality certification protocols will enhance the international comparability of research outcomes [49].

### 3.5. Cross-Disciplinary Innovation Ecosystems

The multi-scale complexity of geographic environments demands interdisciplinary collaboration. Integrated platforms bridging Earth system science, data science, and social science could link remote sensing data with socioeconomic statistics to elucidate connections between ecological conservation and poverty alleviation [50,51]. FAIR principle-driven data sharing (Findable, Accessible, Interoperable, Reusable) and open-source tools (e.g., Google Earth Engine plugin libraries) will democratize multi-scale analytics. Citizen science crowdsourcing models leveraging smartphone geolocation and social media imagery can supplement traditional monitoring gaps (e.g., urban microclimates, species distribution), fostering professional–public collaborative monitoring paradigms [52].

Future advancements in multi-scale geographic environmental monitoring will hinge on the synergistic evolution of theory, technology, and applications. Dynamic models must unravel cross-scale patterns in complex systems, while intelligent technologies must enable precise data acquisition and fusion—ultimately serving global sustainability and human well-being. This trajectory demands concurrent breakthroughs in technological innovation (e.g., Earth observation, digital twins), methodological advancements (e.g., causal inference integrated with machine learning), and global cooperation mechanisms. By bridging Earth system science with societal needs, these efforts will establish a robust foundation for addressing planetary-scale environmental challenges while empowering localized, evidence-based decision-making.

As representatives of all authors, the Guest Editors hope that this Topic supports ongoing research on multi-scale geographic environmental monitoring, including theory, methodology, and applications. In particular, we believe that this Topic can contribute to vital holistic geographic environmental science and provide policy- and decision-makers with a comprehensive view of spatial patterns across ecosystems.

**Author Contributions:** Conceptualization, J.W., Y.W. and Z.Z.; writing—original draft preparation, J.W., Y.W., Y.Z., I.L. and Z.Z. All authors have read and agreed to the published version of the manuscript.

**Funding:** This work was jointly supported by the National Natural Science Foundation of China (42401065 and 42201347), Guangdong Basic and Applied Basic Research Foundation (2023A1515011273), the Specific Innovation Program of the Department of Education of Guangdong Province (2023KTSCX315), and Shenzhen Polytechnic University Research Fund (6025310064K).

**Data Availability Statement:** Not applicable.

**Acknowledgments:** As the Guest Editors, we would like to thank all of the authors who submitted their work to this Topic. Special thanks to all anonymous reviewers involved in the Topic who helped the authors to improve their manuscripts. Thanks also to the editorial staff of *Climate, Drones, Forests, Land, Remote Sensing, and ISPRS International Journal of Geo-Information (IJGI)* for supporting the idea of this Topic: Advances in Multi-Scale Geographic Environmental Monitoring: Theory, Methodology and Applications.

**Conflicts of Interest:** The authors declare no conflicts of interest. The funders had no role in the design of the study; in the collection, analyses, or interpretation of data; in the writing of the manuscript; or in the decision to publish the results.

## References

1. Vitousek, P.M.; Mooney, H.A.; Lubchenco, J.; Melillo, J.M. Human Domination of Earth's Ecosystems. *Science* **1997**, *277*, 494–499. [CrossRef]
2. Golledge, R.G. The Nature of Geographic Knowledge. *Ann. Assoc. Am. Geogr.* **2002**, *92*, 1–14. [CrossRef]
3. Ge, J.; Hu, Y. The Yellow River Civilization and Other River Civilizations of the World. In *A Historical Survey of the Yellow River and the River Civilizations*; Ge, J., Hu, Y., Eds.; Springer Singapore: Singapore, 2021; pp. 85–112.
4. Kunstler, J.H. *The Long Emergency: Surviving the End of Oil, Climate Change, and Other Converging Catastrophes of the Twenty-First Century*; Grove Press: New York, NY, USA, 2007.
5. LeMenager, S.; Shewry, T.; Hiltner, K. *Environmental Criticism for the Twenty-First Century*; Routledge: New York, NY, USA, 2011.
6. Goudie, A.S. *Human Impact on the Natural Environment: Past, Present and Future*; John Wiley & Sons: Hoboken, NJ, USA, 2018.
7. Fu, B. Promoting Geography for Sustainability. *Geogr. Sustain.* **2020**, *1*, 1–7. [CrossRef]
8. Fu, B.; Li, Y. Bidirectional coupling between the Earth and human systems is essential for modeling sustainability. *Natl. Sci. Rev.* **2016**, *3*, 397–398. [CrossRef]
9. Wang, J.; Ding, J.; Ge, X.; Peng, J.; Hu, Z. Monitoring soil salinization on the basis of remote sensing and proximal soil sensing: Progress and perspective. *Natl. Remote Sens. Bull.* **2024**, *28*, 2187–2208. (In Chinese)
10. Wang, J.; Zhen, J.; Hu, W.; Chen, S.; Lizaga, I.; Zeraatpisheh, M.; Yang, X. Remote sensing of soil degradation: Progress and perspective. *Int. Soil Water Conserv. Res.* **2023**, *11*, 429–454. [CrossRef]
11. Montillet, J.P.; Kermarrec, G.; Forootan, E.; Haberreiter, M.; He, X.; Finsterle, W.; Fernandes, R.; Shum, C.K. How Big Data Can Help to Monitor the Environment and to Mitigate Risks due to Climate Change: A review. *IEEE Geosci. Remote Sens. Mag.* **2024**, *12*, 67–89. [CrossRef]
12. Ghamisi, P.; Rasti, B.; Yokoya, N.; Wang, Q.; Hofle, B.; Bruzzone, L.; Bovolo, F.; Chi, M.; Anders, K.; Gloaguen, R.; et al. Multisource and Multitemporal Data Fusion in Remote Sensing: A Comprehensive Review of the State of the Art. *IEEE Geosci. Remote Sens. Mag.* **2019**, *7*, 6–39. [CrossRef]
13. Hu, Z.; Chu, Y.; Zhang, Y.; Zheng, X.; Wang, J.; Xu, W.; Wang, J.; Wu, G. Scale matters: How spatial resolution impacts remote sensing based urban green space mapping? *Int. J. Appl. Earth Obs. Geoinf.* **2024**, *134*, 104178. [CrossRef]
14. Wang, J.; Zhang, S.; Lizaga, I.; Zhang, Y.; Ge, X.; Zhang, Z.; Zhang, W.; Huang, Q.; Hu, Z. UAS-based remote sensing for agricultural Monitoring: Current status and perspectives. *Comput. Electron. Agric.* **2024**, *227*, 109501. [CrossRef]
15. Fei, S.; Guo, Q.; Potter, K. Macrosystems ecology: Novel methods and new understanding of multi-scale patterns and processes. *Landsc. Ecol.* **2016**, *31*, 1–6. [CrossRef]
16. Zhang, Z.; Ding, J.; Wang, J.; Ge, X. Prediction of soil organic matter in northwestern China using fractional-order derivative spectroscopy and modified normalized difference indices. *Catena* **2020**, *185*, 104257. [CrossRef]
17. Dritsas, E.; Trigka, M. Remote Sensing and Geospatial Analysis in the Big Data Era: A Survey. *Remote Sens.* **2025**, *17*, 550. [CrossRef]
18. Li, Y.; Lai, Y.; Lin, Y. The Role of Diversified Geo-Information Technologies in Urban Governance: A Literature Review. *Land* **2024**, *13*, 1408. [CrossRef]
19. Bennett, E.M.; Peterson, G.D.; Gordon, L.J. Understanding relationships among multiple ecosystem services. *Ecol. Lett.* **2009**, *12*, 1394–1404. [CrossRef]
20. Pei, T.; Xu, J.; Liu, Y.; Huang, X.; Zhang, L.; Dong, W.; Qin, C.; Song, C.; Gong, J.; Zhou, C. GIScience and remote sensing in natural resource and environmental research: Status quo and future perspectives. *Geogr. Sustain.* **2021**, *2*, 207–215. [CrossRef]
21. Sheffield, J.; Wood, E.F.; Pan, M.; Beck, H.; Coccia, G.; Serrat-Capdevila, A.; Verbist, K. Satellite Remote Sensing for Water Resources Management: Potential for Supporting Sustainable Development in Data-Poor Regions. *Water Resour. Res.* **2018**, *54*, 9724–9758. [CrossRef]

22. Yang, J.; Gong, P.; Fu, R.; Zhang, M.; Chen, J.; Liang, S.; Xu, B.; Shi, J.; Dickinson, R. The role of satellite remote sensing in climate change studies. *Nat. Clim. Change* **2013**, *3*, 875–883. [CrossRef]
23. Im, J.; Park, H.; Takeuchi, W. Advances in Remote Sensing-Based Disaster Monitoring and Assessment. *Remote Sens.* **2019**, *11*, 2181. [CrossRef]
24. Li, J.; Pei, Y.; Zhao, S.; Xiao, R.; Sang, X.; Zhang, C. A Review of Remote Sensing for Environmental Monitoring in China. *Remote Sens.* **2020**, *12*, 1130. [CrossRef]
25. Pricope, N.G.; Mapes, K.L.; Woodward, K.D. Remote Sensing of Human–Environment Interactions in Global Change Research: A Review of Advances, Challenges and Future Directions. *Remote Sens.* **2019**, *11*, 2783. [CrossRef]
26. Fu, B.; Liu, Y.; Li, Y.; Wang, C.; Li, C.; Jiang, W.; Hua, T.; Zhao, W. The research priorities of Resources and Environmental Sciences. *Geogr. Sustain.* **2021**, *2*, 87–94. [CrossRef]
27. Liu, Y.; Du, J.; Wang, Y.; Cui, X.; Dong, J.; Gu, P.; Hao, Y.; Xue, K.; Duan, H.; Xia, A.; et al. Overlooked uneven progress across sustainable development goals at the global scale: Challenges and opportunities. *Innov.* **2024**, *5*, 100573. [CrossRef]
28. Zhu, C.; Li, Y.; Ding, J.; Rao, J.; Xiang, Y.; Ge, X.; Wang, J.; Wang, J.; Chen, X.; Zhang, Z. Spatiotemporal analysis of AGB and BGB in China: Responses to climate change under SSP scenarios. *Geosci. Front.* **2025**, *16*, 102038. [CrossRef]
29. Chen, H.; Wu, J.; Xu, C. Monitoring Soil Salinity Classes through Remote Sensing-Based Ensemble Learning Concept: Considering Scale Effects. *Remote Sens.* **2024**, *16*, 642. [CrossRef]
30. Qin, S.; Zhang, Y.; Ding, J.; Wang, J.; Han, L.; Zhao, S.; Zhu, C. The Link between Surface Visible Light Spectral Features and Water–Salt Transfer in Saline Soils—Investigation Based on Soil Column Laboratory Experiments. *Remote Sens.* **2024**, *16*, 3421. [CrossRef]
31. Cheng, W.; Ma, C.; Li, T.; Liu, Y. Construction of Ecological Security Patterns and Evaluation of Ecological Network Stability under Multi-Scenario Simulation: A Case Study in Desert–Oasis Area of the Yellow River Basin, China. *Land* **2024**, *13*, 1037. [CrossRef]
32. Liu, B.; Yang, X.; Wang, Z.; Ding, Y.; Zhang, J.; Meng, D. A Comparison of Six Forest Mapping Products in Southeast Asia, Aided by Field Validation Data. *Remote Sens.* **2023**, *15*, 4584. [CrossRef]
33. Papadopoulou, E.E.; Papakonstantinou, A. Combining Drone LiDAR and Virtual Reality Geovisualizations towards a Cartographic Approach to Visualize Flooding Scenarios. *Drones* **2024**, *8*, 398. [CrossRef]
34. Tiškus, E.; Bučas, M.; Vaičiūtė, D.; Gintauskas, J.; Babrauskienė, I. An Evaluation of Sun-Glint Correction Methods for UAV-Derived Secchi Depth Estimations in Inland Water Bodies. *Drones* **2023**, *7*, 546. [CrossRef]
35. Chen, J.; Wang, Y.; Wang, J.; Zhang, Y.; Xu, Y.; Yang, O.; Zhang, R.; Wang, J.; Wang, Z.; Lu, F.; et al. The Performance of Landsat-8 and Landsat-9 Data for Water Body Extraction Based on Various Water Indices: A Comparative Analysis. *Remote Sens.* **2024**, *16*, 1984. [CrossRef]
36. Bu, Z.; Fu, J.; Jiang, D.; Lin, G. Production–Living–Ecological Spatial Function Identification and Pattern Analysis Based on Multi-Source Geographic Data and Machine Learning. *Land* **2023**, *12*, 2029. [CrossRef]
37. Tian, H.; Lu, C.; Pan, S.; Yang, J.; Miao, R.; Ren, W.; Yu, Q.; Fu, B.; Jin, F.-F.; Lu, Y.; et al. Optimizing resource use efficiencies in the food–energy–water nexus for sustainable agriculture: From conceptual model to decision support system. *Curr. Opin. Environ. Sustain.* **2018**, *33*, 104–113. [CrossRef]
38. Gaillard, M.-J.; Sugita, S.; Mazier, F.; Trondman, A.-K.; Broström, A.; Hickler, T.; Kaplan, J.O.; Kjellström, E.; Kokfelt, U.; Kuneš, P. Holocene land-cover reconstructions for studies on land cover-climate feedbacks. *Clim. Past* **2010**, *6*, 483–499. [CrossRef]
39. Chakraborty, T.C.; Venter, Z.S.; Demuzere, M.; Zhan, W.; Gao, J.; Zhao, L.; Qian, Y. Large disagreements in estimates of urban land across scales and their implications. *Nat. Commun.* **2024**, *15*, 9165. [CrossRef]
40. Lenton, T.M.; Abrams, J.F.; Bartsch, A.; Bathiany, S.; Boulton, C.A.; Buxton, J.E.; Conversi, A.; Cunliffe, A.M.; Hebden, S.; Lavergne, T. Remotely sensing potential climate change tipping points across scales. *Nat. Commun.* **2024**, *15*, 343. [CrossRef]
41. Yu, L.; Du, Z.; Li, X.; Zheng, J.; Zhao, Q.; Wu, H.; Weise, D.; Yang, Y.; Zhang, Q.; Li, X.; et al. Enhancing global agricultural monitoring system for climate-smart agriculture. *Clim. Smart Agric.* **2025**, *2*, 100037. [CrossRef]
42. Alvarez-Vanhard, E.; Corpetti, T.; Houet, T. UAV & satellite synergies for optical remote sensing applications: A literature review. *Sci. Remote Sens.* **2021**, *3*, 100019. [CrossRef]
43. Mohammed, M.A.; Lakhan, A.; Abdulkareem, K.H.; Almujally, N.A.; Al-Attar, B.B.S.M.T.; Memon, S.; Marhoon, H.A.; Martinek, R. Edge-Cloud Remote Sensing Data-Based Plant Disease Detection Using Deep Neural Networks With Transfer Learning. *IEEE J. Sel. Top. Appl. Earth Obs. Remote Sens.* **2024**, *17*, 11219–11229. [CrossRef]
44. Al-Shafi, M.; Massarweh, O.; Abushaikha, A.S.; Bicer, Y. A review on underground gas storage systems: Natural gas, hydrogen and carbon sequestration. *Energy Rep.* **2023**, *9*, 6251–6266. [CrossRef]
45. Yabe, T.; Jones, N.K.W.; Rao, P.S.C.; Gonzalez, M.C.; Ukkusuri, S.V. Mobile phone location data for disasters: A review from natural hazards and epidemics. *Comput. Environ. Urban Syst.* **2022**, *94*, 101777. [CrossRef]

46. Yuan, F.; Fan, C.; Farahmand, H.; Coleman, N.; Esmalian, A.; Lee, C.-C.; Patrascu, F.I.; Zhang, C.; Dong, S.; Mostafavi, A. Smart flood resilience: Harnessing community-scale big data for predictive flood risk monitoring, rapid impact assessment, and situational awareness. *Environ. Res. Infrastruct. Sustain.* **2022**, *2*, 025006. [CrossRef]
47. Priyadarshana, T.S.; Martin, E.A.; Sirami, C.; Woodcock, B.A.; Goodale, E.; Martínez-Núñez, C.; Lee, M.B.; Pagani-Núñez, E.; Raderschall, C.A.; Brotons, L. Crop and landscape heterogeneity increase biodiversity in agricultural landscapes: A global review and meta-analysis. *Ecol. Lett.* **2024**, *27*, e14412. [CrossRef]
48. Khan, J.A.; Khan, M.A.; Al-Khalidi, M.; AlHammadi, D.A.; Alasiry, A.; Marzougui, M.; Zhang, Y.; Khan, F. Design of Super Resolution and Fuzzy Deep Learning Architecture for the Classification of Land Cover and Landsliding Using Aerial Remote Sensing Data. *IEEE J. Sel. Top. Appl. Earth Obs. Remote Sens.* **2025**, *18*, 337–351. [CrossRef]
49. Zeng, Y.; Hao, D.; Huete, A.; Dechant, B.; Berry, J.; Chen, J.M.; Joiner, J.; Frankenberg, C.; Bond-Lamberty, B.; Ryu, Y.; et al. Optical vegetation indices for monitoring terrestrial ecosystems globally. *Nat. Rev. Earth Environ.* **2022**, *3*, 477–493. [CrossRef]
50. Li, X.; Feng, M.; Ran, Y.; Su, Y.; Liu, F.; Huang, C.; Shen, H.; Xiao, Q.; Su, J.; Yuan, S.; et al. Big Data in Earth system science and progress towards a digital twin. *Nat. Rev. Earth Environ.* **2023**, *4*, 319–332. [CrossRef]
51. Webster, A.J.; Lin, Y.C.; Scruggs, C.E.; Bixby, R.J.; Crossey, L.J.; Huang, K.; Johnson, A.; de Lancer Julnes, P.; Kremer, C.A.; Morgan, M.; et al. Facilitating convergence research on water resource management with a collaborative, adaptive, and multi-scale systems thinking framework. *Ecol. Soc.* **2025**, *30*, 23. [CrossRef]
52. See, L. A Review of Citizen Science and Crowdsourcing in Applications of Pluvial Flooding. *Front. Earth Sci.* **2019**, *7*, 44. [CrossRef]

International Journal of
*Geo-Information*

MDPI

*Article*

# A System for Analysis and Simulating Hydraulic and Hydrogeological Risks Through WebGIS 3D Digital Platforms

**Mauro Mazzei * and Davide Quaroni**

National Research Council, Istituto di Analisi dei Sistemi ed Informatica, LabGeoInf, Via dei Taurini, 19, I-00185 Rome, Italy; quaroni.davide@gmail.com
* Correspondence: mauro.mazzei@iasi.cnr.it

**Abstract:** The present research activity carried out demonstrated how simulation tools developed through WebGIS 3D digital platforms are capable of producing approximate forecasts of the effects of potentially catastrophic meteorological phenomena that may affect riverbeds in the territories observed. This work presents an analysis and simulation platform with graphic representation of the results in the form of three-dimensional animation. This methodology may represent a useful tool for all bodies and organizations that need to create hypothetical scenarios for the management of emergencies related to flooding events in watercourses, especially in areas of maximum hydrogeological vulnerability in the Italian territory. These scenarios are particularly useful in cases where watercourses are located near inhabited centers, industrial areas or strategic infrastructures, where the risk of material damage and danger to the population is greater. The simulation is based on the morphology of the land adjacent to the bed of an affected watercourse, taking elevation into account to determine the direction of the expansion of the water mass. An important aspect of the platform is the extreme speed of simulation resolution, which allows the tool to be used even in real time. This real-time forecasting approach is crucial for making quick and informed decisions, thus reducing reaction times and improving emergency management on the ground, with a potential positive impact on the safety of the population and the protection of infrastructure.

**Keywords:** WebGIS 3D; hydraulic analysis; hydrogeological analysis; simulation

## 1. Introduction

The Italian territory has more than 1200 watercourses, which represent a fundamental element of the country's natural landscape and hydrographic heritage. This extensive river network not only models the territory but also directly influences the lives of the population living along its riverbanks. From antiquity to the present, Italian rivers have played a crucial role in the development of civilizations, providing resources for agriculture, industry and transport. However, this water wealth also brings with it a significant risk: flood risk.

Floods are considered one of the most destructive natural disasters globally, causing significant human, economic and environmental losses. To address this issue, several global directives and international initiatives have been developed to improve risk management and adaptation to climate change. The objectives are as follows: significantly reduce disaster risk, including the risk of floods, by increasing the resilience of communities and infrastructure; improve the understanding of disaster risk through monitoring and data collection; strengthen risk management and coordination between governments and communities; invest in disaster risk reduction and climate adaptation; and promote sustainable recovery strategies after floods. Some of the main global flood policies and frameworks are as follows: the Sendai Framework for Disaster Risk Reduction; the United Nations Framework Convention on Climate Change (UNFCCC); the European Union Floods Directive (2007/60/EC); the Global Framework for Climate Services (GFCS); Integrated Flood Management (IFM); and the Global Facility for Disaster Reduction and Recovery (GFDRR).

The territory of the Italian State, characterized by a complex morphology ranging from the Alps to the Apennines and the coastal plains, is particularly vulnerable to flooding events. Intense rainfall, especially frequent in certain seasons, together with sometimes inadequate water management can cause floods with serious consequences for infrastructure, the economy and the safety of populations. Floods represent one of the most destructive natural disasters, and in recent decades, an increase in their frequency and intensity has been observed, a phenomenon aggravated by ongoing climate change.

As can be seen from the recent catastrophic events in the Emilia Romagna region, it is essential to provide structural flood risk mitigation measures, as well as to prepare an emergency management plan that limits as much as possible the damage caused by the increasingly frequent floods that occur as a result of climatic events that were once considered exceptional but are now becoming increasingly common and destructive.

The institution that deals with the study of possible measures to mitigate land-related risks, among which flood risk can certainly be counted, is the ISPRA (Istituto Superiore per la Protezione e la Ricerca Ambientale), which, among its many activities, also periodically produces reports on the hazard and health status of the Italian territory. In the hydrological sphere, the most recent is the Report on Flood Hazard Conditions in Italy and Associated Risk Indicators [1] in which hazard levels are defined, the mosaic of the territory according to hazard is outlined, and the most recent data for populations, cultural heritage and installations exposed to flood risk are presented.

Once the risk has been defined, it is necessary for organizations to equip themselves with tools to help create scenarios in the case of climatic events but also to forecast damage to people and structures in the case of events that are already underway. Given the developments of the last few years in computing power and connectivity, it is possible to support emergency management operations with IT tools that exploit internet connection to enable them to be used on portable devices or simple terminals equipped with a browser.

The research activity described in this study aims to create a platform for the real-time simulation of the overflowing of watercourses from their riverbeds, with an interactive three-dimensional visualization of the simulation results, by exploiting the powerful WebGL graphics libraries [2], the Three.js three-dimensional rendering libraries [3] and the two-dimensional maps provided by the Open Layers API [4].

The system uses the polygonal geometries of the waterways provided by OpenStreetMap [5], which generates vector maps for the entire planet and has various layers, including waterways; these geometries are cast in a portion of the surrounding territory, reproduced thanks to the DEM (Digital Elevation Model) raster maps of the TINITALY 1.1 project. 1 [6], which, with a resolution of 10 × 10 m, allow for the accurate modeling of the terrain. This is further enriched by textures from satellite images offered by the COPERNICUS and ESRI services [7], for the better locating of any roads, infrastructure or inhabited centers present in the area of interest.

These data, combined with the input data entered by the user, including the amount of rain falling in an hour (mm/h), the soil permeability index and the time data of the simulation, make it possible to calculate the expansion of the volume of water in the area concerned, making it possible to understand which areas will be most affected by the outflow of water from the riverbed and also having an indication of the heights reached and any built-up areas reached.

The system also proposes the consideration of, albeit with a simplified approach, the contribution given by the water precipitated on the slopes adjacent to the riverbed, which then inevitably fall into it, increasing the water flow involved.

The simulations are performed automatically at any point in the area of interest, which, for the purpose of this research study, was limited to the region of Liguria but can easily be expanded to the entire Italian territory for potential use in real cases; the user can in fact select a point of their choice, and all the data of a 5 × 5 km area around it are automatically downloaded, adapted and used within the simulation, which can also be repeated several times, perhaps with changing the input parameters, to provide a series of diversified scenarios that take into account different starting conditions.

## 2. Materials and Methods

### 2.1. Watercourses

The Italian territory boasts the presence of numerous watercourses (around 1200), which are relatively shorter than those in other European regions due to the country's geographic conformation and the presence of the Apennine Mountain chain that divides the flow of water along the peninsula into two opposing sides. [8]

Among Italian rivers, the longest of those that flow entirely through our territory is the Po, with a total length of 652 km.

The flow and length of the watercourses are due to the characteristics of the soil over which they flow; of particular importance are the steepness of the terrain and its permeability, due to the geological conformation of the underlying soil.

The following considerations can be made about Italian watercourses:

- The rivers with the greatest flow and width are those in the Alpine region, due to the high elevations.
- Rivers flowing along the peninsula, on the other hand, due to the arrangement of the Apennine chain, cross short valleys and generally do not exceed 200 km in length, with a higher average length on the Tyrrhenian side than on the Adriatic and Ionian sides.
- The rivers flowing into the Tyrrhenian Sea are also longer because they initially follow longitudinal Apennine valleys and then flow in cross-river courses with respect to the chain in the sub-Apennine area.

Italian rivers can therefore be divided into 3 main categories: The Alpine rivers, of glacial origin, are subject to flooding in the spring and summer seasons due to the melting of glaciers, with lakes occupying the lower parts of the valleys whose function is to calm the impetus of the rivers and clarify the turbid waters. The Apennine rivers, on the other hand, are subject to sudden autumn flooding due to rainfall. They are supplied by springs that flow at the edges of areas characterized by permeable fissured rocks, due to the lack of snow and glaciers in the Apennines. Finally, Sardinian and Sicilian rivers are usually torrential in character and are therefore rich in water in winter and almost dry in summer. The location within which the flow of river water occurs is called the riverbed. It is created precisely by the erosive action exerted by the water, which acts on the rocky substrate and determines its progressive erosion. For each watercourse, three distinct riverbeds can be identified in the cross-section:

- Flood bed (or ordinary bed): This constitutes the flow channel during ordinary flood periods, usually in spring and autumn. It is delimited laterally by banks or sub-vertical slopes. This riverbed contains coarse materials, deposited as a result of current variations. The presence of tree vegetation is poor.
- Major riverbed (or flood bed): This consists of the maximum area unavailable to the river during flooding. It is elevated and larger than the ordinary flow bed. It

is characterized by the presence of coarse material and sediments, deposited by the water during flooding.

- Leaning bed (or flow channel): This constitutes the channel in which the river current flows during the lean phases. It is not characterized by sufficiently defined banks to allow its immediate distinction from the ordinary channel in which it forms. It often tends to divide and merge seamlessly, depending on the excavation action of the sediments in the ordinary bed.

From a morphological point of view, the largest riverbed is the widest and contains within it the floodplain, which in turn includes the flow channel (see Figure 1).

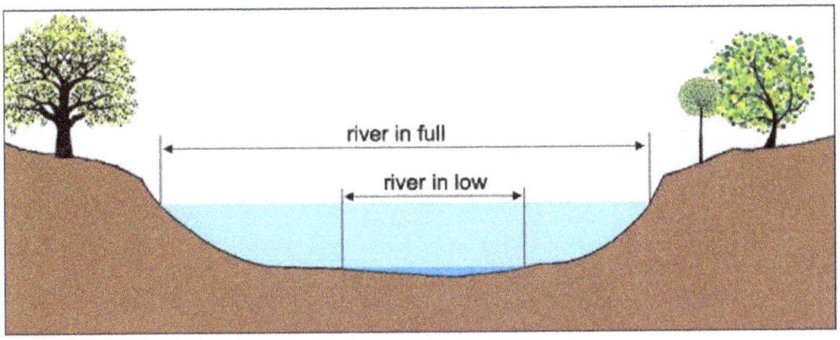


**Figure 1.** River in full and river in low period.

## 2.2. Floods

The main risk associated with watercourses is flooding areas adjacent to the riverbed, which are usually not covered by water. This phenomenon is called flooding, and Italy has established a regulatory reference framework for the assessment and management of flood risks by implementing Directive 2007/60/EC on the assessment and management of flood risks (Floods Directive—FD) through the Legislative Decree 49/2010. [8]

The institute in charge of monitoring the hydraulic hazard of the areas delimited by the District Basin Authorities is the ISPRA (Istituto superiore per la protezione e la ricerca ambientale)—https://www.isprambiente.gov.it/it/istituto—accessed on 6 February 2025 which is a public research body, with legal personality under public law and technical, scientific and administrative autonomy. The ISPRA is under the supervision of the Minister for the Environment and Energy Security (MASE). The institute produces the mosaic of the aforementioned areas, according to the three scenarios of the Legislative Decree 49/2010 (see Figure 2):

- High hazard with a return time between 20 and 50 years (frequent floods);
- Medium hazard with a return time between 100 and 200 years (infrequent floods);
- Low hazard (low probability of floods or extreme event scenarios).

From this mosaic work, it can be seen that the areas of high hydraulic hazard in Italy amount to 16,224 km$^2$ (5.4% of the national territory), the areas of medium hazard amount to 30,194 km$^2$ (10%) and those of low hazard (maximum expected scenario) amount to 42,376 km$^2$ (14%) (mosaic v. 5.0—2020) [9].

One aspect to be considered when forecasting floods and assessing the effects these phenomena may have on the surrounding territory is the hydrogeological risk. This is the risk deriving from slope instability and is closely connected to the geological and morphological conformation of the slopes, which in turn is influenced by the characteristics of the deposits and rocks from which they are formed.

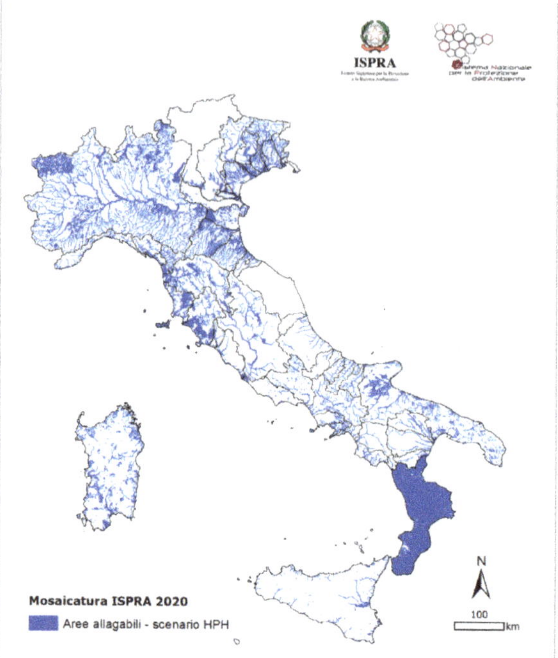


**Figure 2.** Floodable areas per high flood hazard scenario.

Flood risk thus derives from a combination of hydrogeological risk and meteorological, climatic and environmental conditions. A fundamental element is precipitation; rainwater, once it falls on the ground, follows a cycle that depends in turn on geological and geomorphological factors and, not least, the human use of the land [9].

Hydrogeological risk can also be expressed as the combination of hydrogeological hazard (the probability that a landslide or flood of a given magnitude will occur for a given time and area), the vulnerability of built-up areas and the potential damage (often very high as high-value cultural assets may be exposed to landslides and floods).

Another concept that influences potential flood damage is hydrogeological instability, which summarizes the set of processes and actions that cause land and soil degradation; it is well known that the activities that contribute most to this degradation are of a purely 'anthropic' nature and can have very serious consequences, to the point of making a territory particularly exposed to the risk of extreme phenomena. In industrialized and densely populated (and therefore built-up) countries like Italy (see Figure 3) [9], much of the hydrogeological risk depends on the fact that the territory has been extensively modified through the construction of roads, bridges and railways, which interfere with the environment and its natural evolution. Most of the Earth's geological history is characterized by the formation of mountain ranges, which are the most extreme consequence of the movement of the Earth's plates. Mountains, in turn, are subject to erosive phenomena that, over time, lead to the dismantling of slopes and the consequent accumulation of debris towards topographically more depressed areas. The erosive phenomena that lead to the progressive leveling of reliefs can develop through the gradual transfer of rocky debris towards the valley, or its sudden and violent transfer, as in the case of landslides and floods.

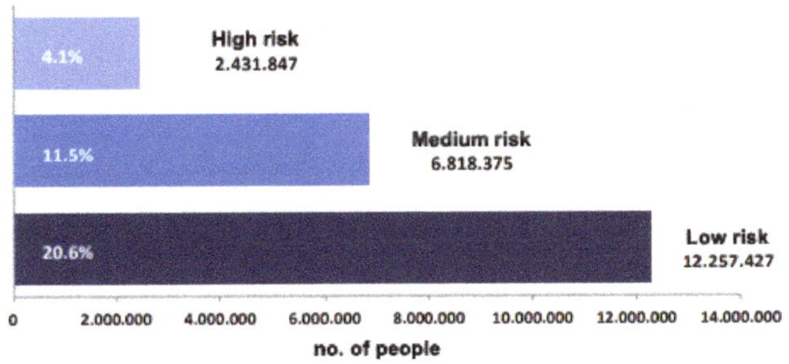


**Figure 3.** Resident populations in flood-prone areas for the three flood probability scenarios at the national level—ISPRA Mosaic, 2020.

The development of hydrogeological events such as floods is relatively simple to describe. They occur when the water of a river can no longer be contained within its banks and invade the surrounding areas. The triggering cause is always the amount of rainfall, which can exceed the absorption capacity of the land; this happens when rainfall is particularly heavy or in cases where land disruption is so severe that the regular flow of rainwater is impossible.

In the case of moderate rainfall, which falls on the land over an even, prolonged period of time, the volume of water is distributed over time and the river flow is sufficient to dispose of the surplus water through the riverbed. On the other hand, when a large amount of rain falls over a very short period of time, the volume of water channeled into the riverbed is greater than the river's disposal capacity. In other words, the volume invading the riverbed is greater than the flow rate and the excess volume is forced out (overflowing) from the embankments.

Moreover, when an area falls prey to hydrogeological instability, the capacity of soils to absorb water can be reduced to zero, thus causing an exponential increase in the volumes of water that pour into rivers during heavy rainfall. Another activity that increases the propensity of an area to be affected by floods is so-called 'sealing': covering land with concrete has the effect of preventing the infiltration of rain into the ground, increasing the volume and speed of water flowing towards rivers. Cementing riverbeds is also responsible for reducing the runoff time of water, which can then flow downstream faster. Finally, floods are made even more destructive by the presence of bridges with too little a span, which is an obstacle to water flowing under them.

### *2.3. River Basins*

A hydrographic basin is a portion of a territory that, due to its orographic conformation, collects rainwater or water from melting glaciers or snowfields, all of which flow towards an impluvial furrow giving rise to a watercourse or towards a basin or depression giving rise to a lake or marshy area (see Figure 4). It differs from a hydrogeological basin in that the latter does not only consider surface water runoff but also infiltration runoff that depends on the stratigraphy and geological conformation of the subsoil [10].

If the water collected is only due to precipitation, we speak of a catchment area. In a valley catchment area, it is possible to identify a place of convergence of water, known as the 'closure section', through which the entire volume of water collected on the surface passes. The altitude at which the closure section is located is the reference altitude in the basin's altitude measurements [10].

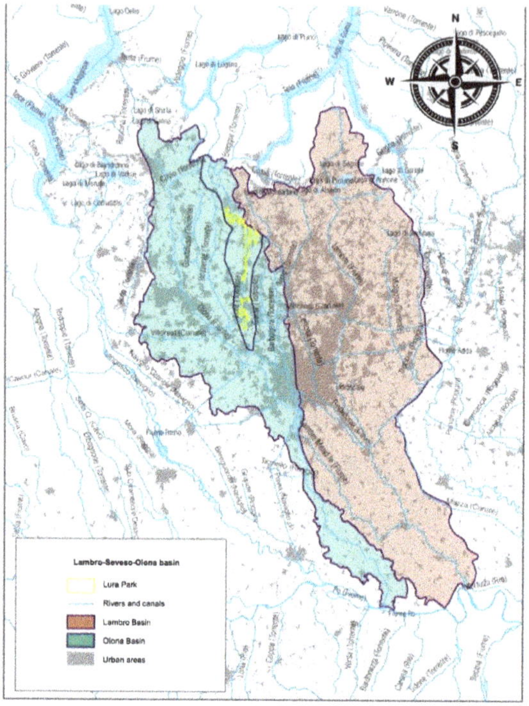


**Figure 4.** A map of the Lambro–Seveso–Olona basin.

The boundary perimeter of each basin, once the closure section has been identified, is recognized by mapping, from the closure section, the line beyond which the water flows over the land, following another path to collect in a different catchment area. This line is called the 'watershed line' or simply the 'watershed'. The watershed can be defined on a topographical map by connecting the peaks of the maximum elevation with a line that is always perpendicular to the contour lines, thus having the direction of the inverse of the maximum gradient [10].

When we consider the water of a hydrographic area that flows underground, the watershed no longer coincides with the surface water identified with topographical methods, through which it is much easier to delineate it, and various aspects must be considered that modify its course and its shape. One of these very important elements to consider is the filtering motion of the water, a motion that develops mainly on horizontal paths, carried out by the water in a saturated soil zone above an impermeable bottom layer. A basin whose waters flow below the earth's surface is also considered a hydrogeological basin; precisely because of the subterranean component of the water, it is much more difficult to identify the watershed, and thus delimit the hydrogeological basin, than the hydrographic area [10].

### 2.4. Basin Shape

Different basins can have very different shapes despite having identical drainage surfaces. The shape can vary from more compact to more elongated, and this difference between basins is therefore assessed by calculating additional parameters besides the simple drainage surface [10].

The first calculation that can be made is the ratio or circularity factor $R_c$, and it is calculated by comparing the area of basin A with that of a circle with the perimeter P, equal to the length of the watershed line of the basin considered:

$$R_c = 4\pi \frac{A}{P^2}$$

The 'uniformity ratio' $R_u$ is defined as the ratio of the perimeter of the basin to the perimeter of the reference circle with the condition that the surface area of the basin is equal to that of the circle:

$$R_u = 4\pi \frac{P}{2\sqrt{\pi A}}$$

The ratio of the area of the basin to that of a square with the side L, equal to the length of the basin, is called the 'shape factor':

$$F = \frac{A}{L^2}$$

The elongation ratio is as follows:

$$E = \frac{2\sqrt{A}}{L\sqrt{\pi}}$$

One of the very important parameters through which the shape of a catchment area is characterized is the Gravelius coefficient (Φ); it represents the ratio between the perimeter P of the catchment area and the perimeter of a circle of an equal area (A). Depending on the value to which the coefficient tends, the shape of the basin appears differently:

- $\Phi = 1$: round shape;
- $1 < \Phi < 1.25$: round–oval shape;
- $1.25 < \Phi < 1.5$: elongated round–oval shape;
- $1.5 < \Phi < 1.75$: elongated oval–oblong rectangular shape.

We measure the compactness of a basin in relation to a circle with the same area with the following formula:

$$K_c = \frac{P}{2\sqrt{\pi A}}$$

*2.5. Altitude Characteristics*

The most interesting antimetrical characteristic is the slope of the basin, and it can be described by the hypsographic curve, which is a monotonically decreasing curve determined by the means of a Cartesian graph in which the ordinate heights and increasing percentages of the total basin area are plotted [10–12].

In order to determine it, it is necessary to have a small-scale cartography with a hydrographic grid and a detailed altimetry (identified with good approximation by contour lines). The most interesting quantities that can be obtained from the hypsographic curve are as follows.

Average elevation: this is the elevation that represents the integral average of the diagram and thus represents the height of a rectangle whose base is the percentage of the total area; its surface area is equivalent to that subtended by the hypsographic curve.

Median altitude: This is the altitude value that corresponds to half of the basin area in the graph of the hypsographic curve. Another widely used parameter is the mean slope, usually calculated according to the Alvar–Horton formula:

$$I_m = \lim_{D_h \to 0} \frac{D_h L_v}{A}.$$

where $L_v$ is the total length of the contour lines of the assigned elevation difference $D_h$.

In practice, this parameter is often estimated approximately by superimposing a square grid on the topographic map of a basin, calculating at each node the difference in elevation between the two nearest contour lines and their distance. Local slopes are thus obtained, the arithmetic means of which can be calculated to obtain the average slope.

### 2.6. Topological Characteristics

The topological representation of a hydrographic network is carried out on the basis of its development on a two-dimensional plane and is based on certain easily classifiable topological properties. Regardless of the scheme adopted, a number of general definitions can be identified

- 'External nodes' (or sources) are defined as those nodes from which a single rod originates.
- 'Internal nodes' (or junctions) are defined as nodes consisting of the points at which more than one rod flows.
- 'Internal branches' are defined as elements connecting internal nodes and junctions or internal nodes themselves.
- 'External branches' are defined as elements connecting an internal node immediately downstream with a source.
- 'Network magnitude' is defined as the total number of sources.
- 'Topological distance' is defined as the number of branches from the source to the network outlet.
- 'Topological level' is defined as the topological distance between the upstream node and the network outlet.
- 'Network diameter' is defined as the maximum topological distance.

### 2.7. Horton–Strahler Sorting System

The direction of water flow and the shape of a water network establish a hierarchical relationship between the various branches that make up the network. The first to systematically propose a quantitative geomorphological analysis of hydrographic networks and their drainage basins was E.R. Horton in 1933 [13]; he based this analysis on the hierarchization of the constituent elements of a network according to certain rules. Horton's classification was taken up in 1952 by Strahler, who proposed a classification procedure essentially based on five simple rules (see Figure 5):

- First-order channels are defined as those originating from sources.
- From the union of two branches of the order 'n' derives one branch of the order 'n + 1'.
- From the union of two branches of different orders, the confluent with the greater order will be the stretch of the channel immediately downstream.
- The succession of two or more branches characterized by the same order 'n' constitutes channels of the same order.
- The channel characterized by the highest order 'N' determines the order of the channel itself.

The Alvar–Horton formula, also known as Horton's Law, describes the hierarchy and organization of the drainage network in river basins. Horton identified empirical laws to characterize river networks, which include the following.

Law of Order of Watercourses: Watercourses can be classified in an order system; when two watercourses of the order n join together, they form a watercourse of the order n+1. This hierarchy represents the structure of the drainage network.

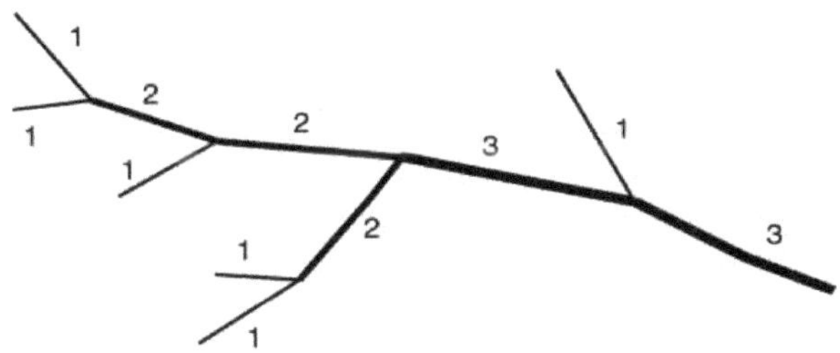


**Figure 5.** Strahler sorting.

Bifurcation Ratio: This is defined as the ratio between the number of watercourses of one order and the number of watercourses of the next highest order.

$$R_b = \frac{N_n}{N_{n+1}}$$

where

$R_b$ is the bifurcation ratio;
$N_n$ is the number of watercourses of the order n;
$N_{n+1}$ is the number of watercourses of the next order.

$$R_l = \frac{L_n}{L_{n+1}}$$

where

$R_l$ is the average length ratio;
$L_n$ is the average length of watercourses of the order n;
$L_{n+1}$ is the average length of watercourses of the next order.
Basin Area Ratio: This defines how the basin area of a certain order compares with that of the next order.

$$R_\alpha = \frac{A_n}{A_{n+1}}$$

where

$R_\alpha$ is the ratio of the basin area;
$A_n$ is the basin area associated with watercourses of the order n;
$A_{n+1}$ is the basin area for the next order.

### 2.8. Corrugation Time

The runoff time is a further morphological and hydrological characteristic and represents the time it takes for a hydraulically disadvantaged water particle, i.e., one falling at the furthest point from the closure section, to reach the section itself. The runoff time ($t_c$), measured in hours, is essentially determined using various analytical formulas.

Giandotti's equation is an empirical formula used in hydrology to estimate the time of concentration (Tc) of a catchment area, i.e., the time required for all rainwater collected at the furthest point of the catchment to reach the outlet or control section. It is useful for calculating the peak runoff during heavy rainfall.

$$t_c = \frac{4 \cdot \sqrt{A} + 1.5 \cdot L}{0.8 \cdot \sqrt{H}}$$

where S is the surface area of the basin, expressed in km$^2$; $L$ is the length of the main stem, expressed in km; Hm is the mean elevation of the basin, expressed in m, referring to the mean sea level; and h0 is the elevation of the closure section, also expressed in m, also referring to the mean sea level. This formula is valid for catchment areas larger than 100 km$^2$.

Viparelli's formula is used in hydraulics to estimate the runoff time in catchment areas, providing a similar measure to Giandotti's formula, but it was specifically developed by Giuseppe Viparelli for hydrological evaluations in small basins under heavy rainfall conditions.

$$t_c = \frac{L}{V}$$

where $L$ indicates the length of the river shaft and $V$ is the velocity of the water particle; the velocity, set at 5.4 km/h, is determined with the aid of an abacus that proposes a relationship between the velocity of the water, the average slope gradients and the type of terrain characterizing the basin.

Pezzoli's formula is an empirical expression used to calculate the runoff time ($t_c$) in catchment areas, particularly in small mountain basins. This formula is mainly used in hydrology to estimate the response time of a basin to intense, short-duration rainfall and is particularly suitable for steeply sloping terrain.

$$t_c = 0.055\frac{L}{\sqrt{p}}$$

where $L$ denotes the length of the main stem and $p$ denotes the average slope of the riverbed. The ridge flow rate can be calculated using empirical formulas such as the rational formula:

$$Q_p = C \cdot i \cdot A$$

where

$Q_p$ is the ridge flow rate (in m$^3$/s);
$C$ is the runoff coefficient, which varies according to permeability and land cover;
$i$ is the precipitation intensity (in mm/h);
$A$ is the basin area (in km$^2$).

Also, the term S denotes the basin area and the term i denotes the rainfall intensity of a duration equal to the runoff time $t_c$, while the term $\varphi$ denotes the runoff coefficient, i.e., the rainfall yield, the value of which depends on the permeability and the type of vegetation cover of the basin. If one wishes to express the flow rate in m$^3$/s, and the surface area of the basin is expressed in km$^2$ and the rainfall intensity i in mm/hour, one must divide the obtained value of Qc by 3.6.

### 2.9. Risk Mitigation Measures

Floods are natural phenomena over which humans have no control (except in the acceleration of climate change that favors increasingly extreme weather events), but what can and must be performed instead is risk mitigation in the event of these events occurring.

Possible mitigation interventions vary depending on whether an area is built-up or non-built-up; in the former case, structural and non-structural interventions are necessary, ranging from engineering works to consolidate unstable slopes to instrumental monitoring and warning networks. An example is the Rercomf landslide monitoring network ARPA Piedmont, Sondrio Geological Monitoring Centre–ARPA Lombardia.

For areas that have not yet been built on, on the other hand, it is fundamental to place new urbanization areas in safe places, with particular attention to strategic buildings such as schools, hospitals and public offices; it is also necessary to implement proper territorial planning, through the application of constraints and land use regulations (PAI) [14].

- The Plan for Hydrogeological Structure is the most effective action for risk reduction in the medium–long term (see Table 1).

**Table 1.** Outline of structural and non-structural measures for hydrogeological risk mitigation.

| | | |
|---|---|---|
| Hydrogeological risk mitigation strategy | - | Knowledge activities: landslide census and mapping (IFFI inventory), mapping of flood areas (Floods Directive) |
| | - | Risk assessment |
| | - | Correct spatial planning (hydrogeological planning) |
| | - | Structural risk reduction interventions for population centers and linear communication infrastructures |
| | - | Communication and dissemination of information to public |
| | - | Land maintenance and good agro-sylvo-pastoral practices |
| | - | Relocations |
| | - | Monitoring networks and warning systems |
| | - | Civil protection emergency plans |

With a view to mitigate hydrogeological instability, in addition to the implementation of structural interventions, cognitive activity on a national scale is strategic, also in accordance with the provisions of Articles 55 and 60 of the Legislative Decree 152/2006.

In addition to risk mitigation policies, physical barriers can be built to protect areas most at risk from flood damage. These include the following:

- Weirs: Erosion by surface water depends on the speed of the watercourse, which can be slowed down by building steps, called weirs, along the watercourse. Between one weir and the next, the water flows with a small gradient, only to jump abruptly to the next step. This reduces the speed of the current and, consequently, also the removal of materials.
- Artificial embankments: In order to prevent flooding, watercourses can be prevented from overflowing by constructing artificial embankments. The design of these defense works must take into account the increase in water masses during floods. For this reason, embankments are erected at a certain distance from the riverbed so as to allow the water to escape if necessary.
- Flood reservoirs and spillways: Flood reservoirs and spillways are built in order to reduce the flow of flooding watercourses. Flood reservoirs are basins into which water is conveyed in order to retain part of the flood wave. Spillways are canals or tunnels that, when opened before the flood wave arrives, drain part of the water into a lake or sea.
- Containment systems: In many rivers, artificial embankments interfere with the normal functioning of river systems. For adequate protection from flood waves, containment systems are essential. Recently, reservoirs have also begun to be used to control and regulate river flows.
- Artificial channels: To contain flood waves along the lower reaches of rivers, artificial channels and riverbed widenings have been dug since ancient times. In some vast stretches of floodplains, they are left clear so that floodwater can drain away completely.
- Reforestation and vegetation control: since deforestation contributes to increasing the risk of floods and landslides, the adequate control of vegetation destruction and

reforestation where necessary can be very important factors in maintaining a territory in equilibrium and preventing land disruption.

## 3. Case Study: Hydraulic and Hydrogeological Risk Simulator

This experimental work involves the creation of a geographical web platform for analyzing and simulating hydrogeological risk on Italian soil in the event of flooding events, using soil elevation and existing watercourse beds as input data. Using this application, it is possible to click on a point of interest, obtain elevation data for a 5 km area around the point itself and simulate the overflow of watercourses in that area in the event of rain, also considering the external contribution due to the slope of the surrounding land.

### *3.1. Area of Interest*

We chose to limit the scope of this work to the area of the Liguria region, given the presence of numerous cartographic data that can be freely downloaded from a dedicated geoportal (https://geoportal.regione.liguria.it/, access on 6 February 2025).

These data include several thematic maps available in WFS and shapefile formats including the hydrographic network, river basins and many other maps useful for monitoring the region's territory.

As far as the geometries of watercourses are concerned, however, it is preferable to start with a polygonal representation of a watercourse, as this allows a starting situation, in which the watercourse is still within its bed, to be as realistic as possible. Since the geoportal does not provide this type of model, we chose to use the free maps of OpenStreetMap.

### *3.2. Type of Application*

The application WebGIS 3d is available and accessible via the project page of LabGeoInf, a research laboratory of the National Research Council in Italy, identified as SmartData—Hydro Simulator. (The web application service is available here: http://150.146.207.228:8080/idro_simul/, access on 6 February 2025) This choice allows it to be easily usable by any user with an internet connection without the need to install local software and makes it possible to exploit the capabilities of modern browsers which, by the means of dedicated libraries such as Three.js, allow the visualization of three-dimensional environments with a large quantity of polygons on the screen, as is often the case in the realization of simulations of this type.

The user interface consists of two different screens:

- The initial screen is a 2D visualization of the region of interest, within which it is possible to move, zoom and obtain information on the features of interest, as well as select an area in which to run the simulation;
- The second screen is a three-dimensional representation of the selected area, where the user can set parameters and start the simulation.

The simulation is displayed on the screen as an animation, in which both the time steps and the total duration are user-selectable parameters. At the end of the simulation, it is possible to compare the initial and final state of the water expanse to assess which surrounding areas might be most affected by flooding.

### *3.3. Input Data—DEM*

The DEM used for this project is TINITALY/1.1, a new version, released in January 2023, of the original TINITALY/01 presented in 2007 (see Figures 6 and 7).

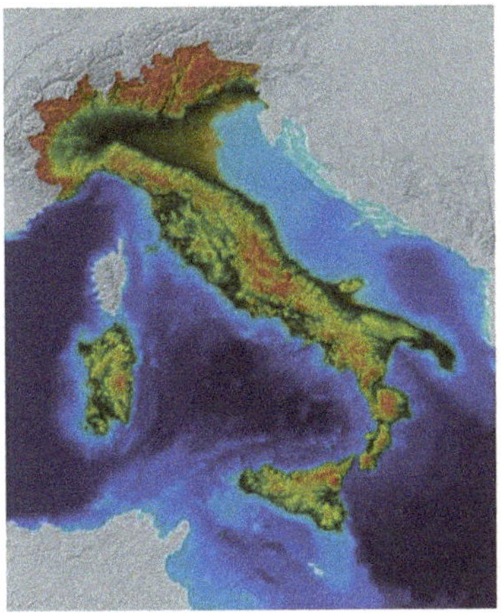

**Figure 6.** DEM Tinitaly/1.1.

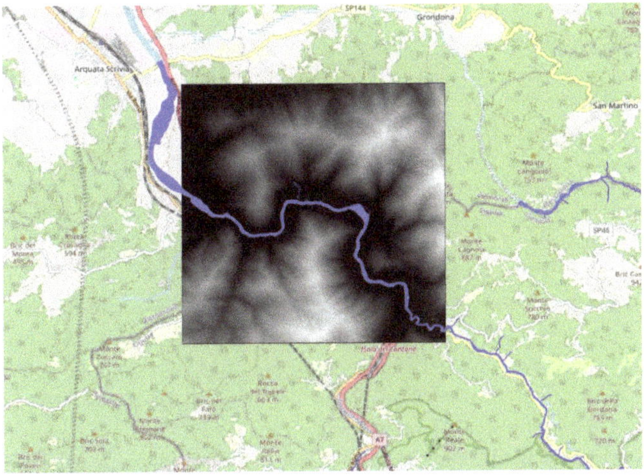


**Figure 7.** Watercourse layer and DEM portion within platform.

This model has a resolution of 10 m and is downloadable in the GeoTIFF format via the WMS, WMTS and WCS. In particular, the WCS (Web Coverage Service) [15] was used in the present study because it allowed us to download a map portion of a specific size. In this case, a size of 500 × 500 pixels was chosen where each pixel covered an area of 10 × 10 m, thus obtaining an elevation map for an area of interest of 5 × 5 km. However, the parameterization of this resolution allows even larger portions of terrain to be downloaded if necessary.

### 3.4. Geometries of Watercourses

For the areas covered by watercourses, it was decided to use map data from OpenStreetMap, a series of open-source maps managed by the OpenStreetMap Foundation and maintained by a large community of contributors. These maps form the basis of thousands of websites and guarantee an accurate and up-to-date representation of the world's territory.

We chose to use such maps because watercourses are represented vectorially in the form of polygons rather than lines, as is the case with several other cartographic sources representing water networks. Polygonal representation makes it possible to simulate as realistically as possible the initial situation of a watercourse and thus the simulation of its spread in the surrounding terrain in the event of a flood [16].

The data were downloaded using the QuickOSM plugin for the QGIS software, which allowed us to download data from the various layers of which the OpenStreetMap map was composed using filters on the type of layer and a bounding box to delimit the area of interest.

Specifically, all geometries were selected in which the key-value pair consisted of natural = water, which characterized the geometries of the water mirror layer in the OpenStreetMap map.

Once downloaded to QGIS in the shapefile format, the geometries were then saved on the PostgreSQL database using the integrated functions of QGIS to be subsequently retrieved in the GeoJSON format by the services of the application for visualization in the form of layers for the OpenLayers Javascript libraries used by the platform.

In addition to the polygonal geometries of the watercourses, it was decided to add a layer to the platform containing the hydrographic network in the geoportal of the region of Liguria. This layer shows the linear geometries of the hydrographic network; the information provided includes the name of the watercourse, the indicative specific surface area, the indicative surface area of the underlying basin (referring to the selected segment), whether it is hydraulically investigated and the relative Strahler order, which measures the branching complexity of the selected stretch.

The geometries were downloaded from the geoportal in the form of shapefiles. As before, they were imported into the QGIS software and then saved to the PostgreSQL database, whose built-in geographical functions were used to filter the results, retaining only sections that intersected the polygonal geometries downloaded from OpenStreetMap.

### 3.5. Simulation Algorithms—Calculation of External Contribution

In order to simulate as realistically as possible the increase in water volumes following floods caused by major meteorological events, it is necessary to take into account not only the amount of rain that falls in the river basin but also that which is deposited in the surrounding areas and then, due to gravity, pours from higher-elevation areas into the riverbed, thus increasing the mass of liquid that could cause the river to flood.

This type of rainwater movement is called surface runoff. In order to take this 'external contribution' into account, while keeping the simulation streamlined and executable in near-real time, it was decided to approach the problem with a simplification: for each pixel of the elevation map, i.e., for each $10 \times 10$ m area within it, the platform selects the preferred path of the water towards neighboring cells, based on the slope, and of all the neighbors, it chooses the one with the steepest downward slope.

The first step is then to calculate, for each cell of the elevation grid, the slope of the four directions with respect to the current cell. These four slopes are then compared with each other to decide which direction the water will take on its path to the lowest point.

In the example in Figure 8, if we had to choose a path for the water from cell A towards the adjacent cells B, C, D and E, the path towards E would be chosen because there is a negative slope towards B and C, while between D and E, E is the one with a greater downward slope than A (see Figure 8).

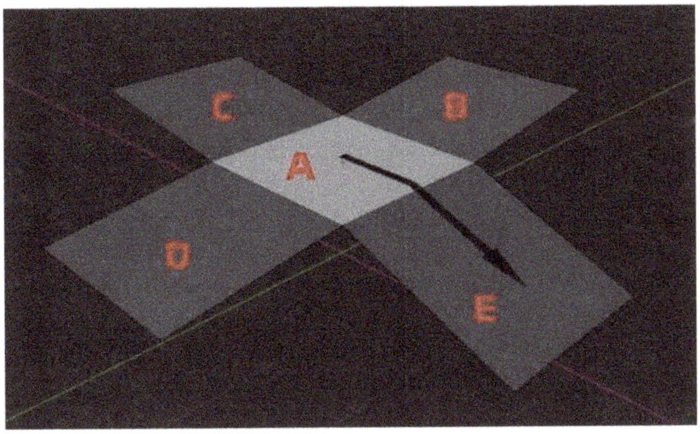


**Figure 8.** Example of runoff based on relative slope.

The path is then continued until a flat area is found, or until an area within the initial riverbed is reached (or the boundaries of the map under consideration). Each pixel affected by the trickle of water thus created is increased by a certain value; repeating the operation on all the pixels of the map will thus result in 'accumulation zones' in which the water tends to stop, forming a static map that is kept in memory. When the dynamic simulation is performed, for each pixel reached by the mass of water simulated, a certain amount of volume is added depending on the external contribution of that pixel, thus taking into account the accumulated rain from the surrounding areas.

Figure 9 shows the complete map calculated for one of the zones of interest in the study area.

**Figure 9.** Static map of external contributions based on surface runoff.

The areas of greatest water accumulation are visible in orange. It can also be seen that at the base of the slopes, water accumulates more, making a strong contribution to the flooding of those areas; the effect diminishes as the elevation increases, until there is no external contribution at the top of the heights.

The starting data for the simulation are as follows:

- The uniform grid containing the elevation data of the area of interest, from which a three-dimensional plane is derived to represent it;
- The initial geometry of the watercourse, represented by several cubes (one for each 10 $\times$ 10 m cell of the DEM), with a very small initial height to simulate the initial body of water.

The execution of the simulation starts after the user has filled in the fields that determine the control parameters through the interface. It is possible to define the amount of water due to rainfall, expressed in mm/m$^2$ per hour of rain; the time step between one frame and the next of the simulation, which can range from 15 to 120 min; and the total time of the simulation duration, in the range of 6 to 48 h. The last parameter to be selected is the soil permeability index, which affects the percentage of water that is considered absorbed by the soil and therefore does not increase the total volume of the overflow. This value can range from 0.0 (no permeability, found in rocky soils) to 1.0 (maximum permeability, which allows the soil to absorb the entire amount of water that falls, a condition that is obviously impossible in real cases).

Once these data are collected, the platform begins by first calculating a static map for external contributions. After that, it raises the geometries of the entire watercourse of the decided step, calculated as follows:

$$((\textit{rainfall amount} + \textit{external contribution}) \times (1 - \textit{permeability index}))$$

$$\times\ (\textit{time step/total time}).$$

An object representing the edge of the watercourse geometry (called the fringe in the platform) is then generated by selecting all the cells that lie on the outer contour of the geometry. This is the area that is compared with the elevation, to be possibly expanded if the top of the cube of each of its cells is at a higher elevation than the direction in which it would like to expand.

In this way, it is possible to obtain an approximate simulation of the expansion in the volume of water as the amount of rain received increases, with an expansion stop in the case of slopes and areas with a higher ground elevation.

The water volumes displayed are also colored in a graduation from white to deep blue as the height of the water volume in a given cell increases to make it even clearer which areas are most affected by flooding.

In Figure 10 is the end result of a simulation that considers a time frame of 48 h from the start of precipitation, assuming a constant amount of precipitation over time.

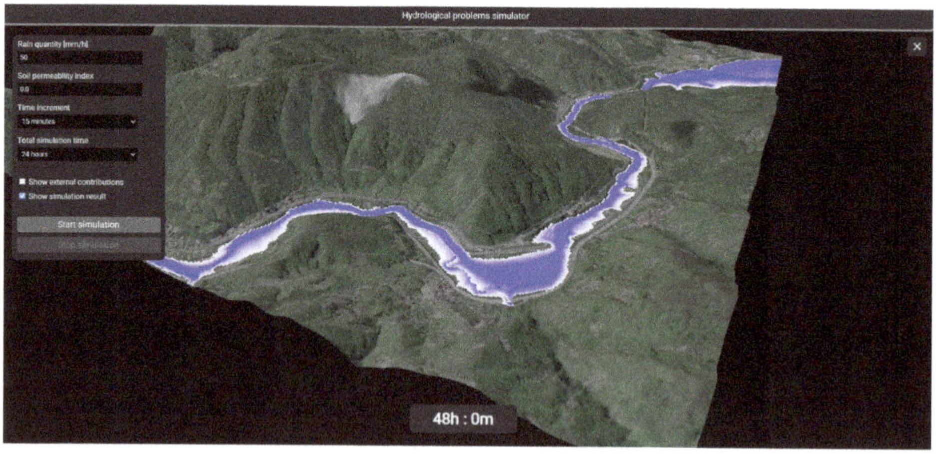


**Figure 10.** The final result of a simulation.

### *3.6. Simulation Platform*

The platform uses the classic client–server scheme (for more details on the platform, see the Supplementary Material), thus allowing for its use by several simultaneous users and publication on the web, possibly integrated within more extensive web portals that allow for different analyses of a territory for the mitigation of various types of risk (see Figure 11).

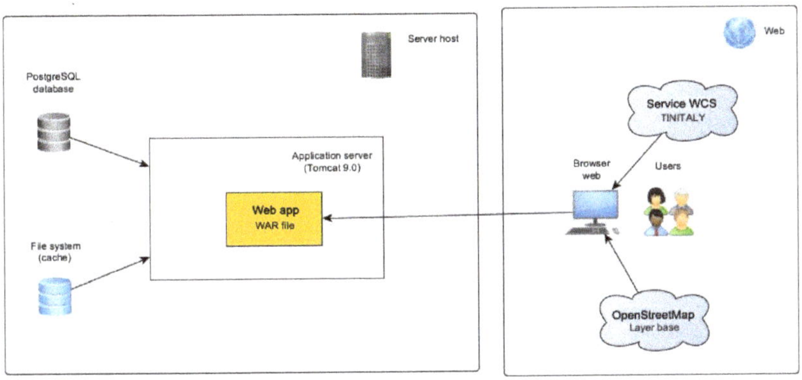


**Figure 11.** System architecture.

## 3.7. Three-Dimensional Rendering

The key part of the platform is the screen that runs the simulation and displays it in 3D. This choice is based on two main reasons: the first is that the three-dimensional X-, Y- and Z-coordinates of the points of the terrain model are necessary for the execution of the simulation, as the expansion of the water volume is decided on the basis of the height of the surrounding terrain in relation to the height of the water; the second is the possibility for the user to rotate and zoom in on the model while the simulation is running, for a better evaluation of the effects of flooding.

The capabilities of modern browsers allow the simulation to run in real time directly on the user's terminal, not overloading the back-end server when running multiple instances of the platform.

The Three.js library provides, in the form of an API, the tools needed to create a 3D Scene complete with polygonal objects, lighting effects, cameras and post-processing within a normal web page, exploiting the potential of the WebGL graphics libraries. Figure 12 shows the conceptual scheme of a Three.js Scene.

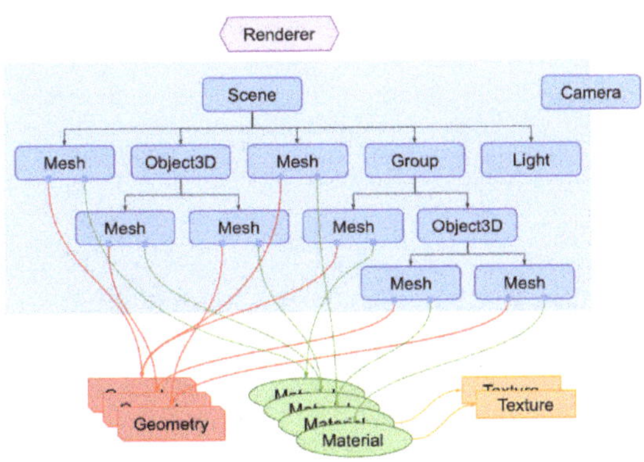


**Figure 12.** A conceptual diagram of a Scene.

Three.js is based on a number of software objects that encapsulate the main functionality of the software: The Renderer is the main object and is responsible for representing everything in a Scene on the screen, framing it as if the user were positioned at the point where the Camera is. This is modeled as a real camera with its own position, orientation and field of view (FOV), framing the Scene (see Figure 13).

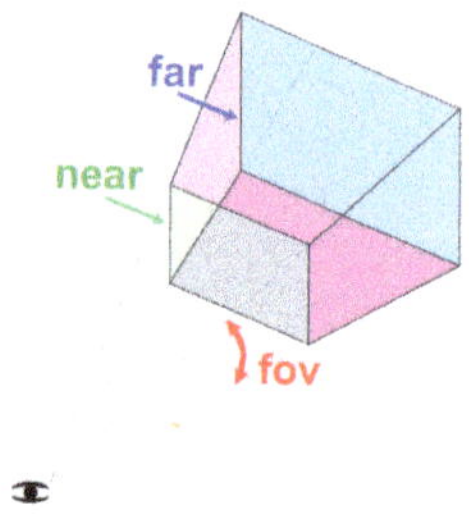


**Figure 13.** Camera-related concepts.

Within the Scene, we find all the objects that make up the structure of the Scene that we frame. The Meshes are the actual 3D objects. These are, in turn, composed of Geometry, the polygonal geometries that describe their appearance, and Material, the materials that define their color, light reflection properties, etc.

Materials may in turn refer to Textures, raster images that are applied to the Geometries, suitably extended or repeated to cover the entire model.

The Scene is then illuminated by Light objects, which may be of different types to simulate various types of light, such as spot, ambient and omnidirectional light and so on.

One of the aspects to which attention must be paid for the correct visualization of three-dimensional space is the co-ordinate system; although they always consist of the three axes X, Y and Z, some applications consider Z as the vertical axis, and others consider Y. Three.js uses the latter convention, as shown in Figure 14.

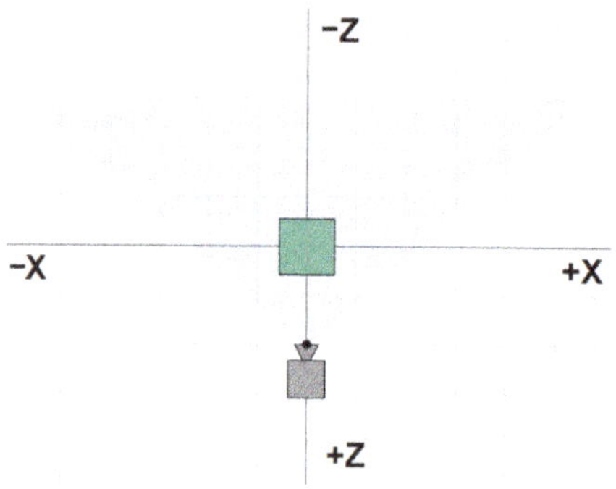


**Figure 14.** Axis directions in Three.js.

Since in geographic systems longitude and latitude values are often traced back to X- and Y-co-ordinates, it would be more convenient, in the case of a geographic application, to use a reference system in which the Z-axis corresponds to the vertical axis, representing the elevation of the terrain; the library allows the reference system to be changed via a simple function: THREE.Object3D.DefaultUp = new THREE.Vector3(0, 0, 1).

This creates a unit vector in the Z-direction and assigns it to the one that is considered 'high' (Up); in this way, the convention is restored to the desired condition.

### 3.8. User Interface

The application consists of two different screens. The initial screen is a classic two-dimensional map showing the watercourses, the hydrographic network and the boundaries of the area of interest. The second is a 3D screen, which can be viewed after selecting an area on which to run the simulation, where it is possible to set parameters and launch the actual simulation.

The two-dimensional map view of the territory is the screen that is displayed when accessing the application URL (see Figure 15) [17].

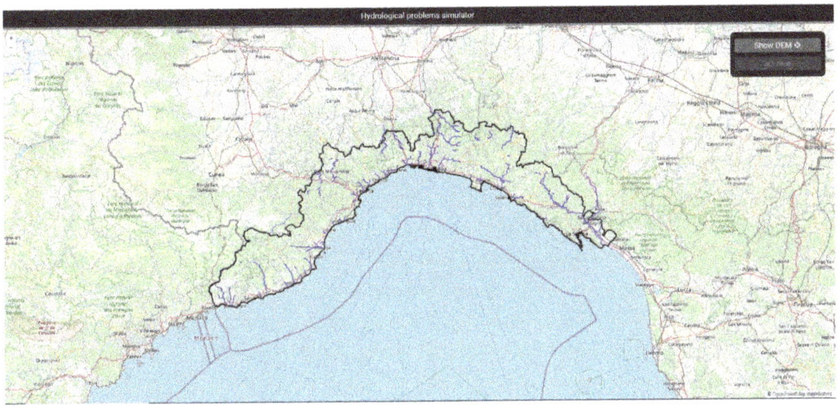


**Figure 15.** User interface—two-dimensional display.

The interface is very simple: the main element is the map, freely navigable and clickable. In addition to the zoom buttons on the left, which are provided directly by the OpenLayers API, there are two buttons for displaying the DEM in a 5 km area around the clicked point and a button for the 3D view, which is activated when the DEM is displayed (see Figure 16).

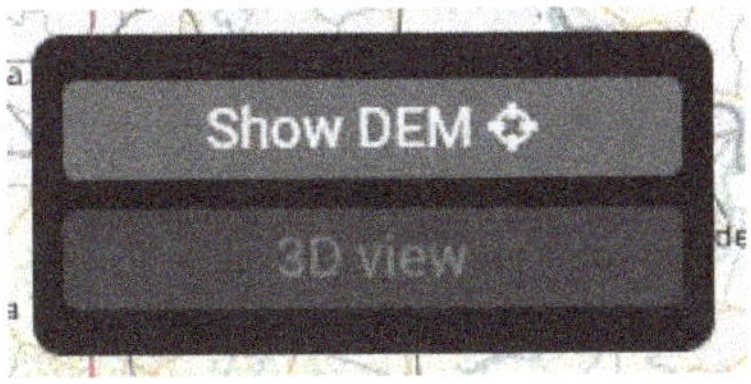


**Figure 16.** Button panel.

On the map, there are, in addition to the OpenStreetMap base layer, two layers: the first displays the watercourses downloaded from OpenStreetMap in vector form, while the second represents the hydrographic network downloaded from the geoportal of the Liguria region.

Clicking on a given point on the map will search for information in the latter layer. If elements are found, these will be colored in yellow and a pop-up window will open containing various information on the selected section of the grid (see Figure 17).

The reported data are as follows:

- The name of the watercourse;
- The Strahler order;
- The indicative specific surface area, expressed in square meters;
- The indicative surface area of the underlying basin, expressed in square meters;
- Whether the watercourse is hydraulically surveyed or not.

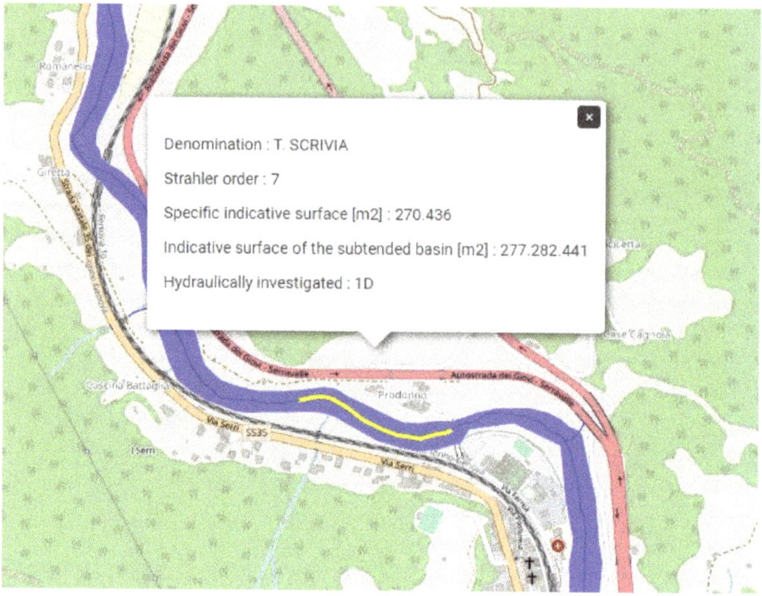


**Figure 17.** Pop-up with details of selected feature.

When clicking on the 'View DEM' button, the user will be asked to click on a point on the map, and once a point has been selected, the TINITALY WCS service will be called up, which returns the GeoTIFF of the elevation for a 5×5 km area around the selected coordinates. The GeoTIFF, in addition to being scanned to send elevation data to the simulation, is also displayed on the map for a better understanding of the area over which it operates (see Figure 18).

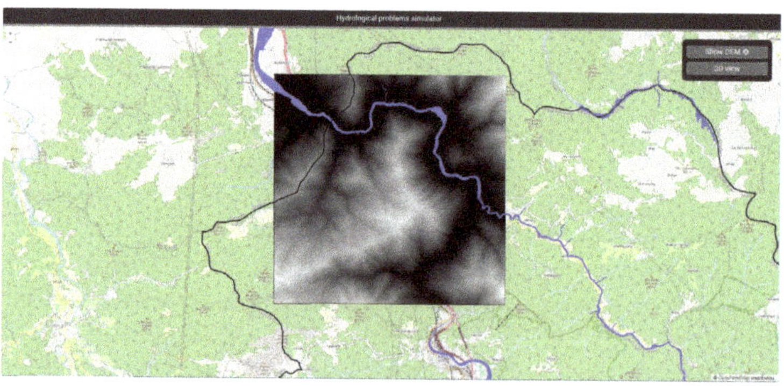

**Figure 18.** Visualization of DEM portion.

At this point, the '3D View' button will be made active and, if there are watercourses within the area, the user can move on to the actual simulation [18,19].

### 3.9. Three-Dimensional Visualization Map

The simulation platform is activated with the '3D View' button, at which point the 3D terrain model, the riverbed of the watercourse in a blue color and all elements of the user interface for interacting with the simulation are displayed, as shown in Figure 19 [20].

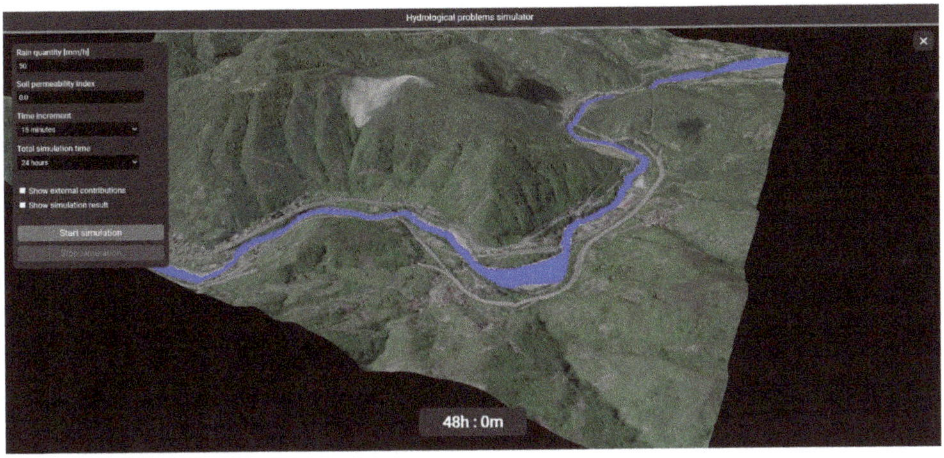


**Figure 19.** User interface—three-dimensional visualization.

In particular, thanks to the panel on the left of the screen, it is possible to select the following (see Figure 20):

- The amount of rain falling in the area in an hour (mm/h);
- The soil permeability index, from 0.0 (completely impermeable) to 1.0 (completely permeable);
- The time increment for each 'frame' of the simulation;
- The total simulation time.

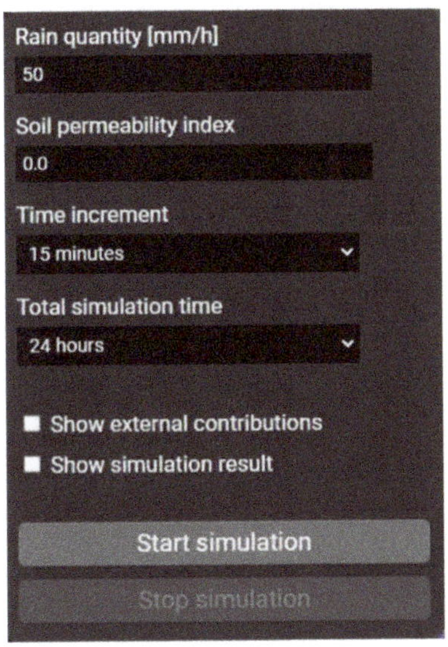


**Figure 20.** Simulation control panel.

In addition to these input parameters, there are two checkboxes: The first allows the user to display, instead of the textured terrain, the map of external contributions, with a graduation from green to red depending on the contribution each cell makes to the volume of water. The second, on the other hand, allows the display of the simulation result to be activated/deactivated after the simulation is completed, as can be seen in Figure 21.

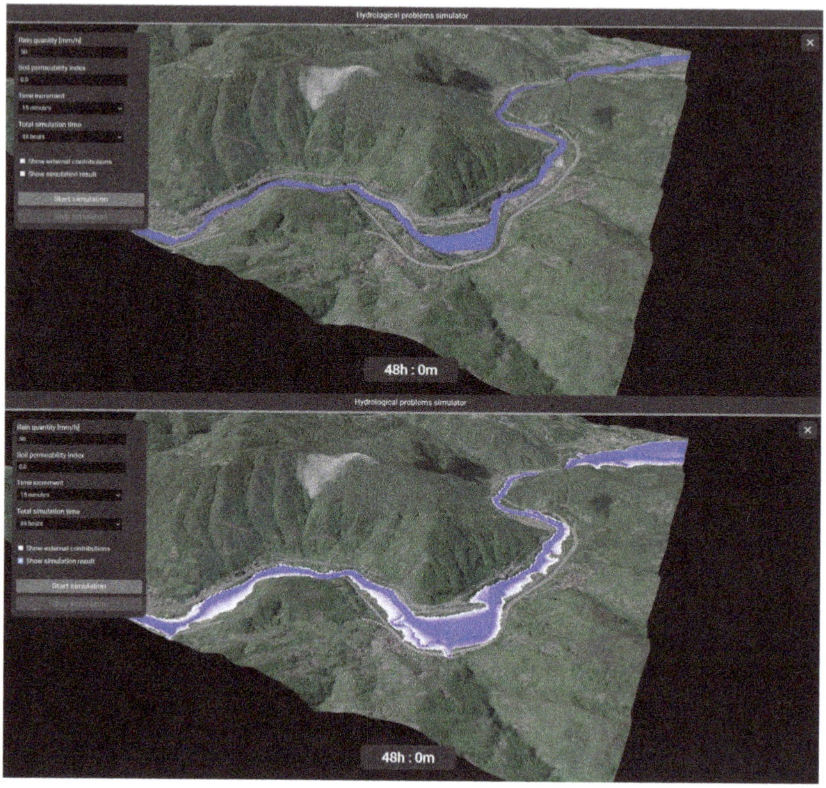


**Figure 21.** A comparison of the situation at the beginning and end of the simulation.

This option allows a better comparison between the initial and final situation and a better understanding of which areas have been flooded compared to the initial riverbed. Below the two checkboxes are the two main buttons: the first allows the simulation to be started, while the second allows it to be interrupted if intermediate effects with respect to the selected time horizon are to be assessed. During the execution of the simulation, the time relative to the current frame is indicated by the timer at the bottom of the screen.

## 4. Conclusions and Future Work

This research activity describes the design and implementation of a tool for the simulation of hydrological risk in the event of rainfall-related flooding in a watercourse, showing the results through a three-dimensional representation of the affected water mass and the surrounding terrain. The simulation is based on the morphology of the terrain adjacent to the affected watercourse bed, taking elevation into consideration to determine the direction of expansion of the water mass.

A distinctive aspect of the platform is its extreme speed in the resolution of simulations, which also allows the tool to be used in real time. This means that during the occurrence of a meteorological event, the platform can be used to monitor the evolution of the situation and predict the effects of any persistence of the phenomenon over time. This real-time forecasting capability is crucial for making quick and informed decisions, thus reducing reaction times and improving emergency management on the ground, with a potential positive impact on the safety of the population and the protection of infrastructure.

A very important feature of the research work performed is the possibility of running simulations on any device with an internet connection, without the need for specific hardware or particularly high resources. This makes the platform accessible from laptops, tablets or even smartphones, offering considerable operational flexibility for users. In this way, the platform can be used not only by emergency operators at the various sites of the agencies involved but also by those who are directly in the field, perhaps during monitoring and control operations in areas at risk.

This feature makes it possible to obtain up-to-date information in real time, even in emergency situations, where quick access to data and the ability to develop up-to-date scenarios can make a difference in managing operations. Mobile use allows, for example, technicians and operators in the field to monitor the evolution of weather and hydrological conditions directly on site, to quickly share simulation results with coordination centers and to make decisions based on concrete and up-to-date data without having to worry about the availability of powerful workstations or computing servers. Moreover, this accessibility reduces operational costs related to the purchase and maintenance of specific equipment, making the platform a particularly cost-effective solution for public authorities, civil protection and land management organizations, even in resource-limited settings.

A further aspect of the simulation that could be worked on is the refinement of the analysis performed; for example, with a point or area description of the permeability of the terrain, one could use these data to determine the amount of water absorbed by the soil instead of relying on a static index entered by the user that remains static for the entire area under consideration. One could also use elevation maps at a higher resolution to model the area, and thus also the riverbed, more accurately; the simulation would also be more accurate as one would be modeling a greater number of cells, each with a smaller size and thus with greater fidelity than in the real case.

It would also be possible to exploit less approximate algorithms for the simulation of surface runoff, i.e., the movement of rainwater on the slopes that flow into the riverbed. In any case, a compromise will have to be found between the fidelity of the simulation and the speed of execution, since excessive complexity in the analyses carried out could compromise the platform's usability in real time, defeating its purpose and leading to the use of more complex software that exploits dynamic fluid simulations to predict the movement of the mass of water involved in the outflow from the riverbed.

**Supplementary Materials:** The following supporting information can be downloaded at https://www.mdpi.com/article/10.3390/ijgi14020073/s1. Additional material describes the functionality of the backend, frontend and simulation algorithm.

**Author Contributions:** Conceptualization, Mauro Mazzei; methodology, Mauro Mazzei; software, Davide Quaroni; validation, Mauro Mazzei and Davide Quaroni; formal analysis, Mauro Mazzei; data curation, Mauro Mazzei and Davide Quaroni; writing—review and editing, Mauro Mazzei and Davide Quaroni; supervision, Mauro Mazzei. All authors have read and agreed to the published version of the manuscript.

**Funding:** This research received no external funding.

**Data Availability Statement:** The sources of the data of the maps are the ISPRA (Institute for Environmental and Environmental Research) the INGV (National Institute of Geophysics and Volcanology) which are used for the project TINITALY/1.1.

**Acknowledgments:** The authors thank Istituto Superiore per la protezione e la Ricerca Ambientale and the l'Istituto Nazionale di Geofisica e Vulcanologia Research Center as sources of documentation necessary for the conduct of this paper.

**Conflicts of Interest:** The authors declare no conflicts of interest.

## References

1. Rapporto Sulle Condizioni di Pericolosità da Alluvione in Italia e Indicatori di Rischio Associati. Available online: https://www.isprambiente.gov.it/files2021/pubblicazioni/rapporti/rapporto_alluvioni_ispra_353_16_11_2021_rev2.pdf (accessed on 1 September 2024).
2. Essential Resources for WebGL Development. Available online: https://www.khronos.org/webgl/ (accessed on 1 September 2024).
3. Threejs.org Documentation, Getting Started. Available online: https://threejs.org/docs/index.html#manual/en/introduction/Creating-a-scene (accessed on 1 September 2024).
4. Openlayers API. Available online: https://openlayers.org/en/latest/apidoc/ (accessed on 1 September 2024).
5. OpenStreetMap. Available online: https://www.openstreetmap.org/ (accessed on 1 September 2024).
6. Tarquini, S.; Isola, I.; Favalli, M.; Battistini, A.; Dotta, G. *TINITALY, a Digital Elevation Model of Italy with a 10 Meters Cell Size (Version 1.1)*; Istituto Nazionale di Geofisica e Vulcanologia (INGV): Rome, Italy, 2023. [CrossRef]
7. ESRI Portal. Available online: https://www.esri.com/en-us/home (accessed on 1 September 2024).
8. Guzzetti, F. Frane e alluvioni, una lunga storia italiana. *Ecoscienza J. ARPA* **2015**, 3, 12–13.
9. *Dissesto Idrogeologico in Italia: Pericolosità e Indicatori di Rischio*; Edizione 2021. Rapporti 356/2021; ISPRA: Rome, Italy, 2021.
10. Çengel, Y.A.; Cimbala, J.M.; Cozzo, G.; Santoro, C. *Meccanica dei Fluidi*; McGraw-Hill: Milan, Italy, 2024.
11. European Commission, Joint Research Centre; De Groeve, T.; Revilla-Romero, B.; Pappenberger, F.; Salamon, P.; Thiemig, V.; Thielen-del Pozo, J.; Hirpa, F. *The Benefit of Continental Flood Early Warning Systems to Reduce the Impact of Flood Disasters*; European Union Publications Office: Luxembourg, 2016.
12. Tazen, F.; Diarra, A.; Kabore, R.F.W.; Ibrahim, B.; Bologo/Traoré, M.; Traoré, K.; Karambiri, H. Trends in flood events and their relationship to extreme rainfall in an urban area of Sahelian West Africa: The case study of Ouagadougou, Burkina Faso. *J. Flood Risk Manag.* **2019**, 12, e12507. [CrossRef]
13. Aich, V.; Liersch, S.; Vetter, T.; Andersson, J.; Müller, E.; Hattermann, F. Climate or Land Use?—Attribution of Changes in River Flooding in the Sahel Zone. *Water* **2015**, 7, 2796–2820. [CrossRef]
14. Massazza, G.; Bacci, M.; Descroix, L.; Ibrahim, M.H.; Fiorillo, E.; Katiellou, G.L.; Panthou, G.; Pezzoli, A.; Rosso, M.; Sauzedde, E.; et al. Recent Changes in Hydroclimatic Patterns over Medium Niger River Basins at the Origin of the 2020 Flood in Niamey (Niger). *Water* **2021**, 13, 1659. [CrossRef]
15. OGC. *OpenGIS® Web Services Architecture Description, Version: 0.1.0*; Whiteside, A., Ed.; OGC 05-042r2; Open Geospatial Consortium: Wayland, MA, USA, 2005; p. 22. Available online: https://www.ogc.org (accessed on 1 September 2024).
16. Giuliani, G.; Nativi, S.; Obregon, A.; Beniston, M.; Lehmann, A. Spatially enabling the Global Framework for Climate Services: Reviewing geospatial solutions to efficiently share and integrate climate data & information. *Clim. Serv.* **2017**, 8, 44–58.
17. Geoportale Della Regione Liguria. Available online: https://geoportal.regione.liguria.it/ (accessed on 1 September 2024).
18. Mazzei, M.; Quaroni, D. Development of a 3D WebGIS Application for the Visualization of Seismic Risk on Infrastructural Work. *ISPRS Int. J. Geo-Inf.* **2022**, 11, 22. [CrossRef]
19. Copernicus, Operational Review of the Record Flood in Niamey, Niger. Sentinel Vision, EVT-719, 27 August 2020. Available online: https://www.sentinelvision.eu (accessed on 10 January 2022).
20. Shen, Y.; Zhang, Y.; Xie, J.; Jiang, R.; Han, H. Study on Rainfall Information Integration System Based on Java Web Start. In Proceedings of the 2015 International Conference on Network and Information Systems for Computers, Wuhan, China, 23–25 January 2015; pp. 431–434.

land

MDPI

Article

# Ecological Management Zoning Identification by Coupling Blue-Green and Gray Infrastructure Networks: A Case Study of Guizhou Province, China

Shuang Song [1,*], Xuanhe Zhang [1], Shaohan Wang [2] and Yue Gong [3]

1 School of Architecture and Urban Planning, Guizhou University, Guiyang 550025, China; ssong@gzu.edu.cn
2 School of Landscape Architecture and Art, Fujian Agriculture and Forestry University, Fuzhou 350002, China; wsh0707@fafu.edu.cn
3 School of Design and Art, Henan University of Technology, Zhengzhou 450001, China; gongyue@haut.edu.cn
* Correspondence: songshuang0916@126.com

**Abstract:** Ecological management zoning is crucial for maintaining regional ecological security and realizing differentiated urban ecological governance. However, the existing zoning methods are overly focused on ecological functional attributes and fail to adequately consider the impacts of human activities, resulting in an insufficiently rational allocation of resources. Taking Guizhou Province as an example, using multi-source data and spatial analysis tools, this study proposed an ecological management zoning framework based on the coupling analysis of the blue-green infrastructure (BGI) network and gray infrastructure (GI) network. The results indicated that (1) the BGI network in the study area included 179 sources, with a total area of 54,228.80 km$^2$, and 232 corridors. (2) There were 53 sources in the GI network, totaling 709.19 km$^2$, and the corridors of the first, second, and third levels were 11,469.31 km, 6703.54 km, and 5341.30 km, respectively. (3) There were 606 barrier points identified, mainly distributed in the central part of the study area, and the total area of the disturbance zone was 1132.50 km$^2$, which had the largest distribution in Qiandongnan, followed by Qiannan. (4) At the county scale, five ecological management zones were identified in the study area based on four indicators, namely, the source area ratio of BGI network, corridor density of BGI network disturbance zone area ratio, and density of barrier point. Then, we proposed targeted optimizations and restorations for each zone. This study organically linked ecological functional attributes and anthropogenic impacts to identify ecological management zones, which will provide new perspectives on synergies between ecological protection and economic development.

**Keywords:** blue-green infrastructure network; ecosystem services; gray infrastructure network; coupling; ecological management zoning

## 1. Introduction

The stability of the ecosystem's structure and function is essential to support the sustainable development of human society. The ecosystem itself is constantly facing various risks when providing multiple services [1]. At the end of the 20th century, with the increasing demand for natural resources for social and economic development, almost 60% of the global ecosystem functions had significantly degraded [2], and the structural conflict between ecology and construction became increasingly inevitable [3–5]. Ecological management zoning for natural resources is the primary way to promote territorial planning, and it is also the mainstream idea of regional management [6]. It has become an urgent task to rationally divide ecological zones and accurately implement policies to maximize ecosystem services [7,8].

Ecological management zoning has methods based on ecological processes and multi-objective pattern optimization [9]. It focuses on the optimization of ecosystem space and the improvement of service capacity within urban zones [10,11]. In essence, ecological management zoning promotes ecological protection and governance by dividing different zones for targeted management, which has important guiding significance for the sustainable development of land space [9,12]. So far, there have been many studies on ecological management zoning. Most of them divided management zones by calculating the ecological compensation funds that should be paid or received by each region [4,13]. Secondly, according to the risk level and ESs bundle, the main functions of different regions were determined to divide the management zone [11,14]. Zoning based on ecological importance and ecological vulnerability has also attracted widespread attention [15]. The topics include single ecosystem, permafrost ecosystem, alpine tundra ecosystem, mangrove ecosystem, grassland ecosystem, mixed ecosystem, etc. [9,16]. However, the above studies cannot portray the interconnections of the elements within the ecosystem and realize overall protection and coordinated management, which is not conducive to enhancing the integrity and connectivity of ecosystems.

Compared with the application of multi-dimensional indicators such as quantity, quality, and function to carry out multi-objective pattern optimization [17], the ecological process method based on the morphological structure of blue-green infrastructure (BGI) is more helpful in identifying the sustainability of ecological patterns [18]. At the same time, it is better to realize the effective control of ecological processes and the optimization and improvement of regional ecological security patterns [18,19]. In this regard, for the goal of sustainable development of ecosystems, in 1986, Forman systematically summarized the two conceptual models of "irreplaceable pattern" and "disaggregation between agglomerations", which laid an important foundation for the construction of a model of urban infrastructure networks known as "source-resistance-surface-corridor" [7]. With the gradual deepening of research, scholars have applied the methods of typology, landscape ecology, operations research, and other disciplines to each core link of infrastructure network construction [20]. The "source-sink" landscape patch identification, network index, effective cost distance model (ECDM), geographic approach to biodiversity protection (GAP), and other modeling methods have appeared one after another [8,21,22]. These methods have been widely applied in studying nature reserves, urban green spaces, and ecological restoration at global, national, and regional levels [11,23]. However, as far as the current urban sustainable development and ecological coordination goals are concerned, there is a lack of research on the interaction between urban development and BGI. This limits the dynamic identification of regional ecological security patterns and the engineering of optimization guidance for ecological space protection.

The BGI network is the natural or semi-natural infrastructure composed of blue-green space and related ancillary systems, such as forests, wetlands, rivers, parks, green roofs, green walls, and bioretention pools [21]. It is considered an effective approach to mitigate and adapt to climate change by providing a series of ecosystem services to enhance the resilience of ecosystems and human well-being [20,24]. As a nature-based solution (NBS), the BGI network has been applied in many cities around the world [1]. In the process of rapid urbanization, the construction of gray infrastructure (GI) has a profound impact on BGI. Its essence is the spatial difference or imbalance between the supply and demand of ecosystem services caused by the change in land use structure, which directly threatens regional ecological security and sustainable development of the social economy [25]. The ecological management zoning based on the coordinated state of the BGI network and GI network can better reflect the complex relationship between the ecosystem and human activities. It is an important basis for ecological restoration, territorial spatial planning, and the allocation of space resources.

As a typical karst landscape distribution area in China, Guizhou Province has sensitive and fragile ecological background conditions, as well as prominent poverty problems [26]. To escape poverty, large-scale production activities have been carried out, causing further damage to the ecological environment [27]. How to construct reasonable ecological management zoning and reduce the impact of anthropogenic activities has become an urgent topic of scientific research. Therefore, we first made a systematic assessment of ecosystem services in Guizhou Province. On this basis, we constructed the BGI network by combining the special geographic environment factors. Then, we extracted the sources and integrated transportation corridors to form the GI network. Finally, the ecological management zones were identified through the coupling analysis of the BGI network and the GI network, and corresponding suggestions were made. This study provides a reference for coordinating ecological protection and socio-economic development and plays an important role in maintaining ecosystem stability, protecting biodiversity, and optimizing ecosystem service functions.

## 2. Materials and Methods

### 2.1. Study Area

Guizhou Province is located in the hinterland of southwest China (103°36′ E~109°35′ E, 24°37′ N~29°13′ N), with a total area of about 176,200 km$^2$, including 9 prefecture-level cities (autonomous prefectures) [28]. The average annual temperature is about 15 °C, and the average annual precipitation is about 1200 mm, which belongs to the typical subtropical monsoon climate [29]. The terrain is high in the west and low in the east. There are many water systems and various landform types [30]. It has developed the largest and deepest karst landform, which is ecologically sensitive and fragile [31]. In recent years, due to unreasonable human activities, great pressure has been exerted on the fragile ecological background conditions, exacerbating the outbreak of regional ecological problems such as soil erosion and rocky desertification [32]. Consequently, the sustainable supply of ecosystem services in the province has been significantly reduced (Figure 1).

### 2.2. Data Collection

The following datasets were used in this study. The basic geographic data (including provincial boundaries, county boundaries, rivers, etc.) were obtained from the National Center for Basic Geographic Information (http://www.ngcc.cn/ (accessed on 2 September 2024)). Land use, Normalized Difference Vegetation Index (NDVI), GDP, and population spatial distribution data were from the Data Center for Resource and Environmental Sciences, Chinese Academy of Sciences (https://www.resdc.cn/ (accessed on 9 September 2024)). Elevation data was provided by the Geospatial Data Cloud (https://www.gscloud.cn/ (accessed on 2 September 2024)). Soil data were from the China Soil Dataset of the World Soil Database (HWSD), which was derived from the Scientific Big Data Center of Cold and Arid Regions (http://bdc.casnw.net/ (accessed on 12 September 2024)). Temperature and precipitation data were provided by the China Weather Network (https://www.cma.gov.cn/ (accessed on 16 September 2024)). Actual surface evapotranspiration data came from the National Tibetan Plateau Science Data Center (https://data.tpdc.ac.cn (accessed on 16 September 2024)). Road data was supported by the National Geographic Information Resources Directory Service System (https://www.webmap.cn/ (accessed on 2 September 2024)).

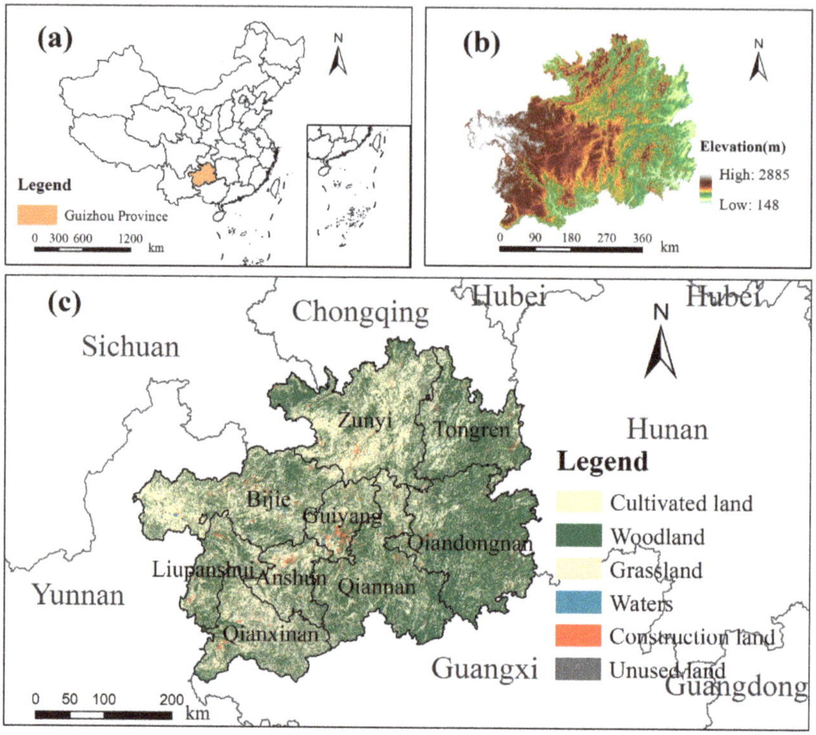


**Figure 1.** Geographical location (**a**), elevation (**b**), and land use types of Guizhou Province (**c**).

*2.3. Methods*

The research framework includes four parts: (1) Extraction of BGI network and GI network. We combine ecosystem service function, principal component analysis, and special geo-environmental factors to determine the sources of BGI network and construct the combined resistance surface. The interaction between sources are simulated using Python 3.9.2 scripting language to determine the direction and extent of corridors. Sources and corridors are synthesized at different levels to extract the GI network in the study area. (2) Coupling the BGI network with the GI network. The conflict areas are extracted as ecological disturbance zones and barrier points, and their distribution characteristics are revealed. (3) Ecological management zoning. We combine the following four indicators: source area ratio of BGI network, corridor density of BGI network disturbance zone area ratio, and density of barrier point. The system clustering algorithm is used to divide the ecological management zone at the county scale. (4) Ecosystem protection and restoration strategies. By comparing the social and ecological conditions of different ecological management areas, the targeted protection and restoration strategies are proposed. The research framework is presented in Figure 2.

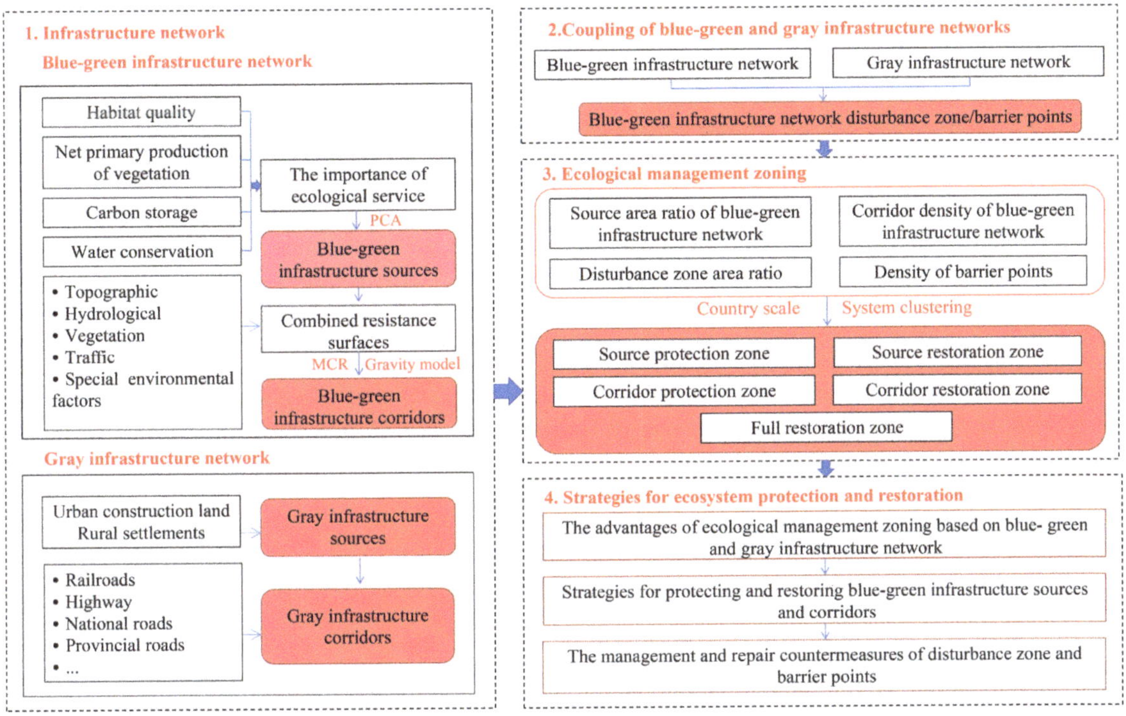


**Figure 2.** The research framework.

2.3.1. Blue-Green Infrastructure Network Identification

Ecosystem service is a comprehensive evaluation of the ability of blue-green patches to transform functions such as energy, matter, and information [33]. There are two methods for the quantitative evaluation of ecological service function: the vegetation net primary productivity (NPP) index and the ecological model [34]. To enhance the accuracy of the source selection results, the study utilized the NPP index while selecting habitat quality, carbon storage, and water conservation, which are crucial for ecological protection, to conduct a comprehensive assessment of the ecosystem service function in the study area.

These two evaluation methods were used as well in the "Technical Guidelines for the Delineation of the Ecological Protection Red Line" issued by China's Ministry of Environmental Protection (MEP) in 2015. Specifically, NPP was assessed with the CASA model, while habitat quality and carbon storage were quantified with the InVEST model. After standardizing the four ecosystem services, the entropy weight method was used to determine the weights, and then the importance of the ecosystem services was calculated [8]. Considering the actual status of karst mountain cities, the blue and green patches larger than 15 km$^2$ with the principal component level in the ecosystem service assessment results were selected as the sources of BGI network in the study area [31].

The ecological flow from one source to another needs to overcome a variety of resistance factors, and there is also a gravitational effect between them [35]. The resistance factors in this study specifically include 3 categories: natural factors, human factors, and special geographical environment factors of karst mountainous cities [23]. A total of 14 factors were selected and the resistance of each factor was set according to the evaluation system (Table 1). The resistance and gravitational force between the sources were calculated using the MCR model and the gravity model, respectively. We utilized the Python scripting language to simulate the interaction between gravitational and resistance forces between sources and processed the overlapping paths to obtain the BGI network [36]. Different levels of BGI network corridors have varying impacts on species migration and the maintenance of stable ecological patterns. Finally, based on the strength of interaction between sources, the BGI network corridors were classified into three levels using the natural breaks method, which provides a basis for the maintenance of blue-green space in the study area.

**Table 1.** Ecological resistance evaluation system.

| Primary Factor | Secondary Factor | Grade | Resistance | Primary Factor | Secondary Factor | Grade | Resistance |
|---|---|---|---|---|---|---|---|
| Natural factors | DEM | >250 m | 9 | Human factors | Land use | Construction land | 9 |
| | | 200–250 m | 7 | | | Unused land | 7 |
| | | 150–200 m | 5 | | | Arable land | 5 |
| | | 100–150 m | 3 | | | Grassland | 3 |
| | | <100 m | 1 | | | Woodland, water | 1 |
| | Slope | >30 | 9 | | Night light intensity index | <0.2 | 9 |
| | | 25–30 | 7 | | | 0.2–0.35 | 7 |
| | | 20–25 | 5 | | | 0.35–0.5 | 5 |
| | | 15–20 | 3 | | | 0.5–0.65 | 3 |
| | | <15 | 1 | | | >0.65 | 1 |
| | NPP | <0.1 | 9 | | Road network density | >1.25 | 9 |
| | | 0.1–0.2 | 7 | | | 0.85–1.25 | 7 |
| | | 0.2–0.3 | 5 | | | 0.52–0.85 | 5 |
| | | 0.3–0.6 | 3 | | | 0.20–0.52 | 3 |
| | | >0.6 | 1 | | | 0–0.20 | 1 |
| | MNDWI | <0.1 | 9 | | Population density | >425 | 9 |
| | | 0.1–0.3 | 7 | | | 300–425 | 7 |
| | | 0.3–0.5 | 5 | | | 175–300 | 5 |
| | | 0.5–0.8 | 3 | | | 50–175 | 3 |
| | | >0.8 | 1 | | | <50 | 1 |
| | Rainfall | >0.056 m | 9 | | GDP | >250 | 9 |
| | | 0.054–0.056 m | 7 | | | 200–250 | 7 |
| | | 0.052–0.054 m | 5 | | | 150–200 | 5 |
| | | 0.050–0.052 m | 3 | | | 100–150 | 3 |
| | | <0.050 m | 1 | | | <100 | 1 |
| | Temperature | <14 °C | 9 | Special geographical environment factors | Rock exposure rate | <15 | 9 |
| | | 14–15 °C | 7 | | | 15–30 | 7 |
| | | 15–16 °C | 5 | | | 30–45 | 5 |
| | | 16–17 °C | 3 | | | 45–60 | 3 |
| | | >17 °C | 1 | | | >60 | 1 |
| | Biodiversity index | <0.3 | 9 | | Soil thickness | <15 | 9 |
| | | 0.3–0.4 | 7 | | | 15–30 | 7 |
| | | 0.4–0.5 | 5 | | | 30–45 | 5 |
| | | 0.5–0.6 | 3 | | | 45–60 | 3 |
| | | >0.6 | 1 | | | >60 | 1 |

### 2.3.2. Gray Infrastructure Network Construction

In this study, construction land was extracted as GI network sources [27]. The representative types of GI network corridors usually include transportation lines, building corridors, and flyovers [25]. Considering the actual situation of the study area, we chose transportation lines as CI network corridors, including railroads, highways, and national, provincial, and county roads. These corridors were superimposed on the sources to form the GI network. The classification of GI network corridors referred to the "Urban Road Engineering Design Code" CJJ 37-90 (2019), with the status of the road network, traffic functions, and service capabilities along the route as the criteria [27]. Among them, railroads, highways, and national highways were categorized as first-level corridors, provincial and county roads as second-level corridors, and township roads as third-level corridors. The width of each level of the GI network corridor was 150 m, 120 m, and 40 m, respectively.

2.3.3. Barrier Point Extraction

The BGI network is a natural network, which plays an important role in conserving biodiversity, protecting water sources, regulating floods, and improving habitats [21]. While the GI network could enhance the living environment, it also threatens the BGI network. It is essential to consider both natural and socio-economic factors to effectively and comprehensively address ecological problems [25]. This study utilized ArcGIS 10.2 to analyze the spatial distribution relationship between the BGI network and GI network. First, BGI network corridors can be disturbed or blocked by GI network corridors, so the intersection was identified as a barrier point. At the same time, GI network corridors intrude into the BGI network sources or the GI network sources intrude into the BGI network corridors, leading to fragmentation, and therefore the overlap zone was recognized as the disturbance zone.

2.3.4. Ecological Management Zoning Based on Systematic Clustering

System clustering, also known as hierarchical clustering, is based on the idea that each sample or variable is considered as a separate cluster [37]. Then, the closest clusters are gradually merged until all clusters are combined into one cluster, which is applicable to various types of data [38]. The specific steps of the systematic clustering algorithm are as follows: (1) Initialization: Each variable is considered as a separate cluster. (2) Calculate distance: The distance between clusters is calculated and the two clusters with the closest distance are selected to merge into a new cluster. (3) Update distance matrix: The distance between the new cluster and all other clusters is recalculated. (4) Repeat merging: Steps 2 and 3 are repeated until all clusters are merged into one cluster. (5) Generate cluster map: A cluster map is generated based on the merging process to help determine the final number of clusters. In this study, we selected four indicators, namely, source area ratio of BGI network, corridor density of BGI network, disturbance zone area ratio, and density of barrier point, to classify ecological management zones by systematic clustering at the county scale.

## 3. Results

### *3.1. Blue-Green Infrastructure Network*

The high values of ecosystem service function were mainly distributed in the mountainous areas of southeastern Guizhou Province, and the low values were concentrated in the central and northwestern parts of the province (Figure 3e). The high values of NPP were mainly concentrated in the southwestern mountainous areas, while low values were mainly distributed in Guiyang, the capital of the province in the central part of the study area, and the surrounding areas (Figure 3a). The high values of habitat quality were mainly distributed in the southern and southeastern mountainous areas, including Tongren and Qiandongnan, as well as the southern part of Qiannan, which shows the best habitat quality (Figure 3b). The high values of carbon storage were mainly distributed in the shallow mountainous areas in the southeast and southwest, with the low values in the northwestern areas (Figure 3c). The overall water conservation was well, with high values mainly located in the eastern and southeastern parts, with the best around the Wuling Mountains and the Qingshui River (Figure 3d).

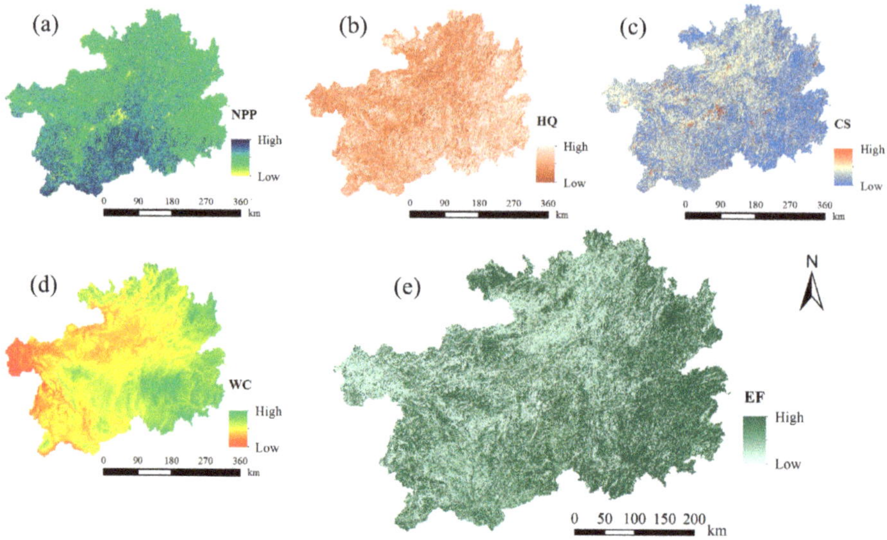


**Figure 3.** Spatial distribution of different ecosystem service functions in Guizhou Province (**a**) NPP (**b**) habitat quality (**c**) carbon storage (**d**) water conservation (**e**) ecosystem service function.

There were 179 sources identified in the BGI network, totaling 54,228.80 km$^2$, accounting for 30.78% of Guizhou Province. The spatial distribution of sources was heterogeneous, with large, highly connected sources concentrated in the southeastern part of the Wuling Mountains. Among them, the sources with the largest area were located in the eastern part of Guizhou Province, including Fanjingshan Mountain, the Qingshui River, and the Wujiang River, which are favorable habitats for many protected species. While the sources in the northwestern plateau were small, dispersed, and poorly connected, even the northern part of Bijie had no source distribution. In the direction of northeast–southwest axis, the sources were very sparse, which greatly affected the connectivity and robustness of the BGI network in the study area.

The low values of combined resistance were located in the southern and eastern parts of the study area, with high vegetation cover, mostly woodland. The high values were in the western part of the area, with high elevations and widespread karst landscapes, the land type mainly being cultivated land. In addition, the central part of the study area was characterized by a scattered distribution of high resistance values, for the land type of urban construction land (Figure 4a). By simulating interactions between sources using the Python scripting language, we extracted 232 BGI network corridors with a total length of 7640.32 km (Figure 4b). The southeastern part of the study area was distributed with many large corridors with good connectivity. They provided an important corridor for the migration and gene exchange of species, and were crucial for the maintenance of regional ecological security and biodiversity. In contrast, the corridors in the northwestern and northeastern parts were relatively sparse. Among them, the first-level corridors were 47, totaling 1840.37 km, mainly distributed in the southeastern part of the study area and dominated by large corridors, including the core corridor connecting the entire study area. There were 110 s-level corridors, with a total of 2932.88 km, concentrated in the central part of the study area. These effectively enriched the hierarchical structure of the BGI network. The third-level corridors were 75, with 2867.07 km, mainly located in the high-altitude western part of the study area. Although the length and connectivity of these corridors may not be as great as that of first-level and second-level corridors, they also play a non-negligible role in the protection of specific species and the maintenance of ecological diversity. Particularly in areas of harsh environments, such as high altitudes, tertiary corridors provide valuable living space for species.

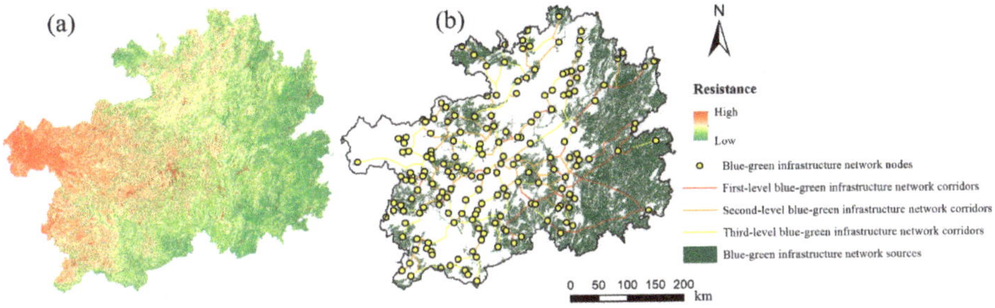


**Figure 4.** Combined resistance surface (**a**) and blue-green infrastructure network (**b**) in Guizhou Province.

### 3.2. Gray Infrastructure Network

The total number of GI sources was 53, with a total area of 709.19 km$^2$, showing the distribution characteristics of single core radiation in Guiyang, with sporadic scattering within the town (Figure 5). This distribution pattern not only reflected the strong radiation power of Guiyang as a regional central city, but also revealed the close connection between gray infrastructure construction and local development needs. The total length of the GI network corridor was 23,514.15 km, which was much larger than the BGI network corridor. Of these, the length of the first-level corridor was 11,469.31 km, accounting for 48.78% of the total length. Most of them were distributed along the rivers, the corridors not only took full advantage of the accessibility provided by the natural waterways, but also exhibited vertical extension, effectively linking the entire study area. The second-level corridors were 6703.54 km in length, mainly in the form of horizontal extension, complementing the primary corridors and connecting the districts and counties. The length of third-level corridors in the GI network was 5341.30 km, accounting for 22.72% of the total length. These corridors, as branches of the first and second levels, formed a crisscrossing network across the study area.

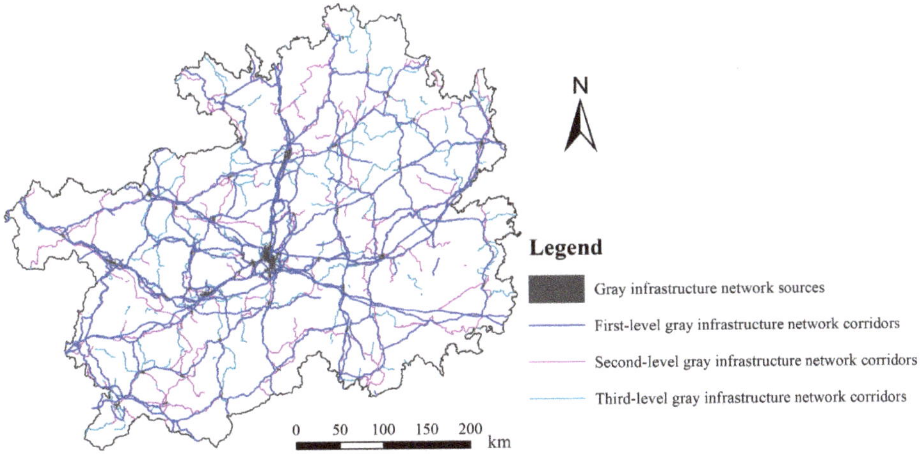


**Figure 5.** Spatial distribution of gray infrastructure network in Guizhou Province.

### 3.3. Barrier Points and Disturbance Zones

There were 606 barrier points in the study area, mainly distributed in the central part, with a relatively sparse distribution in the eastern and western parts of the study area (Figure 6). Of these, the number of first-level barrier points formed with first-level GI network corridors was 318. Qiannan had the highest number of barrier points with 102 points, followed by Guiyang and Bijie with 86 and 84 points. The GI network corridors at different levels have varying impacts on the BGI network sources, forming a total area of the disturbance zone of 1132.50 km$^2$. The disturbance zone formed by the first, second, and third-level GI network corridors was 615.39 km$^2$, 409.99 km$^2$, and 107.12 km$^2$, respectively. Among them, the disturbance zone in Qiandongnan was the largest with 327.48 km$^2$, followed by Qiannan with 281.49 km$^2$.

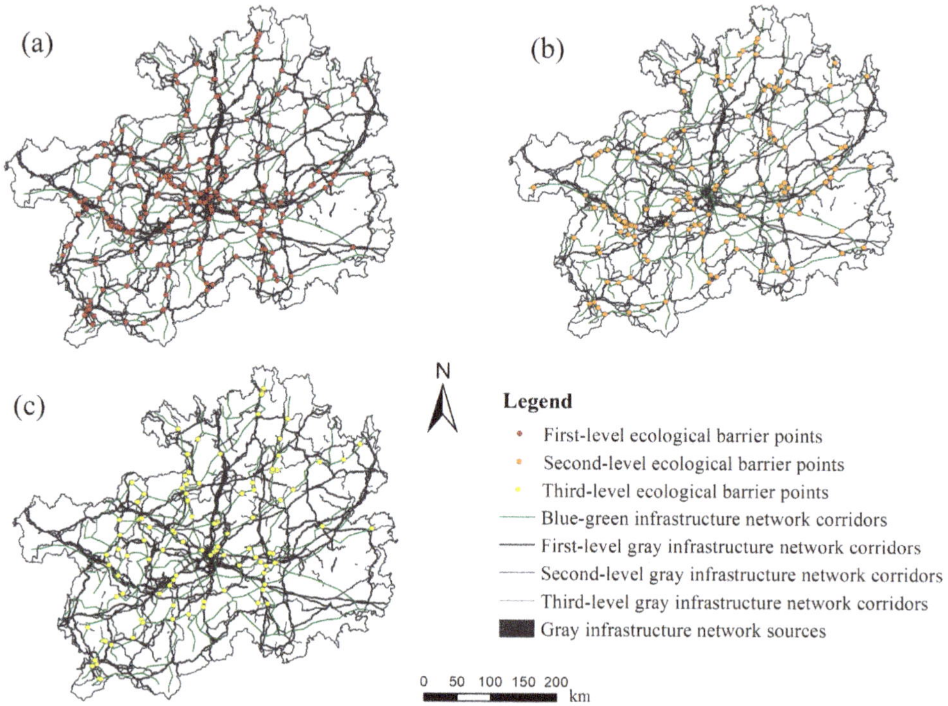


**Figure 6.** Spatial relationship between BGI network and GI network in the study area (**a**) first-level GI network (**b**) second-level GI network (**c**) third-level GI network.

### 3.4. Ecological Management Zone

Based on the results of the system cluster analysis, it was most reasonable to divide the 88 counties in Guizhou Province into five ecological management zones. Namely, they are the source protection zone, source restoration zone, corridor protection zone, corridor restoration zone, and full restoration zone (Figure 7). The spatial pattern of the ecological management zones is shown in Figure 8. The source protection zone included 23 counties, covering 27.05% of Guizhou Province, and mainly distributed in the southeastern part of the province. The zone had a large area of ecological land, with low human activity and well-preserved natural habitats. Therefore, the source area ratio of the BGI network and ecological connectivity in this zone were the largest among the five ecological management zones, while other indicators were lower. The source restoration zone consisted of 20 counties, the area proportion was 27.05%, mainly in the northern and central-southwestern parts. The BGI network sources were relatively small in this zone, with low connectivity and scattered distribution, and human activities posed a great threat to the BGI network resources.

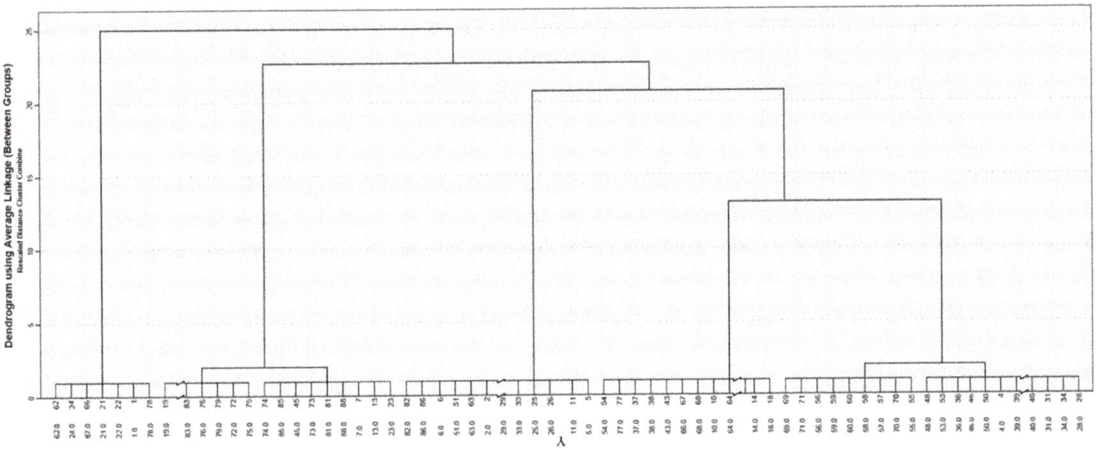


**Figure 7.** Results of system cluster analysis.

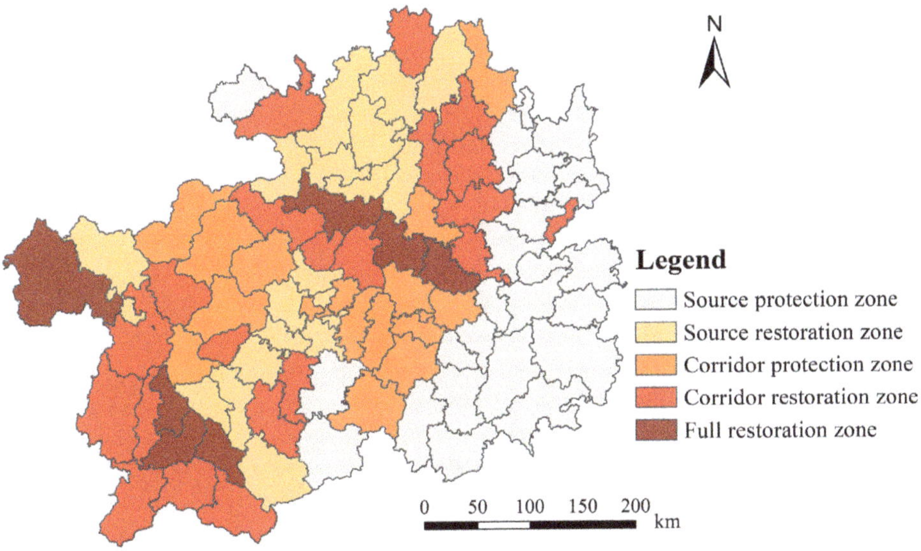


**Figure 8.** Spatial pattern of ecological management zones in Guizhou Province.

The corridor protection zone included 18 counties, covering 18.06% of Guizhou Province, mainly located in the central and northeastern part of the study area. The zone has a large area of medium-sized BGI network corridors that can steadily play an important bridging role. The corridor restoration zone covered 20 counties, accounting for 25.77% of Guizhou Province, mainly in the northeastern and southwestern parts. The BGI network corridors in this zone were sparse and narrow. The full restoration zone included seven counties, covering 9.67% of the total area, mainly in the central and western parts. All elements of the BGI network in this zone were small in size.

### *3.5. Comparison of Natural and Social Conditions of Different Ecological Management Zones*

The blue-green space difference in Guizhou Province is very significant; this further illustrates the need for research on ecological management zoning. Based on the results of the ecological management zoning, we further compared and analyzed the economic and social conditions of each ecological management zone (Figure 9). The source protection zone had the lowest proportion of karst area and the lowest number of nighttime light intensity, GDP, and population density, with the best ecological substrate conditions and the least disturbance by humans relative to the other zones. The source restoration zone had the highest proportion of karst area and relatively high night light intensity and population density, but a low GDP index; combined with the spatial distribution, this zone was mainly concentrated in the light industrial area, which has relatively little direct impact on the environment. The corridor protection zone showed the highest GDP index, and these zones were the main concentration of the coal mining industry. The corridor restoration zone had a relatively low GDP index and a relatively high proportion of karst area, and these zones were primarily dominated by agricultural production. The full restoration zone, with a high proportion of karst area and high nighttime light intensity, was characterized by a large amount of resource-processing substrate, poor substrate quality, and generally strong anthropogenic impact.

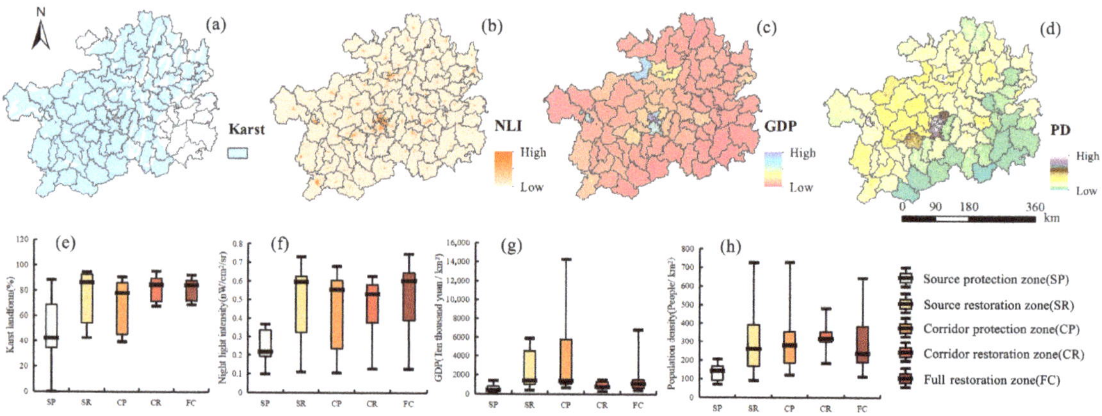


**Figure 9.** The natural and social conditions of different ecological management zones (**a**) karst spatial distribution (**b**) nighttime light intensity spatial distribution (**c**) population density spatial distribution (**d**) population density spatial distribution (**e**) karst statistics (**f**) nighttime light intensity statistics (**g**) nighttime light intensity statistics (**h**) population density statistics.

## 4. Discussion

### *4.1. The Advantages of Ecological Management Zoning Based on Blue-Green and Gray Infrastructure Networks*

Ecological management zoning is an important strategy for implementing differentiated management in the process of ecological construction [18]. By analyzing the distribution of ecological resources, ecosystem services, ecological vulnerability, and other indicators, it is possible to effectively elucidate the matrix conditions and spatial variability of the ecological environment [39]. Ecological management zoning, based on comprehensive consideration of societal and ecological indicators, not only helps in identifying environmental problems accurately but also provides a scientific basis for ecological resource allocation and management. The BGI network combines key point, line, and surface elements, and is an effective landscape planning strategy to achieve biodiversity conservation and ecosystem maintenance. Along with the continuous acceleration of urbanization, the expansion of GI has significantly eroded and negatively affected the ecological integrity of blue-green spaces [2]. In this context, how to maximize ecological benefits while minimizing resource inputs has become a scientific issue that urgently needs to be tackled [6,40]. This study adopted an ecological management zoning approach based on the coupling of the BGI network and GI network, which can clarify the conservation objectives and restoration priorities of different areas. The aim is to develop effective ecological management techniques and strategies to promote optimal integration of BGI network. The results of the study will provide a reference for policymakers to propose strategies that balance ecological protection with economic and social development.

### *4.2. Strategies for Protecting and Restoring Sources and Corridors*

According to the natural and social conditions of each zone, it is possible to propose targeted ecosystem protection and restoration strategies. The source protection zone is the best ecological condition in Guizhou Province, making it a key area for biodiversity conservation and an important guarantee of regional ecological security. Despite the high vegetation cover and good ecosystem conditions, deforestation for the purpose of growing cash crops still exists, posing a serious threat to natural vegetation and wildlife populations [3]. Therefore, the priority within this zone should be to strengthen environmental protection, improve the system of nature reserves, prevent ecological damage, and maintain ecological defense. The source restoration zone is more vulnerable to damage than the source protection zone, as BIN sources were fragmented and not able to form a continuous distribution. To enhance the landscape connectivity within the zone, it is necessary to establish protected reserves where sufficient protection is lacking. Additionally, the zone has the highest proportion of karst landscapes, with thin soils that make it challenging to restore natural vegetation. Widespread rocky desertification and complex geology, as negative feedback, further limit economic development and highlight the urgency of realizing sustainable development [31]. Therefore, irreversible damage to fragile ecosystems caused by human activities should be strictly prevented by gradually expanding the scope of governance.

The corridor protection zone has a relatively high level of economic development. Economic growth will enhance the ability of local governments and communities to invest in ecological protection, promoting the construction of the BGI network [41]. Furthermore, the financial and technical support brought by economic development makes the planning and implementation of BGI network corridors more scientific and effective, thus enhancing biodiversity and ecological service functions [22]. The corridor restoration zone is characterized by a relatively low level of economy and high population density, which leads to greater demand for natural resources and ecosystem services, threatening the stability of the ecosystem. The quality of local ecosystems and the connectivity of the BGI network can be improved by greening both sides of urban roads and rivers and increasing green space at BGI network nodes. Designing greenways around highland lakes can help avoid pollution and damage caused by human activities. There is also a need to increase investment in ecosystem restoration and to rationalize the arrangement of construction land in order to balance economic construction and ecological protection. The full restoration zone is the most vulnerable area of the BGI network and a key area for conservation and protection. Despite the low level of economic development in this zone, the restoration of the BGI network nodes allows for the passage of blocked areas within the corridor and maximizes returns from small investments. Simultaneously, favorable hydrothermal conditions enable ecosystems to recover quickly after disturbance or destruction. Furthermore, considering the density of BGI network corridors in the zone, it is necessary to prioritize the construction of BGI network corridors and to identify alternative relationships between corridors.

### 4.3. The Management and Repair Countermeasures of Barrier Points and Disturbance Zone

According to the distribution of barrier points in the study area, the eastern and southern barrier points were mainly distributed in the urban–rural junction zone, which greatly reduced the ecosystem service function. To address this issue, measures such as installing pipe culverts, culverts under bridges, overpasses, and other ecological corridors can be implemented to ensure the normal passage of organisms under the original transportation system [13,20,42]. This will help minimize the negative impacts of the GI network corridors on the BGI network corridors. Secondly, according to the size of the GI patches, industrial conditions, and intensity of human activities, different management strategies should be formulated [21,43]. If human interference is frequent, it will seriously threaten the regional ecological environment and strict control strategies must be adopted [2]. For the western and northern parts of the study area, where most of the barrier points were generated by mining enterprises and related ancillary facilities [44], stricter control measures should be adopted, and protection and restoration must be carried out in ecologically sensitive and fragile areas. Meanwhile, pollution monitoring and early warning mechanisms should also be established to avoid potential environmental pollution risks and improve the self-recovery capacity of the core area. In addition, corridors with good habitat quality can reduce the resistance of barrier points and improve the survival rate of species migration. Studies [40,43] have shown that corridors of 30–60 m are abundant in herbaceous plants and marginal bird species, which can effectively lower temperature, increase food supply, and filter pollutants. The corridors of 80–100 m feature numerous internal species and higher biodiversity [45,46], and can effectively control the loss of sediment and soil elements [47]. the corridors of 100–200 m are generally considered to be the ideal width range to support the development of a high level of biodiversity [21,48]. They provide ample living space and migration pathways for various organisms, and promote the formation and stabilization of complex relationships within the ecosystem. Considering the specific ecological conditions and conservation needs of the study area, it is recommended that the corridor widths of the first, second, and third levels be set at 200 m, 90 m, and 30 m, respectively [8,9]. This will maximize the ecological service functions of the corridors at each level, and ensure that they play an optimal role in promoting the exchange of species, improving the quality of the environment, and ensuring ecological security.

The disturbance zone was mainly located in the southeastern and northwestern parts of Guizhou, which is the key area for ecological protection and the main battlefield for poverty alleviation [29]. These areas have a high demand for urbanization and development. More than 92.5% of the national territory in Guizhou Province is mountainous and hilly [7]. The competition between construction land and natural space resulting from following the plains urban development pattern has put enormous pressure on the ecosystem in the study area. In the process of new urbanization, it is necessary to ensure that the spatial pattern of the national territory is compatible with the available resources and environment [26]. To address this issue, the techniques and methods for the delineation of bounded land use boundaries have been widely explored [32]. Since 2020, the resource and environmental carrying capacity and suitability evaluation of territorial spatial development has been widely used as an important policy tool in China [22]. However, in the process of preparing municipal-level territorial spatial planning, problems such as neglect of ecosystem integrity, logical paradoxes between it and future planning decisions, and fragmented and discrete spatial identification have gradually emerged, especially in southwestern mountainous cities [39,49]. Therefore, government departments should improve the BGI network system at the county and city levels to further enhance the protection of vegetation in the region, promote ecosystem services, and support sustainable ecological development. The northwestern part of the study area is crucial for the development of energy and coal chemical industries in Guizhou Province [32]. To minimize negative impacts, ecological isolation zones can be established around key and connected BGI network corridors [50,51], and ecological interception methods can be adopted to reduce the impact of external pollutants on the source [52]. It is also necessary to manage GI network corridor facilities, widen green belts, and reduce the number of high-emission vehicles passing through the BGI network sources [53]. Additionally, the number and scale of transportation facilities within ecological sources should be controlled and ecological protection forests need to be constructed [48,54].

### 4.4. Research Limitations and Future Research Directions

Compared with the analysis of ecological processes only focusing on natural elements, this study comprehensively analyzed the spatial connectivity and interference between the BGI network and the GI network, and based on this, cluster analysis was used to divide the ecological management zone and put forward targeted management measures. This can provide full play to the positive role of the BGI network, eliminate the negative role of the GI network, and better maintain the stability and coordinated development of the regional ecosystem. However, there are still some limitations. First, the identification of BGI network sources was only achieved by assessing the importance of ecosystem services at a single time node, without considering the potential trade-offs or synergies between different ecosystem services and the stability of ecosystem service functional outputs. Second, considering the actual situation of the study area with typical karst, this study constructed a resistance surface system consisting of 14 factors such as rock exposure rate, which improved the accuracy to a certain extent, but still deviated from the actual situation; it is necessary to incorporate more regional characterization factors in the future to improve the accuracy continuously. Finally, there are also large differences in resource endowment and socio-economic development within each ecological management zone, and although we point out the key construction directions, the intensity of ecological and environmental management and specific strategies should also be different.

## 5. Conclusions

Zoning that couples the natural conditions with urban development can effectively consider the characteristics of ecologically vulnerable areas. It supports the development of specific ecological restoration programs for overall protection and better promotes the sustainable development of the region. Taking Guizhou Province as the study area, the study extracted the BGI network and GI network and divided the ecological management zones based on four indicators: source area ratio of the BGI network, corridor density of the BGI network, disturbance zone area ratio, and density of barrier points. Then, the natural and social conditions of different zones were compared to propose corresponding ecosystem restoration suggestions. The results showed that the spatial distribution of ecosystem service functions in Guizhou Province was uneven, and improving its regional coordination was the primary task of ecological environment enhancement. The BGI network contained 179 sources and 232 corridors, with obvious spatial heterogeneity. The GI network showed the distribution characteristics of single-core radiation in Guiyang, which was the key object of ecological management. A total of 606 barrier points and 1132.50 km$^2$ of disturbance zone were identified, concentrated in the central and southeastern parts of the study area. Using system clustering, five ecological management zones were divided. Among them, the source protection zone and corridor restoration zone were larger, covering 27.05% and 25.77% of Guizhou Province, respectively. Therefore, the focus of ecological construction in the study area is to enhance the protection of existing high-quality sources and restore shrinking corridors. The results of the study will support the efficient use of ecological resources and help governments and planners to formulate more rational land use and infrastructure development policies.

**Author Contributions:** Conceptualization, S.S., S.W. and X.Z., methodological design, X.Z., S.W. and S.S., data collection, S.S. and Y.G., funding acquisition, S.S., formal analysis, S.S., S.W. and X.Z., writing—original draft, S.S., writing—review and editing, S.S., S.W. and X.Z. All authors have read and agreed to the published version of the manuscript.

**Funding:** This research was funded by the Guizhou Provincial Basic Research Program (Natural Science) (Grant No. Qiankehe Basic-zk [2025] general 588), the Youth Guidance Project of Guizhou Basic Research Program (Natural Sciences) (Grant No. Guizhou Science Foundation [2024] Youth 134), and the Guizhou University Renjihezi (2023 Natural Science) (Grant No. 34) for Shuang Song.

**Data Availability Statement:** The data presented in this study are available on request from the authors. The data are not publicly available due to privacy. We have fully described the process of constructing the BGI network and the GI network, the process of data analysis and processing. If the editors and reviewers have questions about specific data, we will endeavor to provide more detailed explanations and clarifications. Readers interested in detailed information can contact the corresponding authors. All data used in this paper are publicly available through the websites provided in the text.

**Conflicts of Interest:** The authors declare that they have no known competing financial interests or personal relationships that could have appeared to influence the work reported in this paper.

## References

1. Huang, S.; Xiao, X.; Tian, T.; Che, Y. Seasonal influences on preferences for urban blue-green spaces: Integrating land surface temperature into the assessment of cultural ecosystem service value. *Sustain. Cities Soc.* **2024**, *102*, 105237. [CrossRef]
2. Zeng, J.; Cui, X.; Chen, W.; Yao, X. Ecological management zoning based on the supply-demand relationship of ecosystem services in China. *Appl. Geogr.* **2023**, *155*, 102959. [CrossRef]
3. Cheng, Z.; Zhao, T.; Song, T.; Cui, L.; Zhou, X. Assessing the Spatio-Temporal Pattern and Development Characteristics of Regional Ecological Resources for Sustainable Development: A Case Study on Guizhou Province, China. *Land* **2022**, *11*, 899. [CrossRef]
4. Chen, W.; Li, A.; Hu, Y.; Li, L.; Zhao, H.; Han, X.; Yang, B. Exploring the long-term vegetation dynamics of different ecological zones in the farming-pastoral ecotone in northern China. *Environ. Sci. Pollut. Res.* **2021**, *28*, 27914–27932. [CrossRef] [PubMed]
5. Fan, X.; Rong, Y.; Tian, C.; Ou, S.; Li, J.; Shi, H.; Qin, Y.; He, J.; Huang, C. Construction of an Ecological Security Pattern in an Urban-Lake Symbiosis Area: A Case Study of Hefei Metropolitan Area. *Remote Sens.* **2022**, *14*, 2498. [CrossRef]
6. Zhao, H.; He, J.; Liu, D.; Han, Y.; Zhou, Z.; Niu, J. Incorporating ecological connectivity into ecological functional zoning: A case study in the middle reaches of Yangtze River urban agglomeration. *Ecol. Inform.* **2023**, *75*, 102098. [CrossRef]
7. Li, J.; Han, F. Strong ethics and flexible actions, the properties of traditional ecological knowledge (TEK), as key resources for socioecological resilience to the impacts of climate change: A case study of Baojiatun, Yunnan-Guizhou Plateau karst area, southwest China. *Ecol. Soc.* **2022**, *27*, 31. [CrossRef]
8. Song, S.; Gong, Y.; Yu, Y. Integrating pattern, process, and function in urban landscape ecological network planning: A case study of Harbin central city. *Ecol. Indic.* **2024**, *159*, 111671. [CrossRef]
9. Liu, Y.; Jing, Y.; Han, S. Ecological function zoning of Nansi Lake Basin in China based on ecosystem service bundles. *Environ. Sci. Pollut. Res.* **2023**, *30*, 77343–77357. [CrossRef] [PubMed]
10. Huang, H.; Lu, W.; Zuo, G.; Liu, Z.; Liu, C.; Wang, X.; Li, X.; Ma, Y.; Yu, J. Ecological distribution patterns in Chinese seas and adjacent waters: Marine ecological zones. *Sci. Total Environ.* **2023**, *905*, 167259. [CrossRef]
11. Huang, X.; Yan, J.; Zhang, G.; Huang, J.; Wang, Q.; Li, B.; Ma, Y. Marine Ecological Function Zoning and Management Countermeasures: A Case Study of the Sea Area of Zhejiang Province. *Sustainability* **2023**, *15*, 4254. [CrossRef]
12. Ma, L.; Kang, L.; Tian, H.; Xu, D. Quantitative Evaluation of Ecological Cumulative Effects and Zoning Regulations for Prefectural Mining Units in China. *Chin. Geogr. Sci.* **2022**, *32*, 49–63. [CrossRef]
13. Yang, F.; Yang, X.; Wang, Z.; Sun, Y.; Zhang, Y.; Xing, H.; Wang, Q. Spatiotemporal Evolution of Production-Living-Ecological Land and Its Eco-Environmental Response in China's Coastal Zone. *Remote Sens.* **2023**, *15*, 3039. [CrossRef]
14. Zhou, W.; Xi, Y.; Zhai, L.; Li, C.; Li, J.; Hou, W. Zoning for Spatial Conservation and Restoration Based on Ecosystem Services in Highly Urbanized Region: A Case Study in Beijing-Tianjin-Hebei, China. *Land* **2023**, *12*, 733. [CrossRef]

15. Zhang, M.; Bao, Y.; Xu, J.; Han, A.; Liu, X.; Zhang, J.; Tong, Z. Ecological security evaluation and ecological regulation approach of East-Liao River basin based on ecological function area. *Ecol. Indic.* **2021**, *132*, 108255. [CrossRef]
16. Chen, Z.; Zhang, H.; Xu, M.; Liu, Y.; Fang, J.; Yu, X.; Zhang, S. A study on the ecological zoning of the Nantong coastal zone based on the Marxan model. *Ocean. Coast. Manag.* **2022**, *229*, 2292748. [CrossRef]
17. Chen, W.; Gu, T.; Xiang, J.; Luo, T.; Zeng, J.; Yuan, Y. Ecological restoration zoning of territorial space in China: An ecosystem health perspective. *J. Environ. Manag.* **2024**, *364*, 121371. [CrossRef]
18. Jiang, H.; Peng, J.; Zhao, Y.; Xu, D.; Dong, J. Zoning for ecosystem restoration based on ecological network in mountainous region. *Ecol. Indic.* **2022**, *142*, 109138. [CrossRef]
19. Xie, Y.; Wang, S.; Xiang, S.; Wang, Z.; Li, Y.; Wang, Z.; Zhou, M.; Wang, Y.; Gao, M. Ecological zoning and dynamic assessment of effectiveness in the Three Gorges Reservoir Area, China. *Ecol. Modell.* **2024**, *487*, 110563. [CrossRef]
20. Ahmed, S.; Meenar, M.; Alam, A. Designing a Blue-Green Infrastructure (BGI) Network: Toward Water-Sensitive Urban Growth Planning in Dhaka, Bangladesh. *Land* **2019**, *8*, 138. [CrossRef]
21. Aubrechtova, E.; Bydzovska, T.; Horak, J. Blue-green infrastructure and biodiversity: Urbanization and forestation have an important influence on bird diversity in water habitats. *Urban For. Urban Green.* **2024**, *91*, 128151. [CrossRef]
22. Xiang, Q.; Yu, H.; Huang, H.; Li, F.; Ju, L.; Hu, W.; Yu, P.; Deng, Z.; Chen, Y. Assessing the resilience of complex ecological spatial networks using a cascading failure model. *J. Clean. Prod.* **2024**, *434*, 140014. [CrossRef]
23. Song, S.; Wang, S.; Xu, D.; Gong, Y. Elemental evolution characteristics and influencing factors of green infrastructure network in karst mountain cities: A case study of Qianzhong urban agglomeration in Southwest China. *Ecol. Process.* **2024**, *13*, 54. [CrossRef]
24. Well, F.; Ludwig, F. Development of an Integrated Design Strategy for Blue-Green Architecture. *Sustainability* **2021**, *13*, 7944. [CrossRef]
25. Gu, J.; Lyu, X.; Fang, B.; Qiang, H.; Yuan, C. Study on Planning and Design of Blue-Green-Gray Transformation of Lakeside Cities to Deal with the Complex Urban Waterlogging Caused by Extreme Rainstorm. *Land* **2023**, *12*, 289. [CrossRef]
26. Xue, X.; Wang, Z.; Hou, S. NDVI-Based Vegetation Dynamics and Response to Climate Changes and Human Activities in Guizhou Province, China. *Forests* **2023**, *14*, 753. [CrossRef]
27. Wang, L.; Zhao, J.; Ai, D.; Chen, G.; Lin, Y. Integrating risk zoning and multifactor analysis: A strategic approach to ecological carbon sink management. *Ecol. Inform.* **2024**, *82*, 102671. [CrossRef]
28. Cheng, Z.; Zhao, T.; Zhu, Y.; Li, H. Evaluating the Coupling Coordinated Development between Regional Ecological Protection and High-Quality Development: A Case Study of Guizhou, China. *Land* **2022**, *11*, 1775. [CrossRef]
29. Su, Y.; Liao, S.; Ren, J.; Zhao, Z. Research on the decoupling relationship between water resources utilization and economic development at the county scale in Qian'nan Prefecture, Guizhou Province. *Front. Environ. Sci.* **2024**, *12*, 1347652. [CrossRef]
30. Wang, Y.; Zhang, Z.; Chen, X. Spatiotemporal change in ecosystem service value in response to land use change in Guizhou Province, southwest China. *Ecol. Indic.* **2022**, *144*, 109514. [CrossRef]
31. Wang, Q.; Bai, X. Spatiotemporal characteristics and driving mechanisms of land-use transitions and landscape patterns in response to ecological restoration projects: A case study of mountainous areas in Guizhou, Southwest China. *Ecol. Inform.* **2024**, *82*, 102748. [CrossRef]
32. Feng, Q.; Zhou, Z.; Chen, Q.; Zhu, C.; Zhu, M.; Luo, W.; Wang, J. Quantifying the extent of ecological impact from China's poverty alleviation relocation program: A case study in Guizhou Province. *J. Clean. Prod.* **2024**, *444*, 141274. [CrossRef]
33. Yuan, S.; Ren, Z.; Shan, X.; Deng, Q.; Zhou, Z. Seasonal different effects of land cover on urban heat island in Wuhan's metropolitan area. *Urban Clim.* **2023**, *49*, 101547. [CrossRef]
34. Guo, X.; Zhang, Y.; Guo, D.; Lu, W.; Xu, H. How does ecological protection redline policy affect regional land use and ecosystem services? *Environ. Impact Assess. Rev.* **2023**, *100*, 107062. [CrossRef]
35. Xie, Y.; Zhang, L.; Gong, X.; Liu, J.; Liao, X.; Dong, Y. Empirical study of the coupling relationship between biodiversity and environmental geology under different ecological status: Evidence from five typical areas in Guizhou, China. *Environ. Sci. Pollut. Res.* **2022**, *29*, 51398–51410. [CrossRef]
36. Song, S.; Wang, S.-H.; Shi, M.-X.; Hu, S.-S.; Xu, D.-W. Multiple scenario simulation and optimization of an urban green infrastructure network based on complex network theory: A case study in Harbin City, China. *Ecol. Process.* **2022**, *11*, 33. [CrossRef]
37. Leger, J.-B.; Daudin, J.-J.; Vacher, C. Clustering methods differ in their ability to detect patterns in ecological networks. *Methods Ecol. Evol.* **2015**, *6*, 474–481. [CrossRef]
38. Cai, Q.; Liu, J. Hierarchical Clustering of Bipartite Networks Based on Multiobjective Optimization. *IEEE Trans. Netw. Sci. Eng.* **2020**, *7*, 421–434. [CrossRef]
39. Peng, J.; Tang, H.; Su, C.; Jiang, H.; Dong, J.; Xu, D. Regarding reference state to identify priority areas for ecological restoration in a karst region. *J. Environ. Manag.* **2023**, *348*, 119214. [CrossRef]
40. Liu, H.; Tang, D. Ecological zoning and ecosystem management based on landscape ecological risk and ecosystem services: A case study in the Wuling Mountain Area. *Ecol. Indic.* **2024**, *166*, 112421. [CrossRef]

41. Bixler, R.P.; Lieberknecht, K.; Atshan, S.; Zutz, C.P.; Richter, S.M.; Belaire, J.A. Reframing urban governance for resilience implementation: The role of network closure and other insights from a network approach. *Cities* **2020**, *103*, 102726. [CrossRef]
42. Yu, H.; Gu, X.; Liu, G.; Fan, X.; Zhao, Q.; Zhang, Q. Construction of Regional Ecological Security Patterns Based on Multi-Criteria Decision Making and Circuit Theory. *Remote Sens.* **2022**, *14*, 527. [CrossRef]
43. Zhang, Z.; Wang, Q.; Feng, Y.; Sun, Y.; Liu, N.; Yan, S. The spatio-temporal evolution of spatial structure and supply-demand relationships of the ecological network in the Yellow River Delta region of China. *J. Clean. Prod.* **2024**, *471*, 143388. [CrossRef]
44. Ying, B.; Liu, T.; Ke, L.; Xiong, K.; Li, S.; Sun, R.; Zhu, F. Identifying the Landscape Security Pattern in Karst Rocky Desertification Area Based on Ecosystem Services and Ecological Sensitivity: A Case Study of Guanling County, Guizhou Province. *Forests* **2023**, *14*, 613. [CrossRef]
45. Zhao, H.; Jiang, X.; Gu, B.; Wang, K. Evaluation and Functional Zoning of the Ecological Environment in Urban Space-A Case Study of Taizhou, China. *Sustainability* **2022**, *14*, 6619. [CrossRef]
46. Zhang, H.; Li, J.; Tian, P.; Pu, R.; Cao, L. Construction of ecological security patterns and ecological restoration zones in the city of Ningbo, China. *J. Geogr. Sci* **2022**, *32*, 663–681. [CrossRef]
47. Ye, J.; Hou, Z.; Ming, H.; Cheng, Y. Ecological security pattern-based simulation for land use structure change: A case study in Ezhou City, China. *Front. Earth Sci.* **2021**, *15*, 526–542. [CrossRef]
48. Xu, C.; Yang, G.; Wan, R.; Ou, W.; Wang, P. Toward ecological function zoning and comparison to the Ecological Redline Policy: A case study in the Poyang Lake Region, China. *Environ. Sci. Pollut. Res.* **2021**, *28*, 40178–40191. [CrossRef]
49. Guo, K.; Niu, X.; Wang, B. A GIS-Based Study on the Layout of the Ecological Monitoring System of the Grain for Green Project in China. *Forests* **2023**, *14*, 70. [CrossRef]
50. Jiao, M.; Wang, Y.; Hu, M.; Xia, B. Spatial deconstruction and differentiation analysis of early warning for ecological security in the Pearl River Delta, China. *Sustain. Cities Soc.* **2021**, *64*, 102557. [CrossRef]
51. Ju, L.; Luo, M.; Luo, H.; Ma, Z.; Lu, X.; Jiang, G. Hydropower Functional Zoning with Crowdsourced Geospatial Data: A Case Study in Sichuan Province. *Appl. Sci.* **2023**, *13*, 7260. [CrossRef]
52. Li, Y.; Abbas, Z.; Chen, D.; Zhu, Z.; Guo, H.; Zhao, Y. The Eco-Environmental Changes in Typical Coastal Zones of Southern China From 1987 to 2020: A Case Study of Guangdong Coastal Counties. *Front. Environ. Sci.* **2022**, *10*, 874364. [CrossRef]
53. Li, Z.; Yang, F.; Zhong, J.; Zhao, J. Self-Organizing Feature Zoning and Multiple Hotspots Identification of Ecosystem Services: How to Promote Ecological Refined Management of Chengdu-Chongqing Urban Agglomeration. *J. Urban Plan. Dev.* **2023**, *149*, 04022049. [CrossRef]
54. Shao, S.; Yang, Y. Identification of ecological improvement zones in different ecological functional zones in northwest Hubei, China. *Ecol. Indic.* **2023**, *155*, 111032. [CrossRef]

International Journal of *Geo-Information*

*Article*

# Spatiotemporal Changes and Trade-Offs/Synergies of Ecosystem Services in the Qin-Mang River Basin

**Jiwei Zhao [1], Luyao Wang [1,*], Dong Jia [2] and Yaowen Wang [1]**

1 Water Conservancy College, North China University of Water Resources and Electric Power, Zhengzhou 450046, China; zhaojiwei@stu.ncwu.edu.cn (J.Z.); z202210010188@stu.ncwu.edu.cn (Y.W.)
2 College of Civil Engineering and Transportation, North China University of Water Resources and Electric Power, Zhengzhou 450046, China; z202210030407@stu.ncwu.edu.cn
* Correspondence: z202210010075@stu.ncwu.edu.cn; Tel.: +86-152-3716-0597

**Abstract:** The Qin-Mang River Basin is an important biodiversity conservation area in the Yellow River Basin. Studying the spatiotemporal changes in its ecosystem services (ESs) and the trade-offs and synergies (TOSs) between them is crucial for regional ecological protection and high-quality development. This study, based on land use type (LUT), and meteorological and soil data from 1992 to 2022, combined with the InVEST model, correlation analysis, and spatial autocorrelation analysis, explores the impacts of land use/land cover changes (LUCCs) on ESs. The results show that: (1) driven by urbanization and economic development, the expansion of built-up areas has replaced cultivated land and forests, with 35,000 hectares of farmland lost, thereby increasing pressure on ESs; (2) ESs show an overall downward trend, habitat quality (HQ) has deteriorated, carbon storage (CS) remains stable but the area of low CS has expanded, and sediment delivery ratio (SDR) and water yield (WY) fluctuate due to human activities and climate influence; (3) the TOSs of ESs change dynamically, with strong synergies among HQ, CS, and SDR. However, in areas with water scarcity, the negative correlation between HQ and WY has strengthened; (4) spatial autocorrelation analysis reveals that in 1992, significant positive synergies existed between ESs in the northern and northwestern regions, with WY negatively correlated with other services. By 2022, accelerated urbanization has intensified trade-off effects in the southern and eastern regions, leading to significant ecological degradation. This study provides scientific support for the sustainable management and policymaking of watershed ecosystems.

**Keywords:** ecosystem services; trade-offs and synergies; spatial heterogeneity analysis; land use type transition; Qin-Mang River Basin

## 1. Introduction

Ecosystem services (ESs) refer to the direct and indirect benefits provided by natural ecosystems to human society through various pathways. As an essential support system for human survival and sustainable development, these services encompass multiple aspects, including resource supply and ecological regulation, and play an indispensable role in regional ecological security and global ecological stability [1,2]. However, with the intensification of climate change and the rapid development of human activities, approximately 60% of global ecosystems have undergone degradation, posing a severe threat to the stability and sustainability of ecosystems [3,4]. For instance, the shrinkage of wetlands not only leads to the deterioration of water quality but also significantly accelerates the loss of biodiversity [5,6]. This trend underscores the urgency of formulating scientifically sound ecological protection and management policies, the effectiveness of which depends on a comprehensive understanding of the dynamic changes in ESs and their interrelationships [7].

As one of the global biodiversity hotspots, China's ecosystems are under immense pressure from rapid urbanization and economic growth. Land use/land cover changes (LUCCs) have been widely recognized as the main drivers of dynamic changes in ESs [8]. In response to this challenge, the Chinese government has implemented ecological protection policies such as "Grain-for-Green" and "Ecological Protection Red Lines". These measures have alleviated ecological degradation to some extent, but regional imbalances between the supply and demand of ESs remain prominent, especially in ecologically fragile areas and regions undergoing rapid urbanization [9,10]. The Yellow River Basin, as a key ecological barrier in China, maintains a close relationship between the spatiotemporal dynamics of its ESs and human activities, making the study of ecosystem service changes in this area highly significant [11].

The Qin-Mang River Basin is located within the biodiversity conservation area of the Yellow River Basin and serves as a crucial part of the Yellow River wetlands with key ecological functions. Its unique geographical location makes it of vital importance to regional ecological protection. Existing studies primarily focus on the following aspects: water resource protection, biodiversity habitat quality (HQ), and the impacts of LUCCs on ESs. For example, previous studies have assessed the role of water resource management in watershed ecological stability through hydrological models, revealing the impacts of climate change and human activities on the water supply–demand balance [12]. Additionally, studies have explored the effects of LUCCs on HQ using biodiversity indicators, finding that urbanization significantly reduces HQ [13]. However, existing research lacks in-depth exploration of the trade-offs and synergies (TOSs) between ESs, especially regarding the systematic study of spatial heterogeneity and complex interaction mechanisms.

In response to the research gaps mentioned above, this study addresses the following scientific questions: (1) what spatiotemporal changes have occurred in ESs in the Qin-Mang River Basin over the past 30 years? (2) How have the TOSs between different ESs evolved? (3) How have LUCCs affected the spatiotemporal distribution of ESs? To address these questions, this study integrates land use type (LUT) data from 1992 to 2022, combined with the InVEST model, correlation analysis, and spatial autocorrelation analysis, to systematically evaluate the dynamic changes and spatiotemporal heterogeneity of ESs in the Qin-Mang River Basin and reveal the TOSs between these services. This study focuses on four main ESs: HQ, carbon storage (CS), sediment delivery ratio (SDR), and water yield (WY). These services were selected based on their importance for regional ecological security and their close relationship with LUCCs [14,15]. Through quantitative analysis, this study reveals the spatiotemporal dynamic characteristics of these ESs and explores the driving roles of urbanization and climate change on ESs.

The innovations of this study are reflected in the following aspects: (1) a systematic study of the dynamic changes in ESs in the Qin-Mang River Basin, an important ecological barrier region; (2) quantifying the TOSs between ESs using the InVEST model and spatial analysis methods; (3) a detailed analysis of the impact of LUCCs on the spatial heterogeneity and interaction mechanisms of ESs; (4) providing scientific support for optimizing regional land use planning and formulating ecological protection policies, contributing to the high-quality development of the Yellow River Basin. In conclusion, this study not only fills the research gap on ESs in the Qin-Mang River Basin but also provides important theoretical support and practical references for regional ecological protection and management.

## 2. Materials and Methods

### 2.1. Study Area

The Qin-Mang River Basin, located in the northwest of Henan Province, refers to the combined basin of the Qin River and its tributary, the Mang River, with a total area of approximately 306,240 hectares. This basin is a key energy and chemical industry base within the Yellow River Basin and is designated as the core area for the Ecological Protection and Restoration Project for the Mountains, Rivers, Forests, Farmlands, Lakes, Grasslands, and Deserts in the South Taihang region of Henan Province, playing a significant role as an ecological barrier. Geographically, the basin spans from 112°20′ to 113°39′ east longitude and from 34°49′ to 35°21′ north latitude. It connects the North China Plain with the Loess Plateau and is located between the second and third geomorphic steps (Figure 1). The climate in the region is classified as a warm, temperate semi-humid climate, with an average annual temperature of approximately 14.6 °C and an annual precipitation of about 641.7 mm, concentrated primarily in the summer months. The basin is rich in vegetation resources, home to 1609 species of higher plants, and supports several nationally protected species. According to the "Ecological Protection and Restoration Plan for Mountains, Rivers, Forests, Farmlands, Lakes, Grasslands, and Deserts in Henan Province", the Qin-Mang River Basin has been identified as a key area for ecological restoration, involving measures such as forest management, soil and water conservation, and wetland protection. The biodiversity and ESs in this region play a crucial role in ensuring regional ecological security.

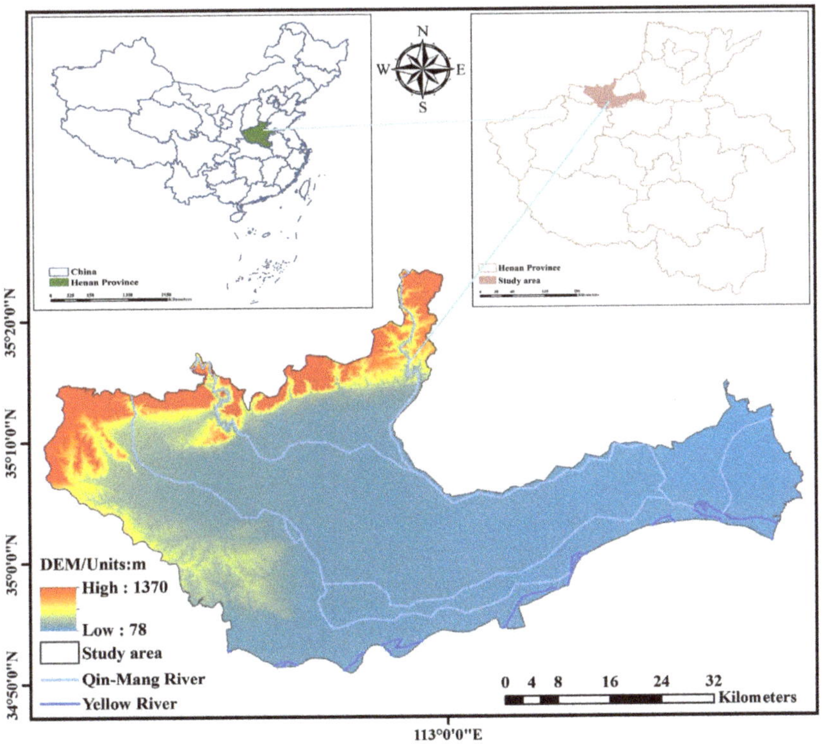


Figure 1. Geographic location and elevation map of the study area.

### 2.2. Data Sources and Preprocessing

This study's analysis is primarily based on LUT, Digital Elevation Model (DEM), administrative boundaries, natural environmental data, and socio-economic data. The selected datasets are intended to comprehensively reflect the ecological and socio-economic characteristics of the Qin-Mang River Basin and provide a reliable foundation for the research. Data sources and preprocessing steps are detailed in Table 1. During the data preprocessing phase, consistency in resolution and uniformity in the coordinate system across different datasets were ensured to enhance the accuracy and validity of the analysis results. All raster data were transformed into the WGS_1984_UTM_zone_49N coordinate system and processed to a uniform spatial resolution. Specific preprocessing steps included data clipping, resampling, reclassification, and spatial analysis operations.

**Table 1.** Main data sources.

| Data Name | Resolution | Data Source | Data Preprocessing |
|---|---|---|---|
| DEM | 90 m | Geospatial Data Cloud (http://www.gscloud.cn/, accessed on 13 July 2024) | The ArcGIS Clip tool was used to extract data for the study area and generate slope and aspect data. |
| LUCCs | 30 m | Resource and Environment Science and Data Center (https://www.resdc.cn/, accessed on 19 July 2024) | The ArcGIS Reclassification tool was used to reclassify land use data into six categories based on the primary land classification standards and then clipped to the study area. |
| Precipitation, Temperature, and Potential Evapotranspiration Data | 1 km | National Meteorological Science Data Centre (http://data.cma.cn, accessed on 30 July 2024) | Monthly data were processed in ArcGIS to obtain annual average data and then clipped to the study area. |
| Soil Type, Vegetation Type, and NDVI | 1 km | The National Tibetan Plateau Data Center (http://data.tpdc.ac.cn/zh-hans/, accessed on 5 August 2024) | Data were clipped to the study area using ArcGIS. |
| Nighttime Light Index, GDP, and Population | 1 km | Resource and Environment Science and Data Center (https://www.resdc.cn/, accessed on 2 August 2024) | Data were clipped to the study area using ArcGIS. |
| Railway, Water System, and Administrative Boundary Vector Data | - | National Catalogue Service for Geographic Information (https://www.webmap.cn/, accessed on 10 July 2024) | The ArcGIS Clip tool was used to extract the study area, and the Euclidean Distance tool was applied to calculate the distance to railways and water systems. |

### 2.3. Research Methods

This study established a comprehensive analysis framework to systematically assess the impact of LUCCs on ESs in the Qin-Mang River Basin from 1992 to 2022, with a focus on revealing the TOSs between ESs and their spatial characteristics. Through the land use transfer matrix, this study quantified LUT transformation characteristics and rates at different time periods, analyzing the driving forces of urbanization and economic development on natural land. Coupled with the InVEST model, this study quantified the spatiotemporal dynamics of key ESs such as HQ, CS, SDR, and WY, providing data support for understanding the impact of LUCCs on ESs. Correlation analysis was used to reveal the TOSs between ESs, and bivariate local Moran's I (BLI) was employed to analyze their spatial clustering and dispersion, identifying key hotspot and coldspot areas. The research framework is shown in Figure 2.

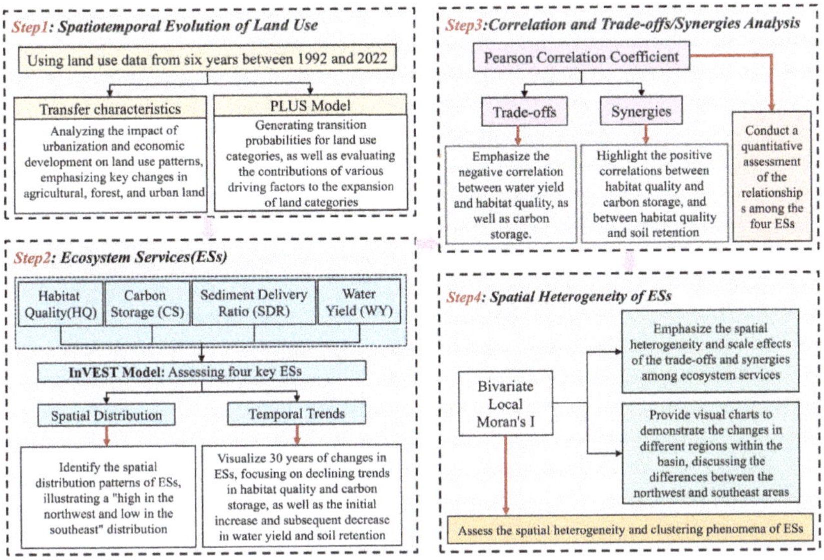


**Figure 2.** Research framework.

2.3.1. Land Use Transfer Matrix

The land use transfer matrix is used to quantify the patterns of LUT and their change rates over different time periods, revealing the spatiotemporal dynamics of land use. In this study, ArcGIS spatial analysis tools were employed to construct the land use transfer matrix and analyze the trends in land use conversion across different time periods. The calculation formula is as follows [16].

$$A_{ij} = \begin{bmatrix} A_{11} & A_{12} & \cdots & A_{1n} \\ A_{21} & A_{22} & \cdots & A_{2n} \\ \cdots & \cdots & \cdots & \cdots \\ A_{n1} & A_{n2} & \cdots & A_{nn} \end{bmatrix} \quad (1)$$

$A_{ij}$ denotes the area of land type $i$ transitioning to land type $j$. The $i$-th row corresponds to the outflow from land type $i$, while the $j$-th column indicates the destination source, representing the inflow to land type $j$.

2.3.2. PLUS Model

The Patch-generating Land Use Simulation (PLUS) model integrates the Land Expansion Analysis Strategy (LEAS) module to assess the spatiotemporal dynamics of LUCCs and simulate the potential for future land use expansion [17]. The LEAS module generates suitability surfaces for LUT transitions using historical data and various driving factors, representing the probability of transitions on a scale from 1 to 255. The input data includes topographic factors (elevation, slope, aspect), socio-economic indicators (GDP, population, nighttime light index), and environmental variables (average annual precipitation, soil types, vegetation types, NDVI, distance to roads, distance to water systems). The data were standardized and adjusted to a consistent spatial resolution to meet the model's requirements. The expansion probability map generated by the LEAS module was validated through visual analysis and found to be consistent with known urbanization and land expansion trends, providing references for land management policies.

#### 2.3.3. InVEST Model

Habitat Quality (HQ)

The HQ module of the InVEST model is commonly used to assess an ecosystem's ability to support species habitats. This module calculates the HQ index, reflecting the impact of different LUTs and management scenarios on HQ. The formula is as follows [18,19].

$$Q_{xj} = H_j \left[ 1 - \left( \frac{D_{xj}^z}{D_{xj}^z + K^z} \right) \right] \tag{2}$$

$Q_{xj}$ represents the HQ index 0f grid cell $x$; $H_j$ is the habitat suitability for the land use type, ranging from 0 to 1; $Z$ is the scaling constant, typically set at 2.5; $K$ is the half-saturation constant, which is 0.5 in this study; $D_{xj}$ denotes the habitat degradation index, indicating the extent of habitat degradation under stress conditions.

Based on previous research and the InVEST model user guide, and in combination with the actual conditions of the study area, a habitat threat factor and threat level evaluation table was constructed (Table 2), as well as a sensitivity evaluation table for LUTs to threat factors (Table 3) [20–22].

**Table 2.** Threat factors and threat levels.

| Threat Factors | Farthest Threat Distance | Threat Degree | Declining Type |
|---|---|---|---|
| Cropland | 4 | 0.4 | Linear |
| Impervious | 8 | 0.8 | Exponential |
| Railways | 7 | 0.7 | Exponential |
| Barren | 5 | 0.5 | Exponential |

**Table 3.** Sensitivity to threat factors.

| LUTs | Habitat Suitability | Cropland | Impervious | Barren |
|---|---|---|---|---|
| Cropland | 0.3 | 0 | 0.8 | 0.4 |
| Forest | 0.9 | 0.7 | 0.8 | 0.3 |
| Grassland | 0.9 | 0.5 | 0.7 | 0.6 |
| Water | 0.8 | 0.4 | 0.7 | 0.4 |
| Impervious | 0 | 0 | 0 | 0 |
| Barren | 0.4 | 0.4 | 0.6 | 0 |

Carbon Storage (CS)

CS is a key indicator of an ecosystem's carbon sequestration and storage capacity, reflecting its role in mitigating global climate change. This module reveals the potential impact of LUCCs and management measures on CS. The formula is as follows [23].

$$C_{total} = C_{above} + C_{below} + C_{soil} + C_{dead} \tag{3}$$

$C_{total}$ represents the total CS (Mg/ha); $C_{above}$ is the CS in aboveground biomass; $C_{below}$ is the CS in belowground biomass; $C_{soil}$ is the CS in soil; $C_{dead}$ is the CS in dead organic matter.

This study constructed a carbon pool density table (Table 4) in consideration of the actual conditions of the study area [24,25].

**Table 4.** Carbon pool density (Mg/ha).

| LUTs | C_Above | C_Below | C_Soil | C_Dead |
|---|---|---|---|---|
| Cropland | 4.85 | 0.92 | 58.2 | 2.84 |
| Forest | 20.92 | 7.53 | 67.3 | 54.2 |
| Grassland | 1.63 | 8.48 | 60.2 | 2.19 |
| Water | 0 | 0 | 62.1 | 0 |
| Impervious | 0 | 0 | 60 | 0 |
| Barren | 0 | 0 | 53.3 | 0 |

Sediment Delivery Ratio (SDR)

SDR is a key indicator in evaluating ESs, mainly measuring the ecosystem's function in preventing soil erosion and maintaining soil quality. The application of this module helps analyze the ecosystem's contribution to maintaining soil health and reducing water and soil loss under different land management scenarios. The formula is as follows [26].

$$RKLS = R \times K \times LS \tag{4}$$

$$USLE = R \times K \times LS \times P \times C \tag{5}$$

$$SDR = RKLS - USLE \tag{6}$$

$RKLS$ indicates potential soil erosion, $USLE$ indicates actual soil erosion, and $SDR$ indicates SDR; $R$ represents the rainfall erosion factor, $K$ indicates the soil erodibility factor, $LS$ denotes the slope length factor, $P$ indicates the soil and water conservation measures factor, and $C$ indicates the vegetation cover and management factor.

Based on previous studies, ecological literature, and standard parameters from the InVEST model, this research constructed vegetation cover factor C and soil conservation factor P parameters (Table 5) [27] for the Qin-Mang River Basin.

**Table 5.** Vegetation coverage factor (C) and soil conservation measure factor (P).

| LUTs | usle_c | usle_p |
|---|---|---|
| Cropland | 0.06 | 0.4 |
| Forest | 0.04 | 1 |
| Grassland | 0.06 | 1 |
| Water | 0 | 0 |
| Impervious | 0 | 0 |
| Barren | 1 | 1 |

Water Yield (WY)

WY is an important indicator for assessing the water supply capacity of a region. The WY evaluation comprehensively considers factors such as average annual precipitation, potential evapotranspiration, effective vegetation water content, and soil depth, reflecting the region's water input, storage, and release processes. The assessment of WY helps researchers to understand the impact of LUCCs on water resources. The calculation formula is as follows [28].

$$Y(x) = (P(x) - AET(x)) \cdot \frac{1}{A} \tag{7}$$

$$AET(x) = P(x) \cdot \frac{PET(x)}{PET(x) + P(x)PET(x)} \tag{8}$$

$Y(x)$ is the average annual WY for cell $x$; $P(x)$ is the annual precipitation for cell $x$; $AET(x)$ is the annual actual evapotranspiration for cell $x$; $A$ is the watershed area used to convert water quantity units to volume; and $PET(x)$ is the potential evapotranspiration for cell $x$; all units are in mm.

Based on previous research, the ecological literature, and standard parameters from the InVEST model, this study constructed biophysical parameters for each LUT (Table 6) [29].

**Table 6.** Biophysical parameters for various LUTs.

| LUTs | Root_Depth | Kc | LULC_Veg |
|---|---|---|---|
| Cropland | 2000 | 0.68 | 1 |
| Forest | 7000 | 0.95 | 1 |
| Grassland | 1700 | 0.62 | 1 |
| Water | 100 | 1 | 0 |
| Impervious | 300 | 0.2 | 0 |
| Barren | 100 | 0.2 | 0 |

#### 2.3.4. Trade-Off and Synergy (TOS) Effect Analysis

The TOS effects of ESs refer to the relationships of mutual influence between different services. The trade-off effect is characterized by an increase in one ES accompanied by a decrease in another, such as agricultural expansion potentially leading to a decline in habitat quality. In contrast, the synergy effect refers to the simultaneous increase or decrease in two or more ESs, such as forest vegetation restoration, which can simultaneously enhance CS and improve soil and water conservation [30]. This study explores the TOS effects between ESs through Pearson correlation analysis and BLI analysis.

##### Pearson Correlation Analysis

Pearson correlation is a widely used statistical method for quantitatively assessing the linear relationship between two variables. The correlation coefficient ranges from −1 to 1, with a value of 1 indicating a perfect positive correlation, −1 indicating a perfect negative correlation, and 0 indicating no correlation. Specifically, a larger positive correlation value suggests the potential for a synergy effect between ESs, while a negative value indicates the possibility of a trade-off effect. All correlation analyses were conducted using Origin software, with the formula as follows [31].

$$r = \frac{\sum_{i=1}^{n} (x_i - \overline{x})(y_i - \overline{y})}{\sqrt{\sum_{i=1}^{n} (x_i - \overline{x})^2} \cdot \sqrt{\sum_{i=1}^{n} (y_i - \overline{y})^2}} \tag{9}$$

$r$ is the Pearson correlation coefficient; $x_i$ and $y_i$ are the $i-th$ data points of the two variables; $\overline{x}, \overline{y}$ are the means of the two variables; and $n$ is the sample size.

##### Bivariate Local Moran's I (BLI)

To further investigate the spatial distribution and interactions of ESs, this study employed BLI to explore spatial autocorrelation and clustering effects. Positive values in the Moran scatter plot indicate the presence of synergy effects between services, suggesting that they tend to increase or decrease together. Negative values, on the other hand, indicate trade-off effects, where the enhancement of one service is associated with the decline of another. The clustering map further reveals spatial aggregation effects and heterogeneity patterns. All calculations were performed using GeoDa software, with the formula as follows [32].

$$I_i = \frac{n(x_i - \overline{x}) \sum_{j=1}^{n} W_{ij}(x_j - \overline{x})}{\sum_{i=1}^{n} (x_i - \overline{x})^2} \tag{10}$$

$I_i$ is the BLI index, $n$ is the total number of spatial units, $x_i$ and $x_j$ represent the ESs quantities of regions $i$ and $j$, $\overline{x}$ is the mean of the observed values, and $\omega_{ij}$ is the spatial weight.

## 3. Results and Analysis

### 3.1. Spatiotemporal Evolution of Land Use/Land Cover Change (LUCC)

LUCC in the Qin-Mang River Basin has significantly impacted the region's ESs over the past 30 years, particularly in terms of HQ and CS. This study analyzed LUT distribution maps from six time periods between 1992 and 2022 (Figure 3) using ArcGIS, revealing the spatiotemporal patterns of LUCC. The results indicate that the dominant LUTs in the basin are cropland and forestland, which together account for over 74% of the total area, while built-up land occupies approximately 10%. Notably, the expansion of built-up land is a significant trend, mainly encroaching upon cropland and forestland, while changes in water bodies are minimal. These changes reflect the profound impact of rapid urbanization and economic growth on the land use pattern of the basin, especially the ecological effects resulting from the expansion of built-up land.

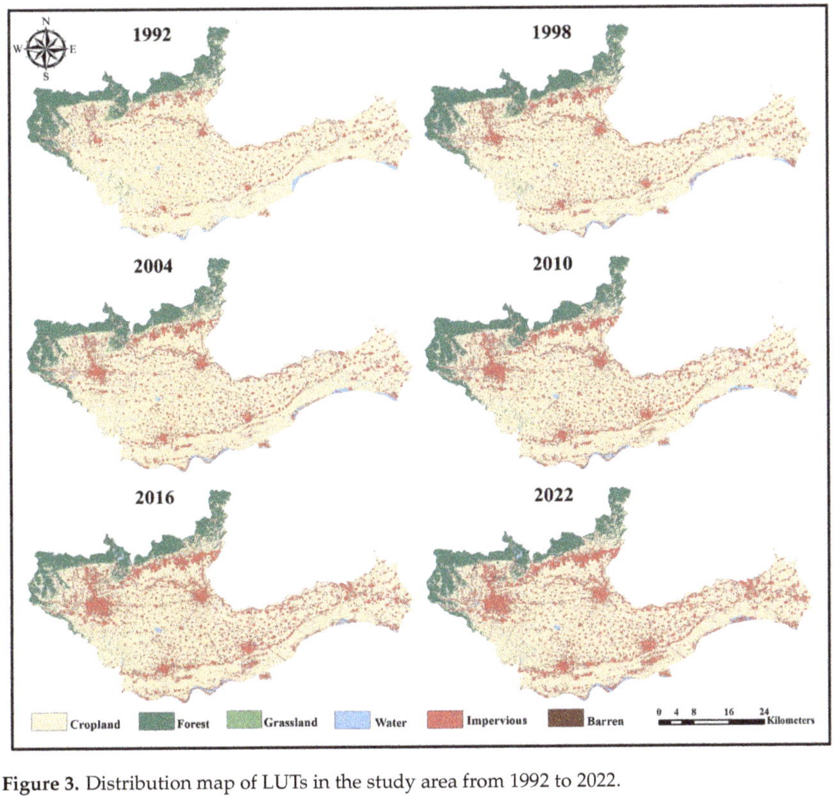


**Figure 3.** Distribution map of LUTs in the study area from 1992 to 2022.

Corresponding to the spatial distribution changes in land use, the temporal evolution of land use patterns also reveals spatial restructuring during the urbanization process. The study shows that the expansion of built-up land is primarily concentrated in the north-eastern and western regions of the basin. This pattern indicates that, with the acceleration of economic growth and urbanization, these areas have transitioned from agricultural land to urban land. The conversion of cropland and forestland has not only altered the land use structure but has also had a profound impact on regional ESs, particularly in terms of HQ and CS, reflecting the immense pressure that urbanization exerts on the ecological environment.

Further analysis of the land use transfer matrix (Table 7) and Figure 4 reveal the processes of LUT transitions between 1992 and 2022. Specifically, cropland decreased by 35,000 ha, mainly being converted into built-up land, reflecting the dominant role of economic development as a driving force behind LUCC. The loss of cropland resources poses a potential threat to food security and reduces its function as a carbon sink. Forestland area decreased by approximately 244.62 ha, particularly in the central and western regions, leading to a significant decline in biodiversity and ecological functions, especially in terms of forest ESs and species habitats. The Sankey diagram (Figure 5) clearly illustrates the conversion trends of different LUTs through the width and direction of the flow, providing a visual representation of the macro dynamics of LUCC.

**Table 7.** Land use transfer matrix for the study area from 1992 to 2022 (ha).

| 1992 | 2022 | | | | | | | |
|---|---|---|---|---|---|---|---|---|
| | **Cropland** | **Forest** | **Grassland** | **Water** | **Impervious** | **Barren** | **Total** | **Transfer Out** |
| Cropland | 195,899.40 | 1633.59 | 115.47 | 1238.85 | 34,946.28 | 4.77 | 233,838.36 | 37,938.96 |
| Forest | 5282.46 | 30,815.46 | 64.98 | 14.67 | 244.62 | 0 | 36,422.19 | 5606.73 |
| Grassland | 1859.31 | 320.76 | 173.52 | 14.58 | 201.06 | 0 | 2569.23 | 2395.71 |
| Water | 1008.72 | 12.96 | 3.78 | 832.77 | 808.83 | 0.09 | 2667.15 | 1834.38 |
| Impervious | 240.12 | 0.18 | 2.70 | 260.82 | 29,670.93 | 0.18 | 30,174.93 | 504.00 |
| Barren | 0 | 0 | 0 | 0 | 0.54 | 0 | 0.54 | 0.54 |
| Total | 204,290.01 | 32,782.95 | 360.45 | 2361.69 | 65,872.26 | 5.04 | 305,672.40 | / |
| Transfer In | 8390.61 | 1967.49 | 186.93 | 1528.92 | 36,201.33 | 5.04 | / | / |

Figure 6 displays the probability of future development for different land types, based on predictions generated by the PLUS model and spatial analysis with ArcGIS, revealing the trends in the expansion of different LUTs. The results show that, in the context of rapid urbanization and economic growth, the probability of expansion of built-up land is high, demonstrating the strong influence of the urbanization process on land use patterns. In contrast, the probability of expansion of cropland and forestland is declining, indicating that these natural land types will continue to face pressure from urbanization. Particularly in areas of rapid economic development, the compression of cropland and forestland will further intensify the impact on ecosystem service functions.

Figure 7a–c further illustrates the contributions of different regional characteristics to the driving factors of newly developed land types. As shown in Figure 7a, in regions with higher GDPs, the expansion of built-up land is most prominent, reflecting the strong driving effect of economic development on land expansion. Figure 7b shows that, in flat terrain areas, cropland expansion is most significant, while regions with steeper slopes face land use bottlenecks due to topographic limitations. Figure 7c reveals that, in areas with higher NDVI, forestland expansion is trending, reflecting the positive land restoration trends in regions with better vegetation coverage, influenced by ecological restoration and environmental protection measures. Overall, the economic, topographic, and ecological characteristics of different regions interact to shape the spatial patterns of LUT transformation.

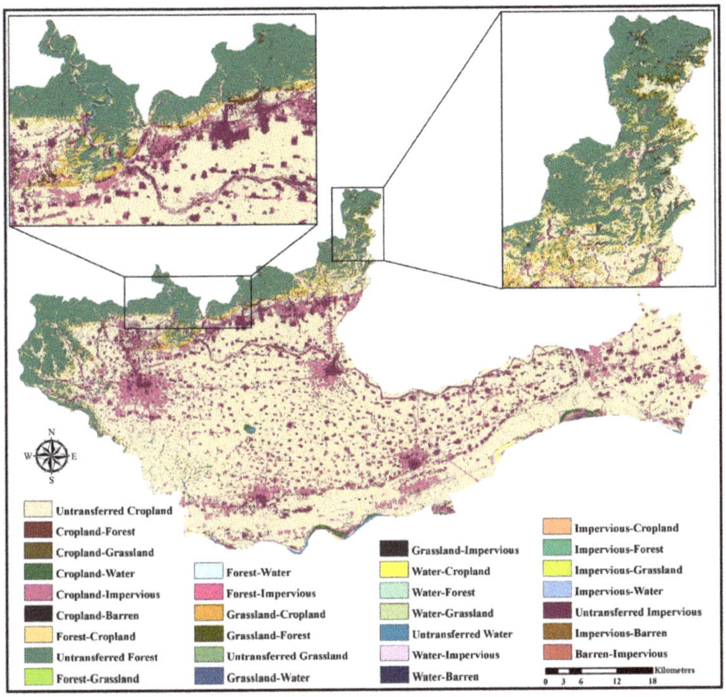


**Figure 4.** LUTs from 1992 to 2022.

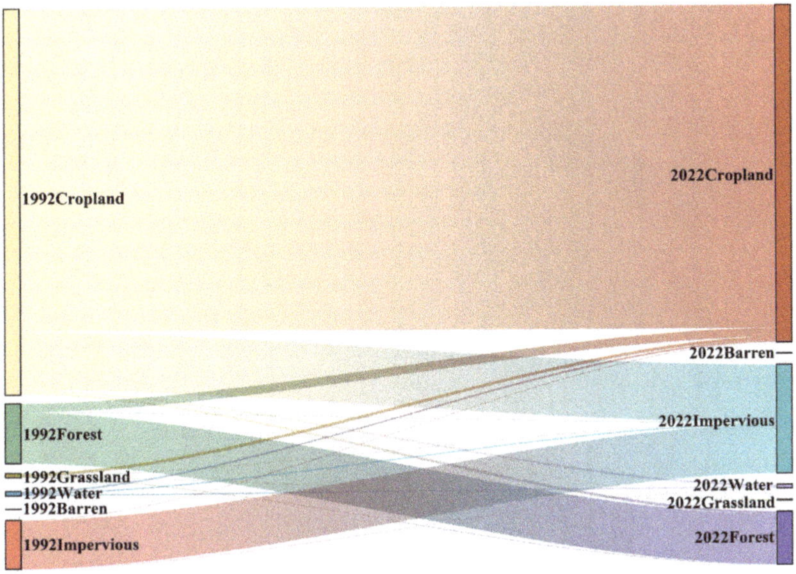


**Figure 5.** Sankey diagram of LUTs from 1992 to 2022.

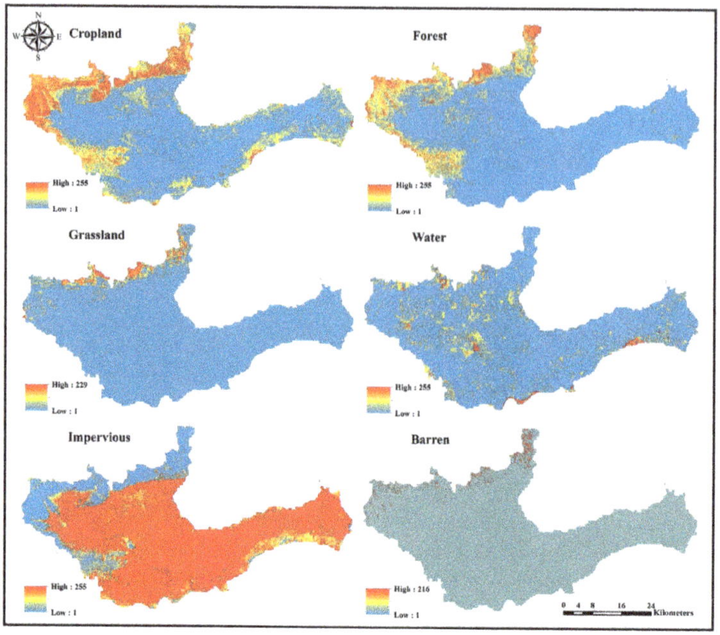


**Figure 6.** Development probabilities of various LUTs.

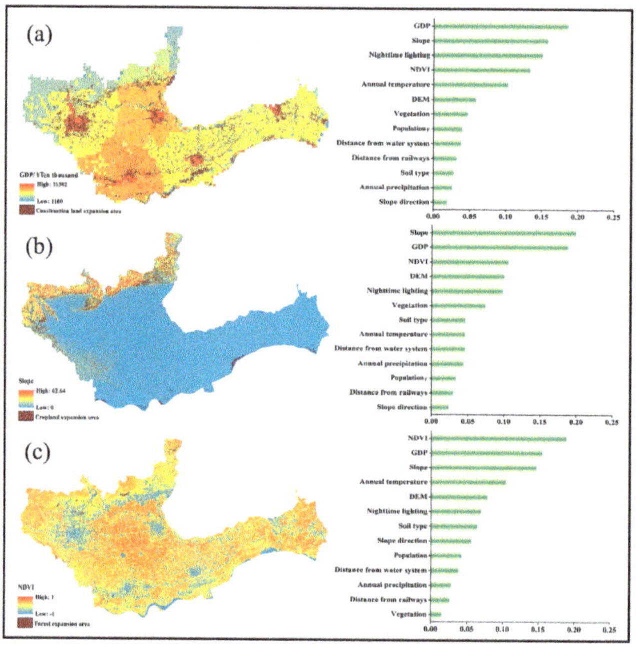


**Figure 7.** (**a**) Relationship between newly added built-up land and GDP level, along with the contribution of expansion driving factors; (**b**) relationship between newly added arable land and slope, along with the contribution of expansion driving factors; (**c**) relationship between newly added forest land and NDVI values, along with the contribution of expansion-driving factors.

*3.2. Spatiotemporal Evolution of Ecosystem Services (ESs)*

3.2.1. Spatiotemporal Evolution of Habitat Quality (HQ)

Based on the InVEST model, the evaluation of HQ in the Qin-Mang River Basin from 1992 to 2022 (Figure 8) indicates a persistent decline in HQ across the basin. The HQ index decreased from 0.34 in 1992 to 0.29 in 2022, with the spatial variation in HQ increasing, as reflected by the rise in standard deviation from 0.16 to 0.19. This change reflects the overall degradation of the ecological environment, particularly in the southern and eastern regions where HQ deterioration is more pronounced. Figure 8 clearly illustrates this trend, with some mountainous areas and nature reserves maintaining relatively good HQ, but with the majority of regions, especially those experiencing rapid urbanization, showing significant declines in HQ.

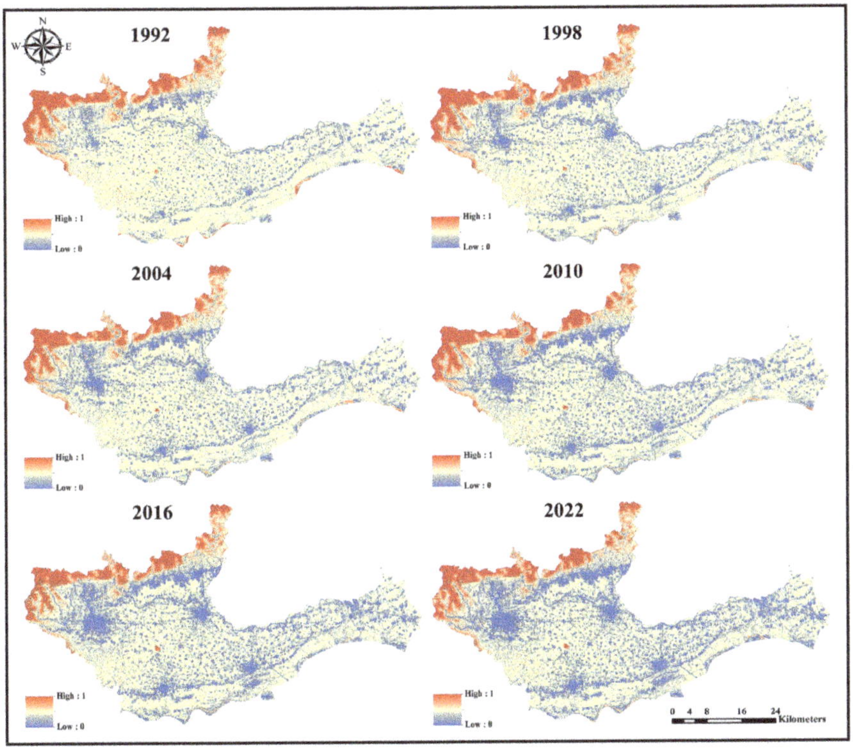


**Figure 8.** Temporal and spatial distribution map of HQ index.

The primary cause of this decline is closely linked to the accelerated urbanization process and the expansion of built-up land. Specifically, urbanization has intensified human interference with natural ecosystems, leading to the disruption of habitat continuity and integrity, especially in ecologically fragile areas where degradation has significantly accelerated. Additionally, LUCCs, particularly the conversion of agricultural and forested lands, have further exacerbated the degradation of HQ. The relationship between the reduction in agricultural land and the expansion of urban areas in the basin reflects the direct impact of human activities on ecosystem health.

3.2.2. Spatiotemporal Evolution of Carbon Storage (CS)

From 1992 to 2022, CS in the Qin-Mang River Basin exhibited significant spatial variation (Figure 9). Overall, CS follows a "high in the northwest, low in the southeast" distribution pattern. The northern and northwestern regions have higher CS, primarily due to the good ecological conditions of the western Henan mountains and the Taihang Mountains, coupled with higher vegetation coverage. In contrast, the southeastern region, characterized by extensive built-up areas and water bodies, has lower CS, weaker vegetation coverage, and a reduced carbon sequestration capacity. This spatial distribution difference is closely related to both the natural conditions and the spatial distribution of human activities in the basin.

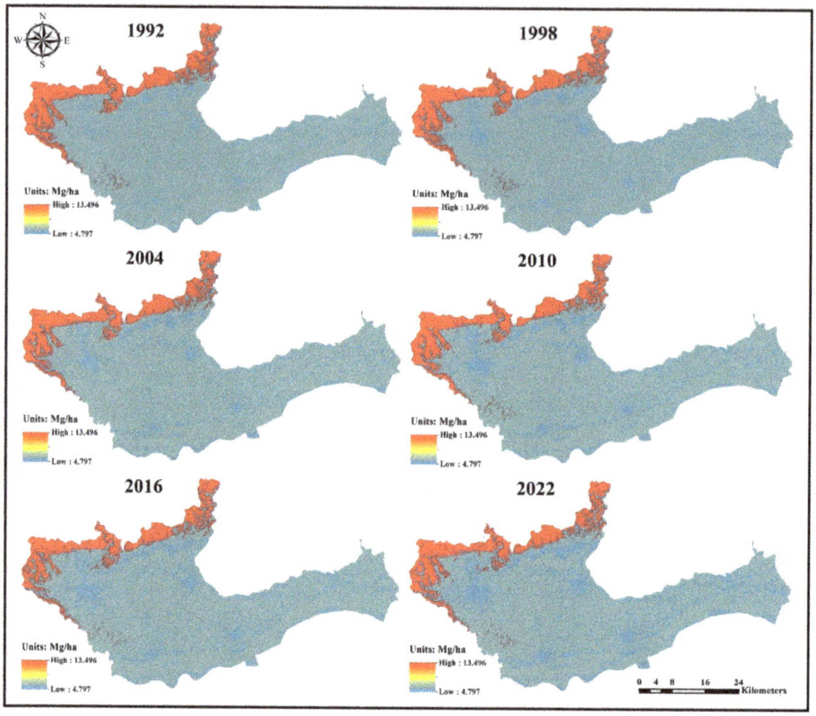


**Figure 9.** Temporal and spatial distribution map of CS.

Time-series analysis shows that, while the CS in high-carbon areas remained stable, the extent of low-carbon areas gradually expanded, particularly in regions with accelerating urbanization. The trend in Figure 9 indicates that urban expansion has led to significant LUCCs, particularly the continuous increase in built-up land, resulting in a reduction in CS in some areas. Notably, in the central and southern regions, as built-up land expanded, areas originally rich in CS were gradually transformed into low-carbon areas. This phenomenon suggests that rapid urbanization not only altered land use patterns but also impacted the spatial distribution of regional CS and ecological functions.

#### 3.2.3. Spatiotemporal Evolution of Sediment Delivery Ratio (SDR)

SDR in the Qin-Mang River Basin exhibited significant spatial variability between 1992 and 2022 (Figure 10). Overall, the distribution follows a "high in the northwest, low in the southeast" pattern. The southern section of the Taihang Mountains in the north, with good vegetation and significant topographical variation, became a high-value zone for SDR. In contrast, the southeastern plains, with a high degree of land development, exhibited weaker SDR. The changes in SDR reflect the influence of different topographies and land use practices on soil conservation functions.

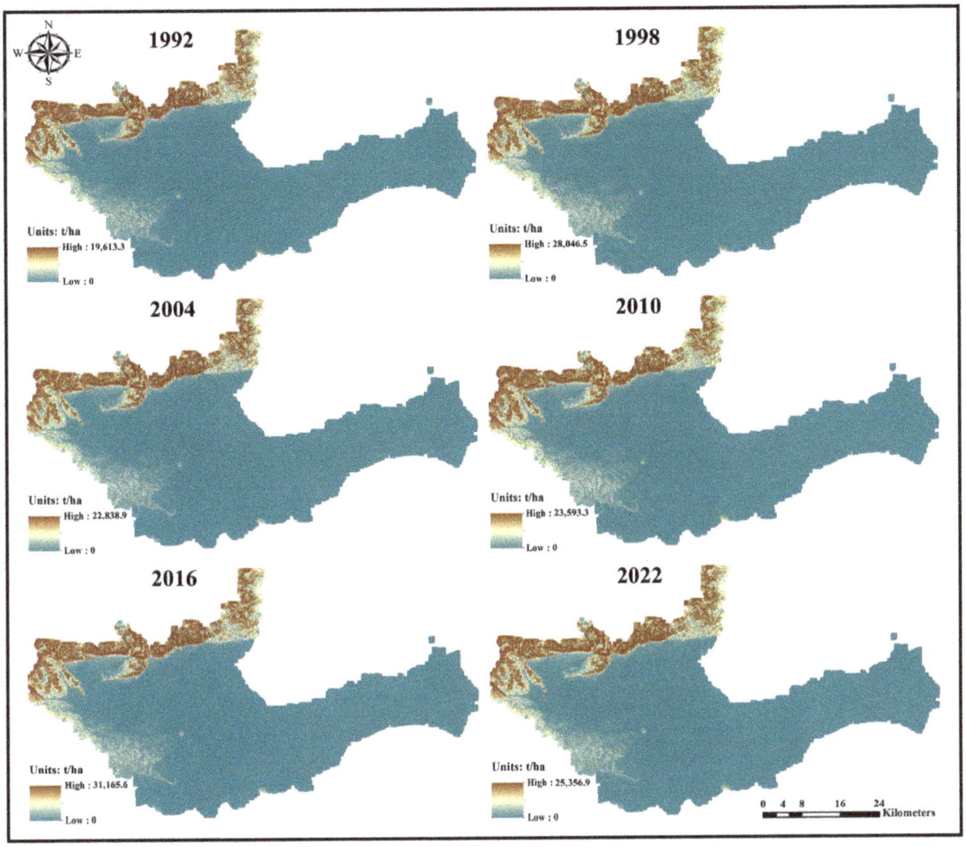


**Figure 10.** Temporal and spatial distribution map of SDR.

Time-series analysis shows that the volume of SDR initially increased and then decreased. From 1992 to 2016, the volume of SDR increased from 19,613.3 t/ha to 31,165.6 t/ha, but by 2022, it had declined to 25,356.9 t/ha. This change is closely related to LUCCs, topographical features, and vegetation cover. Particularly in the southeastern region, rapid urbanization, the reduction in agricultural land, and the expansion of built-up land severely impacted SDR. In contrast, the northwest region, with better vegetation protection, maintained a relatively stable level of SDR.

### 3.2.4. Spatiotemporal Evolution of Water Yield (WY)

The WY from 1992 to 2022 in the Qin-Mang River Basin, calculated using the InVEST model based on average annual precipitation, evapotranspiration, and root limiting layer depth data, shows a "high in the southeast, low in the northwest" distribution pattern (Figure 11). The southeastern region, with abundant precipitation and lower elevation, has a higher WY capacity, while the northwestern mountainous areas, with high evapotranspiration and low precipitation, have a lower WY capacity. Although the overall spatial distribution of WY did not change significantly, the rapid urbanization process, particularly in the central region, has negatively impacted WY services.

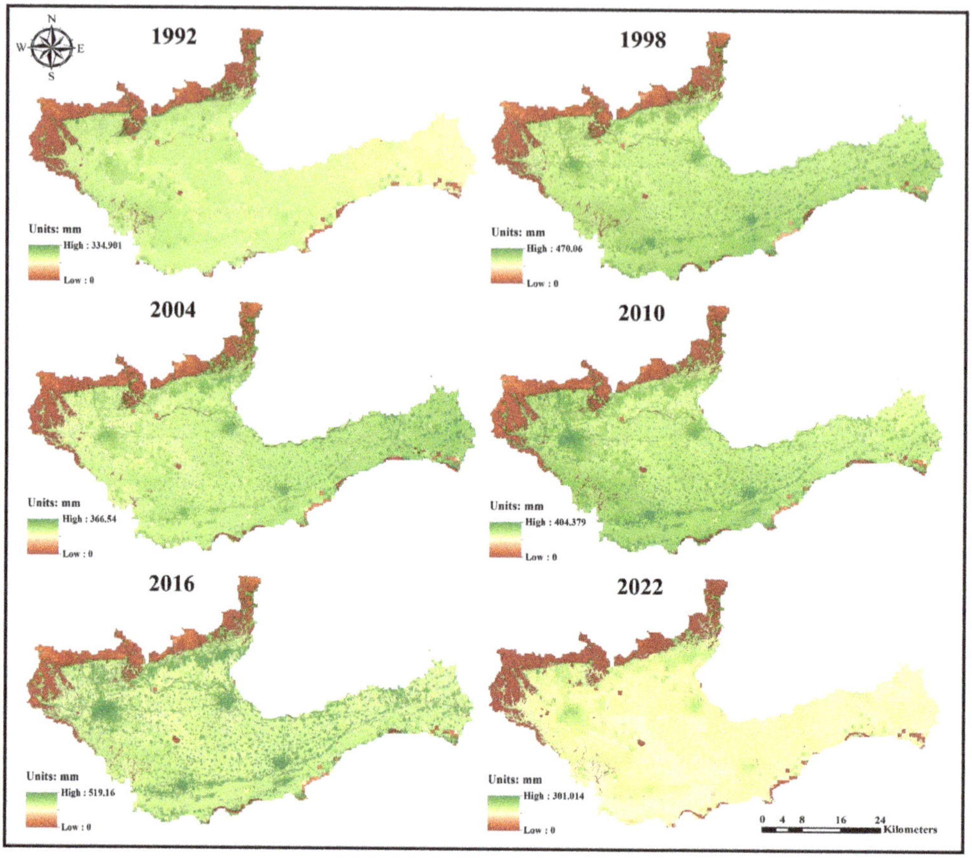


**Figure 11.** Temporal and spatial distribution map of WY.

Time-series analysis (Table 8) shows significant fluctuations in WY across different periods, with the highest yields recorded in 1998 at 796 million cubic meters and 831 million cubic meters in 2016, but with a sharp decline to 400.45 million cubic meters in 2022. This significant drop is closely related to changes in precipitation, increased evapotranspiration, and inadequate water resource management. Figure 11 further illustrates the spatial distribution of these changes and reveals the compressive effect of urban expansion on water resource yields, particularly in areas with reduced agricultural land, where the availability of water resources has been greatly impacted.

**Table 8.** ESs function quantities in the study area from 1992 to 2022.

| Year | HQ | CS/$10^6$ t | SDR/$10^8$ t | WY/$10^6$ $m^3$ |
|---|---|---|---|---|
| 1992 | 0.34 | 23.18 | 4695.05 | 460.39 |
| 1998 | 0.33 | 22.04 | 6924.59 | 796.06 |
| 2004 | 0.32 | 22.83 | 5738.28 | 594.28 |
| 2010 | 0.30 | 22.64 | 5677.55 | 620.46 |
| 2016 | 0.29 | 22.52 | 7723.19 | 831.17 |
| 2022 | 0.29 | 22.62 | 6470.34 | 400.45 |

Table 8 presents the trend in ecosystem service function changes from 1992 to 2022 in the study area. The HQ index decreased from 0.34 to 0.29, reflecting the deterioration of regional ecosystem health, which is closely linked to increasing human activities, urbanization, and LUCCs. CS remained between 22 and 23 million tons throughout the study period, with minor fluctuations, but failed to return to 1992 levels, suggesting that, while the region has some carbon sequestration capacity, its long-term low level may weaken its ability to mitigate climate change. SDR fluctuated significantly from 1992 to 2016, peaking at 69.24 million tons in 1998 and increasing to 77.23 million tons in 2016, but dropped to 64.7 million tons by 2022, reflecting the negative impact of human activities on soil quality. WY fluctuated greatly, reaching 796 million cubic meters in 1998 and 831 million cubic meters in 2016, but falling sharply to 400.45 million cubic meters in 2022, reflecting significant declines caused by changes in precipitation, increased evapotranspiration, and inadequate water resource management.

### 3.3. *Trade-Off and Synergy (TOS) Among Ecosystem Services (ESs)*

#### 3.3.1. Correlation Analysis of TOS in ESs

The changes in the correlation between ESs in the Qin-Mang River Basin from 1992 to 2022 reveal the dynamic processes of functional TOS (Figure 12). The study indicates that, over time, the synergy between ESs gradually weakened in some years, while trade-offs became more pronounced in certain regions. Specifically, in 1992, the correlation coefficient between HQ and CS was 0.86, indicating a significant synergy; the correlation coefficient between HQ and SDR was 0.66, also suggesting a strong synergistic relationship. However, the negative correlation between HQ and WY (−0.20) indicated a potential trade-off between ESs. In 1998, the correlation coefficient between HQ and CS increased to 0.91, suggesting an enhanced synergy. However, the negative correlation between HQ and WY increased to −0.42, reflecting a clear trade-off between high HQ and low WY in some areas. In the same year, the positive correlation between CS and SDR was 0.75, while the negative correlation between CS and WY was −0.52, further illustrating the competition between WY and CS.

By 2004, the correlation coefficient between HQ and CS remained at 0.91, continuing to show a strong synergy. However, the negative correlation between HQ and WY (−0.40) began to strengthen. This trend indicates that, although the synergy between CS and HQ persisted, the negative trade-offs between WY and other ESs were intensifying. In 2010 and 2016, the correlation coefficient between HQ and CS remained stable at 0.91, and the synergy between HQ and SDR remained strong (0.71 and 0.68), but the negative correlation between HQ and WY remained between −0.40 and −0.45, suggesting that changes in WY had a significant impact on the relationships between ESs. By 2022, the correlation coefficient between HQ and CS was 0.90, continuing to show a strong synergy. However, the negative correlation between HQ and WY increased to −0.36, indicating that the competitive relationship between ESs was more pronounced in certain regions. The positive correlation between CS and SDR (0.73) and the negative correlation between CS and WY (−0.41) continued to reflect the trade-offs between WY and other ESs.

Overall, as urbanization progresses and LUCCs increase in the Qin-Mang River Basin, the synergy between ESs has gradually weakened, while trade-offs have become more significant in certain areas. Specifically, between HQ, CS, SDR, and WY, the competitive relationships between ESs have increasingly intensified over time, especially in areas with WY stress. This change not only reveals the mutual constraints between ecosystem service functions but also reflects the profound impact of LUCCs on ESs. To address this challenge, it is urgent to optimize land use patterns, control urbanization, and promote sustainable land management strategies to enhance positive synergies between ESs and mitigate negative trade-offs.

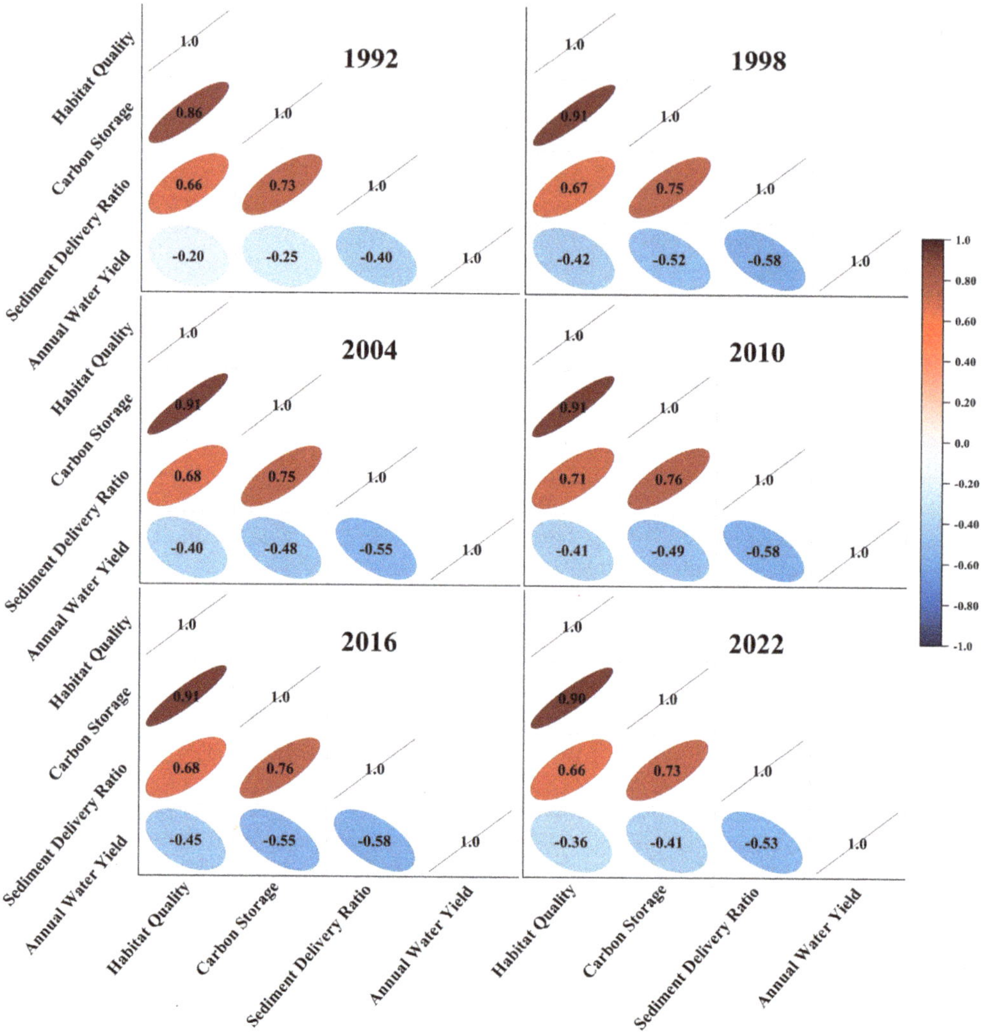


**Figure 12.** Correlation analysis of four ESs in the study area from 1992 to 2022.

### 3.3.2. Spatial Distribution and Heterogeneity of TOSs in ESs

Figure 13 presents the BLI clustering map of ESs in the Qin-Mang River Basin in 1992, while Figure 14 further reveals the spatial relationships between ESs through scatter plots. The results indicate that there is a significant positive spatial autocorrelation between HQ and CS (Moran's I = 0.713), HQ and SDR (Moran's I = 0.630), and CS and SDR (Moran's I = 0.702), while there is a significant negative spatial autocorrelation between HQ and WY (Moran's I = −0.331), CS and WY (Moran's I = −0.407), and SDR and WY (Moran's I = −0.514). All Moran's I values passed the significance test (P values were all less than 0.05), and, through randomization tests, the probability of these clustering patterns being generated was less than 1%, indicating the statistical significance of these spatial autocorrelation relationships.

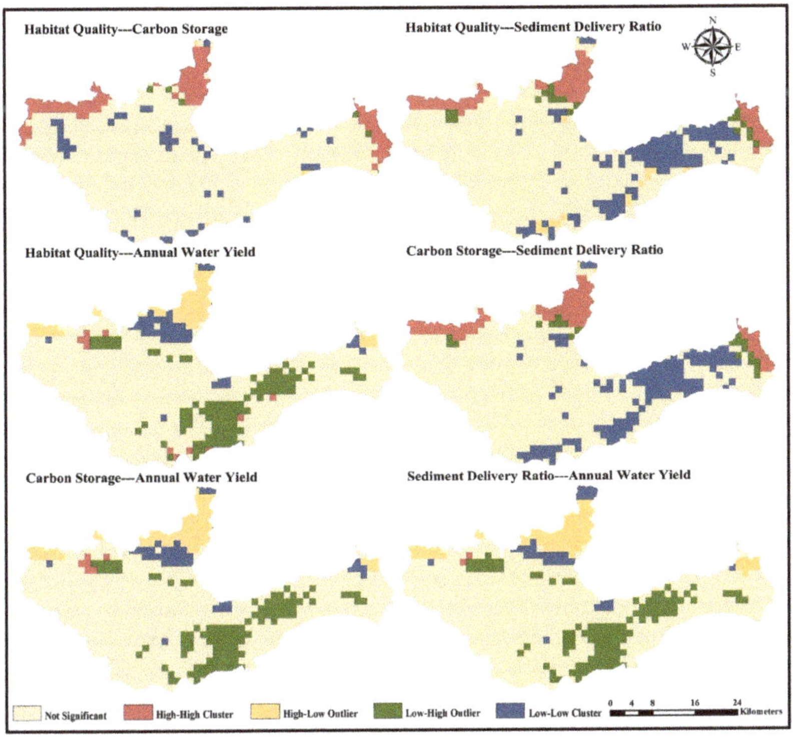


**Figure 13.** Spatial distribution map of TOSs among ESs in 1992.

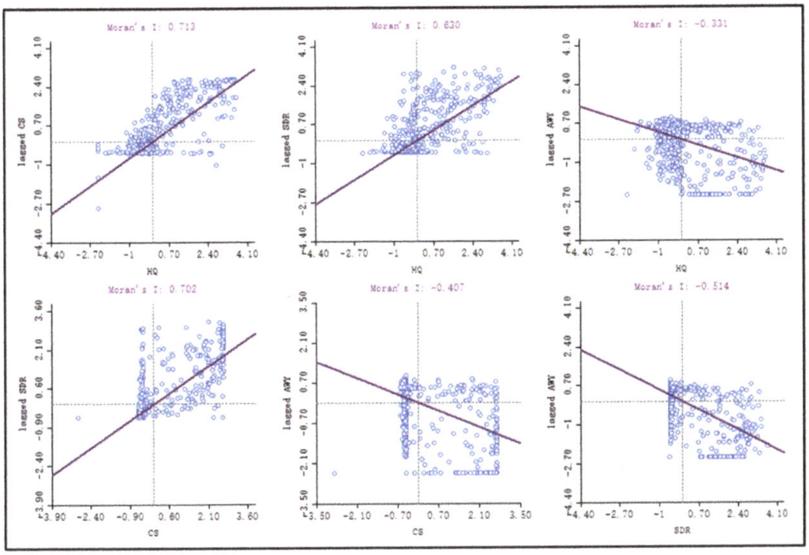


**Figure 14.** BLI scatter plot for 1992.

The spatial analysis in 1992 showed that in the northern and northwestern parts of the study area, HQ, CS, and SDR exhibited significant positive spatial autocorrelation, forming high-value clusters of ecosystem service functions. The strong positive spatial autocorrelation indicates significant synergies between ESs in these regions, particularly the high correlations between HQ and CS and CS and SDR. This phenomenon reflects the relatively favorable ecological conditions in these areas, such as good vegetation coverage and minimal ecological disturbance, which promote the mutual enhancement and coordinated development of ESs. However, there was a significant negative spatial autocorrelation between WY and HQ, CS, and SDR, indicating that, in some regions, higher ecological service functions (e.g., HQ, CS, and SDR) were accompanied by lower WY. This negative correlation reveals the spatial heterogeneity of ecosystem service functions, particularly in areas with low WY, where high-value ecological functions such as HQ and CS tend to concentrate in spatial units corresponding to lower WY.

The BLI clustering map further confirmed this spatial aggregation phenomenon. The high-value regions in the study area were mainly concentrated in the northern mountainous areas, while low-value regions were primarily distributed in the southeast. This spatial pattern highlights the significant spatial heterogeneity of ecosystem service functions, particularly the contrast between high-value and low-value areas. Specifically, the ecosystem service functions in the northern mountainous areas were stronger, reflecting the stability of the natural ecological environment and the synergies between ESs in these areas. In contrast, the southeastern region exhibited weaker ecosystem service functions due to factors such as the expansion of construction land and a decline in vegetation coverage, leading to a significant reduction in the synergies between services.

By 2022, the spatial distribution of ESs in the Qin-Mang River Basin had undergone significant changes (Figure 15). Particularly in the southern and eastern regions, the trade-off phenomena between ESs became more pronounced, with the negative correlations between HQ and WY and between SDR and WY strengthening in these areas. The spatial autocorrelation analysis in 2022 revealed an increase in the number of low-value regions, where ESs experienced severe degradation, the synergies between services were significantly weakened, and trade-off phenomena were significantly exacerbated. The BLI scatter plot showed (Figure 16) that there was a significant positive spatial autocorrelation between HQ and CS (Moran's I = 0.731), HQ and SDR (Moran's I = 0.643), and CS and SDR (Moran's I = 0.706), while significant negative spatial autocorrelation was found between HQ and WY (Moran's I = −0.366), CS and WY (Moran's I = −0.405), and SDR and WY (Moran's I = −0.527). All results had P values less than 0.05, and, through randomization tests, the probability of these clustering patterns being generated was less than 1%, confirming the statistical significance of these spatial autocorrelation relationships.

In the spatial analysis of 2022, the spatial aggregation effect of low-value regions was particularly prominent, reflecting the significant negative impact on ecosystem service functions in these areas. This change is closely related to the accelerating urbanization process, the expansion of construction land, and unsustainable land use patterns. Specifically, with the acceleration of urbanization, the land cover types in the southern and eastern regions have undergone significant changes, and the expansion of construction land and ecological degradation have intensified the competition and conflicts between ESs.

This negative spatial aggregation effect poses a major challenge to the synergies of ESs. In particular, the negative correlations between WY and HQ, CS, and SDR reveal the mutual constraints between ESs in these regions. As ecological degradation intensifies, the ecosystem functions in these areas further deteriorate, leading to the formation of distinct low-value regions in the spatial distribution of ESs. This trend poses a severe challenge to regional ecological stability and sustainable development.

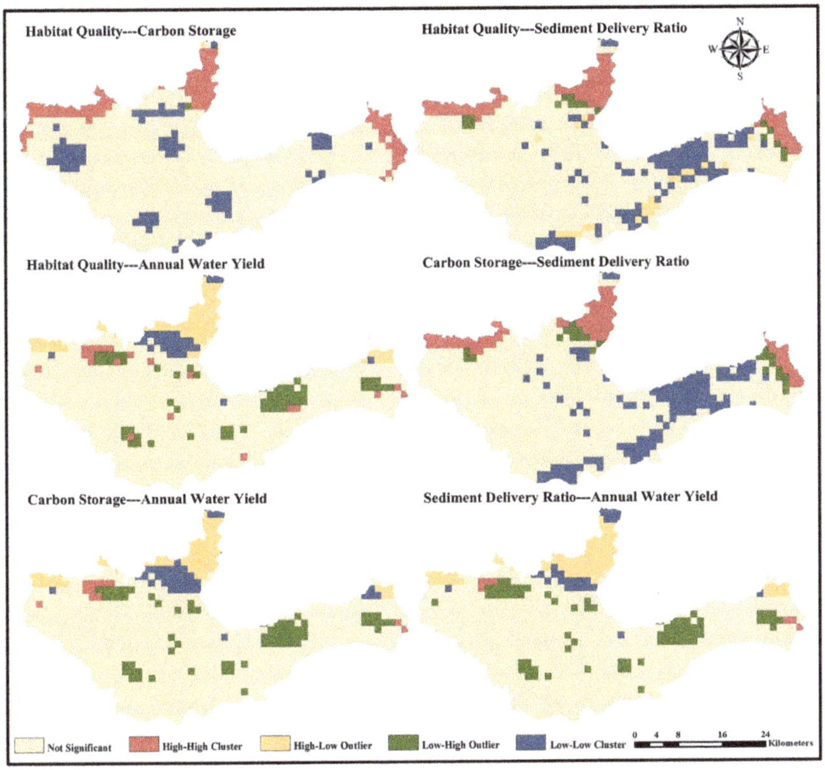


**Figure 15.** Spatial distribution map of TOSs among ESs in 2022.

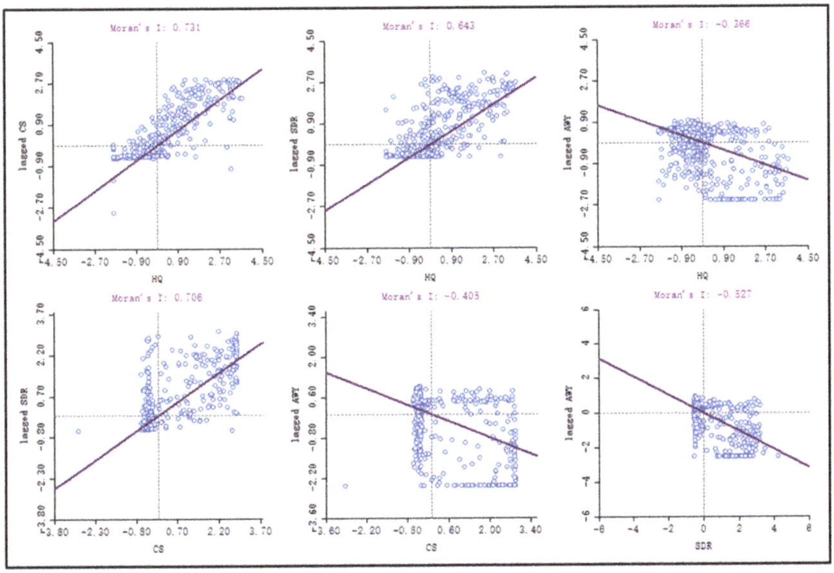


**Figure 16.** BLI scatter plot for 2022.

## 4. Discussion

### 4.1. Interpretation and Comparison of Results

This study demonstrates that LUT transformation in the Qin-Mang River Basin over the past 30 years has been driven by urbanization and economic development, leading to significant changes in the land use pattern of the basin. Although LUCC has stabilized in recent years, the expansion of construction land remains the main trend, which is primarily achieved through the reduction in cultivated land and forest areas. This trend highlights the profound impact of urbanization on the ecological environment, resulting in the weakening of critical ESs.

The expansion of construction land encroaching on cultivated land and forests indicates a close relationship between urbanization and LUT transformation. These transformations have significantly impaired the region's ecosystem service capacity, particularly in terms of HQ, CS, and SDR. The decline in HQ is directly related to the conversion of natural land types to urban areas, and this finding is consistent with research conducted in other urbanized regions [33,34]. This study further confirms the negative impact of urbanization on ecosystem functions, particularly in other regions of the Yellow River Basin, where trends of habitat degradation and decreased CS have also been validated [35]. These studies suggest that urbanization not only leads to the loss of natural habitats but also exacerbates habitat fragmentation, which further affects biodiversity and ecosystem resilience [36].

In terms of spatial distribution, this study found significant regional differences in ESs within the basin, particularly in the northwest and southeast. The northwest region has stronger ecological service functions, while the southeast region is more vulnerable. This disparity suggests that future land management and conservation policies should consider local differences and develop specific strategies [37]. For example, ecological protection measures should be prioritized in the high-ecological-value areas of the northwest, while ecological restoration should be strengthened in the urbanized areas of the southeast to mitigate the negative impacts of urbanization on the ecosystem.

The results of this study further emphasize that policymakers should integrate ecological restoration measures, such as forest rehabilitation and soil and water conservation, into urban planning to achieve a sustainable balance between urbanization and ecosystem integrity. Additionally, considering the importance of spatiotemporal dynamics in land use and ecosystem service management is crucial. The Qin-Mang River Basin, as a key ecological region, plays an irreplaceable role not only in biodiversity conservation but also in maintaining regional ecological stability. Therefore, regional management strategies should place a high priority on the sustainability of ESs to ensure the long-term health and stability of ecosystems.

### 4.2. Mechanisms of Trade-Offs and Synergies (TOSs)

This study delves into the impact of LUCCs on the TOSs between ESs. Correlation analysis reveals that there are complex interactions between Ess, such as HQ, CS, SDR, and WY, and these relationships dynamically change with shifts in land use. Specifically, there is a positive correlation between HQ and CS, indicating that healthy habitats help enhance carbon sequestration, thereby boosting ecosystem CS. This finding is consistent with studies from other regions, where large-scale ecological restoration or forest protection projects typically improve both HQ and CS [38].

However, excessive water resource development significantly affects the synergy and trade-off relationships between ESs. While water resource development may enhance WY services in the short term to support agricultural, industrial, and urban needs, long-term overexploitation leads to habitat degradation, increased soil erosion, and reduced CS, resulting in ecological degradation. When water resource development intersects with

LUCCs, it often exacerbates ecological degradation, such as the drying of water bodies and reduction in wetlands, which affects biodiversity and ecosystem restoration capacity. The negative correlation between water resource development and other ecological functions, such as HQ and WY, has been validated in studies of multiple river basins [39].

As urbanization accelerates, the synergy between ESs weakens, and trade-off relationships become more pronounced. Particularly in areas of construction land expansion and reduction in cultivated land, CS and SDR continue to decline, while WY fluctuates in some regions. This trade-off mechanism is especially evident in areas where urbanization is intensifying, highlighting the pressure urban expansion places on the multifunctionality of ESs. Similar conclusions have been found in both domestic and international studies, where rapid urbanization in the Yangtze River Basin and the Brazilian Amazon has led to negative trade-offs between ESs [40].

This study also analyzes the impact of urbanization on the synergy between WY and CS. As urbanization accelerates, the expansion of built-up areas reduces the amount of green land and forests, which in turn diminishes the capacity for water source conservation, thereby affecting WY. Simultaneously, the decrease in land permeability not only reduces groundwater recharge but also exacerbates soil erosion, further weakening SDR functions [41]. These cumulative effects lead to a gradual reduction in the synergy between water source conservation and CS, particularly in highly urbanized areas such as the southeastern part of the Qin-Mang River Basin, where a clear trade-off relationship emerges. As green land area shrinks and water source conservation zones decrease, the negative trade-off between WY and CS becomes more pronounced. This trend is commonly observed in other watersheds with similar urbanization backgrounds, indicating that the question of how to coordinate the joint enhancement of water source conservation and CS remains a key issue for future urban development.

### 4.3. Spatial Heterogeneity and Scale Effects

This study found that the impact of LUT transformation on ESs exhibits significant spatial heterogeneity. This variability is closely related not only to the type and intensity of land use but also to regional natural environmental conditions and socioeconomic development. The presence of spatial heterogeneity indicates that ESs show marked spatial distribution differences across regions. For example, in the northern mountainous areas of the basin, due to the higher vegetation cover, ESs such as SDR, CS, and HQ are stronger. This is closely related to the lower level of urbanization and relatively intact natural ecosystems in the region. In contrast, in the southeastern areas, urbanization has accelerated, leading to significant expansion of construction land, resulting in a marked decline in services like HQ, SDR, and CS, demonstrating weaker ecological service functions.

Spatial heterogeneity not only reflects the direct impact of different land use patterns on ESs but is also constrained by various factors such as topography, climate conditions, and population density and economic development. For instance, the northern mountainous areas, with their higher altitude and complex terrain, are more resilient to external disturbances, and their ESs remain stable. In contrast, the lowland plain areas in the southern and eastern parts of the basin, characterized by flat terrain and convenient transportation, experience higher levels of urbanization. LUCCs in these areas have more significant impacts on ESs, particularly in terms of habitat fragmentation and ecosystem degradation. This conclusion is consistent with findings from other studies, such as those in the Yangtze River region, where areas with higher urbanization levels also show significant declines in ecosystem service functions [42].

Building on this, this study further explores the impact of temporal and spatial scales on ESs. Through spatiotemporal scale analysis, it was found that the trend of ecosystem service changes is not only closely related to spatial location but is also significantly influenced by changes in the temporal scale. Over shorter time scales (e.g., within 10 years), LUCCs have a more direct impact on ESs, particularly during urbanization, where the expansion of construction land has a significant negative effect on HQ and CS. Meanwhile, over longer time scales (e.g., 30 years or more), changes in ESs show more complex dynamic characteristics, closely related to long-term LUT transformation and policy interventions.

Specifically, over the past 30 years, the HQ and CS in the Qin-Mang River Basin have shown a significant declining trend, which is not only due to intensified urbanization but also the reduction in agricultural land and the decline in forest cover. In the short term, the expansion of construction land may lead to a rapid decline in ESs. However, in the long term, ecological restoration measures (such as converting farmland to forests and wetland restoration) can mitigate the trend of ecological degradation to some extent, as seen in preliminary ecological restoration projects in the central and eastern parts of the basin.

Moreover, scale effects are also particularly pronounced in spatiotemporal scale analysis. At larger scales (regional scale), the impact of LUCCs is more pronounced, and the interrelationships between ESs are more complex. At smaller scales (such as at the local watershed level), changes in ESs are more influenced by local environmental conditions [43]. Therefore, when formulating land use planning and ecological restoration policies, it is essential to consider the influencing factors at different scales and adopt corresponding management measures.

### 4.4. Policy Recommendations and Practical Implications

This study provides theoretical foundations and practical guidance for land use policy and management. As urbanization progresses, the spatiotemporal variability of ESs requires policymakers to carefully consider regional differences and their impacts on ecological functions in planning. In areas with high ecological value (such as the northern mountainous regions of the basin), priority should be given to protecting natural ecosystems, restricting excessive development, and preventing the expansion of construction land as well as irrational agricultural and industrial development to ensure that ecological functions remain intact. Special emphasis should be placed on the protection of key ecological resources, such as forests, grasslands, and wetlands, to maintain regional ecosystem stability.

In the central, southern, and eastern parts of the basin, where urbanization is accelerating, ecological degradation is becoming more severe. There is an urgent need to implement strict ecological red-line protection policies to prevent further exacerbation of ecological degradation. In already degraded areas, large-scale ecological restoration measures, such as afforestation, converting farmland back to forests, and wetland restoration, should be implemented to enhance CS, water conservation capacity, and biodiversity. These efforts can be combined with ecological engineering techniques, such as the construction of ecological barriers and soil moisture management, to further improve ecological restoration outcomes.

This study emphasizes the necessity of region-specific land use policies through spatiotemporal scale analysis. Considering the regional differences in ESs, future land use policies should adopt more precise regional strategies to ensure a reasonable balance between land use and ecological protection needs in different ecological environments. For instance, in regions with higher ecological value, stricter ecological management measures should be implemented. In more urbanized areas, green infrastructure construction and ecological restoration measures should be adopted to gradually achieve a win-win situation for both ecology and economy.

Beyond existing ecological protection and restoration policies, this study also recommends the incorporation of innovative policy tools to enhance ESs. Ecological compensation mechanisms, for example, could incentivize land developers or urban planners to compensate for the loss of ecological functions, thereby assisting in the restoration of the ecological value of affected areas. Furthermore, the implementation of green infrastructure (such as urban greenbelts, ecological parks, and rooftop greening) could effectively enhance urban ecological functions and mitigate the pressures of urbanization on ecosystems. While these policy measures contribute significantly to the enhancement of ESs, local governments may face challenges such as economic constraints, limited management capacity, and public awareness. Therefore, future policy-making should emphasize interdepartmental collaboration and increased public participation to foster a shared commitment to ecological protection, helping to address the challenges posed by rapid urbanization.

### *4.5. Research Limitations and Future Approaches*

Despite this study's in-depth exploration of the impact of LUCCs on ESs, some limitations remain. First, the land use data used in this study have relatively low resolution, which may affect the results of local scale analyses. Second, this study did not explore the impact of other factors, such as climate change, on ESs. Future research could integrate climate change considerations into the assessment framework for ESs.

Future research could focus on the following aspects: first, improving data resolution for more refined spatial analyses; second, examining the combined impacts of climate change and other stressors on ESs; and third, conducting long-term dynamic monitoring to identify long-term trends and patterns in ecosystem service changes.

## 5. Conclusions

This study evaluated the impact of LUT transformation on ESs in the Qin-Mang River Basin using multiple methods, revealing the spatiotemporal dynamics between LUCCs and ESs, along with the driving mechanisms behind them. The main conclusions are as follows:

(1) The LUCCs in the Qin-Mang River Basin have significantly affected ESs. In recent years, the expansion of built-up land has primarily encroached upon arable land and forests, reflecting the close relationship between urbanization and economic development. Over the past 30 years, the area of arable land has decreased by 35,000 hectares, which were converted into built-up land. This change is primarily driven by economic development, posing significant threats to food security and ecosystem service functions. The LUT transition matrix analysis further reveals that the probability of future expansion of built-up land is relatively high, while the probability of expansion for arable land and forests is relatively low, reflecting the immense pressure of urbanization on natural land.

(2) LUCCs have significantly affected ESs, exhibiting certain TOSs. Over the past 30 years, the overall trend of ESs has been in decline. HQ has continuously deteriorated, especially in areas with rapid urbanization. CS has remained stable, but low-CS areas have gradually expanded, indicating that regional carbon sequestration potential still exists. SDR has fluctuated significantly, reflecting the negative impact of human activities on soil quality. WY has also experienced significant fluctuations, influenced by changes in precipitation, increased evapotranspiration, and inadequate water resource management.

(3) Correlation analysis reveals the dynamic evolution of synergies and trade-offs among ESs. HQ and CS have maintained a strong synergy in most years, while the negative correlation between HQ and WY has gradually intensified, especially in areas experiencing water scarcity. This suggests that, with the progress of urbanization and LUCCs, the competition between ESs, particularly the conflict between water resources and other ecological functions, has become more pronounced, reflecting the profound impact of human activities on ESs. As built-up land expands, the negative trade-offs between ESs become more significant.

(4) This study shows that, in 1992, the HQ, CS, SDR, and WY in the northern and northwestern regions of the Qin-Mang River Basin exhibited significant positive spatial autocorrelation, demonstrating strong synergies. However, by 2022, the ESs in the southern and eastern regions have gradually shown clear trade-offs, with the negative correlations between HQ and CS and between SDR and WY strengthening. This change is primarily attributed to the impact of urbanization on the ecosystem's structure and functions, leading to a reduction in natural land and an increase in conflicts between ESs, especially in resource-limited areas where the mutual constraints among ESs become more pronounced.

**Author Contributions:** All authors contributed to the study conception and design. Writing and editing: Jiwei Zhao and Luyao Wang; Chart editing: Luyao Wang, Dong Jia and Yaowen Wang; Preliminary data collection: Luyao Wang. All authors have read and agreed to the published version of the manuscript.

**Funding:** This study was funded by the National Natural Science Foundation of China (51709115), the Henan Provincial Key R&D and Promotion Special Project (Science and Technology Tackling) (182102210066).

**Data Availability Statement:** Data and materials are available from the corresponding author upon request.

**Acknowledgments:** The authors would like to thank the anonymous reviewers and the editors for their profound comments, which helped us improve the quality of this paper.

**Conflicts of Interest:** All authors have read, understood, and have complied as applicable with the statement on "Ethical responsibilities of Authors" as found in the Instructions for Authors.

## References

1. Costanza, R.; De Groot, R.; Sutton, P.; Van der Ploeg, S.; Anderson, S.J.; Kubiszewski, I.; Farber, S.; Turner, R.K. Changes in the global value of ecosystem services. *Glob. Environ. Chang.* **2014**, *26*, 152–158. [CrossRef]
2. Chee, Y.E. An ecological perspective on the valuation of ecosystem services. *Biol. Conserv.* **2004**, *120*, 549–565. [CrossRef]
3. Watson, R.; Baste, I.; Larigauderie, A.; Leadley, P.; Pascual, U.; Baptiste, B.; Demissew, S.; Dziba, L.; Erpul, G.; Fazel, A.M.; et al. *Summary for Policymakers of the Global Assessment Report on Biodiversity and Ecosystem Services of the Intergovernmental Science-Policy Platform on Biodiversity and Ecosystem Services*; IPBES Secretariat: Bonn, Germany, 2019; pp. 22–47. Available online: https://pesquisa.bvsalud.org/portal/resource/pt/biblio-1179448 (accessed on 3 August 2024).
4. Li, J.; Xie, B.; Dong, H.; Zhou, K.; Zhang, X. The impact of urbanization on ecosystem services: Both time and space are important to identify driving forces. *J. Environ. Manag.* **2023**, *347*, 119161. [CrossRef] [PubMed]
5. Davidson, N.C. How much wetland has the world lost? Long-term and recent trends in global wetland area. *Mar. Freshw. Res.* **2014**, *65*, 934–941. [CrossRef]
6. Junk, W.J.; An, S.; Finlayson, C.M.; Gopal, B.; Květ, J.; Mitchell, S.A.; Mitsch, W.J.; Robarts, R.D. Current state of knowledge regarding the world's wetlands and their future under global climate change: A synthesis. *Aquat. Sci.* **2013**, *75*, 151–167. [CrossRef]
7. Díaz, S.; Settele, J.; Brondízio, E.S.; Ngo, H.T.; Agard, J.; Arneth, A.; Balvanera, P.; Brauman, K.A.; Butchart, S.H.; Chan, K.M.; et al. Pervasive human-driven decline of life on Earth points to the need for transformative change. *Science* **2019**, *366*, eaax3100. [CrossRef] [PubMed]
8. Long, H.; Liu, Y.; Hou, X.; Li, T.; Li, Y. Effects of land use transitions due to rapid urbanization on ecosystem services: Implications for urban planning in the new developing area of China. *Habitat Int.* **2014**, *44*, 536–544. [CrossRef]
9. Spake, R.; Lasseur, R.; Crouzat, E.; Bullock, J.M.; Lavorel, S.; Parks, K.E.; Schaafsma, M.; Bennett, E.M.; Maes, J.; Mulligan, M.; et al. Unpacking ecosystem service bundles: Towards predictive mapping of synergies and trade-offs between ecosystem services. *Glob. Environ. Chang.* **2017**, *47*, 37–50. [CrossRef]
10. Pang, X.; Nordström, E.M.; Böttcher, H.; Trubins, R.; Mörtberg, U. Trade-offs and synergies among ecosystem services under different forest management scenarios–The LEcA tool. *Ecosyst. Serv.* **2017**, *28*, 67–79. [CrossRef]
11. Yang, J.; Xie, B.; Tao, W.; Zhang, D. Ecosystem services assessment, trade-off, and bundles in the Yellow River Basin, China. *Diversity* **2021**, *13*, 308. [CrossRef]
12. Qiao, Q.; Zhen, Z.; Liu, L.; Luo, P. The construction of ecological security pattern under rapid urbanization in the Loess Plateau: A case study of Taiyuan city. *Remote Sens.* **2023**, *15*, 1523. [CrossRef]

13. Wu, Y.; Zhu, X.; Gao, W.; Qian, F. The spatial characteristics of coupling relationship between urbanization and eco-environment in the Pan Yangtze River Delta. *Energy Procedia* **2018**, *152*, 1121–1126. [CrossRef]
14. Long, K.; Omrani, H.; Pijanowski, B.C. Impact of local payments for ecosystem services on land use in a developed area of China: A qualitative analysis based on an integrated conceptual framework. *Land Use Policy* **2020**, *96*, 104716. [CrossRef]
15. Zhao, Y.; Liu, Z.; Wu, J. Grassland ecosystem services: A systematic review of research advances and future directions. *Landsc. Ecol.* **2020**, *35*, 793–814. [CrossRef]
16. Pontius, G.R.; Malanson, J. Comparison of the structure and accuracy of two land change models. *Int. J. Geogr. Inf. Sci.* **2005**, *19*, 243–265. [CrossRef]
17. Liang, X.; Guan, Q.; Clarke, K.C.; Liu, S.; Wang, B.; Yao, Y. Understanding the drivers of sustainable land expansion using a patch-generating land use simulation (PLUS) model: A case study in Wuhan, China. *Comput. Environ. Urban Syst.* **2021**, *85*, 101569. [CrossRef]
18. Sharp, R.; Tallis, H.; Ricketts, T.; Guerry, A.; Wood, S.A.; Chaplin-Kramer, R.; Nelson, E.J.; Ennaanay, D.; Wolny, S.; Olwero, N.; et al. *InVEST 3.7.0 User's Guide*; The Natural Capital Project: Standford, CA, USA, 2018. [CrossRef]
19. Yang, K.; Han, Q.; de Vries, B. Urbanization effects on the food-water-energy nexus within ecosystem services: A case study of the Beijing-Tianjin-Hebei urban agglomeration in China. *Ecol. Indic.* **2024**, *160*, 111845. [CrossRef]
20. Guo, F.Y.; Liu, X.H.; Zhang, W.B.; Xing, L.Y.; Wang, R.; Zu, P.Y.M.; Luo, X.; Wang, C.; Zhao, H. Spatiotemporal pattern evolution and driving force analysis of habitat quality in the Yellow River Basin (Henan section) from 2000 to 2040. *Mod. Geol.* **2024**, *38*, 599–611.
21. Chen, M.J.; Bai, Z.K.; Wang, Q.R.; Shi, Z. Habitat quality effect and driving mechanism of land use transitions: A case study of Henan water source area of the middle route of the South-to-North Water Transfer Project. *Land* **2021**, *10*, 796. [CrossRef]
22. Feng, J.M.; Li, C.; Feng, Y.F. Evolution of watershed habitat quality based on ISODATA spatial mosaic clustering: A case study of urban areas surrounding the Yellow River Basin in Henan Province. *Chin. Landsc. Archit.* **2024**, *50*, 89–95.
23. Schulte-Uebbing, L.; de Vries, W. Global-scale impacts of nitrogen deposition on tree carbon sequestration in tropical, temperate, and boreal forests: A meta-analysis. *Glob. Change Biol.* **2018**, *24*, e416–e431. [CrossRef] [PubMed]
24. Li, X.; Wang, Y.; Wu, K.; Feng, Z. Analysis and prediction of carbon balance in production-living-ecological space of Henan Province, China. *Environ. Sci. Pollut. Res.* **2023**, *30*, 75973–75988. [CrossRef] [PubMed]
25. Chuai, X.; Huang, X.; Lai, L.; Wang, W.; Peng, J.; Zhao, R. Land use structure optimization based on carbon storage in several regional terrestrial ecosystems across China. *Environ. Sci. Policy* **2013**, *25*, 50–61. [CrossRef]
26. Didoné, E.J.; Minella JP, G.; Merten, G.H. Quantifying soil erosion and sediment yield in a catchment in southern Brazil and implications for land conservation. *J. Soils Sediments* **2015**, *15*, 2334–2346. [CrossRef]
27. Hu, Y.; Tian, G.; Mayer, A.L.; He, R. Risk assessment of soil erosion by application of remote sensing and GIS in Yanshan Reservoir catchment, China. *Nat. Hazards* **2015**, *79*, 277–289. [CrossRef]
28. Feng, X.M.; Sun, G.; Fu, B.J.; Su, C.H.; Liu, Y.; Lamparski, H. Regional effects of vegetation restoration on water yield across the Loess Plateau, China. *Hydrol. Earth Syst. Sci.* **2012**, *16*, 2617–2628. [CrossRef]
29. Jiang, L.; Wang, Z.; Zuo, Q.; Du, H. Simulating the impact of land use change on ecosystem services in agricultural production areas with multiple scenarios considering ecosystem service richness. *J. Clean. Prod.* **2023**, *397*, 136485. [CrossRef]
30. Briner, S.; Huber, R.; Bebi, P.; Elkin, C.; Schmatz, D.R.; Grêt-Regamey, A. Trade-offs between ecosystem services in a mountain region. *Ecol. Soc.* **2013**, *18*. Available online: https://www.jstor.org/stable/26269360 (accessed on 3 August 2024). [CrossRef]
31. Jopke, C.; Kreyling, J.; Maes, J.; Koellner, T. Interactions among ecosystem services across Europe: Bagplots and cumulative correlation coefficients reveal synergies, trade-offs, and regional patterns. *Ecol. Indic.* **2015**, *49*, 46–52. [CrossRef]
32. Tian, S.; Wu, W.; Shen, Z.; Wang, J.; Liu, X.; Li, L.; Li, X.; Liu, X.; Chen, H. A cross-scale study on the relationship between urban expansion and ecosystem services in China. *J. Environ. Manag.* **2022**, *319*, 115774. [CrossRef] [PubMed]
33. Fang, Z.; Ding, T.; Chen, J.; Xue, S.; Zhou, Q.; Wang, Y.; Huang, Z.; Yang, S. Impacts of land use/land cover changes on ecosystem services in ecologically fragile regions. *Sci. Total Environ.* **2022**, *831*, 154967. [CrossRef] [PubMed]
34. Liu, Z.; Chang, Y.; Pan, S.; Zhang, P.; Tian, L.; Chen, Z. Unfolding the spatial spillover effect of urbanization on composite ecosystem services: A case study in cities of Yellow River Basin. *Ecol. Indic.* **2024**, *158*, 111521. [CrossRef]
35. Ouyang, X.; Tang, L.; Wei, X.; Li, Y. Spatial interaction between urbanization and ecosystem services in Chinese urban agglomerations. *Land Use Policy* **2021**, *109*, 105587. [CrossRef]
36. Di Giulio, M.; Holderegger, R.; Tobias, S. Effects of habitat and landscape fragmentation on humans and biodiversity in densely populated landscapes. *J. Environ. Manag.* **2009**, *90*, 2959–2968. [CrossRef] [PubMed]
37. Wang, K.; Ouyang, X.; He, Q.; Zhu, X. Impact of urban land expansion efficiency on ecosystem services: A case study of the three major urban agglomerations along the Yangtze River economic belt. *Land* **2022**, *11*, 1591. [CrossRef]
38. Di Sacco, A.; Hardwick, K.A.; Blakesley, D.; Brancalion, P.H.; Breman, E.; Cecilio Rebola, L.; Chomba, S.; Dixon, K.; Elliott, S.; Ruyonga, G.; et al. Ten golden rules for reforestation to optimize carbon sequestration, biodiversity recovery and livelihood benefits. *Glob. Change Biol.* **2021**, *27*, 1328–1348. [CrossRef]

39. Fu, B.; Zhang, L.; Xu, Z.; Zhao, Y.; Wei, Y.; Skinner, D. Ecosystem services in changing land use. *J. Soils Sediments* **2015**, *15*, 833–843. [CrossRef]
40. Luo, Q.; Zhou, J.; Zhang, Y.; Yu, B.; Zhu, Z. What is the spatiotemporal relationship between urbanization and ecosystem services? A case from 110 cities in the Yangtze River Economic Belt, China. *J. Environ. Manag.* **2022**, *321*, 115709. [CrossRef]
41. Gong, J.; Dai, X.; Wang, L.; Niu, Z.; Cao, Q.; Huang, C. The Impact of Urbanization on the Supply–Demand Relationship of Ecosystem Services in the Yangtze River Middle Reaches Urban Agglomeration. *Remote Sens.* **2023**, *15*, 4749. [CrossRef]
42. Liu, T.; Yu, L.; Chen, X.; Wu, H.; Lin, H.; Li, C.; Hou, J. Environmental laws and ecological restoration projects enhancing ecosystem services in China: A meta-analysis. *J. Environ. Manag.* **2023**, *327*, 116810. [CrossRef] [PubMed]
43. Seppelt, R.; Lautenbach, S.; Volk, M. Identifying trade-offs between ecosystem services, land use, and biodiversity: A plea for combining scenario analysis and optimization on different spatial scales. *Curr. Opin. Environ. Sustain.* **2013**, *5*, 458–463. [CrossRef]

*land* MDPI

*Article*

# Multi-Scale Mapping of Energy Consumption Carbon Emission Spatiotemporal Characteristics: A Case Study of the Yangtze River Delta Region

**Kangjuan Lv [1], Qiming Wang [2,*], Xunpeng Shi [3], Li Huang [1] and Yatian Liu [4]**

1 SHU-UTS SILC Business School, Shanghai University, Shanghai 201900, China
2 School of Economics, Shanghai University, Shanghai 201900, China
3 Australia-China Relations Institute, University of Technology Sydney, Sydney, NSW 2007, Australia
4 School of Economics and Management, Beijing Jiaotong University, Beijing 100080, China
* Correspondence: wangqiming@shu.edu.cn; Tel.: +86-15000180363

**Abstract:** Climate issues significantly impact people's lives, prompting governments worldwide to implement energy-saving and emission-reducing measures. However, many areas lack carbon emission data at the lower administrative divisions. Additionally, the inconsistency in the standards, scope, and accuracy of carbon dioxide emission statistics across different regions makes mapping carbon dioxide spatial patterns complex. Nighttime light (NTL) data combined with land use data enable the detailed spatial and temporal disaggregation of carbon emission data at a finer administrative level, facilitating scientifically informed policy formulation by the government. Differentiating carbon emission data by sector will help us further identify the carbon emission efficiency in different sectors and help environmental regulators implement the most cost-effective emission-reduction strategy. This study uses integrated remote-sensing data to estimate carbon emissions from fossil fuels (CEFs). Experimental results indicate (1) that the regional CEF can be calculated by combining NTL and Landuse data and has a good fit; (2) the high-intensity CEF area is mainly concentrated in Shanghai and its surrounding areas, showing a concentric circle structure; (3) there are obvious differences in the spatial distribution characteristics of carbon emissions among different departments; (4) hot spot analysis reveals a three-tiered distribution in the Yangtze River Delta, increasing from the west to the east with distinct spatial characteristics.

**Keywords:** energy consumption; carbon emissions; nighttime light; spatiotemporal changes; Yangtze River Delta

## 1. Introduction

Carbon dioxide ($CO_2$) is one of the main factors contributing to global warming. Many countries have set long-term targets for carbon emissions and are attempting to introduce more clean energy to reduce reliance on carbon [1]. Effective regional governance of carbon reduction necessitates an understanding of the spatial distribution of carbon emissions, taking into account the differences in regional endowments. Therefore, accurately identifying the spatiotemporal distribution characteristics and trends of spatial carbon emissions will provide a practical basis for policy formulation [2]. China formally committed in 2020 to peak its carbon emissions by 2030 and achieve carbon neutrality by 2060 [3]. China is still classified as a developing country, and the government faces dual pressures of economic development and carbon reduction. To minimize the economic impact of decarbonization, significant regional disparities in development must be fully considered [4,5]. Estimating regional carbon emissions is highly significant in this context [6].

Understanding the spatial distribution of carbon dioxide emissions enables decision-makers to identify high-emission areas. Human activities are a major source of carbon dioxide emissions and the primary focus of the carbon emission estimation in this paper. County-level areas, as the primary spatial units for grassroots activities, are often used as the most basic units for economic activities [7]. In China, the administrative hierarchy features several levels, with the county serving as a crucial unit of local governance. Unlike U.S. counties, which mainly function as political subdivisions of a state, Chinese counties are directly involved in local administration. U.S. counties primarily handle local services and enforce state laws, while Chinese counties focus on grassroots governance and services. These differences in structure and function are important to understand when studying carbon emissions and economic development at the county level in China, as they reflect the unique governance system that influences local decision-making and policy implementation. We can achieve a more detailed understanding of emission patterns at the county level, enabling targeted analysis that surpasses broader provincial or municipal assessments. However, obtaining county-level carbon emission data is challenging. Due to the lack of statistical data, county-level statistics suffer from issues such as opaque statistical results, inadequate data reliability, and high levels of missing data [8].

The carbon emission estimates of counties provide a basis for the assessment of local governments, which is conducive to improving the awareness of local governments and residents about carbon emissions and promoting stakeholders to actively participate in environmental protection activities. Currently, carbon emission data for China's regions are mostly reflected at the provincial level, and there are few carbon emission estimates for prefecture-level cities or counties [9]. Few studies on county carbon emissions currently exist, and they have limitations in methods, data refinement, and regional coverage [10]. In addition, considering that the proportion of carbon emissions in different industries is only available at the national or provincial level, the aggregated data obscure the differences in carbon emission characteristics between industries in different regions, which is more evident in the county. Therefore, it is necessary to distinguish carbon emissions from different sectors when estimating county carbon emissions. Refined carbon emission data enhance our ability to study how microeconomic activities influence carbon emissions, allowing for a more nuanced analysis of the temporal dynamics of emissions in relation to regional economic activities and land-use patterns. More detailed county carbon emission data will enhance policymakers' understanding of regional emission characteristics, enabling better evaluation of past policies' impact on various sectors. These data will help identify which governance strategies are most effective and differentiate carbon emission efficiency and reduction costs across industries. By considering emissions-reduction costs alongside welfare effects, policymakers can better balance environmental goals with economic growth, allowing for tailored policies in different regions that achieve maximum emission reductions at minimal cost. The compilation and analysis of carbon emission data at the county level facilitate a more precise identification of local emission sources and enable the formulation of effective strategies that align with national emission reduction targets. For enterprises, precise carbon emission data enable the identification of long-term economic impacts from investments in innovation and low-carbon technologies. By leveraging existing measures, companies can mitigate potential future costs associated with environmental regulations. A long-term analysis of sector-specific carbon emissions will allow businesses to accurately determine these costs, highlighting the practical significance of this article.

The advancement of satellite remote-sensing technology has enabled studies to estimate economic development levels [11,12], population density [13,14], and population distribution using satellite data [15]. More and more scholars are paying attention to how to use satellite data to detect carbon dioxide emissions [16]. Existing studies rarely consider the differences in the correlation between carbon emissions from different sectors and night light data when using night light data. Instead, they estimate the total regional carbon emissions directly through night light data. Obviously, due to the resolution and saturation problems of night lights, the simulation accuracy of carbon emissions still needs to be improved [17]. In this regard, by combining land use type data, not only can the carbon emissions of different sectors be estimated, but the heterogeneity of carbon emissions and night lights in different sectors is also taken into account. The spatiotemporal heterogeneity is also incorporated into the model through the Geographically and Temporally Weighted Regression model (GTWR) to achieve the purpose of a more accurate simulation of regional carbon emissions [18].

We propose a multi-scale carbon estimation strategy based on matched data from land use types and nighttime light (NTL-Landuse). The objectives of this study are as follows: (1) relying on the constructed NTL-Landuse, a multi-scale county $CO_2$ estimation strategy is developed to measure county $CO_2$ emissions under different sectors; (2) calculating the value of carbon dioxide emissions in the Yangtze River Delta; (3) introducing spatial statistical methods to examine the traits of various sectors and the overall spatiotemporal patterns of carbon emissions. (4) Evaluating carbon emission trends and formulating targeted emission reduction policies.

The main contributions of this study are as follows: (1) We combine nighttime light (NTL) data with land use information to simulate carbon emissions, improving accuracy over single-method approaches. This method addresses data limitations, offering more precise regional estimates. (2) In contrast to previous studies on total $CO_2$ emissions, we assess carbon emissions by industry sector using the NTL-Landuse framework. This multi-scale approach offers insights into sector-specific emission patterns, allowing for more targeted policy interventions. (3) This study tests the model in the Yangtze River Delta, examining carbon emissions trends and suggesting region-specific reduction policies.

## 2. Literature Review

Many scholars are now concentrating on carbon emissions accounting [19]. Due to different research objectives and purposes, the methods for accounting for carbon emissions also vary [20,21]. According to the approach of carbon accounting, it can generally be divided bottom-up and top-down [22]. The internationally accepted method is the Emission-Factor Approach proposed by the International Panel on Climate Change (IPCC) [23]. Although the IPCC method was initially developed for national-level carbon emission estimation, its theoretical framework is often used for provincial-level administrative divisions and urban [24]. The IPCC method has also been used for carbon emission estimation in various specific sectors such as the steel industry, timber industry, construction, and transportation [25–27]. These studies on carbon emission estimation provide strong practical evidence for regions to achieve policy goals of carbon neutrality and peak carbon emissions. However, the IPCC method relies on the energy balance sheet (EBT), and China only discloses these at the provincial level. Another standard bottom-up method is the carbon mass balance approach, which calculates emissions by tracking elemental inputs and outputs within an organization or system [28]. However, such methods' limitations lie in the extensive manpower, material, and financial resources required for data collection, making it suitable only for estimating carbon emissions in a specific process or system rather than for regional carbon emissions [8].

There are currently two main methods for more refined spatial carbon emission estimation. The first method is based on existing energy balance sheets. For instance,

Chen (2021) estimated industrial carbon emissions in Guangdong Province by compiling energy consumption and cement production data from various prefecture-level cities, depicting the spatiotemporal pattern of industrial carbon emissions in Guangdong Province from 2005 to 2015 [29]. Dong (2018) used input–output tables and IPCC methods to calculate carbon dioxide emissions in four direct-controlled municipalities, namely, Beijing, Tianjin, Shanghai, and Chongqing. Additionally, this study identified the urbanization rate as the primary driver of increased urban carbon dioxide emissions [30]. These studies rely on the collection of field survey data, which are not updated frequently. For example, China's input–output tables are compiled every five years. The second issue is the inevitable presence of various noises and biases in survey data, affecting the research and decision-making based on these data [31]. The third issue is that the field survey method is too costly and lacks sustainability. The second method is to use remote-sensing data for estimation. With the advancement of remote-sensing technology, estimating economic activities through satellite data has emerged as a new method [32]. Among these, NTL, highly correlated with human activities and is often used by scholars as a proxy variable to investigate human activities [33]. Elvidge (1997) laid the groundwork for understanding the correlation between NTL and carbon emissions. Their work demonstrated the potential of using NTL as a proxy variable to measure carbon emissions [34]. Chen (2020) estimated carbon emissions data for Chinese countries from 2000 to 2017 using NTL [10]. Wang (2023) combined multi-source remote-sensing data to estimate carbon emissions at the grid scale in China from 2010 to 2018 and explored potential driving factors for carbon emissions, using Hunan Province as an example [35]. In more detailed research, Zheng (2024) estimated carbon emissions patterns at the "province-city-county-township" four-level scale in Fujian Province using NTL [36]. Zhang (2024) studied carbon emissions at the street level in Xi'an using NTL [37]. Wu (2025) estimated energy-related carbon emissions in the northeast by developing a model linking NTL to emissions. He also applied the Tapio decoupling model to examine the relationship between economic development and carbon emissions, concluding that both follow a three-stage decoupling pattern, with an overall state of decoupling marked by a growth linkage [38]. Lu (2024) utilized high-resolution NTL data obtained from the domestic satellite Luojia 1–01 to estimate electricity consumption in Shenzhen [39].

To further enhance the accuracy of the estimation results, existing studies have made substantial efforts. Some studies have considered incorporating additional data to construct more refined carbon emission estimation models. Meng (2017) further improved the estimation accuracy (R2 = 0.8796) by introducing data such as population density and combining it with NTL to estimate carbon emissions [40]. Wang (2023) combined NTL and $XCO_2$ concentration data to develop a carbon emission and energy consumption estimation model, achieving spatially refined measurements of energy consumption carbon emissions [35]. In terms of model selection, existing studies have identified a strong linear correlation between carbon emissions from human energy consumption, which is why linear regression analysis is often used with regional carbon emission statistics and nighttime light data [41]. Considering that carbon emissions from fossil fuels (CEFs) between cities are not isolated, with one region's emissions being influenced by surrounding cities, a Spatial Dubin Model (SDM) is employed to address the spatial dependence issue in NTL based CEF estimation, and the use of a dynamic SDM model addresses endogeneity problems [42]. Considering the spatial heterogeneity, in addition to the application of the SDM model, some studies have incorporated the Geographically Weighted Regression (GWR) model into this estimation framework, aiming to improve the accuracy of the estimates [43].

Most related research primarily focuses on total carbon dioxide emissions, with little distinction between emissions from different sectors. Shi (2020) has started investigating the relationship between NTL and carbon emissions across various sectors. The study findings indicate that NTL can provide more accurate carbon emission assessments in urban areas with large populations and relatively developed social and economic conditions and that the precision of estimating urban carbon emissions through NTL is higher than that of estimating industrial carbon emissions [44]. Point of Interest (POI) data, as a form of multi-source geographic spatial big data, can be combined with NTL to obtain carbon emission estimates for specific sectors [45]. Wei (2024) effectively measured industrial carbon emissions in the Yellow River Basin by combining NTL and land use data, and further classified industrial carbon emissions using POI data. They categorized industrial carbon emissions into eight sectors and analyzed them individually [46]. Apart from POI, Landuse data are also commonly used to represent regional carbon emissions. Liu (2024) studied county-level Landuse carbon emissions (LUCEs) using changes in China's land use data [47]. Since directly using NTL data results in carbon emission spatialization with high-value areas overly concentrated, making it difficult to discern the internal spatial heterogeneity, combining Landuse data helps to accurately depict carbon emissions [48]. Wei (2021) estimated carbon emissions for various provinces in China by differentiating NTL data under different land use types, further refining the categories into urban, rural, and industrial sectors [49].

In the field of using NTL to measure carbon emissions, existing studies have established a relatively comprehensive research framework. Although these studies have made significant progress, there are still certain limitations. First, current research mostly focuses on depicting the total carbon emissions, and NTL alone cannot differentiate between different categories or industries of carbon emissions. Secondly, aggregating NTL data to simulate carbon emissions masks the spatial heterogeneity of emissions, resulting in overly simplified carbon emission views based on average estimates and failing to capture more effective and refined differences.

The connection and distinction between this study and existing research lie in the fact that this study draws on existing mature approaches, such as the processing of nighttime light data and the calculation of carbon emissions. At the same time, this study attempts to address the issue of sector-specific carbon emission characterization by incorporating Landuse data and solving spatial heterogeneity through the application of the GTWR model. Building on existing research, we believe it is essential to consider both sectoral and spatial heterogeneity when estimating CEF using NTL. Therefore, we propose the research hypotheses of this study. Integrating nighttime light data with land use data will yield more accurate carbon emission estimates by accounting for spatial and sectoral differences.

## 3. Study Area and Research Methods

### 3.1. Study Area

The Yangtze River Delta lies at the intersection of China's eastern coast and the Yangtze River Basin (see Figure 1). It is characterized by flat terrain, predominantly plains, with higher elevations and a combination of mountainous and hilly terrain in the southern region. The climate is hot and rainy in summer and mild and humid in winter, with an annual precipitation of over 800 mm. The area has diverse land use types, high urbanization rates, and a dense population.

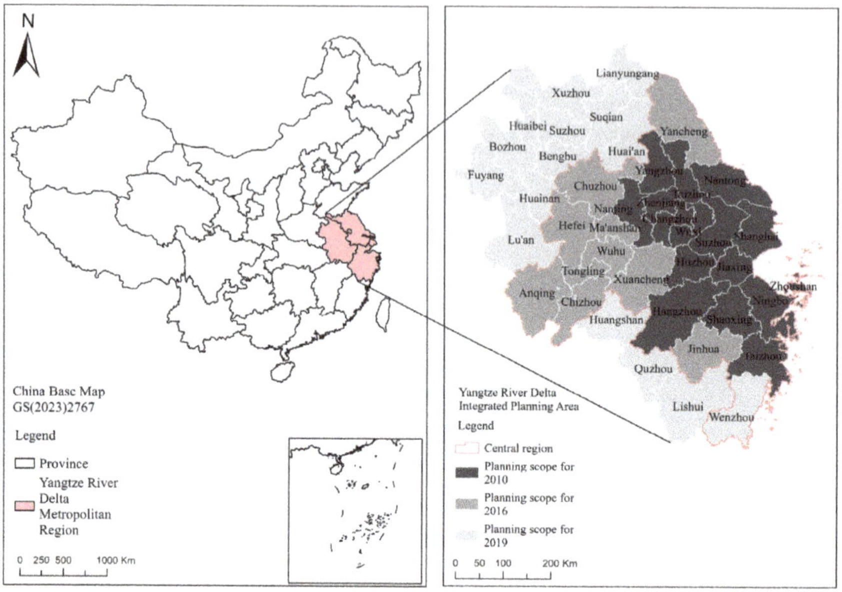


**Figure 1.** Geographical position of the Yangtze River Delta region.

According to data released by the Yangtze River Delta and Yangtze River Economic Belt Research Center, in 2023, the GDP of the three provinces and one city will exceed CNY 30 trillion, a year-on-year increase of 5.7%, accounting for 24.4% of the national total. It is one of China's largest and most robust industrial bases, concentrating on highly competitive manufacturing industry chains and modern service sectors. It is the hub for modern services and advanced manufacturing industries in China, with core industries including electronics, automobiles, and modern finance. The Yangtze River Delta region is a crucial growth hub in China's economic development and a key area for low-carbon transformation. Thus, choosing this region as the study area for this paper is both representative and significant.

*3.2. Data Sources*

We primarily utilized nocturnal light remote-sensing data, land use type data, and energy consumption data. The nocturnal light dataset was sourced from NOAA (https://earthdata.nasa.gov, accessed on 4 January 2025), predominantly employing Defense Meteorological Satellite Program (DMSP-OLS) and Visible infrared Imaging Radiometer (NPP-VIIRS) nocturnal light data. However, due to disparities in spectral resolution, spatial resolution, radiometric resolution, and product update frequency, the two sets of data are incompatible. Therefore, following Wu's (2021) approach, we integrated and calibrated DMSP-OLS annual nocturnal light data and NPP-VIIRS monthly nocturnal light data using a convolutional neural network autoencoder model [50]. This integration yielded DMSP-OLS-like extended time series data, which underwent validation to ensure reasonable accuracy and spatiotemporal consistency. The land use type dataset was obtained from the Chinese Academy of Sciences Resource and Environmental Science Data Center (http://www.resdc.cn, accessed on 10 October 2024), primarily utilizing Landsat remote-sensing image data and employing manual visual interpretation to construct the dataset. We integrated land use types into five categories: agricultural land, urban land, rural land, built-up land, and undeveloped land. Energy consumption data were obtained from the China Energy Statistical Yearbook [51]. It was further categorized into agricultural consumption, transportation consumption, wholesale and retail trade consumption, town resident consumption, rural areas resident consumption, industry consumption, and building consumption.

### 3.3. Research Framework and Research Methods

#### 3.3.1. Research Framework

In this study, we aimed to analyze the spatiotemporal variations of CEF in the Yangtze River Delta region. To achieve this goal, we can be broken down into four main phases: (1) Integration of Nighttime Light Data with Land Use Types. The first step involves integrating NTL with Landuse data for various regions within the Yangtze River Delta. This integration allowed us to correlate the intensity and distribution of lighting with specific land use categories, providing a spatially detailed basis for estimating CEF at a regional level. (2) Calculation of Total and Sectoral CEF Emissions. We calculated the total emissions of CEF across mainland China, as well as the sector-specific emissions, using official statistical data. These emissions were categorized into various sectors, such as industry, transportation, and residential, based on their respective energy consumption patterns and carbon intensity. This step provided a foundational dataset for the analysis of CEF patterns across different spatial units. (3) Selection of Relevant NTL-Landuse Data. The third phase involved selecting appropriate NTL-Landuse data by utilizing sector-specific CEF data as dependent variables. We performed a Pearson correlation analysis to identify the most relevant NTL-Landuse data that were strongly associated with CEF emissions. Once the suitable data were selected, we employed three different regression models—OLS, FE, and GTWR—to fit quantitative relationships between NTL-Landuse data and CEF emissions. These models were assessed using accuracy indicators, such as R-squared and Mean Squared Error (MSE), to determine the most appropriate model for estimation. (4) Spatiotemporal Analysis of CEF Dynamics. Finally, we integrated the estimation results obtained from the best-performing model and applied several analytical techniques to explore the spatiotemporal dynamics of CEF. This included trend analysis to identify temporal changes in CEF, hotspot analysis to detect areas with significant emission concentrations, and the use of standard deviation ellipses to examine the spatial distribution and shifts of emissions over time. These methods enabled us to visualize and interpret the regional variation and trends in CEF across the study period. The detailed process is illustrated in Figure 2.

#### 3.3.2. Correction Processing of Nighttime Light Data

Nighttime light data have been widely applied in various studies. To obtain more valuable long-term time series data, existing research often uses a fusion of DMSP-OLS and NPP-VIIRS datasets. DMSP-OLS suffers from sensor aging issues, resulting in systematic biases and light saturation [52]. These problems are not present in NPP-VIIRS; however, NPP-VIIRS itself has issues with outliers, such as unfiltered transient light sources (e.g., lasers, ships, aircraft) [53]. These distortions affect the accuracy of carbon emission estimation, so it is necessary to calibrate both datasets before their fusion. Additionally, one of the datasets provides annual data, while the other provides monthly data, and their resolutions are inconsistent. Therefore, mutual calibration and fusion of the two datasets are also required to ensure data consistency and accuracy [54].

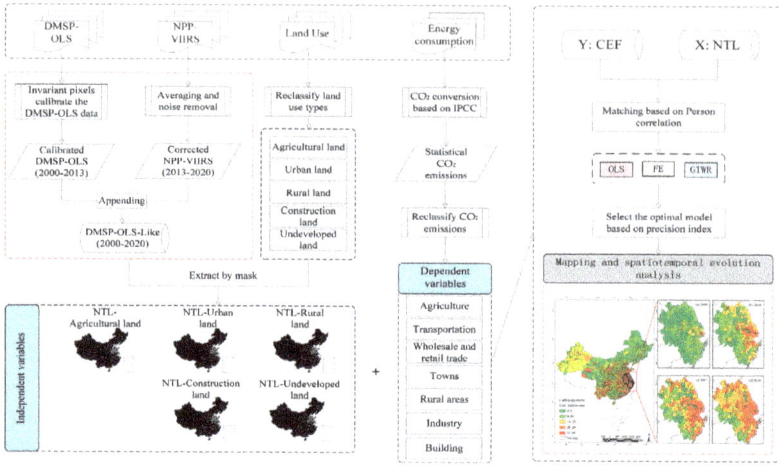


**Figure 2.** Flowchart of the methodology.

Following the common processing methods in the existing literature, we collated and integrated the two sets of data. Firstly, NOAA's F162006 DMSP-OLS data were selected as reference data, using Hegang as the pseudo-invariant calibration region with a quadratic calibration model, following Elvidge's approach (1997) [34]. Secondly, NPP-VIIRS data noise removal is essential due to the lack of filtering in the nighttime light data. Anomalous values in the original images with Digital Number (DN) values less than 0 were reassigned as 0. To account for high DN values likely representing noise (e.g., from airports or fires), Beijing, Shanghai, and Guangzhou were used as reference points, with peak DN values in other regions theoretically not exceeding these cities [55]. By optimizing the monthly data, we further obtain annual data and perform annual calibration based on the yearly data to ensure that there are no abnormal DN values. Thirdly, all spatial data were referenced to the WGS-84 datum, and the spatial resolution was resampled to 1 km [56]. Finally, logarithmic transformation of the NPP-VIIRS data reduced variance in radiance values, and, by selecting 2013 as the common coverage year, a sigmoid model was applied to convert the NPP-VIIRS data into a curve based on DMSP-OLS and the logarithmically transformed NPP-VIIRS data [57].

### 3.3.3. Method for Quantifying the Statistical $CO_2$ Emissions

This study adopts the carbon dioxide emission measurement model proposed by IPCC, which calculates carbon emissions by multiplying the combustion quantity (energy) of each energy source by its corresponding default emission factor and then summing them up to obtain the total carbon emissions [58]. The formula is as follows:

$$CEF_{ij} = C \times \left(AD_{ij} \times NCV_i \times CC_i \times O_{ij}\right) \quad (1)$$

where $i$ represents the type of fuel and $j$ represents the sector. $CEF$ denotes the carbon dioxide emissions produced from fossil fuel combustion. $AD$ represents the physical consumption of fossil energy. $NCV$ denotes the net calorific value, which refers to the heat released when a unit mass of sample burns in excess oxygen under constant volume conditions, with the combustion products composed of oxygen, nitrogen, $CO_2$, sulfur dioxide, gaseous water, and solid ash. *CC* denotes the carbon content per unit of heating value. *O* indicates the percentage of carbon in the energy source converted to $CO_2$. *C* represents the conversion factor, which is 44/12, converting the mass of carbon atoms to the mass of carbon dioxide molecules. According to the IPCC method, the emission factor (EF) can be expressed as ($NCV \times CC \times O \times 44/12$). The emission factor represents the amount of $CO_2$ emitted per unit of activity or energy consumed.

The China Energy Statistical Yearbook provides energy balance sheets by province, listing 27 types of fossil energy consumption across various industrial sectors. Considering the discrepancies between the default net calorific values in the IPCC and actual investigation results in China, the average net calorific value coefficient used in this study is sourced from the "General Principles for Comprehensive Energy Consumption Calculation" [59].

Building upon this foundation, we simulated carbon emissions. Currently, abundant literature has evidenced a significant correlation between *CEF* and *NTL* [60]. Hence, we constructed the following linear regression model:

$$CEF_i^{'} = aNTL_i + b \tag{2}$$

where $CEF_i^{'}$ represents the carbon emissions in the *i* year, *a* is the estimated parameter, and *b* is the constant term. Although there have been continuous improvements in estimating carbon emissions using *NTL* data, uncertainties still exist [61]. Firstly, the presence of light oversaturation and halo effects reduces the authenticity of *NTL* and weakens its correlation with economic activities [62]. Secondly, carbon emission estimates rely solely on a simple linear relationship with *NTL* [63], assuming that the correlation between *NTL* and carbon emissions is consistent across different spatial contexts without considering their spatial and temporal relationships, leading to estimation biases [64]. So, considering potential individual, time, and spatial effects, we further incorporated FE regression estimation and GWTR regression estimation. The optimal model was determined based on precision indicators such as MSE and R2.

In earlier related studies, most researchers adopted basic linear regression models such as OLS. While OLS has the advantage of being simple and straightforward, it is unable to account for individual-specific effects, leading to biases in carbon emissions modeling across individuals and time [65]. Panel estimation models have been widely applied to carbon emission estimation; however, the estimation strategies of panel fixed-effects regression also have certain limitations. For instance, these models typically assume that coefficients remain constant over space and time, which fails to account for localized effects [66]. To address this issue, we further introduce the GTWR model to overcome such limitations. First, GTWR allows the relationship between dependent and independent variables to vary across both space and time. This is particularly important in carbon emission estimation, as the complexity and diversity of carbon emissions result in differences across regions and over time. Second, GTWR permits local estimation of coefficients, which is especially useful in regional studies. Regional factors such as economic structures and policy environments vary across locations, and a single coefficient is insufficient to capture the nuanced spatiotemporal variations in carbon emissions. GTWR, by capturing local effects and spatiotemporal heterogeneity, typically achieves better model fit and higher explanatory power compared to panel fixed-effects models.

The specific regression model for GTWR is as follows:

$$CEF_{ik} = \beta_0(u_i, v_i, t_i) + \beta_k(u_i, v_i, t_i)NTL_{ik} + \varepsilon_i \tag{3}$$

where $CEF_{ik}$ represents the value of the carbon emission for the *k* sector. $(u_i, v_i)$ represents the longitude and latitude coordinates. $t_i$ represents time. $\varepsilon_i$ is the model error term.

$\beta_k(u_i, v_i, t_i)$ is the regression coefficient corresponding to the nighttime light data for the $k$ sector at the $i$ region.

#### 3.3.4. Kernel Density Estimation

Kernel density estimation helps us generate continuous surface density maps, allowing for the intuitive identification of high-density regions of CEF. Kernel density estimation estimates a population distribution from a sample, generating a probability density function to analyze data properties such as clustering and dispersion. The formula is as follows:

$$f(x) = \frac{1}{nh}\sum_{i=1}^{n} K\,[\frac{X_i - x}{h}] \tag{4}$$

$$\lim_{N\to\infty} h(n) = 0 \tag{5}$$

$$\lim_{n\to\infty} nh(N) = N \to \infty \tag{6}$$

where $n$ represents the sample size, $K$ denotes the kernel density function, and $h$ is the bandwidth. In urban and regional analysis, kernel functions compute values per unit area from point or line features, fitting each to a smooth, cone-shaped surface to visualize feature distribution. For point features, kernel density analysis calculates the density of each output raster cell around points, with the surface value peaking at each point and tapering to zero at the search radius. Each cell's density is the sum of all kernel surfaces stacked at its center [67].

#### 3.3.5. Directional Analysis

Directional analysis can quantitatively explain the overall characteristics of CEF in space, such as centrality, distribution, directionality, and spatial form. Common methods for spatial directional analysis include the mean center, standard deviation ellipse (SDE), and standard distance analysis. The SDE method can generate an elliptical space and describe the spatial distribution characteristics through the basic parameters such as the range, center, major axis, minor axis, and azimuth angle of the generated ellipse. Calculating the SDE first requires determining the mean center of the features.

$$\overline{X} = \frac{\sum_{i=1}^{n} x_i}{n},\ \overline{Y} = \frac{\sum_{i=1}^{n} y_i}{n} \tag{7}$$

$$\overline{X_w} = \frac{\sum_{i=1}^{n} w_i x_i}{\sum_{i=1}^{n} w_i},\ \overline{Y_w} = \frac{\sum_{i=1}^{n} w_i y_i}{\sum_{i=1}^{n} w_i} \tag{8}$$

Obtain the calculated average center as the center of the ellipse. The standard deviations of each element's $X$ and $Y$ coordinates are calculated to define the ellipse's major and minor axes and determine its direction. The formula is as follows:

$$SDE_x = \sqrt{\frac{\sum_{i=1}^{n}(x_i - \overline{X})^2}{n}},\ SDE_y = \sqrt{\frac{\sum_{i=1}^{n}(y_i - \overline{Y})^2}{n}} \tag{9}$$

where $SDE_x$ and $SDE_y$ represent the coordinates of the center of the ellipse. To find the angle between the X-axis and true north (0 degrees), calculate accordingly.

$$A = \left(\sum_{i=1}^{n} \tilde{x}_i^2 - \sum_{i=1}^{n} \tilde{y}_i^2\right) \tag{10}$$

$$B = \sqrt{\left(\sum_{i=1}^{n} \tilde{x}_i^2 - \sum_{i=1}^{n} \tilde{y}_i^2\right)^2 + 4\left(\sum_{i=1}^{n} \tilde{x}_i \tilde{y}_i\right)^2} \tag{11}$$

$$C = 2\sum_{i=1}^{n} \tilde{x}_i \tilde{y}_i \quad (12)$$

$$tan\theta = \frac{A + B}{C} \quad (13)$$

$\tilde{x}_i$ and $\tilde{y}_i$ represent the difference between the $i$-th element and the mean center. $\theta$ is the angle. Finally, calculate the lengths:

$$\sigma_x = \sqrt{2}\sqrt{\frac{\sum_{i=1}^{n}\left(\tilde{x}_i cos\theta - \tilde{y}_i sin\theta\right)^2}{n}} \quad (14)$$

$$\sigma_y = \sqrt{2}\sqrt{\frac{\sum_{i=1}^{n}\left(\tilde{x}_i cos\theta + \tilde{y}_i sin\theta\right)^2}{n}} \quad (15)$$

The mean center of the ellipse represents the centroid of the spatial distribution of CEF, while the azimuth angle indicates its primary direction. The major axis highlights areas with a higher concentration of CEF, while the minor axis represents more dispersed areas. If both axes are nearly equal in length, it suggests that the CEF distribution is uniform and does not exhibit clear directional characteristics.

## 4. Results and Analysis

### 4.1. Correlation Analysis of CEF and NTL-Landuse

The vast majority of nighttime light data recorded by satellites from space are generated by human activities and have been proven to be related to economic activities [68,69]. These human economic activities are significant sources of CEF, resulting in a high correlation between NTL and CEF [43]. Based on data availability, this study categorizes energy consumption-related carbon emissions into various sectors: agricultural consumption, transportation consumption, wholesale and retail trade consumption, urban residential consumption, rural residential consumption, industrial consumption, and building consumption. Considering the distinct characteristics of each sector, some sectors have minimal nighttime activities and, thus, cannot be directly captured through NTL data. These are classified as indirect sectors (e.g., agriculture, transportation, wholesale and retail trade). In contrast, sectors with substantial nighttime activities, which can be directly captured by NTL data, are classified as direct sectors (e.g., towns, rural areas, industry, and building).

To construct the estimation model more effectively, it is essential to match CEF from different industry sectors with NTL under various land use types. Apart from the insignificant correlation between NTL corresponding to undeveloped land and CEF emissions from various sectors, significant positive correlations are observed between NTL and CEF under other land use types. However, there are significant differences in the correlation coefficients between NTL and CEF across different land use types (see Table 1).

We need to compare the correlation coefficients between the NTL-Landuse and NTL-Total datasets and the carbon emissions across various sectors to select the appropriate independent variables. From a theoretical perspective, sectoral carbon emissions are closely related to land-use patterns (e.g., agricultural land tends to have more agricultural carbon emissions rather than industrial or construction-related emissions). The correlation between the matched NTL and CEF is shown in Table 2. By comparing with the correlation results of NTL-Total, it can be observed that, except for the agricultural sector, NTL-Landuse has higher correlation coefficients with CEF in all sectors. Although the difference in correlation coefficients between NTL-Landuse and NTL-Total for CEF is small, we still have reasons to believe that using NTL-Landuse can improve the accuracy of CEF estimation. Therefore, using NTL-Landuse as an independent variable is appropriate. Secondly, comparing the correlation coefficients of different sectors with NTL-Landuse data reveals that direct sectors exhibit coefficients ranging between 0.8 and 1, indicating a high linear correlation. This suggests that NTL-Landuse effectively captures the CEF of these sectors. In contrast, indirect sectors have correlation coefficients between 0.6 and 0.8, lower than those of direct sectors but still within a range of significant linear correlation. The existence of differences in correlation coefficients between NTL-Landuse and NTL-Total for CEF also indicates that the aggregated NTL neglects the sectoral heterogeneity when estimating carbon emissions. Therefore, incorporating NTL-Landuse will help identify this sectoral heterogeneity. This is especially important in regions where land use types have distinct structures, as the use of NTL-Landuse will be crucial for identifying CEF. Additionally, relying solely on NTL-Total cannot accurately assess the spatial distribution of sectoral CEF, as both are estimated based on the same nightlight distribution, which can introduce biases when mapping the spatial distribution of sectoral CEF. In such cases, even a slight improvement in accuracy makes the choice of NTL-Landuse for CEF estimation significant for enhancing the precision of identifying emission spatial distribution.

**Table 1.** Pearson correlation coefficient between NTL-Landuse and CEF.

| Type | Total | Indirect Section | | |
|---|---|---|---|---|
| | | Agriculture | Transportation | Wholesale and Retail Trade |
| Agricultural land | 0.8720 * | 0.6554 * | 0.7088 * | 0.7027 * |
| Urban land | 0.8708 * | 0.5840 * | 0.7576 * | 0.7037 * |
| Rural land | 0.8648 * | 0.5797 * | 0.6290 * | 0.6010 * |
| Construction land | 0.7163 * | 0.4485 * | 0.7151 * | 0.6810 * |
| Undeveloped land | 0.0597 | 0.2671 * | −0.0158 | 0.0281 |
| **Type** | **Direct Section** | | | |
| | **Towns** | **Rural areas** | **Industry** | **Building** |
| Agricultural land | 0.8271 * | 0.8052 * | 0.8425 * | 0.7568 * |
| Urban land | 0.8678 * | 0.7453 * | 0.8389 * | 0.7547 * |
| Rural land | 0.7881 * | 0.7870 * | 0.8559 * | 0.6589 * |
| Construction land | 0.7362 * | 0.6814 * | 0.6665 * | 0.7770 * |
| Undeveloped land | −0.0576 | −0.0606 | 0.0754 | 0.0137 |

Note: * denotes $p$-values were all <1%, passing the significance test of 1%.

**Table 2.** NTL-Landuse and NTL-Total correlation comparison.

| | Sector | NTL-Landuse | NTL-Total |
|---|---|---|---|
| Indirect Section | Agriculture | 0.6554 * | 0.6622 * |
| | Transportation | 0.7961 * | 0.7300 * |
| | Wholesale and retail trade | 0.7430 * | 0.7143 * |
| Direct Section | Towns | 0.8945 * | 0.8479 * |
| | Rural areas | 0.8295 * | 0.8078 * |
| | Industry | 0.8681 * | 0.8661 * |
| | Building | 0.8065 * | 0.7708 * |
| | Total | 0.8946 * | 0.8946 * |

Note: * denotes $p$-values were all <1%, passing the significance test of 1%.

In addition, this study incorporates the Spatial Autoregressive Model (SAR) and Spatial Durbin Model (SDM) to examine whether these models can provide more accurate estimations. According to the results of the Akaike Information Criterion (AIC), the AIC value of the GTWR model is significantly smaller than those of the SAR and SDM models, indicating that the GTWR model is the more appropriate choice. Furthermore, the calculation of precision metrics also shows that the GTWR model achieves higher accuracy than both the SAR (Root Mean Squared Error (RMSE) = 0.6301; Mean Absolute Error (MAE) = 0.2602) and SDM (RMSE = 0.6505; MAE = 0.2825) models. This further demonstrates the rationality of using the GTWR model for carbon emission estimation.

### 4.2. CEF Calibration and Validation

We chose three models: OLS, FE, and GTWR. The regression parameters were all logarithmically transformed and considered time effects. In Table 3, by computing the RMSE, R2, and MAE for 30 provinces across the three models, using NTL-Total as the independent variable, the R2 values for OLS, FE, and GTWR were found to be 0.8408, 0.12575, and 0.93133. The corresponding RMSE values were 0.33204, 0.77811, and 0.21807. It can be concluded that the GTWR model outperforms OLS and FE in terms of fitting effect. Furthermore, by comparing the MAE, it is evident that the GTWR model has the smallest value at 0.17324, indicating that the GTWR model estimation results in smaller errors compared to OLS and FE. Similar results were observed when using NTL-Landuse as the independent variable. Additionally, through a comparison between the results obtained using NTL-Total and NTL-Landuse, it was found that both cases effectively reflected CEF. Therefore, it is deemed reasonable to adopt NTL-Landuse as the independent variable and GTWR as the estimation model.

**Table 3.** Performance comparison of three models for estimating CEF.

| Independent Variables | Index | OLS | FE | GTWR |
|---|---|---|---|---|
| NTL-Total | RMSE | 0.33204 | 0.77811 | 0.21807 |
| | R2 | 0.8408 | 0.8751 | 0.93133 |
| | MAE | 0.25601 | 0.64211 | 0.17324 |
| NTL-Landuse | RMSE | 0.33171 | 0.44904 | 0.21858 |
| | R2 | 0.84112 | 0.70884 | 0.93101 |
| | MAE | 0.25223 | 0.34431 | 0.17278 |

### 4.3. Mapping CEF

Using NTL-Landuse as the independent variable input into the GTWR model yielded the estimation results of CEF. To observe the spatial heterogeneity of carbon emissions at the county level, we visualized the calculated CEF using ArcGIS (see Figure 3). The resulting map provides a spatial distribution of CEF at the county level across the study region. This map does not distinguish between emissions from specific industries within these counties but rather illustrates the overall carbon emissions at the county scale. The spatial distribution of CEF shows distinct patterns. Low CEF values are primarily observed in the northeast, northwest, and most rural areas of the central region. In contrast, high CEF values are predominantly concentrated in developed urban areas, where industrial activities, dense populations, and vigorous economic activities contribute to high energy consumption and $CO_2$ emissions. Notably, the Yangtze River Delta region stands out for its high CEF, which results from the concentration of energy-intensive industries, a large population, and a dynamic economy. This is also the reason why we chose the Yangtze River Delta as our research area. By comparing the spatial distribution of CEF in the Yangtze River Delta region in different periods, it can be found that carbon emissions show a clear upward trend. Specifically, CEF in Zhejiang Province increased by 281.19%, in Shanghai Municipality by 83.38%, in Jiangsu Province by 327.93%, and in Anhui Province by 228.85%.

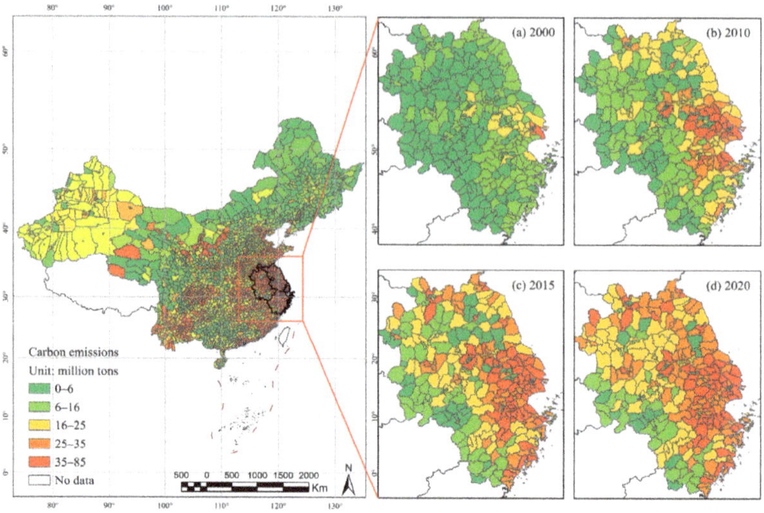


**Figure 3.** Spatial mapping of CEF.

## 4.4. *Temporal Dynamics Characteristics of CEF*

The variation of CEF during the period 2000–2020 is depicted in Figure 4. Overall, over the decade from 2005 to 2015, the CEF of the study area exhibited a rapid upward trend, followed by a subsequent decline in slope, with Shanghai being the first to show a downward trend. When examining the regions individually, apart from Shanghai, the other three provinces all demonstrated a continuous upward trend.

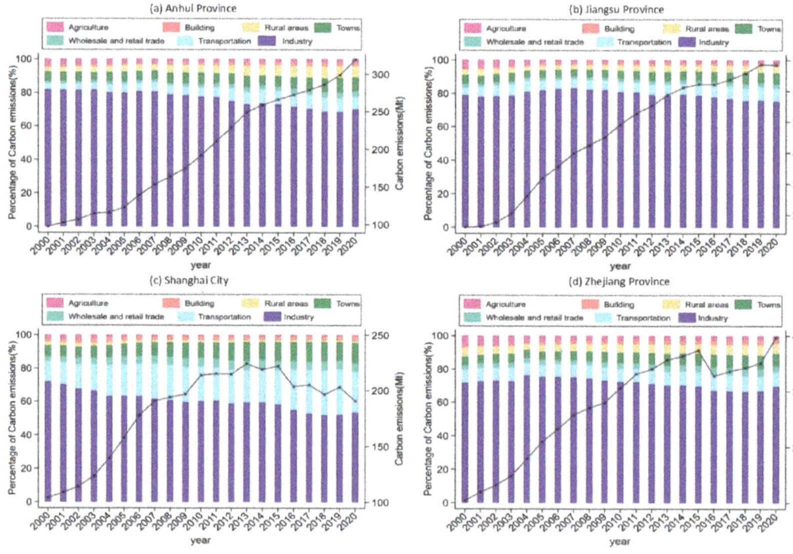


**Figure 4.** Changing trend of CEF in the Yangtze River Delta.

Jiangsu Province experienced the fastest growth in CEF, with a slope of 26.24. In 2007, industrial energy consumption was the primary source of its CEF, accounting for a remarkable 83.18%, the highest among all regions in the study area. By 2020, this proportion had decreased to 75.22%, yet it remained the highest level among the four provinces and municipalities examined. Although the growth of CEF in Jiangsu Province has slowed since the 13th Five-Year Plan period, a clear turning point in emissions has yet to be observed. This is largely due to surrounding areas transferring outdated, energy-intensive, and polluting industries to central and northern Jiangsu. At present, the northern Jiangsu region has formed a structural feature dominated by heavy industries such as chemical industry, coal, and machinery. The industrial sector consumes the most energy among the three major industries, which contributes to the ongoing increase in CEF emissions in the region.

Following Jiangsu Province, Zhejiang Province experienced the next highest growth rate, with a slope of 14.64. Due to the relative scarcity of oil, coal, and electricity resources in Zhejiang Province, it relies heavily on external energy consumption. With the acceleration of urbanization, total energy consumption in Zhejiang Province has continued to increase, with coal consumption reaching 12,758.22 million tons in 2020, accounting for 39% of total primary energy consumption. Although there was a significant decline in Zhejiang Province's CEF in 2016, it resumed growth thereafter, with the growth rate showing an increasing trend year by year.

Anhui Province, another major industrial region, showed a slope of 11.12, with industrial CEF averaging 76.52%. Although there was a certain downward trend in industrial energy consumption carbon emissions during the study period, the decline was relatively weak. Energy consumption carbon emissions from transportation, urban residents, and rural residents have significantly increased, as the growth rate of energy consumption linked to economic activities in the region has outpaced that of industrial energy consumption. During the 13th Five-Year Plan period, the Anhui provincial government introduced a series of policies. However, this has not effectively curbed the overall growth of CEF, as evidenced by the results.

Shanghai Municipality, the most economically developed region in China, also experienced a rapid rise in CEF in the early stages. After 2014, industrial restructuring in Shanghai caused many industrial entities to relocate to nearby cities, resulting in a decrease in industrial CEF, which fell to 53.45% by 2020. This represented the peak of overall CEF since 2013, after which it began to decline. Meanwhile, as urbanization continued to expand, the proportion of CEF from urban residents and transportation increased. Conversely, the proportion of CEF from rural residents declined. Agriculture in Shanghai relies mainly on imports, with minimal local production, thus exerting a relatively minor impact on overall CEF, accounting for only 0.62% in 2020.

### 4.5. Spatial Change Characteristics of CEF

From 2000 to 2020, the CEF in the study area initially exhibited a bimodal distribution (See Appendix A). Apart from the low peak, a small high peak area formed in Shanghai and its surrounding areas. As the scale of CEF emissions continues to expand across regions, the bimodal distribution gradually shifts to a unimodal distribution. The kernel density curve shows a rightward shift, and, although it still exhibits a right-skewed distribution, the skewness value has decreased year by year, indicating a consistent increase in CEF. Meanwhile, the peak of the CEF kernel density has shown a declining trend, suggesting that the data distribution is becoming more dispersed. Specific to the performance in space, Figure 5 clearly shows that Shanghai is developing into a high CEF emission center, which gradually expands to surrounding areas, resulting in a higher proportion of high CEF regions.

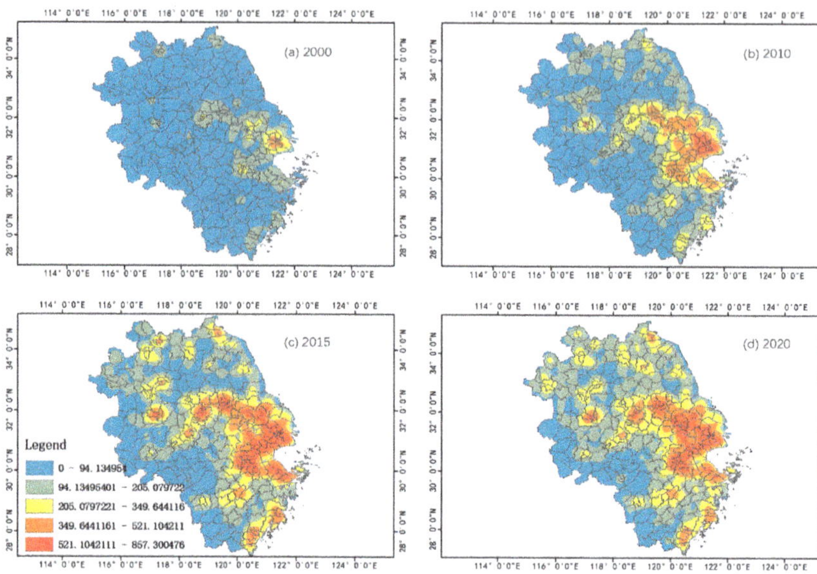


**Figure 5.** Spatial distribution of CEF in the Yangtze River Delta.

In 2000, the overall CEF scale in the study area remained relatively low. However, Shanghai and its surrounding areas exhibited significantly higher CEF emissions levels than other regions during the same period, forming the embryonic stage of future carbon emission centers. As the level of economic development continued to rise, the regional industrialization level improved. By 2010, Shanghai had already demonstrated distinct features of high-density CEF centers, spreading to surrounding areas and displaying characteristics of a central pole. In 2015, CEF in Shanghai peaked and began to decline, with the regional diffusion effect during this phase being significantly greater than the agglomeration effect. The high-density CEF range gradually spread to counties and districts corresponding to Suzhou, Wuxi, Jiaxing, Hangzhou, and other areas. At the same time, scattered small, high-density CEF centers began to emerge in cities such as Nanjing, Hefei, Yangzhou, and Ningbo, forming the basic spatial structure of the carbon emission network. By 2020, the high-density carbon emission areas centered around Shanghai continued to increase and connect with secondary central regions. During this phase, the CEF in Shanghai and Suzhou decreased due to the optimization of their industrial structures and the relocation of inefficient capacity.

The spatial characteristics of CEF emissions across different sectors in the Yangtze River Delta Region generally exhibit similar distribution patterns. Still, there are subtle differences in high-density areas (see Figure 6). Industrial CEF emissions are primarily concentrated in the Minhang District, Jiading District, and Pudong New Area of Shanghai, as well as in surrounding areas such as Suzhou, Hangzhou, and Ningbo, showing distinct spatial agglomeration features. The spatial distribution of transportation CEF emissions is relatively dispersed, mainly forming small high-density areas around important transportation hubs such as Hefei, Nanjing, Yangzhou, and Wuxi. Wholesale and retail trade, construction, and urban resident CEF emissions exhibit similar distributions, mainly concentrated in several large cities and their surrounding areas within the region, without forming obvious spatial agglomerations. Rural resident CEF emissions form a high-density area from Wuxi to Shanghai to Hangzhou, with small high-density areas also present in Hefei and Lianyungang. Agricultural CEF emissions are primarily concentrated in southern

Jiangsu Province, where the high proportion of plains and abundant water resources create contiguous high-density areas. In 2020, Jiangsu's grain production reached 745.8 billion tons, 44.64% of the Yangtze River Delta's total. Agricultural modernization is advanced, with crop cultivation mechanization at 93% and overall agricultural mechanization at 88%, leading the nation.

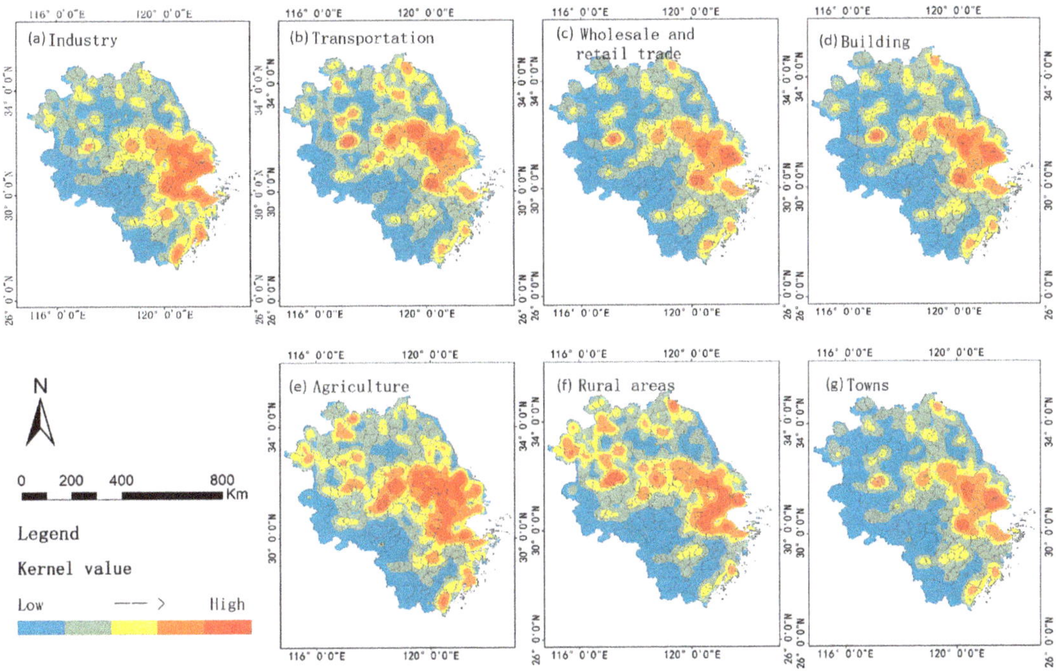


**Figure 6.** Spatial distribution of CEF in various departments in the Yangtze River Delta.

### 4.6. Spatial Correlation Analysis of CEF

Spatial correlation is primarily assessed using the Moran's I index. When Moran's I is greater than 0, it indicates positive spatial autocorrelation within the region; when Moran's I is less than 0, it indicates negative spatial autocorrelation within the region; when Moran's I equals 0, it suggests the data exhibit random spatial distribution characteristics within the region. The overall Moran's I index in the study area can be calculated. In Table 4, the overall Moran's I for CEF increased from 0.315 in 2000 to 0.38 in 2020, indicating a continuous strengthening of the spatial clustering of CEF.

**Table 4.** Global Moran's I values for CEF.

| Year | Sector CEF | Moran's I | *p*-Value |
|---|---|---|---|
| 2000 | Total | 0.315 | 0.001 |
| | Agriculture | 0.415 | 0.001 |
| | Industry | 0.339 | 0.001 |
| | Building | 0.236 | 0.002 |
| | Transportation | 0.216 | 0.001 |
| | Wholesale and retail trade | 0.256 | 0.002 |
| | Towns | 0.262 | 0.001 |
| | Rural areas | 0.240 | 0.001 |

**Table 4.** *Cont.*

| Year | Sector CEF | Moran's I | *p*-Value |
|---|---|---|---|
| 2010 | Total | 0.372 | 0.001 |
| | Agriculture | 0.404 | 0.001 |
| | Industry | 0.394 | 0.001 |
| | Building | 0.302 | 0.001 |
| | Transportation | 0.310 | 0.001 |
| | Wholesale and retail trade | 0.307 | 0.001 |
| | Towns | 0.309 | 0.001 |
| | Rural areas | 0.311 | 0.001 |
| 2015 | Total | 0.366 | 0.001 |
| | Agriculture | 0.402 | 0.001 |
| | Industry | 0.411 | 0.001 |
| | Building | 0.266 | 0.001 |
| | Transportation | 0.281 | 0.001 |
| | Wholesale and retail trade | 0.274 | 0.001 |
| | Towns | 0.286 | 0.001 |
| | Rural areas | 0.278 | 0.001 |
| 2020 | Total | 0.380 | 0.001 |
| | Agriculture | 0.354 | 0.001 |
| | Industry | 0.413 | 0.001 |
| | Building | 0.326 | 0.001 |
| | Transportation | 0.352 | 0.001 |
| | Wholesale and retail trade | 0.335 | 0.001 |
| | Towns | 0.358 | 0.001 |
| | Rural areas | 0.308 | 0.001 |

Looking at different sectors, except for agriculture, which showed a significant decrease in global Moran's I, the other sectors all exhibited a clear upward trend. This is mainly due to the increasing urbanization rate, where the population has transitioned from scattered distribution to concentrated distribution, leading to a concentration of economic activities. Agriculture, constrained by land resources, cannot change with population migration, resulting in a decrease in the clustering degree of agricultural CEF.

To further understand the local differences in CEF, hotspot analysis was conducted on the dataset. Hotspot analysis is a form of local autocorrelation analysis. It calculates a statistic for each feature in the dataset and provides an evaluation of whether its surrounding environment forms a hotspot or a cold spot. A hotspot signifies high values in both the area and its surroundings, while a cold spot indicates low values in both. In general, positive values of GiZScore indicate clusters of high values (hot spots), while negative values indicate clusters of low values (cold spots). CEF hotspots are primarily found in the central and eastern regions, whereas cold spots are mainly in the western areas, with an intermediate buffer zone in between (see Figure 7). After experiencing regional expansion from 2000 to 2015, hotspot areas showed signs of contraction from 2015 to 2020. The number of cold spot areas remained relatively stable at around 35 counties (districts).

In 2000, 23 counties (districts) were identified as hotspots at a 99% confidence level, covering most of Shanghai, Suzhou, Wuxi, and areas like Haiyan County in Jiaxing, and Yuyao and Jiangbei in Ningbo. Only Huangshan District in Huangshan City was identified as a cold spot at the same confidence level. By 2020, hotspots at 99% confidence rose to 26 counties, though, in Shanghai, only Qingpu, Fengxian, and Chongming remained. New hotspots included Chongxing County and Wuxing District in Huzhou and Rugao City in Nantong. Cold spots at this confidence level also grew to 12, encompassing areas like Anqing, Huangshan, Xuancheng, and Longquan in Zhejiang, showing a strengthened spatial clustering of CEF.

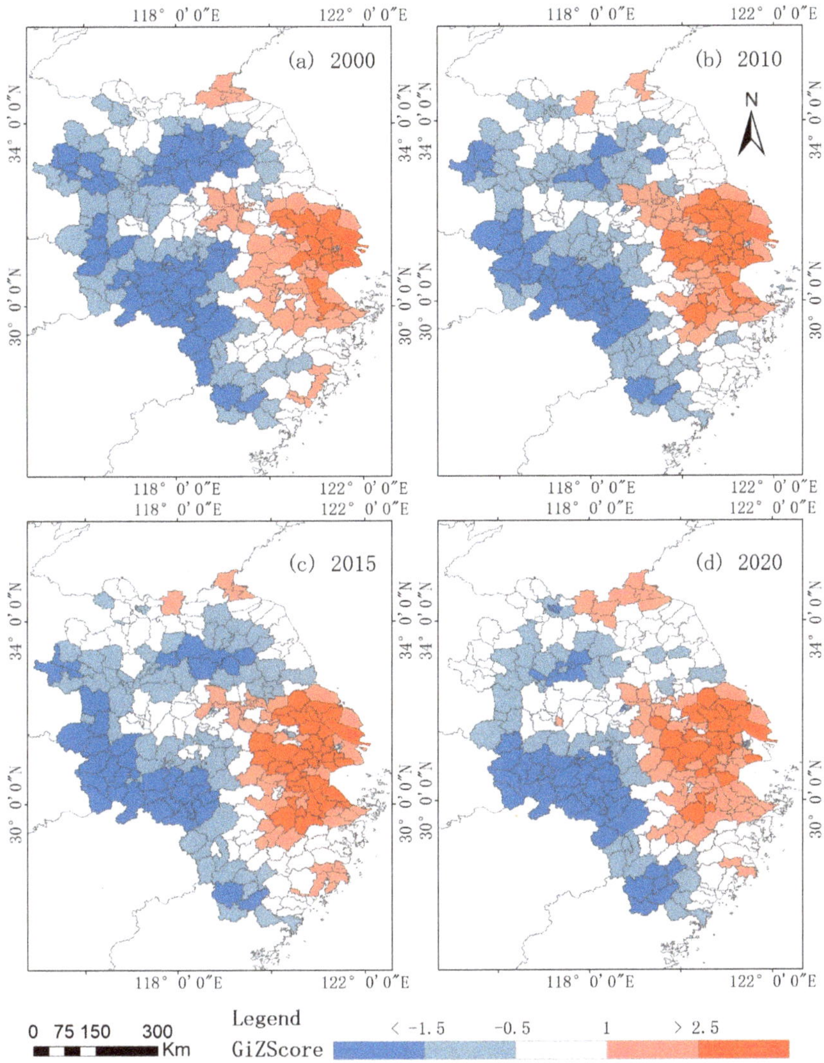


**Figure 7.** Distribution pattern of CEF hot spots in the Yangtze River Delta.

Further calculations of the standard deviation ellipse parameters for CEF in the study area showed a continuous increase in ellipse area from 2000 to 2020, indicating a gradual expansion of CEF coverage. The "Rotation" angle parameter suggests a distribution pattern of CEF in the "southeast–northwest" direction (see Table 5).

**Table 5.** CEF standard deviation ellipse of the Yangtze River Delta.

| Year | Center X | Center Y | X Std Dist | Y Std Dist | Shape-Area | Rotation |
|---|---|---|---|---|---|---|
| 2000 | 119.9208 | 31.2356 | 1.548122 | 2.686693 | 13.54527 | 142.8634 |
| 2010 | 119.7034 | 31.31735 | 1.601379 | 2.696235 | 13.72098 | 139.1775 |
| 2015 | 119.5993 | 31.25213 | 1.631479 | 2.820925 | 14.23804 | 139.3094 |
| 2020 | 119.4639 | 31.36714 | 1.675366 | 2.870402 | 14.52843 | 138.3116 |

The CEF emissions center in the study area, initially located in Wuxi City, Jiangsu Province (31.2356° N, 119.9208° E), gradually shifted northwestward. By 2020, the CEF emissions center had relocated to Changzhou City, Jiangsu Province (31.36714° N, 119.4639° E). The flattening rate decreased from 42.38% to 41.63%, showing a weakening directional orientation in the southeast–northwest direction.

## 5. Discussion

### 5.1. Cross-Validation

While NTL data offer a feasible approach to estimating CEF at the county or sub-regional levels, some errors remain unavoidable, including NTL data, model estimation, and carbon measurement errors, causing deviations from true CEF values. This research does not aim to replace existing carbon estimation methods but to enhance them, focusing on capturing CEF trends and spatial patterns to identify key hotspot and coldspot areas for policymakers. To validate our results, we compare them with established carbon emission databases, specifically the CEADs China City Carbon Emission Database [70], derived from energy balance sheets and industrial energy consumption, and the China County Carbon Emission Database, based on remote-sensing data [10]. Since both databases provide total emissions estimates, this paper also aggregates sector carbon emissions to generate a total and examines the correlation among the various datasets. Ideally, since the errors of each method are independent, they do not have a significant impact on the correlation between the datasets. Therefore, a high correlation should be observed between different datasets.

Table 6 indicates that the correlation between CEF and county CE is 0.7416, city CE_CEADs is 0.7202, and province CE_CEADs is 0.8773. All these correlation coefficients are more significant than 0.7. Therefore, our calculated results are significantly linearly correlated and relatively reliable. Additionally, the CEF estimated for direct departments using NTL data shows a more significant correlation with other datasets than that for indirect departments. This discrepancy highlights the varying performance of NTL in estimating CEF across different sectors, primarily due to the limited nighttime activities of indirect departments, which introduces more noise into the estimates.

**Table 6.** Cross validation results.

| | County CE | City CE_CEADs | Province CE_CEADs |
|---|---|---|---|
| CEF | 0.7416 * | 0.7202 * | 0.8773 * |
| Direct CEF | 0.7641 * | 0.7339 * | 0.8720 * |
| Indirect CEF | 0.6240 * | 0.6405 * | 0.8559 * |

Note: * denotes $p$-values were all <1%, passing the significance test of 1%.

### 5.2. Result Analysis

This study uses nighttime light data to analyze county-level carbon emissions, focusing on the Yangtze River Delta. It measures the spatiotemporal distribution of carbon emissions across counties in the region.

Figure 8 shows the overall characteristics of the area, except for negative growth in CEF emissions in 2016, the CEF emissions in the study area showed positive growth throughout the study period. This conclusion is consistent with that of Wu (2021) [71]. This trend deviates from the "dual carbon" targets set by the central government, indicating significant challenges in achieving these goals. The Yangtze River Delta Region faces relative energy scarcity, relying heavily on external inputs for production activities [72]. The region has many industrial parks that contribute to high energy consumption, especially of fossil fuels, causing an annual rise in CEF emissions.

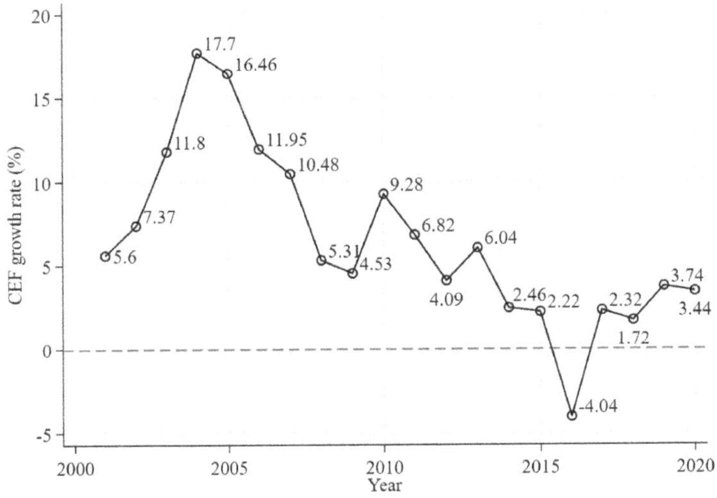


**Figure 8.** CEF growth rate of Yangtze River Delta.

Observing the local characteristics of the region, we identified carbon emission hotspots in the Yangtze River Delta. In the early years, carbon emissions in the Yangtze River Delta were primarily concentrated in Shanghai, but, over time, they gradually expanded to surrounding areas. Meanwhile, other cities in the region, such as Nanjing, Hefei, Yangzhou, and Ningbo, also emerged as secondary growth poles for carbon emissions. In 2000, Shanghai's CEF emissions accounted for 22.34%. By 2020, Shanghai's share had decreased to 11.97%. Jiangsu Province has the highest proportion, with its CEF emissions accounting for 34.39% of the total in 2020 and rising to 43.02% by 2020. This is mainly due to Jiangsu Province's rapid industrial development and its absorption of outdated production capacity transferred from Shanghai. This indicates that, over the past 20 years, carbon emissions in the Yangtze River Delta have shown spillover effects. In addition to the original growth pole, Shanghai, new secondary growth poles have emerged. Therefore, it is essential to focus on the carbon emissions from the CEF centers and implement targeted policy interventions. At the same time, due to the existence of spillover effects, achieving carbon reduction through independent local governance models has become increasingly difficult. Thus, coordinated carbon reduction efforts among urban agglomerations are urgently needed to promote high-quality development and effectively reduce regional CEF emissions.

According to the second law of geography, which states that spatial separation creates differences among geographic features, there is significant heterogeneity in sectoral carbon emissions across regions [73]. The distribution of carbon emissions also varies by sector. Industrial CEF emissions are primarily concentrated in Shanghai's Minhang District, Jiading District, and Pudong New Area, as well as in surrounding areas such as Suzhou, Hangzhou, and Ningbo. Transportation CEF emissions are more spatially dispersed, forming small high-density clusters around key transportation hubs such as Hefei, Nanjing, Yangzhou, and Wuxi. The CEF emissions from wholesale and retail, construction, and urban residential sectors share similar spatial distributions, primarily concentrated in major cities within the region and their surrounding areas. Agricultural CEF emissions are mainly concentrated in southern Jiangsu. Given this sectoral heterogeneity, targeted emission reduction policies can be developed. For instance, regions where emissions are concentrated in the industrial sector may benefit from investments in clean technology and energy efficiency. Similarly, areas with high transportation emissions could focus on improving public transportation infrastructure or incentivizing the adoption of electric vehicles.

### 5.3. Strengths and Limitations

The combination of nighttime light data and land use type produces better regional carbon emissions simulations, achieving a goodness of fit of 0.931, which exceeds estimates from traditional econometric methods [56,65]. Estimates indicate that carbon emissions in the study area rose from 2000 to 2020. Initially showing a bimodal spatial distribution, emissions evolved into a right-skewed pattern centered around Shanghai, with concentric radiation and significant spatial autocorrelation. Hotspots and cold spots form a three-tiered west-to-east pattern, aligning with past findings [74]. Unlike previous models relying solely on nighttime light data for carbon emissions, our model includes land use data, allowing for sector-specific carbon estimates and multiscale analysis of regional emission patterns.

The carbon emission model proposed in this study applies to China and other countries or regions, given the difficulty in obtaining statistical data on carbon emissions. In such cases, the framework proposed in this study can be used for estimation, enabling finer monitoring of carbon emission spatial patterns and providing practical evidence for formulating regional carbon reduction policies.

Although our study provides more accurate multiscale estimates of carbon emissions, there are still limitations. Firstly, the model must demonstrate a better fit for the CEF of indirect sectors. This is primarily due to the weak nighttime activity in these sectors, making it difficult for NTL to capture their CEF directly. Instead, CEF must be estimated indirectly through the correlation between NTL and human activities. However, given that the CEF of indirect sectors constitutes a tiny portion of the total CEF, this does not significantly impact the main research conclusions. Secondly, we focus on using nighttime light data as a proxy for carbon emissions, which serves as a macro indicator of carbon emissions at the regional level. Additionally, considering the heterogeneity across industries and regions, we incorporate land use data and the GTWR model for further identification. However, due to the lack of more granular carbon emission data, we were unable to further refine the analysis at the industry level. Previous studies have used input–output tables to achieve more detailed sectoral carbon emission estimates; however, due to the large time span of input–output tables, the fit is not ideal [75]. In the future, with updated data, subsequent research can consider improving the carbon emission estimates for different sectors, which will provide more accurate cross-industry and cross-region carbon emission estimates. This will help capture the diversity and complexity of carbon emissions across regions and industries more accurately. Finally, the nighttime light data offer carbon emission grid data at a 1 km $\times$ 1 km resolution, failing to capture detailed urban emission patterns [76]. Future research should explore higher-resolution satellite data for more precise carbon emissions analysis.

## 6. Conclusions

This study estimates the CEF of various industries in China. It helps to supplement the existing database. The optimal fitting model is selected based on accuracy indicators to analyze the spatiotemporal patterns of CEF at different scales. The main conclusions are as follows:

(1) The dataset combining nighttime light data with land use type demonstrates a stronger correlation with CEF than nighttime light data alone. Model accuracy metrics show higher R-squared values and lower RMSE and MAE when NTL and land use data are used as independent variables, indicating that the NTL-Landuse approach is

superior for CEF estimation. In model selection, we consider potential time, individual, and spatial effects, with results from OLS, FE, and GTWR models, where GTWR outperforms both OLS and FE.

(2) From 2000 to 2020, the total CEF in the Yangtze River Delta increased, with Shanghai experiencing a significant turning point while the other three provinces continued to rise. The primary driver was growth in industrial CEF, followed by urban population increases and expansions in transportation and wholesale/retail CEF due to urbanization. At the same time, agricultural and rural resident CEF saw smaller gains.

(3) In 2000, the Yangtze River Delta region had low carbon emissions and lacked distinct spatial distribution characteristics. By 2010, Shanghai emerged as a significant growth pole, influencing cities like Suzhou and Jiaxing. By 2015, the high-density CEF area radiating from Shanghai expanded to include Suzhou, Wuxi, Jiaxing, and Hangzhou, with other regions also beginning to show small high-density centers. In 2020, while the total CEF in Shanghai began to decline, its radiating area expanded further, connecting with secondary centers to form large contiguous regions. Although industrial CEF decreased across all provinces during the study period, only Shanghai saw a notable drop in its industrial CEF proportion, while other provinces experienced smaller decreases. Additionally, industrial CEF emissions clustered in adjacent areas of Jiangsu, Zhejiang, and Shanghai, whereas agricultural CEF displayed a dispersed trend. The transportation CEF concentrated around major transport hubs, while the wholesale and retail, construction, and urban resident CEF were predominantly found in larger cities.

(4) The global Moran's I index results indicate that the spatial clustering of CEF in the study area region is continuously strengthening, except for the agricultural CEF, which shows a decline. The spatial pattern of hot and cold spots remains stable, exhibiting a three-level gradient increasing from west to east. The CEF center of gravity has gradually shifted northwest, weakening the characteristics in the southeast–northwest direction.

To achieve regional sustainable development, local governments must implement effective intervention policies. The carbon emission measurements for the Yangtze River Delta region from 2000 to 2020 reveal that carbon emissions exhibit significant spatial spillover, evolving from a single growth pole to a multi-center trend. Since carbon emissions are not confined by administrative boundaries, inter-regional cooperation is crucial. To meet the central government's "dual carbon" goals, the Yangtze River Delta should leverage its resource advantages and promote coordinated governance for pollution and carbon reduction within a regional integration framework. For the existing carbon emission growth pole, Shanghai, it is crucial to focus on and monitor its emissions to prevent further diffusion while also closely observing emerging secondary growth poles such as Nanjing, Hefei, Yangzhou, and Ningbo. Efforts should be made to accelerate the construction of low-carbon cities in these areas to prevent an accelerated increase in carbon emission growth rates.

In addition to requiring close cooperation between regions, different regions should implement appropriate policies based on their unique characteristics to mitigate carbon emissions. Tailored emission reduction measures can be proposed based on the carbon emissions of different sectors. For industrial carbon emissions centers, such as Shanghai, Suzhou, Jiaxing, Hangzhou, Shaoxing, and Ningbo, the promotion of clean energy should be prioritized. Efforts should be made to create green industrial parks and encourage enterprises to transition to greener technologies through tax subsidies and pollution control measures. For regions with high transportation-related carbon emissions, such as Nanjing, Hefei, Suzhou, Hangzhou, Yangzhou, and Wuxi, public transportation infrastructure should be expanded, and the substitution of traditional fuel-powered vehicles with new energy vehicles should be accelerated. Additionally, efforts should be made to reduce reliance on private cars by improving public transportation facilities. In residential areas, subsidies for new energy vehicles should be increased to encourage residents to scrap older, high-emission vehicles. Incentives, such as discounts on public transportation fares, should also be provided to encourage residents to choose more environmentally friendly modes of transportation. For high carbon emission wholesale and retail sectors in regions such as Shanghai, Hangzhou, Wuxi, and Hefei, local governments should encourage the production and sale of goods locally, shorten supply chains, and establish energy-efficient factories and warehouses to reduce emissions along the supply chain. For regions with high carbon emissions in the construction industry, such as Shanghai, Hangzhou, Hefei, and Suzhou, governments should consider introducing green building certification systems. This would encourage the use of low-carbon materials and energy-efficient construction practices to minimize energy consumption and maximize the use of renewable energy, providing the most comfortable indoor environments with the least energy consumption. In southern Jiangsu, where there are significant agricultural carbon emissions, efforts should be made to reduce the proportion of land used for field embankments, increase the level of agricultural mechanization, and minimize unnecessary fertilizer use. This would enhance agricultural productivity and reduce the carbon emission intensity of the agricultural sector. For cities and rural areas with high carbon emissions, such as Shanghai, Hangzhou, Nanjing, Hefei, and Suzhou, sustainable urbanization should be planned carefully. Urban sprawl should be avoided, and efforts should be made to guide residents to concentrate in specific areas to create economies of scale. Green infrastructure should be laid in densely populated areas to reduce carbon emissions while improving residents' quality of life.

**Author Contributions:** K.L.: conceptualization, project administration, funding acquisition. Q.W.: data curation, formal analysis, software, writing—original draft, methodology. X.S.: writing—review and editing, supervision. L.H.: resources, validation. Y.L.: software, visualization. All authors have read and agreed to the published version of the manuscript.

**Funding:** This work was supported by the Major Program of the National Fund of Philosophy and Social Science of China (grant number 23AZD056).

**Data Availability Statement:** The data that have been used are confidential.

**Conflicts of Interest:** The authors declare that they have no known competing financial interests or personal relationships that could have appeared to influence the work reported in this paper.

## Appendix A

**Table A1.** Energy emission factors.

| Energy Type | Emission Factor |
|---|---|
| Raw Coal | 1.4875 |
| Cleaned Coal | 1.8743 |
| Other Washed Coal | 0.595 |
| Briquettes | 0.595 |
| Gangue | 0.595 |
| Coke | 2.023 |
| Coking Oven Gas | 1.19 |
| Blast Furnace Gas | 0.2677 |
| Converter Gas | 0.2677 |
| Other Gas | 0.5751 |
| Crude Oil | 3.0209 |
| Gasoline | 2.9287 |
| Kerosene | 3.0372 |
| Diesel Oil | 3.0998 |

Table A1. Cont.

| Energy Type | Emission Factor |
|---|---|
| Fuel Oil | 3.1744 |
| Naphtha | 3.057 |
| Lubricating | 3.057 |
| Paraffin | 3.057 |
| Solvent Oil | 3.057 |
| Bitumen Asphalt | 3.057 |
| Petroleum Coke | 3.057 |
| Liquefied Petroleum Gas | 3.1052 |
| Refinery Gas | 3.0119 |
| Natural Gas | 1.9802 |
| Liquefied Natural Gas | 3.2153 |
| Heat | 0.095 |

Table A2. Electricity energy emission factors by region.

| Province | Before 2016 | 2016 and After |
|---|---|---|
| Beijing | 0.8292 | 0.6168 |
| Tianjin | 0.8733 | 0.8119 |
| Hebei | 0.9148 | 0.9029 |
| Shanxi | 0.8798 | 0.7399 |
| Inner Mongolia | 0.8503 | 0.7533 |
| Liaoning | 0.8357 | 0.7219 |
| Jilin | 0.6787 | 0.6147 |
| Heilongjiang | 0.8158 | 0.6634 |
| Shanghai | 0.7934 | 0.5641 |
| Jiangsu | 0.73566 | 0.6829 |
| Zhejiang | 0.6822 | 0.5246 |
| Anhui | 0.7913 | 0.7759 |
| Fujian | 0.5439 | 0.391 |
| Jiangxi | 0.7635 | 0.6339 |
| Shandong | 0.9236 | 0.8606 |
| Henan | 0.8444 | 0.7906 |
| Hubei | 0.3717 | 0.3574 |
| Hunan | 0.5523 | 0.4987 |
| Guangdong | 0.6379 | 0.4512 |
| Guangxi | 0.4821 | 0.3938 |
| Hainan | 0.6463 | 0.5147 |
| Chongqing | 0.6294 | 0.4405 |
| Sichuan | 0.2891 | 0.1031 |
| Guizhou | 0.6556 | 0.4275 |
| Yunnan | 0.415 | 0.0921 |
| Shaanxi | 0.8696 | 0.7673 |
| Gansu | 0.6124 | 0.4912 |
| Qinghai | 0.2263 | 0.2602 |
| Ningxia | 0.8184 | 0.6195 |
| Xinjiang | 0.7636 | 0.622 |

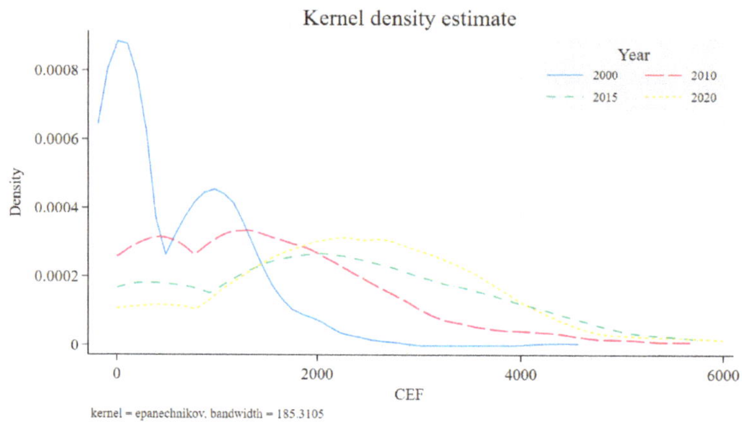


**Figure A1.** Kernel Density Plot.

## References


1. Lin, B.; Li, Z. Towards world's low carbon development: The role of clean energy. *Appl. Energy* **2022**, *307*, 118160. [CrossRef]
2. Guo, B.; Xie, T.; Zhang, W.; Wu, H.; Zhang, D.; Zhu, X.; Ma, X.; Wu, M.; Luo, P. Rasterizing $CO_2$ emissions and characterizing their trends via an enhanced population-light index at multiple scales in China during 2013–2019. *Sci. Total. Environ.* **2023**, *905*, 167309. [CrossRef]
3. Yang, W.-J.; Tan, M.-Z.; Chu, S.-H.; Chen, Z. Carbon emission and financial development under the "double carbon" goal: Considering the upgrade of industrial structure. *Front. Environ. Sci.* **2023**, *10*, 1091537. [CrossRef]
4. Yang, Z.; Shao, S.; Xu, L.; Yang, L. Can regional development plans promote economic growth? City-level evidence from China. *Socio-Econ. Plan. Sci.* **2022**, *83*, 101212. [CrossRef]
5. Sun, X.; Ma, X.; Shi, F.; Han, M.; Xie, H.; He, Y. Decomposition of China's regional carbon emission paths: An analysis of environmental input and output considering regional development differences. *Environ. Sci. Pollut. Res.* **2022**, *29*, 62537–62559. [CrossRef] [PubMed]
6. Zhang, X.; Cai, Z.; Song, W.; Yang, D. Mapping the spatial-temporal changes in energy consumption-related carbon emissions in the Beijing-Tianjin-Hebei region via nighttime light data. *Sustain. Cities Soc.* **2023**, *94*, 104476. [CrossRef]
7. Sun, Y.; Zhao, T.; Xia, L. Spatial-temporal differentiation of carbon efficiency and coupling coordination degree of Chinese county territory and obstacles analysis. *Sustain. Cities Soc.* **2022**, *76*, 103429. [CrossRef]
8. Jing, Q.; Bai, H.; Luo, W.; Cai, B.; Xu, H. A top-bottom method for city-scale energy-related $CO_2$ emissions estimation: A case study of 41 Chinese cities. *J. Clean. Prod.* **2018**, *202*, 444–455. [CrossRef]
9. Sun, C.; Wang, B.; Miao, H. Spatiotemporal dynamics of $CO_2$ emissions: A case study of the "New Yangtze River Delta" in China. *Environ. Sci. Pollut. Res.* **2023**, *30*, 40961–40977. [CrossRef]
10. Chen, J.; Gao, M.; Cheng, S.; Hou, W.; Song, M.; Liu, X.; Liu, Y.; Shan, Y. County-level $CO_2$ emissions and sequestration in China during 1997–2017. *Sci. Data* **2020**, *7*, 391. [CrossRef]
11. Keola, S.; Andersson, M.; Hall, O. Monitoring economic development from space: Using nighttime light and land cover data to measure economic growth. *World Dev.* **2015**, *66*, 322–334. [CrossRef]
12. Watmough, G.R.; Marcinko, C.L.J.; Sullivan, C.; Tschirhart, K.; Mutuo, P.K.; Palm, C.A.; Svenning, J.-C. Socioecologically informed use of remote sensing data to predict rural household poverty. *Proc. Natl. Acad. Sci. USA* **2019**, *116*, 1213–1218. [CrossRef]
13. Doll, C.N.; Pachauri, S. Estimating rural populations without access to electricity in developing countries through night-time light satellite imagery. *Energy Policy* **2010**, *38*, 5661–5670. [CrossRef]
14. Bustos, M.F.A.; Hall, O.; Andersson, M. Nighttime lights and population changes in Europe 1992–2012. *Ambio* **2015**, *44*, 653–665. [CrossRef] [PubMed]
15. Mossoux, S.; Kervyn, M.; Soulé, H.; Canters, F. Mapping population distribution from high resolution remotely sensed imagery in a data poor setting. *Remote Sens.* **2018**, *10*, 1409. [CrossRef]
16. Yang, D.; Luan, W.; Qiao, L.; Pratama, M. Modeling and spatio-temporal analysis of city-level carbon emissions based on nighttime light satellite imagery. *Appl. Energy* **2020**, *268*, 114696. [CrossRef]
17. Han, J.; Meng, X.; Liang, H.; Cao, Z.; Dong, L.; Huang, C. An improved nightlight-based method for modeling urban $CO_2$ emissions. *Environ. Model. Softw.* **2018**, *107*, 307–320. [CrossRef]

18. Fotheringham, A.S.; Crespo, R.; Yao, J. Geographical and temporal weighted regression (GTWR). *Geogr. Anal.* **2015**, *47*, 431–452. [CrossRef]
19. Wang, P.; Li, Y.; Huang, G.; Wang, S.; Suo, C.; Ma, Y. A multi-scenario factorial analysis and multi-regional input-output model for analyzing $CO_2$ emission reduction path in Jing-Jin-Ji region. *J. Clean. Prod.* **2021**, *300*, 126782. [CrossRef]
20. Harris, S.; Weinzettel, J.; Bigano, A.; Källmén, A. Low carbon cities in 2050? GHG emissions of European cities using production-based and consumption-based emission accounting methods. *J. Clean. Prod.* **2020**, *248*, 119206. [CrossRef]
21. Souza Leao, E.B.; Nascimento, L.F.M.; Andrade, J.C.S.; Oliveira, J.A.P. Carbon accounting approaches and reporting gaps in urban emissions: An analysis of the Greenhouse Gas inventories and climate action plans in Brazilian cities. *J. Clean. Prod.* **2020**, *245*, 118930. [CrossRef]
22. Froemelt, A.; Geschke, A.; Wiedmann, T. Quantifying carbon flows in Switzerland: Top-down meets bottom-up modelling. *Environ. Res. Lett.* **2021**, *16*, 014018. [CrossRef]
23. Eggleston, H.S.; Buendia, L.; Miwa, K.; Ngara, T.; Tanabe, K. *2006 IPCC Guidelines for National Greenhouse Gas Inventories*; IPCC: Geneva, Switzerland, 2006.
24. Long, Y.; Sharifi, A.; Huang, L.; Chen, J. Urban carbon accounting: An overview. *Urban Clim.* **2022**, *44*, 101195. [CrossRef]
25. Ahmetoğlu, S.; Tanık, A. Management of carbon footprint and determination of GHG emission sources in construction sector. *Int. J. Environ. Geoinform.* **2020**, *7*, 191–204. [CrossRef]
26. Trofimenko, Y.; Komkov, V.; Trofimenko, K. Forecast of energy consumption and greenhouse gas emissions by road transport in Russia up to 2050. *Transp. Res. Procedia* **2020**, *50*, 698–707. [CrossRef]
27. Petrov, S.P. Application of the IPCC Methodology to Assess the Spatial Distribution of $CO_2$ Emissions in Russian Ferrous Metallurgy. *Stud. Russ. Econ. Dev.* **2024**, *35*, 54–59. [CrossRef]
28. Jiang, M.; Medlyn, B.E.; Drake, J.E.; Duursma, R.A.; Anderson, I.C.; Barton, C.V.M.; Boer, M.M.; Carrillo, Y.; Castañeda-Gómez, L.; Collins, L.; et al. The fate of carbon in a mature forest under carbon dioxide enrichment. *Nature* **2020**, *580*, 227–231. [CrossRef]
29. Chen, L.; Xu, L.; Cai, Y.; Yang, Z. Spatiotemporal patterns of industrial carbon emissions at the city level. *Resour. Conserv. Recycl.* **2021**, *169*, 105499. [CrossRef]
30. Dong, F.; Yu, B.; Hadachin, T.; Dai, Y.; Wang, Y.; Zhang, S.; Long, R. Drivers of carbon emission intensity change in China. *Resour. Conserv. Recycl.* **2018**, *129*, 187–201. [CrossRef]
31. Burke, M.; Driscoll, A.; Lobell, D.B.; Ermon, S. Using satellite imagery to understand and promote sustainable development. *Science* **2021**, *371*, eabe8628. [CrossRef]
32. Bennett, M.M.; Smith, L.C. Advances in using multitemporal night-time lights satellite imagery to detect, estimate, and monitor socioeconomic dynamics. *Remote Sens. Environ.* **2017**, *192*, 176–197. [CrossRef]
33. Zheng, Q.; Seto, K.C.; Zhou, Y.; You, S.; Weng, Q. Nighttime light remote sensing for urban applications: Progress, challenges, and prospects. *ISPRS J. Photogramm. Remote Sens.* **2023**, *202*, 125–141. [CrossRef]
34. Elvidge, C.D.; Baugh, K.E.; Kihn, E.A.; Kroehl, H.W.; Davis, E.R.; Davis, C.W. Relation between satellite observed visible-near infrared emissions, population, economic activity and electric power consumption. *Int. J. Remote Sens.* **1997**, *18*, 1373–1379. [CrossRef]
35. Wang, M.; Wang, Y.; Teng, F.; Ji, Y. The spatiotemporal evolution and impact mechanism of energy consumption carbon emissions in China from 2010 to 2020 by integrating multisource remote sensing data. *J. Environ. Manag.* **2023**, *346*, 119054. [CrossRef] [PubMed]
36. Zheng, Y.; Fan, M.; Cai, Y.; Fu, M.; Yang, K.; Wei, C. Spatio-temporal pattern evolution of carbon emissions at the city-county-town scale in Fujian Province based on DMSP/OLS and NPP/VIIRS nighttime light data. *J. Clean. Prod.* **2024**, *442*, 140958. [CrossRef]
37. Zhang, Y.; Quan, J.; Kong, Y.; Wang, Q.; Zhang, Y.; Zhang, Y. Research on the fine-scale spatial-temporal evolution characteristics of carbon emissions based on nighttime light data: A case study of Xi'an city. *Ecol. Inform.* **2024**, *79*, 102454. [CrossRef]
38. Wu, R.; Wang, R.; Nian, Z.; Gu, J. Spatio-temporal variation and decoupling effects of energy carbon footprint based on nighttime light data: Evidence from counties in northeast China. *Atmos. Pollut. Res.* **2025**, *16*, 102366. [CrossRef]
39. Lu, S.; Xiao, Y.; Lu, Y.; Lin, J. Spatialization of electricity consumption by combining high-resolution nighttime light remote sensing and urban functional zoning information. *Geo-Spat. Inf. Sci.* **2024**, *27*, 1–14. [CrossRef]
40. Meng, X.; Han, J.; Huang, C. An improved vegetation adjusted nighttime light urban index and its application in quantifying spatiotemporal dynamics of carbon emissions in China. *Remote Sens.* **2017**, *9*, 829. [CrossRef]
41. Shi, K.; Chen, Y.; Yu, B.; Xu, T.; Chen, Z.; Liu, R.; Li, L.; Wu, J. Modeling spatiotemporal $CO_2$ (carbon dioxide) emission dynamics in China from DMSP-OLS nighttime stable light data using panel data analysis. *Appl. Energy* **2016**, *168*, 523–533. [CrossRef]
42. Wu, H.; Yang, Y.; Li, W. Dynamic spatiotemporal evolution and spatial effect of carbon emissions in urban agglomerations based on nighttime light data. *Sustain. Cities Soc.* **2024**, *113*, 105712. [CrossRef]
43. Wang, G.; Hu, Q.; He, L.; Guo, J.; Huang, J.; Zhong, L. The estimation of building carbon emission using nighttime light images: A comparative study at various spatial scales. *Sustain. Cities Soc.* **2024**, *101*, 105066. [CrossRef]

44. Shi, K.; Chen, Z.; Cui, Y.; Wu, J.; Yu, B. NPP-VIIRS nighttime light data have different correlated relationships with fossil fuel combustion carbon emissions from different sectors. *IEEE Geosci. Remote Sens. Lett.* **2020**, *18*, 2062–2066. [CrossRef]
45. Xia, B. Spatial Characteristics and Driving Mechanisms of Carbon Neutrality Progress in Tourism Attractions in the Qinghai–Tibet Plateau Based on Remote Sensing Methods. *Remote Sens.* **2024**, *16*, 4481. [CrossRef]
46. Wei, W.; Chen, D.; Zhang, X.; Ma, L.; Xie, B.; Zhou, J.; Wang, M.; Wei, X.; Yang, S.; Yan, P. Estimation of carbon emissions from different industrial categories integrated nighttime light and POI data—A case study in the Yellow River Basin. *J. Environ. Manag.* **2024**, *370*, 122418. [CrossRef]
47. Liu, C.; Hu, S.; Wu, S.; Song, J.; Li, H. County-level land use carbon emissions in China: Spatiotemporal patterns and impact factors. *Sustain. Cities Soc.* **2024**, *105*, 105304. [CrossRef]
48. Gao, F.; Wu, J.; Xiao, J.; Li, X.; Liao, S.; Chen, W. Spatially explicit carbon emissions by remote sensing and social sensing. *Environ. Res.* **2023**, *221*, 115257. [CrossRef] [PubMed]
49. Wei, W.; Zhang, X.; Cao, X.; Zhou, L.; Xie, B.; Zhou, J.; Li, C. Spatiotemporal dynamics of energy-related $CO_2$ emissions in China based on nighttime imagery and land use data. *Ecol. Indic.* **2021**, *131*, 108132. [CrossRef]
50. Wu, Y.; Shi, K.; Chen, Z.; Liu, S.; Chang, Z. Developing improved time-series DMSP-OLS-like data (1992–2019) in China by integrating DMSP-OLS and SNPP-VIIRS. *IEEE Trans. Geosci. Remote Sens.* **2021**, *60*, 4407714. [CrossRef]
51. National Bureau of Statistics of China. China Energy Statistical Yearbook. 2023. Available online: https://www.stats.gov.cn/sj/ndsj/2023/indexeh.htm (accessed on 10 October 2024).
52. Xia, S.; Shao, H.; Wang, H.; Xian, W.; Shao, Q.; Yin, Z.; Qi, J. Spatio-temporal dynamics and driving forces of multi-scale $CO_2$ emissions by integrating DMSP-OLS and NPP-VIIRS data: A case study in Beijing-Tianjin-Hebei, China. *Remote Sens.* **2022**, *14*, 4799. [CrossRef]
53. Ustaoglu, E.; Bovkır, R.; Aydınoglu, A.C. Spatial distribution of GDP based on integrated NPS-VIIRS nighttime light and MODIS EVI data: A case study of Turkey. *Environ. Dev. Sustain.* **2021**, *23*, 10309–10343. [CrossRef]
54. Li, S.; Cao, X.; Zhao, C.; Jie, N.; Liu, L.; Chen, X.; Cui, X. Developing a pixel-scale corrected nighttime light dataset (PCNL, 1992–2021) combining DMSP-OLS and NPP-VIIRS. *Remote Sens.* **2023**, *15*, 3925. [CrossRef]
55. Shi, K.; Yu, B.; Huang, Y.; Hu, Y.; Yin, B.; Chen, Z.; Chen, L.; Wu, J. Evaluating the ability of NPP-VIIRS nighttime light data to estimate the gross domestic product and the electric power consumption of China at multiple scales: A comparison with DMSP-OLS data. *Remote Sens.* **2014**, *6*, 1705–1724. [CrossRef]
56. Zhao, J.; Ji, G.; Yue, Y.; Lai, Z.; Chen, Y.; Yang, D.; Yang, X.; Wang, Z. Spatio-temporal dynamics of urban residential $CO_2$ emissions and their driving forces in China using the integrated two nighttime light datasets. *Appl. Energy* **2019**, *235*, 612–624. [CrossRef]
57. Li, X.; Zhou, Y.; Zhao, M.; Zhao, X. A harmonized global nighttime light dataset 1992–2018. *Sci. Data* **2020**, *7*, 168. [CrossRef]
58. Pachauri, R.K.; Allen, M.R.; Barros, V.R.; Broome, J.; Cramer, W.; Christ, R.; Church, J.A.; Clarke, L.; Dahe, Q.; Dasgupta, P.; et al. Climate Change 2014: Synthesis Report. In *Contribution of Working Groups I, II and III to the Fifth Assessment Report of the Intergovernmental Panel on Climate Change*; IPCC: Geneva, Switzerland, 2014; p. 151.
59. *GB/T 2589-2020*; National Standard of the People's Republic of China. General Principles for Calculation of the Comprehensive Energy Consumption. Standardization Administration of China: Beijing, China, 2020.
60. Yang, X.; Liu, X. Accuracy analysis of inverting provincial-level carbon emissions from night-time light data in China: Comparison based on international carbon emission data. In *IOP Conference Series: Earth and Environmental Science*; IOP Publishing: Bristol, UK, 2020; Volume 601, p. 012046. [CrossRef]
61. Ou, J.; Liu, X.; Wang, S.; Xie, R.; Li, X. Investigating the differentiated impacts of socioeconomic factors and urban forms on $CO_2$ emissions: Empirical evidence from Chinese cities of different developmental levels. *J. Clean. Prod.* **2019**, *226*, 601–614. [CrossRef]
62. Letu, H.; Hara, M.; Tana, G.; Nishio, F. A saturated light correction method for DMSP/OLS nighttime satellite imagery. *IEEE Trans. Geosci. Remote Sens.* **2011**, *50*, 389–396. [CrossRef]
63. Cai, M.; Shi, Y.; Ren, C.; Yoshida, T.; Yamagata, Y.; Ding, C.; Zhou, N. The need for urban form data in spatial modeling of urban carbon emissions in China: A critical review. *J. Clean. Prod.* **2021**, *319*, 128792. [CrossRef]
64. Cui, Y.; Zhang, W.; Wang, C.; Streets, D.G.; Xu, Y.; Du, M.; Lin, J. Spatiotemporal dynamics of $CO_2$ emissions from central heating supply in the North China Plain over 2012–2016 due to natural gas usage. *Appl. Energy* **2019**, *241*, 245–256. [CrossRef]
65. Meng, L.; Graus, W.; Worrell, E.; Huang, B. Estimating $CO_2$ (carbon dioxide) emissions at urban scales by DMSP/OLS (Defense Meteorological Satellite Program's Operational Linescan System) nighttime light imagery: Methodological challenges and a case study for China. *Energy* **2014**, *71*, 468–478. [CrossRef]
66. Zhang, W.; Cui, Y.; Wang, J.; Wang, C.; Streets, D.G. How does urbanization affect $CO_2$ emissions of central heating systems in China? An assessment of natural gas transition policy based on nighttime light data. *J. Clean. Prod.* **2020**, *276*, 123188. [CrossRef]
67. Lotwick, H.W.; Silverman, B.W. Methods for analysing spatial processes of several types of points. *J. R. Stat. Soc. Ser. B Stat. Methodol.* **1982**, *44*, 406–413. [CrossRef]
68. Henderson, J.; Storeygard, V.A.; Weil, D.N. Measuring economic growth from outer space. *Am. Econ. Rev.* **2012**, *102*, 994–1028. [CrossRef] [PubMed]

69. Pinkovskiy, M.; Sala-i-Martin, X. Lights, camera... income! Illuminating the national accounts-household surveys debate. *Q. J. Econ.* **2016**, *131*, 579–631. [CrossRef]
70. Shan, Y.; Guan, Y.; Hang, Y.; Zheng, H.; Li, Y.; Guan, D.; Li, J.; Zhou, Y.; Li, L.; Hubacek, K. City-level emission peak and drivers in China. *Sci. Bull.* **2022**, *67*, 1910–1920. [CrossRef] [PubMed]
71. Wu, W.; Zhang, T.; Xie, X.; Huang, Z. Regional low carbon development pathways for the Yangtze River Delta region in China. *Energy Policy* **2021**, *151*, 112172. [CrossRef]
72. Chen, X.; Di, Q.; Liang, C. The mechanism and path of pollution reduction and carbon reduction affecting high quality economic development—Taking the Yangtze River Delta urban agglomeration as an example. *Appl. Energy* **2024**, *376*, 124340. [CrossRef]
73. Cui, Y.; Khan, S.U.; Deng, Y.; Zhao, M. Spatiotemporal heterogeneity, convergence and its impact factors: Perspective of carbon emission intensity and carbon emission per capita considering carbon sink effect. *Environ. Impact Assess. Rev.* **2022**, *92*, 106699. [CrossRef]
74. Lv, T.; Hu, H.; Zhang, X.; Xie, H.; Fu, S.; Wang, L. Spatiotemporal pattern of regional carbon emissions and its influencing factors in the Yangtze River Delta urban agglomeration of China. *Environ. Monit. Assess.* **2022**, *194*, 515. [CrossRef]
75. Chen, G.; Chen, Z. Carbon emissions and resources use by Chinese economy 2007: A 135-sector inventory and input–output embodiment. *Commun. Nonlinear Sci. Numer. Simul.* **2010**, *15*, 3647–3732. [CrossRef]
76. Hu, K.; Liu, Z.; Shao, P.; Ma, K.; Xu, Y.; Wang, S.; Wang, Y.; Wang, H.; Di, L.; Xia, M.; et al. A review of satellite-based $CO_2$ data reconstruction studies: Methodologies, challenges, and advances. *Remote Sens.* **2024**, *16*, 3818. [CrossRef]

*land*

MDPI

*Article*

# How Do Temporal and Geographical Kernels Differ in Reflecting Regional Disparities? Insights from a Case Study in China

**Chunzhu Wei [1], Xufeng Liu [1], Wei Chen [1], Lupan Zhang [1], Ruixia Chao [1,*] and Wei Wei [2,*]**

[1] School of Geography and Planning, Southern Marine Science and Engineering Guangdong Laboratory (Zhuhai), Sun Yat-sen University, Guangzhou 510275, China; weichzh@mail.sysu.edu.cn (C.W.); liuxf65@mail2.sysu.edu.cn (X.L.); chenw297@mail2.sysu.edu.cn (W.C.); zhanglp28@mail2.sysu.edu.cn (L.Z.)

[2] School of Economic, Guangxi University for Nationalities, Nanning 530006, China

* Correspondence: chaorx3@mail.sysu.edu.cn (R.C.); 20070018@gxmzu.edu.cn (W.W.)

**Abstract:** Rapid economic growth in China has brought about a significant challenge: the widening gap in regional development. Addressing this disparity is crucial for ensuring sustainable development. However, existing studies have largely overlooked the intrinsic spatial and temporal dynamics of regional disparities on various levels. This study thus employed five advanced multiscale geographically and temporally weighted regression models—GWR, MGWR, GTWR, MGTWR, and STWR—to analyze the spatio-temporal relationships between ten key conventional socio-economic indicators and per capita GDP across different administrative levels in China from 2000 to 2019. The findings highlight a consistent increase in regional disparities, with secondary industry emerging as a dominant driver of long-term economic inequality among the indicators analyzed. While a clear inland-to-coastal gradient underscores the persistence of regional disparity determinants, areas with greater economic disparities exhibit pronounced spatio-temporal heterogeneity. Among the models, STWR outperforms others in capturing and interpreting local variations in spatio-temporal disparities, demonstrating its utility in understanding complex regional dynamics. This study provides novel insights into the spatio-temporal determinants of regional economic disparities, offering a robust analytical framework for policymakers to address region-specific variables driving inequality over time and space. These insights contribute to the development of targeted and dynamic policies for promoting balanced and sustainable regional growth.

**Keywords:** social disparity; spatio-temporal non-stationarity; spatial and temporal kernels; spatio-temporal patterns; geographically and temporally weighted regression models

## 1. Introduction

Since the inception of the "Reform and Opening Up" policy in the late 1970s, China has experienced remarkable economic growth, with its real GDP expanding at an average annual rate of 9.8% over the past three decades [1]. However, alongside this economic development, there has been a noticeable increase in rural–urban and regional economic disparities post-2000 [2]. For instance, coastal regions exhibit a per capita GDP double that of inland areas, while Shanghai's per capita GDP surpasses that of inland Guizhou Province by a factor of 10. Such escalating economic inequality can fuel social discontent and political instability within a nation. Moreover, considerable and swiftly escalating inequality may hamper economic growth and jeopardize the prospects of sustainable long-term development [3]. From this standpoint, addressing regional social disparity is essential for advancement and has a beneficial impact on productivity, the economy, and the reinforcement of democracy and citizenship rights [4,5].

Regional social disparity refers to unjust disparities in opportunities, resources, and outcomes among different groups of people. These disparities can in turn contribute to the emergence of spatial disparities, characterized by uneven distributions of various socio-economic factors across varying geographical areas [2,3,6]. Importantly, these spatial disparities often exhibit non-stationarity, which refers to the phenomenon where the statistical properties of a dataset—such as its mean, variance, or relationships between variables—change over space or time rather than remaining constant. In essence, social disparity is a driving force behind the phenomenon of spatial disparity, as these two concepts are intricately interlinked and mutually reinforcing. Therefore, recognizing the spatial disparity and identifying the determinants of social disparity as a foundation for targeted policy interventions to address the unique needs and concerns of disadvantaged regions [7]. This recognition stands as the prerequisite to help implement policies that facilitate social inclusion and provide opportunities for marginalized groups to overcome structural barriers [8].

However, while the widely used global linear regression models are only able to analyze the complicated influence of socio-economic variables in social disparity [9], they neglect the spatial or temporal trends of regional disparity at some scales [10,11]. For example, ordinary least square (OLS) assumes that the regression relationship between independent variables and the explanatory variable is homogeneous in space and time, leading to the inaccuracy or failure of results in exploring disparity patterns and processes. Therefore, a series of OLS-like models, such as the expansion method [12], the distance–decay weighted regression [13], and the geographically weighted regression (GWR) model have been proposed to take spatial heterogeneity into consideration [14–16].

Among these OLS-like models, GWR was one of the widely used methods in regional economic disparity analysis [17]. GWR is widely used in rural [18] and urban [19] analyses, and the research scale ranges from large-scale administrative districts (such as the city agglomeration [20] and provincial levels [21]) to small-scale community studies in non-administrative units. In GWR, predictive variables such as the built environment (e.g., land use patterns, transportation systems, urban design variables) and socio-economic situations (e.g., income, GDP, unemployment rate, education level) are embedded within the GWR framework to demonstrate the social disparity determinants [22].

A critical component of GWR and its extensions is bandwidth optimization, which determines the spatial range (or temporal range in spatio-temporal models) over which observations influence the local regression estimates. In GWR, a single bandwidth is selected to ensure the relationships between variables are adequately smoothed while capturing local variations. However, this uniform bandwidth assumes that all relationships vary at a single spatial scale, which may oversimplify complex socio-economic phenomena [23]. To address this limitation, the multiscale GWR model (MGWR) introduces variable-specific bandwidths, enabling each predictor to adapt to distinct spatial scales. For example, while income disparity might vary significantly over a broad spatial range, unemployment effects could be more localized. MGWR's bandwidth optimization dynamically adjusts to these differences, providing a finer-grained analysis of socio-economic determinants [24–26]. Both GWR and MGWR highlight the importance of spatio-temporal kernels, which serve as weighting functions to determine the influence of observations within a specific spatial or temporal neighborhood. These kernels enable the models to capture variations in social disparity determinants across both space and time, providing a dynamic lens to analyze complex processes. By addressing the non-stationary nature of socio-economic phenomena and introducing spatio-temporal kernels, these models offer powerful tools for exploring how social disparity determinants evolve across different spatial and temporal scales [27,28].

While it is worth noting non-stationarity in a social disparity context, it can also be the case that spatial disparity differs not only in a variety of spaces but at various times (time–non-stationarity). Thus, to improve upon the GWR and MGWR methods, geographically and temporally weighted regression (GTWR) [29,30] and multiple scales of GTWR (MGTWR) models incorporating temporal dimensions alongside spatial bandwidth optimization have been proposed to combine the effects of temporal and spatial variation in the regression model. The time variation considered in the GTWR and MGTWR model is the concept of a time interval instead of the rate of value variation over time. The stepwise strategy applied in the spatio-temporal kernel function of the GTWR and MGTWR model is to fix an optimized spatial bandwidth first, and then it fixes the optimized temporal bandwidth [31,32]. However, this time-isolated bandwidth optimization procedure does not always seem reasonable [33], as it is not able to optimize both the temporal and spatial bandwidths at the same time [34]. At present, GTWR and other improved models of GWR have been used in a large number of case studies in exploring ecological environment and urban construction [35,36], air quality and carbon cycle [37].

To further improve the spatio-temporal kernel function, the recently developed spatio-temporal weighted regression (STWR) model has been proposed by Que [38]. Different from the GTWR and MGTWR, the STWR model treats the time distance as the rate of value variation over a time interval rather than the time interval itself. This model integrates spatial and temporal bandwidth optimization into a unified framework. By weighting observations based on their proximity in both space and time, STWR excels at modeling complex processes that evolve non-uniformly across regions and periods. It is thus more suitable to consider the effects of different variation of observed points over time. In addition, STWR optimized the temporal kernel function in two or three dimensions, which is different with the one-dimension temporal kernel function (e.g., Gaussian kernel or Bi-square kernel) that defined in GTWR and MGTWR. STWR utilizes a weighted average form to calculate the spatio-temporal kernel rather than the multiplication form in GTWR and MGTWR [29,33], which may avoid potential underestimation of combined spatio-temporal effects [38].

Addressing the intricate dynamics of social disparity requires tackling complex, non-linear interactions between temporal and spatial factors. These challenges are further compounded by issues like the modifiable areal unit problem (MAUP) or "neighborhood effect" [39], which arise from variations in geographical definitions and scales. To address these complexities, this study employs advanced spatial and spatio-temporal regression models, including GWR, MGWR, GTWR, MGTWR, and STWR, which effectively account for non-stationarity across spatial and temporal dimensions—critical for analyzing socio-economic data such as GDP per capita [40].

By leveraging spatio-temporal kernels, these models capture localized variations and dynamically adapt to changes over both space and time. Their interpretability and ability to provide geographically and temporally explicit insights set them apart from machine learning techniques like random forests [41], support vector machines [42–44], or neural networks [45,46], which often function as black box approaches. Unlike machine learning models, these regression methods deliver localized coefficients, revealing spatially and temporally varying relationships that offer valuable insights into regional disparities and socio-economic patterns [47]. This interpretability makes these models particularly advantageous for policy-making and academic research, where understanding the underlying mechanisms driving disparities is as crucial as achieving predictive accuracy. Their ability to identify nuanced, context-specific relationships supports the development of targeted interventions and evidence-based strategies to address social inequality effectively [48].

Additionally, socio-economic variables often exhibit unique spatio-temporal patterns, further complicating the analysis of regional disparities. To better understand their dynamics, this study focuses on county-level and prefecture-level cities in China as fundamental spatial units [49]. Using spatio-temporal regression models of GWR, MGWR, GTWR, MGTWR, and STWR, it aims to:

(1) Evaluate and compare the effectiveness of GWR, MGWR, GTWR, MGTWR, and STWR in identifying and quantifying the key determinants of regional economic disparities.
(2) Investigate how these models capture spatio-temporal heterogeneity across different administrative levels, providing deeper insights into the structural factors driving regional social disparities.

This research contributes to refining the analytical framework for studying regional disparities by integrating advanced spatio-temporal modeling techniques with a focus on scale-sensitive and context-specific socio-economic variables.

## 2. Study Area and Data Collection

### 2.1. Study Area

To thoroughly investigate the "neighborhood" effect in various geographically and temporally weighted regression models, this study examines regional disparities in China at two spatial scales: the county level (2357 counties) and the prefecture level (357 cities), as depicted in Figure 1a. The temporal dimension is segmented into four periods: 2000–2005, 2006–2010, 2011–2015, and 2016–2019. Given evidence of increasing rural–urban and regional economic disparities in China since 2000 [2]. we extended the analysis to evaluate social disparities across two dimensions: (1) poverty-stricken versus non-poverty-stricken regions and (2) mega-regions and sub-national regions, represented by the southeastern and northwestern areas divided by the Hu Line. This multi-scale and multi-regional approach ensure sensitivity to the spatio-temporal dynamics of social disparity evolution.

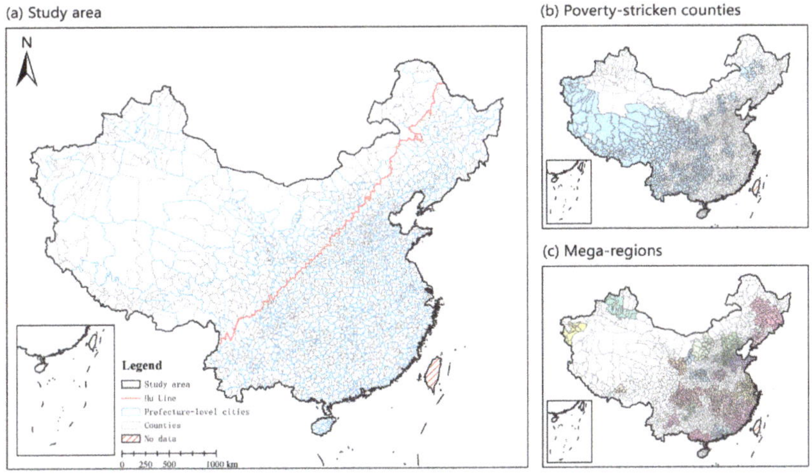


**Figure 1.** County-level and prefecture-level data were used to show the spatio-temporal regression. Other spatial units levels (poverty-stricken vs. non-poverty-stricken regions, mega-regions, and west/east of Hu Line regions) were used to evaluate socio-economic inequity as well as the model's sensitivity in this study.

To address the frequent boundary changes in county-level administrative units, we employed a boundary harmonization strategy. Using the 2019 administrative boundaries of county-level units (2347 counties within 357 cities) as the authoritative reference dataset (http://www.resdc.cn/, accessed on 1 January 2021), we ensured consistent spatial boundaries across all time periods. This harmonization accounted for over 653 boundary changes between 2000 and 2020 [50]. When discrepancies were identified, socio-economic data such as GDP and population were adjusted to a unified base year boundary using spatial overlay techniques in GIS. For counties with boundary modifications, data were recalculated using an area-weighting method or a time series interpolation approach, ensuring comparability across years. To validate these adjustments, the recalibrated data were cross-referenced with secondary datasets to minimize discrepancies and confirm accuracy. These preprocessing methods ensured that the observed temporal trends reflected genuine socio-economic changes rather than artifacts introduced by administrative boundary modifications.

### 2.2. Variables

How to objectively measure socio-economic inequity is a key issue in a regional disparity study. Gross domestic product (GDP) per capita is an indicator that is widely used in socio-economic research [17,27], and the growth rate of real GDP per capita has shown a significant influence on regional development. This indicator is more generic across regions compared with other indicators such as GDP, minimum income, etc. Thus, this study adopted GDP per capita as the dependent variable to depict regional disparities in China.

We selected the independent variables based on a comprehensive literature review and data availability [42,43]. To clarify the rationale behind our selection, we have provided the following explanations: (1) Practitioner number (PN) reflects the active labor force, a key driver of economic productivity and GDP growth [51]. (2) Primary industry (PI) includes agriculture, forestry, and fishing, representing the backbone of rural economies and regional economic transitions [52]. (3) Secondary industry (SI) captures industrial production and urbanization, a major GDP growth driver in emerging economies [53]. (4) Public financial revenue (PFR) indicates government income for funding infrastructure and services, linked to economic growth and decentralization [54]. (5) Public financial expenditure (PFE) reflects government spending on public goods associated with growth and social welfare [55]. (6) Deposit balance (DB) represents household savings and financial stability, influencing inequality and consumption [56]. (7) Loan balance (LB) highlights access to credit and investment capacity, crucial for economic development [57]. (8) Total power of agricultural machinery (TP) indicates mechanization, boosting agricultural productivity and rural economic transformation [58]. (9) Student number (SN) reflects educational access, fostering human capital and long-term growth (Becker, 1964). (10) Hospital bed number (HBN) measures healthcare infrastructure, linked to workforce productivity and social development [59]. (11) Grain yield (GY) indicates agricultural productivity, essential for food security and rural stability [60].

These variables were chosen to explore GDP determinants across spatial and temporal contexts. PN, PI, and GY highlight disparities in employment and agriculture-dependent regions. SI addresses industrialization-driven disparities. PFR and PFE reflect government capacity and investment. DB and LB capture financial inclusion and savings patterns, while TP, SN, and HBN emphasize rural development, education, and healthcare. A detailed description of these variables is provided in Table 1.

**Table 1.** Summary of the variables used.

| Variable Type | Variable Name | Code | Description |
|---|---|---|---|
| Dependent variables | Gross domestic product per capita | GPC | Gross domestic product divided by population |
| Independent variables | Practitioner number | PN | Number of registered practitioners |
| | Primary industry | PI | GDP of primary industry |
| | Secondary industry | SI | GDP of secondary industry |
| | Public financial revenue | PFR | Total public financial revenue of the government |
| | Public financial expenditure | PFE | Total public financial expenditure of the government |
| | Deposit balance | DB | Total saving deposits of urban and rural residents |
| | Loan balance | LB | Balance of various loans of financial institutions at the end of the year |
| | Total power of agricultural machinery | TP | The sum of the rated power of all agricultural machinery power |
| | Student number | SN | Number of students in primary and secondary schools |
| | Hospital bed number | HBN | Number of beds in medical and health institutions |
| | Grain yield | GY | Total grain yield of each county |

The variance inflation factors (VIFs) for 10 of the explanatory variables were all below the acceptable threshold of 10 (ranging from 2.38 to 9.29) across four years of analysis. For public financial revenue (PFR), the VIF exceeded 10 in earlier years but fell below this threshold in 2014 and 2019. PFR was retained in the analysis because it serves as a critical proxy for regional government capacity to invest in development initiatives. It reflects government income from taxes and other sources that fund public services and infrastructure development. Given its historical alignment with economic expansion and fiscal decentralization, PFR provides valuable insights into regional disparities. To ensure comprehensive analysis, we included all 11 variables, as their collective impact is integral to understanding the socio-economic dynamics in this study. This approach balances statistical robustness with the inclusion of theoretically and empirically significant variables, as detailed in the methodology section. The details of the VIFs are shown in Tables S3–S6 in the Supplementary Materials. All the socio-economic statistic data from 2000 to 2019 were collected from the China County Statistical Yearbook (CCSY), published by the National Bureau of Statistics of China. All of these data are publicly available.

## 3. Methodology

As shown in Figure 2, the research methodology in this study involved two major steps: (1) OLS, GWR, MGWR, GTWR, MGTWR, and STWR model regression; the relationship between the dependent and independent variables was established using the six regression models; and (2) comparison between the results of the OLS, GWR, MGWR, GTWR, MGTWR, and STWR models. A calibration procedure was implemented to optimize the bandwidth of the regression models according to $R^2$ and the Akaike information criterion (AIC), and then the spatio-temporal heterogeneity of different variables was explored across two different spatial domains.

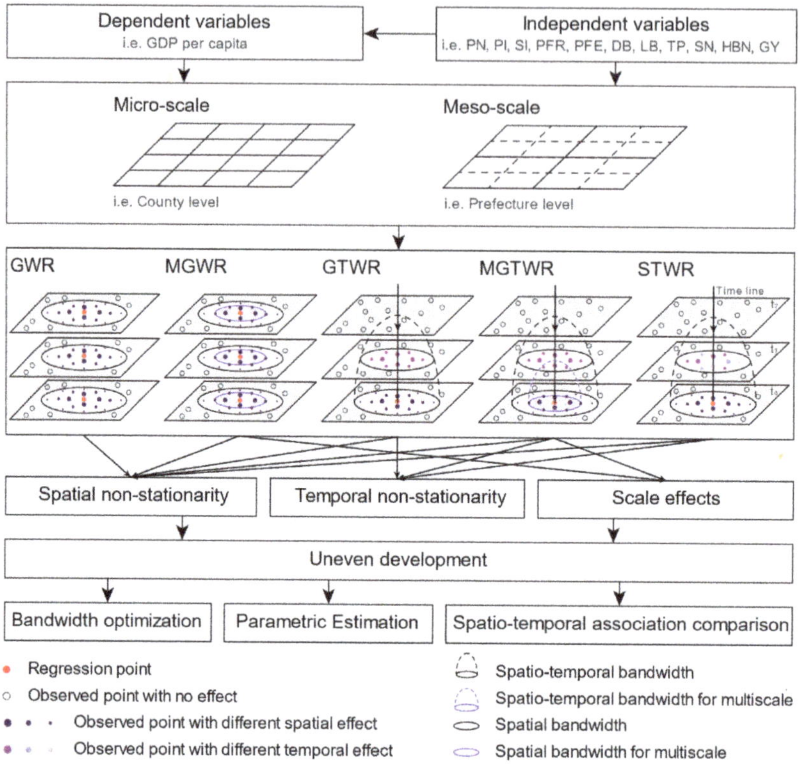


**Figure 2.** Workflow of the methodology.

### 3.1. GWR Model

Traditionally, the relationship between the dependent and independent socio-economic variables has been built using ordinary least squares (OLS) regression, i.e.,

$$y_i = \beta_0 + \sum_{k=1}^{m} \beta_k x_{ik} + \varepsilon \quad (1)$$

where $\beta_0$ represents the intercept value, and $\beta_k$ denotes the $k$th coefficient, and $k$ is assumed to be 11 in this study. $x_{ik}$ represents the $k$th independent variable, and $y_i$ represents the dependent variable.

Based on the OLS model, the GWR methodology was proposed, which used local regression framework to explain spatial non-stationarity by estimating various parameters locally in space [14]. The parameters were estimated locally, using distance-weighted subsampling at neighboring locations, which can be expressed as:

$$y_i = \beta_0(u_i, v_i) + \sum_{k=1}^{m} \beta_k(u_i, v_i) x_{ik} + \varepsilon_i \quad (2)$$

where $y_i$ is a response variable of point $i$ at location i with the coordinate $(u_i, v_i)$, and $x_{ik}$ represents the $k$th in $m$ predictive variables of point $i$. $\beta_0(u_i, v_i)$ is the intercept value, $\beta_k(u_i, v_i)$ denotes the estimated parameter for the $k$th predictor variables, and $\varepsilon_i$ is the error term of point $i$. $\beta_k(u_i, v_i)$ in the GWR model is estimated using the predictive variables of point $i$'s neighbors in the scope defined by the space distance to the point $i$ and can be expressed by Equation (3):

$$\beta_k(u_i, v_i) = [X^T W(u_i, v_i) X]^{-1} X^T W(u_i, v_i) y \quad (3)$$

where $X$ is the independent variable matrix, $X^T$ is the transpose of $X$, $y$ is the independent variable, and $W(u_i, v_i)$ denotes an $n \times n$ diagonal geographic weighting matrix for point $i$, which is formed through a kernel function, such as Gaussian and bi-square. Here $\beta_k(u_i, v_i)$ denotes the impact of $x_{ik}$ on $y_i$ (GDP per capita in this case), and higher $\beta_k(u_i, v_i)$ signifies larger impact on regional economic disparity.

### 3.2. MGWR Model

MGWR is an extension of GWR. While GWR assumes consistent local relationships across models at the same spatial scale [24], MGWR allows the conditional relationships between the response variable and the different predictor variables to vary at different spatial scales. This adaptability is facilitated by aligning each explanatory variable's neighborhood with its own spatial extent, a principle that underpins MGWR's enhanced coefficient estimation within local regression models. This innovation is reflected in the bandwidths, which indicate the range over which data is borrowed and can vary by parameter surface. The expression of MGWR Equation (4) is refined from Equation (2).

$$y_i = \beta_0(u_i, v_i) + \sum_{k=1}^{m} \beta_{bwk}(u_i, v_i) x_{ik} + \varepsilon_i \quad (4)$$

Assuming that there are $n$ observations, for observation $i \in \{1, 2, \ldots, n\}$ at location $(u_i, v_i)$, bwk in $\beta_{bwk}$ indicates the bandwidth used for the calibration of the jth conditional relationship.

### 3.3. GTWR Model

To further reveal the influence of spatial–temporal heterogeneity of independent variables, the GTWR model is used to explore regional differences and periodic laws of influencing effects [33,61]. The base formula of the GTWR model is shown as follows:

$$y_i = \beta_0(u_i, v_i, t_i) + \sum_{k=1}^{m} \beta_k(u_i, v_i, t_i) x_{ik} + \varepsilon_i \quad (5)$$

where $y_i$ is the response variable at point $i$ with spatial and temporal coordinates $(u_i, v_i, t_i)$. In GTWR, $\beta_k(u_i, v_i, t_i)$ is optimized using the predictive variables of point $i$' s neighbors in the scope defined by the space–time distance to the point $i$, and it can be obtained from Equation (6), where $W(u_i, v_i, t_i)$ is a diagonal spatio-temporal weighting matrix specific to location $i$:

$$\beta_k(u_i, v_i, t_i) = \left[X^T W(u_i, v_i, t_i) X\right]^{-1} X^T W(u_i, v_i, t_i) y \quad (6)$$

The elements in the spatio-temporal weighting matrix can be represented as:

$$W(u_i, v_i, t_i) = \begin{cases} W_{ij}^S \times W_{ij}^T, \ 0 \le \Delta t < b^{S\tau} \\ 0, \ otherwise \end{cases}$$
$$W_{ij}^T = exp\left(\frac{d_{ij}^{T\,2}}{h_T^2}\right), \ W_{ij}^S = exp\left(\frac{d_{ij}^{S\,2}}{h_S^2}\right) \quad (7)$$
$$d_{ij}^T = \sqrt{(t_i - t_j)^2} \ (t_i > t_j) \ d_{ij}^S = \sqrt{(u_i - u_j)^2 + (v_i - v_j)^2}$$

where $W_{ij}^T$ and $W_{ij}^S$ are the time-weighted and space-weighted elements between observations $i$ and $j$, respectively. $h_T$ and $h_S$ are the spatial–temporal bandwidth. $d_{ij}^S$ is defined as the spatial distance between observations $i$ and $j$; $v$ is the bandwidth in time. $\tau$ is used for adjusting the inconsistency of the time distance and space distance, and therefore $b^{S\tau}$ is the spatial bandwidth $b^S$ at a certain time stage $\tau$, and $b^\tau$ denotes the temporal bandwidth. The remaining steps of GTWR are similar to those in GWR.

### 3.4. The MGTWR Model

The calibration of the MGTWR model is based on a back-fitting algorithm, and the MGTWR model is regarded as a generalized additive form [25]:

$$y_i = \beta_0(u_i, v_i, t_i) + \sum_{k=1}^{m} \beta_{bwk}(u_i, v_i, t_i)x_{ik} + \varepsilon_i = \sum_{k=1}^{m} f_{ik} + \varepsilon_i \quad (8)$$

The MGTWR model regression coefficients bwk for the kth covariate are calibrated using the specific bandwidth bwk. The back-fitting algorithm first initializes the additive term vector $\mathrm{f_k} = [\mathrm{f_{1,k}}, \mathrm{f_{2,k}}, \ldots, \mathrm{f_{n,k}}]$ using the GTWR model. The residual term $\hat{\varepsilon}$ at this stage can be obtained from $\hat{\varepsilon} = \mathrm{y} - \sum_{\mathrm{k}=1}^{\mathrm{m}} \mathrm{f_k}$. The first covariate generates the optimal bandwidth bw1 and updates the first additive term vector $\mathrm{f_k}$ as well as $\hat{\varepsilon}$. Then this process continues on to the next covariate until all optimal bandwidths bwk are updated.

### 3.5. STWR Model

The spatio-temporal weighted regression model (STWR) is a space–time regression model similar to GTWR to explore local spatio-temporal non-stationarity. The most notable improvement of STWR compared with GTWR is that it is designed to have a weighted average form in the spatio-temporal weighting matrix [38]. The distances are calculated as:

$$W_{ij}^{ST} = (1 - \alpha)W_{ij}^{S} + \alpha W_{ij}^{T} \quad (9)$$

where $\alpha \in [0, 1]$ is an adjustable factor to scale the space and time effect. This improves on the potential weight underestimation and accelerates the calculation through kernel functions. $W_{ij}^S$ is the same as in the GTWR model and $W_{ij}^T$ is not the time interval, but the rate of value variation between an observed point and a regression point through a time interval, shown as follows:

$$W_{ij}^{T} = \begin{cases} \left[\frac{2}{1+\exp\left(-d_{ij}^{T}/b^{T}\right)} - 1\right], \ 0 < \Delta t < b^{T} \\ 0, \ otherwise \end{cases} \quad (10)$$

$$d_{ij}^{T} = \frac{\left|(y_i - y_j)/y_j\right|}{t_i - t_j} \quad (11)$$

where $\Delta t$ is the time interval $t_i - t_j$ between point $i$ and point $j$, and $b^T$ is the temporal bandwidth selected from time intervals. $d_{ij}^T$ represents spatial distance and temporal distance between point $i$ and point $j$, respectively, $t_i$ is the timestamp of point $i$, and $t_j$ is the timestamp of point $j$ $(t_j < t_i)$; $y_i$ and $y_j$ are dependent variables. Thus, $d_{ij}^T$ is actually the rate of value variation from time $t_j$ to $t_i$, rather than the time interval, as in GTWR.

### 3.6. Model Evaluation

The above algorithms were implemented in the package of GWmodel and STWR v1.0 model in Python [38,62]. To avoid randomness within different regression models as well as to ensure the stability of the model results, we repeated the simulation experiment 200 times for each model for the purposes of selecting the minimum Akaike information criterion (AICc) to define the bandwidth of the model. Then several corresponding indicators, namely the mean absolute error (MAE), the root mean square error (RMSE), the coefficient of determination ($R^2$), and the adjusted $R^2$ ($Radj^2$), to quantitatively evaluate the similarity between the observations and the estimated values of the resulting response variable.

High $R^2$ and low RMSE, MAE, and AICc indicate a good fit between the model and the observation.

## 4. Results

### 4.1. Statistical Evaluation of the Six Regression Models

Before using the regression models, the spatial–temporal non-stationarity of explanatory variables was evaluated using the interquartile range (IQR). The IQR measures the spread of the middle half of the data, and a larger IQR value indicates a larger data variability. Comparison of the IQR of the regression coefficients from the GWR, MGWR, GTWR, MGTWR, and STWR models with twice the standard error (SE) from the OLS model is shown in Table 2. The extra local variation in the rightmost column indicated that all 11 explanatory variables exhibit spatio-temporal non-stationarities.

**Table 2.** The comparison of interquartile range of the regression coefficients from the GWR, MGWR, GTWR, MGTWR, and STWR models.

| Variable | IQR (GWR) | IQR (MGWR) | IQR (GTWR) | IQR (MGTWR) | IQR (STWR) | 2 × SE (OLS) | Extra-Local Variation |
|---|---|---|---|---|---|---|---|
| PN | 0.19 | 0.19 | 0.19 | 0.11 | 0.20 | 0.14 | Yes |
| PI | 0.14 | 0.20 | 0.16 | 0.09 | 0.17 | −0.13 | Yes |
| SI | 0.69 | 0.05 | 0.86 | 0.57 | 0.80 | 0.10 | Yes |
| PFR | 0.69 | 0.04 | 0.74 | 0.65 | 0.51 | 1.48 | Yes |
| PFE | 0.4 | 0.07 | 0.43 | 0.27 | 0.34 | 0.33 | Yes |
| DB | 0.63 | 0.04 | 0.78 | 0.47 | 0.75 | −0.08 | Yes |
| LB | 0.70 | 0.03 | 0.76 | 0.40 | 0.67 | 0.19 | Yes |
| TP | 0.12 | 0.12 | 0.15 | 0.09 | 0.16 | −0.04 | Yes |
| SN | 0.25 | 0.13 | 0.30 | 0.17 | 0.28 | 0.01 | Yes |
| HBN | 0.15 | 0.16 | 0.13 | 0.12 | 0.10 | −0.34 | Yes |
| GY | 0.18 | 0.10 | 0.22 | 0.13 | 0.14 | −0.09 | Yes |

Table 3 presents a comparative analysis of the average fitted results obtained from five regression models at two levels. At the county level, the statistical results of the STWR model exhibited the highest $R^2$ values of 0.97 along with the lowest AICc value for the fitting results at the county level, and relatively high $R^2$ values of 0.95 with reasonable RMSE and MAE values at the prefecture-level. However, at the prefecture-level, the GTWR model outperformed another four models with the highest $R^2$ and $Radj^2$, as well as the lowest RMSE and MAE values.

Comparing the performances of STWR and MGTWR models with GTWR revealed that, based on $R^2$, $Radj^2$, and AICc values, they exhibited similar capabilities, without one model clearly surpassing the others. However, when assessing the RMSE and MAE values, STWR and MGTWR did not outperform GTWR. Therefore, these results indicate that the STWR, GTWR, and MGTWR models offer advantages over OLS, GWR, and MGWR in the fitting process. Still, they do not distinctly outperform each other based on different validation methods.

**Table 3.** Statistical evaluation of the OLS, GWR, MGWR, GTWR, MGTWR, and STWR models.

| Level | Model | $R^2$ | $R_{adj}^2$ | AICc | RMSE | MAE |
|---|---|---|---|---|---|---|
| County | OLS | 0.68 | 0.68 | −7668.20 | 0.03 | 0.02 |
| | GWR | 0.93 | 0.90 | −9631.55 | 0.02 | 0.01 |
| | MGWR | 0.93 | 0.90 | −9278.35 | 0.02 | 0.01 |
| | GTWR | 0.97 | 0.95 | −49,311.54 | 0.02 | 0.01 |
| | MGTWR | 0.94 | 0.92 | −49,579.53 | 0.02 | 0.01 |
| | **STWR** | **0.97** | **0.95** | **−11,380.14** | **0.02** | **0.01** |
| Prefecture | OLS | 0.70 | 0.69 | −808.76 | 0.13 | 0.09 |
| | GWR | 0.86 | 0.81 | −898.94 | 0.05 | 0.03 |
| | MGWR | 0.86 | 0.81 | −853.78 | 0.05 | 0.03 |
| | **GTWR** | **0.96** | **0.94** | **−5625.34** | **0.12** | **0.08** |
| | **MGTWR** | **0.95** | **0.92** | **−4317.74** | **0.12** | **0.08** |
| | **STWR** | **0.95** | **0.92** | **−1228.29** | **0.20** | **0.14** |

### 4.2. Spatial Dynamics of the Beta Parameter for the 11 Variables in Response to GDP per Capita

During the study period, the mean nominal GDP per capita exhibited a consistent increasing trend across both county- and prefecture-level regions (Figure 3b), with the most significant growth observed between 2005 and 2015. After adjusting for the GDP deflator coefficients provided by the National Bureau of Statistics (2000: 0.35%; 2005: 1.78%; 2010: 3.18%; 2015: 1.44%; 2019: 2.90%), the trend of real GDP per capita mirrored that of nominal GDP per capita (Figure 3b). However, regional disparities in GDP per capita widened significantly after 2005.

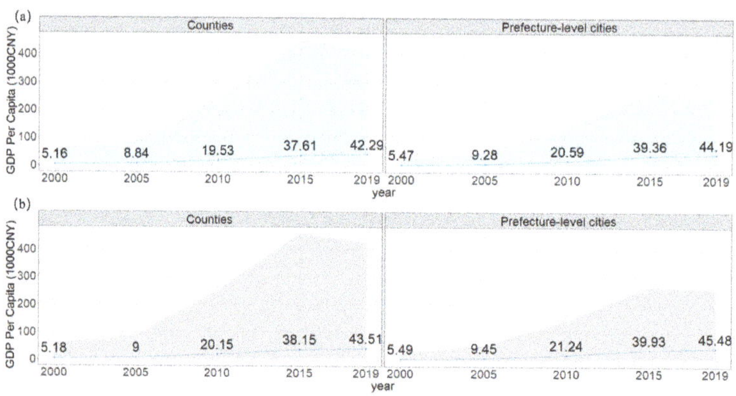


**Figure 3.** The regional differences based on real GDP per capita (**a**) and nominal GDP per capita (**b**) from 2000 to 2019. The blue dots represent the regional mean value, and the shaded area represents the variance.

Specifically, regional differences in GDP per capita increased modestly during the first five years but grew rapidly over the subsequent 15 years. The growth rates of mean GDP per capita were 740.1% and 728.2% at the county and prefecture levels, respectively. Although the regional per capita differences at the prefecture level were initially less than half of those at the county level during the first decade, these disparities continued to expand after 2015. In contrast, per capita differences at the county level began to narrow during this period.

Among the 11 independent variables, secondary industry (SI) has always had the most positive influences and made the biggest contribution to the GDP per capita over the past 20 years. While the relationships between GDP per capita and the other variables vary dramatically—for example, in 2000, SI, PN and SN were three of the most important variables related to the GDP per capita at the prefecture level, but the variables' combination changed to SI, PN, and DB in 2010 and SI, PFR, and HBN in 2019. Therefore, we only selected the SI, PN, and HBN variables in 2019 as an example to further demonstrate the spatial pattern variances of their intercept coefficients in response to GDP per capita.

As shown in the spatio-temporal regression results in Figures 4 and 5, based on five models, SI maintained a positive and significant influence on GDP per capita using different models across county-level and prefecture-level regions, and the highest value was observed in the northwestern regions of China. Following SI, PFE also showed a gradient spatial pattern from the southeastern parts to the northwestern parts of China. Unlike SI and PFE, the spatial pattern of HBN is more diverse across the study regions.

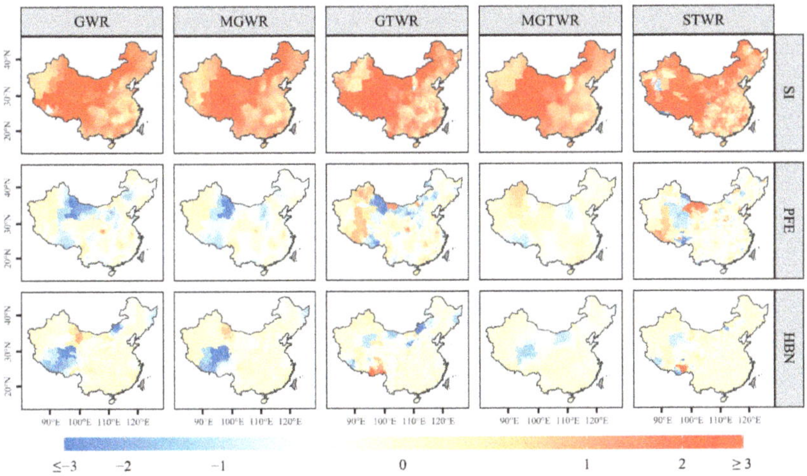


**Figure 4.** The spatial patterns of the $\beta_k$ parameter for the predictive variables in 2019 at the county level.

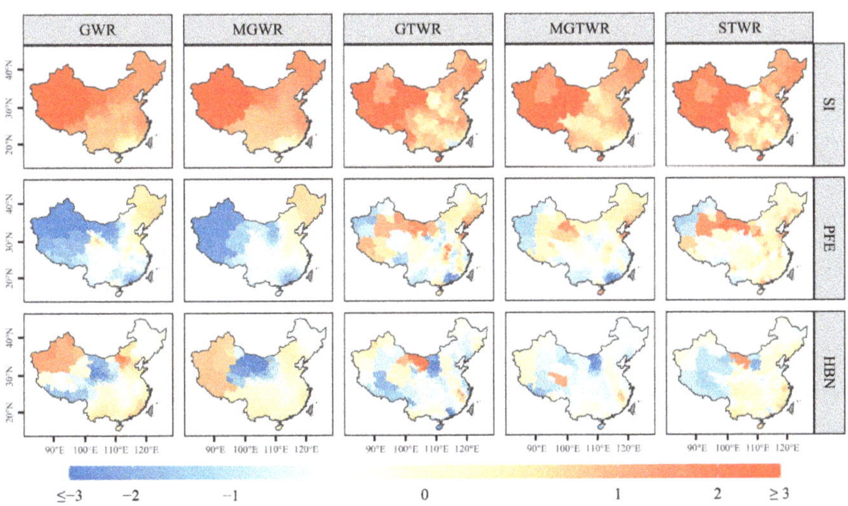


**Figure 5.** The spatial patterns of the $\beta_k$ parameter for the predictive variables in 2019 at the prefecture level.

In term of the models' performance, the spatial pattern of intercept coefficients based on the STWR model was more consistent with that based on the GTWR model and MGTWR model. Comparatively, the spatial patterns of intercept coefficients based on GWR and MGWR were less fragmented than those of STWR, GTWR, and MGTWR. In addition, when compared with the results based on county-level and city-level regions, the variances in PFE and HBN at the prefecture level are less significant than those at the county level in the southeastern part of China as well as the western part of China, and most of the intercept coefficients changed from the negative value to the positive values across these regions. Generally, the spatial patterns of intercept coefficients based on GTWR, MTGWR, and STWR are more sensitive to the variables, as well as the spatial units.

### *4.3. Spatial Heterogeneity of Regional Disparity Influencing Variables*

Figure 6 provides a deeper insight into the spatio-temporal patterns of SI intercept coefficients over five different years, using various regression models. Notably, when it comes to the SI variable, GTWR sometimes falls short in detecting changes in parameter fitting as effectively as GWR. For instance, in 2010, significant variations were observed in SI parameters within the region at the intersection of Qinghai and Tibet Provinces. While GTWR employs an adjustment factor τ to balance the influence of time and space, it tends to over-emphasize the temporal effects in most Chinese counties while neglecting spatial fitting within local neighborhoods.

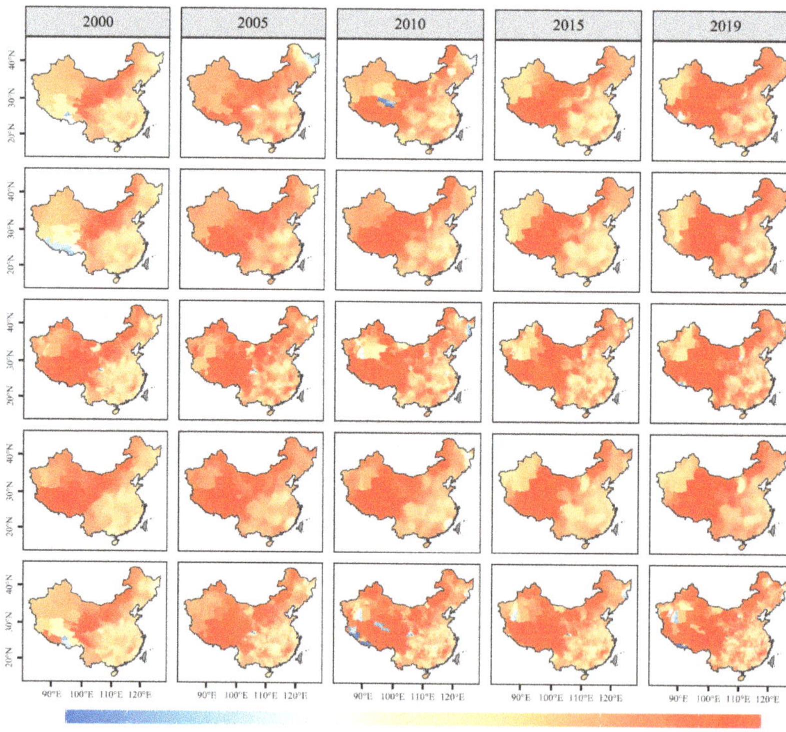


**Figure 6.** Spatio-temporal pattern of the the $\beta_k$ parameter for SI for different models at the county level.

Furthermore, regarding the spatio-temporal patterns of SI intercept coefficients, the variations within the MGWR and MGTWR models are relatively insignificant. This suggests that the utilization of time intervals as temporal weights through the adjustment factor c in MGTWR is not as impactful as the rate of variation in SI attribute values. In contrast, STWR excels in capturing the combined effects of a non-stationary spatio-temporal process involving the observed SI variable, effectively capturing local SI influences on GPC in both space and time. This results in more pronounced spatio-temporal variations when compared to MGWR and MGTWR.

## 5. Discussion

### *5.1. The Optimization of Bandwith Selection in Different Models*

The optimization of the spatio-temporal kernel function in STWR is distinguished by the introduction of a scale parameter, which adjusts the relative contributions of spatial and temporal kernels to the regression points. This parameter effectively quantifies the relative strength of spatial and temporal influences, enabling a more nuanced analysis of spatio-temporal interactions. However, the advantages of this optimization may not be immediately evident when temporal variations are minimal [34]. For example, as illustrated in Table 4, during the initial period (2000–2005), the $R^2$ values across the five models are similar. This is because models such as STWR, GTWR, and MGTWR lack prior temporal observations to leverage, and any observed $R^2$ differences are likely due to variations in their spatial bandwidth search ranges.

As the time intervals extend (e.g., from 5 years to 20 years), significant differences begin to surface. Models like GWR and MGWR demonstrate a consistent decline in $R^2$ values across the time stages, indicating reduced model fit over longer periods. In contrast, STWR maintains consistently high $R^2$ values, reflecting its robustness in capturing spatio-temporal dynamics. Although MGTWR occasionally achieves higher $R^2$ values than STWR, its performance comes with a trade-off in model complexity, as evidenced by substantially higher AICc values. AICc penalizes models with excessive complexity, balancing model fit against overfitting risks.

**Table 4.** Bandwidth comparison of the five models based on the bi-square kernel.

| Level | Year | Model | $R^2$ | AICc |
|---|---|---|---|---|
| County | 1999 | MGWR | 0.95 | −10,622.3 |
| | | MGTWR | 0.95 | −10,347.9 |
| | | GWR | 0.95 | −11,088.7 |
| | | GTWR | 0.95 | −11,078.8 |
| | | STWR | 0.96 | −11,032.9 |
| | 2004 | MGWR | 0.94 | −9208.77 |
| | | MGTWR | 0.95 | −20,219.7 |
| | | GWR | 0.94 | −9560.87 |
| | | GTWR | 0.95 | −20,724.8 |
| | | STWR | 0.96 | −10,285.9 |
| | 2009 | MGWR | 0.94 | −9286.07 |
| | | MGTWR | 0.95 | −29,307.6 |
| | | GWR | 0.93 | −9572.26 |
| | | GTWR | 0.95 | −30,821.3 |
| | | STWR | 0.96 | −10,114.4 |

**Table 4.** *Cont.*

| Level | Year | Model | $R^2$ | AICc |
|---|---|---|---|---|
| County | 2014 | MGWR | 0.92 | −8829.88 |
| | | MGTWR | 0.96 | −35,262.5 |
| | | GWR | 0.92 | −9044.22 |
| | | GTWR | 0.95 | −40,490.7 |
| | | STWR | 0.93 | −9904.19 |
| | 2019 | MGWR | 0.93 | −8444.73 |
| | | MGTWR | 0.94 | −49,579.5 |
| | | GWR | 0.93 | −8891.7 |
| | | GTWR | 0.97 | −49,311.5 |
| | | STWR | 0.97 | −11,380.1 |
| Prefecture | 1999 | MGWR | 0.89 | −782.94 |
| | | MGTWR | 0.92 | −738.84 |
| | | GWR | 0.92 | −872.11 |
| | | GTWR | 0.92 | −873.75 |
| | | STWR | 0.93 | −871.55 |
| | 2004 | MGWR | 0.85 | −968.72 |
| | | MGTWR | 0.93 | −1722.67 |
| | | GWR | 0.83 | −956.22 |
| | | GTWR | 0.93 | −2055.85 |
| | | STWR | 0.92 | −1131.59 |
| | 2009 | MGWR | 0.85 | −760.3 |
| | | MGTWR | 0.95 | −2556.34 |
| | | GWR | 0.89 | −873.24 |
| | | GTWR | 0.95 | −3221.36 |
| | | STWR | 0.98 | −1286.9 |
| | 2014 | MGWR | 0.82 | −915.43 |
| | | MGTWR | 0.95 | −3388.8 |
| | | GWR | 0.79 | −899.6 |
| | | GTWR | 0.95 | −4317.06 |
| | | STWR | 0.97 | −1238.52 |
| | 2019 | MGWR | 0.88 | −841.5 |
| | | MGTWR | 0.97 | −5434.55 |
| | | GWR | 0.89 | −893.54 |
| | | GTWR | 0.96 | −5625.34 |
| | | STWR | 0.95 | −1228.29 |

Furthermore, our comprehensive examination of parameter bandwidths across the five regression models highlights an important distinction: while STWR may not offer the same degree of flexibility in adjusting parameter bandwidths as the MGWR and MGTWR models [63,64], it consistently exhibited smaller bandwidths for all 11 parameters compared to their counterparts in other models. For instance, as demonstrated in Table 5, regardless of whether we considered county or city scales, the bandwidths for all parameters in STWR were 15 and 9, respectively. These values were significantly smaller than those observed in other models. These smaller bandwidths indicate that STWR optimizes fewer initial neighbors, focusing on a more confined local area for spatio-temporal analysis.

**Table 5.** Bandwidth of the five models based on the bi-square kernel.

| Level | Model | PN | PI | SI | PFR | PFE | DB | LB | TP | SN | HBN | GY |
|---|---|---|---|---|---|---|---|---|---|---|---|---|
| County | MGWR | 43 | 43 | 43 | 43 | 43 | 43 | 43 | 43 | 43 | 43 | 47 |
| | MGTWR | 54 | 54 | 54 | 54 | 54 | 54 | 54 | 54 | 54 | 54 | 54 |
| | | | | GWR: 81 | | GTWR: 67 | | STWR: 15 | | | | |
| Prefecture | MGWR | 43 | 43 | 46 | 46 | 43 | 43 | 43 | 44 | 44 | 44 | 51 |
| | MGTWR | 22 | 20 | 17 | 24 | 14 | 19 | 21 | 29 | 17 | 20 | 23 |
| | | | | GWR: 83 | | GTWR: 70.1 | | STWR: 9 | | | | |

This reduced bandwidth demonstrates STWR's capability to effectively address spatio-temporal non-stationarity within localized contexts. By focusing on smaller neighborhoods, STWR enhances the robustness of its regression outcomes, allowing it to capture finer-grained variations in the spatio-temporal dynamics of the data. This localized focus not only improves the interpretability of the results but also highlights STWR's ability to maintain stability and accuracy in regions with complex or heterogeneous spatio-temporal patterns. Moreover, while MGTWR may occasionally achieve a higher fit, the findings emphasize that STWR offers a more optimal balance between goodness of fit and model parsimony. By avoiding excessive model complexity, STWR ensures greater reliability and interpretability across varying temporal scales. This balance underscores STWR's practicality and effectiveness as a tool for spatio-temporal analysis, particularly in applications where achieving both precision and simplicity is essential.

### 5.2. The Kernel Function of Different Spatio-Temporal Regression Model

It is worth noting that the STWR model optimizes the weight matrix for spatial and temporal intervals rather than the spatial kernel function. Traditional kernel density estimation methods, such as the bi-square kernel and Gaussian kernel, incorporate a distance decay function that assigns more weight to observations closer to a regression point and less weight to those farther away [16,65]. Therefore, we also compared the performance of the bi-square kernel and Gaussian kernel in specifying spatial weighting. Since the Gaussian model has not been fully implemented in the STWR model, we only compared its results in four other models. As shown in Tables 6 and S1, all the models achieved higher R-squared values but also higher AIC information criteria when using the Gaussian kernel [25,39]. The Gaussian kernel demonstrates higher uncertainty than the bi-square kernel.

Furthermore, we compared the bandwidth differences between the Gaussian and bi-square kernels (Table 6 vs. Table S2) and found that the bandwidths for all 11 parameters were more diverse with the Gaussian kernel. This may be because the bi-square weight assigns a weight of zero to observations outside the bandwidth, effectively eliminating their influence on the local regression estimate. As a result, the bi-square kernel is less sensitive to detecting variations in the magnitudes of observation values. However, it excels in capturing local spatial heterogeneity effects from observations to the regression point.

**Table 6.** The comparison of bi-square and Gaussian-based models.

| | | Gaussian | | Bi-Square | |
|---|---|---|---|---|---|
| Level | Model | $R^2$ | AICc | $R^2$ | AICc |
| County | OLS | 0.68 | −7668.20 | 0.68 | −7668.20 |
| | GWR | 0.90 | −9131.82 | 0.93 | −9631.55 |
| | MGWR | 0.82 | −8659.29 | 0.93 | −9278.35 |
| | GTWR | 0.73 | −39,537.20 | 0.97 | −49,311.54 |
| | MGTWR | 0.88 | −45,273.30 | 0.94 | −49,579.53 |
| | STWR | / | / | 0.97 | −11,380.14 |
| Prefecture | OLS | 0.70 | −808.76 | 0.70 | −808.76 |
| | GWR | 0.83 | −868.43 | 0.86 | −898.94 |
| | MGWR | 0.75 | −836.39 | 0.86 | −853.78 |
| | GTWR | 0.78 | −4420.36 | 0.96 | −5625.34 |
| | MGTWR | 0.88 | −4990.48 | 0.95 | −4317.74 |
| | STWR | / | / | 0.95 | −1228.29 |

### *5.3. The First Law of Geography and the MAUP Issue in the Spatio-Temporal Regressions*

Tobler's First Law Of Geography [66], which states that "everything is related to everything else, but near things are more related than distant things," serves as the foundational principle for spatial–temporal geographic weighted regression. This principle underpins the analysis of spatial heterogeneity and temporal dynamics, enabling a more nuanced understanding of geographic patterns. The findings depicted in Figure 7 further affirm this fundamental principle by illustrating the coefficient consistency of parameters across various spatial–temporal geographic weighted regression models. Consistent with Zhang [67] and Li et al. [68], this homogeneity arises from uniform socio-economic conditions, smaller GDP per capita differences, and limited industrial diversification. Similar trends are discussed by Song et al. [69] and Chen [70], who note minimal disparities in western China due to slower economic integration and less diverse economic structures.

Conversely, regions with greater GDP per capita disparities, such as the Mega-region and areas east of the Hu Line, show lower spatial consistency due to significant economic heterogeneity and rapid industrialization. Kanbur and Zhang [71] and Gao et al. [72] highlight how diversification and income disparities in these regions create fragmented spatial patterns. Advanced models like GTWR, MGTWR, and STWR, which incorporate spatial and temporal heterogeneity, effectively capture these dynamics. Wu et al. [73] emphasize the value of temporal dimensions in understanding socio-economic processes, while Han et al. [74] and Wang et al. [75] demonstrate the efficacy of these models in regions undergoing rapid transformation, such as China's eastern Mega-region.

The modifiable areal unit problem (MAUP) further impacts the fitting outcomes of spatial–temporal models [76,77]. Parameters like HBN, GY, and PI, which correlate positively with GDP at the prefecture level, may exhibit negative correlations at the county level (Figures S1–S20). Openshaw [78] and Fotheringham and Wong [79] stress that MAUP-induced variability necessitates multi-scale regression approaches. Multi-scale analyses improve reliability by exploring the effects of spatial and temporal heterogeneity at varying scales [25,80]. Recent advancements, such as the One4All-ST framework [81], mitigate scale inconsistencies through flexible, unified spatio-temporal modeling.

Rational parameter selection also plays a crucial role. Indicators like HBN and PI offer insights into human development and infrastructure investment, while GY captures economic growth dynamics, making them essential for analyzing spatial–temporal variations [75,82]. This aligns with Jiang and Yao [83], who underscore the importance of selecting indicators reflective of regional disparities in heterogeneous development landscapes like China. Therefore, adopting multi-scale models, flexible frameworks, and carefully selected parameters is crucial for addressing MAUP and enhancing the interpretability of spatial–temporal analyses. These approaches deepen our understanding of the interplay between spatial proximity, temporal dynamics, and socio-economic disparities while refining the application of Tobler's First Law across varying scales.

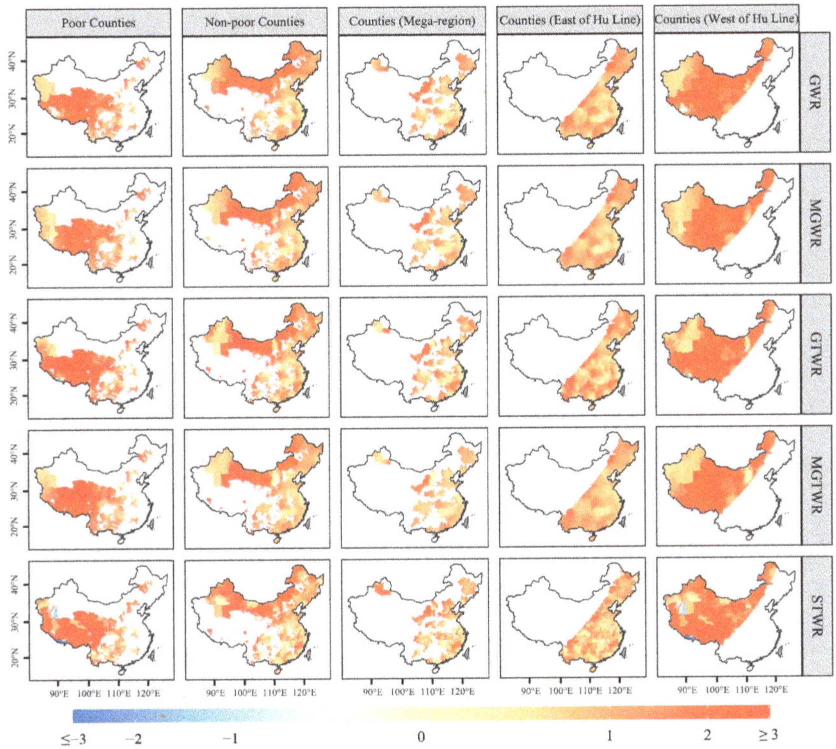


**Figure 7.** Spatio-temporal pattern of the $\beta_k$ parameter for SI for different models across different regions at the county level in 2019.

## 5.4. Suggestions for Regional Disparity Alleviation

China has made great leaps in reducing poverty and reaching the benchmarks laid out in the Sustainable Development Goals (SDGs) [84,85], which has lifted more than 500 million of its citizens out of extreme poverty over the last three decades. Despite these achievements, regional disparities persist, necessitating a deeper understanding of the spatial and temporal dynamics of inequality to inform effective policy-making. This study, using the GWR, MGWR, GTWR, MGTWR and STWR models from 2000 to 2019, offers timely insights into the socio-economic variables influencing regional disparities across various scales (all counties, poverty-stricken and non-poverty-stricken counties, mega-regions, and sub-national levels). The results highlight several critical issues and policy implications.

Equalization of Public Services. One significant finding is the role of the public financial revenue (PFR) variable, which emerged as the second most important factor contributing to regional disparities from 2000 to 2020. This reflects the success of initiatives like the "China Rural Poverty Alleviation and Development (CRPAD) Program" [86,87] in developing healthcare, education, and technology services in the central and western regions. However, other public service variables, such as SN and HBN, demonstrated weak relationships with economic growth, indicating potential misallocation of government resources. A similar point was also raised by Blanchard [88]. Therefore, this misallocation, characterized by overinvestment in manufacturing and underinvestment in domestic service industries, particularly at the county level, underscores the need for more balanced resource distribution. Insufficient equalization of public services remains a significant challenge for achieving high-quality economic development.

Rebalancing the Economy to Promote Domestic Consumption. When time interval variables were considered in the GTWR and STWR models, deposit balance (DB) emerged as a more significant factor than PFR in poverty-stricken regions and areas west of the Hu Line. This finding suggests that rising inter-regional inequality has contributed to high personal savings rates in these areas [89,90], highlighting the need to transition from high savings to higher domestic consumption [91]. Rebalancing the economy is essential for addressing these disparities, particularly in inland poverty-stricken regions and small cities, where fostering local demand could boost economic growth.

Monitoring County-Level Disparities and Addressing the MAUP Issue. The study confirmed Tobler's First Law of Geography, showing that adjacent counties tend to share more similar socio-economic determinants than distant ones. Patterns such as the positive congestive SI pattern in Xinjiang and the negative congestive DB pattern in the Pearl River Delta mega-regions highlight the importance of continuous monitoring and tailored interventions that account for the unique socio-economic and geographic characteristics of these regions. However, the analysis also revealed significant MAUP effects, particularly in regions with high inequities, such as Xinjiang and Inner Mongolia. For instance, the SI factor in Shaanxi varied in significance between county and mega-region scales, underscoring the need for cross-scale evaluations. Policies designed without considering scale variations may lead to inaccurate conclusions and ineffective interventions, making multi-scale analyses essential for designing robust and accurate policies to address regional disparities effectively [92,93].

Strategic Focus on High-Disparity Regions: Regions with pronounced inequities, such as Xinjiang and Inner Mongolia, require tailored strategies to address their unique challenges. These areas exhibit stronger MAUP effects and greater resource allocation needs, reinforcing the importance of targeted development efforts. Focusing on technical efficiency improvements and creating sustainable development models for these regions can significantly bridge the gap between high-growth and high-inequity areas, fostering equitable economic development [94–96].

### *5.5. Research Limitation*

We should note that there is not one ideal dataset that can comprehensively represent the complexities of socially and economically meaningful regional inequity. Due to the data limitation of the China County Statistical Yearbook (CCSY) published by the National Bureau of Statistics of China, this study can only obtain 11 variables to delineate the regional variations. Meanwhile, to explore regional disparities, it is essential to address inflationary distortions in socio-economic data. In this study, the data collected from CCSY were adjusted using the same inflation coefficient, ensuring their robustness in demonstrating their contribution to nominal GDP per capita. However, if the goal is to analyze the variance of socio-economic data across different spatial and temporal scales, it would be advisable to account for inflation coefficients to achieve more accurate and comparable results. Moreover, different combinations of socio-economic indicators may lead to various alternative representations of regional disparity and different spatio-temporal patterns in their relationships [90,93]. Therefore, it is crucial to integrate diverse socio-economic variables, such as remote sensing data, open access geographical data, etc., to study regional disparities and social perceptions of inequity in the future. The framework outlined in this study transforms static statistical data into multiple scales of geographical units but lacks ground truth data for validation. Therefore, it is advisable to employ composite indicators based on multiple dimensions of socio-economic and environmental features to further scrutinize the uncertainty and sensitivity of indicators in identifying inequities across different scales.

## 6. Conclusions

To better analyze the spatio-temporal dynamics of regional economic disparities in China, this study builds upon existing spatial and temporal regression models (e.g., GWR, MGWR, GTWR, and MGTWR) and advances the framework with the spatio-temporal weighted regression (STWR) model proposed by Que [34]. By introducing a "time distance" perspective, the study explores the intertwined spatial and temporal complexities of regional economic disparities. The contributions of this research can be summarized as follows:

(1) Extending Traditional Spatial Regression by Incorporating Temporal Dynamics. Unlike models such as GWR, MGWR, GTWR, and MGTWR, which focus on spatially varying relationships at specific time steps, STWR incorporates a spatio-temporal kernel function. This kernel optimizes the weighting of observations based on their spatial and temporal proximity, enabling a more accurate representation of regional economic disparities where spatial and temporal non-stationarity coexist. This enhancement ensures a deeper understanding of the dynamic interactions driving disparities over time and space.

(2) Improving the Interpretability of Regional Disparities. By employing county-level spatial units and analyzing disparities across diverse regions—poverty-stricken and non-poverty-stricken areas, mega-regions, and sub-national regions delineated by the Hu Line—STWR ensures robust estimation of regional disparities. The flexibility of spatial units enhances the model's ability to reflect nuanced differences across various regions, making the results more interpretable and actionable for policymakers and stakeholders.

(3) Encouraging Data-Driven Policymaking. This study successfully applies STWR to analyze a range of socio-economic variables, including primary and secondary industries, public financial revenue, and deposit balances, to assess regional economic disparities. By integrating spatial and temporal kernels, STWR provides a holistic analysis of socio-economic inequities, offering a comprehensive understanding of the factors driving disparities. This approach highlights specific spatial and temporal dimensions of inequality, enabling more targeted resource allocation and intervention strategies.

This study also emphasizes the practical applications of GTWR, MGTWR, and STWR models, which provide significant advancements in analyzing the spatial and temporal impacts of China's "accurate poverty targeting" [97] initiatives and offer deeper insights into regional economic disparities. The key contributions are as follows:

(1) Impact of Industrialization on Regional Disparities. Post-2000, heavy industry has intensified the rural–urban divide, particularly between mega- and non-mega-regions and areas west and east of the Hu Line. These findings highlight structural challenges in regional development, underscoring the need for tailored policies to address industrial dependency and promote balanced growth.

(2) Policy-Driven Transformations Post-2010. The analysis of the public financial revenue (PFR) variable highlights the Chinese government's focused efforts to improve public

services, ecological development, and capacity building, with a stronger impact observed in poverty-stricken areas. These results validate the effectiveness of targeted poverty alleviation strategies and provide a blueprint for scaling similar efforts.

(3) Targeted Interventions to Address Disparities. Spatio-temporal insights identify specific counties where targeted investments in healthcare, education, and technology can effectively reduce inequities. The findings highlight the importance of sustained and region-specific policy efforts, particularly in poverty-stricken areas and regions west of the Hu Line. Enhancing technical efficiency and implementing long-term strategies can significantly contribute to equitable economic development and growth [89,90].

It is crucial to acknowledge that the effective application of GTWR, MGTWR, and STWR models hinges on the presence of spatio-temporal non-stationarity in parameters related to social disparities. Although socio-economic indicators may exhibit variations within spatial bandwidths and spatio-temporal kernels, relying solely on a combined measurement of spatial and temporal distances can lead to oversimplification or misinterpretation. Calculating distances in three dimensions—time and two-dimensional space—remains a significant methodological challenge. To address this, future research should focus on optimizing spatio-temporal kernel functions and experimenting with varying spatial and temporal bandwidths to more accurately capture the complexities of spatio-temporal heterogeneity.

**Supplementary Materials:** The following supporting information can be downloaded at: https://www.mdpi.com/article/10.3390/land14010059/s1.

**Author Contributions:** C.W.: conceptualization, writing. X.L. and W.C.: methodology, software application. L.Z.: data collection, data analysis. R.C. and W.W.: review, editing, manuscript revision. All authors contributed to the study. All authors have read and agreed to the published version of the manuscript.

**Funding:** This research was financially supported by the National Natural Science Foundation of China, NSFC Young Scientist Fund, No. 42201205; National Nature Science Foundation of China, NSFC General Fund, No. 42471443.and the Southern Marine Science and Engineering Guangdong Laboratory (Zhuhai) project No. 311021018.

**Data Availability Statement:** Data will be made available on request.

**Conflicts of Interest:** The authors have no competing interests to declare that are relevant to the content of this article.

## References

1. Pan, J.; Hu, Y. Spatial Identification of Multi-Dimensional Poverty in Rural China: A Perspective of Nighttime-Light Remote Sensing Data. *J. Indian Soc. Remote Sens.* **2018**, *46*, 1093–1111. [CrossRef]
2. Kanbur, R.; Zhang, X. Fifty Years of Regional Inequality in China: A Journey Through Central Planning, Reform, and Openness. *Rev. Dev. Econ.* **2005**, *9*, 87–106. [CrossRef]
3. Cingano, F. *Trends in Income Inequality and Its Impact on Economic Growth*; OECD: Paris, France, 2014.
4. United Nations. The Inefficiency of Inequality. Available online: https://www.un-ilibrary.org/content/books/9789210586283 (accessed on 7 June 2024).
5. Wang, Y.; Chen, X.; Sun, P.; Liu, H.; He, J. Spatial-Temporal Evolution of the Urban-Rural Coordination Relationship in Northeast China in 1990–2018. *Chin. Geogr. Sci.* **2021**, *31*, 429–443. [CrossRef]
6. Su, S.; Gong, Y.; Tan, B.; Pi, J.; Weng, M.; Cai, Z. Area Social Deprivation and Public Health: Analyzing the Spatial Non-Stationary Associations Using Geographically Weighed Regression. *Soc. Indic. Res.* **2017**, *133*, 819–832. [CrossRef]
7. Zhang, G.; Chen, Y.; Wang, G.; Zhou, C. Spatial-Temporal Evolution and Influencing Factors of Digital Financial Inclusion: County-Level Evidence from China. *Chin. Geogr. Sci.* **2023**, *33*, 221–232. [CrossRef]
8. Wei, C.; Cabrera Barona, P.; Blaschke, T. A New Look at Public Services Inequality: The Consistency of Neighborhood Context and Citizens' Perception across Multiple Scales. *ISPRS Int. J. Geo-Inf.* **2017**, *6*, 200. [CrossRef]

9. Wan, G. Accounting for Income Inequality in Rural China: A Regression-Based Approach. *J. Comp. Econ.* **2004**, *32*, 348–363. [CrossRef]
10. Anselin, L.; Getis, A. Spatial Statistical Analysis and Geographic Information Systems. *Ann. Reg. Sci.* **1992**, *26*, 19–33. [CrossRef]
11. Cabrera-Barona, P.; Wei, C.; Hagenlocher, M. Multiscale Evaluation of an Urban Deprivation Index: Implications for Quality of Life and Healthcare Accessibility Planning. *Appl. Geogr.* **2016**, *70*, 1–10. [CrossRef]
12. Casetti, E. Bayesian Regression and the Expansion Method. *Geogr. Anal.* **1992**, *24*, 58–74. [CrossRef]
13. Gutiérrez, J.; Cardozo, O.D.; García-Palomares, J.C. Transit Ridership Forecasting at Station Level: An Approach Based on Distance-Decay Weighted Regression. *J. Transp. Geogr.* **2011**, *19*, 1081–1092. [CrossRef]
14. Brunsdon, C.; Fotheringham, A.S.; Charlton, M.E. Geographically Weighted Regression: A Method for Exploring Spatial Nonstationarity. *Geogr. Anal.* **1996**, *28*, 281–298. [CrossRef]
15. Xu, Z.; Cai, Z.; Wu, S.; Huang, X.; Liu, J.; Sun, J.; Su, S.; Weng, M. Identifying the Geographic Indicators of Poverty Using Geographically Weighted Regression: A Case Study from Qiandongnan Miao and Dong Autonomous Prefecture, Guizhou, China. *Soc. Indic. Res.* **2019**, *142*, 947–970. [CrossRef]
16. Shen, X.; Zhou, Y.; Jin, S.; Wang, D. Spatiotemporal Influence of Land Use and Household Properties on Automobile Travel Demand. *Transp. Res. Part D Transp. Environ.* **2020**, *84*, 102359. [CrossRef]
17. Wei, Y.D. Spatiality of Regional Inequality. *Appl. Geogr.* **2015**, *61*, 1–10. [CrossRef]
18. Yuan, X.; Li, Y.; Song, Y.; Lu, H.; Wang, Y.; Ge, B.; Wang, J. Spatial Distribution Characteristics and Driving Factors of 777 Traditional Villages in Yunnan Province: A Study Based on GWR Model and Geodetector. *Land* **2024**, *13*, 2004. [CrossRef]
19. Zhang, X.; Du, L.; Song, X. Identification of Urban Renewal Potential Areas and Analysis of Influential Factors from the Perspective of Vitality Enhancement: A Case Study of Harbin City's Core Area. *Land* **2024**, *13*, 1934. [CrossRef]
20. Shi, T.; Xu, H.; Bai, X. Spatiotemporal Evaluation and Driving Factor Screening for Regulating and Supporting Ecosystem Service Values in Suzhou–Wuxi–Changzhou Metropolitan Area's Green Space. *Land* **2024**, *13*, 1191. [CrossRef]
21. Li, B.; Lu, Y.; Li, Y.; Zuo, H.; Ding, Z. Research on the Spatiotemporal Distribution Characteristics and Accessibility of Traditional Villages Based on Geographic Information Systems—A Case Study of Shandong Province, China. *Land* **2024**, *13*, 1049. [CrossRef]
22. Lu, B.; Brunsdon, C.; Charlton, M.; Harris, P. Geographically Weighted Regression with Parameter-Specific Distance Metrics. *Int. J. Geogr. Inf. Sci.* **2017**, *31*, 982–998. [CrossRef]
23. Wei, C.; Cabrera-Barona, P.; Blaschke, T. Local Geographic Variation of Public Services Inequality: Does the Neighborhood Scale Matter? *Int. J. Environ. Res. Public Health* **2016**, *13*, 981. [CrossRef] [PubMed]
24. Fotheringham, A.S.; Yang, W.; Kang, W. Multiscale Geographically Weighted Regression (MGWR). *Ann. Am. Assoc. Geogr.* **2017**, *107*, 1247–1265. [CrossRef]
25. Wu, C.; Ren, F.; Hu, W.; Du, Q. Multiscale Geographically and Temporally Weighted Regression: Exploring the Spatiotemporal Determinants of Housing Prices. *Int. J. Geogr. Inf. Sci.* **2019**, *33*, 489–511. [CrossRef]
26. Tong, Z.; Kong, Z.; Jia, X.; Zhang, H.; Zhang, Y. Multiscale Impact of Environmental and Socio-Economic Factors on Low Physical Fitness among Chinese Adolescents and Regionalized Coping Strategies. *Int. J. Environ. Res. Public Health* **2022**, *19*, 13504. [CrossRef] [PubMed]
27. Soranno, P.A.; Cheruvelil, K.S.; Bissell, E.G.; Bremigan, M.T.; Downing, J.A.; Fergus, C.E.; Filstrup, C.T.; Henry, E.N.; Lottig, N.R.; Stanley, E.H.; et al. Cross-Scale Interactions: Quantifying Multi-Scaled Cause–Effect Relationships in Macrosystems. *Front. Ecol. Environ.* **2014**, *12*, 65–73. [CrossRef]
28. Rollinson, C.R.; Finley, A.O.; Alexander, M.R.; Banerjee, S.; Dixon Hamil, K.-A.; Koenig, L.E.; Locke, D.H.; DeMarche, M.L.; Tingley, M.W.; Wheeler, K.; et al. Working across Space and Time: Nonstationarity in Ecological Research and Application. *Front. Ecol. Environ.* **2021**, *19*, 66–72. [CrossRef]
29. Huang, B.; Wu, B.; Barry, M. Geographically and Temporally Weighted Regression for Modeling Spatio-Temporal Variation in House Prices. *Int. J. Geogr. Inf. Sci.* **2010**, *24*, 383–401. [CrossRef]
30. Wei, Q.; Zhang, L.; Duan, W.; Zhen, Z. Global and Geographically and Temporally Weighted Regression Models for Modeling PM2.5 in Heilongjiang, China from 2015 to 2018. *Int. J. Environ. Res. Public Health* **2019**, *16*, 5107. [CrossRef] [PubMed]
31. Hu, J.; Zhang, J.; Li, Y. Exploring the Spatial and Temporal Driving Mechanisms of Landscape Patterns on Habitat Quality in a City Undergoing Rapid Urbanization Based on GTWR and MGWR: The Case of Nanjing, China. *Ecol. Indic.* **2022**, *143*, 109333. [CrossRef]
32. Zhang, J.; Dong, Z. Assessment of Coupling Coordination Degree and Water Resources Carrying Capacity of Hebei Province (China) Based on WRESP2D2P Framework and GTWR Approach. *Sustain. Cities Soc.* **2022**, *82*, 103862. [CrossRef]
33. Fotheringham, A.S.; Crespo, R.; Yao, J. Geographical and Temporal Weighted Regression (GTWR). *Geogr. Anal.* **2015**, *47*, 431–452. [CrossRef]
34. Que, X.; Ma, C.; Ma, X.; Chen, Q. Parallel Computing for Fast Spatiotemporal Weighted Regression. *Comput. Geosci.* **2021**, *150*, 104723. [CrossRef]

35. Wu, Y.; Kan, H.; Deng, A. Spatio-Temporal Correlation and Optimization of Urban Development Characteristics and Carbon Balance in Counties: A Case Study of the Anhui Province, China. *Land* **2024**, *13*, 810. [CrossRef]
36. Yang, L.; Xu, Y.; Zhu, J.; Sun, K. Spatiotemporal Evolution and Influencing Factors of the Coupling Coordination of Urban Ecological Resilience and New Quality Productivity at the Provincial Scale in China. *Land* **2024**, *13*, 1998. [CrossRef]
37. Wang, X.; Liu, Y. Enhancing Agricultural Ecological Efficiency in China: An Evolution and Pathways under the Carbon Neutrality Vision. *Land* **2024**, *13*, 187. [CrossRef]
38. Que, X.; Ma, X.; Ma, C.; Chen, Q. A Spatiotemporal Weighted Regression Model (STWR v1.0) for Analyzing Local Nonstationarity in Space and Time. *Geosci. Model Dev.* **2020**, *13*, 6149–6164. [CrossRef]
39. Mansour, S.; Al Kindi, A.; Al-Said, A.; Al-Said, A.; Atkinson, P. Sociodemographic Determinants of COVID-19 Incidence Rates in Oman: Geospatial Modelling Using Multiscale Geographically Weighted Regression (MGWR). *Sustain. Cities Soc.* **2021**, *65*, 102627. [CrossRef] [PubMed]
40. Johnston, R.J. Local Effects in Voting at a Local Election. *Ann. Assoc. Am. Geogr.* **1974**, *64*, 418–429. [CrossRef]
41. Hanson, H.A.; Martin, C.; O'Neil, B.; Leiser, C.L.; Mayer, E.N.; Smith, K.R.; Lowrance, W.T. The Relative Importance of Race Compared to Health Care and Social Factors in Predicting Prostate Cancer Mortality: A Random Forest Approach. *J. Urol.* **2019**, *202*, 1209–1216. [CrossRef]
42. Luo, Y.; Su, S. SpatioTemporal Random Forest and SpatioTemporal Stacking Tree: A Novel Spatially Explicit Ensemble Learning Approach to Modeling Non-Linearity in Spatiotemporal Non-Stationarity. *Int. J. Appl. Earth Obs. Geoinf.* **2025**, *136*, 104315. [CrossRef]
43. Bindajam, A.A.; Mallick, J.; Talukdar, S.; Shahfahad; Shohan, A.A.A.; Rahman, A. Modeling the Spatiotemporal Heterogeneity of Land Surface Temperature and Its Relationship with Land Use Land Cover Using Geo-Statistical Techniques and Machine Learning Algorithms. *Environ. Sci. Pollut. Res.* **2023**, *30*, 106917–106935. [CrossRef] [PubMed]
44. Georganos, S.; Grippa, T.; Niang Gadiaga, A.; Linard, C.; Lennert, M.; Vanhuysse, S.; Mboga, N.; Wolff, E.; Kalogirou, S. Geographical Random Forests: A Spatial Extension of the Random Forest Algorithm to Address Spatial Heterogeneity in Remote Sensing and Population Modelling. *Geocarto Int.* **2021**, *36*, 121–136. [CrossRef]
45. Liu, R.; Yuan, H.; Chen, W.; Hu, Q.; Zhou, M.; Bao, L. Spatio-Temporal Heterogeneity and Scenario Prediction of Influencing Factors of Transportation Carbon Emissions in the Yangtze River Economic Belt, China. *Environ. Res. Commun.* **2024**, *6*, 115022. [CrossRef]
46. Tang, B.; Ma, K.; Liu, Y.; Wang, X.; Tang, S.; Xiao, Y.; Cheke, R.A. Managing Spatio-Temporal Heterogeneity of Susceptibles by Embedding It into an Homogeneous Model: A Mechanistic and Deep Learning Study. *PLoS Comput. Biol.* **2024**, *20*, e1012497. [CrossRef]
47. Chen, R.; Wang, C.; Que, X.; Liao, F.H.; Ma, X.; Wang, Z.; Li, Z.; Wen, K.; Lai, Y.; Xu, X. Exploring Urban Heat Distribution and Thermal Comfort Exposure Using Spatiotemporal Weighted Regression (STWR). *Buildings* **2024**, *14*, 1883. [CrossRef]
48. Wang, Z.; Que, X.; Li, M.; Liu, Z.; Shi, X.; Ma, X.; Fan, C.; Lin, Y. Spatiotemporally Weighted Regression (STWR) for Assessing Lyme Disease and Landscape Fragmentation Dynamics in Connecticut Towns. *Ecol. Inform.* **2024**, *84*, 102870. [CrossRef]
49. Yang, Z.; Zhang, X.; Lei, J.; Duan, Z.; Li, J. Spatio-Temporal Pattern Characteristics of Relationship Between Urbanization and Economic Development at County Level in China. *Chin. Geogr. Sci.* **2019**, *29*, 553–567. [CrossRef]
50. Jianghui, W.J. Types, Characteristics and Tendency of County-Level Administrative Division Adjustment in China since 2000. *Trop. Geogr.* **2018**, *38*, 799–809. [CrossRef]
51. Todaro, M.P.; Smith, S.C. *Economic Development*, 13th ed.; Pearson: Hoboken, NJ, USA, 2020; ISBN 978-1-292-29115-4.
52. Lewis, W.A. Economic Development with Unlimited Supplies of Labour. *Manch. Sch.* **1954**, *22*, 139–191. [CrossRef]
53. Banelienė, R. Industry Impact on GDP Growth in Developed Countries under R&D Investment Conditions. *J. Small Bus. Strategy (Arch. Only)* **2021**, *31*, 66–80.
54. Zhang, J. Data Analysis of Fiscal Expenditure and GDP Based on Financial Budget Performance Evaluation Indicators. *Discret. Dyn. Nat. Soc.* **2022**, *2022*, 1141618. [CrossRef]
55. Barro, R.J. Government Spending in a Simple Model of Endogenous Growth. *J. Political Econ.* **1990**, *98*, S103–S125. [CrossRef]
56. Bai, Y. An Empirical Analysis of the Relationship Between Chinese GDP and Deposit Savings. In Proceedings of the 7th International Conference on Economic Management and Green Development, Oxford, UK, 6 August 2023; Li, X., Yuan, C., Kent, J., Eds.; Springer Nature: Singapore, 2024; pp. 873–884.
57. Rachuba, J. GDP Growth as a Bank Loan Quality Determinant. *J. Bank. Financ. Econ.* **2020**, *14*, 21–37. [CrossRef]
58. Chen, X.; Shuai, C.; Zhang, Y.; Wu, Y. Decomposition of Energy Consumption and Its Decoupling with Economic Growth in the Global Agricultural Industry. *Environ. Impact Assess. Rev.* **2020**, *81*, 106364. [CrossRef]
59. Bloom, D.; Canning, D. Health as Human Capital and Its Impact on Economic Performance. *Geneva Pap. Risk Insur. Issues Pract.* **2003**, *28*, 304–315. [CrossRef]
60. Strydom, N.; Struweg, J. Malthus Revisited: Long-Term Trends in South African Population Growth and Agricultural Output. *Agrekon* **2016**, *55*, 34–61. [CrossRef]

61. Wu, B.; Li, R.; Huang, B. A Geographically and Temporally Weighted Autoregressive Model with Application to Housing Prices. *Int. J. Geogr. Inf. Sci.* **2014**, *28*, 1186–1204. [CrossRef]
62. Lu, B.; Harris, P.; Charlton, M.; Brunsdon, C. The GWmodel R Package: Further Topics for Exploring Spatial Heterogeneity Using Geographically Weighted Models. *Geo-Spat. Inf. Sci.* **2014**, *17*, 85–101. [CrossRef]
63. Hong, Z.; Mei, C.; Wang, H.; Du, W. Spatiotemporal Effects of Climate Factors on Childhood Hand, Foot, and Mouth Disease: A Case Study Using Mixed Geographically and Temporally Weighted Regression Models. *Int. J. Geogr. Inf. Sci.* **2021**, *35*, 1611–1633. [CrossRef]
64. Hong, Z.; Wang, J.; Wang, H. Introducing Bootstrap Test Technique to Identify Spatial Heterogeneity in Geographically and Temporally Weighted Regression Models. *Spat. Stat.* **2022**, *51*, 100683. [CrossRef]
65. Li, Z.; Fotheringham, A.S. Computational Improvements to Multi-Scale Geographically Weighted Regression. *Int. J. Geogr. Inf. Sci.* **2020**, *34*, 1378–1397. [CrossRef]
66. Tobler, W.R. A Computer Movie Simulating Urban Growth in the Detroit Region. *Econ. Geogr.* **1970**, *46*, 234–240. [CrossRef]
67. Zhang, W. Rethinking Regional Disparity in China. *Econ. Plan.* **2001**, *34*, 113–138. [CrossRef]
68. Li, Z.; He, S.; Su, S.; Li, G.; Chen, F. Public Services Equalization in Urbanizing China: Indicators, Spatiotemporal Dynamics and Implications on Regional Economic Disparities. *Soc. Indic. Res.* **2020**, *152*, 1–65. [CrossRef]
69. Song, L.; Wang, P.; Xiang, K.; Chen, W.-Q. Regional Disparities in Decoupling Economic Growth and Steel Stocks: Forty Years of Provincial Evidence in China. *J. Environ. Manag.* **2020**, *271*, 111035. [CrossRef]
70. Chen, A. Reducing China's Regional Disparities: Is There a Growth Cost? *China Econ. Rev.* **2010**, *21*, 2–13. [CrossRef]
71. Kanbur, R.; Wan, G.; Zhang, X. Introduction: Growing Inequality in China. *J. Asia Pac. Econ.* **2005**, *10*, 405–407. [CrossRef]
72. Gao, J.; Liu, Y.; Chen, J.; Cai, Y. Demystifying the Geography of Income Inequality in Rural China: A Transitional Framework. *J. Rural Stud.* **2022**, *93*, 398–407. [CrossRef]
73. Wu, S.; Wang, Z.; Du, Z.; Huang, B.; Zhang, F.; Liu, R. Geographically and Temporally Neural Network Weighted Regression for Modeling Spatiotemporal Non-Stationary Relationships. *Int. J. Geogr. Inf. Sci.* **2021**, *35*, 582–608. [CrossRef]
74. Han, L.; Zhang, X.; Zhou, W.; Shen, M.; Huang, Y.; Li, W.; Qian, Y. Transformation of China's Urbanization and Eco-Environment Dynamics: An Insight with Location-Based Population-Weighted Indicators. *Environ. Sci. Pollut. Res.* **2021**, *28*, 16558–16567. [CrossRef] [PubMed]
75. Wang, H.; Zhang, B.; Liu, Y.; Liu, Y.; Xu, S.; Zhao, Y.; Chen, Y.; Hong, S. Urban Expansion Patterns and Their Driving Forces Based on the Center of Gravity-GTWR Model: A Case Study of the Beijing-Tianjin-Hebei Urban Agglomeration. *J. Geogr. Sci.* **2020**, *30*, 297–318. [CrossRef]
76. Iyanda, A.E.; Osayomi, T. Is There a Relationship between Economic Indicators and Road Fatalities in Texas? A Multiscale Geographically Weighted Regression Analysis. *GeoJournal* **2021**, *86*, 2787–2807. [CrossRef]
77. Fotheringham, A.S.; Sachdeva, M. Scale and Local Modeling: New Perspectives on the Modifiable Areal Unit Problem and Simpson's Paradox. *J. Geogr. Syst.* **2022**, *24*, 475–499. [CrossRef]
78. Openshaw, S. Ecological Fallacies and the Analysis of Areal Census Data. *Environ. Plan A* **1984**, *16*, 17–31. [CrossRef] [PubMed]
79. Fotheringham, A.S.; Wong, D.W.S. The Modifiable Areal Unit Problem in Multivariate Statistical Analysis. *Environ. Plan A* **1991**, *23*, 1025–1044. [CrossRef]
80. Yan, J.; Wu, B.; Zheng, H. Multiscale Cooperative Optimization in Multiscale Geographically Weighted Regression Models. *Int. J. Geogr. Inf. Sci.* **2024**, 1–20. [CrossRef]
81. Chen, L.; Fang, J.; Liu, T.; Cao, S.; Wang, L. A Unified Model for Spatio-Temporal Prediction Queries with Arbitrary Modifiable Areal Units. In Proceedings of the 2024 IEEE 40th International Conference on Data Engineering (ICDE), Utrecht, The Netherlands, 13–16 May 2024; pp. 1352–1365.
82. Jiang, Y.; Shi, C. Estimating Sustainability and Regional Inequalities Using an Enhanced Sustainable Development Index in China. *Sustain. Cities Soc.* **2023**, *99*, 104555. [CrossRef]
83. Jiang, Y. Trade Integration and Regional Inequality: A Theoretical Framework with Empirical Implications for China. *J. Chin. Econ. Bus. Stud.* **2016**, *14*, 365–384. [CrossRef]
84. Liu, X.; Yuan, M. Assessing Progress towards Achieving the Transport Dimension of the SDGs in China. *Sci. Total Environ.* **2023**, *858*, 159752. [CrossRef] [PubMed]
85. Cao, M.; Chen, M.; Zhang, J.; Pradhan, P.; Guo, H.; Fu, B.; Li, Y.; Bai, Y.; Chang, L.; Chen, Y.; et al. Spatio-Temporal Changes in the Causal Interactions among Sustainable Development Goals in China. *Humanit. Soc. Sci. Commun.* **2023**, *10*, 450. [CrossRef]
86. Zhou, Y.; Liu, Z.; Wang, H.; Cheng, G. Targeted Poverty Alleviation Narrowed China's Urban-Rural Income Gap: A Theoretical and Empirical Analysis. *Appl. Geogr.* **2023**, *157*, 103000. [CrossRef]
87. Ge, Y.; Hu, S.; Song, Y.; Zheng, H.; Liu, Y.; Ye, X.; Ma, T.; Liu, M.; Zhou, C. Sustainable Poverty Reduction Models for the Coordinated Development of the Social Economy and Environment in China. *Sci. Bull.* **2023**, *68*, 2236–2246. [CrossRef]
88. Blanchard, O.; Bean, C.; Münchau, W. European Unemployment: The Evolution of Facts and Ideas. *Econ. Policy* **2006**, *21*, 7–59. [CrossRef]

89. Fleisher, B.; Li, H.; Zhao, M.Q. Human Capital, Economic Growth, and Regional Inequality in China. *J. Dev. Econ.* **2010**, *92*, 215–231. [CrossRef]
90. van Treeck, T.; Sturn, S. *Income Inequality as a Cause of the Great Recession? A Survey of Current Debates*; International Labour Organization: Geneva, Switzerland, 2012.
91. Lewin, P.A.; Watson, P.; Brown, A. Surviving the Great Recession: The Influence of Income Inequality in US Urban Counties. *Reg. Stud.* **2018**, *52*, 781–792. [CrossRef]
92. Ezcurra, R.; Pascual, P.; Rapún, M. The Dynamics of Regional Disparities in Central and Eastern Europe during Transition. *Eur. Plan. Stud.* **2007**, *15*, 1397–1421. [CrossRef]
93. Voskamp, I.M.; Visscher, M.N.; Vreugdenhil, C.; Van Lammeren, R.J.A.; Sutton, N.B. Spatial, Infrastructural and Consumer Characteristics Underlying Spatial Variability in Residential Energy and Water Consumption in Amsterdam. *Sustain. Cities Soc.* **2021**, *72*, 102977. [CrossRef]
94. Feldman, M.P.; Florida, R. The Geographic Sources of Innovation: Technological Infrastructure and Product Innovation in the United States. *Ann. Assoc. Am. Geogr.* **1994**, *84*, 210–229. [CrossRef]
95. Doloreux, D. What We Should Know about Regional Systems of Innovation. *Technol. Soc.* **2002**, *24*, 243–263. [CrossRef]
96. Zhao, X.; Qian, Y. Does Digital Technology Promote Green Innovation Performance? *J. Knowl. Econ.* **2024**, *15*, 7568–7587. [CrossRef]
97. Tan, X.; Wang, Z.; An, Y.; Wang, W. Types and Optimization Paths Between Poverty Alleviation Effectiveness and Rural Revitalization: A Case Study of Hunan Province, China. *Chin. Geogr. Sci.* **2023**, *33*, 966–982. [CrossRef]

land

Article

# Impact of Urban Neighborhood Morphology on PM2.5 Concentration Distribution at Different Scale Buffers

**Zhen Wang [1,2], Kexin Hu [1,2], Zheyu Wang [1,2], Bo Yang [3] and Zhiyu Chen [2,4,*]**

1 School of Architecture and Urban Planning, Huazhong University of Science and Technology, Wuhan 430074, China; wangz@hust.edu.cn (Z.W.); m202277093@hust.edu.cn (K.H.); d202481686@hust.edu.cn (Z.W.)
2 Hubei Engineering and Technology Research Center of Urbanization, Wuhan 430074, China
3 School of Architecture and Urban Planning, University of Arizona, Tucson, AZ 85721, USA; boyang17@arizona.edu
4 Center-South Architecture Design Institution (CSADI), Wuhan 430071, China
* Correspondence: chenzhiyu@csadi.cn

**Abstract:** PM2.5 air pollution is a critical global health issue. This paper introduces an innovative framework to explore the multi-scale relationship between urban morphology and PM2.5 concentrations. An enhanced Land Use Regression (LUR) model integrates geographic, architectural, and visual factors, enabling analysis from neighborhood to regional scales. A stratified sampling strategy, combined with standardized mobile monitoring and fixed-site data, establishes a robust and verifiable data collection methodology. Cross-validation (CV $R^2 > 0.70$) further confirms the model's reliability and robustness. The nested buffer analysis reveals scale-dependent effects of urban morphology on PM2.5 concentrations, providing quantitative evidence for planning interventions. Quantitative analysis shows land use ($\beta = 0.42$, $p < 0.01$), visual factors ($\beta = 0.38$, $p < 0.01$), and building density ($\beta = 0.35$, $p < 0.01$) in descending order of influence. Geographic factors are significant at the regional scale (2000–3000 m) while architectural parameters dominate at the neighborhood scale (50–500 m), informing both macro-scale spatial optimization and micro-scale design. This framework, through standardized parameters and reproducible procedures, supports cross-regional and cross-scale air quality assessments, providing quantitative metrics for urban planning, neighborhood optimization, and public space design.

**Keywords:** air quality; pollutant dispersion; urban neighborhood morphology; Land Use Regression (LUR); different scale buffers

## 1. Introduction

Air pollution is a major global public health issue, with exposure to outdoor air pollutants leading to adverse health outcomes [1–3]. An increase of 10 $\mu g/m^3$ in the daily average PM2.5 concentration correlates with a roughly 2% rise in hospital admissions for respiratory diseases and mortality rates [4,5]. Cardiovascular mortality and morbidity rates also escalate with higher PM2.5 levels [6–8]. According to the World Health Organization, approximately 4.2 million premature deaths annually are attributed to fine particulate matter (PM2.5) exposure [9]. Ambient air pollution is identified as a critical problem in high-density Asian cities [10]. In such environments, densely packed urban neighborhoods can hinder air circulation, restricting pollutant dispersion [11,12]. Current research and practice increasingly focus on improving urban pollutant distribution and diffusion levels [13–15], with enhancing diffusion rates being an effective approach to reducing pollution concentration [16]. The comprehensive analysis of how urban neighborhood morphology

 https://doi.org/10.3390/land14010007

affects PM2.5 spatial distribution could significantly improve pollutant diffusion, thus reducing health risks associated with exposure [17].

Numerous studies have examined the impact of urban morphology on the spatial distribution of PM2.5 concentrations. It has been established that factors such as building density [18], street canyon depth [19,20], and the aerodynamic effects of street trees [21] significantly influence PM2.5 concentrations, although their effects vary across different cities. However, a notable limitation of current research is the insufficiently comprehensive exploration of influencing factors, particularly the complex interactions between key factors and the lack of a systematic theoretical framework for understanding these effects across varying urban forms and climatic conditions [22]. Key factors influencing PM2.5 concentration distribution include topography [23], building morphology, podium-level porosity, building density, height standard deviation [24], building setback distance [25], and accessibility [26]. These factors exhibit different impacts on various urban scales, making them crucial considerations for air quality optimization. Despite this, the existing research often lacks an in-depth discussion of scale effects, with significant differences in how key factors affect PM2.5 distribution across different scales. At the local neighborhood scale, building morphology and porosity significantly influence air flow and pollutant dispersion [24], while at the city or regional scale, factors like topography and accessibility have more pronounced effects [23,26]. This cross-scale variability leads to a lack of systematic analysis and comprehensive understanding of the interactions between key factors at different scales, thereby limiting the applicability and practical implications of research findings.

Discussions on the spatial scales of urban morphology's impact on PM2.5 distribution reveal that different spatial scales significantly affect pollutant concentration and distribution characteristics. Research indicates that urban scale, building morphology, and model resolution play critical roles in influencing PM2.5 concentrations [27,28]. Finer scales, such as 50–500 m, generally provide more accurate data on pollutant distribution while coarser scales may obscure important local effects [29]. Additionally, scale effects vary across different models, particularly in the treatment of deposition processes at micro (10–500 m) and macro (5–100 km) scales [30]. These differences reflect the profound impact of model choice and spatial resolution on result accuracy [31–33]. Nevertheless, existing studies often lack a systematic analysis of the effects of influencing factors across different scales, particularly with regard to scale effects within varying buffer zones. This limitation impedes a thorough understanding of how urban morphology affects PM2.5 concentrations across multiple scales and affects the formulation of precise and effective air quality management strategies.

In addition, the LUR model is a generalized model for simulating the spatial variability of atmospheric pollutant concentrations at the urban scale, which was first proposed by a small-scale air quality and health variability study in 1997 [34]. When using the Land Use Regression (LUR) model to analyze the spatial distribution of urban PM2.5, urban environmental and meteorological monitoring stations are often used to obtain the relevant pollutant and meteorological data, and the number of urban environmental monitoring stations built in China is limited and the distribution spacing is large, so this approach is commonly used to analyze the characteristics of pollutant distributions at larger scales [35–37]. Usually, multiple fixed monitoring stations are placed within a city, and the measured data from these stations are used to construct an LUR model to estimate the air pollutant concentration distribution in the whole city [38–40]. However, the majority of existing studies have predominantly concentrated on the macro scale, but the PM2.5 concentrations at the neighborhood scale also have significant variability, and the neighborhood pattern is an important influence on the differences in the distribution of PM2.5 concentrations [41,42]. Meanwhile, one of the main limitations of the LUR model is the need for a large number of fixed real measurement stations for sampling data. An insufficient number of fixed monitoring stations leads to a large granularity in the acquired data [43–45]. Mobile measurements provide an alternative to the current situation wherein most cities and regions do not have sufficient numbers of fixed monitoring stations that are capable of monitoring specific pollutants. Mobile measurements allow the use of a limited number of monitoring devices and can greatly improve the accuracy of measurements in the high-resolution range with a good spatial coverage of samples, allowing the spatial distribution of air pollutant concentrations to be measured at the urban neighborhood scale [46,47].

In summary, research on the relationship between urban morphology and air quality is incomplete, focusing primarily on single-layer or conventional indicators. There has been inadequate attention to three-dimensional structures, visual factors, and boundary configurations. Additionally, studies have often remained at broader regional or singular local scales, with limited comparison of urban neighborhood morphology's impact on PM2.5 concentration distribution across varying buffer sizes, preventing a comprehensive quantification of scale effects and the exploration of compact city development impacts on air quality. Based on the above, this paper proposes an innovative research framework encompassing three levels, geographic, architectural, and visual influencing factors, to systematically reveal the mechanisms by which urban neighborhood morphology affects PM2.5 distribution. By introducing neighborhood-level visual factors and supplementing 3D morphology data, this paper provides a more comprehensive perspective by examining the visible factors of buildings, vegetation, the sky, and impermeable surfaces, offering insights into the nuanced local-scale impacts of urban neighborhood morphology from a human visual angle. Addressing the limitations of previous single-scale studies, this research employs a multi-scale approach, establishing 11 buffer zones (50–3000 m), obtaining PM2.5 data from mobile measurements and utilizing Land Use Regression (LUR) models for regression analysis to assess key factors affecting PM2.5 concentrations across scales. The buffer size used in this paper is more similar to that in a typical land use regression method, which is helpful to reveal the significant influence of key factors on the spatial distribution of PM2.5 concentrations at a specific buffer scale and provides support for an in-depth understanding of the scale effect and a comprehensive understanding of the scale effect of PM2.5 concentration.

## 2. Methodology

### 2.1. Research Area

This paper aims to investigate residents' exposure risks to PM2.5 in the central urban area of Wuhan. Using ArcGIS, a $500 \times 500$ m$^2$ grid was established to acquire urban data, including road intensity, building density, and building height. The spatial distribution maps of population density, road density, building density, and average building height in the central Wuhan are shown in Figure 1. Regarding population density, based on the 2021 national bureau of statistics of China, Jiang'an, Jianghan, Qiaokou, and Wuchang districts in the central urban area have population densities exceeding 10,000 people/square kilometer, with Hongshan district having the highest and Qingshan district the lowest, showing an "inner high, outer low" distribution. Road intensity information, extracted from OpenStreetMap data, indicates higher road density in Hankou, with lower to medium density prevailing outside the third ring road. Building density and height exhibit an "inner ring high, outer ring low" trend in the central developed area, mainly featuring medium-to-high-rise and high-rise buildings, with the city center dominated by medium-to-high-rise structures.

By comprehensively analyzing population density, road intensity, building density, average building height, and PM2.5 concentrations, we find that Wuchang district shows slightly higher annual PM2.5 concentrations, posing elevated population exposure risks, especially in Hongshan and Wuchang districts, each with a population exceeding one million. In comparison to Hankou and Hanyang, Wuchang district was deemed more suitable for an in-depth exploration of the relationship between urban green spaces and PM2.5 concentrations. Consequently, we selected the Wuchang Riverside and Optics Valley areas as typical neighborhood study regions.

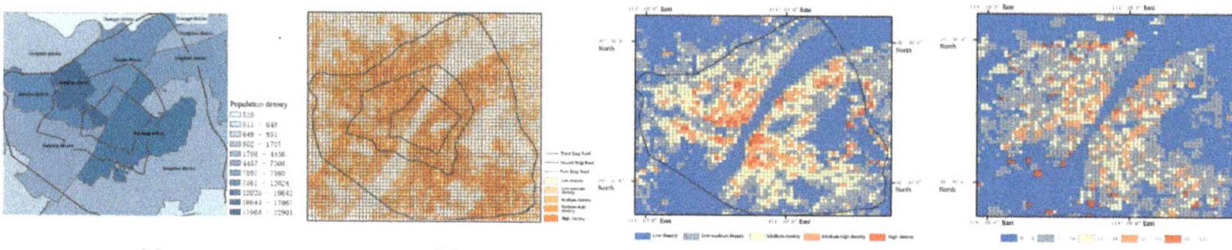

(a) (b) (c) (d)

**Figure 1.** Spatial distribution of population density, road density, building density, and average building height in Wuhan's central city: (**a**) spatial distribution of population density in central Wuhan, 2020; (**b**) spatial distribution of road intensity in central Wuhan (500 × 500 m$^2$); (**c**) building density in central Wuhan (500 × 500 m$^2$); (**d**) average building height in central Wuhan (500 × 500 m$^2$).

### 2.2. Data Sources

#### 2.2.1. PM2.5 Concentration Data

Research on PM2.5 concentrations often relies on long-term data from national environmental monitoring stations, focusing on regional and urban-scale spatial distributions in areas such as Beijing–Tianjin–Hebei, Fujian, and Anhui provinces in China [48–50]. However, Wuhan's existing 22 monitoring stations are insufficient for neighborhood-scale studies. This research combines mobile monitoring with fixed-point measurements [51–53], employing driving, cycling, and walking methods across typical neighborhoods in Hongshan, Wuchang, and Qingshan districts to obtain PM2.5 spatial distributions at the neighborhood level. Comparative analysis of data from mobile monitoring and stationary environmental stations reveals that at the neighborhood scale, stationary station data show minimal variation in simulated PM2.5 concentrations. The Land Use Regression (LUR) model constructed with national environmental station data exhibits lower accuracy for PM2.5 distribution at this scale. In contrast, mobile measurements enhance accuracy and spatial resolution for neighborhood-scale PM2.5 concentration studies.

#### 2.2.2. Impact Factors for the Three Tiers

(1) Geographical Influencing Factors: Wuhan's land cover data are sourced from the National Geographical Information Resource Catalog Service System (https://www.webmap.cn (accessed on 1 December 2022), utilizing 30 m global land cover data, with GlobeLand30 being a project under China's National 863 Program. Selected factors include cultivated land, forest, grassland, water bodies, and artificial surfaces. Land-use data, obtained from the Planning Bureau's master plan, encompass municipal and East Lake scenic area plans. Industrial land, park green spaces, residential land, and commercial land were chosen as influencing factors. Urban road network information from OpenStreetMap (https://www.openstreetmap.org (accessed on 5 December 2023) was saved in Shapefile format using ArcGIS. Road intensity is represented by road length, focusing on second-, third-, and fourth-tier roads, along with bus routes, as primary influencing factors.

(2) Building Morphology Factors: To explore air pollution distribution at the neighborhood scale, our study utilized building information provided by Baidu Maps to obtain Wuhan's building height and boundary information. Using ArcGIS, the sampled points of

three typical neighborhood mobile measurement paths were buffered to calculate building density, average building height, and floor area ratio.

(3) Visual Influencing Factors: Street-view images were sourced from the Baidu Maps Open Platform, accessed using Python through the Baidu Maps API. This allowed the batch retrieval of street-view images for desired path points. The study primarily acquired visual factors for buildings, vegetation, sky, and impervious surfaces from the street-view images at 0°, 90°, 180°, and 270° angles for each path point.

The specific classifications of the impact factors for the three tiers are shown in Table 1.

**Table 1.** Impact factors for the three tiers (source: self-mapping).

| **Impact Factor** | | | **Impact Factor** | |
|---|---|---|---|---|
| Geographic Impact Factor (GIF) | Land cover | Cropland<br>Forests<br>Grassland<br>Water<br>Artificial surfaces | Architectural Form Factor (AFF) | Building density<br>Average building height<br>Floor area ratio |
| | Land use properties | Industrial land<br>Parks and green spaces<br>Residential<br>Commercial | Visual Impact Factor (VIF) | Architectural visibility factor<br>Plant visibility factor<br>Sky visibility factor<br>Impervious surface Visibility factor |
| | Road intensity | Secondary roads<br>Tertiary roads<br>Fourth-class roads<br>Public transport routes | | |

### 2.3. Overview of the Methodology

Methodically investigating the influence of urban neighborhood morphology on PM2.5 distribution across three dimensions—geographical influencing factors, building morphology factors, and visual influencing factors—this study utilized a hybrid approach of mobile and fixed monitoring to gather spatial PM2.5 concentration data within typical neighborhoods. Varied scale buffers were created and a land regression model was applied to scrutinize the disparities in the impacts of three-tier influencing factors on the spatial dispersion of PM2.5 concentrations. This research delves into the scale effects of these influencing factors across the three dimensions. The specific technical route is illustrated in Figure 2.

#### 2.3.1. Obtaining PM2.5 Data by Mobile Monitoring

Elevated concentrations of PM2.5 are noted in the industrial zones of Qingshan, Jiangbei, Hankou, and Hanshi in Wuhan, while regions with more lakes, such as Hanyang and the southern part of Wuchang, demonstrate lower PM2.5 levels [14]. Selected monitoring stations include Wuchang Ziyang, Qingshan Ganghua, and Hongshan Geological University, forming dynamic monitoring routes that, along with station data, establish a foundation for mobile measurements. Path planning integrates considerations of urban green spaces, land use properties, and floor area ratios. The diverse business types, urban green spaces, and building morphology distinctions in these three districts facilitate an in-depth exploration of the relationship between neighborhood spatial morphology and PM2.5. Please refer to Table 2 for specifics, where construction intensity and land types are categorized based on regional artificial surfaces and surrounding land characteristics.

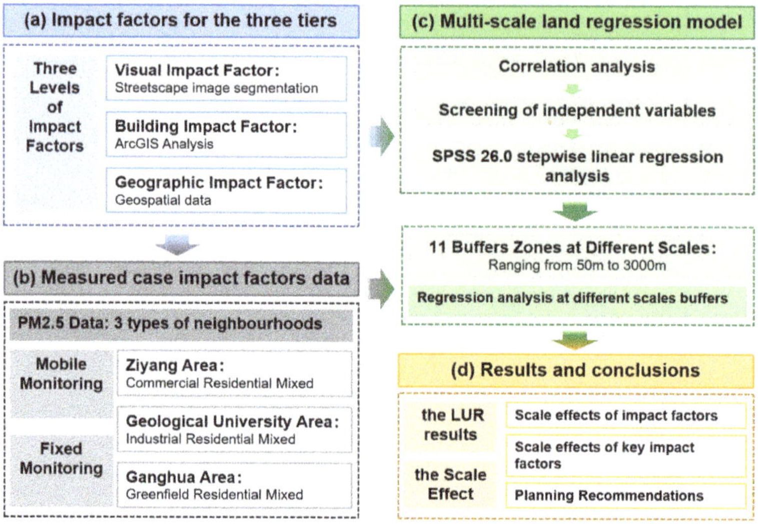


**Figure 2.** Technological roadmap.

**Table 2.** Mobile survey neighborhood surrounding information (source: self-mapping).

| | Environmental Monitoring Site | Construction Intensity | Site Type | Urban Green Space | Road Density | Building Heights | Building Density |
|---|---|---|---|---|---|---|---|
| Ziyang District | Wuchang Ziyang Station | High Strength | Commercial Residential Mixed | Decentralized | Medium–high Density | Multi-story, medium–high rise | High density |
| Ganghua District | Qingshan Ganghua Station | Medium–high strength | Industrial Residential Mixed | Decentralized | Medium–high Density | High-rise | Medium–high density |
| Geological University District | Hongshan Geological University Station | Medium–high strength | Greenfield Residential Mixed | Centralized | High density | Mid-rise | Medium–high density |

Wuchang Ziyang station, situated in the old city area, features a plethora of commercial and residential buildings characterized by multi-story and medium-to-high-density structures. The surrounding Ziyang Park, Shouyi Square, and universities contribute to the dispersed greenery. Qingshan Ganghua station is positioned in an industrial area undergoing ecological enhancement, marked by medium-to-high-density multi-story structures. The adjacent residential areas, universities, science parks, and Qingshan Riverside display variations in greenery. Hongshan Geological University station, located in the university district, boasts high green coverage and medium-to-high-density mid-rise structures. The three national monitoring sites are shown in Figure 3a.

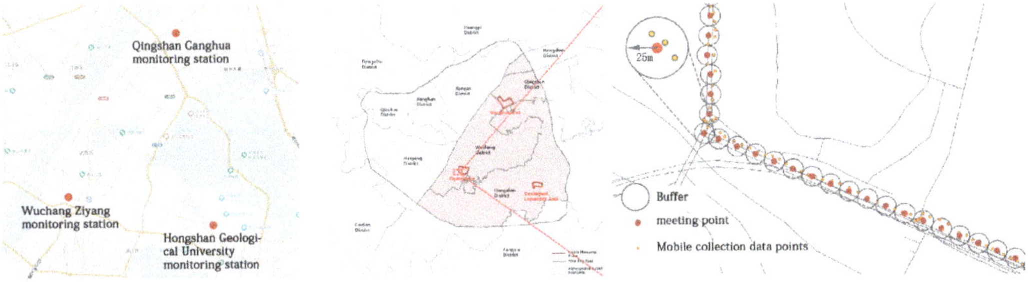


(a) (b)

**Figure 3.** Study area: (**a**) national monitoring sites; (**b**) data aggregation process.

The mobile measurement paths span the boundaries, with Ziyang and Geological University paths covering 4.6 km and 4.2 km, respectively, while Ganghua is slightly longer at 6.7 km. From 27 to 30 July 2022, research teams conducted on-site mobile measurements around Hongshan Geological University, Wuchang Ziyang, and Qingshan Ganghua monitoring stations. Four times daily, covering a total length of 4.2–6.3 km per session lasting 20–30 min, electric bikes or shared bicycles were utilized. Equipment included particulate matter detectors, GPS, and temperature–humidity recorders. Riding speeds ranged from 12 to 15 km/h, with six volunteers managing two measurement points each, simultaneously conducting mobile and fixed-point measurements.

Through 12 mobile measurements, data for the dependent variable in the Land Use Regression (LUR) model were collected. In the three typical districts, Ziyang collected 232–256 data points per session and Geological University gathered 237–271 data points, with path lengths of 4.6 km, 6.7 km, and 4.2 km. Aggregation points were established at 50 m intervals, with 88 points for Ziyang, 132 for Ganghua, and 85 for Geological University. Average values of mobile measurement data within the 50 m diameter range of each aggregation point were used as particle data. The data aggregation process is shown in Figure 3b.

### 2.3.2. Street View Segmentation by Machine Learning

With the advent of Street View imagery, the trend of using methods based on Street View image segmentation for SVF measurement has attracted more and more attention [54]. Street View imagery is captured by cameras mounted on vehicles, tricycles, bicycles, and pedestrians with wide spatial coverage in urban areas [55]. The method of using Google Street View images to convert conformal projection into hemispherical view images to explore the urban sky visibility factor can explore the correlation between the urban morphological index and urban microclimate on a large scale [56,57] and can also evaluate urban street greening, which can be effectively evaluated in combination with the Green Landscape Index (GLI) [58,59]. The use of Street View imagery to generate hemispherical images allows for SVF measurements at scale, and Street View imagery covers the entire world, providing unprecedented data support for research [60,61]. This paper uses street view image segmentation to obtain visual influencing factors, supplements the three-dimensional neighborhood morphology data, and examines the visual factors of buildings, plants, sky, and impervious ground, which provides a more three-dimensional perspective for the study, and deeply excavates the subtle influence of urban neighborhood morphology on the local scale of human perspective.

The image segmentation process and processing results are shown in Figure 4. This study utilized the ADE20K scene parsing dataset to enhance an open-source project for image segmentation, focusing on the proportions of buildings, vegetation, sky, and impervious surfaces. The segmentation achieved an accuracy of 80%. Mobile collection routes and surrounding street-view images were obtained, and path points were delineated using OpenStreetMap and ArcGIS. The Wuchang Ziyang district acquired 2693 path points, with 88 lacking image data; Qingshan Ganghua district obtained 2625 path points, with 40 lacking image data; Hongshan Geological University district gathered 4337 path points, with 231 lacking image data. Each point comprised four perspective images, totaling 37,184 street-view images.

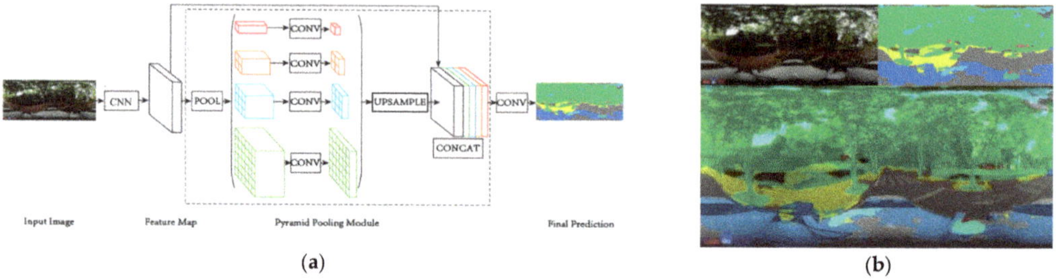


**Figure 4.** Image segmentation process and processing results: (**a**) image segmentation process; (**b**) outcome of the process.

### 2.3.3. Regression Analysis by LUR

Land Use Regression (LUR) models are extensively applied to simulate pollutant spatial distribution at urban and neighborhood scales. At the city scale, LUR models were used to obtain high-resolution spatial distribution maps of air pollution in Tianjin [62], Chongqing [63], Wuhan [50], Shenzhen [51], and Beijing [64]. At the regional scale, LUR models were established in Beijing–Tianjin–Hebei [48], Hubei, Fujian [65], and Anhui [66] to explore the spatial distribution of regional air pollution. On the neighborhood scale, researchers used monitoring equipment to drive cars, bicycles, walks, and drones to obtain neighborhood-scale pollutant data through mobile measurements to explore the risk of air pollution exposure [67]. Recent research has focused on enhancing spatial resolution by incorporating factors such as building morphology, emission sources, and land use properties to construct neighborhood-scale LUR models [68]. This paper builds on these advancements, integrating data across three levels of influencing factors, utilizing multi-source information to construct a comprehensive LUR model, and providing an in-depth exploration of PM2.5 spatial distribution at the neighborhood scale.

Before modeling, insignificant and inconsistently influential factors with the predicted impact were filtered out. The average PM2.5 concentrations during morning (9:00–11:00) and afternoon (16:30–18:30) in three representative neighborhoods served as dependent variables. The remaining variables after screening were employed as independent variables in a stepwise linear regression conducted using SPSS 26.0. The modeling procedure followed Wu et al.'s methodology [13,14].

The analysis of the correlation between the 208 groups of predicted influencing factors and the PM2.5 concentration groups of dependent variables in the morning and afternoon periods was carried out, and then the direction of the influencing factors was predicted, and the variables that were inconsistent with the predicted direction were screened out. The correlation coefficients of the influencing factors of the remaining land cover, land use properties, road strength, building form, and visual impact factors were sorted to find out the most relevant variables. Finally, the remaining influencing factors of each group and the PM2.5 concentrations of the dependent variable in the morning and afternoon periods were grouped for stepwise linear regression, and an LUR model was established for each period of the three typical neighborhoods.

The formulated expression for the established LUR model is as follows:

$$y = \beta 0 + \beta 1X1 + \beta 2X2 + \ldots\ldots \beta nXn \tag{1}$$

$y$ represents the PM2.5 concentration during each time period, $\beta 0$ is the intercept, and $X1, X2 \ldots\ldots$ Xn are the predicted influential factors, each with its respective coefficient ($\beta 1, \beta 2 \ldots\ldots \beta$n).

## 3. Analysis

### 3.1. Land Regression Modeling Assessment

This study utilized the Variance Inflation Factor (VIF) to detect multicollinearity. The VIF serves as an indicator for assessing the degree of multicollinearity in multiple linear regression, reflecting the ratio of the variance of the estimated regression coefficients to the variance when assuming nonlinear correlation among the independent variables. In the morning regression model for the Wuchang district, seven factors were included, and the VIF for each factor was below 10, indicating the absence of severe collinearity issues in the regression model.

Model stability was validated using Leave-One-Out Cross-Validation (LOOCV), where n variables were divided into n-1 training sets and one validation set. The model was constructed using the training set, and the results of the LUR for each time period were introduced. The simulation values were calculated in each iteration and compared with the monitoring values, repeated $n$ times to obtain the Pearson correlation coefficient ($R^2$), Root Mean Squared Error (*RMSE*), and Mean Absolute Error (*MAE*). In this context, $x$ represents the predicted value, $y$ denotes the true value, and $m$ is the sample size.

The *RMSE* is a measure of the deviation between predicted and true values, representing the square root of the ratio of the squared differences between predicted and true values to the number of data collection instances (m). In practical measurements, the number of data collection instances (m) is always limited, and the true value can only be replaced by the optimal predicted value. The Formula (2) for calculating the Root Mean Squared Error (*RMSE*) is as follows:

$$RMSE = \sqrt{\frac{\sum_{i=1}^{m} (x_i - y_i)^2}{m}} \tag{2}$$

The Mean Absolute Error (*MAE*) is the average of the absolute values of the deviations between individual predicted values and the true arithmetic meaning. The *MAE* helps to avoid the problem of errors canceling each other out, providing an accurate reflection of the actual size of prediction errors. The Formula (3) for calculating the Mean Absolute Error (*MAE*) is as follows:

$$MAE = \frac{\sum_{i=1}^{m} |x_i - y_i|}{m} \tag{3}$$

The Root Mean Squared Error (*RMSE*) for the morning and afternoon regression models across all regions ranges from 1.63 to 7.783 $\mu g/m^3$ while the Mean Absolute Error (*MAE*) falls between 1.312 and 6.374 $\mu g/m^3$. Comparatively, previous studies had reported *RMSE* values ranging from 1.33 to 14.2 $\mu g/m^3$ [8–13], indicating the favorable performance of the Land Use Regression models established in this study.

In the morning regression model for the Wuchang district, seven influencing factors were considered, resulting in a well-adjusted $R^2$ of 0.883, indicating a good fit. In the afternoon regression model for the same area, eight influencing factors were included, yielding an adjusted $R^2$ of 0.906, indicating excellent model fitting.

### 3.2. Spatial Distribution of PM2.5 Concentrations

The spatial distribution of PM2.5 concentrations in the Wuchang District, Ziyang District, Ganghua District, and Geological University District during the morning and afternoon is shown in Figure 5. The distribution of PM2.5 concentrations in the Wuchang district over two periods indicates lower levels in the northwest and central lake regions, suggesting that water bodies play a role in limiting PM2.5 diffusion. Conversely, higher concentrations are observed in the central urban area for both periods, likely due to high population density and traffic volume. In the morning, the southeastern region shows elevated PM2.5 levels, significantly influenced by industrial activities. In the afternoon, there is an increase in the concentrations in the eastern and northern areas, with the southwest region's PM2.5 dispersing beyond the third ring road and localized high values appearing in the south.

In the Wuchang Jiang'an district, PM2.5 concentrations are slightly lower than in inland areas. A comparison of PM2.5 levels around three typical neighborhoods shows that the Qingshan Ganghua and Kepu Park areas have the lowest concentrations, followed by the Wuchang Ziyang and Ziyang Park areas, with the Geological University district having the highest concentration. Notably, the typical neighborhoods within the Qingshan industrial zone have lower PM2.5 levels than the Geological University district, which has a significantly higher greening rate.

In the Ziyang district, the distribution of PM2.5 shows that the southern Ziyang Lake effectively reduces PM2.5 concentrations in its surroundings during both morning and afternoon while high-rise buildings on the west side lead to localized increases in concentration. Notably, PM2.5 levels in the west and north areas are significantly higher than in nearby low-rise residential areas, with building morphology being a primary influencing factor. In the morning, the southern area exhibits lower PM2.5 concentrations, associated with high greenery and a low traffic volume. Certain northern areas also show reduced PM2.5 levels due to high green coverage and minimal traffic activity. In the afternoon, PM2.5 concentrations fluctuate between 25 and 43 $\mu g/m^3$, mainly influenced by factors such as the floor area ratio, artificial surfaces, and impervious ground. The central residential area's floor area ratio is lower, and there is minimal variation in concentrations across other regions.

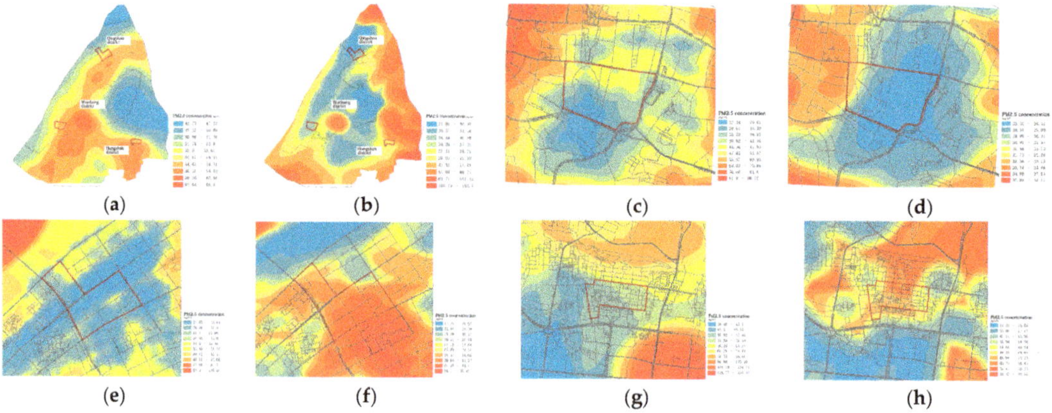

**Figure 5.** Spatial distribution of PM2.5 concentrations: (**a**) Wuchang district (morning); (**b**) Wuchang district (afternoon); (**c**) Ziyang district (morning); (**d**) Ziyang district (afternoon); (**e**) Ganghua district (morning); (**f**) Ganghua district (afternoon); (**g**) Geological University (morning); (**h**) Geological University (afternoon).

In the Ganghua district, morning PM2.5 concentrations range from 21 to 130 μg/m$^3$ while other areas show levels between 17 and 38 μg/m$^3$. The central and surrounding regions have lower concentrations due to their high greenery, with multiple parks and schools contributing to good environmental quality. In the north, PM2.5 concentrations are 38–45 μg/m$^3$ in newly constructed high-rise residential areas, where the building height affects PM2.5 distribution. In the afternoon, PM2.5 levels decrease over the Yangtze River while increasing inland, with higher concentrations in the southeastern residential areas due to high building density. The northeast school area benefits from dispersed greenery, which helps alleviate pollution.

In the Geological University district, PM2.5 concentrations are similar in the morning and afternoon, with some localized variations. The southeast industrial park is a high-concentration area, significantly impacting its surroundings in the morning, with high PM2.5 levels in the eastern scenic and mountainous areas. In the afternoon, pollutants shift from the scenic area to urban construction zones, and the northwest school area experiences lower concentrations due to the reduced university population during summer holidays. The concentrated natural ecological environment of the scenic and mountainous areas hinders PM2.5 dispersion, particularly in the morning. Within the campus, areas like the western side have low PM2.5 levels due to high road canopy coverage. High-concentration zones are found in the eastern and western residential areas within the campus and in external communities, with on-campus residential areas notably influenced by the northern scenic area and off-campus high-density residential areas forming high-concentration regions.

### *3.3. Analysis of PM2.5 Correlation with Influencing Factors*

#### 3.3.1. Establishment of Different Scale Buffers

Before constructing the LUR model, it is necessary to establish buffers with PM2.5 concentration aggregation points as centers. Eleven buffers, ranging from 50 m to 3000 m, are created with different distance scales, including building, neighborhood, city, and regional scales. Considering that visual impact factors primarily concern human-scale observations and are not suitable for larger scales, eight buffer zones are analyzed within the range of 50 m to 1000 m. This analysis focuses on the proportions of visual impact factors, such as buildings, vegetation, the sky, and impervious surfaces. Furthermore, detailed discussions on the statistical methods and variations in various predictive factors are conducted, providing foundational data for multi-scale urban air quality research. The classification of independent variables and their data sources is provided in Table 3.

**Table 3.** Classification of independent variables and data sources (source: self-mapping).

| Impact Factor | Description | Data Source | Spatial Resolution | Unit Measure | Computation Method |
|---|---|---|---|---|---|
| Geographic Impact Factor (GIF) | Includes land cover, land use properties, and road intensity to measure geographic influence on environmental and spatial characteristics. | GIS Data, National Geographic Information Resources | 50–3000 m | m$^2$ | Based on spatial analysis, remote sensing, and traffic data processing |

**Table 3.** *Cont.*

| Impact Factor | | Description | Data Source | Spatial Resolution | Unit Measure | Computation Method |
|---|---|---|---|---|---|---|
| Land cover | Cropland | Areas covered by agricultural crops | National Geographic Information Resources Catalogue Service System | 50, 100, 150, 200, 300, 400, 500, 1000, 1500, 2000, 3000 | $m^2$ | Based on spectral data classification |
| | Forests | Areas dominated by trees and forest ecosystems. | | | | Based on vegetation index(e.g., NDVI) analysis |
| | Grassland | Areas dominated by herbaceous plants, typically used for grazing. | | | | Based on vegetation index analysis |
| | Water | Areas covered by water bodies, including lakes, rivers, and oceans. | | | | Water body classification |
| | Artificial surfaces | Surfaces altered by human activities, such as buildings, roads, and parking lots. | | | | Based on surface classification algorithms |
| Land use properties | Industrial Land | Land used for industrial activities, such as factories and manufacturing zones. | Wuhan City Planning One Map + ArcGIS | 50, 100, 150, 200, 300, 400, 500, 1000, 1500, 2000, 3000 | $m^2$ | Based on land use classification algorithms |
| | Parks and Green Spaces | Areas for public recreation and greenery typically include parks and gardens. | | | | Based on green space lists and spatial analysis |
| | Residential | Land used for residential buildings, including urban, suburban, and rural areas. | | | | Based on land use classification analysis |
| | Commercial | Land used for commercial activities, such as shops, offices, etc. | | | | Based on land use classification analysis |
| Road intensity | Secondary roads | Roads connecting arterial roads with secondary streets. | OpenStreetMap | 50, 100, 150, 200, 300, 400, 500, 1000, 1500, 2000, 3000 | m | Road classification analysis |
| | Tertiary roads | Smaller roads providing access to individual properties or neighborhoods. | | | | Road classification analysis |
| | Fourth-class roads | Smaller roads, often unpaved, are found in rural or remote areas. | | | | Road classification analysis |
| | Public transport routes | Routes for public transportation systems (buses, metro, trains, etc.). | | | | Based on transport network analysis |
| Architectural Form Factor (AFF) | Building Density | The number of buildings per unit area, indicating the concentration of buildings in an area. | Baidu Architecture + ArcGIS | 50, 100, 150, 200, 300, 400, 500, 1000, 1500, 2000, 3000 | % | Building base area/Plot area × 100% |
| | Average Building Height | The average height of buildings in an area, reflecting the scale and intensity of development. | | | / | Height measurement and digital modeling analysis |
| | Floor Area Ratio (FAR) | The ratio of total building floor area to the land area it occupies. | | | m | Total floor area/Plot area calculation |

**Table 3.** *Cont.*

| Impact Factor | | Description | Data Source | Spatial Resolution | Unit Measure | Computation Method |
|---|---|---|---|---|---|---|
| Visual Impact Factor (VIF) | Architectural Visibility Factor | Measures the visibility of buildings from different vantages points. | Baidu Street View + Image Segmentation | 50, 100, 150, 200, 300, 400, 500, 1000 | % | Based on building features and sightline analysis |
| | Plant Visibility Factor | Measures the visibility of plants or green spaces in a given area. | | | | Based on plant visibility analysis and segmentation |
| | Sky Visibility Factor | Measures the extent of sky visibility from a given point. | | | | Sky visibility analysis |
| | Impervious Surface Visibility Factor | Measures the visibility of impervious surfaces (such as roads, buildings) in the landscape. | | | | Impervious surface visibility analysis |

3.3.2. Influencing Factors and PM2.5 Correlation Test

In this study, we delineated 11 buffer zones to encompass various geographical impact factors (including land cover, land use properties, and road intensity), architectural morphology factors (such as building density, building height, and floor area ratio), and visual impact factors (comprising proportions of buildings, vegetation, the sky, and impervious surfaces in street view). This approach yielded a total of 208 sets of predictive variables. Through a meticulous correlation analysis of PM2.5 concentration data obtained from mobile measurements in typical urban neighborhoods with these predictive variables, we explored the intricate relationships among the three hierarchical impact factors and PM2.5 concentrations.The analysis of the correlation between PM2.5 and different influencing factors is shown in Figure 6.

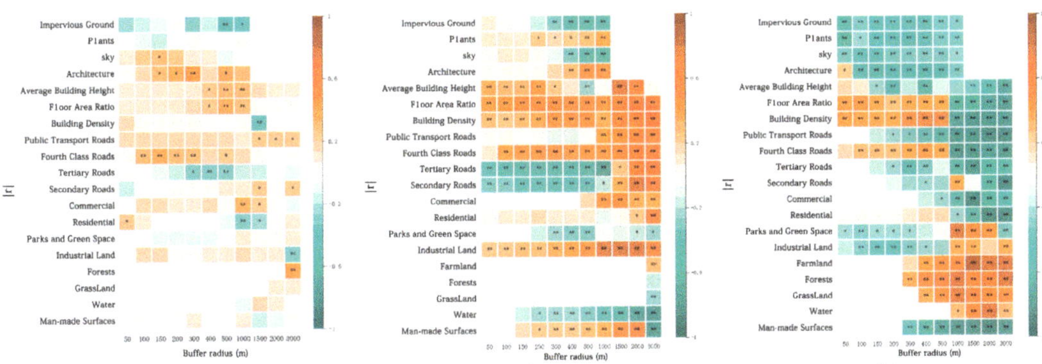


(**a**) (**b**) (**c**)

**Figure 6.** Analysis of PM2.5 correlation with different influencing factors (*/**indicates the strength of correlation): (**a**) PM2.5 correlation with different influencing factors of Ziyang district; (**b**) PM2.5 correlation with different influencing factors of Ganghua district; (c) PM2.5 correlation with different influencing factors of Geological University district.

(1) Ziyang District

As illustrated in Figure 6a, the daily average PM2.5 concentrations in the Ziyang district exhibit minimal sensitivity to artificial surfaces, water bodies, grasslands, and forests, reflecting a relatively low proportion of natural land cover in the area. The most pronounced correlation between artificial surfaces, water bodies, and PM2.5 concentrations is observed within a 1500 m radius buffer zone, characterized by a coefficient of −0.187. Regarding grasslands, the strongest correlation with the PM2.5 concentrations is found within a 3000 m radius buffer zone, showcasing a coefficient of −0.125.

Residential and commercial land use properties display significant correlations with PM2.5 concentrations across various buffer zones. In the 1000 m and 1500 m radius buffer zones, residential land use properties exhibit negative correlations with coefficients of −0.324 and −0.257, respectively. Conversely, commercial land use properties show positive correlations with coefficients of 0.379 and 0.215. The densely populated commercial areas feature mid-to-high-density multi-story and high-rise buildings. Parks and green spaces demonstrate negative correlations within the 50–500 m radius buffer zone, with coefficients ranging from −0.004 to −0.117. However, in the 1000–3000 m radius buffer zone, positive correlations are observed, with coefficients ranging from 0.04 to 0.123. While smaller-scale parks like Ziyang Park aid in mitigating environmental pollution, expanding parks and green spaces on a larger scale may impede PM2.5 dispersion.

Fourth-level roads exhibit significant correlations with PM2.5 concentrations across five buffer zones, with the strongest correlation observed in the 150 m radius buffer zone (coefficient: 0.337). Other road types, including second level and fourth-level roads, as well as bus routes, show associations with PM2.5 concentrations in 2~3 buffer zones. The dense distribution of second-level roads, fourth-level roads, and bus routes, coupled with increased traffic and exhaust emissions, contributes to environmental pollution.

Regarding building morphology factors, the plot ratio and average building height demonstrate positive correlations with PM2.5 concentrations, particularly within the 500 m and 1000 m radius buffer zones, with the plot ratio exhibiting the highest correlation coefficient of 0.343 within the 500 m radius buffer zone. Building density reveals a maximum correlation coefficient of 0.301 within the 1000 m radius buffer zone. The relationship between building density and PM2.5 concentration is positive within the 400 m radius buffer zone and negative beyond this range. The most significant negative correlation is observed within the 1500 m radius buffer zone, with a coefficient of −0.375.

Concerning visual impact factors, the visual proportions of street buildings and the sky correlate positively with PM2.5 concentration, displaying a noticeable trend. The visual factor of buildings demonstrates the most significant correlation within the 300 m radius buffer zone, with a coefficient of 0.413. The visual proportion of the sky exhibits a negative correlation within the 500–1000 m radius buffer zone, suggesting that larger scales facilitate PM2.5 dispersion. The visual factor of vegetation consistently shows negative correlation coefficients below 0.2. Street trees contribute to alleviating the PM2.5 concentration to some extent. The visual factor of impervious surfaces, primarily urban roads, correlates with road width, showing the strongest association within the 500 m radius buffer zone, with a coefficient of −0.407.

(2) Ganghua District

As illustrated in Figure 6b, artificial surfaces and water bodies in land cover show significant correlations with the PM2.5 concentration. Artificial surfaces display a positive correlation within the 3000 m radius buffer zone, with a correlation coefficient reaching −0.738. In contrast, water bodies exhibit a negative correlation within the 3000 m radius buffer zone, with a correlation coefficient of −0.738. Situated along the Yangtze River, this region is notably influenced by water bodies, which help impede PM2.5 dispersion. Forests and grasslands contribute to mitigating environmental pollution while croplands demonstrate a notable positive correlation with the PM2.5 concentration, showing a correlation coefficient of 0.316, consistent with existing research findings.

The findings unveil a positive correlation between industrial land use properties and PM2.5 concentration across all 11 buffer zones, notably pronounced in the Qing Shan Industrial Zone, where the maximum correlation within the 1500 m buffer zone reaches −0.794. Park and green land use properties display a negative correlation, particularly within the 400 m buffer zone, with a correlation coefficient of −0.287, contributing to pollution alleviation. Commercial and residential land use properties exhibit a positive correlation, especially in buffer zones beyond 1000 m. Road intensity demonstrates a positive correlation with the PM2.5 concentration in all buffer zones, with fewer secondary and tertiary roads and bus routes within the 500 m buffer zone. However, in areas beyond 1000 m, traffic flow is positively correlated with the PM2.5 concentration, with the strongest correlation coefficient of 0.709 observed within the 3000 m buffer zone.

Building morphology factors exhibit a significant positive correlation with the PM2.5 concentration. In the 500–1000 m radius buffer zone, where structures are predominantly multi-storied, the average building height decreases, resulting in a negative correlation. The strongest association is observed within the 3000 m radius buffer zone, with a coefficient of 0.702. The building density, plot ratio, and average building height display correlation coefficients exceeding 0.6 within the 1000–3000 m radius buffer zone, indicating a strong correlation and highlighting the substantial impact of the built environment on environmental quality.

Among visual factors, the building and plant correlate positively with the PM2.5 concentration, with the strongest correlation being within the 1000 m buffer zone, presenting coefficients of 0.581 and −0.598, respectively. The visibility factor of the sky shows a positive correlation within the 200 m buffer zone, turning negative beyond 300 m, with the most significant correlation observed within the 1000 m radius buffer zone, presenting a coefficient of −0.598. The visibility factor of impervious surfaces follows a similar pattern, exhibiting a significant association with the PM2.5 concentration within the 400 m, 500 m, and 1000 m radius buffer zones.

(3) Geological University District

As illustrated in Figure 6c, the correlation among artificial surfaces, water bodies, grasslands, forests, and arable land within this area diverges from the previous two testing zones. This testing area is distinguished by the concentrated distribution of water bodies, grasslands, forests, and arable land around the East Lake Natural Scenic Area. The results indicate that the extensive concentrated distribution of natural ecological environments impedes the dispersion of PM2.5 within the city.

Regarding land use properties, parks and green spaces within the 1000 m radius buffer zone exhibit a significantly positive correlation with the PM2.5 concentration, with a coefficient of 0.744, underscoring the significant impact of East Lake water on the PM2.5 concentration. Industrial land use properties within the 1500 m buffer zone demonstrate the strongest correlation with the PM2.5 concentration, with a coefficient of 0.399. Residential land use properties exhibit the strongest correlation within the 3000 m buffer zone, with a coefficient of −0.835. Fourth-level roads exhibit the strongest correlation within the 500 m buffer zone, with a coefficient of −0.782. Other roads within the 1000 m buffer zone display an increased proportion of second-level roads, significantly positively correlating with the PM2.5 concentration. The strongest correlation within the 3000 m buffer zone is observed with a coefficient of −0.808. This area is proximate to the city's third ring road, characterized by lower traffic flow, more surrounding parks and green spaces, and a lower road network density, showcasing significant differences from the other two areas.

Building Form Factors: Building density and plot ratio display notable positive correlations with the PM2.5 concentration within the 500 m radius buffer zone whereas they demonstrate a negative correlation within the 1000 m radius buffer zone. Both factors reveal their strongest correlation within the 3000 m radius buffer zone, with correlation coefficients of −0.776 and −0.787, respectively. Beyond the 500 m radius buffer zone, the influence of natural ecological land becomes more pronounced. This area, except for the campus interior, mainly consists of high-density residential and commercial zones, rendering this urban layout unfavorable for mitigating environmental pollution.

Visual Impact Factors: All Visual Impact Factors within this area showcase a negative correlation with the PM2.5 concentration. The area boasts a high proportion of roads within the campus, exerting a significant impact on visual impact factor results. Roads within the campus are adorned with tall plane trees, resulting in high canopy closure, low sky openness, and low building density. In contrast, the external campus environment differs, suggesting that the internal campus environment is less conducive to PM2.5 dispersion.

Each of the three areas exhibits distinct characteristics, resulting in differences in correlation results. The Ziyang district demonstrates weaker correlations with the predicted influencing factors, with road intensity being the only factor strongly associated with the PM2.5 concentration. The Qing Shan Ganghua and Science Park area is influenced by industrial land, and water areas alleviate environmental pollution on a larger scale. The Geological University Area is influenced by the surrounding ecological environment, where extensive urban green spaces may have a negative impact on the surrounding environment. Road intensity in all three areas is significantly correlated with the PM2.5 concentration at multiple buffer scales, with fourth-level roads being a key factor. Within the building form factors, building density and plot ratio exhibit positive correlations with the PM2.5 concentration, emphasizing the need for urban planning to consider the environmental impact of form. Among visual impact factors, the non-permeable ground visual factor exhibits a negative correlation with the PM2.5 concentration, highlighting the strong association between the road width and PM2.5 concentration.

## 4. Results

### *4.1. Scaling Effects of Geographic Influence Factors*

(1) Land cover

Artificial surfaces and water bodies exert an influence on the PM2.5 concentration across all four scales. Cultivated land and forests impact the PM2.5 concentration at the urban and regional scales while at the building and neighborhood scales, the correlation is not significant. Grassland significantly influences the PM2.5 concentration only at the regional scale. The absence of a significant correlation at smaller scales might be attributed to the uniform distribution of land cover within the buffer zone. Cultivated land, forests, grasslands, water bodies, and artificial surfaces demonstrate more pronounced impacts on the PM2.5 concentration at the regional and urban scales. The frequency distribution of land cover correlations is shown in Figure 7.

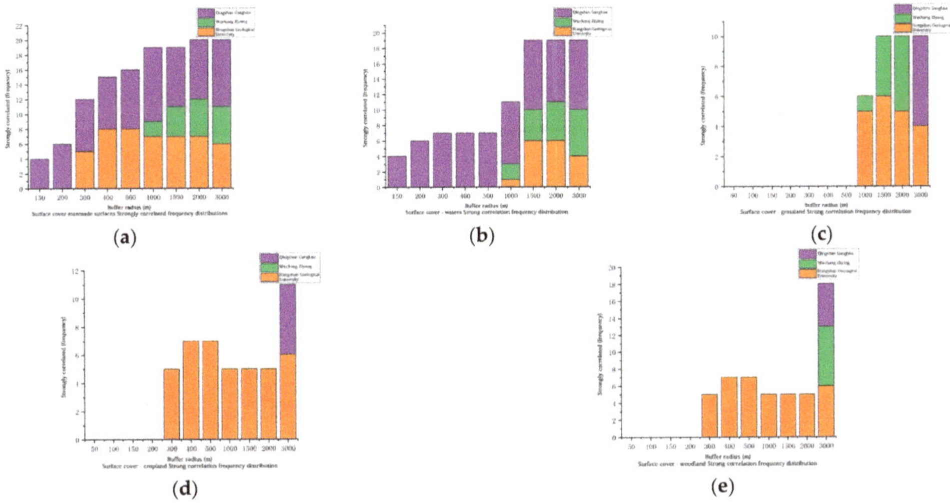


**Figure 7.** Frequency distribution of land cover correlations: (**a**) artificial surface correlations; (**b**) water correlations; (**c**) grassland correlations; (**d**) forest correlations; (**e**) cropland correlations.

(2) Land use properties

Industrial land, park greenery, residential land, and commercial land all influence the PM2.5 concentration across all four scales, with particularly notable effects at the regional and urban scales. The impact of industrial land remains consistent across different scales, being most pronounced in the 150 m buffer zone. Park greenery demonstrates a gradual increase in impact across various scales, with significance observed in the 150 m and 400 m buffer zones, and more prominently at the urban and regional scales within the 1500 m and 3000 m buffer zones. Residential land has a comparable impact at the building and neighborhood scales, with the most notable effect observed in the 50 m buffer zone. Commercial land exhibits its strongest influence in the 1000 m buffer zone, with significant effects on the building and neighborhood scales within the 500 m buffer zone.The frequency distribution of correlations for land use properties is shown in Figure 8.

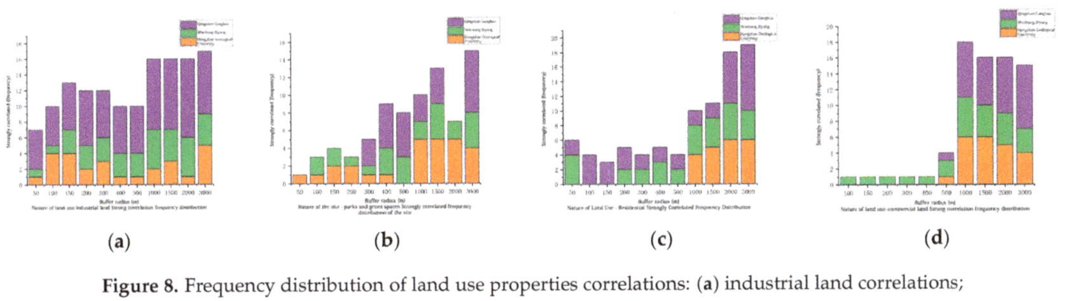


**Figure 8.** Frequency distribution of land use properties correlations: (**a**) industrial land correlations; (**b**) park and green space correlations; (**c**) residential correlations; (**d**) commercial correlations.

(3) Road intensity

The correlation of secondary roads and bus routes with the PM2.5 concentration strengthens as the spatial scale increases. For tertiary roads, there is an increasing–decreasing–increasing trend at the building scale and a decreasing–increasing trend at the neighborhood scale. The most significant trend is observed in the 300 m buffer zone, with consistent effects at the urban and regional scales, peaking at the 2000 m buffer zone. For fourth-class roads, except in the 50 m buffer zone, the impact intensity remains consistent across other buffer zones, with the maximum influence observed in the 3000 m radius buffer zone. Bus routes exhibit the strongest influence at the building scale in the 150 m buffer zone, at the neighborhood scale in the 300 m and 400 m buffer zones, and at the urban scale in the 1000 m and 2000 m buffer zones, aligning with the regional scale at 2000 m. The frequency distribution of correlations for road intensity is shown in Figure 9.

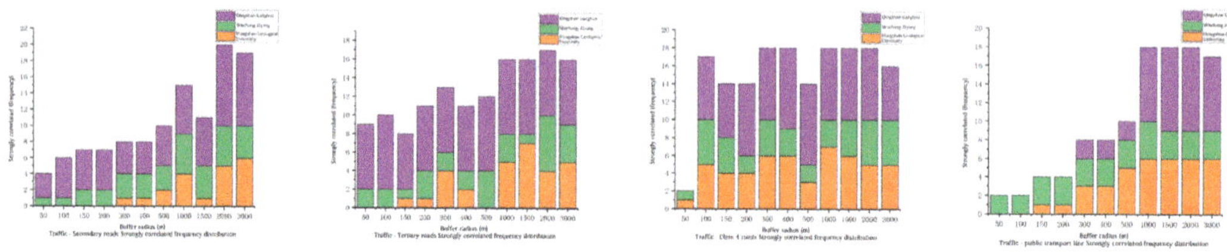

(**a**) (**b**) (**c**) (**d**)

**Figure 9.** Frequency distribution of road intensity correlations: (**a**) secondary road correlations; (**b**) tertiary road correlations; (**c**) fourth-class road correlations; (**d**) public transport route correlations.

### *4.2. Scale Effects of Architectural Form Factors*

Building morphology factors exhibit differences from other predictive factors. Building density increases with scale, and at the neighborhood scale, the impact is greater within the 400 m radius buffer zone than at that scale. The plot ratio is most significant at the neighborhood scale, particularly in the 300 m radius buffer zone, showing an increasing–decreasing trend. Its influence gradually rises at the building, urban, and regional scales across various buffer zones, peaking in the 150 m, 1500 m, and 3000 m radius buffer zones. Building height shows considerable fluctuations in its impact, remaining relatively constant with scale changes. The 500 m radius buffer zone has the smallest effect while the 200 m buffer zone is most significant at the building scale. At the urban and regional scales, significance is observed at the 1000 m and 2000 m radius buffer zones, respectively. The frequency distribution of correlations for architectural form is shown in Figure 10.

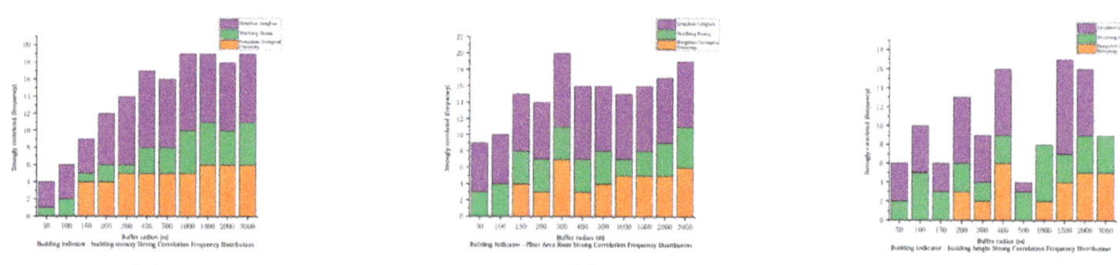

(**a**) (**b**) (**c**)

**Figure 10.** Frequency distribution of architectural form correlations: (**a**) building density correlations; (**b**) Floor Area Ratio correlations; (**c**) average building height correlations.

### *4.3. Scale Effects of Visual Impact Factors*

Figure 11 illustrates that visual impact factors strengthen with an increasing scale. The effects of building and sky visual factors are most pronounced within the 500 m radius buffer zone while plant and impervious area visual effects peak within the 1000 m radius buffer zone. In the Wuchang Ziyang and Ziyang Park area, the correlation frequency with the PM2.5 concentration is low, indicating a minimal impact of visual factors. Building visual factors exhibit effects at multiple scales whereas the scale impact of sky visual factors is less significant. In the Qingshan Ganghua and Science Park area, visual impact factors significantly affect the PM2.5 concentration, showing correlated effects in all eight buffer zones, with each factor exhibiting the strongest impact within different scale ranges. For the Hongshan Geological University and campus area, the impact effects across various buffer zones are relatively consistent, suggesting that the influence of visual impact factors on the PM2.5 concentration in this region is less affected by scale variations.

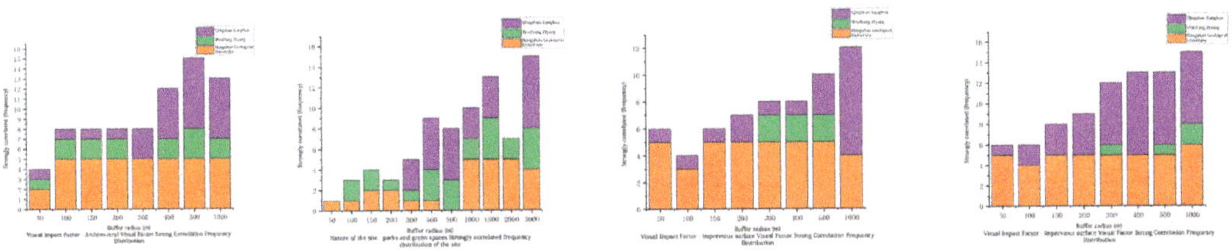

(a) (b) (c) (d)

**Figure 11.** Frequency distribution of significant correlations for visual impact factors: (**a**) building correlations for Visual Impact Factors; (**b**) sky correlations for Visual Impact Factors; (**c**) plant correlations for Visual Impact Factors; (**d**) impervious surface correlations for Visual Impact Factors.

## 5. Discussion

### *5.1. Scaling Effects of Key Impact Factors*

To delve into the primary influencing factors affecting the neighborhood-level PM2.5 concentration, we undertook bivariate correlation analysis, juxtaposing the daily average PM2.5 concentration data gathered from mobile measurements against geographical, architectural, and visual impact factors. Across 11 distinct buffer zones, we scrutinized the absolute values of correlation coefficients for each category of influencing factors, as delineated in Figure 12 to unravel their impacts on PM2.5 concentration. The findings were as follows:

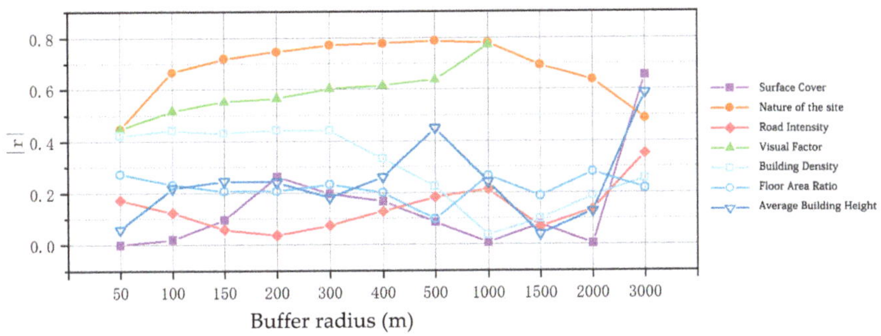


**Figure 12.** Comparison of absolute values of correlation coefficients between PM2.5 concentrations and multi-scale buffer impact factors.

(1) Land Cover: The correlation between land cover and the PM2.5 concentration diminishes beyond the 3000 m buffer zone ($|r| < 0.3$), but it intensifies with scale in the 200–1000 m and 1500–3000 m radius buffer zones. Artificial surfaces predominate within the 50–1000 m buffer zone, resulting in smaller correlation coefficients. In larger areas with increased water bodies, grasslands, forests, and cultivated land, the impact of land cover on the PM2.5 concentration strengthens in larger-scale buffer zones.

(2) Land Use Properties: Land use properties exhibit a robust correlation with the PM2.5 concentration in multiple buffer zones ($|r| > 0.5$) and demonstrate an inverted "U"

shape. The PM2.5 concentration is significantly influenced by land use properties, with anthropogenic environmental impacts outweighing natural geographic factors.

(3) Road Intensity: Road intensity strongly correlates with the PM2.5 concentration within the 3000 m buffer zone ($|r| > 0.6$), moderately correlates within the 200 m buffer zone, and weakly correlates at other scales. The PM2.5 concentration is significantly influenced by road intensity at larger scales.

(4) Building Form Factors: The correlation with volume ratio (floor area ratio) is weak ($0.1 < |r| < 0.3$). Building density shows a strong correlation within the 50–300 m buffer zone, with a stable trend. Building height exhibits increasing correlation within the 50–500 m buffer zone, with a slight peak at 500 m, followed by a decline. The correlation impact significantly increases within the 1500–3000 m buffer zone. Diverse layouts of building density and height within the 500 m–1000 m range are beneficial for improving environmental quality.

(5) Visual Impact Factors: Visual impact factors exhibit a strong correlation with the PM2.5 concentration ($|r| > 0.5$), strengthening with increasing scale. This indicates a significant association between street quality and the PM2.5 concentration.

### 5.2. Innovation Point

(1) This paper introduces a novel three-tier research framework that encompasses geographical, architectural, and visual aspects, providing a comprehensive understanding of the mechanisms underlying PM2.5 distribution influenced by building neighborhood morphology. By addressing the partiality often observed in urban air quality research, this framework offers a more holistic perspective. The incorporation of visual factors, including the analysis of visual impact factors such as buildings, vegetation, the sky, and impermeable surfaces, allows for a deeper exploration of how urban neighborhood morphology affects the PM2.5 concentration at the local scale. This approach contributes to offering more specific and practical recommendations for improving air quality for urban residents.

(2) The paper focuses on comprehensively considering multi-scale influencing factors on PM2.5 concentration by establishing 11 scale buffer zones and utilizing a Land Use Regression model to analyze influencing factors across different scales. The findings unveil the scale effects of PM2.5 concentration and identify key factors with significant impacts at each scale. This research serves as a reference for selecting appropriate scales in subsequent related studies. Additionally, it provides targeted guidance for updating regulations pertaining to urban planning at different scales.

### 5.3. Significance of Guidance

(1) Scalable Planning and Architectural Morphology Optimization

A multi-scale integrated planning approach is proposed. At the regional scale, the focus should be on optimizing the balance between water bodies, green spaces, and agricultural land, particularly in high-density urban development areas. Increasing the area of water bodies can significantly reduce PM2.5 concentrations. At the urban scale, the land use structure should be optimized to prevent the proximity of industrial zones to residential areas and ensure the rational distribution of green spaces. Scattered green spaces, including micro-parks, can help mitigate the urban heat island effect. At the neighborhood scale, one may avoid the concentration of high-density building clusters by adopting low-density, mixed-use designs. Increasing the distance between buildings improves air circulation. The integration of resilient building design and green infrastructure is crucial. Green building principles, such as using energy-efficient, low-carbon materials and incorporating green roofs and vertical greening, can enhance air quality and reduce PM2.5

buildup. Additionally, ensuring spatial permeability within building clusters and avoiding large concentrations of buildings with similar heights can help prevent the localized accumulation of PM2.5.

(2) Street Landscape Optimization Based on Visual Impact

To optimize the visual impact, street width and building layout should be adjusted to avoid the "street canyon" effect. Open street designs that ensure air flow should be prioritized. Green belts and tree planting along streets should be enhanced, particularly with tree species that help absorb particulate matter and improve air quality. The continuity of the street greening system must be ensured, and urban green spaces should be integrated into the larger city ecosystem. Building facades should be designed for transparency and permeability, avoiding closed-off structures. Increasing the amount of light and air permeability in building facades enhances visual comfort and contributes to reducing the PM2.5 concentration in an area.

(3) Efficient Transportation System and Pollution Source Dispersion

To optimize transportation systems and reduce pollution, it is essential to avoid locating major traffic arteries near dense residential areas. Underground transportation systems or circular traffic networks should be developed to minimize traffic congestion. Smart traffic management systems, utilizing big data and AI technologies, can regulate traffic flow and reduce pollution caused by traffic jams. Additionally, promoting public transport and non-motorized transportation options, such as cycling and walking, can reduce private car usage, thus reducing pollution sources.

(4) Innovative Water Body and Green Space Layout

Integrating green infrastructure is key. Expanding urban water bodies, particularly in dense areas and on the urban periphery, can alleviate heat island effects and improve the overall environmental quality. Water bodies help reduce PM2.5 levels and regulate microclimates. A multi-layered green space system should be constructed, incorporating vertical greening, such as via rooftop gardens and green walls, particularly in high-density urban zones. This approach will enhance connectivity and diversity in green spaces, which can significantly reduce PM2.5 concentrations and improve urban resilience.

## 6. Conclusions

This paper has analyzed the spatial distribution of PM2.5 concentrations in Wuhan through multi-scale buffer analysis, quantifying the impact of geographical, architectural, and visual factors. The results show that land use, building density, and visual factors significantly affect PM2.5 concentrations at varying scales.

Key findings include the following. Land use properties significantly impact PM2.5 concentrations at larger scales (2000 m–3000 m), with water bodies and green spaces reducing PM2.5 by up to 20%. Building density affects PM2.5 at the neighborhood scale (300 m–500 m), with high-density areas showing concentrations up to 15% higher. Visual factors, such as street width and green belts, improve air quality at the urban scale (1000 m–1500 m), with street greenery reducing PM2.5 by about 10%. The multi-scale buffer analysis and Land Use Regression models employed in this study offer a comprehensive understanding of spatial effects on PM2.5 concentrations. However, limitations in data accuracy and scale selection exist, and future studies could enhance precision by refining models and incorporating high-resolution data. The method is highly replicable and can be applied to PM2.5 analysis in other cities. It is particularly effective in large-scale urban environments, where multi-scale assessments are needed. The method can be expanded to analyze other environmental factors such as noise and heat island effects. Integration with remote sensing and sensor data will further enhance its applicability and accuracy. Based on the research findings, the following measures can effectively reduce PM2.5 concentrations and improve urban quality. In urban planning, professionals must avoid locating industrial areas near high-density residential zones, increase green spaces and the numbers of water bodies, and enhance their connectivity. In building design, they must increase spacing, adopt low-density mixed-use layouts, and reduce pollution in high-density areas. They must improve facade transparency to enhance airflow. In street design, they must optimize the street width to avoid "street canyon" effects, and increase greenery, especially in high-density areas.

Future research will make significant advancements in PM2.5 concentration analysis by integrating mobile measurement technologies, machine learning algorithms, and multi-objective optimization techniques. The use of mobile measurement tools, such as buses, subways, and trams, will address the spatial and temporal limitations of traditional fixed monitoring stations, offering the more comprehensive coverage of PM2.5 concentrations across urban environments. This approach will allow for the precise tracking of dynamic pollution changes and spatial distribution, enhancing data accuracy and flexibility. By combining this with GIS analysis, the real-time tracking of pollutant dispersion and evolving sources will become possible. Meanwhile, machine learning algorithms such as Support Vector Machines (SVMs), Random Forests (RFs), and Gradient Boosting Trees (GBDTs) will replace traditional regression methods, enabling the analysis of complex nonlinear relationships and interactions, thereby improving prediction accuracy. These algorithms can handle large datasets effectively, continuously updating models to adapt to environmental changes and provide more accurate PM2.5 predictions. Additionally, multi-objective optimization algorithms like Genetic Algorithms (GAs), Particle Swarm Optimization (PSO), and Non-dominated Sorting Genetic Algorithm II (NSGA-II) will facilitate finding optimal trade-offs in urban planning and air quality management, offering precise decision support for balancing environmental and development needs. As these technologies evolve, dynamic monitoring systems will become more intelligent, utilizing big data and cloud platforms to provide real-time PM2.5 concentration monitoring and actionable insights, advancing air quality management towards more refined and intelligent solutions.

**Author Contributions:** Conceptualization, Z.W. (Zhen Wang) and Z.C.; Methodology, Z.C.; Software, Z.C.; Validation, Z.C., Z.W. (Zhen Wang) and K.H.; Formal Analysis, Z.C.; Investigation, Z.C.; Resources, Z.C.; Data Curation, Z.C. and K.H.; Writing—Original Draft Preparation, K.H.; Writing—Review and Editing, K.H. and Z.W. (Zheyu Wang); Visualization, K.H. and Z.W. (Zheyu Wang); Supervision, B.Y.; Project Administration, Z.W. (Zhen Wang); Funding Acquisition, Z.W. (Zhen Wang). All authors have read and agreed to the published version of the manuscript.

**Funding:** This study was supported by the National Natural Science Foundation Project of China (Grant Number: 52078229) and Hubei Province Housing and Urban-Rural Construction Science and Technology Program (Project Approval No. 2023-79).

**Data Availability Statement:** Data will be made available on request.

**Acknowledgments:** We would like to thank Yunchen Dong, Yezi Yang, Xuying Zhao, Chenxi Lv, Jiaxuan Wang, Tianzi Lian, Zhicong Wu, Qian Wang, and Zihan Zhou for their great help in the research. All individuals included in this section have consented to the acknowledgement.

**Conflicts of Interest:** The authors declare no conflict of interest.

## References


1. Beelen, R.; Raaschou-Nielsen, O.; Stafoggia, M.; Andersen, Z.J.; Weinmayr, G.; Hoffmann, B.; Wolf, K.; Samoli, E.; Fischer, P.; Nieuwenhuijsen, M.; et al. Effects of long-term exposure to air pollution on natural-cause mortality: An analysis of 22 European cohorts within the multicenter ESCAPE project. *Lancet* **2014**, *383*, 785–795. [CrossRef] [PubMed]
2. Hoek, G.; Krishnan, R.M.; Beelen, R.; Peters, A.; Ostro, B.; Brunekreef, B.; Kaufman, J.D. Long-term air pollution exposure and cardio-respiratory mortality: A review. *Environ. Health* **2013**, *12*, 43. [CrossRef] [PubMed]
3. Krewski, D.; Jerrett, M.; Burnett, R.T.; Ma, R.; Hughes, E.; Shi, Y.; Turner, M.C.; Pope, C.A., III; Thurston, G.; Calle, E.E.; et al. Extended follow-up and spatial analysis of the American Cancer Society study linking particulate air pollution and mortality. *Res. Rep. Health Eff. Inst.* **2009**, *140*, 5–114.
4. Wong, T.W.; Tam, W.S.; Yu, T.S.; Wong, A.H.S. Associations between daily mortalities from respiratory and cardiovascular diseases and air pollution in Hong Kong, China. *Occup. Environ. Med.* **2002**, *59*, 30–35. [CrossRef] [PubMed]
5. Wong, G.W.K.; Ko, F.W.S.; Lau, T.S.; Li, S.T.; Hui, D.; Pang, S.W.; Leung, R.; Fok, T.F.; Lai, C.K.W. Temporal relationship between air pollution and hospital admissions for asthmatic children in Hong Kong. *Clin. Exp. Allergy* **2001**, *31*, 565–569. [CrossRef]
6. Pope III, C.A.; Burnett, R.T.; Thun, M.J.; Calle, E.E.; Krewski, D.; Ito, K.; Thurston, G.D. Lung cancer, cardiopulmonary mortality, and long-term exposure to fine particulate air pollution. *JAMA* **2002**, *287*, 1132–1141. [CrossRef]
7. Pope III, C.A.; Dockery, D.W. Health effects of fine particulate air pollution: Lines that connect. *J. Air Waste Manag. Assoc.* **2006**, *56*, 709–742. [CrossRef]
8. Raaschou-Nielsen, O.; Andersen, Z.J.; Beelen, R.; Samoli, E.; Stafoggia, M.; Weinmayr, G.; Hoffmann, B.; Fischer, P.; Nieuwenhuijsen, M.J.; Brunekreef, B.; et al. Air pollution and lung cancer incidence in 17 European cohorts: Prospective analyses from the European Study of Cohorts for Air Pollution Effects (ESCAPE). *Lancet Oncol.* **2013**, *14*, 813–822. [CrossRef]
9. World Health Organization. 2 May 2018. 9 out of 10 People Worldwide Breathe Polluted Air, but More Countries Are Taking Action. Available online: https://www.who.int/news/item/02-05-2018-9-out-of-10-people-worldwide-breathe-polluted-air-but-more-countries-are-taking-action (accessed on 2 December 2022).
10. Schwela, D.; Haq, G.; Huizenga, C.; Han, W.J.; Fabian, H. *Urban Air Pollution in Asian Cities: Status, Challenges and Management*; Taylor & Francis: Abingdon, UK, 2012.
11. Landsberg, H.E. *The Urban Climate*; Academic press: London, UK, 1981.
12. Ng, E. Policies and technical guidelines for urban planning of high-density cities—Air ventilation assessment (AVA) of Hong Kong. *Build. Environ.* **2009**, *44*, 1478–1488. [CrossRef]
13. Fernando, H.J.S.; Lee, S.M.; Anderson, J.; Princevac, M.; Pardyjak, E.; GrossmanClarke, S. Urban fluid mechanics: Air circulation and contaminant dispersion in cities. *Environ. Fluid Mech.* **2001**, *1*, 107–164. [CrossRef]
14. Carruthers, D. Dispersion in cities. In *Environmental Medicine*; Ayres, J.G., Harrison, R.M., Nichols, G.L., CBE, R.L.M., Eds.; CRC Press: Boca Raton, FL, USA, 2010; p. 320.
15. Vardoulakis, S.; Fisher, B.E.; Pericleous, K.; Gonzalez-Flesca, N. Modelling air quality in street canyons: A review. *Atmos. Environ.* **2003**, *37*, 155–182. [CrossRef]
16. Heinrich, J.; Gehring, U.; Cyrys, J.; Brauer, M.; Hoek, G.; Fischer, P.; Bellander, T.; Brunekreef, B. Exposure to traffic related air pollutants: Self-reported traffic intensity versus GIS modelled exposure. *Occup. Environ. Med.* **2005**, *62*, 517–523. [CrossRef] [PubMed]
17. Zannetti, P. *Air Pollution Modeling: Theories, Computational Methods, and Available Software*; Computational mechanics publications: New York, NY, USA, 1990.
18. Robins, A.; MacDonald, R. Review of flow and dispersion in the vicinity of groups of buildings, Annexure B, University of Surrey, 2001.
19. Li, C.; Wang, Z.; Li, B.; Peng, Z.-R.; Fu, Q. Investigating the relationship between air pollution variation and urban form. *Build. Environ.* **2019**, *147*, 559–568. [CrossRef]
20. Li, Z.; Zhang, H.; Wen, C.-Y.; Yang, A.-S.; Juan, Y.-H. Effects of height-asymmetric street canyon configurations on outdoor air temperature and air quality. *Build. Environ.* **2020**, *183*, 107195. [CrossRef]
21. Lin, X.; Chamecki, M.; Yu, X. Aerodynamic and deposition effects of street trees on $PM_{2.5}$ concentration: From street to neighborhood scale. *Build. Environ.* **2020**, *185*, 107291. [CrossRef]
22. Han, J.; Lee, J. A comparative study of air pollution levels in different urban street configurations in Pangyo New Town, South Korea. *Build. Environ.* **2023**, *243*, 110695. [CrossRef]
23. Huang, C.; Hu, T.; Duan, Y.; Li, Q.; Chen, N.; Wang, Q.; Zhou, M.; Rao, P. Effect of urban morphology on air pollution distribution in high-density urban neighborhoods based on mobile monitoring and machine learning. *Build. Environ.* **2022**, *219*, 109173. [CrossRef]
24. Wu, S.; Chen, C.; Wang, Y.; Yu, Z.; Cao, J.; Zhang, R.; Song, H. Differential effects of valley city morphology on mesoscale flow field characteristics. *Build. Environ.* **2021**, *205*, 108283. [CrossRef]

25. Liu, M.; Chen, H.; Wei, D.; Wu, Y.; Li, C. Nonlinear relationship between urban form and street-level $PM_{2.5}$ and CO based on mobile measurements and gradient boosting decision tree models. *Build. Environ.* **2021**, *205*, 108265. [CrossRef]
26. Li, Z.; Zhang, H.; Juan, Y.-H.; Wen, C.-Y.; Yang, A.-S. Effects of building setback on thermal comfort and air quality in the street canyon. *Build. Environ.* **2022**, *208*, 108627. [CrossRef]
27. Zhang, A.; Xia, C.; Li, W. Relationships between 3D urban form and ground-level fine particulate matter at street neighborhood level: Evidence from fifteen metropolises in China. *Build. Environ.* **2022**, *211*, 108745. [CrossRef]
28. Li, S.; Zou, B.; Ma, X.; Liu, N.; Zhang, Z.; Xie, M.; Zhi, L. Improving air quality through urban form optimization: A review study. *Build. Environ.* **2023**, *243*, 110685. [CrossRef]
29. Likhvar, V.N.; Pascal, M.; Markakis, K.; Colette, A.; Hauglustaine, D.; Valari, M.; Klimont, Z.; Medina, S.; Kinney, P. A multi-scale health impact assessment of air pollution over the 21st century. *Sci. Total Environ.* **2015**, *514*, 439–449. [CrossRef] [PubMed]
30. Shi, Y.; Xie, X.; Fung, J.C.-H.; Ng, E. Identifying Critical Building Morphological Design Factors of Street-Level Air Pollution Dispersion in High-Density Built Environment Using Mobile Monitoring. *Build. Environ.* **2018**, *128*, 248–259. [CrossRef]
31. Tiwari, A.; Kumar, P.; Baldauf, R.; Zhang, K.M.; Pilla, F.; Di Sabatino, S.; Brattich, E.; Pulvirenti, B. Considerations for evaluating green infrastructure impacts in microscale and macroscale air pollution dispersion models. *Sci. Total Environ.* **2019**, *672*, 410–426. [CrossRef]
32. Santiago, J.L. Estimates of population exposure to atmospheric pollution and health-related externalities in a real city: The impact of spatial resolution on the accuracy of results. *Sci. Total Environ.* **2022**, *819*, 152062. [CrossRef]
33. Gong, S.; Liu, Y.; He, J.; Zhang, L.; Lu, S.; Zhang, X. Multi-scale analysis of the impacts of meteorology and emissions on $PM_{2.5}$ and O3 trends at various regions in China from 2013 to 2020 1: Synoptic circulation patterns and pollution. *Sci. Total Environ.* **2022**, *815*, 152770. [CrossRef]
34. Zhang, J.; Wan, Y.; Tian, M.; Li, H.; Chen, K.; Xu, X.; Yuan, L. Comparing multiple machine learning models to investigate the relationship between urban morphology and $PM_{2.5}$ based on mobile monitoring. *Build. Environ.* **2024**, *248*, 111032. [CrossRef]
35. Briggs, D.J.; Collins, S.; Elliott, P.; Fischer, P.; Kingham, S.; Lebret, E.; Pryl, K.; VAN Reeuwijk, H.; Smallbone, K.; VAN DER Veen, A. Mapping Urban Air Pollution Using GIS: A Regression-Based Approach. *Int. J. Geogr. Inf. Sci.* **1997**, *11*, 699–718. [CrossRef]
36. Brauer, M.; Hoek, G.; van Vliet, P.; Meliefste, K.; Fischer, P.; Gehring, U.; Heinrich, J.; Cyrys, J.; Bellander, T.; Lewne, M. Estimating long-term average particulate air pollution concentrations: Application of traffic indicators and geographic information systems. *Epidemiology* **2003**, *14*, 228–239. [CrossRef]
37. Ross, Z.; Jerrett, M.; Ito, K.; Tempalski, B.; Thurston, G.D. A land use regression for predicting fine particulate matter concentrations in the New York City region. *Atmos. Environ.* **2007**, *41*, 2255–2269. [CrossRef]
38. Liu, C.; Henderson, B.H.; Wang, D.; Yang, X.; Peng, Z.-R. A land use regression application into assessing spatial variation of intra-urban fine particulate matter ($PM_{2.5}$) and nitrogen dioxide ($NO_2$) concentrations in City of Shanghai, China. *Sci. Total Environ.* **2016**, *565*, 607–615. [CrossRef] [PubMed]
39. Chunli, F.; Runkui, L. Analysis of spatial and temporal changes of air pollution in Beijing from 2013 to 2019 based on land use regression model. *J. Environ. Sci.* **2021**, *41*, 1231–1238.
40. Wu, J.S.; Liao, X.; Peng, J.; Huang, X.L. Modeling of spatial differentiation of $PM_{2.5}$ concentration and influencing factors in Chongqing. *Environ. Sci.* **2015**, *36*, 759–767.
41. Hong, J.; Xin, G. Research on the influence of spatial morphology and $PM_{2.5}$ diffusion during winter heating period in settlements in severe cold areas. *Build. Sci.* **2020**, *36*, 64–73.
42. Fei, D.; Ming, C.; Min, W.; Shengwei, Z.; Fan, F. Research on the influence of urban neighborhood morphology on $PM_{10}$ and PM $_{2.5}$—Taking Wuhan as an example. *China Gard.* **2020**, *36*, 109–114.
43. Zhang, J.; Dai, C.; Cui, P. Research on the coupling mechanism and prediction of $PM_{2.5}$ distribution and neighborhood morphology in Harbin neighborhood. *Build. Sci.* **2022**, *38*, 78–86.
44. Hankey, S.; Marshall, J.D. Land Use Regression Models of On-Road Particulate Air Pollution (Particle Number, Black Carbon, $PM_{2.5}$, Particle Size) Using Mobile Monitoring. *Environ. Sci. Technol.* **2015**, *49*, 9194–9202. [CrossRef]
45. Messier, K.P.; Chambliss, S.E.; Gani, S.; Alvarez, R.; Brauer, M.; Choi, J.J.; Hamburg, S.P.; Kerckhoffs, J.; LaFranchi, B.; Lunden, M.M. Mapping air pollution with Google Street View cars: Efficient approaches with mobile monitoring and land use regression. *Environ. Sci. Technol.* **2018**, *52*, 12563–12572. [CrossRef]
46. Gu, P.; Li, H.Z.; Ye, Q.; Robinson, E.S.; Apte, J.S.; Robinson, A.L.; Presto, A.A. Intracity variability of particulate matter exposure is driven by carbonaceous sources and correlated with land-use variables. *Environ. Sci. Technol.* **2018**, *52*, 11545–11554.
47. Lim, C.C.; Kim, H.; Vilcassim, M.R.; Thurston, G.D.; Gordon, T.; Chen, L.-C.; Lee, K.; Heimbinder, M.; Kim, S.-Y. Mapping urban air quality using mobile sampling with low-cost sensors and machine learning in Seoul, South Korea. *Environ. Int.* **2019**, *131*, 105022. [CrossRef] [PubMed]
48. Li, X.; Yang, T.; Zeng, Z.; Li, X.; Zeng, G.; Liang, J.; Xiao, R.; Chen, X. Underestimated or overestimated? Dynamic assessment of hourly $PM_{2.5}$ exposure in the metropolitan area based on heatmap and micro-air monitoring stations. *Sci. Total Environ.* **2021**, *779*, 146283. [CrossRef] [PubMed]

49. Xu, G.; Jiao, L.; Xiao, F.; Zhang, X. Land use regression model to simulate the spatial distribution of $PM_{2.5}$ concentration in Beijing-Tianjin-Hebei. *Arid. Zone Resour. Environ.* **2016**, *30*, 116–120.
50. Yu, L.; Yang, W. Study on spatial and temporal distribution of $PM_{2.5}$ mass concentration in Fujian Province based on land use regression model. *J. Subtrop. Resour. Environ.* **2020**, *15*, 29–39.
51. Mingliang, Y.; Zongjiu, Z. Simulation analysis of spatial distribution of $PM_{2.5}$ concentration based on LUR model. *J. Atmos. Environ. Opt.* **2022**, *17*, 347–359.
52. Zhan-Yong, W.; Ming, C.; Zhong-Ren, P.; Ya, G. Spatial and temporal distribution of roadside $PM_{2.5}$ and CO concentrations based on mobile observations. *China Environ. Sci.* **2017**, *37*, 4428–4434.
53. Jie, S.; Suhong, Z.; Yinong, P.; Rongping, L.; Jianbin, X. Study on spatial and temporal patterns of urban $PM_{2.5}$ pollution based on mobile monitoring—A case study of Guangzhou city centre. *Trop. Geogr.* **2020**, *40*, 229–242.
54. Qingyu, L.; Tingqiao, H.; Nan, C.; Qi, W.; Pinhua, R.; Yusen, D.; Chenyu, H. An empirical study on the distribution characteristics of atmospheric particulate matter mass concentration in a high-density neighbourhood. *Environ. Pollut. Prev.* **2022**, *44*, 206–211+217.
55. Carrasco-Hernandez, R.; Smedley, A.R.; Webb, A.R. Using urban canyon geometries obtained from Google Street View for atmospheric studies: Potential applications in the calculation of street level total shortwave irradiances. *Energy Build.* **2015**, *86*, 340–348. [CrossRef]
56. Gong, F.Y.; Zeng, Z.C.; Zhang, F.; Li, X.J.; Ng, E.; Norford, L.K. Mapping sky, tree, and building view factors of street canyons in a high-density urban environment. *Build. Environ.* **2018**, *134*, 155–167. [CrossRef]
57. Zeng, L.Y.; Lu, J.; Li, W.Y.; Li, Y.C. A fast approach for large-scale sky view factor estimation using street view images. *Build. Environ.* **2018**, *135*, 74–84. [CrossRef]
58. Wei, G.; Menghan, J.; Mei, Z.; Yong, G.; Haohan, M. Review of Progress and Quantitative Measurement Methods of Research on street Space. *Urban Plan.* **2021**, 1–9.
59. Middel, A.; Lukasczyk, J.; Maciejewski, R.; Demuzere, M.; Roth, M. Sky View Factor footprints for urban climate modeling. *Urban Clim.* **2018**, *25*, 120–134. [CrossRef]
60. Middel, A.; Lukasczyk, J.; Zakrzewski, S.; Arnold, M.; Maciejewski, R. Urban form and composition of street canyons: A human-centric big data and deep learning approach. *Landsc. Urban Plan.* **2019**, *183*, 122–132. [CrossRef]
61. Li, X.; Zhang, C.; Li, W.; Ricard, R.; Meng, Q.; Zhang, W. Assessing street-level urban greenery using Google Street View and a modified green view index. *Urban For. Urban Green.* **2015**, *14*, 675–685. [CrossRef]
62. Li, C.; Zhipeng, B.; Di, S.; Yan, Y.; Huamin, L.; Quan, L. Spatial distribution of air pollutant concentrations in Tianjin using LUR model. *China Environ. Sci.* **2009**, *29*, 685–691.
63. Limin, J.; Gang, X.; Suli, Z.; Ming, M.; Ting, D.; Jiangyue, L. Simulation of spatial distribution of $PM_{2.5}$ concentration in Wuhan based on LUR. *J. Wuhan Univ.* **2015**, *40*, 1088–1094.
64. Maodan, X.; Jiansheng, W. Impacts of land use and landscape pattern on $PM_{2.5}$ concentration—A case study of Shenzhen City. *J. Peking Univ.* **2017**, *53*, 160–170.
65. Li, X.; Ratti, C.; Seiferling, I. Quantifying the shade provision of street trees in urban landscape: A case study in Boston, USA, using Google Street View. *Landsc. Urban Plan.* **2018**, *169*, 81–91. [CrossRef]
66. Wanying, S.; Zhen, Y.; Pingping, W.; Qiyan, D.; Xingming, L. Simulation of atmospheric $PM_{2.5}$ concentration distribution and population exposure study based on LUR model—A case study of Hubei Province. *J. Cent. China Norm. Univ.* **2019**, *53*, 451–458.
67. Hatzopoulou, M.; Weichenthal, S.; Barreau, G.; Goldberg, M.; Farrell, W.; Crouse, D.; Ross, N. A web-based route planning tool to reduce cyclists' exposures to traffic pollution: A case study in Montreal, Canada. *Environ. Res.* **2013**, *123*, 58–61. [CrossRef] [PubMed]
68. Okokon, E.O.; Yli-Tuomi, T.; Turunen, A.W.; Taimisto, P.; Pennanen, A.; Vouitsis, I.; Samaras, Z.; Voogt, M.; Keuken, M.; Lanki, T. Particulates and noise exposure during bicycle, bus and car commuting: A study in three European cities. *Environ. Res.* **2017**, *154*, 181–189. [CrossRef] [PubMed]

forests

MDPI

Article

# Phenological Response of an Evergreen Broadleaf Tree, *Quercus acuta*, to Meteorological Variability: Evaluation of the Performance of Time Series Models

**Jeongsoo Park *, Minki Hong and Hyohyemi Lee**

Ecological Observation Team, National Institute of Ecology, Seocheon 33657, Republic of Korea; mkhong0321@nie.re.kr (M.H.); hyohyemi@nie.re.kr (H.L.)
* Correspondence: jspark@nie.re.kr; Tel.: +82-41-950-5618

**Abstract:** Phenological events are key indicators for the assessment of climate change impacts on ecosystems. Most previous studies have focused on identifying the timing of phenological events, such as flowering, leaf-out, leaf-fall, etc. In this study, we explored the characteristics of the green chromatic coordinate (GCC) values of the evergreen broadleaf tree (*Quercus acuta* Thunb.), which is a widely used index that serves as a proxy for the seasonal and physiological responses of trees. Additionally, we estimated their relationship with meteorological variables using time series models, including time series decomposition and a seasonal autoregressive integrated moving average with exogenous regressors (SARIMAX). Our results showed that the GCC values and the meteorological variables, which were collected at daily intervals, exhibited a strong autocorrelation and seasonality. This suggests that time series analysis methods are more suitable than ordinary least squares (OLS) regression methods for the fulfillment of statistical assumptions. The time series analysis results highlighted a strong association between precipitation and GCC variation in evergreen broadleaf trees, particularly during the dry season. These results improve our understanding of the response of plant phenology to climate change.

**Keywords:** phenology; time series; *Quercus acuta*; meteorological factors; SARIMAX; climate change

## 1. Introduction

The importance of plant phenological events in the context of climate change is undeniable [1], particularly in regions with distinct seasonality, as phenological mismatches can have severe consequences for agriculture and ecosystems [2]. It is well known that the timing of biological events is strongly controlled by meteorological variations. A previous study reported that the flowering timing of 90% of 243 studied plants was significantly related to temperature [3], and other researchers also argued that temperature is one of the most important factors for plant phenology [4,5]. However, few studies have assessed the importance of multiple meteorological factors such as temperature, precipitation, and solar radiation on plant phenological responses.

Traditional plant phenologic observations, including flowering, leaf-out, leaf-falling, etc., have been recorded by repeated visual observations. Capturing the exact date of key events is relatively labor-intensive and costly. Recently, repeated photography taken with ground-based cameras has been used to document changes in phenological variation. Specifically, the calculated RGB (red, green, and blue) bands can be manipulated in numerous ways to detect patterns and classify objects. RGB images can be captured inexpensively and effectively estimate percent plant cover and biomass [6], and the phenocams can serve as a bridge between satellite remote sensing and field observation methods in plant phenology research [7,8]. The interpretation of time series RGB bands can be improved by incorporating meteorological data collected alongside them [9], and time series statistical models can help identify phenological triggers hidden within these datasets.

Time series data consists of sequential observations taken at equally spaced intervals. Time series analysis is based on the fundamental assumption that past patterns will continue to repeat in the future [10]. The most commonly used method for time series analysis is an autoregressive integrated moving average model (ARIMA). This time series model is frequently applied in various fields such as meteorology, epidemiology, limnology, etc. [11–14], but it is seldom applied in the terrestrial ecological field [15]. In particular, ground-based phenology datasets have rarely been analyzed with time series statistical methods.

The primary objective of this study is to evaluate the applicability of ground-based digital photography and time series analysis as tools for estimating relationships between tree phenological responses and meteorological factors. The first specific objective is to demonstrate the utility of time series models for ground-based phenocam datasets. The second specific objective is to identify which meteorological variables, namely temperature, precipitation, and solar radiation, have an influence on the GCC variability of evergreen broad-leaved forests.

## 2. Data and Methods

### *2.1. Study Site and Data Collection*

Mt. Sangwang is part of Wando-gun, Jeolla province (34.3° N, 126.6° E) (Figure 1). Mt. Sangwang is located southeast of the Korean Peninsula and belongs to the humid subtropical climate zone (Cfa) according to the Köppen climate classification. Our study site supports one of the largest evergreen broadleaf forests in the Korean Peninsula. Dominant tree species are *Quercus acuta* Thunb., *Castanopsis sieboldii* (Makino) Hatus., and *Quercus glauca* Thunb. Its annual average temperature is 13.6°C and the annual average precipitation was nearly 1310 mm over the study period. It has more than 60% of its precipitation during the summer (monsoon season), and winter and spring suffer from drought (Figure 2). Our site is located 240 m above sea level.

**Figure 1.** Locations of tree phenological monitoring site using phenocam and weather station on Mt. Sangwang. Right picture is the phenocam image with regions of interest (ROI: red polygon).

The phenocam used in this study was positioned to capture the southwest-facing canopy of a *Quercus acuta* forest from the ecological research tower of the National Institute of Forest Science (34°21′32.6″ N, 126°40′22.4″ E). The camera, a Nikon D5600 DSLR, was connected to a modem for remote data transmission, and a plastic housing was used to protect the camera and other electronic equipment from harsh environmental conditions.

To obtain optimal daily green chromatic coordinate (GCC) values for phenological analysis, we selected one image per day throughout the study period from September 2017 to August 2023. Typically, to avoid shadows and minimize light scatter, this was taken between 12:00 and 13:00. Regions of interest (ROI) were defined for the evergreen broadleaf tree, *Q. acuta,* as described by [16,17]. The distance from the camera to the ROI is 8 m. The vegetation index of the GCC was calculated by extracting the relative brightness of the red, green, and blue color channels (RGB), a process which is widely used to identify phenological phase changes [18,19].

$$GCC = \frac{DN_G}{DN_R + DN_G + DN_B}$$

where $DN$ is the digital number and $R$, $G$, and $B$ denote the red, green, and blue channels. The GCC value acts as a proxy for understanding tree physiological processes, including chlorophyll content, photosynthetic activity, and carbon assimilation. These processes were performed using the R package 'phenopix' [20].

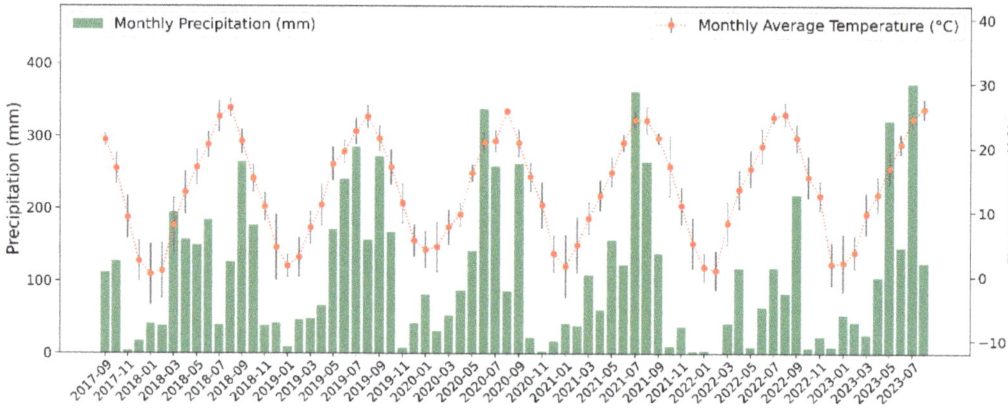


**Figure 2.** Temperature and precipitation over the study period. The vertical bar is the total monthly precipitation and each point is the mean monthly temperature. The vertical grey lines are the standard deviation of the temperature.

The daily meteorological data was provided by the weather station of the Korea Meteorological Administration [21]. Meteorological factors included mean temperature (Tmean), maximum temperature (Tmax), minimum temperature (Tmin), range temperature (Trange), mean humidity (Hmean), maximum humidity (Hmax), minimum humidity (Hmin), precipitation (precip), and solar radiation (radiation) (Table 1). The meteorological observation station was located 4.4 km away from the phenocam observation site. (Figure 1).

**Table 1.** Summary statistics of the daily observed variables during the sampling period from September 2017 to August 2023.

| Variables (Unit) | Mean | Std | Min | Max | Ljung–Box Test | | ADF Test | |
|---|---|---|---|---|---|---|---|---|
| | | | | | Statistics | *p*-Value | Statistics | *p*-Value |
| GCC | 0.38 | 0.022 | 0.33 | 0.46 | 47,040 | 0.000 | −3.80 | 0.003 |
| Tmean (°C) | 13.69 | 8.38 | −8.08 | 28.5 | 48,094 | 0.000 | −2.05 | 0.265 |
| Tmax (°C) | 16.75 | 8.08 | −5.47 | 32.6 | 46,227 | 0.000 | −2.51 | 0.113 |
| Tmin (°C) | 11.05 | 8.82 | −11.4 | 26.4 | 48,873 | 0.000 | −2.54 | 0.105 |

Table 1. Cont.

| Variables (Unit) | Mean | Std | Min | Max | Ljung–Box Test | | ADF Test | |
|---|---|---|---|---|---|---|---|---|
| | | | | | Statistics | $p$-Value | Statistics | $p$-Value |
| Trange (°C) | 5.70 | 1.96 | 0.36 | 14.0 | 5492 | 0.000 | −4.86 | 0.000 |
| Hmean (%) | 75.4 | 14.3 | 28.9 | 100 | 10,533 | 0.000 | −3.83 | 0.003 |
| Hmax (%) | 92.0 | 9.86 | 38 | 100 | 4278 | 0.000 | −4.73 | 0.000 |
| Hmin (%) | 57.1 | 18.2 | 13 | 100 | 10,975 | 0.000 | −3.60 | 0.006 |
| Precipitation (mm) | 3.59 | 13.3 | 0 | 165.5 | 124 | 0.001 | −38.91 | 0.000 |
| Solar radiation (MJ/m$^2$) | 14.67 | 7.28 | 0.91 | 30.6 | 7310 | 0.000 | −4.49 | 0.000 |

### 2.2. Statistical Analysis

To estimate the relative importance of meteorological variables on the greenness of evergreen broad-leaved forests, we applied several time series analysis methods. Before statistical analysis, the gap-filling process was carried out based on the spline interpolation method (cubic spline) due to missing 23% of the GCC values [22]. We chose the cubic spline interpolation method because this method is generally more robust for unequally spaced points compared to many other interpolation methods [23]. Missing values are primarily caused by mechanical failures in data collection or storage. We also standardized all the collected variables (Z-Score normalization) to compare these variables on the same scale. To examine the characteristics of the data, the Ljung–Box test was used to assess autocorrelation, and the ADF test was conducted to evaluate stationarity in our datasets [24].

As time series data have many underlying patterns, it is necessary to split the time series data into trend, season, and residual components using seasonal-trend decomposition using LOESS (STL) [25]. After decomposing the time series data, we extracted the trend component to compare the distance between the GCC index and meteorological variables using the dynamic time warping (DTW) method [26]. By focusing on the underlying trends rather than short-term fluctuations or noise, the trend component distance helps explore how two different time series evolve together over a long-term period [27].

The basic time series models, ARIMA models, can be divided into autoregressive (AR) and moving average (MA) terms, which are defined as ARIMA ($p$, $d$, $q$), where $p$ and $q$ are the orders of AR and MA models, respectively, and $d$ represents the degree of series integration (differencing) [28,29]. The SARIMA model can analyze seasonal lags by incorporating seasonal parameters $(P, D, Q)^S$. Additionally, the SARIMAX model can incorporate exogenous variables to reflect the structural changes of the analyzed process. Its general equation can be defined as

$$Y(t) = c + \varphi 1Y(t-1) + \ldots + \varphi pY(t-p) + \theta 1\epsilon(t-1) + \ldots + \theta q\epsilon(t-q) + \varphi a(Y(t-s) - Y(t-s-m)) + \beta X(t) + \varepsilon(t)$$

where $Y(t)$ is the value of the time series at time $t$, $c$ is the constant term, $\varphi 1$ to $\varphi p$ represent the autoregressive parameters, $\theta 1$ to $\theta q$ represent the moving average parameters, $\beta X(t)$ represents the exogenous variables at time t and $\beta$ is a vector of coefficients for those variables, $\varepsilon(t)$ is the error term at time $t$, $s$ is the seasonal period, and $m$ is the number of seasonal periods.

Our SARIMAX model incorporates meteorological factors as exogenous variables to estimate the effects of environmental variables on the GCC variations. Owing to computation expenses and to minimize the effects of outliers, the weekly mean time series values were used for the SARIMAX model. In this model, the periodicity parameter S was set to 52 weeks (1 year). The procedure for defining the structure of SARIMAX was as follows: (1) Multicollinearity was investigated before entering the exogenous variables into the SARIMAX model. (2) An augmented Dickey–Fuller test (ADF) was performed to verify the stationary and the periodical terms were estimated using the autocorrelation function (ACF) and partial ACF (PACF) diagrams of GCC data. (3) The parameters p, P, q, and Q range from 0 to 4, and the parameters d and D range from 0 to 1 and were constructed to fit the time series of the GCC, and then the lowest Akaike information criterion (AIC) value was selected considering the optimal SARIMAX model. (4) Forecast accuracy was evaluated using the root mean square error (RMSE) and mean absolute percent error (MAPE). We considered the lower values of these statistics to be better for forecasting accuracy. (5) The selected optimal model was diagnosed with the Ljung–Box test for the independence of residuals, and the Jarque–Bera test for the normal distribution of residuals. All time series analyses were performed using the "statsmodels" package in Python 3.6.3. [30].

## 3. Results

### 3.1. Time Series Data Decomposition

The summary statistics for the daily GCC values and meteorological factors, as well as the results of the Ljung–Box test and the ADF test, are described in Table 1. The mean GCC value was $0.38 \pm 0.02$ and was highly autocorrelated over time ($p < 0.001$) according to the results of the Ljung–Box test. The Ljung–Box test also showed significant autocorrelation in all meteorological factors. The ADF test showed that the temperature variables (Tmean, Tmax, Tmin) were statistically non-stationary.

To explore the time series data in more detail, the GCC time series was decomposed into three components: trend, seasonal, and residual (Figure 3). The trend component had the lowest value in early 2022 and gradually increased. The seasonal component showed a distinct seasonal periodicity of the GCC, typically peaking annually in summer.

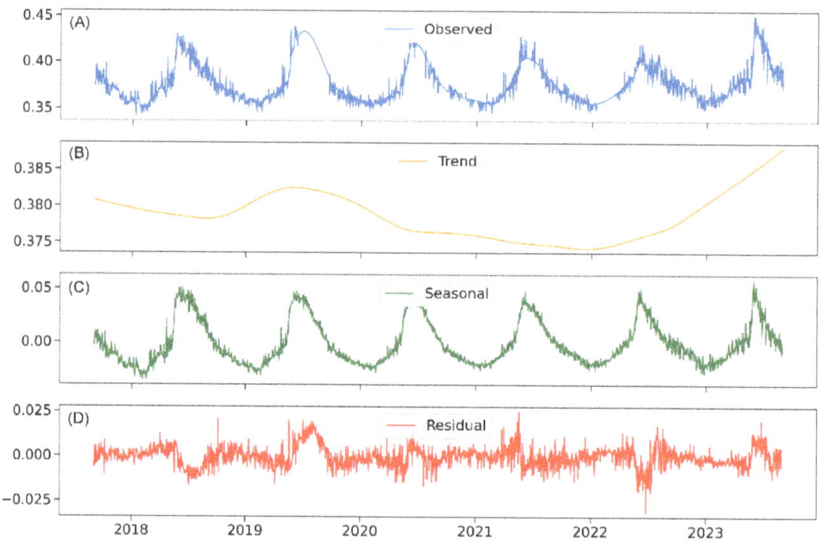


**Figure 3.** Decomposition of GCC time series value. (**A**) presents the observed time series of GCC. (**B–D**) depict the decomposed components: the trend component, seasonal component, and residual, respectively.

This study calculated paired dynamic time warping (DTW) distances to compare the similarity of time series datasets (Figure 4). Although the observed GCC time series had higher similarities with temperature variables, the trend component, with seasonal and random variations removed to reveal the underlying long-term pattern, showed greater similarity to humidity and precipitation variables than temperature (Figure A1).

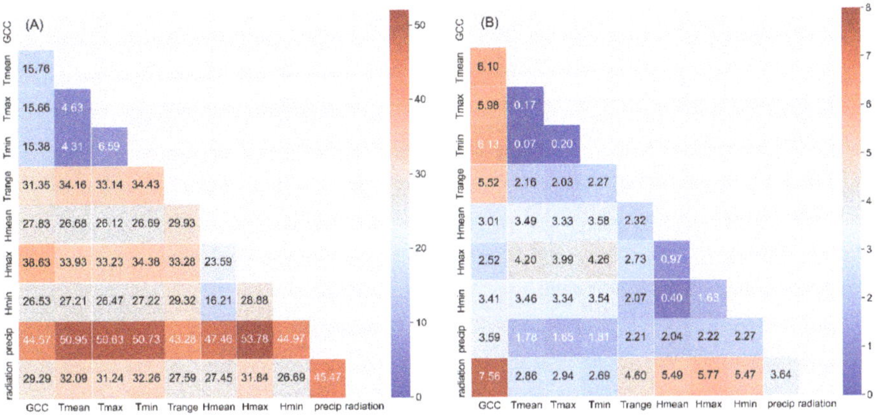

**Figure 4.** Heat map of the paired dynamic time warping (DTW) distance matrix for the observed GCC time series (**A**) and the trend component of the GCC (**B**). A lower DTW distance between the two time series variables indicates a greater similarity in their temporal changes.

### 3.2. SARIMAX

To estimate the relationship between weekly GCC values and meteorological factors, the SARIMAX time series model was developed. The time series GCC data was tested for stationary by an augmented Dickey–Fuller (ADF) test. The test statistics confirmed the stationary after the first difference (−10.86, $p < 0.001$). The best-fitted SARIMAX model developed for the GCC was selected: SARIMA (0, 1, 1) $(0, 1, 1)_{52}$ with four meteorological exogenous variables (AIC = −65.8, BIC = −42.2) (Table 2).

**Table 2.** The best fitted SARIMAX (0, 1, 1) $(0, 1, 1)_{52}$ model parameter's estimate and their significance for the GCC on Mt. Sangwang.

| Model Parameter | Estimate | SE | Prob > \|Z\| | Confidence Interval | |
|---|---|---|---|---|---|
| | | | | 0.025 | 0.975 |
| Tmax | 0.095 | 0.056 | 0.091 | −0.015 | 0.205 |
| Hmax | 0.039 | 0.024 | 0.096 | −0.007 | 0.086 |
| Solar radiation | 0.041 | 0.027 | 0.135 | −0.013 | 0.094 |
| Precipitation | 0.062 | 0.024 | 0.009 | 0.015 | 0.108 |
| Moving average, lag 1 | −0.160 | 0.052 | 0.002 | −0.262 | −0.058 |
| Seasonal moving average, seasonal lag 1 | −0.580 | 0.093 | 0.000 | −0.763 | −0.397 |
| Model variance ($\sigma^2$) | 0.037 | 0.004 | 0.000 | 0.030 | 0.044 |
| AIC = −65.8, BIC = −42.2, Ljung–Box (Prob (Q) = 0.95), Jarque–Bera (Prob (JB) = 0.07) | | | | | |

Our SARIMAX model includes the moving average term (lag 1), especially the seasonal moving average term (52-week moving average), which significantly accounted for seasonal variations of the GCC time series. To identify the importance of meteorological factors, we chose Tmax, Hmax, solar radiation, and precipitation as the optimal exogenous variables considering multicollinearity. Among the meteorological variables, Tmax had the highest positive feature importance, with a weight coefficient of 0.095, but it also had a larger standard error (SE) compared to other exogenous variables. Precipitation was the next important exogenous factor with a weight coefficient of 0.062 and a significant positive confidence interval. However, solar radiation and Hmax were not relevant to the prediction of the GCC within a 95% confidence interval.

Figure 5 displays observed and predicted values for visualization of the performance of our SARIMAX (0, 1, 1)(0, 1, 1)$_{52}$ model. This model was evaluated based on its RMSE and MAPE. When comparing the model performance with and without exogenous factors, the model with exogenous factors shows a slightly better fit to the observed data. The SARIMAX model with exogenous factor has the RMSE value of 0.246, suggesting that the model's predictions are 0.246 units away from the actual values on average. Model performance was validated using an independent test dataset to ensure robustness. The plot demonstrates that the model's predicted values closely align with the observed values, providing evidence of its accuracy.

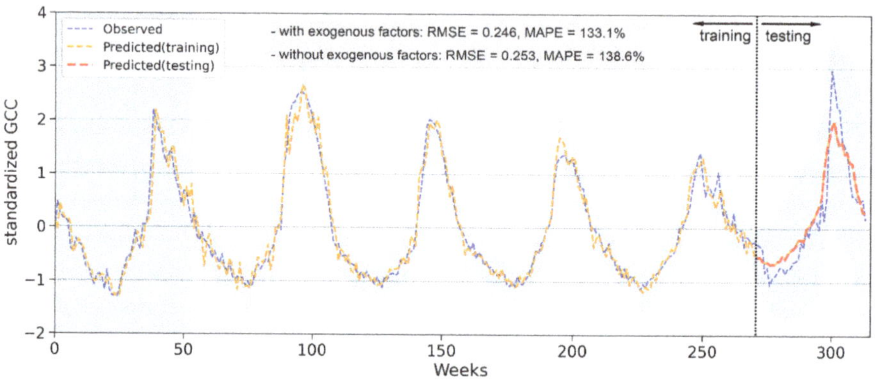


**Figure 5.** Observed data and predicted model for the GCC from September 2017 to August 2023 using SARIMAX model with exogenous factors. The grey shade area represents the 95% confidence interval.

To interpret the results of our model for the GCC, the diagnostic residual plots for the model were generated (Figure A2). Our model showed randomly scattered residuals, indicating that this model correctly captured the underlying patterns in the data. In addition, the normality of the residuals of our model was revealed by a histogram and a Q-Q plot. Finally, the correlogram of our model's residuals shows no significant correlation at any lag, indicating the removal of any autocorrelation from the data. The developed SARIMAX model was tested for independence and normal distribution of the residuals using the Ljung–Box test ($p = 0.95$) and the Jarque–Bera test ($p = 0.07$), which showed the adequacy of the fit of the estimated model.

## 4. Discussion

This study explored the effects of meteorological factors on the GCC value of *Q. acuta* using time series analysis. Daily or weekly changes in the GCC and meteorological variables are recorded as a sequence of data points over an interval time, meaning that past values can influence future ones due to trends, seasonality, or cyclical patterns in the data [27]. These temporal dependencies should be considered in these time series datasets for the validity of the model [31]. Other researchers have emphasized the importance of time series analysis in the ecological field [15]. However, these methods are rarely applied to terrestrial ecology, particularly to ground-based phenocam datasets.

Most of the GCC value was focused on identifying the seasonal key events, such as the start of season (SOS), end of season (EOS), and peak of season (POS). The timing of these events is a vital mechanism for maintaining species coexistence, since plants are the base of the food chain [32]. Furthermore, it can determine how ecosystems respond to the impact of climate change [33], but it takes at least 10 or 20 years to identify the effect of climate on the tree's phenological response because these seasonal patterns occur once a year. Additionally, the change of the GCC for the evergreen tree is vague, so detecting seasonal key events is more difficult in the evergreen tree compared to the deciduous tree.

To overcome these weaknesses, we applied a time series statistical method to identify the effect of meteorological factors on GCC variations of the evergreen tree. Time series statistical methods, such as ARIMA, SARIMA, SARIMAX, etc., are commonly applied to analyze or predict meteorological variables due to their data properties [34]. The satellite remote sensing dataset is also frequently explored by time series methods. For example, the ARIMA model was conducted to predict the vegetation health index and to estimate the relation with meteorological variables [35], and to evaluate the danger of forest fires for past and future climatic periods [36]. However, time series methods have not been applied to ground-based phenocam data. The Ljung–Box test and the ACF plots showed strong autocorrelation in the GCC index and meteorological factors, indicating that ordinary least squares (OLS) regression is not suitable, as it could result in unreliable statistical inferences and spurious relationships. However, numerous previous phenocam studies have conducted data analyses using ordinary least squares regression methods [37–39]. Our SARIMAX model demonstrated excellent predictions of GCC values and good diagnostic results using phenology data from September 2017 to August 2023. This study demonstrates that time series statistical methods are a valuable approach for analyzing ground-based phenocam data.

When comparing the long-term trends of GCC values with meteorological variables, the time series trends of humidity and precipitation showed greater similarity to GCC variations than those of temperature and solar radiation. Specifically, precipitation at our study site from October 2021 to February 2022 was only 55 mm, just 24% of the average rainfall. During the same period, the GCC values of *Q. acuta* were also significantly low. Our SARIMAX model also confirmed the relevance of precipitation for the GCC fluctuation. Precipitation was the only statistically significant exogenous variable, suggesting that variability in precipitation may have a greater impact on the phenological variability of *Q. acuta* than temperature. A previous study also suggested that variation of precipitation is a critical driver in phenological shifts in drylands [40]. Prolonged drought can affect groundwater availability, potentially triggering an unavoidable physiological response in trees or even leading to tree mortality [41]. The results of satellite-derived monitoring also show that the interannual variation in annual rainfall distribution influenced the length of the growing season [42]. Deep soil water can offer a relatively stable water source for trees, but the shallow soil causes water to drain quickly after rainfall at our site [43]. Consequently, prolonged droughts lasting several months can put significant stress on the trees. In particular, evergreen broad-leaved trees may be more sensitive to drought than deciduous broad-leaved trees during winter and early spring, as they continue to lose water through transpiration via their stomata [44,45], even in the drought season on the Korean Peninsula.

While our study provided statistical evidence of the impact of drought on the GCC value of *Q. acuta*, a limitation is that it did not establish a causal relationship between precipitation and GCC variation. To better understand the effects of meteorological variables on the phenological responses of trees, it is crucial to measure physiological indicators such as sap flow, photosynthetic rate, and osmotic potential. Our study site is situated at the boundary between the temperate and warm–temperate climate zones, where evergreen broad-leaved trees must adapt to low temperatures and drought during the winter season. At our site, *Q. acuta* may prioritize accumulating soluble carbohydrates, amino acids, and water-soluble proteins in their cells to regulate osmosis over enhancing photosynthesis. Investigating the combined effects of low temperatures and drought could be an important research topic. Additionally, monitoring soil moisture contents, which is directly influenced by precipitation, is crucial for estimating causal relationships.

## 5. Conclusions

In order to clarify the influence of meteorological variables on the phenological variation in the evergreen broad-leaved forests, we used ground-based digital photographic images and analyzed the collected data by means of time series methods. The results showed strong autocorrelation and seasonality in the GCC value collected at daily intervals. This means that unreliable statistical inferences and spurious relationships can be made using the ordinary least squares (OLS) regression method. It is crucial to carefully examine the characteristics of the data and to use appropriate statistical models to satisfy basic statistical assumptions. We recommend using time series analysis on phenocam data sets that are highly autocorrelated and seasonal. Our time series analysis highlighted the influence of dry season precipitation on the GCC variation of evergreen broad-leaved trees on the Korean Peninsula. Our results help to understand how plant phenology responds to climate change and improve the accuracy of phenological modeling.

**Author Contributions:** Conceptualization, J.P. and H.L.; methodology, J.P.; software, M.H. and J.P.; validation, J.P. and M.H.; formal analysis, J.P.; investigation, J.P.; resources, M.H.; data curation, J.P. and M.H.; writing—original draft preparation, J.P. and M.H.; writing—review and editing, J.P.; visualization, J.P.; supervision, J.P. and H.L.; project administration, J.P. and H.L.; funding acquisition, J.P. and H.L. All authors have read and agreed to the published version of the manuscript.

**Funding:** This research was funded by the National Institute of Ecology (grant no. No. NIE-B-2024-02).

**Data Availability Statement:** The data presented in this study are available on request from the corresponding author.

**Conflicts of Interest:** The authors declare no conflicts of interest.

## Appendix A

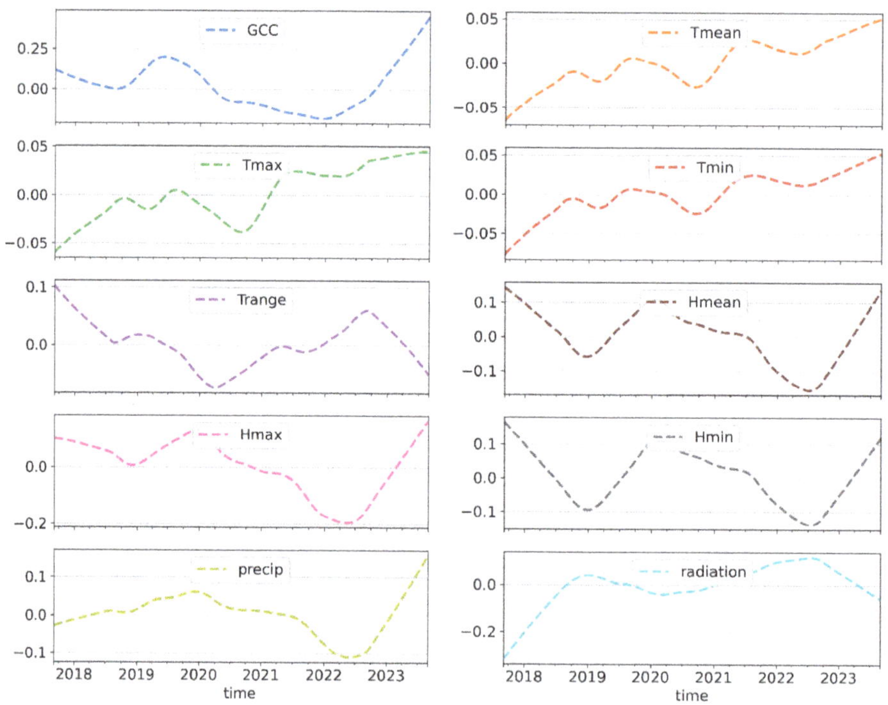


**Figure A1.** The variability of the trends in the GCC and nine meteorological factors. The trend components have been estimated after decomposing the individual time series data using the seasonal-trend decomposition with LOESS (STL).

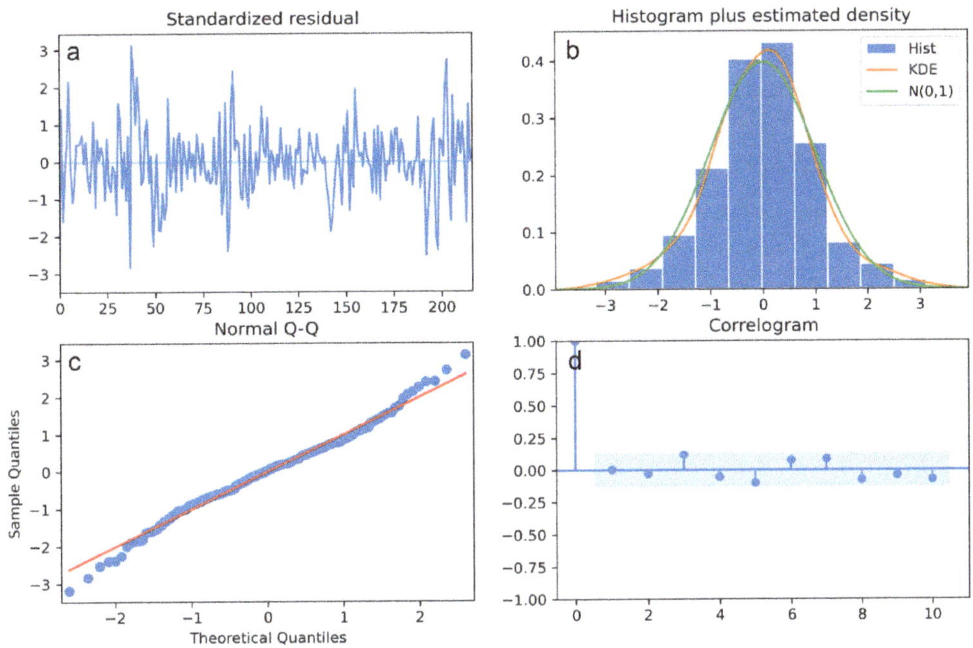


**Figure A2.** The diagnostic plot for the residuals of our model. The residuals are the difference between the predictions of the model one step ahead and the actual values of the time series. (**a**) The standardized residual. (**b**) Residual distribution, the orange line (KDE) shows a smoothed version of this histogram and the green line shows a normal distribution. (**c**) The Q-Q plot compares the distribution of the residuals with the normal distribution. (**d**) The correlogram plot is the autocorrelation function of the residuals. A total of 95% of the correlations should be insignificant for lags greater than zero (within the blue shading).

## References


1. Morisette, J.T.; Richardson, A.D.; Knapp, A.K.; Fisher, J.I.; Graham, E.A.; Abatzoglou, J.; Wilson, B.E.; Breshears, D.D.; Henebry, G.M.; Hanes, J.M.; et al. Tracking the rhythm of the seasons in the face of global change: Phenological research in the 21st century. *Front. Ecol. Environ.* **2009**, *7*, 253–260. [CrossRef]
2. Beard, K.H.; Kelsey, K.C.; Leffler, A.J.; Welker, J.M. The Missing Angle: Ecosystem Consequences of Phenological Mismatch. *Trends Ecol. Evol.* **2019**, *34*, 885–888. [CrossRef]
3. Fitter, A.H.; Fitter, R.S.R.; Harris, I.T.B.; Williamson, M.H. Relationships between first flowering date and temperature in the flora of a locality in central England. *Funct. Ecol.* **1995**, *1*, 55–60. [CrossRef]
4. Sparks, T.H.; Jeffree, E.P.; Jeffree, C.E. An examination of the relationship between flowering times and temperature at the national scale using long-term phenological records from the UK. *Int. J. Biometeorol.* **2000**, *44*, 82–87. [CrossRef]
5. Menzel, A.; Sparks, T.H.; Estrella, N.; Koch, E.; Aasa, A.; Ahas, R.; Alm-Kübler, K.; Bissolli, P.; Braslavská, O.G.; Briede, A.; et al. European phenological response to climate change matches the warming pattern. *Glob. Change Biol.* **2006**, *12*, 969–1976. [CrossRef]
6. Luscier, J.D.; Thompson, W.L.; Wilson, J.M.; Gorham, B.E.; Dragut, L.D. Using digital photographs and object-based image analysis to estimate percent ground cover in vegetation plots. *Front. Ecol. Environ.* **2006**, *4*, 408–413. [CrossRef]
7. Browning, D.; Karl, J.; Morin, D.; Richardson, A.; Tweedie, C. Phenocams Bridge the Gap between Field and Satellite Observations in an Arid Grassland Ecosystem. *Remote Sens.* **2017**, *9*, 1071. [CrossRef]
8. Klosterman, S.T.; Hufkens, K.; Gray, J.M.; Melaas, E.; Sonnentag, O.; Lavine, I.; Mitchell, L.; Norman, R.; Friedl, M.A.; Richardson, A.D. Evaluating remote sensing of deciduous forest phenology at multiple spatial scales using PhenoCam imagery. *Biogeosciences* **2014**, *11*, 4305. [CrossRef]
9. Crimmins, M.A.; Crimmins, T.M. Monitoring Plant Phenology Using Digital Repeat Photography. *Environ. Manag.* **2008**, *41*, 949–958. [CrossRef] [PubMed]

10. Raman, R.K.; Mohanty, S.K.; Bhatta, K.S.; Karna, S.K.; Sahoo, A.K.; Mohanty, B.P.; Das, B.K. Time series forecasting model for fisheries in Chilika lagoon (a Ramsar site, 1981), Odisha, India: A case study. *Wetlands Ecol. Manag.* **2018**, *26*, 677–687. [CrossRef]
11. Bai, L.; Lu, K.; Dong, Y.; Wang, X.; Gong, Y.; Xia, Y.; Wang, X.; Chen, L.; Yan, S.; Tang, Z.; et al. Predicting monthly hospital outpatient visits based on meteorological environmental factors using the ARIMA model. *Sci. Rep.* **2023**, *13*, 2691. [CrossRef]
12. He, J.; Wei, X.; Yin, W.; Wang, Y.; Qian, Q.; Sun, H.; Xu, Y.; Magalhaes, R.J.S.; Guo, Y.; Zhang, W. Forecasting Scrub Typhus Cases in Eight High-Risk Counties in China: Evaluation of Time-Series Model Performance. *Front. Environ. Sci.* **2022**, *9*, 783864. [CrossRef]
13. Mulla, S.; Pande, C.B.; Singh, S.K. Times Series Forecasting of Monthly Rainfall using Seasonal Auto Regressive Integrated Moving Average with EXogenous Variables (SARIMAX) Model. *Water Resour. Manag.* **2024**, *38*, 1825–1846. [CrossRef]
14. Guo, Y.; Lai, X.; Gan, M. Cyanobacterial biomass prediction in a shallow lake using the time series SARIMAX models. *Ecol. Inf.* **2023**, *78*, 102292. [CrossRef]
15. Boero, F.; Kraberg, A.C.; Krause, G.; Wiltshire, K.H. Time is an affliction: Why ecology cannot be as predictive as physics and why it needs time series. *J. Sea Res.* **2015**, *101*, 12–18. [CrossRef]
16. Richardson, A.D.; Braswell, B.H.; Hollinger, D.Y.; Jenkins, J.P.; Ollinger, S.V. Near-surface remote sensing of spatial and temporal variation in canopy phenology. *Ecol. Appl.* **2009**, *19*, 1417–1428. [CrossRef] [PubMed]
17. Alberton, B.; Almeida, J.; Helm, R.; Torres, R.D.S.; Menzel, A.; Morellato, L.P.C. Using phenological cameras to track the green up in a cerrado savanna and its on-the-ground validation. *Ecol. Inf.* **2014**, *19*, 62–70. [CrossRef]
18. Richardson, A.D.; Jenkins, J.P.; Braswell, B.H.; Hollinger, D.Y.; Ollinger, S.V.; Smith, M.L. Use of digital webcam images to track spring green-up in a deciduous broadleaf forest. *Oecologia* **2007**, *152*, 323–334. [CrossRef] [PubMed]
19. Toomey, M.; Friedl, M.A.; Frolking, S.; Hufkens, K.; Klosterman, S.; Sonnentag, O.; Baldocchi, D.D.; Bernacchi, C.J.; Biraud, S.C.; Bohrer, G.; et al. Greenness indices from digital cameras predict the timing and seasonal dynamics of canopy-scale photosynthesis. *Ecol. Appl.* **2015**, *25*, 99–115. [CrossRef]
20. Filippa, G.; Cremonese, E.; Migliavacca, M.; Galvagno, M.; Forkel, M.; Wingate, L.; Tomelleri, E.; di Cella, U.M.; Richardson, A.D. Phenopix: AR package for image-based vegetation phenology. *Agric. For. Meteorol.* **2016**, *220*, 141–150. [CrossRef]
21. Korea Meteorological Administration (KMA). Available online: https://www.weather.go.kr (accessed on 2 August 2024).
22. De Boor, C. *A Practical Guide to Splines*, revised ed.; Springer: Berlin/Heidelberg, Germany, 2001.
23. Kvasov, B.I. *Methods of Shape-Preserving Spline Approximation*; World Scientific: Singapore, 2000.
24. Chatfield, C.; Xing, H. *The Analysis of Time Series: An Introduction with, R*; Chapman and hall/CRC: Boca Raton, FL, USA, 2019.
25. Cleveland, R.B.; Cleveland, W.S.; McRae, J.E.; Terpenning, I. STL: A seasonal-trend decomposition. *J. Off. Stat.* **1990**, *6*, 3–73.
26. Bankó, Z.; Abonyi, J. Correlation based dynamic time warping of multivariate time series. *Expert Syst. Appl.* **2012**, *17*, 12814–12823. [CrossRef]
27. Brockwell, P.J.; Davis, R.A. (Eds.) *Introduction to Time Series and Forecasting*; Springer: New York, NY, USA, 2002.
28. Yin, R. Forecasting short-term timber prices with univariate ARIMA models. *South. J. Appl. For.* **1999**, *23*, 53–58. [CrossRef]
29. Banaś, J.; Utnik-Banaś, K. Evaluating a seasonal autoregressive moving average model with an exogenous variable for short-term timber price forecasting. *For. Policy Econ.* **2021**, *131*, 102564. [CrossRef]
30. Seabold, S.; Perktold, J. Statsmodels: Econometric and Statistical Modeling with Python. *SciPy* **2010**, *7*, 92–96.
31. Mo, H.H.; Cho, K.; Shin, K.I. A Review of Time Series Analysis for Environmental and Ecological Data. *Korean J. Environ. Biol.* **2016**, *34*, 365–373. [CrossRef]
32. Jorge, C.; Silva, J.M.N.; Boavida-Portugal, J.; Soares, C.; Cerasoli, S. Using Digital Photography to Track Understory Phenology in Mediterranean Cork Oak Woodlands. *Remote Sens.* **2021**, *13*, 776. [CrossRef]
33. Kramer, K.; Leinonen, I.; Loustau, D. The importance of phenology for the evaluation of impact of climate change on growth of boreal, temperate and Mediterranean forests ecosystems: An overview. *Int. J. Biometeorol.* **2000**, *44*, 67–75. [CrossRef] [PubMed]
34. Kaur, J.; Parmar, K.S.; Singh, S. Autoregressive models in environmental forecasting time series: A theoretical and application review. *Environ. Sci. Pollut. Res.* **2023**, *30*, 19617–19641. [CrossRef]
35. Vilanova, R.S.; Delgado, R.C.; da Silva Abel, E.L.; Teodoro, P.E.; Silva Junior, C.A.; Wanderley, H.S.; Capristo-Silva, G.F. Past and future assessment of vegetation activity for the state of Amazonas-Brazil. *Remote Sens. Appl. Soc. Environ.* **2020**, *17*, 100278. [CrossRef]
36. Jesus, C.S.L.D.; Delgado, R.C.; Wanderley, H.S.; Teodoro, P.E.; Pereira, M.G.; Lima, M.; Silva Junior, C.A.D. Fire risk associated with landscape changes, climatic events and remote sensing in the Atlantic Forest using ARIMA model. *Remote Sens. Appl. Soc. Environ.* **2022**, *26*, 100761. [CrossRef]
37. Yu, L.; Liu, T.; Bu, K.; Yan, F.; Yang, J.; Chang, L.; Zhang, S. Monitoring the long term vegetation phenology change in Northeast China from 1982 to 2015. *Sci. Rep.* **2017**, *7*, 14770. [CrossRef] [PubMed]
38. Xie, Y.; Civco, D.L.; Silander, J.A. Species-specific spring and autumn leaf phenology captured by time-lapse digital cameras. *Ecosphere* **2018**, *9*, e02089. [CrossRef]
39. Chen, X.; Zhang, Y. The impact of vegetation phenology changes on the relationship between climate and net primary productivity in Yunnan, China, under global warming. *Front. Plant Sci.* **2023**, *14*, 1248482. [CrossRef] [PubMed]
40. Currier, C.M.; Sala, O.E. Precipitation versus temperature as phenology controls in drylands. *Ecology* **2022**, *103*, e3793. [CrossRef]
41. Piayda, A.; Dubbert, M.; Rebmann, C.; Kolle, O.; e Silva, F.C.; Correia, A.; Pereira, J.S.; Werner, C.; Cuntz, M. Drought impact on carbon and water cycling in a Mediterranean Quercus suber L. woodland during the extreme drought event in 2012. *Biogeosciences* **2014**, *11*, 7159–7178. [CrossRef]

42. Medeiros, R.; Andrade, J.; Ramos, D.; Moura, M.; Pérez-Marin, A.M.; dos Santos, C.A.; da Silva, B.B.; Cunha, J. Remote Sensing Phenology of the Brazilian Caatinga and Its Environmental Drivers. *Remote Sens.* **2022**, *14*, 2637. [CrossRef]
43. Korean Soil Information System (KSIS). Available online: https://soil.rda.go.kr/ (accessed on 20 September 2024).
44. Chapin, F.S., III; Matson, P.A.; Vitousek, P.M. *Principles of Terrestrial Ecosystem Ecology*; Springer Science & Business Media: Berlin/Heidelberg, Germany, 2011.
45. De Souza, B.C.; Carvalho, E.C.D.; Oliveira, R.S.; de Araujo, F.S.; de Lima, A.L.A.; Rodal, M.J.N. Drought response strategies of deciduous and evergreen woody species in a seasonally dry neotropical forest. *Oecologia* **2020**, *194*, 221–236. [CrossRef] [PubMed]

*forests*

MDPI

*Article*

# Characteristics of Drought Events and Their Impact on Vegetation Dynamics in the Arid Region of Northwest China

**Guixiang Zhou [1,2], Junqiang Yao [1,3,4,*], Jing Chen [1,3,4], Yaning Chen [5], Chuan Wang [5,6], Yinxue Mo [1,7], Chenzhi Ma [1,7], Yuhui Yang [2], Moyan Li [2] and Peng Zheng [2]**

1 Institute of Desert Meteorology, China Meteorological Administration, Urumqi 830002, China; yangq@idm.cn (G.Z.); chenjing@idm.cn (J.C.); moyinxue@stu.xju.edu.cn (Y.M.); machenzhi@stu.xju.edu.cn (C.M.)

2 College of Geography and Tourism, Xinjiang Normal University, Urumqi 830054, China; yangyuhui@xjnu.edu.cn (Y.Y.); limy525@xjnu.edu.cn (M.L.); zhengpeng@stu.xjnu.edu.cn (P.Z.)

3 Key Laboratory of Tree-Ring Physical and Chemical Research, China Meteorological Administration, Urumqi 830002, China

4 Field Scientific Experiment Base of Akdala Atmospheric Background, China Meteorological Administration, Akdala 836499, China

5 State Key Laboratory of Desert and Oasis Ecology, Key Laboratory of Ecological Safety and Sustainable Development in Arid Lands, Xinjiang Institute of Ecology and Geography, Chinese Academy of Sciences, Urumqi 830011, China; chenyn@ms.xjb.ac.cn (Y.C.); wangchuan201@mails.ucas.ac.cn (C.W.)

6 University of Chinese Academy of Sciences, Beijing 100049, China

7 College of Ecology and Environment, Xinjiang University, Urumqi 830046, China

* Correspondence: yaojq@idm.cn; Tel.: +86-0991-2641466

**Abstract:** The arid region of Northwest China (ARNC) is responsive to global climate change, and drought events have occurred frequently in recent decades. However, studies about the effect of meteorological and drought stress on vegetation change in the ARNC are still insufficient. In this study, we analyzed the spatiotemporal trends of meteorological factors (temperature, TMP; precipitation, PRE; standardized precipitation evapotranspiration index, SPEI), drought stress factors (vapor pressure deficit, VPD; soil moisture, SM), and vegetation (normalized difference vegetation index, NDVI) during 1982–2021. We also investigated the characteristics of drought events by the run theory, including drought times, drought duration, drought severity, and drought intensity. The impacts of meteorological and drought stress factors on the vegetation were explored using Pearson correlation analysis and the structural equation model (SEM). We found that the annual and growing season TMP, PRE, VPD, SM, and NDVI showed an increasing trend in the ARNC during 1982–2021. In contrast, SPEI exhibited a decreasing trend in the annual and growing season. In addition, the characteristics of the drought events varied significantly in the ARNC. The drought events primarily occurred in the Tarim River Basin, Turpan-Hami Basin, and the Hexi Corridor. The Pearson correlation analysis and SEM results consistently demonstrated that TMP and SM exerted greater impacts on vegetation growth than PRE, VPD, and SPEI. The factors that determine vegetation change were TMP and PRE. Exploration of meteorological and drought stress factors that influence vegetation change is essential for comprehending the influence of dominant factors on vegetation change.

**Keywords:** arid region; drought stress; NDVI; SPEI; vegetation change; VPD

## 1. Introduction

Drylands are among the regions most sensitive to global warming [1]. The global demand for atmospheric evaporation also increases and drought events occur frequently, increasing the risk of future climate change [2–4]. Drought events occur frequently and have significant influence on economies, societies, and ecosystems [5–8].

Drought is generally considered a complex phenomenon due to insufficient precipitation or excessive evapotranspiration, and it can be divided into four types: meteorological, agricultural, hydrological, and socio-economic drought [7–12]. Meteorological droughts are precipitation deficits that occur over time [13,14], and usually occur on a temporal scale of 1 to 3 months. Both the ecological carbon and water cycles are significantly affected by drought [15]. Drought can weaken ecosystem functions by reducing vegetation productivity, eventually leading to widespread vegetation death, or an increase in plant diseases and insect pests, indirectly leading to vegetation death [16]. The hydraulic imbalance hypothesis considers that drought affects vegetation root water absorption, resulting in xylem embolism due to a soil water deficit and eventually causing vegetation death [17]. One study found that continuous climate warming in the future could reduce vegetation productivity [18]. The trend of green vegetation is mainly caused by the fertilization effect of carbon dioxide [19].

Soil moisture (SM) and the vapor pressure deficit (VPD) are the main constraints to drought in ecosystems; they are coupled with land–atmosphere interactions that may limit vegetation change and carbon uptake in terrestrial ecosystems [3,20,21]. The effects of SM and VPD on vegetation change remain insufficient because of the complexity of land–atmosphere interactions and biological processes [3,22–24]. A high VPD can limit photosynthesis and reduce change in vegetation [25,26]. A recent related study found that SM is regarded as a determining factor in vegetation change [22].

Vegetation has an indispensable role in regulating the surface material cycle and energy conversion [8,23,27–29]. In the arid region of Northwest China (ARNC), vegetation change is driven by precipitation and temperature rather than human disturbance [30]. The normalized difference vegetation index (NDVI) is a dynamic indicator that can better describe the characteristics of vegetation change [31,32]. The global inventory modeling and mapping studies NDVI3g (GIMMS NDVI3g) is currently the longest available NDVI dataset [33–35]. A research team from Peking University proposed a new machine learning model to generate the latest version of the NDVI product (PKU GIMMS NDVI). The model employed the global Landsat NDVI sample and data integration methodology, utilizing both advanced very high resolution radiometer (AVHRR) data and moderate resolution imaging spectroradiomter (MODIS) data as the base data [36].

The palmer drought severity index (PDSI), standardized precipitation index (SPI), and standardized precipitation evapotranspiration index (SPEI) are indicators used to analyze a drought event [5,37]. The SPEI is a common indicator used to monitor vegetation change [6,38,39]. The SPEI has better advantages than the PDSI and SPI for assessing drought conditions [5,40–42]. The SPEI combines the multi-scala of the SPI and the sensitivity of the PDSI to reveal the evaporative demand changes [13,42].

The ARNC is highly susceptible to global warming [43]. The "warming and humidification" phenomenon was significant in the ARNC since 1980s, with the rise in temperature and precipitation [43]. At the same time, the drought frequency has increased, having serious impacts on the vegetation ecology [42]. Nevertheless, the VPD is a limiting factor of vegetation greening [26]. Based on hydraulic imbalance, the VPD would be the main physical constraint. However, studies about the effect of meteorological and drought stress on vegetation change in the ARNC are still insufficient so far, and the dominant factors that affect vegetation change remain controversial. Therefore, this study investigated the spatiotemporal characteristics of drought events in the ARNC during 1982–2021 and explored the effects of meteorological and drought stress factors on vegetation change.

## 2. Materials and Methods

### 2.1. Study Area

The ARNC is geographically located on the Asian–European continent, between 73° E–107° E and 34° N–50° N, covering an area of $2.50 \times 10^7$ km$^2$. It encompasses the entirety of Xinjiang, the majority of Gansu Province, and portions of the Inner Mongolia Autonomous Region and Ningxia Hui Autonomous Region (Figure 1) [43]. It has mountain ranges such as the Tianshan Mountains, Kunlun Mountains, Altai Mountains, Qilian Mountains, and the Pamir Plateau, as well as the Tarim Basin, Jungar Basin, Turpan-Hami Basin, and Hexi Corridor. Here, a unique temperate continental climate formed based on the complex mountain–basin topography, which is characterized by large temperature variations, low precipitation, and intense evaporation. The land use types include crop land, forest, grass land, water, residential land, deserts such as the Gobi, etc. Furthermore, the region is mainly characterized by sparse vegetation, scarce water resources, and an extremely fragile ecological environment [42,43].

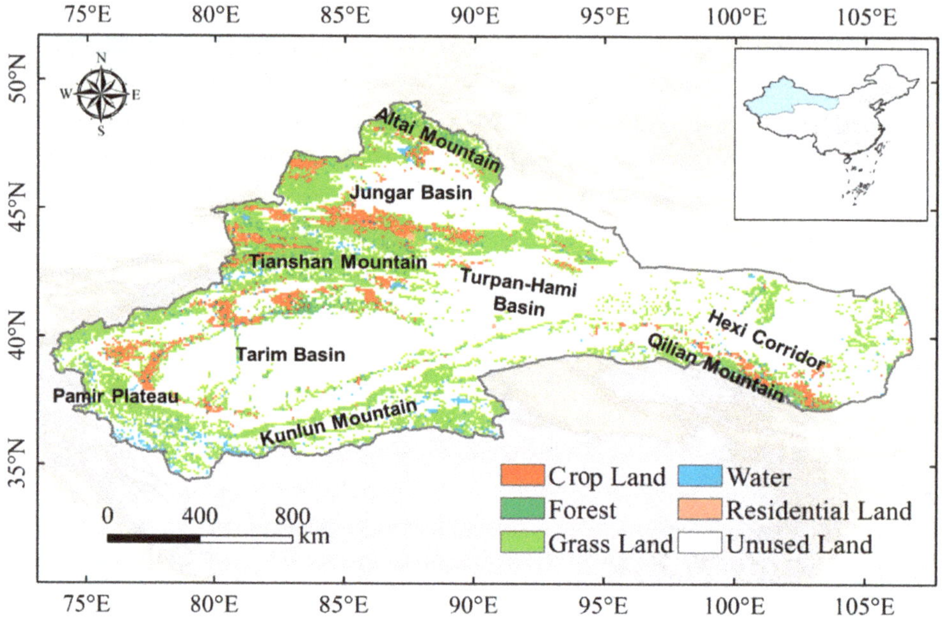


**Figure 1.** Geographical location of study area. The land use type data were derived from the Resource and Environment Data Cloud Platform (http://www.resdc.cn/; accessed on 3 January 2024).

## 2.2. Data Sources

The temperature (TMP), precipitation (PRE), and actual vapor pressure (AVP) data were obtained from the Climate Research Unit (CRU) time-series (TS) data version 4.06 (https://crudata.uea.ac.uk/cru/data/hrg/; accessed on 3 September 2024). The CRU TS4.06 dataset was derived by the angular distance weighting interpolation of monthly climate anomalies from extensive networks of weather station observations [44]. The SPEI is used popularly to evaluate drought severity under warming conditions. It is calculated using the monthly PRE and potential evapotranspiration (PET) data from the CRU data [38]. The FAO-56 Penman–Monteith method has been used for computing PET. The SPEI data were derived from the global standardized precipitation evapotranspiration index database (SPEIbase v2.10) (https://spei.csic.es/database.html; accessed on 21 January 2024), and the temporal and spatial resolution are monthly and $0.5^\circ \times 0.5^\circ$, respectively. The surface SM dataset was acquired from the global land evaporation Amsterdam model (GLEAM, version 3.6) (https://www.gleam.eu/; accessed on 24 December 2023) [45].

The NDVI was obtained from the PKU GIMMS NDVI dataset (version 1.2), which was calculated using the AVHRR and MODIS NDVI data (https://zenodo.org/records/8253971; accessed on 8 December 2023). The temporal and spatial resolution are half month and $1/12^\circ$, respectively [36]. The annual and seasonal NDVI were calculated from the monthly maximum NDVI dataset. The NDVI value is less than 0.1 in the region without vegetation cover [34].

The bilinear interpolation was employed to resample the TMP, PRE, VPD, SM, and SPEI so that their spatial resolution was consistent with the NDVI, based on the MATLAB R2021b software [28]. The land use dataset for 2020 for China was derived from the Resources and Environment Science and Data Center (https://www.resdc.cn/; accessed on 3 January 2024). We analyzed vegetation changes in the crop land, forest, and grassland from China's multi-period land use land cover remote sensing monitoring dataset (http://www.resdc.cn/; accessed on 3 January 2024).

## 2.3. Methodologies

### 2.3.1. Characteristics of Drought Events

Drought is a multi-scale phenomenon that varies markedly in its impact on ecosystems and can have different response times. Therefore, drought index must be associated with a multi-timescale for monitoring drought events. The SPEI is an effective tool to monitor drought events at multi-timescale under warming conditions. In general, the 6-month SPEI can be used to describe agro-ecological drought [46]. We used the six-month scale SPEI data to identify drought events that were highly correlated with vegetation changes in arid regions [3,42]. The SPEI data were obtained from the Consejo Superior de Investigaciones Científicas (CSIC) (https://digital.csic.es/handle/10261/202305; accessed on 3 September 2024). The SPEI is classified into four categories, and the SPEI setting thresholds are less than to $-0.5$, $-1.0$, $-1.5$, and $-2.0$, which represent mild, moderate, severe, and extreme droughts, respectively [45]. Run theory was employed to define the drought events that affect vegetation change. Run theory is broadly utilized to recognize drought events. We extracted drought events using run theory and set the threshold to $-1$ (SPEI $< -1$), which was defined as a drought event if it was below this threshold and lasted longer than two months [47]. The characteristics of drought events include the drought times (DT), drought duration (DD), drought severity (DS), and drought intensity (DI). The DT is the sum of the number of drought events. DD is the number of months from the beginning to the end of the drought. DS is the absolute value of the cumulative SPEI, and the DI is the absolute value of the average SPEI during the drought event [47]. The annual mean and growing season (from April to October) mean values were obtained for drought event characteristics. The annual characteristics of drought events were identified for the period 1980–2021. The above analysis was carried out in MATLAB R2021b software.

### 2.3.2. Characterization of VPD and SM

Drought stress can restrict vegetation growth, with the indicators typically being VPD and SM. Severe drought stress is often correlated with a reduced SM and increased VPD [21,24,48]. The surface SM is directly represented by the soil moisture data of the GLEAM products [3]. The VPD is calculated as the saturated vapor pressure (SVP) minus the AVP, which is used to indicate atmospheric aridity [24]. We calculated the SVP and VPD from the TMP and AVP data in the CRU 4.06 datasets. SVP and VPD were calculated using Equations (1) and (2) [49,50]:

$$SVP = 6.112 \times exp\left(\frac{17.67 \times T_{as}}{T_{as} + 243.5}\right) \tag{1}$$

$$VPD = SVP - AVP \tag{2}$$

In this formula, SVP represents saturation vapor pressure (kPa), AVP is the actual vapor pressure (kPa), Tas represents surface temperature (°C), and VPD represents vapor pressure deficit (kPa).

2.3.3. Trend Analysis

We employed Theil–Sen analysis and Mann–Kendall (MK) method to test the long-time series trends [51,52]. The non-parametric statistical methods can be effectively used for trend testing, such as Theil–Sen slope analysis [51]. Similarly, the MK test is a non-parametric statistic where the sample is not constrained by a particular distribution pattern and is not disturbed by outliers [52,53]. It is widely used to study an element variations over long periods [54]. The above analysis was carried out in MATLAB R2021b software.

2.3.4. Pearson Correlation Analysis and Structural Equation Model

The influences of meteorological and drought stress factors on vegetation growth were explored using Pearson correlation analysis and structural equation model (SEM). We also employed Pearson correlation analysis to reveal the effect of drought on vegetation change [40]. SEM is an analytical technique that enables the quantification of the dynamic interrelationships between multiple variables. It can be employed to elucidate both direct and indirect effects between related variables [55]. Thus, the SEM was employed to assess the effect of meteorological and drought stress factors on vegetation change. All variables were standardized prior to SEM analysis. The fit indices included a small $\chi^2$ value, a $p$-value greater than 0.05, a goodness of fit index (GFI) value greater than 0.90, and a root mean square error of approximation (RMSEA) less than 0.08 [56]. The SEM data modeling was carried out in SPSS Amos 24.0 (https://spssanalysis.com/sem-in-spss-amos/; accessed on 3 December 2023).

## 3. Results

### 3.1. Spatiotemporal Changes in Drought Events in the ARNC

3.1.1. Spatial Trends of the Drought Events

In general, the spatial characteristics of annual and growing season drought events differed significantly in the ARNC (Figure 2). However, their spatial characteristics also exhibited some similarity, with the spatial distribution of DD and DI being more similar.

The annual DT was 0.5 times/year, and it was greater than 0.5 times/year in the Tarim Basin, Turpan-Hami Basin, and west of the Alxa Plateau (Figure 2a). The annual DD was 2.5 mon/year. The Tarim Basin and the Hexi Corridor were completed over a period of more than 3.2 mon/year (Figure 2c). The annual DS was 4.0. The DS in the Tarim Basin and the Hexi Corridor was more than 5.0 (Figure 2e). The annual DI was 0.7, and it was more than 0.8 in the western Tianshan, northern Tarim Basin, and the Qilian Mountains (Figure 2g).

The growing season DT was 0.3 times/year and it was greater than 0.3 times/year in the Tarim Basin, Turpan-Hami Basin, and west of the Alxa Plateau (Figure 2b). The growing season DD was 1.6 mon/year. The Tarim Basin and north Hexi corridor lasted for more than 2.1 mon/year (Figure 2d). The growing season DS was 2.5. The DS in the Tarim Basin and the northern Hexi Corridor was more than 3.4 (Figure 2f). The growing season DI was 0.5; it was more than 0.5 in the western Tianshan, northern Tarim Basin, and the Qilian Mountains (Figure 2h).

3.1.2. Change Trends of the Drought Events

The annual DT of moderate drought changed rapidly, followed by that of mild, severe, and extreme droughts (Figure 3a). The change in annual DD was large, and the annual DD of mild, moderate, and severe drought showed significantly increasing trends, with rates of 0.09 mon/year, 0.08 mon/year, and 0.06 mon/year, respectively (Figure 3b). The variation in annual DS was large, and the annual DS of mild, moderate, and severe droughts showed significant increasing trends, with rates of 0.14/year, 0.14/year, and 0.11/year, respectively. (Figure 3c). The annual DI of moderate, severe, and extreme droughts changed negatively, and only mild droughts showed a slowly increasing trend (Figure 3d).

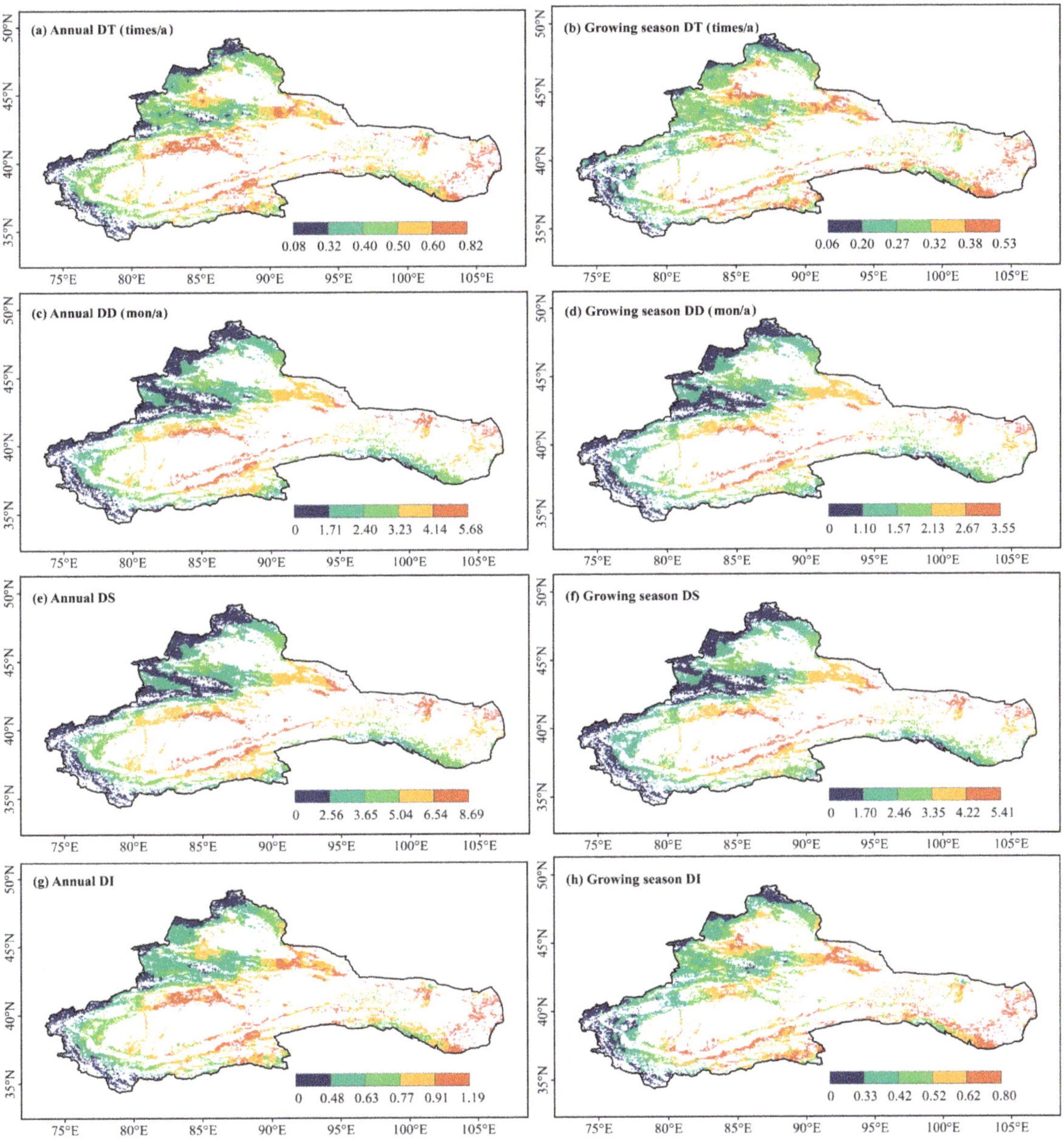


**Figure 2.** Spatial distribution of the annual (**left**) and growing season (**right**) drought events in the ARNC during 1982–2021. The following characteristics of drought events are included: drought times (DT), drought duration (DD), drought severity (DS), and drought intensity (DI).

The growing season characteristics of drought events during 1982–2021 in the ARNC were identified. The growing season DT also changed rapidly. The growing season DT of mild, moderate, and severe droughts showed significant increasing trends, with rates of 0.012 times/year, 0.011 times/year, and 0.008 times/year, respectively (Figure 4a). The growing season DD of mild, moderate, and severe droughts showed significant increasing trends, with rates of 0.058 mon/year, 0.060 mon/year, and 0.039 mon/year, respectively (Figure 4b). The variation in growing season DS was large, and the rates of mild, moderate, and severe droughts in DS showed significant increasing trends of 0.11/year, 0.109/year and 0.073/year, respectively. (Figure 4c). The growing season DI of mild droughts showed increasing trends, while that of severe and extreme droughts changed negatively (Figure 4d).

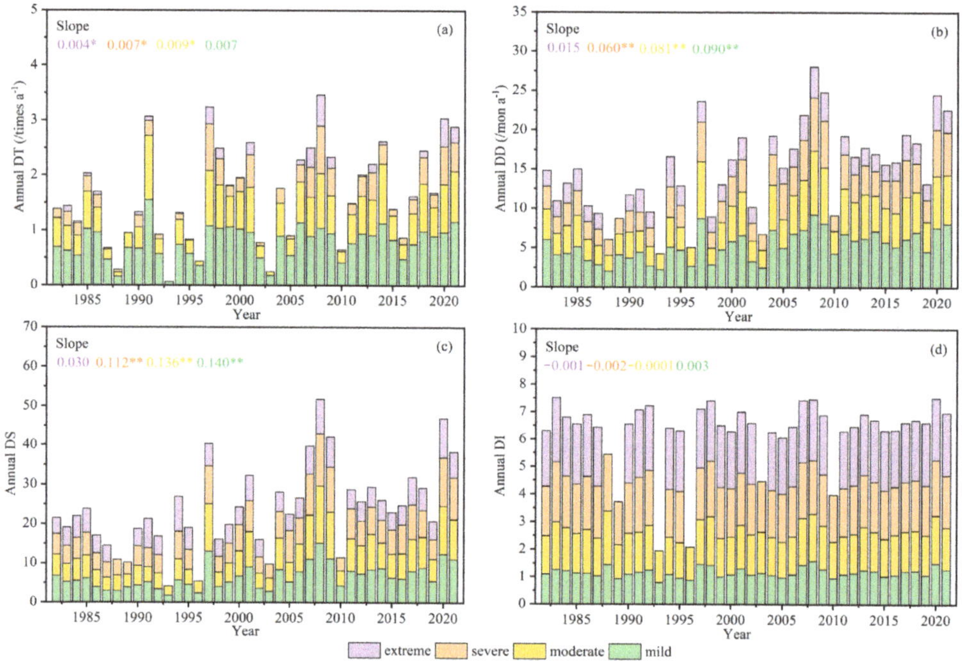


**Figure 3.** Change trends of the characteristics of annual drought events ((**a**) DT; (**b**) DD; (**c**) DS; (**d**) DI) in different categories in the ARNC during 1982–2021. The following characteristics of drought events are included: drought times (DT), drought duration (DD), drought severity (DS), and drought intensity (DI). The SPEI thresholds are less than −0.5, −1.0, −1.5, and −2.0, which represent mild, moderate, severe, and extreme droughts, respectively. The numbers (e.g., 0.004) represent the changing trends of different drought categories. * indicates that $p$ is less than 0.05; ** indicates that $p$ is less than 0.01.

### *3.2. Spatiotemporal Changes of NDVI and Meteorological and Drought Stress Factors in the ARNC*

#### 3.2.1. Spatiotemporal Changes of NDVI in the ARNC

In terms of temporal trends, the annual NDVI increased slowly from 1982 to 2021 at a rate of 0.0001/year (Figure 5a). The growing season NDVI also rose slowly at a rate of 0.0005/year during 1982–2021 (Figure 5b).

In terms of annual spatial changes from 1982 to 2021, the highest NDVI values were primarily concentrated within the Ili River Basin (Figure 6a). The lowest NDVI values were primarily concentrated within the Tarim Basin and the northern Hexi Corridor. For the NDVI changes, the 66% grid indicates a stable trend, while the 20% grid shows a significant increasing trend (Figure 6b). In terms of growing season spatial changes from 1982 to 2021, the highest NDVI values were also primarily concentrated within the Ili River Basin (Figure 6c). In addition, the oasis zone at the edge of the Tarim Basin also has higher NDVI values. The regional NDVI was 60% that of the grid, indicating a stable trend. The regional NDVI was 28% of the grid, indicating a significant increasing trend (Figure 6d).

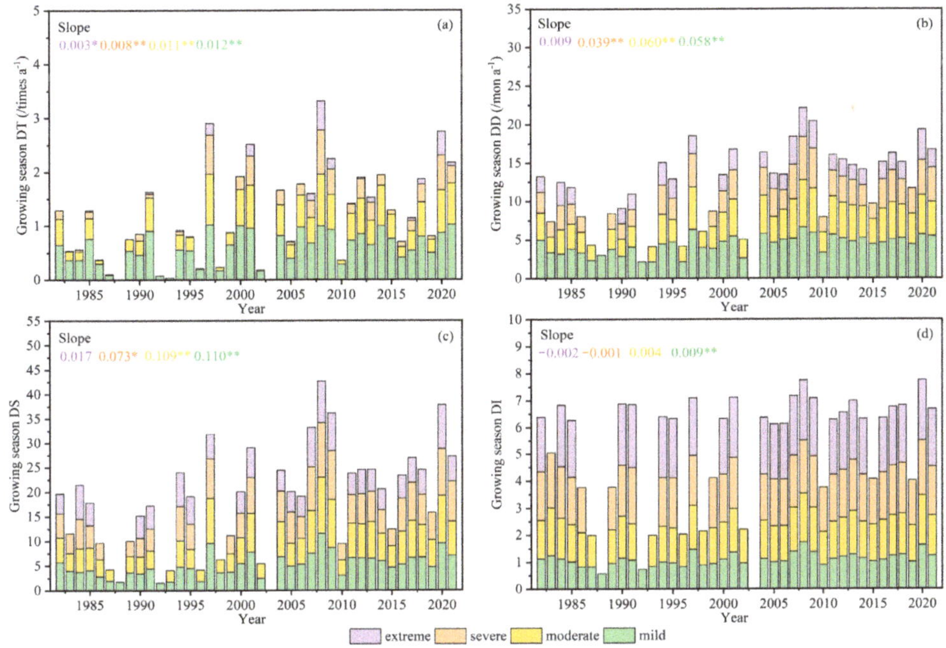


**Figure 4.** Change trends of the characteristics of growing season drought events ((**a**) DT; (**b**) DD; (**c**) DS; (**d**) DI) in different categories in the ARNC during 1982–2021. The following characteristics of drought events are included: drought times (DT), drought duration (DD), drought severity (DS), and drought intensity (DI). The SPEI thresholds are less than −0.5, −1.0, −1.5, and −2.0, which represent mild, moderate, severe, and extreme droughts, respectively. The numbers (e.g., 0.003) represent the changing trends of different drought categories. * indicates that $p$ is less than 0.05; ** indicates that $p$ is less than 0.01.

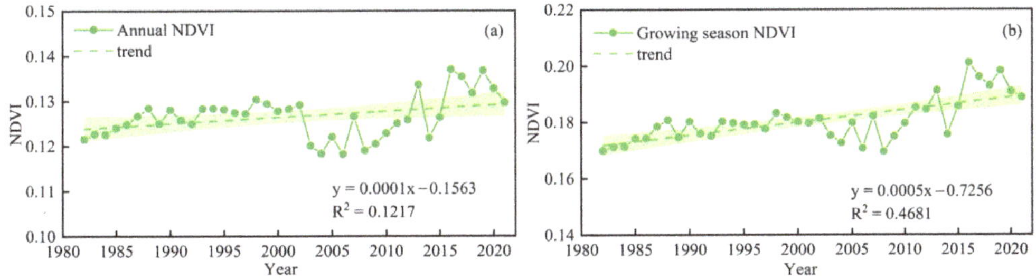


**Figure 5.** The annual (**a**) and growing season (**b**) change trends of the NDVI in the ARNC from 1982 to 2021. The shaded portion is the 95% confidence interval, based on Theil–Sen analysis.

### 3.2.2. Spatiotemporal Changes of Meteorological and Drought Stress Factors

The annual variation trends of TMP and PRE in the ARNC increased significantly with a rate of 0.04 °C/year and 0.36 mm/year during 1982–2021, respectively (Figure 7a,b). In contrast, the annual VPD and SM showed slowly increasing trends from 1982 to 2021 with rates of 0.0025 kPa/year and 0.0008 $m^3m^{-3}$/year, respectively (Figure 7c,d). In contrast, the annual SPEI showed slowly decreasing trends from 1982 to 2021 at a rate of −0.02/year (Figure 7e).

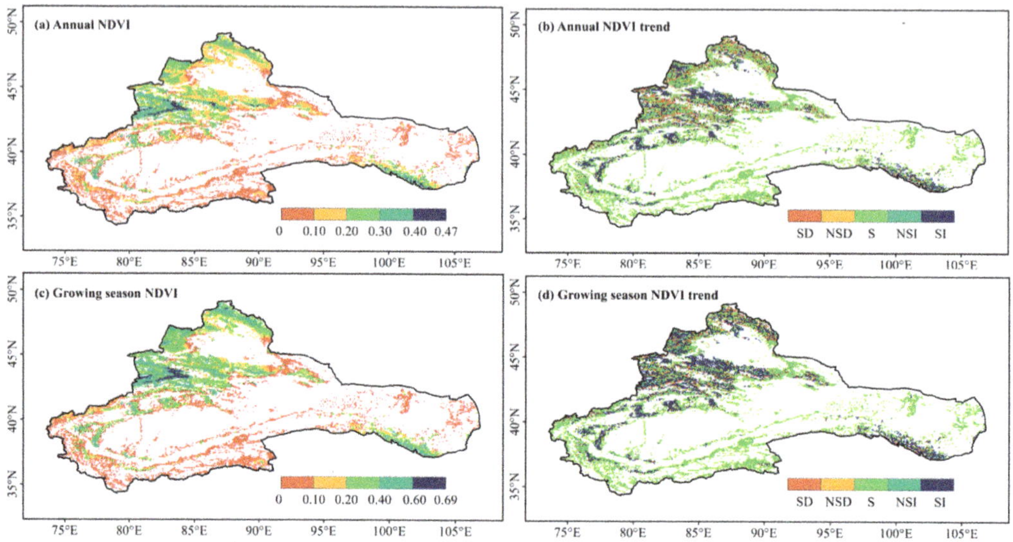


**Figure 6.** The spatial distribution patterns and trends of the annual and growing season NDVI in the ARNC from 1982 to 2021 (S represents a stable trend; SD represents a significant decreasing trend; NSD represents a non-significant decreasing trend; NSI represents a non-significant increasing trend; SI represents a significant increasing trend, based on Theil–Sen analysis and the Mann–Kendall test).

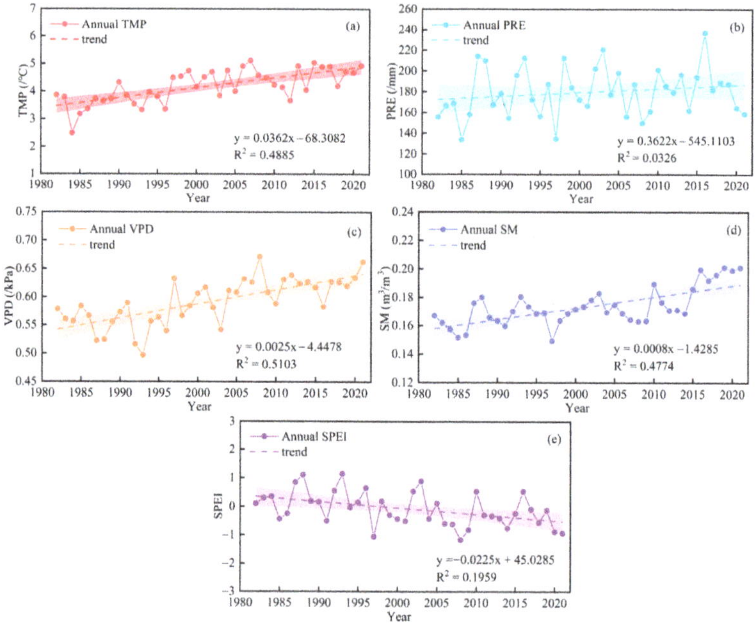


**Figure 7.** The change trends of the annual TMP (**a**), PRE (**b**), VPD (**c**), SM (**d**), and SPEI (**e**) in the ARNC between 1982 and 2021. The shaded portion represents the 95% confidence interval, based on Theil–Sen analysis and the Mann–Kendall test. The following meteorological and drought stress factors are included: temperature (TMP), precipitation (PRE), vapor pressure deficit (VPD), soil moisture (SM), and standardized precipitation evapotranspiration index (SPEI).

From 1982 to 2021, the growing season variation trends of TMP and PRE in the ARNC increased significantly at rates of 0.04 °C/year and 0.28 mm/year, respectively (Figure 8a,b). The growing season variation trends of VPD and SM also showed a slowly increasing trend from 1982 to 2021 at rates of 0.004 kPa/year and 0.0008 $m^3m^{-3}$/year, respectively (Figure 8c–e). In contrast, the growing season SPEI showed slowly decreasing trends from 1982 to 2021 at a rate of −0.03/year (Figure 8e). In addition, Figures 7 and 8 illustrate that the SPEI exceeds zero for the period from 1982 to 1997, indicating a transition from non-arid to arid conditions.

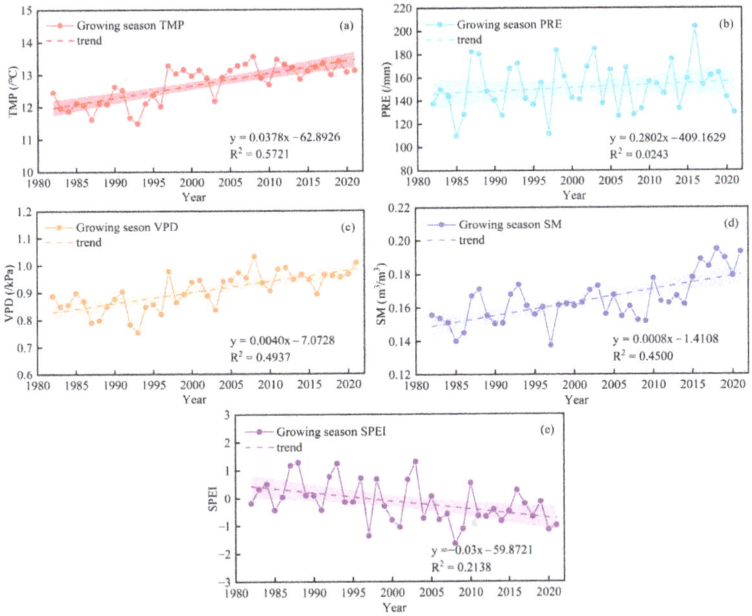


**Figure 8.** The temporal trends of growing season TMP (**a**), PRE (**b**), VPD (**c**), SM (**d**) and SPEI (**e**) in the ARNC between 1982 and 2021. The shaded portion represents the 95% confidence interval, based on Theil–Sen analysis and the Mann–Kendall test. The following meteorological and drought stress factors are included: temperature (TMP), precipitation (PRE), vapor pressure deficit (VPD), soil moisture (SM), and standardized precipitation evapotranspiration index (SPEI).

In terms of annual spatial changes, the highest TMP values were primarily concentrated within the Tarim Basin and Jungar Basin, and the lowest TMP values were primarily concentrated within the south Kunlun Mountains and central Tianshan mountains (Figure 9a). The vast majority of regional TMP indicated a significant increasing trend (Figure 9b). The highest PRE values were mainly concentrated within the Ili River Basin, and the lowest PRE values were primarily concentrated within the Tarim Basin and the northern Hexi Corridor (Figure 9c). The PRE was 69% of the grid, indicating a non-significant increasing trend. In contrast, the PRE was 27% of the grid, indicating a non-significant decreasing trend. (Figure 9d). The highest VPD values were primarily concentrated within the Tarim Basin. The lowest VPD values were primarily concentrated within the Kunlun, Tianshan, and Qilian Mountains (Figure 9e). The regional VPD was 93% of the grid, indicating a significant increasing trend (Figure 9f). The highest SM values were primarily concentrated within the Altay Mountains and central Tianshan Mountains, and the lowest SM values were primarily concentrated within the Tarim Basin and the northern Hexi Corridor (Figure 9g). In general, the regional SM was 57% of the grid, indicating a significant increasing trend (Figure 9h). The highest SPEI values in the region were primarily concentrated within the Altay, Tianshan, and western Kunlun Mountains, the lowest SPEI values were mainly concentrated within the Tarim Basin (Figure 9i). The regional SPEI was 50% of the grid, indicating a significant decreasing trend. The regional SPEI was 39% of the grid, indicating a non-significant decreasing trend (Figure 9j).

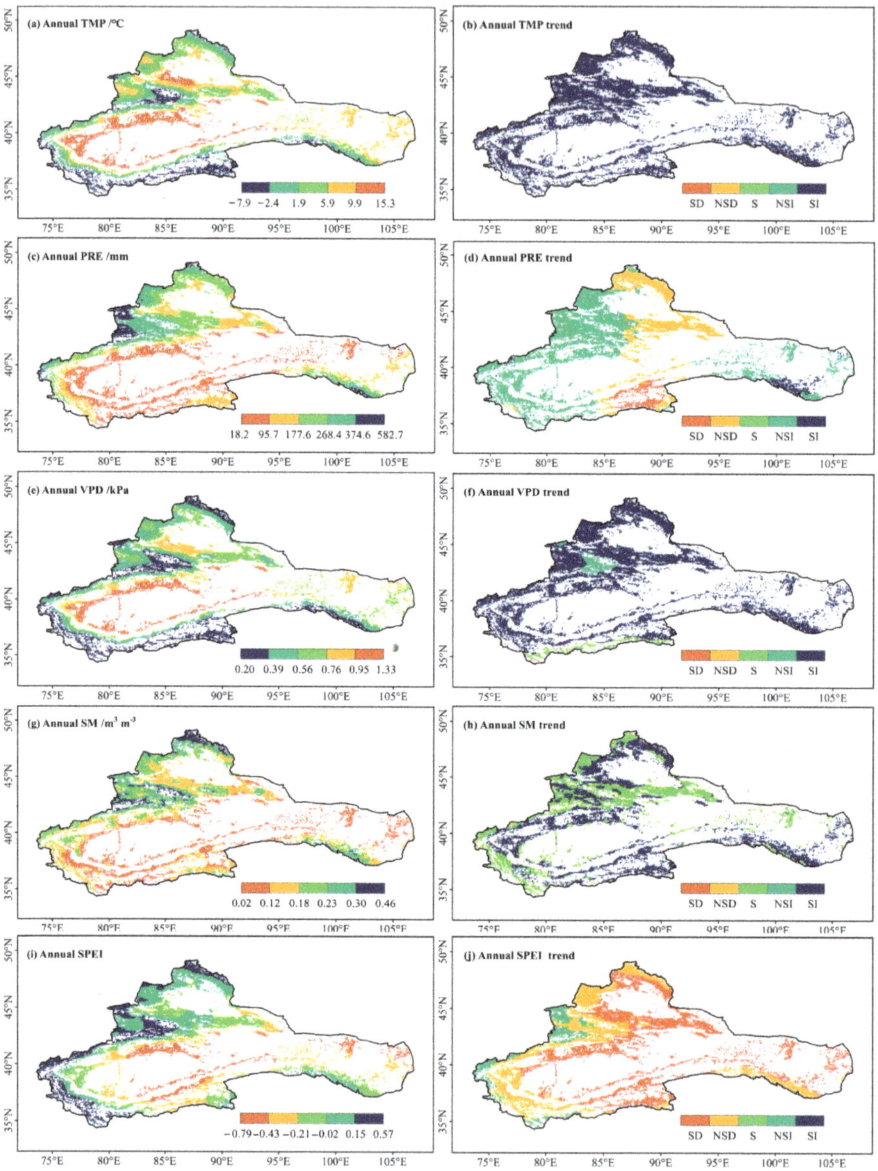


**Figure 9.** Spatial distribution patterns of the annual TMP, PRE, VPD, SM, and SPEI in the ARNC from 1982 to 2021 (S represents a stable trend; SD represents a significant decreasing trend; NSD represents a non-significant decreasing trend; NSI represents a non-significant increasing trend; and SI represents a significant increasing trend, based on Theil–Sen analysis and the Mann–Kendall test). The following meteorological and drought stress factors are included: temperature (TMP), precipitation (PRE), vapor pressure deficit (VPD), soil moisture (SM), and standardized precipitation evapotranspiration index (SPEI).

The spatial pattern of growing season TMP is similar to annual TMP (Figure 10a). The regional TMP also indicated a significant increasing trend (Figure 10b). The highest PRE values were primarily concentrated within the western Ili River Basin and Qilian Mountains, and the lowest PRE values were primarily concentrated within the Tarim Basin and the northern Hexi Corridor (Figure 10c). The PRE was 63% of the grid, indicating a non-significant increasing trend. In contrast, the PRE was 32% of the grid, indicating a non-significant decreasing trend (Figure 10d). The spatial pattern of the growing season VPD is similar to the annual VPD (Figure 10e). The regional VPD was 97% of the grid, indicating a significant increasing trend (Figure 10f). The highest SM values were primarily concentrated within the Altay Mountains and Tianshan Mountains, and the lowest SM values were primarily concentrated within the Tarim Basin and the northern Hexi Corridor (Figure 10g). The regional SM was 60% of the grid, indicating a significant increasing trend (Figure 10h). The highest SPEI values were primarily concentrated within the Kunlun Mountains, and the lowest SPEI values were in the Tarim Basin and the northern Hexi Corridor (Figure 10i). The SPEI was 55% of the grid, indicating a significant decreasing trend, and 39% of the grid, indicating a non-significant decreasing trend (Figure 10j).

### 3.3. *Impacts of Drought and Meteorological and Drought Stress Factors on Vegetation Change*

Figure 11 shows the spatial pattern of the correlation coefficient between the annual NDVI and meteorological and drought stress factors. The annual NDVI more strongly correlated with TMP (0.72) than PRE (0.71) (Figure 11a,b). The highest correlation coefficient between the annual NDVI and SM (0.91) was greater than that of the VPD (0.83), indicating that the annual NDVI was strongly influenced by SM (Figure 11c,d). The annual NDVI showed the highest correlation coefficient with the SPEI (0.79) in the ARNC (Figure 11e). The annual NDVI was significantly affected by the TMP in the Tianshan and Altay Mountains. The annual NDVI was significantly associated with PRE in the Ili River Basin. The annual NDVI was strongly influenced by the VPD in the Qilian and Tianshan Mountains. The SM change was susceptible to change in the edge of the Tarim Basin and the northern Hexi Corridor. The annual NDVI was greatly affected by the SPEI in the Ili River Basin.

Figure 12 shows the spatial pattern of the correlation coefficient between the growing season NDVI and meteorological and drought stress factors. The growing season NDVI was more strongly associated with TMP (0.75) than PRE (0.73) (Figure 12a,b). The highest correlation coefficient between the growing season NDVI and SM (0.92) was greater than the VPD (0.79), indicating that the growing season NDVI was strongly influenced by SM (Figure 12c,d). The growing season NDVI exhibited the highest correlation coefficient for SPEI (0.78) in the ARNC (Figure 12e). The growing season NDVI was influenced by TMP in the Tianshan and Qilian Mountains. The growing season NDVI was influenced by PRE in the Ili River Basin. The growing season NDVI was negatively affected by the VPD in the northwest ARNC. It was also significantly affected by SM in the edge of the Tarim Basin and the northern Hexi Corridor. The growing season NDVI was significantly affected by the SPEI in the Ili River Basin.

The SEM elucidated the effects of TMP, PRE, VPD, SM, and SPEI on the NDVI (Figure 13a). The results revealed that the annual SPEI had a positive influence on PRE (0.38). Furthermore, the NDVI had a negative effect with the annual SPEI (−0.49). The standardized path coefficient from the annual NDVI to SM (0.55) was higher compared to the VPD (−0.56). It indicated that the NDVI was most extensively influenced by SM, TMP, and PRE annually, as shown by the standardized path coefficients from the SM (0.55), TMP (0.20), and PRE (0.14) to the NDVI.

For the growing season, the SPEI exhibited a positive influence on PRE (0.43). Moreover, there was a negative impact of the SPEI (−0.24) on the change in the growing season NDVI. Notably, the standardized path coefficient from the growing season NDVI to SM (0.70) was greater than that of the VPD (−0.59). It indicated that the NDVI was most extensively impacted by SM and TMP in the growing season, as shown by the standardized path coefficients from the SM (0.70), TMP (0.65) and PRE (0.03) to the NDVI(Figure 13b).

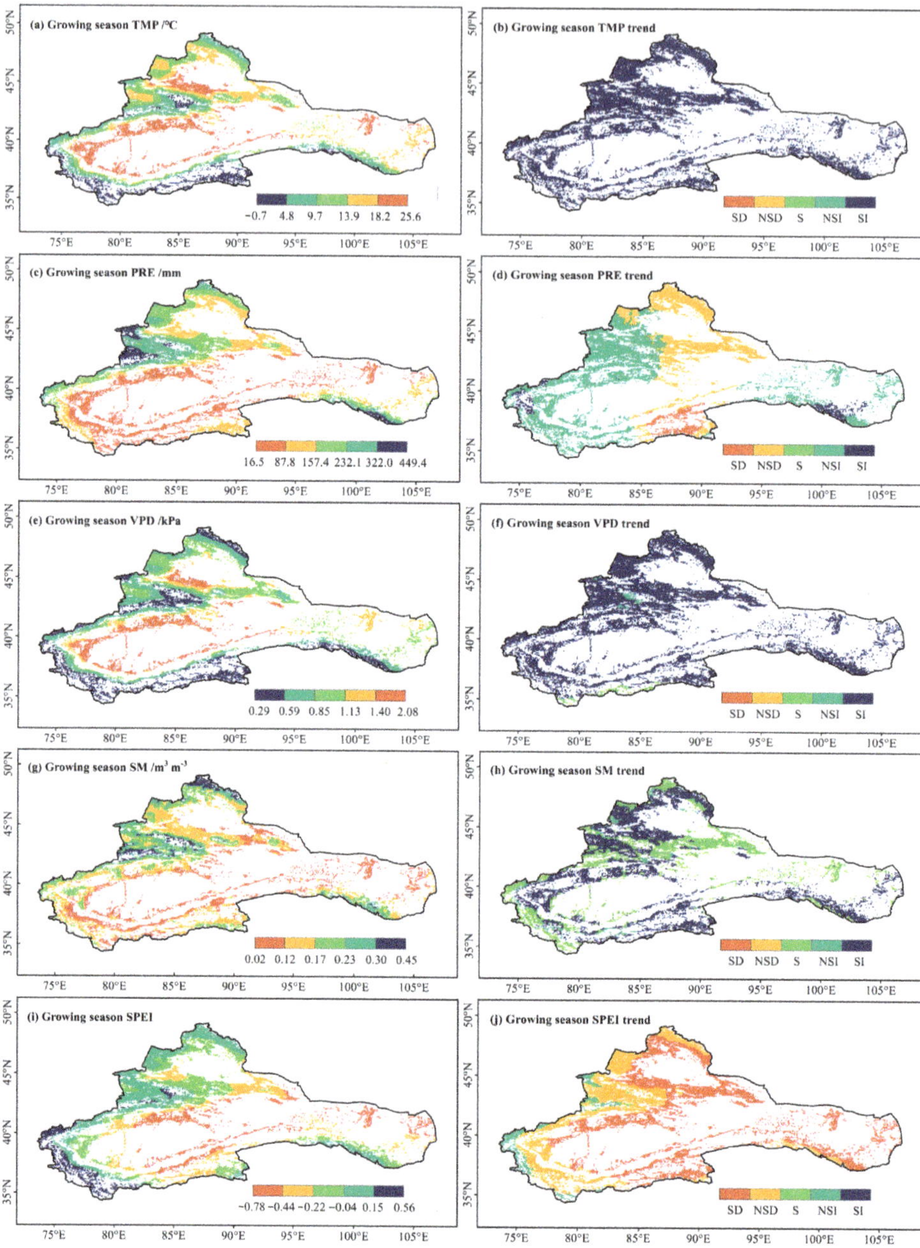


**Figure 10.** Spatial patterns of the growing season TMP, PRE, VPD, SM, and SPEI in the ARNC from 1982 to 2021 (S represents a stable trend; SD represents a significant decreasing trend; NSD represents a non-significant decreasing trend; NSI represents a non-significant increasing trend; SI represents a significant increasing trend, based on Theil–Sen analysis and the Mann–Kendall test). The following meteorological and drought stress factors are included: temperature (TMP), precipitation (PRE), vapor pressure deficit (VPD), soil moisture (SM), and standardized precipitation evapotranspiration index (SPEI).

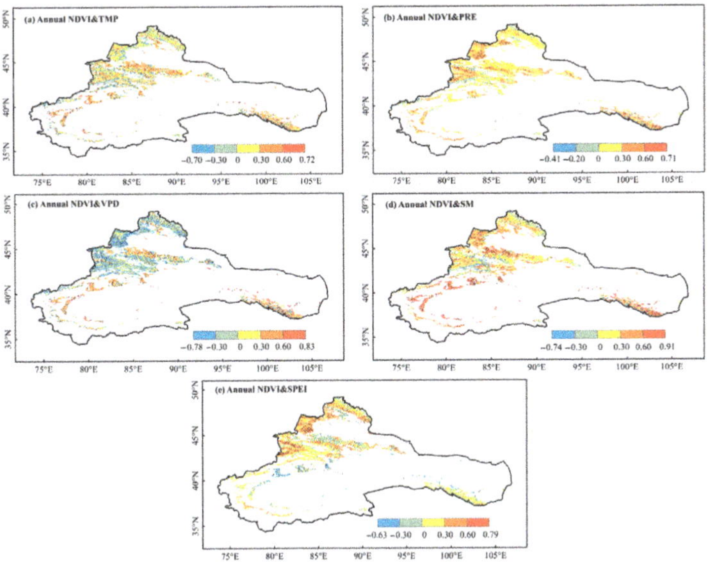


**Figure 11.** Spatial patterns of the correlation between the annual NDVI and TMP (**a**), PRE (**b**), VPD (**c**), SM (**d**), and SPEI (**e**) in the ARNC from 1982 to 2021. The following meteorological and drought stress factors are included: temperature (TMP), precipitation (PRE), vapor pressure deficit (VPD), soil moisture (SM), and standardized precipitation evapotranspiration index (SPEI).

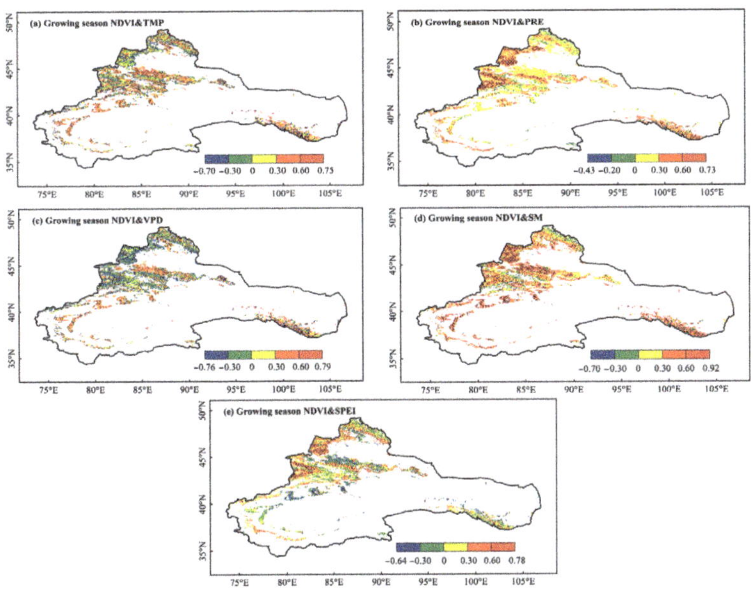


**Figure 12.** Spatial patterns of the correlation between the growing season NDVI and TMP (**a**), PRE (**b**), VPD (**c**), SM (**d**), and SPEI (**e**) in the ARNC from 1982 to 2021. The following meteorological and drought stress factors are included: temperature (TMP), precipitation (PRE), vapor pressure deficit (VPD), soil moisture (SM), and standardized precipitation evapotranspiration index (SPEI).

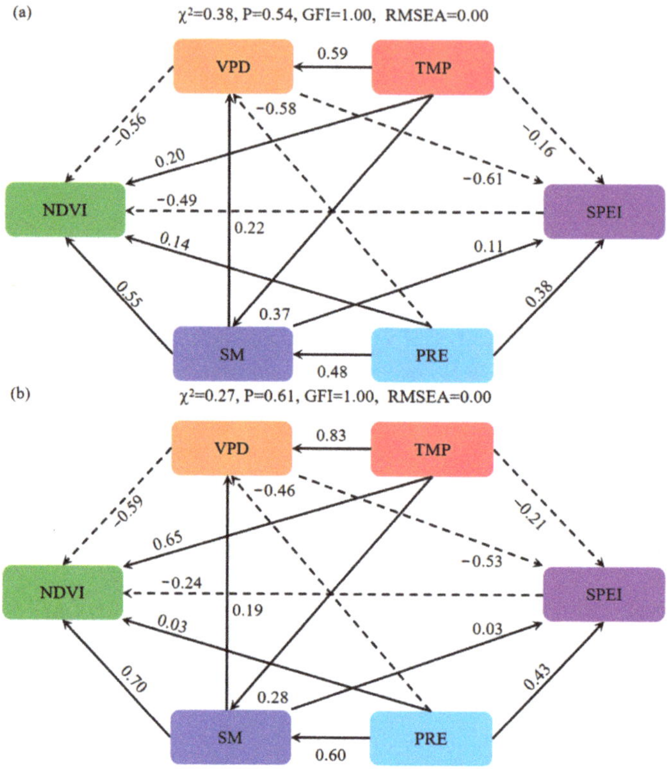


**Figure 13.** Structural equation models showing the relationship between the TMP, PRE, VPD, SM, SPEI, and NDVI annually (**a**) and during growing season (**b**) on the ARNC. The solid arrows indicate a positive correlation between variables, while the dashed arrows indicate a negative correlation between variables. The following meteorological and drought stress factors are included: temperature (TMP), precipitation (PRE), vapor pressure deficit (VPD), soil moisture (SM), and standardized precipitation evapotranspiration index (SPEI).

The NDVI was positively and directly influenced by TMP, PRE, and SM, and the SM directly affected the NDVI, with the largest standardized path coefficient from the SM to the NDVI (Figure 13a,b). Similarly, the TMP and SM affected the NDVI directly between the annual and growing season. Our PCA and SEM results consistently demonstrated that TMP and SM exerted greater impacts on vegetation growth than PRE, the VPD, and the SPEI (Figure 13a,b). Through this comprehensive analysis, it is evident that the TMP and SM play dominant roles in the regulation of vegetation growth.

## 4. Discussion

Many studies have indicated that the climate change in the ARNC was exhibiting a "warming and wetting" trend, and the wetting trend has accelerated, with significant increases in temperature and precipitation, leading to vegetation greening [57,58]. However, some studies have also pointed out that the ARNC has a drying trend and its shift during 1951–2016 [59], and that more frequent and more severe droughts will occur in the ARNC under the 2 °C temperature rise scenario [60]. In this study, we found that the TMP and PRE also showed increasing trends, while the SPEI showed a declining trend during 1982–2021. We also used the run theory to define the drought event characteristics, including DT, DD, DS, and DI [47]. For the four drought categories, the DT, DD, DS, and DI of drought events show consistent spatial patterns and change trends in the ARNC during 1982–2021. The findings provide new insights into the understanding of drought event changes in the ARNC.

In terms of vegetation changes, previous studies have primarily used the GIMMS3g-NDVI data spanning the period from 1982 to 2015 [30–32]. In this study, we used a new version of the GIMMS NDVI product (PKU GIMMS NDVI, which is generated by a new machine learning model and data integration methodology based on the Landsat, AVHRR and MODIS data [36]) to examine the NDVI trends in the ARNC over an extended period from 1982 to 2021, thereby providing a more thorough comprehension. Based on the PKU GIMMS NDVI data, we diagnosed that the NDVI was relatively low during 2003–2008 (Figure 5). During this period, the climate showed significant changes characterized by a noticeable increase in temperature, which resulted in a higher VPD [61]. The significant rise in temperature and the increase in the VPD enhanced evaporation, leading to reduced soil moisture. Additionally, the substantial decrease in precipitation also directly contributed to lower soil moisture. These factors combined likely exacerbated drought conditions, negatively affecting the normal growth of vegetation and resulting in lower NDVI values.

We explored the influence of meteorological and drought stress factors on vegetation changes during 1982–2021. As drought stress factors, the VPD and SM affect terrestrial ecosystems to varying degrees under land–atmosphere interactions [62,63]. However, the VPD and SM remain controversial in exploring the effects of vegetation change [48,64]. Some argue that the VPD is a dominant factor for vegetation change [65,66], while others believe that the SM plays an important role in vegetation productivity [22]. We found that the TMP and SM were the dominant environmental drivers affecting vegetation growth in the ARNC. The TMP indirectly affects the NDVI by affecting VPD, while PRE affects the NDVI by affecting SM. He et al. [67] also suggested that the PRE is the dominant factor influencing vegetation change in the ARNC, which can influence vegetation change by controlling SM.

The ecological resilience of vegetation to withstand potential future drought risks has emerged as a prominent concern in current research on ecosystem sustainability [68,69]. How drought events affect future vegetation change remains unclear. Future studies will discuss spatial patterns and mechanisms of vegetation drought resilience.

It is important to maintain the spatial resolution of TMP, PRE, VPD, SM, and SPEI data consistent with the PKU GIMMS NDVI. The resampling technique may have affected the quality and accuracy of the data [53]. Consequently, the characteristics of drought events identified at different timescales were also different. Moreover, there is increased uncertainty in identifying vegetation ecosystem stress factors under changing meteorological and drought stress factors. The mechanisms of interaction between meteorological and drought stress factors and vegetation growth may be more diverse and complex than previously discussed. In this study, we explored the influences of the TMP, PRE, VPD, SM, and SPEI on the NDVI. In the future, more excellent physical models are needed to further investigate and better describe the effects of meteorological and drought stress factors on vegetation change.

## 5. Conclusions

In this study, we investigated the spatiotemporal trends of the vegetation, meteorological, and drought stress factors in the ARNC during 1982–2021. We found that the annual and growing season vegetation NDVI, the meteorological factors (TMP, PRE), and the drought stress factors (the VPD, SM) showed increasing trends, while the SPEI showed a declining trend during 1982–2021. For the annual (growing season) NDVI changes, the 66% (60%) grid indicates a stable trend, while the 20% (28%) grid shows a significant increasing trend in the ARNC. For the four drought categories, the times, duration, severity, and intensity of drought events show consistent spatial patterns and change trends in the ARNC during 1982–2021.

Based on Pearson correlation analysis and the SEM method, we explored the effects of meteorological and drought stress factors on vegetation changes during 1982–2021. We found that the NDVI was influenced by the TMP, PRE, VPD, and SM. The TMP and SM exerted greater impacts on vegetation change than PRE, the VPD, and the SPEI. TMP and SM were the dominant environmental drivers affecting vegetation growth in the ARNC. The TMP indirectly affects the NDVI by affecting the VPD, while PRE affects the NDVI by affecting SM. It is noteworthy that there is a close relationship between the VPD and SM, and their coupled effect plays a key role in vegetation changes in arid regions. Future research should focus on the impact of the coupling of the VPD and SM on vegetation changes.

**Author Contributions:** Conceptualization, J.Y., Y.C. and G.Z.; methodology, G.Z.; software, G.Z., J.C., C.W. and C.M.; validation, G.Z., J.Y. and J.C.; writing—original draft preparation, G.Z.; writing—review and editing, G.Z., P.Z., Y.M., M.L. and Y.Y.; funding acquisition, J.Y. All authors have read and agreed to the published version of the manuscript.

**Funding:** This research was funded by the Shanghai Cooperation Organization (SCO) Science and Technology Partnership and International S&T Cooperation Program (2023E01022), the Science and Technology Youth Top-Notch Talent Support Program (Tianshan Talents) of Xinjiang (2022TSY-CCX0005), the National Natural Science Foundation of China (W2412135, U2344224), the Key Laboratory Opening Foundation of Xinjiang Uygur Autonomous Region (2023D04048), and the "Tianchi Talents" (Xinjiang) Plan Project (Moyan Li).

**Data Availability Statement:** The data presented in this study are available upon request from the corresponding author.

**Conflicts of Interest:** The authors declare no conflicts of interest.

## References

1. Berg, A.; McColl, K.A. No projected global drylands expansion under greenhouse warming. *Nat. Clim. Chang.* **2021**, *11*, 331–337. [CrossRef]
2. Jiang, L.L.; Liu, W.L.; Liu, B.; Yuan, Y.; Bao, A.M. Monitoring vegetation sensitivity to drought events in China. *Sci. Total Environ.* **2023**, *893*, 164917. [CrossRef]
3. Yu, X.J.; Zhang, L.X.; Zhou, T.J.; Zhang, X. Long-term changes in the effect of drought stress on ecosystems across global drylands. *Sci. China Earth Sci.* **2022**, *66*, 146–160. [CrossRef]
4. Ding, G.; Chen, S.; Sun, Y.; Ma, S.; Chen, J. Holocene Hydroclimatic Variations in the Asian Drylands: Current Understanding and Future Perspectives. *J. Earth Sci.* **2024**, *35*, 292–295. [CrossRef]
5. Geng, G.P.; Zhou, H.K.; Wang, T. Assessing the relationship between drought and vegetation dynamics in northern China during 1982–2015. *Theor. Appl. Climatol.* **2022**, *148*, 467–479. [CrossRef]
6. Jiang, W.X.; Wang, L.C.; Zhang, M.; Yao, R.; Chen, X.X.; Gui, X.; Sun, J.; Cao, Q. Analysis of drought events and their impacts on vegetation productivity based on the integrated surface drought index in the Hanjiang River Basin, China. *Atmos. Res.* **2021**, *254*, 105536. [CrossRef]
7. West, H.; Quinn, N.; Horswell, M. Remote sensing for drought monitoring & impact assessment: Progress, past challenges and future opportunities. *Rem. Sens Environ.* **2019**, *232*, 111291.
8. Won, J.; Seo, J.; Kim, S. A copula model integrating atmospheric moisture demand and supply for vegetation vulnerability mapping. *Sci. Total Environ.* **2022**, *812*, 151464. [CrossRef]
9. Dikshit, A.; Pradhan, B.; Huete, A.; Park, H.J. Spatial based drought assessment: Where are we heading? A review on the current status and future. *Sci. Total Environ.* **2022**, *844*, 157239. [CrossRef]
10. Mishra, A.K.; Singh, V.P. A review of drought concepts. *J. Hydrol.* **2010**, *391*, 202–216. [CrossRef]
11. Hao, Z.C.; Singh, V.P.; Xia, Y.L. Seasonal Drought Prediction: Advances, Challenges, and Future Prospects. *Rev. Geophys.* **2018**, *56*, 108–141. [CrossRef]
12. Shi, M.Q.; Yuan, Z.; Shi, X.L.; Li, M.X.; Chen, F.; Li, Y. Drought assessment of terrestrial ecosystems in the Yangtze River Basin, China. *J. Clean. Prod.* **2022**, *362*, 132234. [CrossRef]
13. Ding, Y.B.; Xu, J.T.; Wang, X.W.; Cai, H.J.; Zhou, Z.Q.; Sun, Y.N.; Shi, H.Y. Propagation of meteorological to hydrological drought for different climate regions in China. *J. Environ. Manag.* **2021**, *283*, 111980. [CrossRef]
14. Wu, J.F.; Yao, H.X.; Chen, X.H.; Wang, G.X.; Bai, X.Y.; Zhang, D.J. A framework for assessing compound drought events from a drought propagation perspective. *J. Hydrol.* **2022**, *604*, 127228. [CrossRef]
15. Zhang, Y.; Keenan, T.F.; Zhou, S. Exacerbated drought impacts on global ecosystems due to structural overshoot. *Nat. Ecol. Evol.* **2021**, *5*, 1490–1498. [CrossRef]

16. Piao, S.L.; Zhang, X.P.; Chen, A.P.; Liu, Q.; Lian, X.; Wang, X.H.; Peng, S.S.; Wu, X.C. The impacts of climate extremes on the terrestrial carbon cycle: A review. *Sci. China Earth Sci.* **2019**, *49*, 1321–1334. [CrossRef]
17. Zhou, G.Y.; Zhou, L.Y.; Shao, J.J.; Zhou, X.H. Effects of extreme drought on terrestrial ecosystems: Review and prospects. *Chin. J. Plant Ecol.* **2020**, *44*, 515–525. [CrossRef]
18. Zhang, Y.C.; Piao, S.L.; Sun, Y.; Rogers, B.M.; Li, X.Y.; Lian, X.; Liu, Z.H.; Chen, A.P.; Peñuelas, J. Future reversal of warming-enhanced vegetation productivity in the Northern Hemisphere. *Nat. Clim. Chang.* **2022**, *12*, 581–586. [CrossRef]
19. Piao, S.L.; Wang, X.H.; Park, T.; Chen, C.; Lian, X.; He, Y.; Bjerke, J.W.; Chen, A.P.; Ciais, P.; Tømmervik, H.; et al. Characteristics, drivers and feedbacks of global greening. *Nat. Rev. Earth Environ.* **2020**, *1*, 14–27. [CrossRef]
20. Konings, A.G.; Williams, A.P.; Gentine, P. Sensitivity of grassland productivity to aridity controlled by stomatal and xylem regulation. *Nat. Geosci.* **2017**, *10*, 284–288. [CrossRef]
21. Liu, L.B.; Gudmundsson, L.; Hauser, M.; Qin, D.H.; Li, S.C.; Seneviratne, S.I. Soil moisture dominates dryness stress on ecosystem production globally. *Nat. Commun.* **2020**, *11*, 4892. [CrossRef] [PubMed]
22. Cheng, Y.M.; Liu, L.; Cheng, L.; Fa, K.Y.; Liu, X.C.; Huo, Z.L.; Huang, G.H. A shift in the dominant role of atmospheric vapor pressure deficit and soil moisture on vegetation greening in China. *J. Hydrol.* **2022**, *615*, 128680. [CrossRef]
23. Lin, X.N.; Niu, J.Z.; Berndtsson, R.; Yu, X.X.; Zhang, L.; Chen, X.W. NDVI Dynamics and Its Response to Climate Change and Reforestation in Northern China. *Remote Sens.* **2020**, *12*, 4138. [CrossRef]
24. Zhou, S.; Zhang, Y.; Williams, A.P.; Gentine, P. Projected increases in intensity, frequency, and terrestrial carbon costs of compound drought and aridity events. *Sci. Adv.* **2019**, *5*, eaau5740. [CrossRef]
25. Grossiord, C.; Buckley, T.N.; Cernusak, L.A.; Novick, K.A.; Poulter, B.; Siegwolf, R.T.W.; Sperry, J.S.; McDowell, N.G. Plant responses to rising vapor pressure deficit. *New Phytol.* **2020**, *226*, 1550–1566. [CrossRef]
26. Yuan, W.P.; Zheng, Y.; Piao, S.L.; Ciais, P.; Lombardozzi, D.; Wang, Y.P. Increased atmospheric vapor pressure deficit reduces global vegetation growth. *Sci. Adv.* **2019**, *5*, eaax1396. [CrossRef]
27. Du, Z.Q.; Zhao, J.; Pan, H.H.; Wu, Z.T.; Zhang, H. Responses of vegetation activity to the daytime and nighttime warming in Northwest China. *Environ. Monit. Assess.* **2019**, *191*, 721. [CrossRef]
28. Gao, J.B.; Jiao, K.W.; Wu, S.H. Investigating the spatially heterogeneous relationships between climate factors and NDVI in China during 1982 to 2013. *J. Geogr. Sci.* **2019**, *29*, 1597–1609. [CrossRef]
29. Li, P.; Kou, X.; Wang, Y.; Niu, L. Building a More Sustainable Chinese Loess Plateau. *J. Earth Sci.* **2024**, *35*, 283–287. [CrossRef]
30. Shang, J.X.; Zhang, Y.; Peng, Y.; Huang, Y.H.; Zhu, L.; Wu, Z.Y.; Wang, J.; Cui, Y.X. Climate change drives NDVI variations at multiple spatiotemporal levels rather than human disturbance in Northwest China. *Environ. Sci. Pollut. Res.* **2022**, *29*, 13782–13796. [CrossRef]
31. Liu, Y.; Tian, J.Y.; Liu, R.H.; Ding, L.Q. Influences of Climate Change and Human Activities on NDVI Changes in China. *Remote Sens.* **2021**, *13*, 4326. [CrossRef]
32. Lu, Q.Q.; Tao, J.; Li, L.D.; Yong, L.Z. The response characteristics of NDVI with different vegetation cover types to temperature and precipitation in China. *Ecol. Environ. Sci.* **2020**, *29*, 23–34.
33. Han, H.; Ma, G.; Ta, Z.; Zhao, T.; Li, P.; Li, X. Evolution of Vegetation Growth Season on the Loess Plateau under Future Climate Scenarios. *Forests* **2024**, *15*, 1526. [CrossRef]
34. Guo, W.; He, H.; Li, X.; Zeng, W. Greater Greening Trend in the Loess Plateau of China Inferred from Long-Term Remote Sensing Data: Patterns, Causes and Implications. *Forests* **2022**, *13*, 1630. [CrossRef]
35. Zhang, H.; Li, L.; Zhao, X.; Chen, F.; Wei, J.; Feng, Z.; Hou, T.; Chen, Y.; Yue, W.; Shang, H.; et al. Changes in Vegetation NDVI and Its Response to Climate Change and Human Activities in the Ferghana Basin from 1982 to 2015. *Remote Sens.* **2024**, *16*, 1296. [CrossRef]
36. Li, M.; Cao, S.; Zhu, Z.; Wang, Z.; Myneni, R.B.; Piao, S. Spatiotemporally consistent global dataset of the GIMMS Normalized Difference Vegetation Index (PKU GIMMS NDVI) from 1982 to 2022. *Earth Syst. Sci. Data* **2023**, *15*, 4181–4203. [CrossRef]
37. Wen, X.; Tu, Y.H.; Tan, Q.F.; Li, W.Y.; Fang, G.H.; Ding, Z.Y.; Wang, Z.N. Construction of 3D drought structures of meteorological drought events and their spatio-temporal evolution characteristics. *J. Hydrol.* **2020**, *590*, 125539. [CrossRef]
38. Vicente-Serrano, S.M.; Beguería, S.; López-Moreno, J.I. A Multiscalar Drought Index Sensitive to Global Warming: The Standardized Precipitation Evapotranspiration Index. *J. Clim.* **2010**, *23*, 1696–1718. [CrossRef]
39. Wang, H.J.; Li, Z.; Cao, L.; Feng, R.; Pan, Y.P. Response of NDVI of Natural Vegetation to Climate Changes and Drought in China. *Land* **2021**, *10*, 966. [CrossRef]
40. Luo, N.N.; Mao, D.H.; Wen, B.L.; Liu, X.T. Climate Change Affected Vegetation Dynamics in the Northern Xinjiang of China: Evaluation by SPEI and NDVI. *Land* **2020**, *9*, 90. [CrossRef]
41. Alsafadi, K.; Bashir, B.; Mohammed, S.; Abdo, H.G.; Mokhtar, A.; Alsalman, A.; Cao, W. Response of Ecosystem Carbon–Water Fluxes to Extreme Drought in West Asia. *Remote Sens.* **2024**, *16*, 1179. [CrossRef]
42. Wan, Y.; Han, H.; Mao, Y.; He, B.-J. Responses of Climatic Drought to Vegetation Cover Dynamics: A Case Study in Yunnan, China. *Forests* **2024**, *15*, 1689. [CrossRef]
43. Chen, Y.N.; Li, Z.; Fan, Y.T.; Wang, H.J.; Deng, H.J. Progress and prospects of climate change impacts on hydrology in the arid region of northwest China. *Environ. Res.* **2015**, *139*, 11–19. [CrossRef]
44. Harris, I.; Osborn, T.J.; Jones, P.; Lister, D. Version 4 of the CRU TS monthly high-resolution gridded multivariate climate dataset. *Sci. Data* **2020**, *7*, 109. [CrossRef]

45. Miralles, D.G.; Holmes, T.R.H.; de Jeu, R.A.M.; Gash, J.H.; Meesters, A.G.C.A.; Dolman, A.J. Global land-surface evaporation estimated from satellite based observations. *Hydrol. Earth Syst. Sci.* **2011**, *15*, 453–469. [CrossRef]
46. Potop, V.; Možný, M.; Soukup, J. Drought at various time scales in the lowland regions and their impact on vegetable crops in the Czech Republic. *Agric. For. Meteorol.* **2012**, *156*, 121–133. [CrossRef]
47. Liu, H.; Jiang, L.L.; Liu, B.; Liu, R.; Xiao, Z.L. Characteristics of drought in China and its effect on vegetation change in recent 40 years. *Acta Ecol. Sin.* **2023**, *43*, 7936–7949.
48. Yu, T.; Jiapaer, G.; Bao, A.M.; Zheng, G.X.; Zhang, J.F.; Li, X.; Yuan, Y.; Huang, X.R.; Umuhoza, J. Disentangling the relative effects of soil moisture and vapor pressure deficit on photosynthesis in dryland Central Asia. *Ecol. Indic.* **2022**, *137*, 108698. [CrossRef]
49. Jiang, P.; Hu, L.Q.; Xu, T.T. Spatiotemporal variations of vapor pressure deficit in Xinjiang in recent 60 years. *Arid Land Geogr.* **2023**, *46*, 1–10.
50. Zhang, T.Y.; Chen, Z.; Zhang, W.K.; Jiao, C.C.; Yang, M.; Wang, Q.F.; Han, L.; Fu, Z.; Sun, Z.Y.; Li, W.H.; et al. Long-term trend and interannual variability of precipitation-use efficiency in Eurasian grasslands. *Ecol. Indic.* **2021**, *130*, 108091. [CrossRef]
51. Gocic, M.; Trajkovic, S. Analysis of changes in meteorological variables using Mann-Kendall and Sen's slope estimator statistical tests in Serbia. *Glob. Planet. Chang.* **2013**, *100*, 172–182. [CrossRef]
52. Jiang, W.G.; Yuan, L.H.; Wang, W.J.; Cao, R.; Zhang, Y.F.; Shen, W.M. Spatio-temporal analysis of vegetation variation in the Yellow River Basin. *Ecol. Indic.* **2015**, *51*, 117–126. [CrossRef]
53. Wang, H.J.; Li, Z.; Niu, Y.; Li, X.C.; Cao, L.; Feng, R.; He, Q.N.; Pan, Y.P. Evolution and Climate Drivers of NDVI of Natural Vegetation during the Growing Season in the Arid Region of Northwest China. *Forests* **2022**, *13*, 1082. [CrossRef]
54. Yao, J.; Qu, L.; Chen, Y.; Mao, W.; Chen, J. Hydro-climatological drivers of the unprecedented flooding in August 2022 along the Tarim River, China. *J. Hydrol.* **2024**, *639*, 131630. [CrossRef]
55. Liu, H.H.; Liu, Y.; Chen, Y.; Fan, M.G.; Chen, Y.; Gang, C.C.; You, Y.F.; Wang, Z.N. Dynamics of global dryland vegetation were more sensitive to soil moisture: Evidence from multiple vegetation indices. *Agric. For. Meteorol.* **2023**, *331*, 109327. [CrossRef]
56. Wang, T.; Yang, D.; Zheng, G.; Shi, R. Possible negative effects of earlier thaw onset and longer thaw duration on vegetation greenness over the Tibetan Plateau. *Agric. For. Meteorol.* **2022**, *326*, 109192. [CrossRef]
57. Zhang, Q.; Yang, J.; Wang, W.; Ma, P.; Lu, G.; Liu, X.; Yu, H.; Fang, F. Climatic warming and humidification in the arid region of Northwest China: Multi-scale characteristics and impacts on ecological vegetation. *J. Meteorol. Res.* **2021**, *35*, 113–127. [CrossRef]
58. Zhang, Q.; Zhu, B.; Yang, J.; Ma, P.; Liu, X.; Lu, G.; Wang, Y.; Yu, H.; Liu, W.; Wang, D. New characteristics about the climate humidification trend in Northwest China. *Chin. Sci. Bull.* **2021**, *66*, 3757–3771. [CrossRef]
59. Ma, Z.; Fu, C.; Yang, Q.; Zheng, Z.; Lv, M.; Li, M.; Duan, Y.; Chen, L. Drying trend in northern China and its shift during 1951–2016. *Chin. J. Atmos. Sci.* **2018**, *42*, 951–961.
60. Lu, D.Y.; Zhu, X.F.; Liu, T.T.; Zhang, S.Z. Changes in meteorological drought characteristics in China under the 2 °C temperature rise scenario. *Arid Land Geogr.* **2023**, *46*, 1227–1237.
61. Li, M.; Yao, J.; Guan, J.; Zheng, J. Vegetation Browning Trends in Spring and Autumn over Xinjiang, China, during the Warming Hiatus. *Remote Sens.* **2022**, *14*, 1298. [CrossRef]
62. Lian, X.; Piao, S.; Chen, A.; Huntingford, C.; Fu, B.; Li, L.Z.; Roderick, M.L. Multifaceted characteristics of dryland aridity changes in a warming world. *Nat. Rev. Earth Environ.* **2021**, *2*, 232–250. [CrossRef]
63. Zhou, S.; Williams, A.P.; Berg, A.M.; Cook, B.I.; Zhang, Y.; Hagemann, S.; Gentine, P. Land–Atmosphere feedbacks exacerbate concurrent soil drought and atmospheric aridity. *Proc. Natl. Acad. Sci. USA* **2019**, *116*, 18848–18853. [CrossRef]
64. Zhang, Y.; Zhang, Y.J.; Lian, X.; Zheng, Z.T.; Zhao, G.; Zhang, T.; Xu, M.J.; Huang, K.; Chen, N.; Li, J.; et al. Enhanced dominance of soil moisture stress on vegetation growth in Eurasian drylands. *Natl. Sci. Rev.* **2023**, *10*, nwad108. [CrossRef]
65. Guan, L.; Chen, Y.; Dong, X. Impacts of High Temperature and Vapor Pressure Deficit on the Maize Opened Spikelet Ratio and Pollen Viability. *Agronomy* **2024**, *14*, 2510. [CrossRef]
66. Zhong, Z.Q.; He, B.; Wang, Y.P.; Chen, H.W. Disentangling the effects of vapor pressure deficit on northern terrestrial vegetation productivity. *Sci. Adv.* **2023**, *9*, eadf3166. [CrossRef]
67. He, B.; Chen, A.F.; Wang, H.L.; Wang, Q.F. Dynamic Response of Satellite-Derived Vegetation Growth to Climate Change in the Three North Shelter Forest Region in China. *Remote Sens.* **2015**, *7*, 9998–10016. [CrossRef]
68. Wang, T.Y.; Wang, P.; Wu, Z.N.; Wang, H.L.; Yu, Z.L.; Xu, H.S.; Yin, J.; Yin, L.H.; Yu, J.J.; Yan, D.H. Progress in the study of ecological resilience of vegetation under drought stress. *Adv. Earth Sci.* **2023**, *38*, 790–801.
69. Zhang, Y.; Liu, X.H.; Jiao, W.Z.; Wu, X.C.; Zeng, X.M.; Zhao, L.J.; Wang, L.X.; Guo, J.Q.; Xing, X.Y.; Hong, Y.X. Spatial Heterogeneity of Vegetation Resilience Changes to Different Drought Types. *Earth's Future* **2023**, *11*, e2022EF003108. [CrossRef]

land

Article

# Navigating Ecological–Economic Interactions: Spatiotemporal Dynamics and Drivers in the Lower Reaches of the Jinsha River

Zhongyun Ni [1], Yinbing Zhao [1,2,3,*], Jingjing Liu [1], Yongjun Li [1], Xiaojiang Xia [1] and Yang Zhang [1,4]

1 College of Geography and Planning, Chengdu University of Technology, Chengdu 610059, China; nizhongyun2012@mail.cdut.edu.cn (Z.N.)
2 State Key Laboratory of Urban and Regional Ecology, Research Center for Eco-Environmental Sciences, Chinese Academy of Sciences, Beijing 100085, China
3 Human Geography Research Center of Qinghai-Tibet Plateau and Its Eastern Edge, Chengdu University of Technology, Chengdu 610059, China
4 School of Architecture, Southeast University, Nanjing 210096, China
* Correspondence: zhaoyinbing06@cdut.cn

**Abstract:** The lower reaches of the Jinsha River, serving as a vital ecological barrier in southwestern China and playing a crucial role in advancing targeted poverty alleviation efforts, remain under-explored in terms of the coupling between ecological and economic development, creating a gap in understanding the region's sustainable development potential. This study combines the remote sensing ecological index (RSEI) derived from MODIS data and the biodiversity richness index (BRI) based on land use data to create the ecological environment index (EEI) using a weighted approach. It also develops the economic development index (EDI) from economic data using the entropy weight method. By integrating the EEI and EDI, the study calculates key metrics, including the ecological–economic coupling degree (EECD), coupling coordination degree (EECCD), and relative development degree (EERDD), and examines their spatiotemporal changes from 2000 to 2020. Additionally, the study applies a geographic detector model to identify the spatial drivers of the EEI, an obstacle factor diagnosis model to pinpoint the main barriers to EDI, and a neural network model to uncover the underlying forces shaping EECCD. The results indicate that: (I) From 2000 to 2020, the overall improvement rate of the ecological and economic subsystems was greater than that of the ecological–economic coupling system. The entire region is still in the Running-In Stage, and the coordination level has been upgraded from near imbalance to marginal coordination. About 85% of the counties' EERDDs are still in the EDI Behind EEI Stage. (II) The structural composition of the EEI shows a pattern of low Dry Hot Valley Area and high in other areas, mainly driven by natural factors, although human activities had a notable impact on these interactions. (III) Originating from an impact model primarily driven by economic factors and supplemented by ecological factors, both EDI and EECCD exhibit a pattern of high in the south and low in the north, with improvements spreading northward from the urban area of Kunming. The development gradient differences between 24 poverty-stricken counties and 16 non-poverty-stricken counties have been reduced. (IV) For the six types of ecological–economic coupling development zones, it is essential to adopt localized approaches tailored to the differences in resource and environmental characteristics and development stages. Key efforts should focus on enhancing ecological protection and restoration, increasing financial support, implementing ecological compensation mechanisms, and promoting innovative models for sustainable development.

**Keywords:** ecological environment index; economic development index; coupling coordination degree; geographic detector; obstacle recognition; neural network model; natural and anthropogenic drivers; development gradient; targeted poverty alleviation

## 1. Introduction

The ecological and economic environments exhibit a mutual relationship of influence and constraint, wherein they foster and propel each other's growth to a certain degree. Engaging in research on the coupled development of the ecological economy holds immense significance for advancing harmonious coexistence between humanity and nature, as well as bolstering the overall competitiveness of regions [1,2]. To characterize the relationship between the ecological and economic environments, researchers often use both single and composite indicators, such as the environmental Kuznets curve [3,4], human development index (HDI) [5,6], green GDP [7,8], ecological footprint [9], eco-efficiency index [10,11], sustainability index [12,13], natural capital index [14,15], sustainable development indicators (SDIs) [16], ecosystem service valuation (ESV) [17,18], coupling degree [19], coupling coordination degree (CCD) [20,21], and relative development degree [22]. These indicators provide a variety of evaluation perspectives, offering researchers nuanced insights into the ecological–economic interplay. The CCD is widely employed to assess the coordinated development between systems or among elements within a system. However, there remains a gap in optimizing the selection of evaluation indicators, improving data accuracy, and refining data processing methods, all of which are essential to further enhance the precision and reliability of CCD analyses.

Based on the above indicators, different analysis models were constructed to further explore the coupling mechanism between the ecological and economic environments. System dynamics models simulate the impacts of various economic conditions on ecological–economic systems [23,24]. The input–output model analyzes industrial linkages and resource usage, guiding industrial optimization and resource allocation strategies [25]. The grey correlation degree model quantifies the influence of multiple factors on ecological–economic development through ranking analysis, offering policy insights [26,27]. The ecological footprint model compares human resource consumption with ecological carrying capacity to guide sustainable development efforts [28,29]. Moreover, the coupling coordination model, which constructs a comprehensive evaluation index system for ecological–economic systems, calculates the degree of coupling and coordination to assess development trends and levels [30,31]. Additionally, various methods have been employed to analyze the causes or driving forces of changes in economic subsystems, ecological environment subsystems, or ecological–economic coupling systems, including the fixed effects regression model [32], the ordinary least squares regression model [33], the geographically weighted regression model [34], and obstacle factor analysis [35]. Simultaneously, neural network models are gaining increasing recognition for their advanced pattern recognition capabilities in analyzing changes and complex interactions within ecological and economic systems [36,37]. However, these models present significant gaps in both dynamic change monitoring and model interpretability. Consequently, the accuracy of ecological–economic coupling evaluations and predictions remains constrained, undermining the scientific foundation for effective policy formulation.

Research on the coupling development of ecological and economic systems spans multiple spatial and temporal scales, each reflecting different ecological, regional, and economic dynamics. Key scales include the global scale [38,39], which addresses global ecological–economic interactions like carbon emissions and resource use patterns; the national or regional scale [40,41], focusing on the impacts of large-scale policies on sustainability; the county or city level [42], examining localized ecological and economic interrelations driven by regional policies and practices; and the urban scale, where studies focus on urban growth, resource management, and environmental preservation [43]. Research at the watershed or environmental region scale [44,45] closely aligns with the comprehensive and dynamic nature of natural ecological processes, providing a practical framework for addressing the challenges of economic development more effectively. In China, significant focus has been placed on studying the ecological and economic coupling within major regions such as the Yellow River Basin and the middle and lower reaches of the Yangtze River Basin [46–49]. However, there has been relatively limited research in areas like the lower reaches of the Jinsha River, located in the upper Yangtze River Basin, despite the ecological and economic significance of these regions.

The lower reaches of the Jinsha River are vital for biodiversity conservation in the Hengduan Mountains and soil conservation in the dry hot valleys of Sichuan and Yunnan. Despite its ecological fragility, the region is rich in species diversity. As a key ecological barrier in the Yangtze River Basin, safeguarding the environment here is crucial for ensuring the ecological security of the entire Yangtze River Economic Belt and, by extension, the nation [50]. Moreover, this region is abundant in mineral, tourism, agriculture, forestry, and hydropower resources. The combined installed capacity of the four hydropower stations—Wudongde, Baihetan, Xiluodu, and Xiangjiaba—equals that of two Three Gorges Dams [51]. As resource development and ecological protection efforts have advanced, the region has experienced industrial upgrades and transformation, with poverty alleviation efforts in Wumeng and Daliang Mountains further complicating the interaction between its ecological and economic environments [52]. While past research has predominantly focused on the ecological environment [53,54], the coupling effects between ecology and economy remain underexplored. This gap limits the support needed to consolidate poverty alleviation achievements and foster regional development.

This study investigates the ecological and economic development challenges in the lower reaches of the Jinsha River by integrating a wide range of data sources, including ecological environment, terrain and climate, land resources and utilization, economic indicators, urbanization and population dynamics, infrastructure, and accessibility. Leveraging this integration, the study establishes a comprehensive evaluation index system to assess ecological–economic coupling. It thoroughly examines the spatiotemporal evolution patterns of ecological and economic subsystems, as well as the broader ecological–economic coupling system. Additionally, it explores the driving mechanisms behind these changes and offers strategic recommendations for sustainable future development.

## 2. Study Area and Data Sources

### 2.1. Study Area

Located in southwestern China, the lower reaches of the Jinsha River (LR-Jinsha) refer specifically to the basin stretching from the confluence of the Yalong River in Panzhihua City to the confluence of the Minjiang River in Yibin (Figure 1a). This study focuses on the counties (districts) within the administrative boundaries of the lower Jinsha River basin. The research area is geographically defined by coordinates 100°57′–104°38′ E and 24°29′–28°53′ N, covering an area of 81,035 km$^2$. It spans seven cities (prefectures) in Sichuan and Yunnan provinces, comprising a total of 40 counties (districts), including 16 A-class regions (non-poverty areas) and 24 B-class regions (lifted out of poverty by 2020) (Figure 1b). This regional classification is primarily based on factors such as geographical location, resource availability, economic development levels, poverty status, poverty alleviation goals, policy orientation, and regional development strategies. The 16 A-class regions include Anning, Chenggong, Dongqu, Fumin, Guandu, Huidong, Huili, Jinning, Ningnan, Panlong, Renhe, Shuifu, Songming, Wuhua, Xishan, and Yuanmou. The 24 B-class regions include Butuo, Daguan, Dayao, Dongchuan, Huize, Jinyang, Leibo, Ludian, Luquan, Meigu, Mouding, Pingshan, Puge, Qiaojia, Suijiang, Wuding, Xundian, Yanjin, Yao'an, Yiliang, Yongren, Yongshan, Zhaojue, and Zhaoyang.

The research area features significant variations in altitude and terrain, with steeper slopes in the central and western regions and gentler slopes in the eastern region. The landform is divided into five zones: the Wumeng Mountain Area, Sichuan Basin, Southwest Sichuan Mountain Area, Dry Hot Valley Area, and Plateau Basin Valley Area (Figure 2a). The region predominantly falls within the subtropical monsoon climate zone, characterized by uneven precipitation distribution. The annual average precipitation exceeds 800 mm, and the average annual temperature is approximately 13–14 °C. It is further categorized into three subzones: North Subtropical Zone, Central Subtropical Zone, and South Subtropical Zone (Figure 2a).

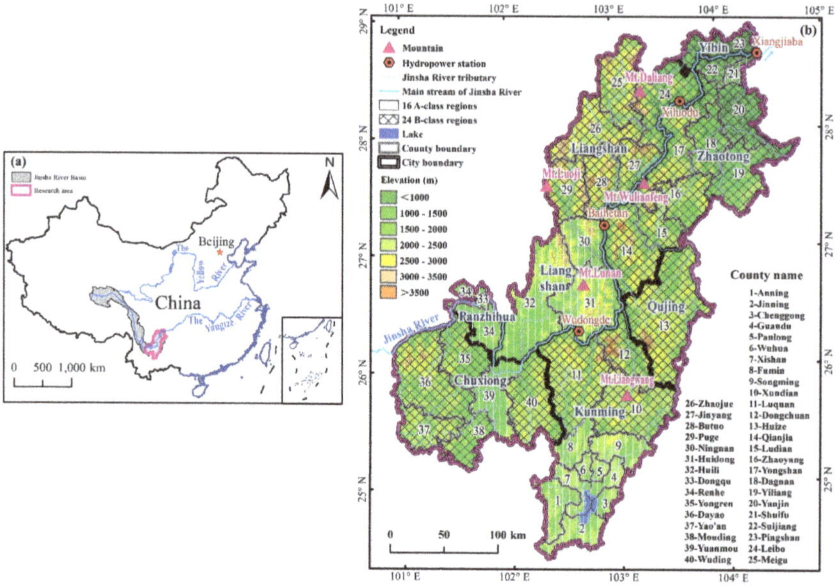


**Figure 1.** Location of the study area. (**a**) Location of the study area in China; (**b**) administrative division of the study area.

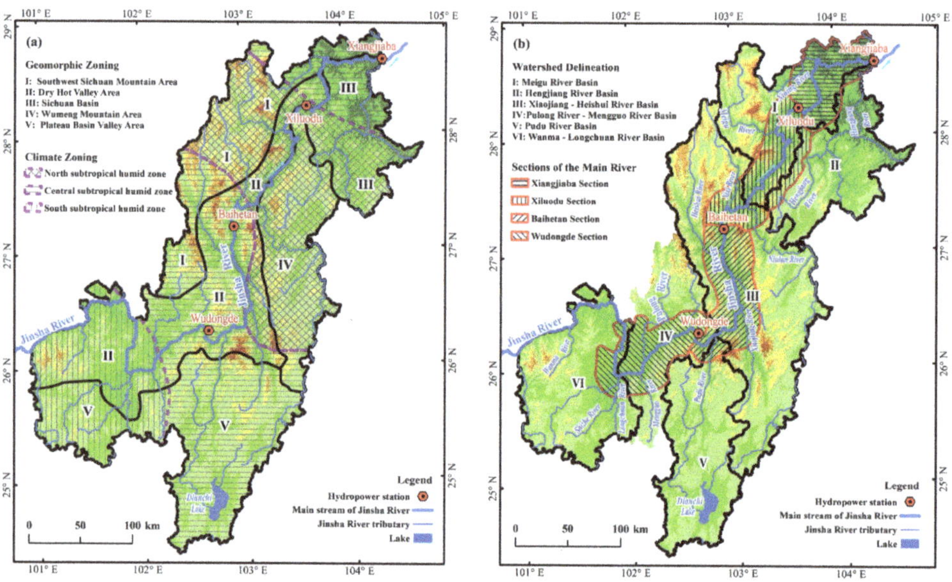


**Figure 2.** Subdivision of the research area: (**a**) Geomorphic and climatic zoning; (**b**) watershed zoning and main stream segmentation.

The water system in the study area is divided into six sub-basins: Meigu River Basin, Hengjiang River Basin, Xiaojiang Heishui River Basin, Pulong River Mengguo River Basin, Pudu River Basin, and Wanma Longchuan River Basin (Figure 2b). Additionally, based on the locations of four major cascade hydropower stations downstream on the Jinsha River, the river is segmented into four sections: Xiangjiaba Section, Xiluodu Section,

Baihetan Section, and Wudongde Section (Figure 2b). According to the "National Ecological Function Zoning" issued by China in 2015, the research area is recognized as an important zone for soil conservation, biodiversity preservation, agricultural production, and urban agglomeration. It is classified as both extremely important and relatively important for ecological protection. This complex ecological and economic landscape forms the basis for the integrated development of ecology and economy in the region.

### 2.2. Data Sources

To identify the ecological–economic coupling system and its dynamic changes, this study primarily gathers time-series data for five key periods: 2000, 2005, 2010, 2015, and 2020. The collected data encompass evaluations of ecological environment quality, assessments of economic development levels, and the driving factors influencing ecological environment changes, economic growth, and the causes of ecological–economic coupling (Table 1). The selection of these specific years is guided by considerations such as the research objectives, data availability, policy and economic contexts, and the evolving dynamics of the ecological–economic coupling system.

**Table 1.** Calculation and analysis data of ecological–economic coupling system.

| No. | Data | Type | Data Sources | Spatial Resolution |
|---|---|---|---|---|
| 1 | Gross value of primary industry [①] | Statistical | Sichuan Statistical Yearbook (2001–2021) and Yunnan Statistical Yearbook (2001–2021) | County |
| 2 | Second gross domestic product [①] | Statistical | | County |
| 3 | Gross value of the tertiary industry [①] | Statistical | | County |
| 4 | Urbanization rate [①] | Statistical | | County |
| 5 | Per capita cropland area [①] | Statistical | China Regional Economic Statistical Yearbook (2001–2021) | County |
| 6 | Total GDP [①] | Statistical | | County |
| 7 | Per capita GDP [①] | Statistical | | County |
| 8 | Total retail sales of consumer goods [①] | Statistical | | County |
| 9 | Nighttime light index [①③] | Grid | Institute of Geographic Sciences and Resources, Chinese Academy of Sciences [a] | 1 km |
| 10 | Cropland area [②] | Grid | Land use dataset created by Yang and Huang (2021) from Wuhan University | 1 km |
| 11 | Woodland area [②] | Grid | | |
| 12 | Grassland area [②] | Grid | | |
| 13 | Water area [②] | Grid | | |
| 14 | Construction land area [②] | Grid | | |
| 15 | Unused land area [②] | Grid | | |
| 16 | Normalized Difference Vegetation Index (NDVI) [②] | Grid | Calculated from Moderate-Resolution Imaging Spectroradiometer (MODIS) data, sourced from the USGS [b] | 1 km |
| 17 | Wetness Index (WET) [②] | Grid | | |
| 18 | Normalized Difference Built-up and Bareness Index (NDBSI) [②] | Grid | | |
| 19 | Land Surface Temperature (LST) [②] | Grid | | |
| 20 | Annual average temperature [③] | Grid | National Qinghai Tibet Plateau Scientific Data Center [c] | 1 km |
| 21 | Annual precipitation [③] | Grid | | |
| 22 | Elevation [③] | Grid | Geographic Spatial Data Cloud [d] | 30 m |
| 23 | Slope [③] | Grid | | |
| 24 | Aspect [③] | Grid | | |
| 25 | Land use type [③] | Grid | Land use dataset created by Yang and Huang (2021) from Wuhan University | 30 m |
| 26 | Cropland density [③] | Grid | | 1 km |
| 27 | Construction land density [③] | Grid | | |
| 28 | Gross domestic product density [③] | Grid | Institute of Geographic Sciences and Resources, Chinese Academy of Sciences [a] | 1 km |

**Table 1.** *Cont.*

| No. | Data | Type | Data Sources | Spatial Resolution |
|---|---|---|---|---|
| 29 | Population density ③ | Grid | School of Geography and Environmental Sciences at the University of Southampton [e] | 100 m |
| 30 | Euclidean distance of road ③ | Grid | Calculated from road vector data, sourced from National Geomatics Center of China [f] | 1 km |

[a] https://www.resdc.cn/ (accessed on 26 July 2024); [b] https://earthexplorer.usgs.gov/ (accessed on 26 July 2024); [c] https://data.tpdc.ac.cn/ (accessed on 26 July 2024); [d] http://www.gscloud.cn/ (accessed on 26 July 2024); [e] https://www.worldpop.org/ (accessed on 26 July 2024); [f] http://www.ngcc.cn (accessed on 26 July 2024); Symbol ① indicates that the data are used to analyze obstacles to the economic development index. Symbols ① and ② indicate that the data are used to analyze the Ecological–Economic Coupling Coordination Degree via the neural network model. Symbol ③ represents the driving forces for analyzing the ecological environment index using the geographic detector model.

#### 2.2.1. Basic Geographic Data

In addition to the thematic data listed in Table 1, the basic geographic data used in this article—such as administrative divisions, place names, water systems, mountain ranges, and power stations—are all sourced from the National Geomatics Center of China. Geomorphic divisions [55], climate divisions [56], and other relevant data are also referenced. To ensure the consistency and accuracy of the research data, multi-source heterogeneous spatial data are uniformly processed: the coordinate system used is WGS1984, with the Universal Transverse Mercator (UTM) projection, specifically using UTM 48N Zone. Statistical data are assigned to the corresponding county-level units for use as vector data, while grid data are standardized in spatial resolution using resampling or interpolation methods as required for evaluation and analysis.

#### 2.2.2. Statistical Data

Statistical data are primarily used for evaluating and analyzing the driving factors of economic development levels, assessing the ecological–economic coupling and coordination degree, and analyzing the degree of obstacles. These data mainly come from the Sichuan Statistical Yearbook, Yunnan Statistical Yearbook, China Regional Economic Statistical Yearbook, and local government annual national economic and social development bulletins. To ensure the accuracy of the spatial representation of economic data, all data from the periods 2000, 2005, 2010, 2015, and 2020 were standardized to the spatial scale of counties and districts, totaling 40 evaluation units in the study area. Data on the gross value of primary industry, secondary gross domestic product, gross value of the tertiary industry, urbanization rate, per capita cropland area, total GDP, per capita GDP, and total retail sales of consumer goods were all processed using this method.

#### 2.2.3. Grid Data

The five data generation methods for cropland area, woodland area, grassland area, water area, construction land area, and unused land area are as follows: using a 2 km × 2 km grid, the area of each grid corresponding to the land use type was calculated, then the statistical value was assigned to the center point of the grid. The inverse distance weighted method was applied to interpolate and generate raster data with a spatial resolution of 1 km. Similarly, the two data generation methods for cropland density and construction land density are as follows: using a 1 km × 1 km grid, the area of cropland or construction land corresponding to each grid was calculated, then the inverse distance weighted method was applied to interpolate and generate raster data with a spatial resolution of 1 km. The four data types of NDVI (Normalized Difference Vegetation Index), WET (Wetness Index), NDBSI (Normalized Difference Built-up and Bareness Index), and LST (Land Surface Temperature) were calculated from MODIS (Moderate-Resolution Imaging Spectroradiometer) data. The elevation data came directly from the Digital Elevation Model (DEM), while the slope and aspect data are derived from the DEM. The data for annual average temperature, annual precipitation, gross domestic product density, and population density were all downloaded and uniformly processed to a spatial resolution of 1 km.

2.2.4. MODIS Data

As a crucial tool for earth science research and application development, the Google Earth Engine (GEE) data processing platform offers extensive data storage and processing capabilities [57]. Users can access a wide variety of remote sensing data stored on the GEE platform in real time, eliminating the need to download and store large amounts of data. This makes data analysis and processing both convenient and efficient. Additionally, GEE provides several effective cloud removals processing solutions, helping users handle obstructions such as clouds and shadows in remote sensing images, thereby enhancing image quality and usability. Due to the vast area of the study region and its complex meteorological conditions, the availability of cloud-free and low-cloud Landsat satellite data is limited. As a result, this study utilizes MODIS data in place of Landsat satellite data. Utilizing this platform, through band operation technology, we achieved minimum cloud amount fusion, obtaining same-season MODIS datasets for five different years—2000, 2005, 2010, 2015, and 2020—as data sources for remote sensing ecological index extraction (Figure A1).

## 3. Methodology

The technical workflow of this study is presented in Figure 3: (I) We extracted four key components—NDVI, WET, NDBSI, and LST—from the MODIS dataset, which were synthesized to form the remote sensing ecological index (RSEI). (II) Land use data were used to calculate the area proportions of different land types, which were then utilized to construct the biodiversity richness index (BRI). (III) The integration of the RSEI and BRI produced the ecological environment index (EEI), while a geographic detector was applied to analyze the driving forces, leveraging additional datasets. (IV) The economic development index (EDI) was derived from economic dataset, revealing spatiotemporal patterns, and the obstacle degree method was employed to identify key drivers. (V) The EEI and EDI data were used to calculate the ecological–economic coupling degree (EECD), the ecological–economic coupling coordination degree (EECCD), and the ecological–economic relative development degree (EERDD), with the spatiotemporal patterns of these indices analyzed. A neural network model was specifically applied to explore the drivers behind the EECCD. (VI) Tailored development strategies were then proposed based on the characteristics of ecological–economic coupling in the region.

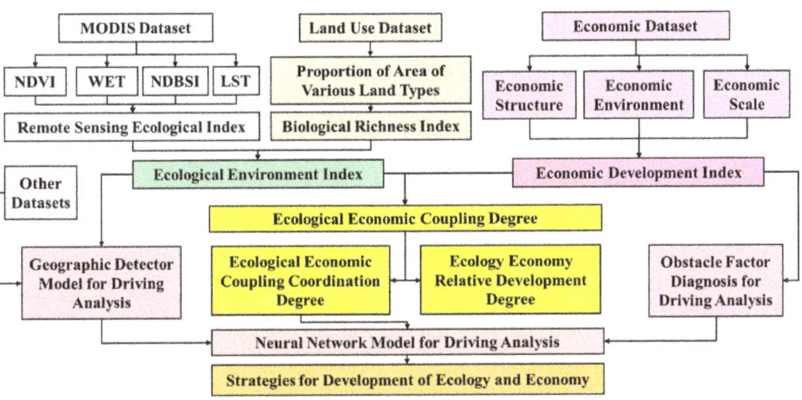


**Figure 3.** Technical workflow of this study.

### 3.1. Construction of Evaluation Index System

Following the principles of scientific rigor, representativeness, and comprehensiveness in indicator selection, this study comprehensively examined the current situation of the research area and constructed an ecological–economic coupling system indicator evaluation system based on previous experience (Table 2). The economic subsystem is characterized by three components: economic structure, economic environment, and economic scale [58–60]. Three indicators—total primary industry output, total secondary industry output, and total tertiary industry output—were selected to reflect the economic structure [60], and these indicators were used to generate the economic structure index (EcoESTI). Three indicators—urbanization rate, per capita cropland area, and nighttime light index—were selected to reflect the economic environment [61–63], and these indicators were used to generate the economic environment index (EcoEEI). Three indicators—total GDP, per capita GDP, and total retail sales of consumer Goods—were selected to reflect the economic scale, and these indicators were used to generate the economic scale index (EcoESCI). The ecological subsystem is characterized by the RSEI and the biological richness index (BRI). The RSEI includes four indicators: NDVI, WET, NDBSI, and LST. The biological richness index includes six indicators: cropland area, woodland area, grassland area, water area, construction land area, and unused land area.

**Table 2.** Evaluation index system for ecological–economic coupling system.

| System | Subsystem | First Level Indicator | Second Level Indicator | Code | Weight |
|---|---|---|---|---|---|
| Ecological Economic Coupling System | Economic Subsystem | Economic Structure | Total Primary Industry Output | $C_{11}$ | Determined by Entropy Weight Method |
| | | | Total Secondary Industry Output | $C_{12}$ | |
| | | | Total Tertiary Industry Output | $C_{13}$ | |
| | | Economic Environment | Urbanization Rate | $C_{21}$ | |
| | | | Per Capita Cropland Area | $C_{22}$ | |
| | | | Nighttime Light Index | $C_{23}$ | |
| | | Economic Scale | Total GDP | $C_{31}$ | |
| | | | Per Capita GDP | $C_{32}$ | |
| | | | Total Retail Sales of Consumer Goods | $C_{33}$ | |
| | Ecological Subsystem | Biological Richness Index | Cropland Area | $E_{11}$ | Determined by Analytic Hierarchy Process |
| | | | Woodland Area | $E_{12}$ | |
| | | | Grassland Area | $E_{13}$ | |
| | | | Water Area | $E_{14}$ | |
| | | | Construction Land Area | $E_{15}$ | |
| | | | Unused land Area | $E_{16}$ | |
| | | Remote Sensing Ecological Index | NDVI | $E_{21}$ | Determined by Principal Component Analysis Method |
| | | | WET | $E_{22}$ | |
| | | | NDBSI | $E_{23}$ | |
| | | | LST | $E_{24}$ | |

### 3.2. Entropy Weight Method

The entropy weight method is an objective weighting technique that assigns weights according to the amount of information each indicator contains. It depends on the degree of data dispersion, which helps to mitigate the overlap of information between indicator variables and reduce human subjectivity. Consequently, it provides an objective reflection of the utility value of the indicators [64]. Due to the inconsistency in the units of different indicators representing economic development levels, direct calculations are not feasible. Therefore, the raw data must first be standardized to eliminate the effects of differing dimensions and the positive or negative orientations of various indicators. Since a standardized value of "0" can affect subsequent calculations, the positive and negative indicators are calculated separately using Formulas (1) and (2), respectively:

$$x'_{ij} = \frac{X_{ij} - min(X_j)}{max(X_j) - min(X_j)} + a \tag{1}$$

$$x'_{ij} = \frac{max(X_i) - X_{ij}}{max(X_j) - min(X_j)} + a \tag{2}$$

where $x'_{ij}$ represents the standardized value; $X_{ij}$ is the raw value of the system indicator; $max(X_j)$ is the maximum value of the $j$-th indicator; $min(X_j)$ is the minimum value of the $j$-th indicator; and $a = 1$.

After completing data standardization, the entropy weight method was used to determine the weights of each indicator. The calculation steps are as follows: calculate the information entropy $e_j$ of the indicators (Formula (3)), calculate the weight $w_j$ of each indicator (Formula (4)), and then determine the economic development level index $F(x)$ using the comprehensive index method after obtaining the weights of the various economic development indicators (Formula (5)).

$$e_j = -k\sum_{i=1}^{m} p_{ij}\ln p_{ij} \tag{3}$$

where $p_{ij} = \frac{x'_{ij}}{\sum_{i=1}^{m} x'_{ij}}$, $k = \frac{1}{\ln m}$.

$$w_j = \frac{1 - e_j}{\sum\limits_{j=1}^{m}(1 - e_j)} \tag{4}$$

$$F(x) = w_j x'_{ij} \tag{5}$$

Multiply the weights of the obtained economic indicators with the normalized economic indicators (Table A1), then add them up to obtain the economic development index (EDI).

### 3.3. Remote Sensing Ecological Index

The remote sensing ecological index (RSEI), introduced by Chinese scholar Xu [65,66], offers a scientific approach to characterizing ecological attributes. It serves as a comprehensive indicator that leverages remote sensing technology and data analysis methods to assess the quality of the regional ecological environment effectively. This index comprises four components: moisture, greenness, dryness, and heat. In this paper, MODIS data are utilized to calculate these various components of the RSEI. The formula is as follows:

$$RSEI = f\ (Moisture, Greenness, Dryness, Heat) \tag{6}$$

Moisture is calculated using the wetness component from the Tasseled Cap Transformation to represent soil moisture, as this component is specifically designed to capture key attributes of soil and plant moisture. Soil moisture conditions are identified as the primary characteristic expressed in this wetness component. The wet component of MODIS data can be calculated as follows [66]:

$$Wet = 0.1147B_1 + 0.2489B_2 + 0.2408B_3 + 0.3132B_4 - 0.3122B_5 - 0.6416B_6 - 0.5087B_7 \tag{7}$$

where $B_1$ to $B_7$ represent bands 1 to 7, corresponding to the *red*, $NIR_1$, *blue*, *green*, $NIR_2$, short-wavelength infrared 1 ($SWIR_1$), and short-wavelength infrared 2 ($SWIR_2$) bands of the MODIS image, respectively.

Greenness is represented by the Normalized Difference Vegetation Index (NDVI) [67]. Vegetation, being the most sensitive indicator of ecological conditions, is utilized to monitor vegetation changes and synthesize RSEI. The calculation formula is as follows [68]:

$$NDVI = \frac{NIR - Red}{NIR + Red} \tag{8}$$

where $NIR$ and $Red$ are the near-infrared ($NIR_1$) and red bands of the MODIS image (MOD09A1), respectively.

Dryness is typically represented by the bare Soil Index (SI), which indicates the scarcity of vegetation in the ecological environment [69]. However, a significant portion of the study area is covered by construction land, which can also contribute to soil "dryness". The Index-based Built-up Index (IBI) reflects the condition of the construction land itself to some extent [70]. Therefore, the dryness component can be represented as the average of SI and IBI, termed the Normalized Difference Built-up and Bareness Index (NDBSI) [71], and is calculated using the following formula:

$$BI = \frac{(SWIR + Red) - (NIR + Blue)}{(SWIR + Red) + (NIR + Blue)} \tag{9}$$

$$IBI = \frac{\frac{2SWIR}{SWIR+NIR} - \frac{NIR}{NIR+Red} - \frac{Green}{Green+SWIR}}{\frac{2SWIR}{SWIR+NIR} + \frac{NIR}{NIR+Red} + \frac{Green}{Green+SWIR}} \tag{10}$$

$$NDBSI = \frac{SI + IBI}{2} \tag{11}$$

where $Red$, $Green$, $Blue$, $NIR$, and $SWIR$ represent the red, green, blue, near-infrared, and shortwave infrared bands of MODIS images, respectively.

The heat component is represented by Land Surface Temperature (LST). We use the MOD11A2 Daytime Land Surface Temperature (DLST) dataset, derived from the MODIS series, to integrate LST data within the GEE platform [72].

The RSEI is synthesized based on its four components: WET, NDVI, NDBSI, and LST. RSEI synthesis can be performed using various methods such as Principal Component Analysis (PCA) [73], the Analytic Hierarchy Process (AHP) [74], and the entropy weight method (EWM) [75]. In this paper, the well-established PCA method is employed. To facilitate the measurement and comparison of RSEI across different time periods, a normalization method is applied to process the RSEI for each period. The final RSEI values range from [0, 1], with values closer to 1 indicating better ecological conditions.

### 3.4. Biological Richness Index

Although RSEI provides a multi-dimensional assessment of ecological environment quality, it should be supplemented with additional indices or components, such as the Biological Richness Index (BRI), particularly in regions experiencing significant changes in terrain, climate, and land use types. The BRI measures the richness and scarcity of local species by calculating the differences in species numbers across various land use types. It is a key indicator for evaluating the quality of the regional ecological environment [76]. The calculation formula is as follows:

$$BRI = \frac{A_{bri}(0.11 \times CLA + 0.35 \times WLA + 0.21 \times GLA + 0.28 \times WA + 0.04 \times CSLA + 0.01 \times ULA)}{} / Area \tag{12}$$

where $BRI$ stands for the biological richness index, $A_{bri}$ is the normalization coefficient, and $CLA$, $WLA$, $GLA$, $WA$, $CSLA$, and $ULA$ represent the areas of cropland, woodland, grassland, wetland, construction land, and unused land, respectively. $Area$ denotes the total area of the region.

### 3.5. Ecological Environment Index

The ecological environment index (EEI) is a comprehensive framework of indicators designed to assess the quality of the ecological environment within a given area. By integrating multiple environmental factors and ecological variables into an assessment system, the EEI aims to translate the state of ecological environmental quality into quantifiable and comparable indicators through scientific algorithms and models. Building on previous research, the EEI is synthesized by combining RSEI and BRI. The calculation formula is as follows:

$$EEI = \alpha \times RSEI + \beta \times BRI \tag{13}$$

where $\alpha$ represents the weight of $RSEI$, and $\beta$ represents the weight of $BRI$. The weight of $RSEI$ and $BRI$ is determined using the Analytic Hierarchy Process, with $\alpha$ set to 0.85 and $\beta$ set to 0.15.

To better observe and analyze the EEI across different periods, the results were classified into five levels at 0.2 intervals: Poor (0.0–0.2), Fair (0.2–0.4), Moderate (0.4–0.6), Good (0.6–0.8), and Excellent (0.8–1.0).

To further investigate changes in EEI this study applies the Theil–Sen median analysis and the Mann–Kendall significance test to evaluate the trends and their statistical significance. The Theil–Sen median analysis, a robust non-parametric method, is utilized to estimate the rate of change in long-term time series data, including meteorological, hydrological, and vegetation data [77]. The calculation formula is as follows:

$$\beta_{EEI} = Median\left(\frac{EEI_j - EEI_i}{j - i}\right),\ 2000 \le i \le j \le 2020 \tag{14}$$

where $\beta_{EEI}$ indicates the rate of change in EEI during the study period. The variables $i$ and $j$ represent the start and end years, respectively, covering the period from 2000 to 2020. $EEI_i$ and $EEI_j$ correspond to the EEI values in the years $i$ and $j$, respectively.

The Mann–Kendall significance test is a non-parametric, rank-based method commonly used to detect turning points in long-term time-series data for meteorological, hydrological, and vegetation variables [78]. Based on the significance levels derived from both the $\beta_{EEI}$ and the Mann–Kendall test, the trends in $\beta_{EEI}$ are classified into four categories: significant increase ($\beta > 0, p < 0.05$), insignificant increase ($\beta > 0, p \ge 0.05$), significant decrease ($\beta < 0, p < 0.05$), and insignificant decrease ($\beta < 0, p \ge 0.05$).

### 3.6. Coupling Coordination Degree

The intricate nature of the relationship between the ecological and economic systems transcends mere interactions, necessitating a deeper exploration. To this end, this article employs a sophisticated coupling coordination model for an exhaustive analysis. Drawing upon the interconnectedness between these two pivotal systems, we have devised a model that quantifies the extent of their coupling, specifically focusing on the harmony between economic development and ecological environmental protection. The calculation formula is as follows:

$$C = \sqrt{\frac{u_1 \times u_2}{[(u_1 + u_2)/2]^2}} = 2\frac{\sqrt{u_1 \times u_2}}{u_1 + u_2} \tag{15}$$

where $u_1$ represents the comprehensive evaluation index of the economic development subsystem, $u_2$ represents the comprehensive evaluation index of the ecological environment subsystem, and C denotes the ecological–economic coupling degree (EECD), with a value range of (0, 1]. A higher EECD value indicates a stronger degree of correlation between the two systems. When EECD reaches 1, it signifies that the two systems have achieved the optimal coupling state. The classification of coupling degrees is shown in Table 3.

**Table 3.** Classification criteria for coupling degree.

| EECD | Coupling Stage |
|---|---|
| (0.0, 0.3] | Low-Level Coupling Stage |
| (0.3, 0.5] | Antagonistic Stage |
| (0.5, 0.8] | Running-In Stage |
| (0.8, 1.0] | High-Level Coupling Stage |

As the EECD merely quantifies the strength of correlation between entities without elucidating the directional nature of their impact, it cannot solely determine whether the influence between two systems is positive or negative. In essence, relying solely on the coupling degree is insufficient to ascertain whether the impact is enhancing or detracting. To gain a more precise and objective understanding of the positive interaction dynamics between the two systems, we employ the following ecological–economic coupling coordination degree (EECCD) model, with its calculation formula detailed as follows:

$$EECCD = \sqrt{C \times T} \tag{16}$$

$$T = \alpha u_1 + \beta u_2 \tag{17}$$

where C represents the EECD, and T is the comprehensive evaluation index of the overall benefits of the two systems. $\alpha$ and $\beta$ are undetermined coefficients, and the sum of the two is 1. Considering the balance between ecosystem and economic development system in evaluation, this article sets $\alpha = \beta = 0.5$. EECCD is a quantitative indicator of the degree of coupling and coordination between economic development and ecological environment protection. Its value directly reflects the degree of coupling and coordination between the two, that is, the quality of interaction and coordination between systems. The specific criteria for dividing EECCD values are detailed in Table 4.

**Table 4.** Classification of different coordination types.

| EECCD | Coordination Level | EECCD | Coordination Level |
|---|---|---|---|
| (0.0, 0.1] | Extreme Imbalance | (0.5, 0.6] | Marginal Coordination |
| (0.1, 0.2] | Severe Imbalance | (0.6, 0.7] | Elementary Coordination |
| (0.2, 0.3] | Moderate Imbalance | (0.7, 0.8] | Moderate Coordination |
| (0.3, 0.4] | Mild Imbalance | (0.8, 0.9] | Effective Coordination |
| (0.4, 0.5] | Near Imbalance | (0.9, 1] | Optimal Coordination |

*3.7. Relative Development Degree*

While the EECCD offers valuable insights into the intensity of the interconnection between ecological environment quality and economic development, it falls short in distinguishing the nuances between different systems. To address this limitation, this article introduces an ecological–economic relative development degree (EERDD) model, which effectively characterizes the dynamics of leadership and lag between ecological environment quality and economic development. The specific calculation formula for this model is presented in Equation (18).

$$EERDD = \frac{u_1}{u_2} \tag{18}$$

where $EERDD$ denotes the relative development degree, $u_1$ represents the Economic Development Index (EDI), and $u_2$ signifies the Ecological Environment Index (EEI). Drawing upon existing research findings [22], the value of EERDD serves as a criterion for categorizing regional economic development types. Specifically, when EERDD > 1.2, it signifies that the region's economic development is ahead of its projected pace, indicating a favorable economic advancement. Conversely, when 0.8 < EERDD $\leq$ 1.2, it indicates a harmonious synchronization between economic development and the ecological environment. Lastly, when EERDD $\leq$ 0.8, it reveals that the region's economic development lags behind its ecological environment, suggesting a potential need for economic rejuvenation or ecological preservation measures.

### 3.8. Geographic Detector Model

As a statistical toolkit, geographic detectors not only provide a quantitative assessment of the spatial heterogeneity of geographic elements but also facilitate a deeper analysis of driving mechanisms [79]. This article uses the EEI in the LR-Jinsha as the dependent variable ($Y$) and considers the factors influencing its changes as independent variables ($X$). By applying the geographic detector model, the study aims to uncover the potential driving mechanisms behind the spatial distribution patterns of ecological environment quality in the LR-Jinsha. Geographic detectors encompass both single-factor detection and interactive detection methods.

Through single-factor detection, the explanatory power of each driving factor $X$ on the spatial differentiation of ecological environment quality $Y$ is determined, allowing for a quantitative analysis of the relative importance of each factor. The calculation formula is as follows:

$$q = 1 - \frac{\sum_{h=1}^{L} N_h \sigma_h^2}{N_h \sigma^2} \tag{19}$$

where the $q$ value measures the correlation between the ecological environment quality ($Y$) and the driving factor detection and evaluation indicators (X), with a range from 0 to 1. A higher $q$ value indicates a stronger explanatory power. $L$ represents the number of layers of the driving factor $X$; $N_h$ and $N$ are the number of sample units in the $h$-th layer and the entire region, respectively; and $\sigma_h^2$ and $\sigma^2$ denote the variances of ecological environment quality values for the h-th layer and the entire region, respectively.

The primary aim of interaction detection is to analyze the pairwise interactions between driving factors to determine whether these interactions influence the explanatory power of ecological environment quality. Specifically, it assesses whether interactions strengthen, weaken, or have no impact on the relationship between the factors and ecological environment quality. These interactions can be categorized into five distinct scenarios, each with its specific outcome and corresponding range of $q$-values, as detailed in Table 5.

**Table 5.** Five different situations of interaction detection.

| Interaction Type | $q$-Value Range |
|---|---|
| Nonlinear attenuation | $q(X_1 \cap X_2) < \mathrm{Min}(q(X_1), q(X_2))$ |
| Single-factor nonlinear attenuation | $\mathrm{Min}(q(X_1), q(X_2)) < q(X_1 \cap X_2) < \mathrm{Max}(q(X_1), q(X_2))$ |
| Dual-factor enhancement | $\mathrm{Max}(q(X_1), q(X_2)) < q(X_1 \cap X_2) < q(X_1) + q(X_2)$ |
| Independent | $q(X_1 \cap X_2) = q(X_1) + q(X_2)$ |
| Nonlinear enhancement | $q(X_1 \cap X_2) > q(X_1) + q(X_2)$ |

Using a geographic detector, the independent impact of various factors on ecological environment quality and their interactive effects were investigated. Based on the study of the ecological environment quality index system in the research area, four factors—climate, topography, land use, and socio-economic factors—were further analyzed. Due to the requirement of geographic detectors for input data to be categorical variables, continuous variables were discretized. Following the principle of maximizing the $q$-value, two grading strategies were selected: quantile and natural breaks, with a range of 5–12 grading levels. The grading strategies are shown in Table 6, and the results of each factor's processing are provided in Table A1.

To ensure the scientific validity of the research, 32,513 sampling points were randomly set at intervals of 5 km in the research area. After excluding sample points and abnormal data in larger water areas such as the lower reaches of the Jinsha River and Dianchi Lake, 31,882 valid sample points were retained. Finally, the independent variables ($X_1$–$X_{13}$) and the dependent variable ($Y$—EEI) were assigned to the valid sample points, and subsequent correlation calculations and analyses were performed.

**Table 6.** Grading of ecological environment index (EEI) driving factors.

| Factor Type | Driving Factors | Variable Code | Unit | Grading Method | Number of Grades |
|---|---|---|---|---|---|
| Climatic | Annual average temperature | $X_1$ | °C | Natural breaks | 9 |
| | Annual precipitation | $X_2$ | mm | Natural breaks | 9 |
| Topographical | Elevation | $X_3$ | m | Quantile | 9 |
| | Slope | $X_4$ | ° | Natural breaks | 9 |
| | Aspect | $X_5$ | Dimensionless | - | 10 |
| Land use | Land use type | $X_7$ | Dimensionless | - | 6 |
| | Cropland density | $X_8$ | Dimensionless | Natural breaks | 12 |
| | Construction land density | $X_9$ | Dimensionless | Quantile | 12 |
| Socio-economic | Nighttime light index | $X_{10}$ | Dimensionless | Quantile | 10 |
| | Gross domestic product density | $X_{11}$ | CNY 10,000/km$^2$ | Natural breaks | 10 |
| | Population density | $X_{12}$ | Person/km$^2$ | Quantile | 10 |
| | Euclidean distance of road | $X_{13}$ | m | Natural breaks | 12 |

### 3.9. Obstacle Factor Diagnosis

The obstacle factor diagnosis model is utilized to identify the main barriers to the economic development index in the study area. In this model, the factor contribution degree ($F_{ij}$) quantifies the impact of a single factor on the entire system, representing its weight in achieving the overall objective. The indicator deviation degree ($I_{ijt}$) measures the gap between a specific indicator and the standards of an economic development system. The degree of impact of the obstacle ($O_{ijt}$) assesses the influence of $F_{ij}$ and $I_{ijt}$ on the development of economy [80].

$$F_{ij} = W_i \times W_{ij} \tag{20}$$

$$I_{ijt} = 1 - P_{ijt} \tag{21}$$

$$O_{ijt} = I_{ijt} \times \frac{F_{ij}}{\sum_{i=1}^{n} F_{ij} \times I_{ijt}} \times 100\% \tag{22}$$

where $n$ represents the total number of indicators. $W_i$ is the weight of the $i$-th classification indicator, while $W_{ij}$ denotes the weight of the $j$-th individual indicator under the $i$-th classification indicator, $P_{ijt}$ is the normalized value of the single indicator after range normalization. $O_{ijt}$ indicates the degree of obstacle posed by the $j$-th individual indicator within the $i$-th classification indicator in the $t$-th year. The weights $W_i$ and $W_{ij}$ are determined using the economic development index system, ensuring the accuracy and scientific validity of the evaluation.

### 3.10. Neural Network Model

A neural network model was employed to identify the driving forces of EECCD. This model simulates the connections and mechanisms of human brain neurons by processing information through a highly interconnected network structure. It comprises multiple layers: an input layer, one or more hidden layers, and an output layer. Each layer contains multiple neurons, also known as nodes or units, that transmit information via weighted connections. By incorporating hidden layers and activation functions, the multilayer perceptron (a type of feedforward neural network) can approximate complex nonlinear functions, enabling it to effectively solve intricate problems (Figure 4) [81].

In this study, a multilayer perceptron neural network model was constructed, with 14 factors as independent variables: total primary industry output, total secondary industry output, total tertiary industry output, urbanization rate, per capita cropland area, nighttime light index, total GDP, per capita GDP, total retail sales of consumer goods, BRI, NDVI, WET, NDSBI, and LST. The EECCD was used as the dependent variable. The influence coefficients of the 14 factors were obtained through analysis.

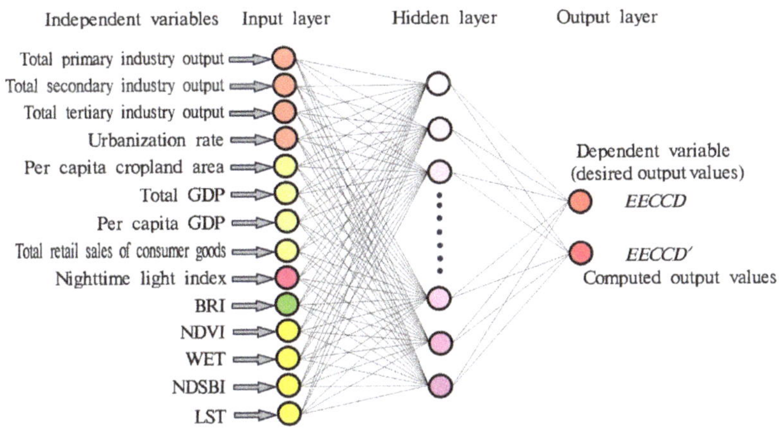


**Figure 4.** Three-layered feed-forward neural network.

# 4. Results

## *4.1. Spatiotemporal Dynamics of Ecological Environment*

### 4.1.1. Spatiotemporal Dynamics in the RSEI

From 2000 to 2020, both the RSEI and its four components exhibited a trend of ecological improvement. The WET component showed an overall upward trend, with a slight decline occurring only between 2010 and 2015, resulting in a net increase of 0.22 (Figures 5a and A3). The NDVI component rose from 0.657 in 2000 to 0.688 in 2020, marking a total increase of 0.031 (Figures 5b and A4). In contrast, the NDBSI component declined from 0.569 in 2000 to 0.555 in 2020, a total decrease of 0.014 (Figures 5c and A5). Similarly, the LST component decreased from 0.553 in 2000 to 0.483 in 2020, a total reduction of 0.070 (Figures 5d and A6).

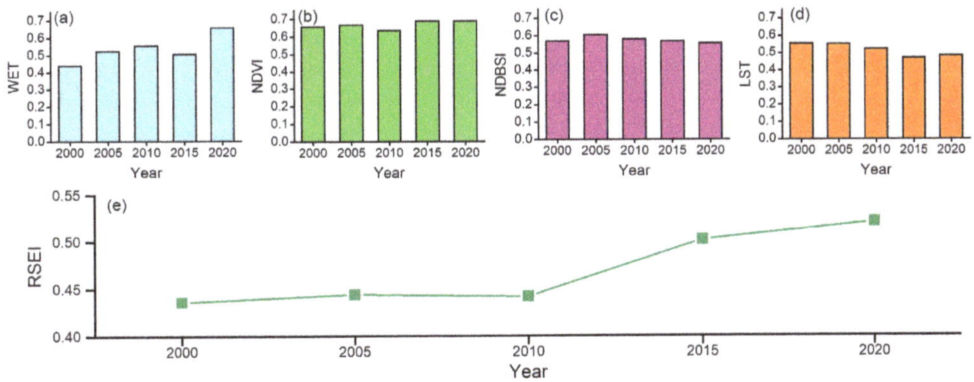


**Figure 5.** Changes in RSEI and its four components in the entire region from 2000 to 2020: (**a**) WET Variation; (**b**) NDVI Variation; (**c**) NDBSI Variation; (**d**) LST Variation; (**e**) RSEI Variation.

Throughout the period, the RSEI in the study area demonstrated an overall trend of improvement (Figure 5e). Taking 2010 as a pivotal year, two distinct stages can be identified: a phase of low-speed and fluctuating improvement from 2000 to 2010, followed by a phase of rapid and sustained improvement from 2010 to 2020. Specifically, the RSEI value was 0.436 in 2000, rose to 0.444 in 2005, slightly decreased to 0.442 in 2010, then quickly climbed to 0.502 in 2015, and further increased to 0.521 in 2020. From 2000 to 2020, the overall RSEI in the study area increased by 0.085, reflecting a growth rate of 19.50%.

The spatial distribution of the RSEI displayed strong geomorphological zoning characteristics (Figure 6a), with average RSEI values ranked from low to high as follows: the Dry Hot Valley Area, the Wumeng Mountain Area, the Plateau Basin Valley Area, the Southwest Sichuan Mountain Area, and the Sichuan Basin. Across the period from 2000 to 2020, the changes in the RSEI revealed four distinct levels of spatial variability (Figure 6b). Specifically, areas experiencing significant reductions comprised merely 0.15% of the total region, while those with non-significant decreases accounted for 15.02%. Conversely, non-significantly improved regions encompassed 76.38% of the area, and significantly improved zones made up 8.45%. The RSEI variation stages in these five sub-regions were generally consistent with those of the entire region, yet each sub-region exhibited unique patterns of change. The sub-regions were ranked by the intensity of RSEI change from highest to lowest as follows: the Wumeng Mountain Area (22.89%), the Plateau Basin Valley Area (19.44%), the Sichuan Basin (18.70%), the Dry Hot Valley Area (17.62%), and the Southwest Sichuan Mountain Area (11.27%).

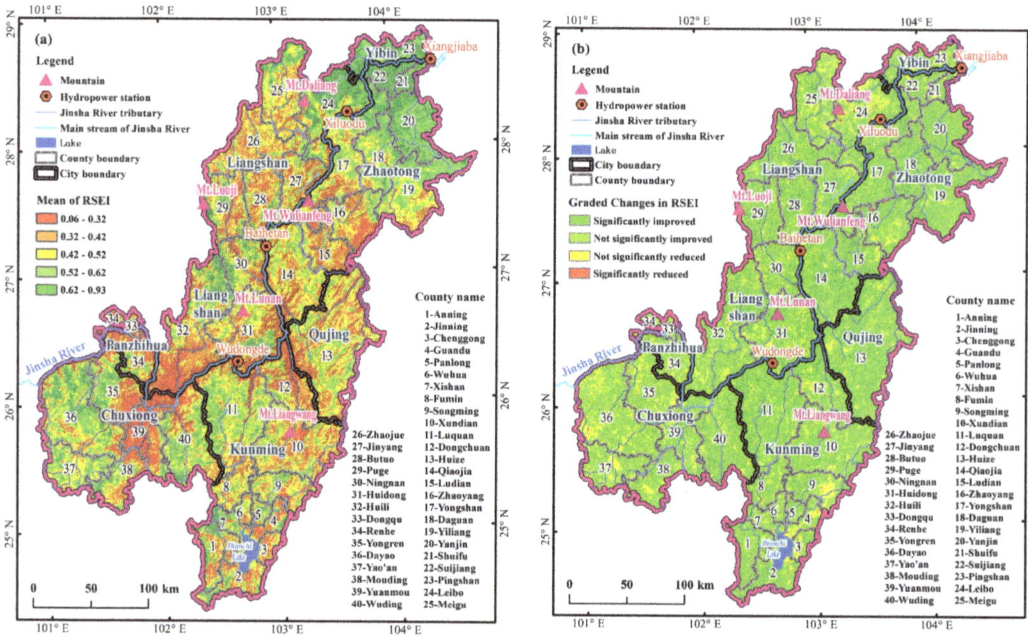


**Figure 6.** Distribution of RSEI. (**a**) Mean RSEI from 2000 to 2020; (**b**) Classification of RSEI changes from 2000–2020.

4.1.2. Spatiotemporal Dynamics in the BRI

The macro spatial distribution of the BRI remained largely unchanged between 2000 and 2020, showing an overall improving trend. The BRI is primarily influenced by land use types (Figure A7), resulting in a mosaic pattern of various BRI values (Figure 7a). High-value areas are predominantly found in forests, grasslands, and wetlands, while low-value areas are mainly located in cropland, construction land, and unused land. During this period, the BRI exhibited two distinct phases: a relatively stable phase from 2000 to 2005, where no significant changes were observed, and a stable improvement phase from 2005 to 2020. During this latter phase, the BRI increased from 0.426 in 2005 to 0.681 in 2010, continued to 0.691 in 2015, and reached 0.699 in 2020. Analysis of BRI changes reveals that areas with significant reduction and minor reduction accounted for 0.26% and 10.79% of the total region, respectively, primarily in areas where forests, grasslands, and wetlands were converted. Conversely, areas with minor improvement and significant improvement made up 68.76% and 20.18% of the total area, respectively, with significant improvements mainly occurring in regions where other land types transitioned to forests, grasslands, and wetlands (Figure 7b).

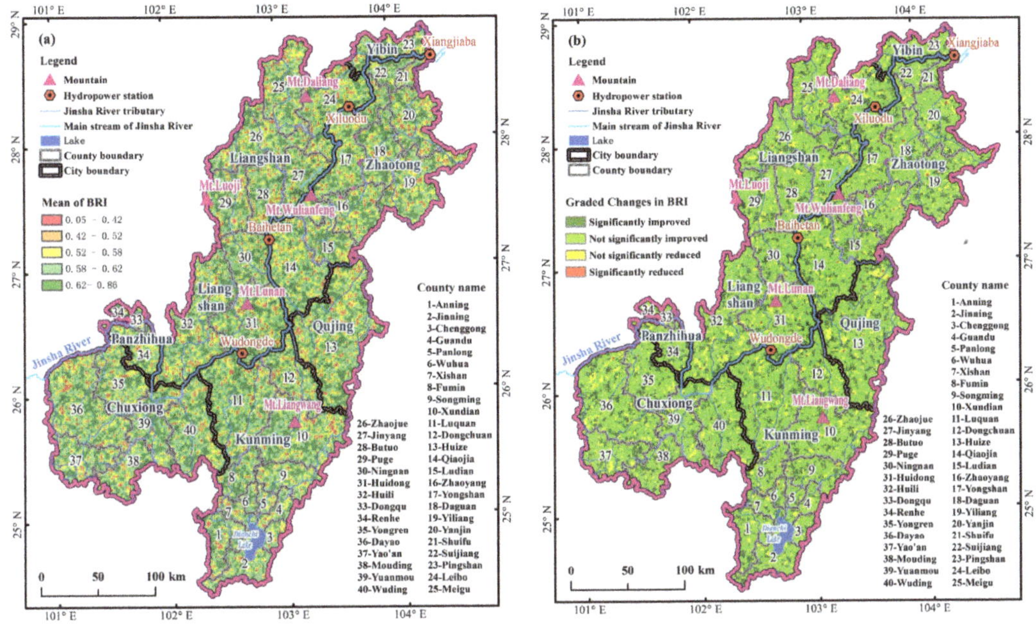


**Figure 7.** Distribution of BRI. (**a**) Mean BRI from 2000 to 2020; (**b**) classification of BRI changes from 2000 to 2020.

The changes in the BRI are primarily driven by adjustments in land use structure (Figure 8). In 2000, the study area was composed of approximately 55.98% forest land, 27.89% cropland, 15.09% grassland, 0.75% wetland, 0.27% construction land, and 0.02% unused land. By 2020, forest land expanded to 58.37%, cropland increased to 28.61%, grassland decreased to 11.47%, wetland increased to 0.92%, and construction land increased to 0.61%, while unused land remained unchanged at 0.02%. Over the two decades, forest land consistently increased, primarily due to the conversion of cropland and grassland driven by reforestation initiatives. Grassland saw a steady decline, with much of it converting to cropland and wetlands. Cropland remained relatively stable, with notable transitions involving construction land, wetland, forest land, and grassland. Construction land expanded significantly due to urban and rural development, infrastructure projects, and targeted poverty alleviation, mainly at the expense of cropland and grassland. Wetlands experienced growth, largely due to the construction of large-scale hydropower stations and associated water conservancy facilities. These projects included the expansion of reservoir areas at the Xiangjiaba (96 km$^2$), Xiluodu (134 km$^2$), Baihetan (1040 km$^2$), and Wudongde (127 km$^2$) hydropower stations. Unused land consistently diminished, predominantly transforming into cropland and grassland.

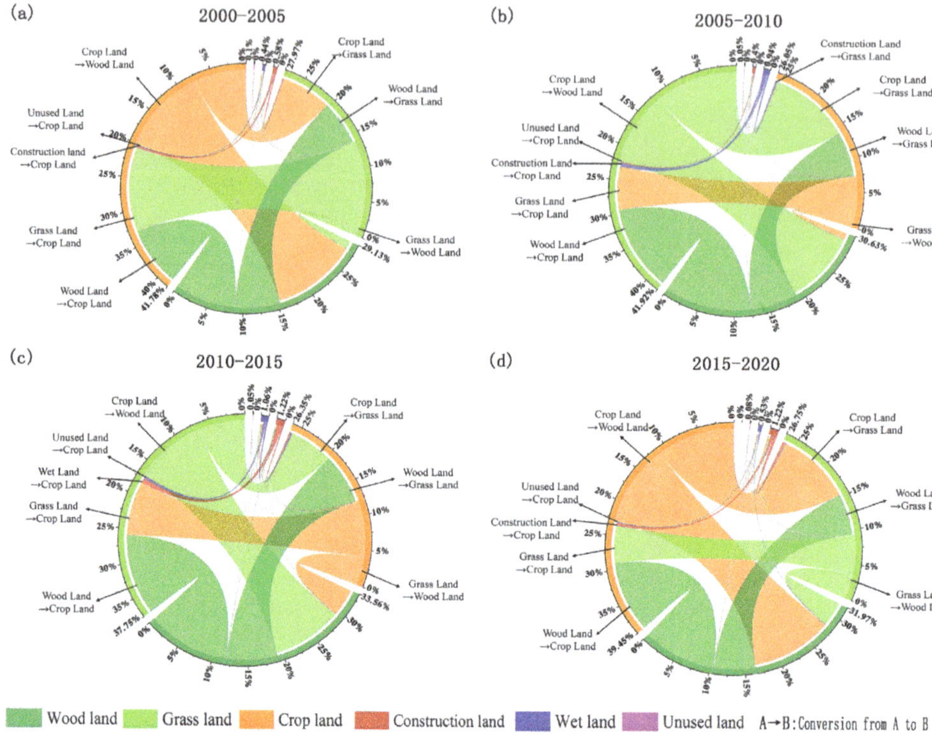


**Figure 8.** Transfer relationships among various land use types in the study area from 2000 to 2020.

#### 4.1.3. Spatiotemporal Dynamics in the EEI

The EEI in the study area shows a continuous improvement trend, driven by the influence of two components: the RSEI and the BRI. From 2000 to 2020, the EEI increased by approximately 19.95%. The change can be divided into two phases: a slow improvement phase from 2000 to 2010, during which the EEI rose from 0.431 in 2000 to 0.434 in 2005, and further to 0.437 in 2010; and a rapid growth phase from 2010 to 2020, where the EEI increased by 0.068 to reach 0.505 in 2015 and maintained a growth rate of 0.012, reaching 0.517 by 2020.

Based on the five-level classification of EEI (Excellent, Good, Moderate, Fair, and Poor) (Figure 9a), the Poor category accounts for approximately 1.43% of the total area, mainly distributed in the southern part of the study area, near regions like Huili, Yongren, Yuanmou, Luquan, and Dongchuan, close to the main stream of the Jinsha River. The Fair category occupies around 31.38% of the area, primarily located in the central valleys and ridges of the study area. The Moderate category, which dominates the region, covers about 50.50% of the total area and is mainly found in the central and southern parts of the study area. The Good and Excellent categories are primarily concentrated in the northern and western parts, together accounting for 16.69% of the area.

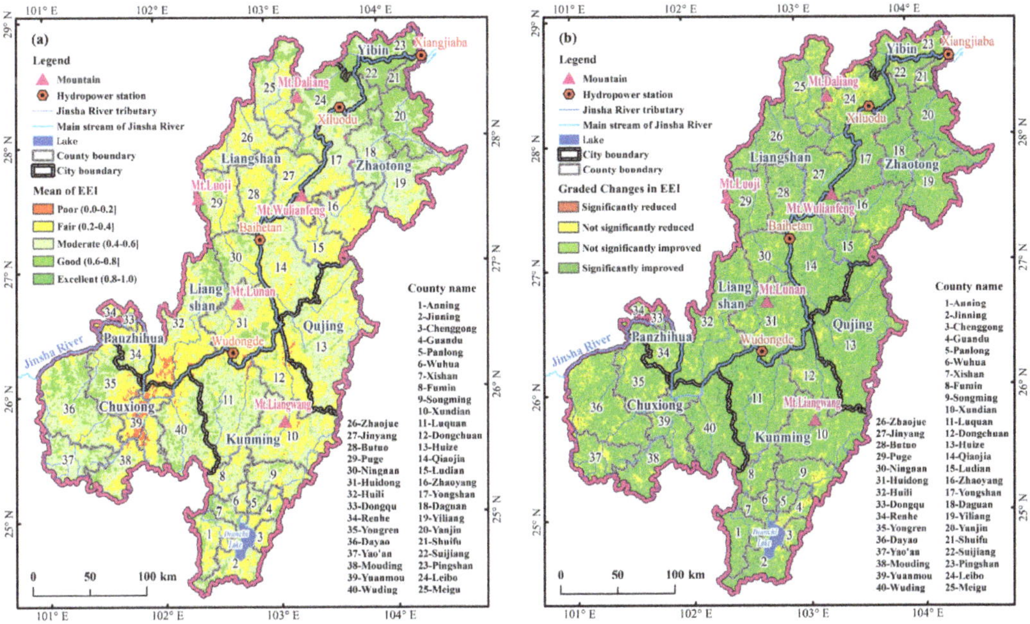


**Figure 9.** Distribution of EEI. (**a**) Mean EEI from 2000 to 2020; (**b**) classification of EEI changes from 2000 to 2020.

Based on the significance test results of EEI changes, the study area primarily exhibited non-significant and significant improvements in ecological environmental quality (Figure 9b). Significant improvement areas, which account for 8.45% of the total area, are mainly located in the border regions of the Wanma–Longchuan River Basin and Pulong–Mengguo River Basin, the northern part of the Xiaojiang–Heishui River Basin, the Southwest Sichuan Mountain Area, and the central Hengjiang–Luoze River Basin. These improvements are widespread and present in both high and low EEI zones. Non-significant improvement areas dominate the study area, making up 76.38%, and are concentrated in the northeastern low-altitude regions, the Pudu River Basin, and the southern Wumeng Mountain Area. These regions are characterized by forest and grassland, with EEI values generally ranging from moderate to high. Areas showing significant degradation are minimal, covering only 0.15% of the total area, and are sparsely located in the western Wanma–Longchuan River Basin, southeastern Kunming, and northern Meigu River Basin, primarily in forested areas. Non-significant degradation zones make up 15.02% of the study area, typically appearing in patchy distributions in southwestern Chuxiong, northern Liangshan, and around the Dianchi Basin in Kunming, with predominant land uses being cultivation and construction.

The structural composition of the EEI has undergone continuous optimization from 2000 to 2020 (Figure 10). During this period, the proportion of the Poor category decreased from 3% to 1%, while the Fair category declined significantly from 40% in 2000 to 20% in 2020. Meanwhile, the Moderate category showed an increase from 45% in 2000 to 50% in 2020. The Good category also saw notable growth, rising from 11% in 2000 to 28% in 2020. The Excellent category, however, remained relatively stable, fluctuating between 1% and 2% throughout the two decades.

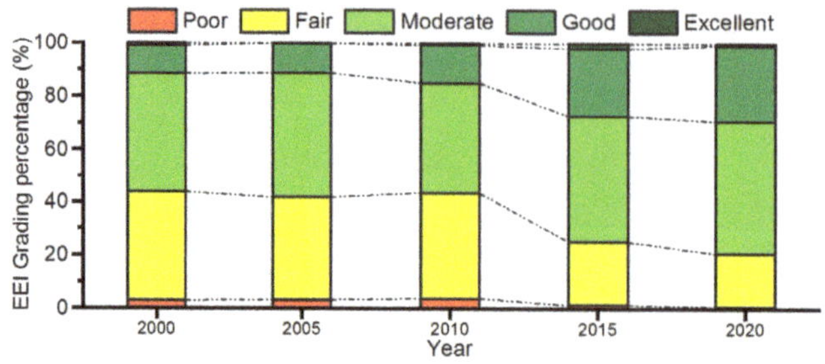


**Figure 10.** Changes in the percentage composition of five EEI grades from 2000 to 2020.

At the geomorphic zoning scale, the average EEI values and their respective improvement rates are as follows (Figure 11): Sichuan Basin (0.5734, 22%), Southwest Sichuan Mountain Area (0.4957, 18.57%), Plateau Basin Valley Area (0.4743, 20.58%), Wumeng Mountain Area (0.4298, 27.15%), and Dry Hot Valley Area (0.4177, 18.60%). Notably, significant variations in the EEI are observed within both the Southwest Sichuan Mountain Area and the Dry Hot Valley Area.

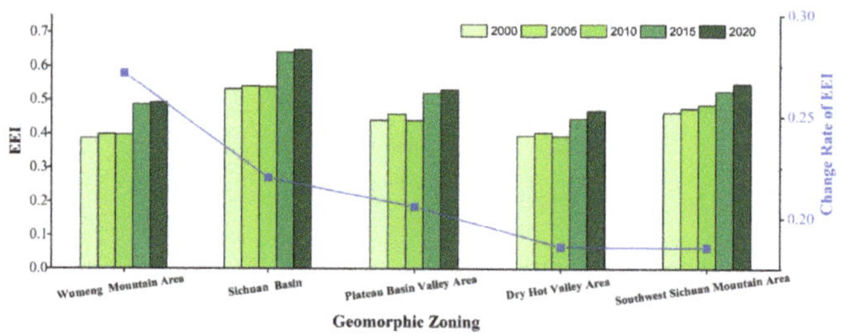


**Figure 11.** EEI changes in five geomorphic zones from 2000 to 2020.

At the sub-watershed scale (Figure 12), the average EEI values and their respective improvement rates are as follows: Meigu River Basin (0.5331, 10.08%), Hengjiang River Basin (0.5095, 24.26%), Pudu River Basin (0.4751, 22.31%), Wanma-Longchuan River Basin (0.4531, 13.72%), Pulong-Mengguo River Basin (0.4317, 26.06%), and Xiaojiang-Heishui River Basin (0.4267, 23.75%). Notably, considerable variation in EEI is also observed within both the Wanma-Longchuan River Basin and the Pulong-Mengguo River Basin.

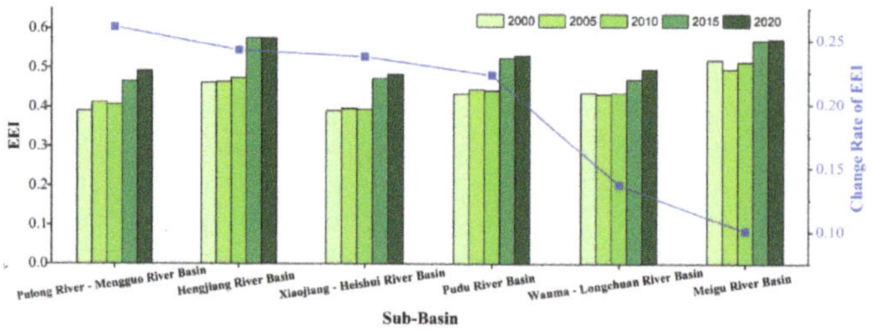


**Figure 12.** Changes in EEI of six sub-basins from 2000 to 2020.

The distribution and variation of EEI across different sections of the main stream show a clear gradient pattern (Figure 13). In ascending order of elevation, the average EEI is 0.6343 in the Xiangjiaba section, 0.4493 in the Xiluodu section, 0.4041 in the Baihetan section, and 0.3376 in the Wudongde section. From 2000 to 2020, EEI demonstrated a trend of smaller increases in high-value areas and larger increases in low-value areas. Specifically, the growth rates of EEI were 6.95% in the Xiangjiaba section, 18.70% in the Xiluodu section, 19.81% in the Baihetan section, and 25.76% in the Wudongde section.

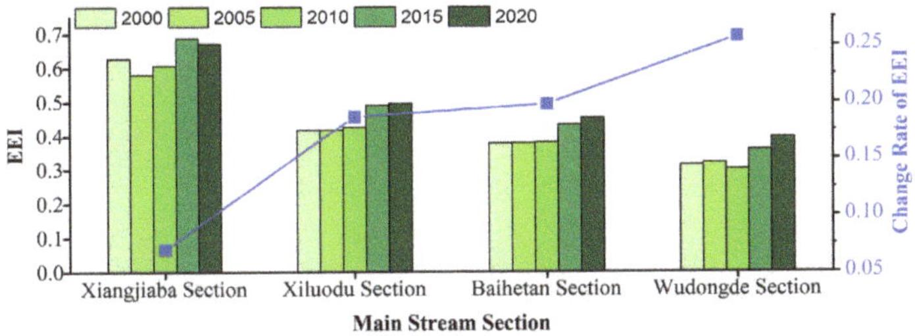


**Figure 13.** EEI changes in four sections of the main stream from 2000 to 2020.

From 2000 to 2020, the county-level EEI exhibited an overall improvement trend, with a noticeable shift in EEI levels from Fair to Moderate and Good (Figure 14, Table A2). In 2000, four counties—Suijiang, Shuifu, Pingshan, and Leibo—were classified as Good, while 22 counties were classified as Moderate. Fourteen counties, including Huidong, Chenggong, Renhe, Butuo, Zhaoyang, Ludian, Dongchuan, and Yuanmou, were classified as Fair. By 2020, the counties classified as Good had expanded to include Suijiang, Yanjin, Shuifu, Pingshan, Daguan, and Leibo. The number of counties classified as Moderate increased to 33, while the Fair classification was reduced to just one county: Yuanmou.

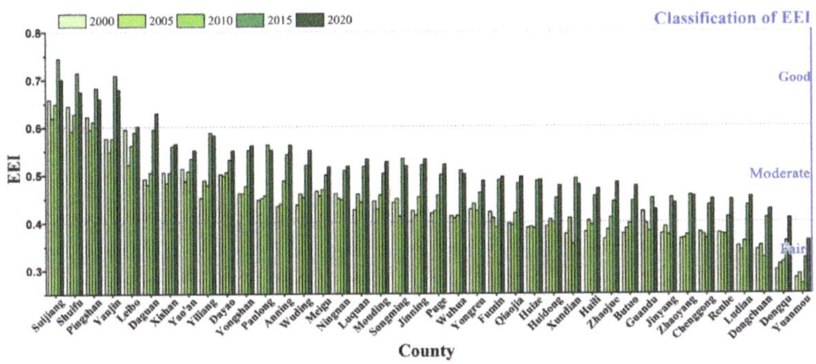


**Figure 14.** Changes in EEI from 2000 to 2020.

## 4.2. Spatiotemporal Dynamics of Economic Development

### 4.2.1. Spatiotemporal Dynamics in the Economic Structure Index

Using the economic structure index (EcoESTI) to analyze the spatiotemporal changes in the economic structure subsystem, the study area exhibited a consistent upward trend in EcoESTI from 2000 to 2020, with the most significant growth occurring between 2015 and 2020 (Figure 15, Table A3). The region's average EcoESTI increased from 0.0680 in 2000 to 0.075 in 2005, remained steady at 0.075 in 2010, rose slightly to 0.076 in 2015, and reached 0.092 in 2020. This reflects a total increase of 0.0270 over the two decades, representing a growth rate of 39.74%.

Between 2000 and 2020, eight counties—Guandu, Wuding, Xishan, Panlong, Dongqu, Anning, Zhaoyang, and Huili—achieved an average EcoESTI value exceeding 0.1000, indicating strong development potential in agriculture, forestry, mineral resources, water resources, and tourism. These counties are primarily located in the southern and central parts of the study area. In contrast, nine counties—Leibo, Yongren, Puge, Zhaojue, Jinning, Meigu, Daguan, Butuo, and Suijiang—recorded an average EcoESTI value below 0.050, highlighting significant opportunities to improve resource endowments, infrastructure, and industrial structures. These counties are mainly concentrated in the northern part of the study area.

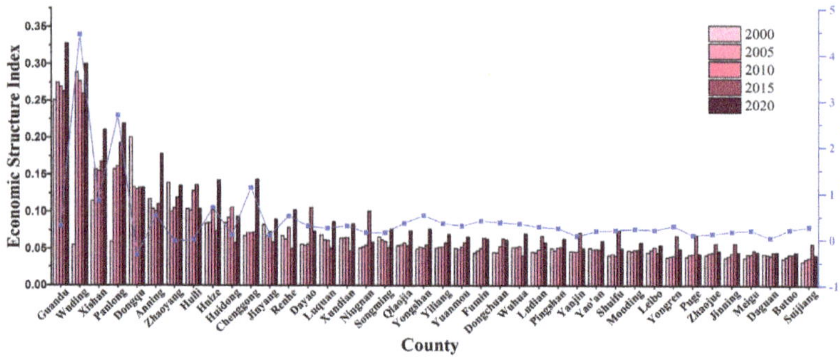


**Figure 15.** Changes in economic structure index (EcoESTI) from 2000 to 2020.

4.2.2. Spatiotemporal Dynamics in the Economic Environment Index

The average economic environment index (EcoEEI) for the region in 2000, 2005, 2010, 2015, and 2020 was 0.1041, 0.0917, 0.0983, 0.0958, and 0.0978, respectively, indicating a slight decline followed by a rebound in 2020 (Figure 16, Table A4). Spatially, the EcoEEI displays a clear north-south differentiation, with lower values in the north and higher values in the south, roughly divided by a line formed by Ningnan, Huidong, and Dongchuan. From 2000 to 2020, ten counties—Dongqu, Wuding, Panlong, Guandu, Chenggong, Xishan, Anning, Renhe, Jinyang, and Songming—consistently maintained an average EcoEEI above 0.1000. In contrast, five counties—Zhaojue, Pingshan, Qiaojia, Yiliang, and Yanjin—had an average EcoEEI below 0.060.

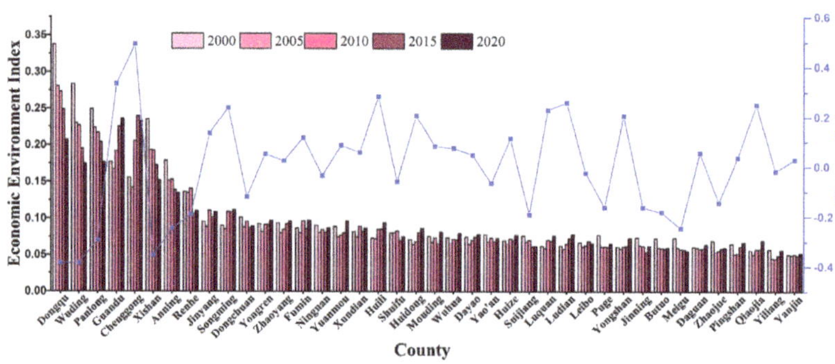


**Figure 16.** Changes in economic environment index (EcoEEI) from 2000 to 2020.

Between 2000 and 2020, 22 counties experienced positive EcoEEI growth, with five counties—Chenggong, Guandu, Huili, Ludian, and Qiaojia—achieving growth rates exceeding 25%, largely due to urbanization and infrastructure development that offset the impact of reduced per capita cropland. Conversely, 18 counties saw a negative EcoEEI growth rate, with five counties—Anning, Panlong, Xishan, Wuding, and Dongqu—experiencing declines of more than 25%, primarily due to factors such as reforestation, grassland restoration, and the reduction of per capita cropland caused by construction projects.

4.2.3. Spatiotemporal Dynamics in the Economic Scale Index

The economic scale index (EcoESCI) in the research area exhibited a continuous growth trend from 2000 to 2020 (Figure 17, Table A5). The average EcoESCI values for the region were 0.0793, 0.0922, 0.0963, 0.1073, and 0.1094 in 2000, 2005, 2010, 2015, and

2020, respectively, representing an overall growth rate of 37.96% over the two decades. The spatial distribution of EcoESCI revealed a pattern of higher values in the central region and lower values in the northern and southern regions.

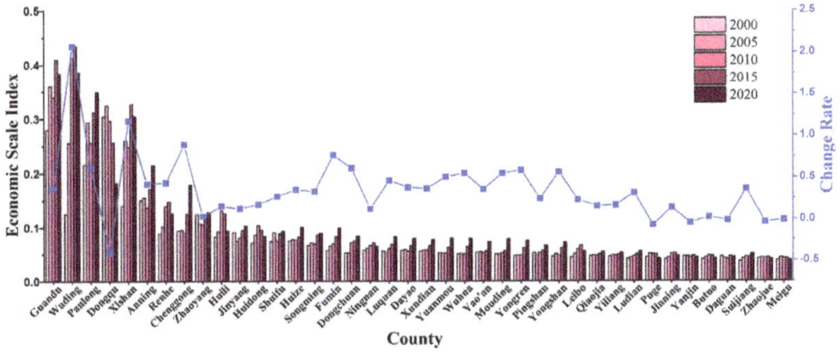


**Figure 17.** Changes in economic scale index (EcoESCI) from 2000 to 2020.

Between 2000 and 2020, ten counties, including Guandu, Wuding, Panlong, Dongqu, Xishan, Anning, Renhe, Chenggong, Zhaoyang, and Huili, achieved an average EcoESCI value greater than 0.1000. Conversely, eleven counties, including Qiaojia, Yiliang, Ludian, Puge, Jinning, Yanjin, Butuo, Daguan, Suijiang, Zhaojue, and Meigu, had an average EcoESCI value below 0.0500.

Over the same period, 36 counties experienced positive EcoESCI growth. Notably, Wuding and Xishan counties recorded significant growth rates of 207.46% and 117.77%, respectively. In contrast, counties located in the Daliangshan and Wumeng mountainous areas, such as Meigu, Daguan, Zhaojue, Yanjin, Puge, and Dongqu, exhibited growth rates below 0, with Dongqu experiencing a notable decline of −40.31%.

#### 4.2.4. Spatiotemporal Dynamics in the EDI

The EDI in the study area, influenced by the combined effects of the economic structure, economic environment, and economic scale subsystems, exhibited a consistent growth trend from 2000 to 2020 (Figure 18, Table A6). The average EDI values for the entire region were 0.152, 0.159, 0.171, 0.186, and 0.203 in 2000, 2005, 2010, 2015, and 2020, respectively. This reflects an overall increase of 0.051, or approximately 33.55%, over the two decades. Analyzing the distribution of EDI levels across forty counties, in 2000, twenty-six counties had an EDI value below 0.1, eight counties had an EDI between 0.1 and 0.25, and six counties had an EDI above 0.25. By 2020, the number of counties with an EDI value below 0.1 decreased to eleven, while those with an EDI between 0.1 and 0.25 increased to twenty-two, and those with an EDI above 0.25 increased to seven.

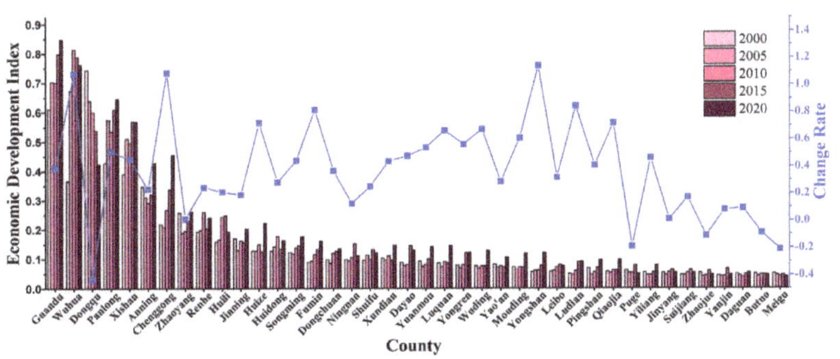


**Figure 18.** Changes in the EDI from 2000 to 2020.

Throughout the period of EDI change, counties with higher EDI rankings displayed variability in their rankings, while those with lower rankings exhibited lagging or declining trends. Counties with mid-range EDI rankings generally maintained stable growth. From 2000 to 2020, a total of 35 counties experienced positive EDI growth rates, with Yongshan, Chenggong, and Wuhua showing growth rates exceeding 100%. Conversely, Yanjin, Jinyang, and Zhaoyang had growth rates below 10%. Five counties experienced a decline in EDI, including Butuo, Zhaojue, Puge, and Meigu in Liangshan Prefecture, and Dongqu in Panzhihua, with Dongqu experiencing the most significant decrease at 43.15%.

The EDI exhibits a distinct spatial differentiation between the northern and southern regions (Figure 19). In the study area, a spatial differentiation pattern of EDI has formed along the line connecting Puge, Butuo, Leibo, and Ludian. South of this line lies the high EDI value zone and relatively high-growth area, while the area along the line and to the north, mainly encompassing the Daliangshan and Wumeng mountainous regions, is characterized by low EDI values and slow development. Additionally, EDI values and growth rates gradually decrease as one moves from the central areas of Zhaoyang in Zhaotong, Huili, Dongqu, and Renhe in Panzhihua, and Anning, Chenggong, Guandu, and Xishan in Kunming towards the surrounding regions.

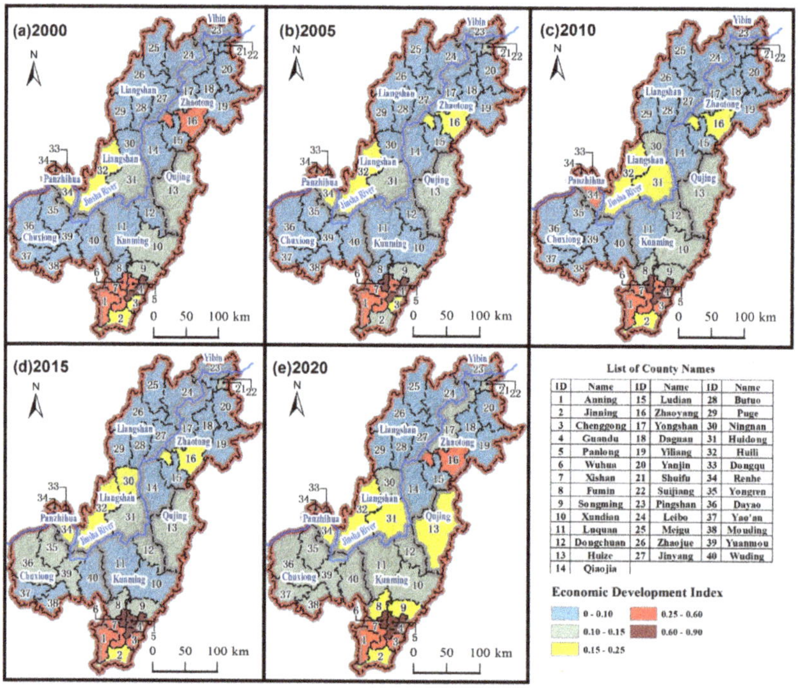


**Figure 19.** Distribution of EDI changes from 2000 to 2020.

Moran's index analysis of the EDI data from five periods between 2000 and 2020 shows an increase from 0.801 in 2000 to 1.543 in 2020, indicating a growing positive spatial correlation of EDI among the counties in the study area, with similar values becoming more spatially clustered (Figure A8).

### *4.3. Spatiotemporal Dynamics of Ecology–Economy Coupling*

#### 4.3.1. Spatiotemporal Dynamics in the EECD

During the period from 2000 to 2020, the overall ecological–economic coupling degree (EECD) in the study area remained at a high level, showing a slight decline followed by a modest increase (Figure 20, Table A7). The average EECD across the study area remained relatively stable, with a slight overall increase from 0.7668 in 2000 to 0.7883 in 2020. The number of counties in the High-Level Coupling Stage increased gradually, rising from 17 counties in 2000 to 20 counties by 2015, where it stabilized through 2020. The number of counties in the Running-In Stage fluctuated slightly, decreasing from 22 counties in 2000 to 20 counties by 2015, where it also stabilized through 2020. The Antagonistic Stage was nearly absent, with only one county (Suijiang) in this stage in 2000, and none in subsequent years. Additionally, sample data from 40 counties show that the EECD metric is only able to roughly depict the High-Level Coupling Stage and Running-In Stage, falling short in capturing the nuanced differences in ecological–economic interactions between counties.

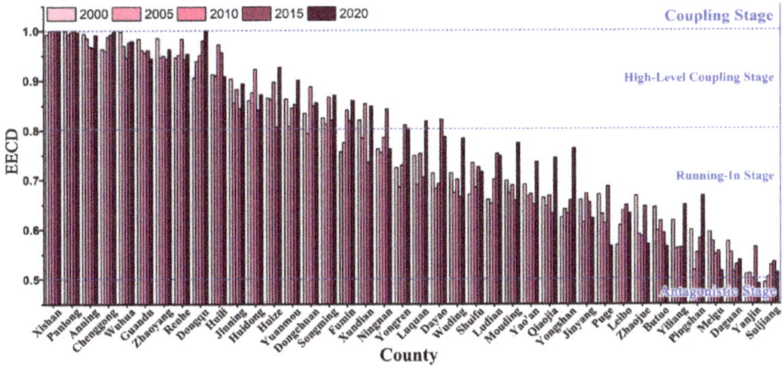


**Figure 20.** Changes in EECD from 2000 to 2020.

#### 4.3.2. Spatiotemporal Dynamics in the EECCD

From 2000 to 2020, the average EECCD in the study area showed an overall increasing trend. The average EECCD was 0.472 in 2000, slightly decreased to 0.468 in 2005, and then increased to 0.481 in 2010. During the period from 2000 to 2010, the entire area was predominantly in the near imbalance stage. By 2015, the EECCD surged to 0.514 and further increased to 0.531 by 2020, with the entire area predominantly entering the marginal coordination stage from 2015 to 2020. From 2000 to 2020, the EECCD increased by 0.059, representing an approximate growth of 12.50%.

The average EECCD and its rate of change across 40 counties from 2000 to 2020 did not exhibit a linear negative correlation (Figure 21, Table A8), indicating the varying impacts of economic development history and stages among these counties. In the 24 B-class regions, the average EECCD was 0.4205 in 2000, slightly declined to 0.4077 in 2005, rebounded to 0.4197 in 2010, surged to 0.4524 in 2015, and continued to rise to 0.4715 in 2020, marking an overall increase of 12.13% from 2000 to 2020. During the same period, the 16 A-class regions exhibited a steady upward trajectory, with the average EECCD increasing from 0.5501 in 2000 to 0.5591 in 2005, 0.5736 in 2010, 0.6064 in 2015, and 0.6191 in 2020, reflecting a total increase of 12.54%. Among the top twenty counties in terms of EECCD growth rate, twelve were poverty-alleviated counties, with six counties—Yongshan, Ludian, Huize, Qiaojia, Wuding, and Luquan—ranking within the top 10 at 1st, 4th, 5th, 7th, 8th, and 9th, respectively. Conversely, eight of the counties ranked 31st to 40th were also poverty-alleviated counties, underscoring the disparity in EECCD changes among these regions.

The coordination levels of counties in the study area have consistently improved over time (Figures 22 and 23, Table A9). The number of counties classified as Mild Imbalance and Near Imbalance decreased from 30 in 2000 to 18 in 2020. However, four poverty-alleviated counties—Meigu, Zhaojue, Jinyang, and Butuo—remain in the Mild Imbalance category. In contrast, the number of counties under Marginal Coordination increased from four in 2000 to fifteen in 2020, marking a critical transitional phase in the development of ecology–economy coupling. The number of counties classified as Elementary Coordination and Moderate Coordination remained stable, at six and seven counties each. In 2000, Guandu was the only county classified under Moderate Coordination. Starting in 2005, this category expanded to include Panlong, Wuhua, and Xishan in Kunming, a classification that persisted through 2020. By then, Anning in Kunming was also included, bringing the total number of counties under Moderate Coordination to five.

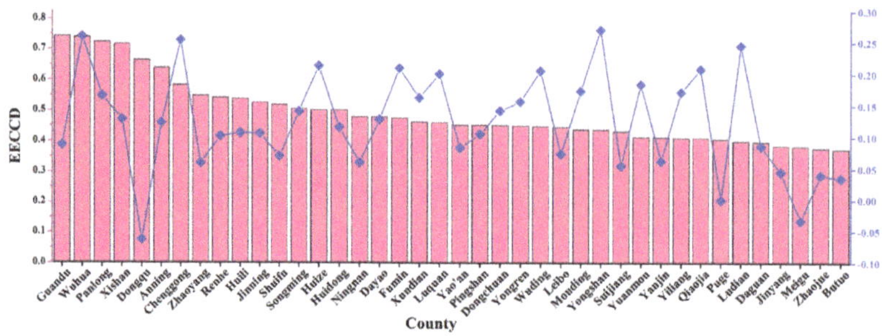


**Figure 21.** Comparison of average EECCD and its change rate from 2000 to 2020.

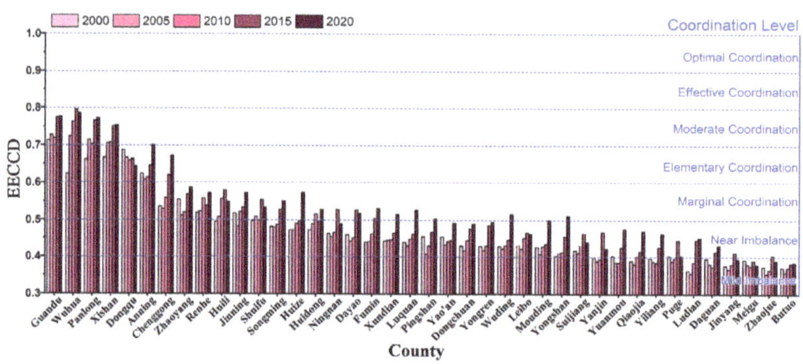


**Figure 22.** EECCD changes and comparison with coordination level in 40 counties from 2000 to 2020.

From 2000 to 2020, the distribution of the average EECCD in the study area displayed a spatial pattern with higher values in the south and lower values in the north (Figure 24a). The southern counties of Guandu (0.742), Wuhua (0.739), Panlong (0.723), and Xishan (0.717) in Kunming were classified as having a Moderate Coordination level, representing the highest average EECCD values in the region during this period. Dongqu (0.644) in Panzhihua and Anning (0.639) in Kunming were at the Elementary Coordination level. In the northern part of the study area, the counties of Butuo (0.373), Zhaojue (0.377), Meigu (0.383), Jinyang (0.385) in Liangshan Prefecture, and Daguan (0.398) in Zhaotong were classified as Mild Imbalance level, indicating the lowest EECCD values. Nine counties, including Chenggong and Huidong, were at the Marginal Coordination level, while twenty counties, such as Ningnan, Puge, and Ludian, were at the Near Imbalance level, predominantly located in the southwestern, central-northern, and northern regions. Regarding the Jinsha River main stream segmentation, EECCD values were relatively high throughout most areas, except for regions east and north of the Baihetan Hydropower Station, with particularly high values near Panzhihua. The hot spot analysis revealed a prominent hot spot area centered around Kunming and cold spot areas in Zhaojue and Butuo, whereas the remaining counties did not display significant clustering of either hot spots or cold spots (Figure 24b).

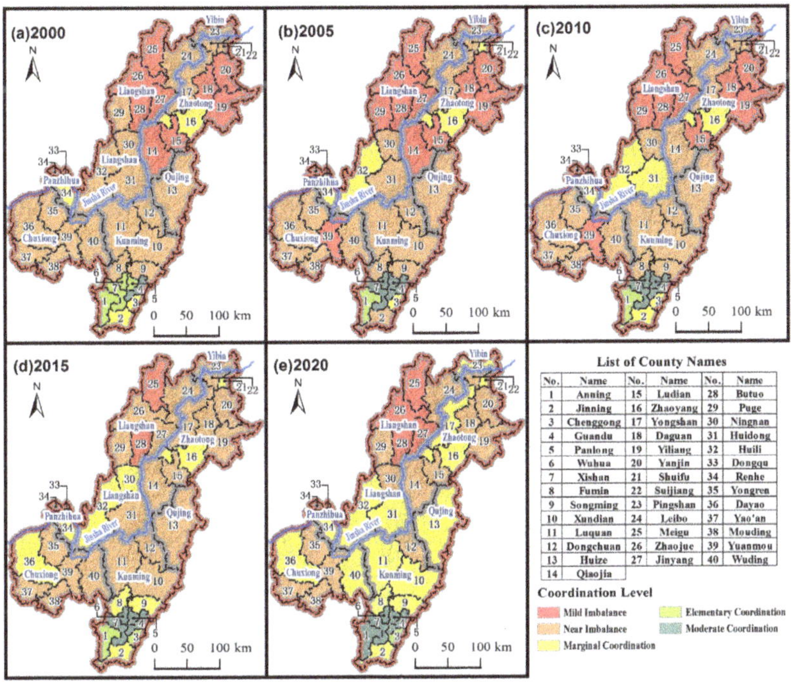


Figure 23. Spatial distribution of EECCD from 2000 to 2020.

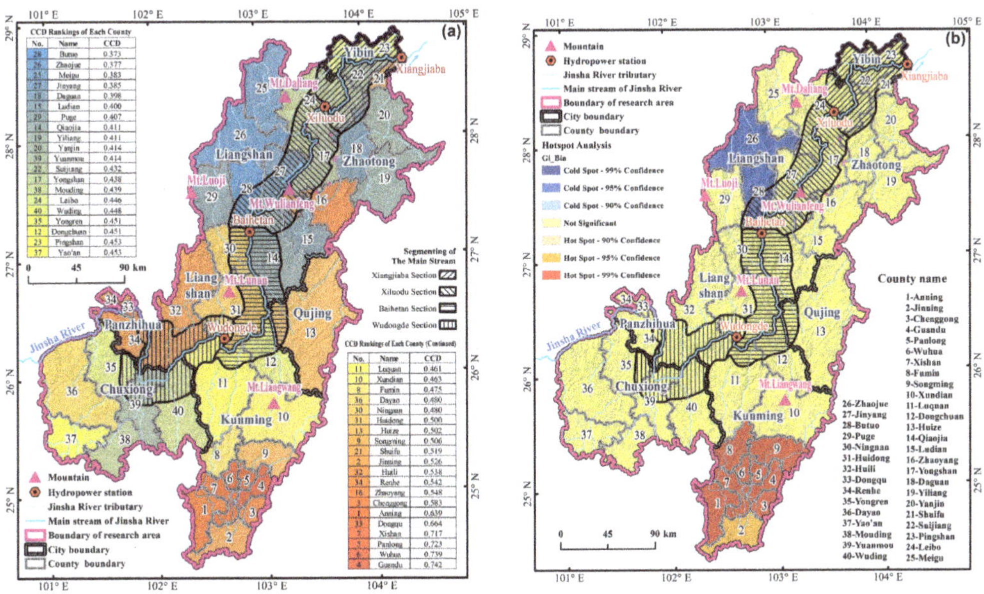


Figure 24. Spatial distribution of EECCD. (a) Average EECCD from 2000 to 2020; (b) hot spot analysis of average EECCD from 2000 to 2020.

4.3.3. Spatiotemporal Dynamics in the EERDD

From 2000 to 2020, although the ecological–economic relative development degree (EERDD) growth in the 24 B-class regions lagged behind that of the 16 A-class regions, a positive trend was still apparent. The average EERDD in the 24 B-class regions for 2000, 2005, 2010, 2015, and 2020 was 0.1801, 0.1554, 0.1746, 0.1705, and 0.2004, respectively, demonstrating an initial decline followed by a subsequent increase, culminating in an overall growth of 11.27%. In comparison, the average EERDD in the 16 A-class regions during the same period was 0.6128, 0.7381, 0.7889, 0.7039, and 0.7254, respectively, showing a pattern of increase, decline, and then increase, with a total growth of 18.37%. Overall, the EERDD development in these 16 counties was significantly better than in the 24 B-class regions.

According to the EERDD stage criteria, economic development in most counties within the study area continues to lag behind ecological environment improvements (Figure 25, Table A10). From 2000 to 2020, only two to four counties consistently maintained an EERDD classified as the "EDI Leads EEI Stage", with just one county remaining in this stage by 2020. Dongqu was in the "EDI Leads EEI Stage" from 2000 to 2015 but shifted to the "EDI and EEI Synced Stage" by 2020. Wuhua remained in the "EDI Leads EEI Stage" from 2005 to 2020. Panlong entered the "EDI Leads EEI Stage" in 2005, downgraded to the "EDI and EEI Synced Stage" in 2015, but returned to the "EDI Leads EEI Stage" in 2020. During this period, only one to four counties were classified under the "EDI and EEI Synced Stage", while the remaining counties, including all 24 B-class regions, were in the "EDI Behind EEI stage".

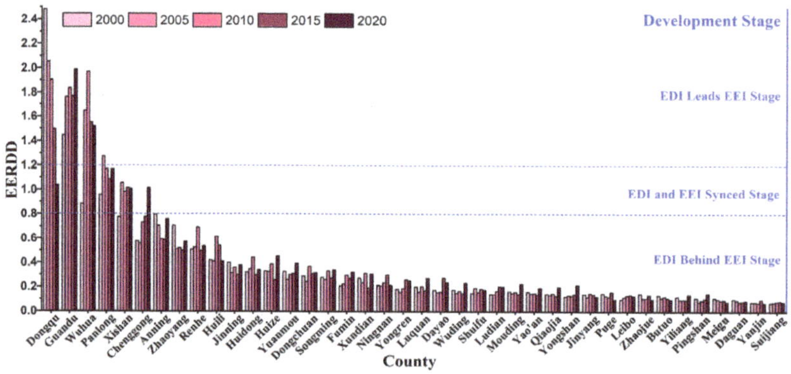


**Figure 25.** Ranking of ecological–economic relative development degree (EERDD) averages from 2000 to 2020.

Analyzing the EERDD growth rate from 2000 to 2020 reveals that counties with stronger economic foundations exhibited higher improvement rates (Figure 26, Table A11). Among the twenty-four B-class regions, eight—Yanjin, Jinyang, Daguan, Zhaoyang, Butuo, Meigu, Zhaojue, and Puge—experienced negative growth rates. Similarly, among the sixteen non-poverty-stricken counties, four—Huili, Jinning, Anning, and Dongchuan—also recorded negative growth rates. A negative growth rate indicates that economic development is relatively lagging, suggesting that the improvement rate of EEI exceeded the growth rate of EDI. Notably, among the 24 B-class regions, Yongshan, Ludian, Qiaojia, Yongren, Huize, and Mouding ranked among the top 10 counties in terms of EERDD growth rate.

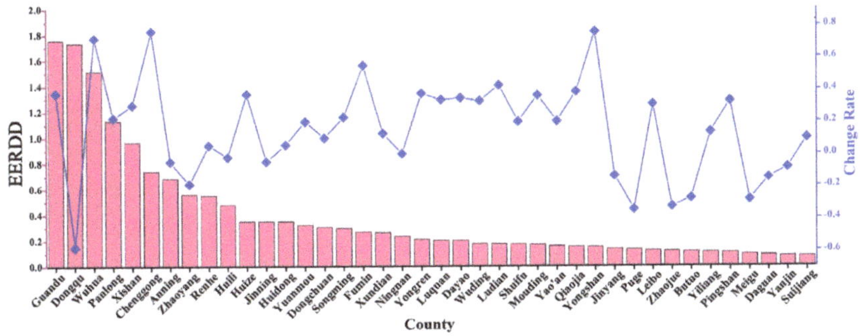


**Figure 26.** Comparison of average EERDD and its change rate in 40 counties from 2000 to 2020.

## 5. Discussion

### *5.1. Analysis of Spatiotemporal Dynamics in Ecological Environment*

#### 5.1.1. Single-Factor Driving Force Analysis

The primary factors driving the ecological environment index, ranked by their impact, include climatic, topographic, land use, and socioeconomic variables (Figure 27). These factors, listed from highest to lowest influence, are annual average temperature, aspect, elevation, land use type, nighttime light index, annual precipitation, cropland density, population density, GDP density, slope, Euclidean distance from roads, and construction land density. Natural conditions are the dominant factors influencing the spatial differentiation of the ecological environment index in the study area, with significant variation in the driving forces among different natural factors. Annual average temperature emerges as the strongest driver, with a $q$-value as high as 0.496, followed by slope ($q$-value of 0.378) and elevation ($q$-value of 0.304). The $q$-values for annual precipitation and slope are both below 0.2, indicating a relatively weak driving effect on the ecological environment index.

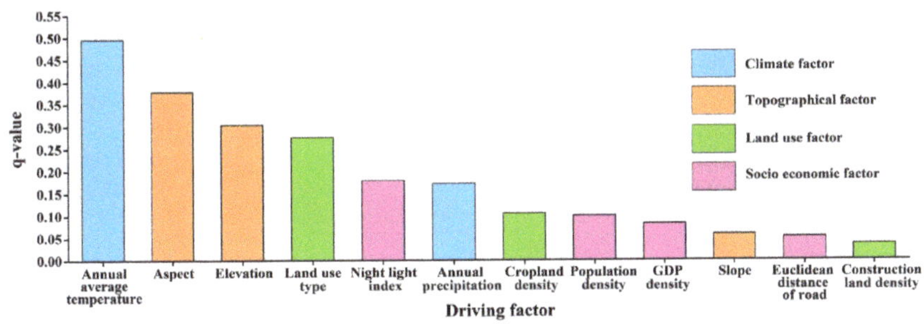


**Figure 27.** Ranking of single-factor driving force analysis results for changes in EEI.

Climatic Factors

Annual precipitation and annual average temperature are key indicators for evaluating hydrothermal conditions, and their impacts on the ecological environment of the study area vary significantly [82]. Among these factors, the influence of annual average temperature and its variations is more pronounced. High annual average temperatures predominantly occur in the dry-hot valley region of the LR-Jinsha [83], particularly in the border areas of Panzhihua, Chuxiong Prefecture, and Liangshan Prefecture. These elevated temperatures pose challenges for improving the ecological environment index. Compared to precipitation, temperature in the LR-Jinsha exerts a more substantial impact on the ecological environment. This is mainly because temperature fluctuations directly affect vegetation growth, soil moisture levels, and ecosystem stability, while the effects of changes in precipitation are relatively indirect and complex. Additionally, the combined effects of global warming and human activities have heightened the sensitivity of the ecological environment to temperature changes in the region [84].

Topographical Factors

Among the topographic factors, the $q$-values of aspect and elevation both exceed 0.3, indicating their significant driving influence on the ecological environment index in the study area. In contrast, the slope factor has a comparatively minor impact, with a $q$-value of only 0.057. Aspect plays a critical role in determining the amount of solar radiation received by different areas [85]. In the mountainous regions of the LR-Jinsha, southern-facing slopes tend to be warmer and drier, while northern-facing slopes are cooler and wetter. This variation affects key ecological elements such as soil moisture, vegetation coverage, and evapotranspiration rates, all of which are vital for shaping the region's ecological environment [86]. Elevation also leads to temperature fluctuations, influencing the distribution of vegetation, soil types, and biodiversity [87]. The wide elevation range in the LR-Jinsha creates diverse microclimates, which have a substantial impact on the area's ecological dynamics. While slope is important for processes like soil erosion, water runoff, and land stability, its broader influence on climate and ecological variables is more localized. Compared to altitude and aspect, slope has a more limited impact on the region's ecological landscape [88].

Anthropogenic Factors

From 2000 to 2020, the influence of human activities on the ecological environment exhibited a downward trend. Among the contributing factors, land use, which captures the effects of various human activities [89], had a notably strong impact, with a $q$-value of 0.276. To address the region's fragile ecological conditions and mitigate the adverse effects of human activity, national and local governments launched large-scale initiatives, such as returning farmland to forests and enhancing ecological protection [90,91]. These efforts led to a marked increase in forest land and water bodies, both of which possess essential ecological source properties. Additionally, these projects facilitated the optimization of the quantity and spatial distribution of construction land.

The $q$-value of the nighttime light index, which reflects the activity of the local economy and the intensity of human activities [92], is 0.179, ranking second. The driving force of factors such as population density, GDP density, Euclidean distance of roads, and construction land density gradually weakened, with their $q$-values not reaching 0.1. This may be due to the limited accuracy of the relevant data and the concentrated reflection of human activities in a relatively confined space, which contrasts with the larger spatial scope of the study area, leading to a decrease in explanatory power.

5.1.2. Two-Factor Interaction Analysis

The Two-Factor Interaction Analysis presents a better explanatory effect [93]. As illustrated in Figure 28, the interaction between annual average temperature ($X_1$) and nighttime light index ($X_{10}$) is the most significant, with a $q$-value of 0.671. This is followed by the interaction between annual average temperature and elevation ($X_3$), with a $q$-value of 0.635, and the interaction between annual average temperature and construction land density ($X_9$), with a $q$-value of 0.165. These findings indicate that temperature is a crucial factor affecting the ecological environment index of the study area [94]. Comparative analysis shows that in univariate analysis, the impact of construction land density ($X_9$) and nighttime light index ($X_{10}$) on the ecological environment is relatively small. However, when considering the interaction effect, the $q$-value of the interaction between these factors significantly increases with the annual average temperature.

The interaction between natural elements and human activity elements exhibits a dual-factor enhancement effect, significantly driving the quality of the ecological environment [95]. Specifically, the interaction between annual precipitation ($X_2$) and aspect ($X_5$) with other factors exerts a significant influence on the ecological environment. This is primarily due to their capacity to enhance soil moisture, support vegetation growth, and promote biodiversity, leading to an overall positive effect on the ecosystem [96]. In the interaction between annual average temperature and other factors, the $q$-values of interaction with Euclidean distance of roads ($X_{13}$) and cropland density ($X_8$) are slightly lower at 0.485 and 0.496, respectively. However, the $q$-values of interaction with other human activity factors are all over 0.515. The interaction between slope ($X_4$) and land use type ($X_7$) is the strongest among all interactions, with a $q$-value as high as 0.576. This result reveals the combined effect of slope and human activity-related factors, which has a more profound impact on the spatial analysis of EEI in the study area [97].

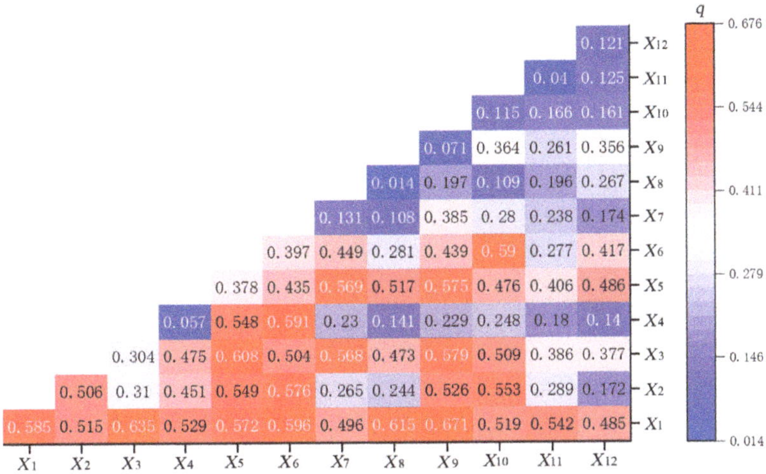

**Figure 28.** Detection results of the interaction between ecological environment driving factors.

## 5.2. Analysis of Spatiotemporal Dynamics in Economic Development

### 5.2.1. Analysis of Influencing Factors at the Subsystem Level

The obstacle factor diagnosis method was applied to calculate the factor contribution degrees of three subsystems—economic structure, economic environment, and economic scale—to the EEI in the 24 B-class and 16 A-class regions, respectively. The analysis revealed that the factor contribution degree of the economic structure increased, the economic environment's contribution degree decreased, and the economic scale's contribution degree fluctuated before stabilizing (Figure 29).

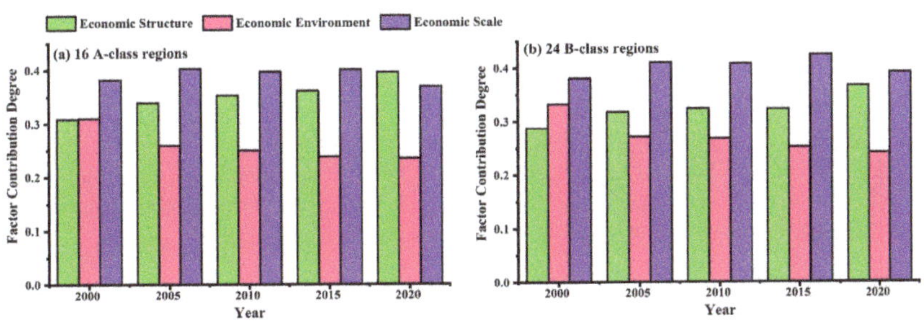


**Figure 29.** Factor contribution degrees of three subsystems—economic structure, economic environment, and economic scale—to EEI. (**a**) 16 A-class regions; (**b**) 24 B-class regions.

#### Economic Structure

Both types of regions showed an increase in the factor contribution degree of economic structure. In the B-class regions, it rose from 28.72% in 2000 to 36.69% in 2020, while in the A-class regions, it increased from 30.87% in 2000 to 39.61% in 2020. The economic structure factor contribution in the 16 A-class regions was generally higher than in the 24 B-class regions, and the rate of increase was also greater. This reflects a common challenge across both regions: optimizing and upgrading industrial structures [98]. Traditional industries, such as agriculture and low-end manufacturing, still constitute a significant proportion of the economic structure and lack support from high value-added and high-tech industries. This leads to lagging industrial optimization and represents a major obstacle to economic development [99]. In the 16 A-class regions, where the economic foundation is relatively strong, the traditional industrial structure no longer meets the needs of high-quality development. Thus, economic structural optimization has become a more significant barrier to further development [100]. Additionally, in some counties, there is a large gap in urban and rural economic structures and a low level of urbanization, hindering effective industrial upgrading and structural adjustment.

Economic Environment

The obstacles related to the economic environment showed a continuous downward trend. The factor contribution degree of the economic environment to economic development levels in both types of regions decreased steadily over time. In B-class regions, it fell from 33.23% in 2000 to 24.14% in 2020; in A-class regions, it dropped from 30.95% in 2000 to 23.49% in 2020. However, the decline in the 16 A-class regions was slightly less pronounced than in the 24 B-class regions. This trend can be attributed to improvements in infrastructure, policy environment, and business environment optimization. The implementation of poverty alleviation and rural revitalization policies has significantly enhanced infrastructure—such as transportation, energy, and communication—as well as public services like education and healthcare in the 24 B-class regions [101]. These improvements reduce the barriers posed by the economic environment to economic development. The government has also provided more policy support to the 24 B-class regions, including infrastructure investment, tax incentives, and talent recruitment, which have effectively improved the economic environment. In the 16 A-class regions, local governments have prioritized optimizing the business environment through various measures such as introducing one-stop government services, streamlining administrative approval processes, refining business environment policies, enhancing the efficiency of government services, and innovating service models [102]. These initiatives have collectively contributed to reducing barriers within the economic environment.

Economic Scale

The contribution of economic scale exhibited fluctuations but showed a general trend towards stabilization over time. In both A- and B-class regions, the contribution of economic scale to the EDI remained consistently significant, although with some variation. In the 24 B-class regions, the contribution primarily ranged between 38% and 42% throughout the study period, while in the 16 A-class regions, it ranged from 36.90% to 40.24%. This trend reflects the relatively modest economic base in both types of regions. Notably, the limited economic scale and fragile industrial foundation in the 24 B-class regions present substantial constraints on expanding economic growth. For instance, by 2020, the average GDP and total retail sales of consumer goods in these 24 regions represented only 25% of the average in China's western region, with per capita GDP standing at just 44% of the national average.

Additionally, the economic growth models across both types of counties are relatively homogeneous, often relying on a few key industries. When these industries face growth bottlenecks or slow down, they have a more substantial impact on the overall economic scale, as seen in regions such as Dongchuan and Butuo. In areas with less favorable geographic conditions, such as Leibo, Jinyang, Meigu, Puge, and Zhaojue, a shrinking demographic dividend and limited resources further hinder the expansion and development of economic scale.

Overall Analysis

Compared to the 16 A-class regions, the economic environment factor contribution degree in the 24 B-class regions has decreased more significantly, whereas the economic structure factor contribution degree is more prominent in the 16 A-class regions. This suggests that the 24 B-class regions are more reliant on improvements in infrastructure and the business environment, while the 16 A-class regions more urgently require economic structural transformation and optimization. The reasons for this include the following: First, the 24 B-class regions are in the early stages of economic development, where the role of infrastructure and environmental improvements is relatively substantial. The 16 A-class regions, having reached a certain level of economic development, rely more on industrial upgrading and economic restructuring for higher-quality development [103]. Second, government support policies for the 24 B-class regions have primarily focused on infrastructure, public services, and improving the business environment, whereas for the 16 A-class regions, the emphasis has been on guiding economic structural optimization and fostering innovation-driven development [104].

### 5.2.2. Analysis of Influencing Factors at the Indicator Level

Using the obstacle factor diagnosis method, we analyzed the impact of nine economic factors on the EDI and found notable differences between the 16 A-class and 24 B-class regions (Figure 30). In the 16 A-class regions, the average influence of each factor on EDI, ranked from highest to lowest, is as follows: social consumption (17.05%), human activity intensity (15.75%), tertiary industry (16.72%), total GDP (13.73%), secondary industry (13.31%), urbanization rate (5.91%), per capita cropland (4.16%), per capita GDP (8.24%), and primary industry (5.93%). In contrast, in the 24 B-class regions, the factors' average influence is ranked as: human activity intensity (17.65%), social consumption (16.57%), tertiary industry (15.31%), total GDP (13.12%), secondary industry (12.51%), urbanization rate (7.04%), per capita GDP (10.62%), primary industry (4.58%), and per capita cropland (2.22%).

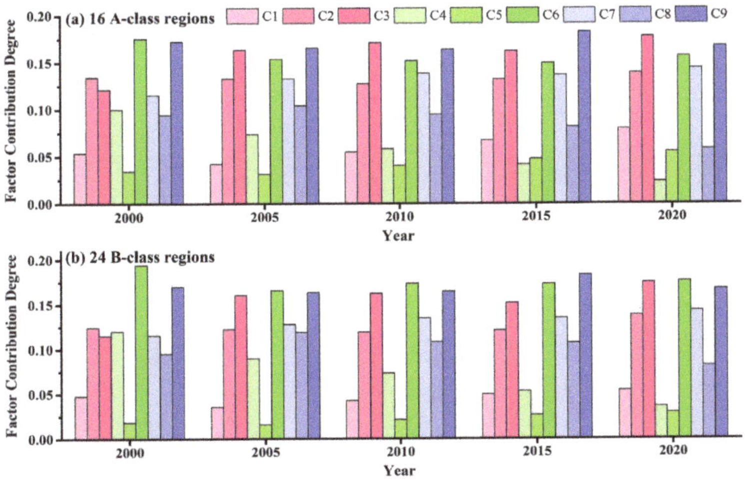


**Figure 30.** Contribution of 9 economic factors to EEI: (**a**) 16 A-class regions; (**b**) 24 B-class regions. C1 denotes total output value of primary industry, C2 denotes total output value of secondary industry, and C3 denotes total output value of tertiary industry. C4 represents urbanization rate, C5 represents per capita cultivated land area, and C6 represents nighttime light index. C7 indicates total GDP, C8 indicates GDP per capita, and C9 indicates total retail sales of consumer goods.

The influence of the four factors—human activity intensity, social consumption, tertiary industry, and total GDP—is similar in both types of regions, indicating that these factors play a crucial role in the economic development across the LR-Jinsha. This suggests that economic activities, consumer spending capacity, and the growth of the service industry significantly impact the economic landscape of the entire region. As the economy develops and social consumption levels increase, the role of the tertiary industry and human activity intensity in driving regional economic growth becomes increasingly prominent [105,106].

There are notable differences in the influence of certain factors between the two types of regions. In the 24 B-class regions, the urbanization rate and secondary industry exert a stronger influence, reflecting that these regions are still in the early stages of urbanization and industrialization. Here, enhancing the level of urbanization and establishing a robust industrial base play a crucial role in driving economic growth. Over time, the marginal effect of urbanization has gradually diminished, which explains the recent downward trend in its impact [107]. Conversely, in the 16 A-class regions, the primary industry and per capita cropland exert a more pronounced influence. This indicates a stronger reliance on natural or agricultural resources for economic development, highlighting the dependence of these regions on the agricultural economy and related sectors. This pattern is particularly evident in regions such as Huidong, Huili, Renhe, Ningnan, and Yuanmou, where agriculture plays a pivotal role in supporting local economies.

The impact of per capita GDP differs between the two types of regions. In the 24 B-class regions, the influence of per capita GDP is generally higher than in the 16 A-class regions. The fluctuations in per capita GDP in the 24 B-class regions suggest that economic development and wealth accumulation per capita in these areas may be influenced by factors such as resource allocation, infrastructure development, and policy support [108].

## 5.3. *Analysis of Spatiotemporal Dynamics in Ecology–Economy Coupling*

### 5.3.1. Analysis in 16 A-Class Regions

The impact factors of EECCD in 16 A-class regions exhibited significant dynamic changes from 2000 to 2020 (Figure 31). The influence of the secondary industry and the nighttime light index, which indicates human activity intensity, weakened over this period. Conversely, the influence of factors such as the tertiary industry, WET, and per capita cropland area gradually increased. These changes in the weight of economic restructuring and ecological factors reflect adjustments in development strategies and environmental policies during the ecological–economic coupling process in these regions.

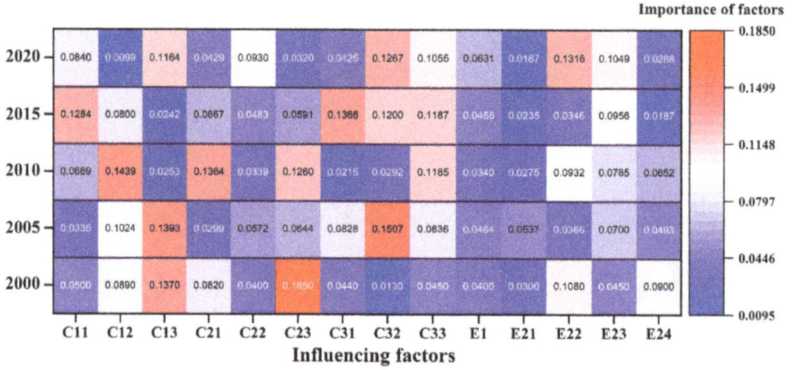


**Figure 31.** The importance of influencing factors on EECCD in 16 A-class regions from 2000 to 2020. C11 denotes total output value of primary industry, C12 denotes total output value of secondary industry, and C13 denotes total output value of tertiary industry. C21 represents urbanization rate, C22 represents per capita cultivated land area, and C23 represents nighttime light index. C31 indicates total GDP, C32 indicates GDP per capita, and C33 indicates total retail sales of consumer goods. E1 stands for BRI, E21 stands for NDVI, E22 stands for WET, E23 stands for NDSBI, and E24 represents LST.

In terms of economic structure, from 2000 to 2020, the impact of the total output value of the primary industry on the coupling and coordination degree of ecological economy fluctuated upward, while the impact of the secondary industry first rose and then fell. The impact of the tertiary industry once decreased but later rebounded, showing the impact of changes in industrial structure on EECCD. The impact of the primary industry on EECCD showed a fluctuating upward trend from 2000 to 2020, reaching a peak of 0.1284 in 2015, indicating that the impact of the ecological–economic coupling of agricultural activities was gradually increasing, but there was a certain decline in 2020. The influence of the secondary industry gradually increased in 2005 and 2010, reaching a peak of 0.1439 in 2010, but then significantly decreased to 0.0099 in 2020. This reflects the gradual shift of the economic center of gravity in these regions from industry to other industries. The influence of the tertiary industry was relatively high in 2000 and 2005, and then significantly decreased in 2010 and 2015, but it rebounded to 0.1164 in 2020, reflecting that the service industry has become an important factor affecting the coupling of ecological economy to a certain extent.

In terms of economic environment, from 2000 to 2020, the impact of urbanization rate on the coupling of ecological economy fluctuated significantly, while the impact of per capita cropland area first decreased and then increased. The explanatory power of nighttime light index as an indicator of economic activity gradually weakened. The impact of urbanization rate on the coupling degree was 0.0820 in 2000, but it dropped to a minimum of 0.0299 in 2005, then reached a peak of 0.1364 in 2010, and then declined again. This indicates that there are significant fluctuations in the impact of urbanization on the coupling of ecological and economic development. The impact of per capita cropland area showed a trend of first decreasing and then increasing, reaching a peak of 0.0930 in 2020, which may be related to the rational use and optimal management of agricultural resources in the region. The intensity of human activities had the greatest impact in 2000, reaching 0.1850, and then showed a general downward trend. This indicates that, over time, the explanatory power of nighttime light index as a proxy for economic activity on the coupling of ecological and economic development has gradually weakened.

From the perspective of economic indicators, the impact of the three factors on the coupling of ecological economy fluctuates. The increase in economic aggregate does not fully and continuously improve EECCD, while per capita income and consumption have a more significant impact in specific years. The impact of total GDP on the coupling of ecological economy fluctuates, reaching a peak of 0.1366 in 2015, but then declining. the increase in economic aggregate does not necessarily continuously improve the coupling degree of ecological economy. The impact of per capita GDP is more significant in 2005 and 2020, which are 0.1507 and 0.1267 respectively, reflecting that the per capita income level has a more significant impact on EECCD at specific stages. The impact of total retail sales of consumer goods began to increase significantly in 2005, reaching 0.0836, and remained at a high level of 0.1187 and 0.1055 in 2015 and 2020, indicating that the impact of consumption on the coupling of ecological and economic development is gradually increasing. From 2000 to 2020, the average total retail sales of consumer goods in 16 counties increased by 17.43 times.

From the perspective of BRI, its impact on EECCD remained relatively stable throughout the entire time period, but slightly increased to 0.0631 in 2020. This indicates that the importance of land use and its derivative BRI is gradually increasing. Especially in urban and rural construction land, cultivated land, and their peripheral areas within the district, such as Panzhihua [109] and Kunming City [110].

The impact of each RSEI component on the coupling of ecology and economy showed different trends, indicating that the changes in the four factors had different effects on the coupling of ecology and economy in different periods. The impact of the NDVI has gradually weakened, from 0.0300 in 2000 to 0.0187 in 2020, indicating that the impact of vegetation coverage on the coupling of ecological economy has gradually decreased. The impact of WET reached a peak of 0.1080 in 2000, then declined and rose again to 0.1316 in

2020, indicating that the impact of WET on ecological economy has significantly changed over different time periods. The impact of the NDSBI is gradually increasing, indicating that land use change has an increasingly important impact on EECCD. The impact of LST was higher in 2000, at 0.0900, and then gradually decreased. Over time, the direct impact of changes in LST on EECCD has weakened.

5.3.2. Analysis in 24 B-Class Regions

From 2000 to 2020, the impact of various economic factors on EECCD in the 24 B-class regions exhibited significant variations across different years (Figure 32). The influence of the secondary industry has gradually increased, while the marginal effect of urbanization has decreased. The growth of economic aggregates and per capita income continues to be the primary driving force. Although the role of ecological environment indicators has risen, their overall impact remains relatively low.

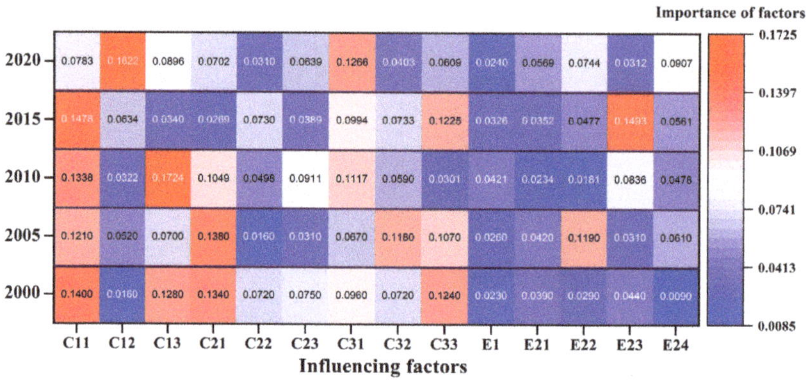


**Figure 32.** Importance of influencing factors on EECCD in 24 B-class regions from 2000 to 2020. C11 denotes total output value of primary industry, C12 denotes total output value of secondary industry, and C13 denotes total output value of tertiary industry. C21 represents urbanization rate, C22 represents per capita cultivated land area, and C23 represents nighttime light index. C31 indicates total GDP, C32 indicates GDP per capita, and C33 indicates total retail sales of consumer goods. E1 stands for BRI, E21 stands for NDVI, E22 stands for WET, E23 stands for NDSBI, and E24 represents LST.

From the perspective of the economic structure's components on EECCD, the roles of the three industrial sectors have been evolving over time. The primary industry had a significant impact on EECCD from 2000 to 2015, but its influence declined sharply by 2020, reflecting a shift towards an upgraded regional industrial structure and a reduced role of the primary sector. In contrast, the impact of the secondary industry on EECCD has been increasing, particularly in 2020, when it became one of the main influencing factors. This underscores the growing importance of industrialization in promoting EECCD. Meanwhile, the impact of the tertiary industry peaked in 2010 but then gradually weakened, likely due to the saturation of the service sector and adjustments or transformations in the economic structures of various counties.

From the perspective of the economic environment's impact on EECCD, the marginal effect of the urbanization rate is diminishing, the influence of per capita cropland area remains limited, and the relationship between economic activity—represented by the nighttime light index—and EECCD is complex. From 2000 to 2010, the impact of the urbanization rate on EECCD was relatively stable but then significantly decreased, indicating either a diminishing marginal effect of urbanization or an increase in environmental problems associated with urbanization. The impact of per capita cultivated land area has consistently been low and fluctuating, reflecting the limited role of agricultural resources on EECCD and constraints from factors such as land use changes and agricultural policies. The nighttime light index, an indirect measure of economic activity, peaked in 2010 and then declined, highlighting the intricate relationship between economic activity intensity and EECCD, influenced by economic structure, energy efficiency, and policy regulation.

The impact of economic scale on EECCD has been stable and rising, with total GDP and per capita GDP consistently driving EECCD forward. The influence of total GDP peaked in 2020, while per capita GDP reached its highest point in 2005 and then slightly declined but remained high. In contrast, the impact of total retail sales of consumer goods, which was relatively stable from 2000 to 2015, declined in 2020, likely due to changes in consumption patterns and policy regulation [111].

From 2000 to 2020, the impact of the BRI on EECCD remained relatively low and exhibited significant fluctuations, indicating its limited contribution to ecological and economic coupling coordination. This could be attributed to the gradual harmonization of land use changes, ecological protection efforts, ecological restoration, and human activity intensity with ecological and economic development during the initiative's implementation. These areas likely underwent a period of adjustment where the BRI's influence on ecological sustainability and economic growth became more coordinated over time, leading to fluctuating but overall moderate impacts [112].

The four components of the RSEI currently have a low but increasing impact on EECCD, reflecting the effectiveness of ecological protection and restoration efforts, as well as the influence of climate warming. While the impacts of the NDVI, the WET, and the NDSBI are relatively low, they are trending upward, indicating that the role of the ecological environment in EECCD has strengthened following the implementation of ecological protection and restoration projects. Although the impact of LST is also relatively low, it is rising, likely due to global warming and regional human activities, which could further affect both the ecological environment and economic development [113].

### 5.3.3. Comparative Analysis

Overall, there are distinct differences in the impact of ecological and economic factors on the EECCD between the 24 B-class regions and the 16 A-class regions from 2000 to 2020 (Figure 33). The poverty-alleviated counties are more reliant on the primary industry, economic scale, and urbanization, with ecological and environmental factors playing a relatively minor role. In contrast, the 16 A-class regions are more influenced by economic factors such as consumption, human activity intensity, and the tertiary industry, while ecological and environmental factors have a comparatively more prominent impact.

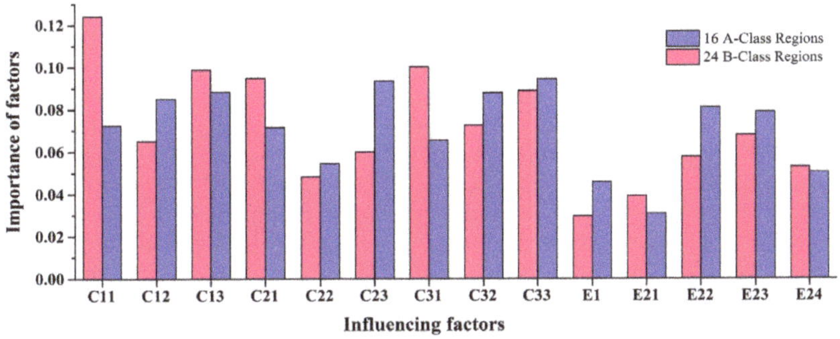


**Figure 33.** Comparison of importance of 16 A-Class regions and 24 B-Class regions from 2000 to 2020. C11 denotes total output value of primary industry, C12 denotes total output value of secondary industry, and C13 denotes total output value of tertiary industry. C21 represents urbanization rate, C22 represents per capita cultivated land area, and C23 represents nighttime light index. C31 indicates total GDP, C32 indicates GDP per capita, and C33 indicates total retail sales of consumer goods. E1 stands for BRI, E21 stands for NDVI, E22 stands for WET, E23 stands for NDSBI, and E24 represents LST.

The impact of economic structure on the coupling of ecological economy differs significantly between the 24 B-class regions and the 16 A-class regions. In the poverty-alleviated counties, the total primary industry output has the greatest influence on ecological–economic coupling, with a value of 0.1242, which is substantially higher than that in non-poor counties. This suggests that agriculture remains a crucial factor supporting economic development in these areas, with agricultural output making a significant contribution to the coupling of ecology and economy. The impact of total GDP is the second highest in poverty-alleviated counties, at 0.1001, indicating that overall economic growth is more vital in these regions due to their relatively weaker economic foundation [114]. The overall economic growth has a notable effect on enhancing the coupling and coordination of the ecological economy in these counties. In contrast, in the 16 A-class regions, the impact of total tertiary industry output (0.0884) and total secondary industry output (0.0850) is more prominent, reflecting a greater reliance on the service and industrial sectors for promoting economic development and ecological–economic coupling. Meanwhile, the influence of the primary industry in the 16 A-class regions is comparatively smaller, at 0.0726, indicating a shift in their economic structure towards the secondary and tertiary sectors, driven by modern industrial growth [115].

The influence of consumption and human activity intensity on the EECCD is notable. In the 24 B-class regions, the effects of total retail sales of consumer goods (0.0889) and nighttime light index (0.0600) are relatively minor. This suggests that consumption and overall human activity in these areas are less dynamic compared to the 16 A-class regions, resulting in a limited impact on EECCD. Conversely, in the 16 A-class regions, total retail sales of consumer goods (0.0943) and nighttime light index (0.0933) are the highest, indicating that consumption levels and comprehensive human activities play a crucial role in the EECCD of these regions. This reflects a greater dependence on consumption and market vitality in the economies of these areas [116].

Economic efficiency and industrial transformation positively impact the EECCD [117]. In the 24 B-class regions, the effects of per capita GDP (0.0725) and total secondary industry output (0.0652) are relatively modest. This is attributed to the weaker economic efficiency and industrial base in these counties, where economic growth is more reliant on the expansion of the primary industry and, to some extent, the overall economic scale. Conversely, in the 16 A-class regions, the impact of per capita GDP (0.0879) is more pronounced, indicating that economic development in these regions is more focused on enhancing economic efficiency and per capita income, such as Dongchuan, Zhaoyang, and Wuding. Additionally, the substantial effect of total secondary industry output (0.0850) in these 16 counties suggests that industrialization plays a significant role in contributing to EECCD [118].

There are differences in the roles of the RSEI and the BRI in influencing EECCD. In the 24 B-class regions, the impacts of WET (0.0576), NDSBI (0.0678), and LST (0.0529) are relatively small, indicating that changes in ecological factors have limited effects on EECCD, with economic factors playing a more dominant role. The influence of the BRI (0.0295) and the NDVI (0.0393) is the lowest in both types of counties, suggesting that improvements in natural ecological indicators have a less significant impact on EECCD in these areas. In contrast, in the 16 A-class regions, WET (0.0808) and NDSBI (0.0788) have a more substantial impact on EECCD, highlighting the importance of ecological environmental governance, as changes in ecological indicators contribute more to EECCD coordination. While the effects of the NDVI (0.0307) and the BRI (0.0458) remain relatively small, they are slightly higher than in poverty-alleviated counties, reflecting the positive impact of improved vegetation cover and biodiversity on EECCD in these regions [119].

The impacts of urbanization and land use on EECCD differ across regions [120]. In the 24 B-class regions, the urbanization rate (0.0948) plays a more significant role, as these areas are still experiencing rapid urbanization, which strongly drives economic growth. Being in the accelerating stage of urbanization, these regions benefit from increased economic development and industrial restructuring, alongside improvements in resource allocation efficiency and ecological governance. In contrast, in the 16 A-class regions, the effects of urbanization rate (0.0716) and per capita cropland area (0.0545) are less pronounced, suggesting that urbanization and agricultural land use have a relatively limited impact on EECCD in these areas.

### 5.4. Implications and Limitations

#### 5.4.1. Implications

To foster coordinated ecological and economic development in the LR-Jinsha, future efforts should focus on strengthening infrastructure construction, enhancing the transportation network, and improving internal and external connectivity to reduce geographical constraints. Financial, tax, and monetary support must be provided to solidify the achievements of targeted poverty alleviation. Increased investment in education and the promotion of science and technology are critical. It is also essential to further develop and utilize resources such as tourism, minerals, water conservancy, photovoltaics, agriculture, and traditional Chinese medicine, while continuously advancing ecological engineering. By leveraging ecological economics and ecological compensation, the industrial structure can be optimized. Based on the spatial–temporal evolution of economic development levels, ecological environment indices, and the EECCD, the development paths in the study area can be categorized into six types (Figure 34).

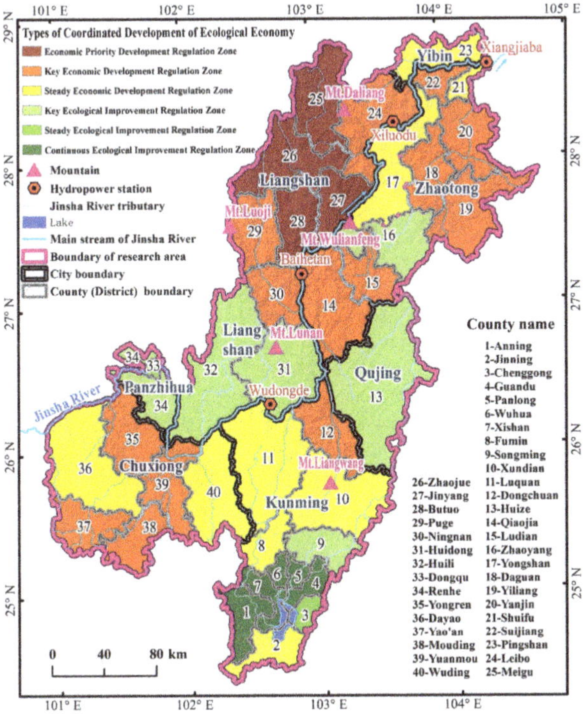


**Figure 34.** Coordinated development path of ecological environment and economic environment.

Economic Priority Development Regulation Zone: Situated in the southwestern mountainous region of Sichuan, primarily encompassing Meigu, Zhaojue, Butuo, and Jinyang counties in Liangshan Prefecture, these areas should prioritize economic growth while improving the synergy between ecological quality and economic development. Emphasis should be placed on advancing industrial transformation and upgrading, reinforcing the modernization of traditional industries, fostering the development of emerging industries, and promoting infrastructure and industrial park construction.

Key Economic Development Regulation Zone: The most extensive category, encompassing counties such as Mouding, Yaoan, Yongren, and Yuanmou in Chuxiong Prefecture; Leibo, Ningnan, and Puge in Liangshan Prefecture; and Daguan, Ludian, Qiaojia, Suijiang, Yanjin, and Yiliang in Zhaotong City. These regions are characterized by lower levels of economic development but possess significant growth potential. Strengthening economic growth while aligning it with ecological improvements is crucial. To achieve this, it is essential to enhance infrastructure development, improve the investment environment, optimize the industrial structure, and foster the growth of unique local industries. Moreover, regional cooperation and coordinated development are key strategies for unlocking the full potential of these areas.

Steady Economic Development Regulation Zone: Includes Dayao and Wuding counties in Chuxiong Prefecture, Fumin, Jinning, Luquan, and Xundian counties in Kunming City, as well as Shuifu and Yongshan counties in Zhaotong City. These areas are transitioning from an eco-economic imbalance to a more coordinated state. To further advance this transition, steady economic growth is essential to foster a more balanced relationship between ecological and economic factors. Key priorities include optimizing the industrial structure, enhancing ecological protection and restoration efforts, and promoting regional coordination and development strategies.

Key Ecological Improvement Regulation Zone: Covers counties like Songming, Huidong, Huili, and Renhe District, along with Huize County and Zhaoyang District. This zone prioritizes ecological protection and is essential for enhancing the overall ecological quality of the Jinsha River basin. Further efforts are needed to strengthen ecological protection and restoration, optimize industrial and energy structures, improve the ecological environment governance system, and promote regional cooperation and collaborative governance.

Steady Ecological Improvement Regulation Zone: Located in Chenggong District of Kunming City and Dong District of Panzhihua City, where positive eco-economic coordination trends are observed. Steady ecological improvements are necessary to further advance towards coordination. Further efforts are needed to strengthen ecological environment protection and governance, optimize industrial and energy structures, and promote coordinated development of ecological economy.

Continuous Ecological Improvement Regulation Zone: Includes Anning City, Guandu, Panlong, Wuhua, and Xishan districts in Kunming City. These areas exhibit a high degree of internal coordination and therefore necessitate ongoing efforts to enhance ecological quality. Key priorities include strengthening water resource protection and governance, improving air quality, implementing ecological restoration and greening initiatives, advancing pollution control measures, and promoting the development of a green economy.

Meanwhile, the regions designated for prioritized economic development and regulation must balance economic growth with ecological preservation, addressing challenges such as underdeveloped infrastructure and a lack of industrial diversity. Regions marked for steady economic development regulation need to accelerate the transformation of coordination mechanisms and ensure effective policy implementation. Areas with a focus on ecological enhancement face issues like fragile ecosystems and funding shortages. Regions aimed at stable ecological improvement must work to raise public environmental awareness and expedite progress in coordination efforts. Finally, areas striving for continuous ecological advancement should innovate ecological development models to meet higher environmental quality standards.

### 5.4.2. Limitations

The coupling development of ecological and economic systems has diverse and complex roles and feedback mechanisms. This article constructs a dataset and uses various methods to explore the spatiotemporal changes and driving forces of the ecological and economic coupling system in the LR-Jinsha region, which has different resource endowments, environmental conditions, development history, and development models. However, there are still some aspects that need improvement. This paper is limited by the availability and consistency of data, and the selection of economic factors can be further increased, such as the proportion of R&D funds in GDP, the proportion of fixed assets investment in GDP, trade balance, labor productivity, employment rate, unemployment rate, human development index, etc. As economic data mostly belongs to statistical data, obtaining raster data or rasterizing data is one of the directions for improvement. The length of research can be extended to before and after 2000–2020, and simulation prediction research can be conducted on this basis to enhance decision support for research results. For ecological and economic coupled development systems that combine natural and social attributes, the potential of machine learning and other methods can be further explored to identify the relationships and processes between complex systems. Furthermore, additional research is required to refine the methodological framework for analyzing the spatiotemporal coupling mechanisms between ecology and economy. Such efforts aim to offer a distinctive ecological and economic perspective for understanding the processes and impacts of the human–environment coupling system.

## 6. Conclusions

This article integrates multiple datasets and methodological models to identify the spatiotemporal changes and driving factors of the ecological environment subsystem, economic environment subsystem, and the ecological–economic coupling system. Here are the key findings:

From 2000 to 2020, the overall improvement rates of the ecological and economic subsystems surpassed that of the ecological–economic coupling system, with a 19.95% increase in the EEI, a 33.55% rise in the EDI, and a 12.50% improvement in the EECCD. The entire region remains in the Running-In Stage, with the coordination level progressing from Near Imbalance to Marginal Coordination. Furthermore, 85% of the counties' EERDD are still in the EDI Behind EEI stage.

The structural composition of the EEI exhibits a pattern of being low in the Dry Hot Valley area and high in other regions, primarily influenced by natural factors, although human activities also significantly impact these interactions. Key natural drivers include temperature, slope, and elevation, while land use and nighttime light reflect human influence. Interaction analysis highlights the combined effects of both natural and human factors, with temperature and nighttime light demonstrating dual-factor enhancement effects. Originating from an impact model primarily driven by economic factors and supplemented by ecological factors, both the EDI and EECCD show a pattern of being higher in the south and lower in the north, with improvements gradually expanding northward from the urban area of Kunming. A-class regions shifted their focus from secondary to tertiary industries and ecological considerations, while B-class regions continued to rely on primary industries and urbanization. Economic restructuring and urbanization played a crucial role, with ecological factors gradually gaining importance. The development gradient gap between the 24 poverty-stricken counties and the 16 non-poverty-stricken counties has narrowed. Simultaneously, policies such as green development, ecological protection, industrial upgrading and transformation, regional coordinated development, targeted poverty alleviation, and poverty alleviation consolidation have played—and will continue to play—a positive role in fostering sustainable development.

Based on their spatiotemporal evolution and corresponding strategies, the development paths of the 40 counties in the study area can be classified into six categories: Economic Priority, Key Economic, Steady Economic, Key Ecological, Steady Ecological, and Continuous Ecological Improvement Regulation Zones. Meanwhile, areas leading in economic development may face challenges such as outdated infrastructure, a limited industrial structure, and the need to balance economic growth with ecological sustainability. Similarly, regions prioritized for ecological protection grapple with issues like fragile ecosystems, insufficient funding, and low public awareness of environmental conservation. It is recommended to increase investments in transportation and water conservancy infrastructure, strengthen coordinated regional economic development, and promote a diversified industrial structure. Furthermore, efforts should focus on enhancing ecological protection and restoration, boosting financial support, implementing ecological compensation mechanisms, and fostering innovative models for sustainable ecological development.

**Author Contributions:** Z.N. wrote the paper with contributions from all co-authors. Y.Z. (Yinbing Zhao) conceived and designed the research, and provided funding. J.L. and Y.L. contributed to the data processing. X.X. and Y.Z. (Yang Zhang) contributed to the methodology. All authors have read and agreed to the published version of the manuscript.

**Funding:** This research was supported by the Soft Science Project of Sichuan Provincial Department of Science and Technology (2022JDR0061), the Open Foundation of the State Key Laboratory of Urban and Regional Ecology of China (SKLURE2022-2-6), the Philosophy and Social Sciences Research Fund Project of Chengdu University of Technology in 2021 (YJ2021-YB023),and the Key Laboratory of Supported by Open Foundation of the Research Center for Human Geography of Tibetan Plateau and Its Eastern Edge (Chengdu University of Technology) (RWDL2022-ZD002 and RWDL2021-ZD002).

**Data Availability Statement:** Data will be made available on reasonable request.

**Conflicts of Interest:** The authors declare no conflicts of interest.

## Abbreviations of Nouns

| | |
|---|---|
| LR-Jinsha | The lower reaches of the Jinsha River |
| 16 A-class regions | Sixteen counties in the study area are classified as non-impoverished |
| 24 B-class regions | By 2020, 24 counties in the study area had been lifted out of poverty |
| MODIS | Moderate-Resolution Imaging Spectroradiometer |
| NDVI | Normalized Difference Vegetation Index |
| NDBSI | Normalized Difference Built-up and Bareness Index |
| LST | Land Surface Temperature |
| WET | Wetness Index |
| RSEI | Remote sensing ecological index |
| BRI | Biodiversity richness index |
| EEI | Ecological environment index |
| EDI | Economic development index |
| EcoESTI | Economic structure index |
| EcoEEI | Economic environment index |
| EcoESCI | Economic scale index |
| EECD | Ecological–economic coupling degree |
| EECCD | Ecological–economic coupling coordination degree |
| EERDD | Ecological–economic relative development degree |

## Appendix A. Supplementary Data

**Table A1.** Weights of EDI evaluation indicators for the five periods.

| Indicator \ Year | Total Primary Industry Output | Total Secondary Industry Output | Total Tertiary Industry Output | Urbanization Rate | Nighttime Light Index | Per Capita Cropland Area | Total GDP | Per Capita GDP | Total Retail Sales of Consumer Goods |
|---|---|---|---|---|---|---|---|---|---|
| 2000 | 0.059 | 0.119 | 0.111 | 0.122 | 0.181 | 0.042 | 0.115 | 0.092 | 0.159 |
| 2005 | 0.047 | 0.117 | 0.153 | 0.094 | 0.157 | 0.037 | 0.123 | 0.116 | 0.157 |
| 2010 | 0.054 | 0.114 | 0.154 | 0.080 | 0.164 | 0.043 | 0.129 | 0.106 | 0.157 |
| 2015 | 0.066 | 0.115 | 0.143 | 0.061 | 0.163 | 0.046 | 0.129 | 0.106 | 0.171 |
| 2020 | 0.066 | 0.131 | 0.162 | 0.046 | 0.163 | 0.053 | 0.135 | 0.086 | 0.157 |

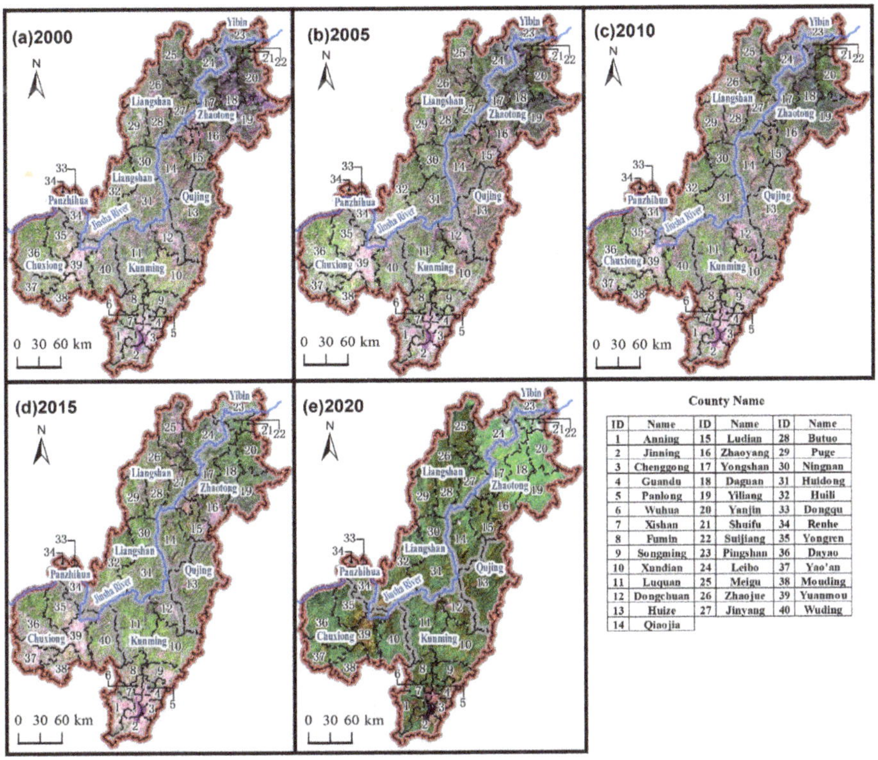


**Figure A1.** MODIS data for extracting remote sensing ecological indices from five periods.

**Table A2.** Changes in ecological environment index (EEI) from 2000 to 2020.

| County | EEI | | | | | County | EEI | | | | |
|---|---|---|---|---|---|---|---|---|---|---|---|
| | 2000 | 2005 | 2010 | 2015 | 2020 | | 2000 | 2005 | 2010 | 2015 | 2020 |
| Suijiang | 0.6584 | 0.6194 | 0.6491 | 0.7447 | 0.7010 | Puge | 0.4192 | 0.4237 | 0.4575 | 0.4997 | 0.5213 |
| Shuifu | 0.6453 | 0.5919 | 0.6278 | 0.7144 | 0.6745 | Wuhua | 0.4140 | 0.4088 | 0.4143 | 0.5088 | 0.5013 |
| Pingshan | 0.6216 | 0.5949 | 0.6115 | 0.6830 | 0.6600 | Yongren | 0.4277 | 0.4392 | 0.4244 | 0.4625 | 0.4865 |
| Yanjin | 0.5766 | 0.5484 | 0.5760 | 0.7090 | 0.6794 | Fumin | 0.4217 | 0.4092 | 0.3903 | 0.4886 | 0.4950 |
| Leibo | 0.5952 | 0.5218 | 0.5614 | 0.5888 | 0.6017 | Qiaojia | 0.3985 | 0.3952 | 0.4190 | 0.4816 | 0.4949 |
| Daguan | 0.4918 | 0.4793 | 0.5045 | 0.5947 | 0.6299 | Huize | 0.3891 | 0.3914 | 0.3880 | 0.4866 | 0.4889 |
| Xishan | 0.5054 | 0.4836 | 0.5044 | 0.5599 | 0.5649 | Huidong | 0.3924 | 0.4061 | 0.3999 | 0.4510 | 0.4770 |
| Yao'an | 0.5135 | 0.4872 | 0.5069 | 0.5341 | 0.5509 | Xundian | 0.3753 | 0.4086 | 0.3548 | 0.4922 | 0.4791 |
| Yiliang | 0.4521 | 0.4884 | 0.4789 | 0.5880 | 0.5833 | Huili | 0.3801 | 0.4039 | 0.3951 | 0.4552 | 0.4706 |
| Dayao | 0.5011 | 0.4983 | 0.5055 | 0.5313 | 0.5504 | Zhaojue | 0.3649 | 0.3851 | 0.4103 | 0.4434 | 0.4833 |
| Yongshan | 0.4615 | 0.4608 | 0.4765 | 0.5517 | 0.5616 | Butuo | 0.3769 | 0.3871 | 0.3984 | 0.4447 | 0.4762 |
| Panlong | 0.4478 | 0.4511 | 0.4566 | 0.5633 | 0.5520 | Guandu | 0.4227 | 0.3980 | 0.3819 | 0.4507 | 0.4270 |
| Anning | 0.4352 | 0.4396 | 0.4878 | 0.5427 | 0.5631 | Jinyang | 0.3763 | 0.3912 | 0.3736 | 0.4520 | 0.4413 |
| Wuding | 0.4371 | 0.4604 | 0.4527 | 0.5204 | 0.5510 | Zhaoyang | 0.3652 | 0.3668 | 0.3740 | 0.4574 | 0.4549 |
| Meigu | 0.4664 | 0.4565 | 0.4697 | 0.4997 | 0.5167 | Chenggong | 0.3791 | 0.3747 | 0.3658 | 0.4362 | 0.4486 |
| Ningnan | 0.4608 | 0.4509 | 0.4473 | 0.5091 | 0.5182 | Renhe | 0.3781 | 0.3759 | 0.3742 | 0.4108 | 0.4476 |
| Luquan | 0.4275 | 0.4598 | 0.4425 | 0.5173 | 0.5330 | Ludian | 0.3502 | 0.3415 | 0.3600 | 0.4353 | 0.4536 |
| Mouding | 0.4453 | 0.4279 | 0.4579 | 0.5027 | 0.5267 | Dongchuan | 0.3429 | 0.3509 | 0.3249 | 0.4094 | 0.4272 |
| Songming | 0.4419 | 0.4500 | 0.4130 | 0.5343 | 0.5190 | Dongqu | 0.3001 | 0.3109 | 0.3154 | 0.3598 | 0.4081 |
| Jinning | 0.4251 | 0.4150 | 0.4533 | 0.5202 | 0.5322 | Yuanmou | 0.2814 | 0.2903 | 0.2695 | 0.3233 | 0.3615 |

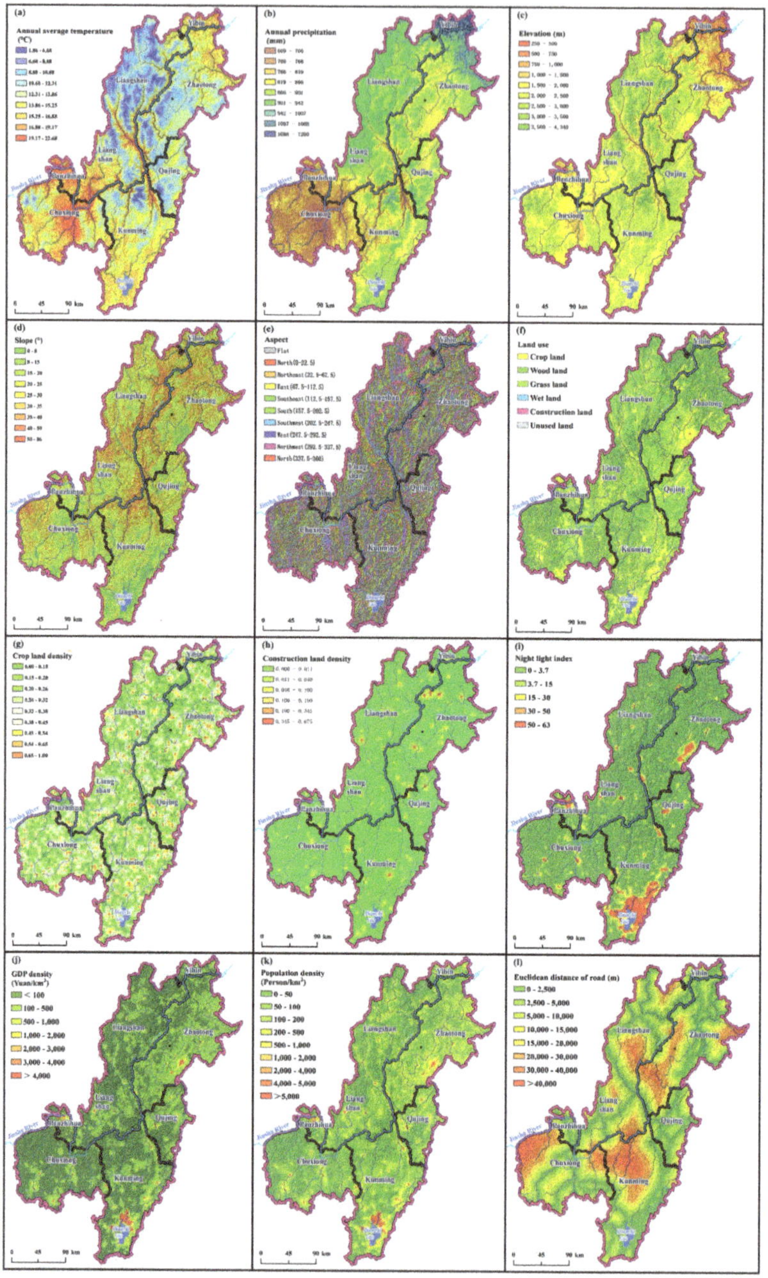


**Figure A2.** The driving factors of the ecological environment index (EEI):(**a**) Annual average temperature; (**b**) Annual precipitation; (**c**) Elevation; (**d**) Slope; (**e**) Aspect; (**f**) Land use; (**g**) Crop land density; (**h**) Construction land density; (**i**) Night light index; (**j**) GDP density; (**k**) Population density; (**l**) Euclidean distance of road.

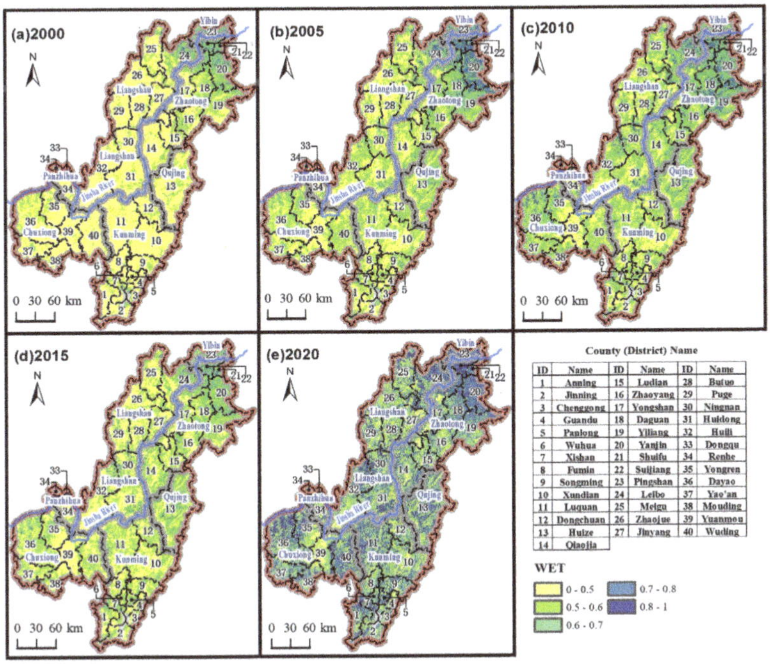


**Figure A3.** Spatial distribution of WET from 2000 to 2020.

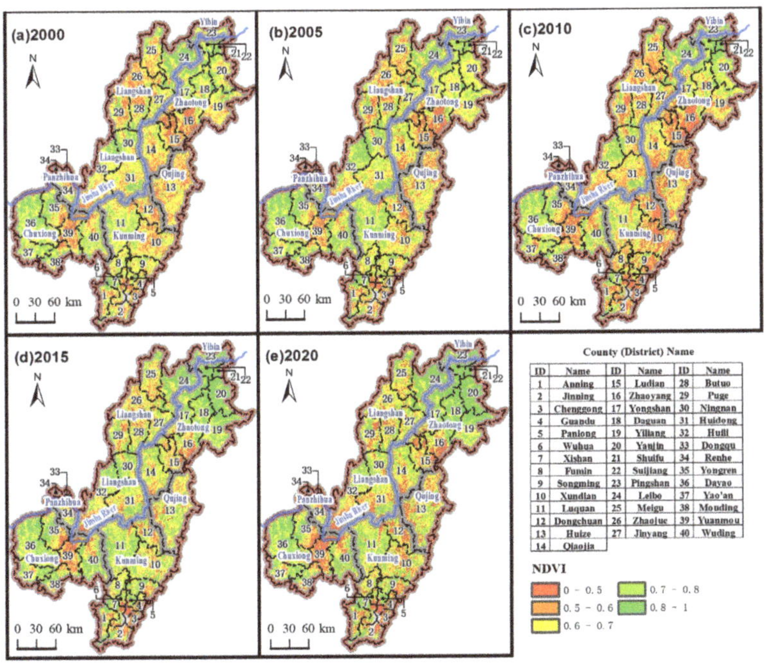


**Figure A4.** Spatial distribution of NDVI from 2000 to 2020.

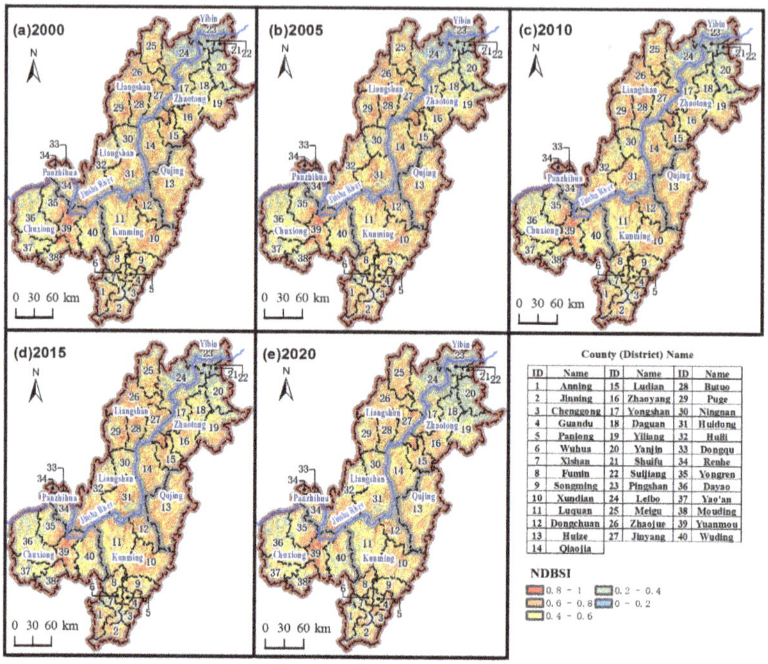


**Figure A5.** Spatial distribution of NDBSI from 2000 to 2020.

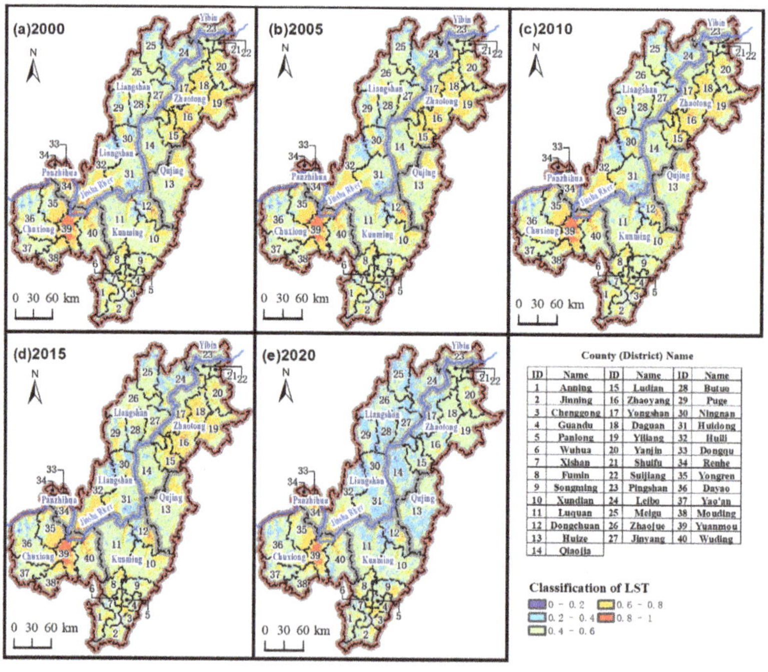


**Figure A6.** Spatial distribution of LST from 2000 to 2020.

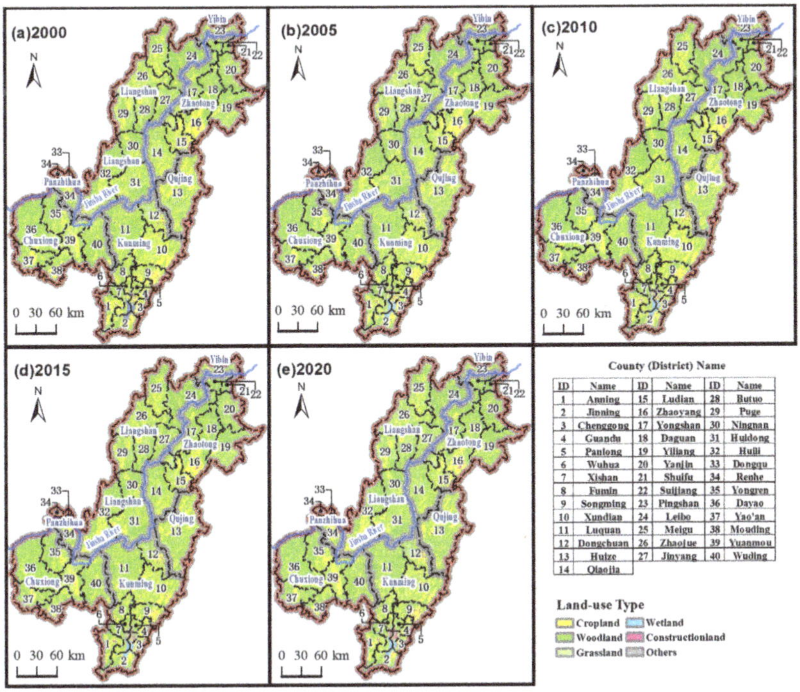


**Figure A7.** Spatial distribution of land use types from 2000 to 2020.

**Table A3.** Changes in economic structure index (EcoESTI) from 2000 to 2020.

| Name | EcoESTI | | | | | Change Rate | Name | EcoESTI | | | | | Change Rate |
|---|---|---|---|---|---|---|---|---|---|---|---|---|---|
| | 2000 | 2005 | 2010 | 2015 | 2020 | | | 2000 | 2005 | 2010 | 2015 | 2020 | |
| Guandu | 0.2514 | 0.2752 | 0.2692 | 0.2627 | 0.3275 | 0.3026 | Yiliang | 0.0513 | 0.0526 | 0.0530 | 0.0581 | 0.0699 | 0.3624 |
| Wuding | 0.0552 | 0.2883 | 0.2769 | 0.2595 | 0.2999 | 4.4362 | Yuanmou | 0.0509 | 0.0494 | 0.0530 | 0.0587 | 0.0665 | 0.3070 |
| Xishan | 0.1143 | 0.1573 | 0.1554 | 0.1679 | 0.2103 | 0.8408 | Fumin | 0.0447 | 0.0477 | 0.0509 | 0.0650 | 0.0633 | 0.4158 |
| Panlong | 0.0594 | 0.1573 | 0.1616 | 0.1927 | 0.2187 | 2.6813 | Dongchuan | 0.0454 | 0.0450 | 0.0536 | 0.0637 | 0.0624 | 0.3762 |
| Dongqu | 0.2004 | 0.1335 | 0.1302 | 0.1329 | 0.1328 | −0.3371 | Wuhua | 0.0519 | 0.0521 | 0.0538 | 0.0410 | 0.0702 | 0.3529 |
| Anning | 0.1172 | 0.1042 | 0.1010 | 0.1105 | 0.1778 | 0.5174 | Ludian | 0.0459 | 0.0456 | 0.0489 | 0.0674 | 0.0587 | 0.2814 |
| Zhaoyang | 0.1391 | 0.1001 | 0.1050 | 0.1196 | 0.1354 | −0.0268 | Pingshan | 0.0509 | 0.0477 | 0.0516 | 0.0524 | 0.0636 | 0.2515 |
| Huili | 0.1039 | 0.1015 | 0.1283 | 0.1360 | 0.1039 | −0.0006 | Yanjin | 0.0467 | 0.0462 | 0.0463 | 0.0721 | 0.0506 | 0.0837 |
| Huize | 0.0843 | 0.0853 | 0.1024 | 0.0740 | 0.1428 | 0.6935 | Yao'an | 0.0511 | 0.0493 | 0.0501 | 0.0491 | 0.0608 | 0.1885 |
| Huidong | 0.0851 | 0.0922 | 0.1061 | 0.0581 | 0.0937 | 0.1008 | Shuifu | 0.0417 | 0.0434 | 0.0412 | 0.0747 | 0.0506 | 0.2127 |
| Chenggong | 0.0677 | 0.0716 | 0.0715 | 0.0726 | 0.1437 | 1.1228 | Mouding | 0.0477 | 0.0466 | 0.0482 | 0.0491 | 0.0588 | 0.2341 |
| Jinyang | 0.0824 | 0.0679 | 0.0729 | 0.0590 | 0.0899 | 0.0919 | Leibo | 0.0456 | 0.0483 | 0.0522 | 0.0449 | 0.0553 | 0.2133 |
| Renhe | 0.0678 | 0.0629 | 0.0792 | 0.0503 | 0.1023 | 0.5105 | Yongren | 0.0386 | 0.0398 | 0.0406 | 0.0680 | 0.0503 | 0.3046 |
| Dayao | 0.0564 | 0.0551 | 0.0561 | 0.1059 | 0.0732 | 0.2990 | Puge | 0.0394 | 0.0420 | 0.0434 | 0.0686 | 0.0434 | 0.1037 |
| Luquan | 0.0689 | 0.0620 | 0.0615 | 0.0510 | 0.0863 | 0.2535 | Zhaojue | 0.0417 | 0.0441 | 0.0447 | 0.0572 | 0.0471 | 0.1293 |
| Xundian | 0.0642 | 0.0647 | 0.0655 | 0.0474 | 0.0838 | 0.3049 | Jinning | 0.0381 | 0.0400 | 0.0437 | 0.0572 | 0.0449 | 0.1795 |
| Ningnan | 0.0511 | 0.0529 | 0.0549 | 0.1011 | 0.0593 | 0.1611 | Meigu | 0.0379 | 0.0426 | 0.0426 | 0.0477 | 0.0456 | 0.2027 |
| Songming | 0.0659 | 0.0622 | 0.0604 | 0.0517 | 0.0766 | 0.1625 | Daguan | 0.0434 | 0.0418 | 0.0407 | 0.0451 | 0.0451 | 0.0384 |
| Qiaojia | 0.0546 | 0.0555 | 0.0581 | 0.0545 | 0.0746 | 0.3661 | Butuo | 0.0369 | 0.0395 | 0.0420 | 0.0426 | 0.0448 | 0.2144 |
| Yongshan | 0.0500 | 0.0527 | 0.0511 | 0.0558 | 0.0764 | 0.5279 | Suijiang | 0.0326 | 0.0360 | 0.0376 | 0.0566 | 0.0417 | 0.2773 |

**Table A4.** Changes in economic environment index (EcoEEI) from 2000 to 2020.

| Name | EcoEEI | | | | | Change Rate | Name | EcoEEI | | | | | Change Rate |
|---|---|---|---|---|---|---|---|---|---|---|---|---|---|
| | 2000 | 2005 | 2010 | 2015 | 2020 | | | 2000 | 2005 | 2010 | 2015 | 2020 | |
| Dongqu | 0.3375 | 0.2808 | 0.2735 | 0.2490 | 0.2074 | −0.3854 | Mouding | 0.0757 | 0.0677 | 0.0737 | 0.0658 | 0.0819 | 0.0820 |
| Wuding | 0.2836 | 0.2299 | 0.2265 | 0.1954 | 0.1745 | −0.3847 | Wuhua | 0.0744 | 0.0662 | 0.0718 | 0.0711 | 0.0799 | 0.0750 |
| Panlong | 0.2496 | 0.2234 | 0.2169 | 0.2039 | 0.1764 | −0.2932 | Dayao | 0.0749 | 0.0646 | 0.0704 | 0.0748 | 0.0784 | 0.0467 |
| Guandu | 0.1771 | 0.1668 | 0.1919 | 0.2253 | 0.2361 | 0.3333 | Yao'an | 0.0780 | 0.0683 | 0.0736 | 0.0685 | 0.0728 | −0.0659 |
| Chenggong | 0.1559 | 0.1420 | 0.2057 | 0.2401 | 0.2324 | 0.4907 | Huize | 0.0697 | 0.0662 | 0.0723 | 0.0701 | 0.0776 | 0.1133 |
| Xishan | 0.2354 | 0.1933 | 0.1927 | 0.1725 | 0.1520 | −0.3543 | Suijiang | 0.0764 | 0.0682 | 0.0709 | 0.0616 | 0.0618 | −0.1909 |
| Anning | 0.1788 | 0.1516 | 0.1530 | 0.1390 | 0.1346 | −0.2474 | Luquan | 0.0620 | 0.0591 | 0.0708 | 0.0695 | 0.0763 | 0.2294 |
| Renhe | 0.1364 | 0.1346 | 0.1407 | 0.1062 | 0.1105 | −0.1899 | Ludian | 0.0623 | 0.0579 | 0.0652 | 0.0730 | 0.0785 | 0.2592 |
| Jinyang | 0.0958 | 0.0893 | 0.1109 | 0.1022 | 0.1086 | 0.1341 | Leibo | 0.0676 | 0.0617 | 0.0630 | 0.0691 | 0.0660 | −0.0240 |
| Songming | 0.0904 | 0.0857 | 0.1093 | 0.1089 | 0.1119 | 0.2380 | Puge | 0.0777 | 0.0616 | 0.0614 | 0.0613 | 0.0652 | −0.1614 |
| Dongchuan | 0.1018 | 0.0881 | 0.0959 | 0.0877 | 0.0896 | −0.1197 | Yongshan | 0.0609 | 0.0585 | 0.0616 | 0.0622 | 0.0733 | 0.2046 |
| Yongren | 0.0927 | 0.0819 | 0.0919 | 0.0919 | 0.0974 | 0.0511 | Jinning | 0.0741 | 0.0629 | 0.0619 | 0.0542 | 0.0622 | −0.1612 |
| Zhaoyang | 0.0940 | 0.0803 | 0.0850 | 0.0932 | 0.0963 | 0.0237 | Butuo | 0.0731 | 0.0597 | 0.0591 | 0.0587 | 0.0598 | −0.1818 |
| Fumin | 0.0872 | 0.0802 | 0.0966 | 0.0858 | 0.0974 | 0.1170 | Meigu | 0.0736 | 0.0598 | 0.0576 | 0.0572 | 0.0556 | −0.2446 |
| Ningnan | 0.0906 | 0.0811 | 0.0848 | 0.0811 | 0.0874 | −0.0347 | Daguan | 0.0606 | 0.0597 | 0.0582 | 0.0584 | 0.0640 | 0.0561 |
| Yuanmou | 0.0889 | 0.0746 | 0.0773 | 0.0802 | 0.0966 | 0.0858 | Zhaojue | 0.0698 | 0.0541 | 0.0558 | 0.0585 | 0.0598 | −0.1431 |
| Xundian | 0.0821 | 0.0749 | 0.0894 | 0.0821 | 0.0868 | 0.0577 | Pingshan | 0.0649 | 0.0511 | 0.0518 | 0.0616 | 0.0673 | 0.0359 |
| Huili | 0.0735 | 0.0723 | 0.0851 | 0.0855 | 0.0943 | 0.2830 | Qiaojia | 0.0561 | 0.0507 | 0.0575 | 0.0576 | 0.0701 | 0.2496 |
| Shuifu | 0.0803 | 0.0807 | 0.0833 | 0.0701 | 0.0756 | −0.0586 | Yiliang | 0.0576 | 0.0456 | 0.0453 | 0.0493 | 0.0565 | −0.0181 |
| Huidong | 0.0716 | 0.0641 | 0.0684 | 0.0808 | 0.0862 | 0.2051 | Yanjin | 0.0506 | 0.0497 | 0.0510 | 0.0492 | 0.0520 | 0.0285 |

**Table A5.** Changes in economic scale index (EcoESCI) from 2000 to 2020.

| Name | EcoESCI | | | | | Change Rate | Name | EcoESCI | | | | | Change Rate |
|---|---|---|---|---|---|---|---|---|---|---|---|---|---|
| | 2000 | 2005 | 2010 | 2015 | 2020 | | | 2000 | 2005 | 2010 | 2015 | 2020 | |
| Guandu | 0.2803 | 0.3613 | 0.3401 | 0.4097 | 0.3840 | 0.3696 | Xundian | 0.0557 | 0.0566 | 0.0573 | 0.0652 | 0.0762 | 0.3688 |
| Wuding | 0.1255 | 0.2559 | 0.4132 | 0.4341 | 0.3858 | 2.0745 | Yuanmou | 0.0522 | 0.0513 | 0.0518 | 0.0619 | 0.0786 | 0.5062 |
| Panlong | 0.2161 | 0.2943 | 0.2563 | 0.3130 | 0.3495 | 0.6175 | Wuhua | 0.0505 | 0.0510 | 0.0506 | 0.0635 | 0.0784 | 0.5529 |
| Dongqu | 0.3045 | 0.3248 | 0.2966 | 0.2563 | 0.1818 | −0.4031 | Yao'an | 0.0531 | 0.0538 | 0.0521 | 0.0555 | 0.0723 | 0.3611 |
| Xishan | 0.1401 | 0.2598 | 0.2478 | 0.3274 | 0.3052 | 1.1777 | Mouding | 0.0499 | 0.0502 | 0.0515 | 0.0561 | 0.0774 | 0.5499 |
| Anning | 0.1510 | 0.1550 | 0.1370 | 0.1708 | 0.2146 | 0.4216 | Yongren | 0.0471 | 0.0478 | 0.0479 | 0.0607 | 0.0748 | 0.5886 |
| Renhe | 0.0881 | 0.1014 | 0.1398 | 0.1474 | 0.1265 | 0.4367 | Pingshan | 0.0525 | 0.0479 | 0.0526 | 0.0563 | 0.0655 | 0.2459 |
| Chenggong | 0.0944 | 0.0959 | 0.0896 | 0.1251 | 0.1786 | 0.8915 | Yongshan | 0.0454 | 0.0497 | 0.0482 | 0.0600 | 0.0710 | 0.5642 |
| Zhaoyang | 0.1241 | 0.1079 | 0.1058 | 0.1169 | 0.1285 | 0.0355 | Leibo | 0.0444 | 0.0504 | 0.0583 | 0.0658 | 0.0548 | 0.2335 |
| Huili | 0.0816 | 0.0929 | 0.1313 | 0.1263 | 0.0943 | 0.1558 | Qiaojia | 0.0465 | 0.0474 | 0.0463 | 0.0490 | 0.0537 | 0.1564 |
| Jinyang | 0.0913 | 0.0744 | 0.0798 | 0.0951 | 0.1032 | 0.1314 | Yiliang | 0.0450 | 0.0480 | 0.0475 | 0.0485 | 0.0526 | 0.1679 |
| Huidong | 0.0701 | 0.0854 | 0.1037 | 0.0952 | 0.0825 | 0.1772 | Ludian | 0.0413 | 0.0438 | 0.0464 | 0.0496 | 0.0543 | 0.3161 |
| Shuifu | 0.0732 | 0.0898 | 0.0745 | 0.0873 | 0.0933 | 0.2743 | Puge | 0.0445 | 0.0503 | 0.0492 | 0.0489 | 0.0413 | −0.0717 |
| Huize | 0.0742 | 0.0766 | 0.0759 | 0.0812 | 0.1005 | 0.3534 | Jinning | 0.0407 | 0.0436 | 0.0505 | 0.0513 | 0.0465 | 0.1404 |
| Songming | 0.0660 | 0.0707 | 0.0686 | 0.0852 | 0.0879 | 0.3325 | Yanjin | 0.0457 | 0.0457 | 0.0444 | 0.0464 | 0.0437 | −0.0423 |
| Fumin | 0.0561 | 0.0649 | 0.0689 | 0.0818 | 0.0994 | 0.7697 | Butuo | 0.0399 | 0.0437 | 0.0469 | 0.0465 | 0.0409 | 0.0261 |
| Dongchuan | 0.0515 | 0.0525 | 0.0708 | 0.0748 | 0.0830 | 0.6118 | Daguan | 0.0449 | 0.0425 | 0.0402 | 0.0452 | 0.0443 | −0.0131 |
| Ningnan | 0.0567 | 0.0603 | 0.0660 | 0.0706 | 0.0639 | 0.1274 | Suijiang | 0.0365 | 0.0413 | 0.0448 | 0.0443 | 0.0497 | 0.3630 |
| Luquan | 0.0555 | 0.0529 | 0.0596 | 0.0673 | 0.0812 | 0.4636 | Zhaojue | 0.0415 | 0.0431 | 0.0430 | 0.0438 | 0.0402 | −0.0318 |
| Dayao | 0.0566 | 0.0582 | 0.0556 | 0.0650 | 0.0783 | 0.3844 | Meigu | 0.0388 | 0.0437 | 0.0423 | 0.0412 | 0.0385 | −0.0062 |

**Table A6.** Changes in the economic development index (EDI) from 2000 to 2020.

| Name | EDI | | | | | Change Rate | Name | EDI | | | | | Change Rate |
|---|---|---|---|---|---|---|---|---|---|---|---|---|---|
| | 2000 | 2005 | 2010 | 2015 | 2020 | | | 2000 | 2005 | 2010 | 2015 | 2020 | |
| Guandu | 0.6110 | 0.7026 | 0.7009 | 0.7977 | 0.8482 | 0.3880 | Yuanmou | 0.0924 | 0.0750 | 0.0816 | 0.1010 | 0.1423 | 0.5405 |
| Wuhua | 0.3656 | 0.6734 | 0.8162 | 0.7891 | 0.7608 | 1.0809 | Luquan | 0.0866 | 0.0738 | 0.0913 | 0.0880 | 0.1442 | 0.6649 |
| Dongqu | 0.7442 | 0.6383 | 0.6000 | 0.5385 | 0.4227 | −0.4320 | Yongren | 0.0788 | 0.0691 | 0.0797 | 0.1207 | 0.1231 | 0.5629 |
| Panlong | 0.4282 | 0.5746 | 0.5343 | 0.6098 | 0.6451 | 0.5065 | Wuding | 0.0770 | 0.0691 | 0.0757 | 0.0758 | 0.1289 | 0.6738 |
| Xishan | 0.3905 | 0.5101 | 0.4953 | 0.5682 | 0.5680 | 0.4544 | Yao'an | 0.0825 | 0.0712 | 0.0753 | 0.0733 | 0.1063 | 0.2881 |
| Anning | 0.3474 | 0.3105 | 0.2905 | 0.3205 | 0.4276 | 0.2309 | Mouding | 0.0737 | 0.0642 | 0.0729 | 0.0711 | 0.1186 | 0.6095 |
| Chenggong | 0.2183 | 0.2091 | 0.2666 | 0.3381 | 0.4552 | 1.0849 | Yongshan | 0.0566 | 0.0606 | 0.0605 | 0.0780 | 0.1212 | 1.1399 |
| Zhaoyang | 0.2577 | 0.1881 | 0.1954 | 0.2296 | 0.2606 | 0.0112 | Leibo | 0.0579 | 0.0602 | 0.0730 | 0.0800 | 0.0764 | 0.3195 |
| Renhe | 0.1926 | 0.1985 | 0.2592 | 0.2042 | 0.2400 | 0.2457 | Ludian | 0.0498 | 0.0470 | 0.0601 | 0.0900 | 0.0919 | 0.8454 |
| Huili | 0.1592 | 0.1664 | 0.2439 | 0.2477 | 0.1930 | 0.2123 | Pingshan | 0.0687 | 0.0464 | 0.0557 | 0.0703 | 0.0967 | 0.4084 |
| Jinning | 0.1697 | 0.1313 | 0.1632 | 0.1565 | 0.2022 | 0.1913 | Qiaojia | 0.0575 | 0.0533 | 0.0615 | 0.0611 | 0.0988 | 0.7186 |
| Huize | 0.1287 | 0.1278 | 0.1500 | 0.1253 | 0.2213 | 0.7201 | Puge | 0.0621 | 0.0536 | 0.0535 | 0.0788 | 0.0503 | −0.1895 |

**Table A6.** *Cont.*

| Name | EDI 2000 | 2005 | 2010 | 2015 | 2020 | Change Rate | Name | EDI 2000 | 2005 | 2010 | 2015 | 2020 | Change Rate |
|---|---|---|---|---|---|---|---|---|---|---|---|---|---|
| Huidong | 0.1271 | 0.1413 | 0.1774 | 0.1342 | 0.1630 | 0.2827 | Yiliang | 0.0542 | 0.0460 | 0.0454 | 0.0559 | 0.0793 | 0.4636 |
| Songming | 0.1226 | 0.1184 | 0.1378 | 0.1460 | 0.1768 | 0.4418 | Jinyang | 0.0533 | 0.0462 | 0.0556 | 0.0627 | 0.0539 | 0.0103 |
| Fumin | 0.0884 | 0.0925 | 0.1159 | 0.1326 | 0.1605 | 0.8160 | Suijiang | 0.0457 | 0.0454 | 0.0529 | 0.0625 | 0.0534 | 0.1689 |
| Dongchuan | 0.0989 | 0.0854 | 0.1199 | 0.1264 | 0.1354 | 0.3688 | Zhaojue | 0.0533 | 0.0410 | 0.0430 | 0.0596 | 0.0474 | −0.1123 |
| Ningnan | 0.0987 | 0.0939 | 0.1052 | 0.1527 | 0.1111 | 0.1253 | Yanjin | 0.0433 | 0.0414 | 0.0414 | 0.0676 | 0.0466 | 0.0774 |
| Shuifu | 0.0956 | 0.1136 | 0.0987 | 0.1322 | 0.1198 | 0.2530 | Daguan | 0.0493 | 0.0437 | 0.0387 | 0.0488 | 0.0538 | 0.0900 |
| Xundian | 0.1024 | 0.0958 | 0.1115 | 0.0949 | 0.1473 | 0.4387 | Butuo | 0.0502 | 0.0426 | 0.0475 | 0.0479 | 0.0458 | −0.0884 |
| Dayao | 0.0882 | 0.0776 | 0.0816 | 0.1456 | 0.1304 | 0.4783 | Meigu | 0.0506 | 0.0458 | 0.0421 | 0.0461 | 0.0400 | −0.2111 |

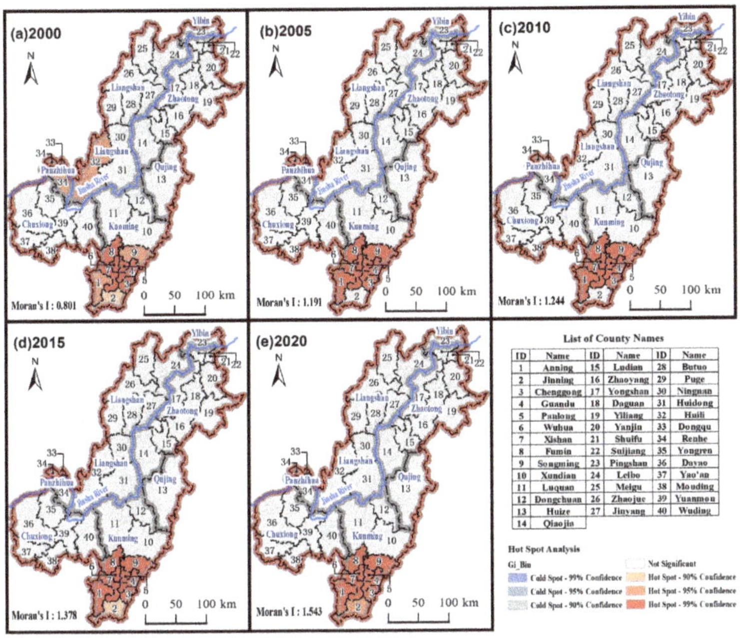


**Figure A8.** Hotspot analysis of economic development index (EDI).

**Table A7.** Changes in ecological–economic coupling degree (EECD) from 2000 to 2020.

| Name | EECD 2000 | 2005 | 2010 | 2015 | 2020 | Name | EECD 2000 | 2005 | 2010 | 2015 | 2020 |
|---|---|---|---|---|---|---|---|---|---|---|---|
| Xishan | 0.9918 | 0.9996 | 1.0000 | 1.0000 | 1.0000 | Luquan | 0.7486 | 0.6906 | 0.7532 | 0.7050 | 0.8188 |
| Panlong | 0.9998 | 0.9927 | 0.9969 | 0.9992 | 0.9970 | Dayao | 0.7136 | 0.6828 | 0.6918 | 0.8218 | 0.7871 |
| Anning | 0.9937 | 0.9851 | 0.9673 | 0.9663 | 0.9906 | Wuding | 0.7138 | 0.6736 | 0.7008 | 0.6662 | 0.7840 |
| Chenggong | 0.9631 | 0.9590 | 0.9876 | 0.9919 | 1.0000 | Shuifu | 0.6706 | 0.7350 | 0.6853 | 0.7259 | 0.7158 |
| Wuhua | 0.9981 | 0.9697 | 0.9452 | 0.9764 | 0.9786 | Ludian | 0.6603 | 0.6521 | 0.7004 | 0.7536 | 0.7486 |
| Guandu | 0.9833 | 0.9609 | 0.9556 | 0.9606 | 0.9439 | Mouding | 0.6981 | 0.6735 | 0.6884 | 0.6589 | 0.7747 |
| Zhaoyang | 0.9850 | 0.9467 | 0.9495 | 0.9434 | 0.9624 | Yao'an | 0.6908 | 0.6670 | 0.6713 | 0.6515 | 0.7365 |
| Renhe | 0.9457 | 0.9511 | 0.9834 | 0.9419 | 0.9533 | Qiaojia | 0.6638 | 0.6474 | 0.6680 | 0.6322 | 0.7448 |
| Dongqu | 0.9050 | 0.9387 | 0.9505 | 0.9800 | 0.9998 | Yongshan | 0.6240 | 0.6410 | 0.6325 | 0.6587 | 0.7641 |
| Huili | 0.9123 | 0.9092 | 0.9716 | 0.9554 | 0.9084 | Jinyang | 0.6595 | 0.6147 | 0.6718 | 0.6543 | 0.6228 |
| Jinning | 0.9032 | 0.8545 | 0.8823 | 0.8433 | 0.8933 | Puge | 0.6703 | 0.6313 | 0.6124 | 0.6860 | 0.5666 |
| Huidong | 0.8597 | 0.8753 | 0.9228 | 0.8409 | 0.8714 | Leibo | 0.5685 | 0.6091 | 0.6383 | 0.6489 | 0.6324 |
| Huize | 0.8643 | 0.8616 | 0.8968 | 0.8070 | 0.9263 | Zhaojue | 0.6671 | 0.5899 | 0.5862 | 0.6463 | 0.5702 |

**Table A7.** *Cont.*

| Name | EECD 2000 | 2005 | 2010 | 2015 | 2020 | Name | EECD 2000 | 2005 | 2010 | 2015 | 2020 |
|---|---|---|---|---|---|---|---|---|---|---|---|
| Yuanmou | 0.8627 | 0.8079 | 0.8446 | 0.8518 | 0.9004 | Butuo | 0.6443 | 0.5978 | 0.6172 | 0.5926 | 0.5659 |
| Dongchuan | 0.8337 | 0.7936 | 0.8875 | 0.8492 | 0.8550 | Yiliang | 0.6183 | 0.5610 | 0.5626 | 0.5632 | 0.6492 |
| Songming | 0.8246 | 0.8121 | 0.8662 | 0.8211 | 0.8706 | Pingshan | 0.5987 | 0.5181 | 0.5530 | 0.5817 | 0.6678 |
| Fumin | 0.7570 | 0.7755 | 0.8404 | 0.8196 | 0.8600 | Meigu | 0.5945 | 0.5755 | 0.5493 | 0.5561 | 0.5163 |
| Xundian | 0.8207 | 0.7845 | 0.8531 | 0.7363 | 0.8482 | Daguan | 0.5756 | 0.5536 | 0.5147 | 0.5294 | 0.5383 |
| Ningnan | 0.7623 | 0.7554 | 0.7851 | 0.8425 | 0.7625 | Yanjin | 0.5096 | 0.5108 | 0.5000 | 0.5639 | 0.4902 |
| Yongren | 0.7249 | 0.6854 | 0.7297 | 0.8103 | 0.8030 | Suijiang | 0.4927 | 0.5047 | 0.5281 | 0.5345 | 0.5131 |

**Table A8.** Average EECCD and its change rate from 2000 to 2020.

| County | Average EECCD | Change Rate | County | Average EECCD | Change Rate |
|---|---|---|---|---|---|
| Guandu | 0.7418 | 0.0884 | Yao'an | 0.4530 | 0.0837 |
| Wuhua | 0.7386 | 0.2596 | Pingshan | 0.4526 | 0.1055 |
| Panlong | 0.7234 | 0.1662 | Dongchuan | 0.4512 | 0.1422 |
| Xishan | 0.7166 | 0.1289 | Yongren | 0.4510 | 0.1565 |
| Dongqu | 0.6642 | −0.0626 | Wuding | 0.4484 | 0.2056 |
| Anning | 0.6386 | 0.1234 | Leibo | 0.4462 | 0.0743 |
| Chenggong | 0.5832 | 0.2537 | Mouding | 0.4390 | 0.1737 |
| Zhaoyang | 0.5484 | 0.0596 | Yongshan | 0.4378 | 0.2711 |
| Renhe | 0.5420 | 0.1021 | Suijiang | 0.4324 | 0.0552 |
| Huili | 0.5380 | 0.1069 | Yuanmou | 0.4144 | 0.1841 |
| Jinning | 0.5260 | 0.1062 | Yanjin | 0.4136 | 0.0630 |
| Shuifu | 0.5186 | 0.0703 | Yiliang | 0.4114 | 0.1717 |
| Songming | 0.5056 | 0.1411 | Qiaojia | 0.4110 | 0.2082 |
| Huize | 0.5016 | 0.2135 | Puge | 0.4066 | 0.0000 |
| Huidong | 0.5004 | 0.1163 | Ludian | 0.4000 | 0.2452 |
| Ningnan | 0.4800 | 0.0606 | Daguan | 0.3984 | 0.0861 |
| Dayao | 0.4796 | 0.1285 | Jinyang | 0.3852 | 0.0452 |
| Fumin | 0.4754 | 0.2096 | Meigu | 0.3832 | −0.0332 |
| Xundian | 0.4628 | 0.1625 | Zhaojue | 0.3772 | 0.0401 |
| Luquan | 0.4610 | 0.2005 | Butuo | 0.3732 | 0.0350 |

**Table A9.** EECCD in 40 counties from 2000 to 2020.

| County | 2000 | 2005 | 2010 | 2015 | 2020 | County | 2000 | 2005 | 2010 | 2015 | 2020 |
|---|---|---|---|---|---|---|---|---|---|---|---|
| Guandu | 0.7130 | 0.7270 | 0.7190 | 0.7740 | 0.7760 | Pingshan | 0.4550 | 0.4080 | 0.4290 | 0.4680 | 0.5030 |
| Wuhua | 0.6240 | 0.7240 | 0.7630 | 0.7960 | 0.7860 | Yao'an | 0.4540 | 0.4320 | 0.4420 | 0.4450 | 0.4920 |
| Panlong | 0.6620 | 0.7140 | 0.7030 | 0.7660 | 0.7720 | Dongchuan | 0.4290 | 0.4160 | 0.4440 | 0.4770 | 0.4900 |
| Xishan | 0.6670 | 0.7050 | 0.7070 | 0.7510 | 0.7530 | Yongren | 0.4280 | 0.4170 | 0.4290 | 0.4860 | 0.4950 |
| Dongqu | 0.6870 | 0.6670 | 0.6600 | 0.6630 | 0.6440 | Wuding | 0.4280 | 0.4220 | 0.4300 | 0.4460 | 0.5160 |
| Anning | 0.6240 | 0.6080 | 0.6140 | 0.6460 | 0.7010 | Leibo | 0.4310 | 0.4210 | 0.4500 | 0.4660 | 0.4630 |
| Chenggong | 0.5360 | 0.5290 | 0.5590 | 0.6200 | 0.6720 | Mouding | 0.4260 | 0.4070 | 0.4270 | 0.4350 | 0.5000 |
| Zhaoyang | 0.5540 | 0.5120 | 0.5200 | 0.5690 | 0.5870 | Yongshan | 0.4020 | 0.4090 | 0.4120 | 0.4550 | 0.5110 |
| Renhe | 0.5190 | 0.5230 | 0.5580 | 0.5380 | 0.5720 | Suijiang | 0.4170 | 0.4100 | 0.4310 | 0.4640 | 0.4400 |
| Huili | 0.4960 | 0.5090 | 0.5570 | 0.5790 | 0.5490 | Yanjin | 0.3970 | 0.3880 | 0.3930 | 0.4680 | 0.4220 |
| Jinning | 0.5180 | 0.4830 | 0.5220 | 0.5340 | 0.5730 | Yuanmou | 0.4020 | 0.3840 | 0.3850 | 0.4250 | 0.4760 |
| Shuifu | 0.4980 | 0.5090 | 0.4990 | 0.5540 | 0.5330 | Qiaojia | 0.3890 | 0.3810 | 0.4010 | 0.4140 | 0.4700 |
| Songming | 0.4820 | 0.4800 | 0.4880 | 0.5280 | 0.5500 | Yiliang | 0.3960 | 0.3870 | 0.3840 | 0.4260 | 0.4640 |
| Huize | 0.4730 | 0.4730 | 0.4910 | 0.4970 | 0.5740 | Puge | 0.4020 | 0.3880 | 0.3960 | 0.4450 | 0.4020 |
| Huidong | 0.4730 | 0.4890 | 0.5160 | 0.4960 | 0.5280 | Ludian | 0.3630 | 0.3560 | 0.3840 | 0.4450 | 0.4520 |
| Ningnan | 0.4620 | 0.4540 | 0.4660 | 0.5280 | 0.4900 | Daguan | 0.3950 | 0.3810 | 0.3740 | 0.4130 | 0.4290 |
| Dayao | 0.4590 | 0.4430 | 0.4510 | 0.5270 | 0.5180 | Jinyang | 0.3760 | 0.3670 | 0.3800 | 0.4100 | 0.3930 |
| Fumin | 0.4390 | 0.4410 | 0.4610 | 0.5050 | 0.5310 | Meigu | 0.3920 | 0.3800 | 0.3750 | 0.3900 | 0.3790 |
| Xundian | 0.4430 | 0.4450 | 0.4460 | 0.4650 | 0.5150 | Zhaojue | 0.3740 | 0.3550 | 0.3650 | 0.4030 | 0.3890 |
| Luquan | 0.4390 | 0.4290 | 0.4480 | 0.4620 | 0.5270 | Butuo | 0.3710 | 0.3580 | 0.3710 | 0.3820 | 0.3840 |

**Table A10.** Ranking of ecological–economic relative development degree (EERDD).

| County | EERDD 2000 | EERDD 2005 | EERDD 2010 | EERDD 2015 | EERDD 2020 | Average Value | Ranking | County | EERDD 2000 | EERDD 2005 | EERDD 2010 | EERDD 2015 | EERDD 2020 | Average Value | Ranking |
|---|---|---|---|---|---|---|---|---|---|---|---|---|---|---|---|
| Dongqu | 2.4802 | 2.0527 | 1.9020 | 1.4969 | 1.0358 | 1.7935 | 1 | Luquan | 0.2027 | 0.1606 | 0.2064 | 0.1701 | 0.2706 | 0.2021 | 21 |
| Guandu | 1.4456 | 1.7655 | 1.8355 | 1.7701 | 1.9863 | 1.7606 | 2 | Dayao | 0.1761 | 0.1557 | 0.1614 | 0.2740 | 0.2370 | 0.2008 | 22 |
| Wuhua | 0.8832 | 1.6470 | 1.9701 | 1.5507 | 1.5175 | 1.5137 | 3 | Wuding | 0.1762 | 0.1501 | 0.1673 | 0.1456 | 0.2340 | 0.1746 | 23 |
| Panlong | 0.9563 | 1.2738 | 1.1701 | 1.0825 | 1.1686 | 1.1303 | 4 | Shuifu | 0.1482 | 0.1918 | 0.1572 | 0.1850 | 0.1776 | 0.1720 | 24 |
| Xishan | 0.7728 | 1.0548 | 0.9820 | 1.0149 | 1.0054 | 0.9660 | 5 | Ludian | 0.1422 | 0.1376 | 0.1670 | 0.2067 | 0.2026 | 0.1712 | 25 |
| Chenggong | 0.5759 | 0.5582 | 0.7289 | 0.7751 | 1.0147 | 0.7305 | 6 | Mouding | 0.1655 | 0.1500 | 0.1592 | 0.1414 | 0.2252 | 0.1683 | 26 |
| Anning | 0.7984 | 0.7062 | 0.5956 | 0.5906 | 0.7594 | 0.6900 | 7 | Yao'an | 0.1607 | 0.1461 | 0.1486 | 0.1372 | 0.1930 | 0.1571 | 27 |
| Zhaoyang | 0.7055 | 0.5127 | 0.5223 | 0.5020 | 0.5728 | 0.5631 | 8 | Qiaojia | 0.1442 | 0.1350 | 0.1467 | 0.1269 | 0.1996 | 0.1505 | 28 |
| Renhe | 0.5095 | 0.5281 | 0.6926 | 0.4971 | 0.5361 | 0.5527 | 9 | Yongshan | 0.1227 | 0.1315 | 0.1270 | 0.1413 | 0.2157 | 0.1477 | 29 |
| Huili | 0.4189 | 0.4120 | 0.6173 | 0.5441 | 0.4102 | 0.4805 | 10 | Jinyang | 0.1417 | 0.1181 | 0.1489 | 0.1388 | 0.1221 | 0.1339 | 30 |
| Jinning | 0.3993 | 0.3163 | 0.3599 | 0.3008 | 0.3799 | 0.3512 | 11 | Puge | 0.1481 | 0.1264 | 0.1170 | 0.1577 | 0.0965 | 0.1291 | 31 |
| Huidong | 0.3238 | 0.3480 | 0.4437 | 0.2977 | 0.3418 | 0.3510 | 12 | Leibo | 0.0973 | 0.1154 | 0.1301 | 0.1358 | 0.1270 | 0.1211 | 32 |
| Huize | 0.3308 | 0.3266 | 0.3865 | 0.2575 | 0.4527 | 0.3508 | 13 | Zhaojue | 0.1462 | 0.1065 | 0.1049 | 0.1344 | 0.0980 | 0.1180 | 33 |
| Yuanmou | 0.3283 | 0.2584 | 0.3026 | 0.3124 | 0.3936 | 0.3191 | 14 | Butuo | 0.1333 | 0.1101 | 0.1193 | 0.1077 | 0.0962 | 0.1133 | 34 |
| Dongchuan | 0.2885 | 0.2435 | 0.3691 | 0.3089 | 0.3170 | 0.3054 | 15 | Yiliang | 0.1198 | 0.0942 | 0.0949 | 0.0951 | 0.1360 | 0.1080 | 35 |
| Songming | 0.2774 | 0.2630 | 0.3337 | 0.2733 | 0.3406 | 0.2976 | 16 | Pingshan | 0.1105 | 0.0780 | 0.0910 | 0.1029 | 0.1466 | 0.1058 | 36 |
| Fumin | 0.2096 | 0.2260 | 0.2970 | 0.2715 | 0.3243 | 0.2657 | 17 | Meigu | 0.1086 | 0.1003 | 0.0896 | 0.0922 | 0.0773 | 0.0936 | 37 |
| Xundian | 0.2728 | 0.2344 | 0.3143 | 0.1929 | 0.3074 | 0.2644 | 18 | Daguan | 0.1003 | 0.0912 | 0.0768 | 0.0820 | 0.0853 | 0.0871 | 38 |
| Ningnan | 0.2142 | 0.2083 | 0.2351 | 0.2998 | 0.2144 | 0.2344 | 19 | Yanjin | 0.0750 | 0.0754 | 0.0718 | 0.0954 | 0.0686 | 0.0772 | 39 |
| Yongren | 0.1842 | 0.1573 | 0.1878 | 0.2611 | 0.2531 | 0.2087 | 20 | Suijiang | 0.0694 | 0.0734 | 0.0816 | 0.0839 | 0.0762 | 0.0769 | 40 |

**Table A11.** Average ecological–economic relative development degree (EERDD) and its change rate in 40 counties from 2000 to 2020.

| County | Average EERDD | Change Rate | County | Average EERDD | Change Rate |
|---|---|---|---|---|---|
| Guandu | 1.7597 | 0.3740 | Luquan | 0.2034 | 0.3353 |
| Dongqu | 1.7374 | −0.5824 | Dayao | 0.2024 | 0.3459 |
| Wuhua | 1.5152 | 0.7182 | Wuding | 0.1761 | 0.3277 |
| Panlong | 1.1300 | 0.2220 | Ludian | 0.1746 | 0.4249 |
| Xishan | 0.9672 | 0.3010 | Shuifu | 0.1721 | 0.1987 |
| Chenggong | 0.7421 | 0.7620 | Mouding | 0.1697 | 0.3608 |
| Anning | 0.6873 | −0.0488 | Yao'an | 0.1576 | 0.2007 |
| Zhaoyang | 0.5605 | −0.1882 | Qiaojia | 0.1517 | 0.3838 |
| Renhe | 0.5509 | 0.0523 | Yongshan | 0.1500 | 0.7583 |
| Huili | 0.4800 | −0.0207 | Jinyang | 0.1336 | −0.1385 |
| Huize | 0.3513 | 0.3688 | Puge | 0.1285 | −0.3482 |
| Jinning | 0.3508 | −0.0485 | Leibo | 0.1211 | 0.3052 |
| Huidong | 0.3495 | 0.0554 | Zhaojue | 0.1171 | −0.3297 |
| Yuanmou | 0.3226 | 0.1992 | Butuo | 0.1124 | −0.2784 |
| Dongchuan | 0.3052 | 0.0986 | Yiliang | 0.1084 | 0.1344 |
| Songming | 0.2975 | 0.2277 | Pingshan | 0.1065 | 0.3264 |
| Fumin | 0.2676 | 0.5470 | Meigu | 0.0932 | −0.2878 |
| Xundian | 0.2616 | 0.1269 | Daguan | 0.0868 | −0.1490 |
| Ningnan | 0.2353 | 0.0007 | Yanjin | 0.0778 | −0.0856 |
| Yongren | 0.2104 | 0.3741 | Suijiang | 0.0771 | 0.0980 |

## References


1. Tian, Y.; Huang, P.; Zhao, X. Spatial analysis, coupling coordination, and efficiency evaluation of green innovation: A case study of the Yangtze River Economic Belt. *PLoS ONE* **2020**, *15*, e0243459. [CrossRef] [PubMed]
2. An, S.; Zhang, S.; Hou, H.; Zhang, Y.; Xu, H.; Liang, J. Coupling coordination analysis of the ecology and economy in the Yellow River Basin under the background of high-quality development. *Land* **2022**, *11*, 1235. [CrossRef]
3. Ekins, P. The Kuznets curve for the environment and economic growth: Examining the evidence. *Environ. Plan. A* **1997**, *29*, 805–830. [CrossRef]
4. Leal, P.H.; Marques, A.C. The evolution of the environmental Kuznets curve hypothesis assessment: A literature review under a critical analysis perspective. *Heliyon* **2022**, *8*, e11521. [CrossRef]
5. Sagar, A.D.; Najam, A. The human development index: A critical review. *Ecol. Econ.* **1998**, *25*, 249–264. [CrossRef]
6. Phillips, J. Determining sustainability using the Environmental Performance Index and Human Development Index–An alternative approach to the Environmental Human Index through a holistic quantitative dynamic framework. *Sci. Total Environ.* **2023**, *884*, 163752. [CrossRef]
7. Zheng, X.; Chen, Y. A better strategy: Using green GDP to measure economic health. *Front. Environ. Sci.* **2024**, *12*, 1459764. [CrossRef]

8. Zheng, Y.; Chen, M. Promoting Green GDP for more balanced development. *Envtl. Pol'y L.* **2007**, *37*, 416.
9. Ahmed, Z.; Asghar, M.M.; Malik, M.N.; Nawaz, K. Moving towards a sustainable environment: The dynamic linkage between natural resources, human capital, urbanization, economic growth, and ecological footprint in China. *Resour. Policy* **2020**, *67*, 101677. [CrossRef]
10. Desli, E.; Gkoulgkoutsika, A.; Sdrolia, E.; Zarotiadis, G. Eco-efficiency: A methodological framework and assessment. *Clean. Environ. Syst.* **2021**, *3*, 100049. [CrossRef]
11. Coluccia, B.; Valente, D.; Fusco, G.; De Leo, F.; Porrini, D. Assessing agricultural eco-efficiency in Italian Regions. *Ecol. Indic.* **2020**, *116*, 106483. [CrossRef]
12. Babcicky, P. Rethinking the foundations of sustainability measurement: The limitations of the Environmental Sustainability Index (ESI). *Social. Indic. Res.* **2013**, *113*, 133–157. [CrossRef]
13. Cucchiella, F.; D'Adamo, I.; Gastaldi, M.; Koh, S.L.; Rosa, P. A comparison of environmental and energetic performance of European countries: A sustainability index. *Renew. Sustain. Energy Rev.* **2017**, *78*, 401–413. [CrossRef]
14. Mora, F. The use of ecological integrity indicators within the natural capital index framework: The ecological and economic value of the remnant natural capital of México. *J. Nat. Conserv.* **2019**, *47*, 77–92. [CrossRef]
15. Czúcz, B.; Molnár, Z.; Horváth, F.; Nagy, G.G.; Botta-Dukát, Z.; Török, K. Using the natural capital index framework as a scalable aggregation methodology for regional biodiversity indicators. *J. Nat. Conserv.* **2012**, *20*, 144–152. [CrossRef]
16. Han, X.; Cao, T.; Yan, X. Comprehensive evaluation of ecological environment quality of mining area based on sustainable development indicators: A case study of Yanzhou Mining in China. *Environ. Dev. Sustain.* **2021**, *23*, 7581–7605. [CrossRef]
17. Li, F.; Wang, F.; Liu, H.; Huang, K.; Yu, Y.; Huang, B. A comparative analysis of ecosystem service valuation methods: Taking Beijing, China as a case. *Ecol. Indic.* **2023**, *154*, 110872. [CrossRef]
18. Kang, N.; Hou, L.; Huang, J.; Liu, H. Ecosystem services valuation in China: A meta-analysis. *Sci. Total Environ.* **2022**, *809*, 151122. [CrossRef]
19. Li, L.; Fan, Z.; Feng, W.; Yuxin, C.; Keyu, Q. Coupling coordination degree spatial analysis and driving factor between socio-economic and eco-environment in northern China. *Ecol. Indic.* **2022**, *135*, 108555. [CrossRef]
20. Wang, Y.; Song, C.; Cheng, C.; Wang, H.; Wang, X.; Gao, P. Modelling and evaluating the economy-resource-ecological environment system of a third-polar city using system dynamics and ranked weights-based coupling coordination degree model. *Cities* **2023**, *133*, 104151. [CrossRef]
21. Shi, T.; Yang, S.; Zhang, W.; Zhou, Q. Coupling coordination degree measurement and spatiotemporal heterogeneity between economic development and ecological environment----Empirical evidence from tropical and subtropical regions of China. *J. Clean. Prod.* **2020**, *244*, 118739. [CrossRef]
22. Jiang, H.-N.; Zhang, W.-Z.; Yu, J.-H.; Zhang, J.-W. Spatial coupling of innovative milieu and industrial structure transformation of resource-based cities in Shanxi province. *J. Nat. Resour.* **2020**, *35*, 269–283. [CrossRef]
23. Guan, D.; Gao, W.; Su, W.; Li, H.; Hokao, K. Modeling and dynamic assessment of urban economy–resource–environment system with a coupled system dynamics–geographic information system model. *Ecol. Indic.* **2011**, *11*, 1333–1344. [CrossRef]
24. Fang, W.; An, H.; Li, H.; Gao, X.; Sun, X.; Zhong, W. Accessing on the sustainability of urban ecological-economic systems by means of a coupled emergy and system dynamics model: A case study of Beijing. *Energy Policy* **2017**, *100*, 326–337. [CrossRef]
25. Moffatt, I.; Hanley, N. Modelling sustainable development: Systems dynamic and input–ouput approaches. *Environ. Model. Softw.* **2001**, *16*, 545–557. [CrossRef]
26. Long, X.; Wu, S.; Wang, J.; Wu, P.; Wang, Z. Urban water environment carrying capacity based on VPOSR-coefficient of variation-grey correlation model: A case of Beijing, China. *Ecol. Indic.* **2022**, *138*, 108863. [CrossRef]
27. Ke, S.; Qiao, D.; Chen, Z. Analysis of the influencing factors on China's forestry economic growth based on grey correlation model. *For. Econ. Rev.* **2019**, *1*, 17–31. [CrossRef]
28. Wang, Z.; Yang, L.; Yin, J.; Zhang, B. Assessment and prediction of environmental sustainability in China based on a modified ecological footprint model. *Resour. Conserv. Recycl.* **2018**, *132*, 301–313. [CrossRef]
29. Sabir, S.; Gorus, M.S. The impact of globalization on ecological footprint: Empirical evidence from the South Asian countries. *Environ. Sci. Pollut. Res.* **2019**, *26*, 33387–33398. [CrossRef]
30. Zuo, Z.; Guo, H.; Cheng, J.; Li, Y. How to achieve new progress in ecological civilization construction?–Based on cloud model and coupling coordination degree model. *Ecol. Indic.* **2021**, *127*, 107789. [CrossRef]
31. Tomal, M. Evaluation of coupling coordination degree and convergence behaviour of local development: A spatiotemporal analysis of all Polish municipalities over the period 2003–2019. *Sustain. Cities Soc.* **2021**, *71*, 102992. [CrossRef]
32. Zhu, X.; Zuo, X.; Li, H. The dual effects of heterogeneous environmental regulation on the technological innovation of Chinese steel enterprises—Based on a high-dimensional fixed effects model. *Ecol. Econ.* **2021**, *188*, 107113. [CrossRef]
33. Cheng, J.; Zhang, X.; Gao, Q. Analysis of the spatio-temporal changes and driving factors of the marine economic–ecological–social coupling coordination: A case study of 11 coastal regions in China. *Ecol. Indic.* **2023**, *153*, 110392. [CrossRef]
34. Zhang, Q.; Shen, J.; Sun, F. Spatiotemporal differentiation of coupling coordination degree between economic development and water environment and its influencing factors using GWR in China's province. *Ecol. Model.* **2021**, *462*, 109794. [CrossRef]
35. Lv, L.; Chen, C.; Wang, Z. Spatiotemporal differentiation and the obstacle factors influencing the coupling coordination between economic development and water pollution control capability in the Yangtze River Economic Belt. *Environ. Sci. Pollut. Res.* **2022**, *29*, 75681–75698. [CrossRef]

36. Sun, Q.; Sui, Y.-J. Agricultural Green Ecological Efficiency Evaluation Using BP Neural Network–DEA Model. *Systems* **2023**, *11*, 291. [CrossRef]
37. Wu, M.; Wei, Y.; Lam, P.T.; Liu, F.; Li, Y. Is urban development ecologically sustainable? Ecological footprint analysis and prediction based on a modified artificial neural network model: A case study of Tianjin in China. *J. Clean. Prod.* **2019**, *237*, 117795. [CrossRef]
38. Chang, C.-P.; Dong, M.; Sui, B.; Chu, Y. Driving forces of global carbon emissions: From time-and spatial-dynamic perspectives. *Econ. Model.* **2019**, *77*, 70–80. [CrossRef]
39. Duan, C.; Zhu, W.; Wang, S.; Chen, B. Drivers of global carbon emissions 1990–2014. *J. Clean. Prod.* **2022**, *371*, 133371. [CrossRef]
40. Sueyoshi, T.; Yuan, Y. Social sustainability measured by intermediate approach for DEA environmental assessment: Chinese regional planning for economic development and pollution prevention. *Energy Econ.* **2017**, *66*, 154–166. [CrossRef]
41. Wu, X.; Liu, S.; Cheng, F.; Hou, X.; Zhang, Y.; Dong, S.; Liu, G. A regional strategy for ecological sustainability: A case study in Southwest China. *Sci. Total Environ.* **2018**, *616*, 1224–1234. [CrossRef] [PubMed]
42. Han, Z.; Jiao, S.; Zhang, X.; Xie, F.; Ran, J.; Jin, R.; Xu, S. Seeking sustainable development policies at the municipal level based on the triad of city, economy and environment: Evidence from Hunan province, China. *J. Environ. Manag.* **2021**, *290*, 112554. [CrossRef] [PubMed]
43. Artmann, M.; Inostroza, L.; Fan, P. Urban sprawl, compact urban development and green cities. How much do we know, how much do we agree? *Ecol. Indic.* **2019**, *96*, 3–9. [CrossRef]
44. Yuan, L.; Li, R.; He, W.; Wu, X.; Kong, Y.; Degefu, D.M.; Ramsey, T.S. Coordination of the industrial-ecological economy in the Yangtze River Economic Belt, China. *Front. Environ. Sci.* **2022**, *10*, 882221. [CrossRef]
45. Liu, L.; Yang, Y.; Liu, S.; Gong, X.; Zhao, Y.; Jin, R.; Duan, H.; Jiang, P. A comparative study of green growth efficiency in Yangtze River Economic Belt and Yellow River Basin between 2010 and 2020. *Ecol. Indic.* **2023**, *150*, 110214. [CrossRef]
46. Song, Y.-S.; Du, C.; Yang, C.; Qin, C. Ecological Environmental Quality Evaluation of Yellow River Basin. *Procedia Eng.* **2012**, *28*, 754–758. [CrossRef]
47. Zhang, Z.; Li, H.; Cao, Y. Research on the coordinated development of economic development and ecological environment of nine Provinces (regions) in the Yellow River Basin. *Sustainability* **2022**, *14*, 13102. [CrossRef]
48. Zhang, S.; Yang, P.; Xia, J.; Qi, K.; Wang, W.; Cai, W.; Chen, N. Research and analysis of ecological environment quality in the Middle Reaches of the Yangtze River Basin between 2000 and 2019. *Remote Sens.* **2021**, *13*, 4475. [CrossRef]
49. Zhang, Y.; Sun, M.; Yang, R.; Li, X.; Zhang, L.; Li, M. Decoupling water environment pressures from economic growth in the Yangtze River Economic Belt, China. *Ecol. Indic.* **2021**, *122*, 107314. [CrossRef]
50. Sun, J.; Zhou, L.; Zong, H. Landscape pattern vulnerability of the eastern Hengduan Mountains, China and response to elevation and artificial disturbance. *Land* **2022**, *11*, 1110. [CrossRef]
51. Qin, Y.; Wang, F.; Zhang, S.; Tu, Y.; Chen, C.; Yang, Y. Impacts of cascade hydropower development on aquatic environment in middle and lower reaches of Jinsha River, China: A review. *Environ. Sci. Pollut. Res.* **2024**, *31*, 54363–54380. [CrossRef] [PubMed]
52. Liu, Y.; Xu, Y. A geographic identification of multidimensional poverty in rural China under the framework of sustainable livelihoods analysis. *Appl. Geogr.* **2016**, *73*, 62–76. [CrossRef]
53. Huang, X.-R.; Gao, L.-Y.; Yang, P.-P.; Xi, Y.-Y. Cumulative impact of dam constructions on streamflow and sediment regime in lower reaches of the Jinsha River, China. *J. Mt. Sci.* **2018**, *15*, 2752–2765. [CrossRef]
54. Zhang, X.-F.; Yan, H.-C.; Yue, Y.; Xu, Q.-X. Quantifying natural and anthropogenic impacts on runoff and sediment load: An investigation on the middle and lower reaches of the Jinsha River Basin. *J. Hydrol. Reg. Stud.* **2019**, *25*, 100617. [CrossRef]
55. Li, B.; Pan, B.; Cheng, W.; Han, J.; Qi, D.; Zhu, C. Research on geomorphological regionalization of China. *Acta Geogr. Sin.* **2013**, *68*, 291–306. [CrossRef]
56. Zheng, J.; Bian, J.J.; Ge, Q.; Hao, Z.; Yin, Y.; Liao, Y. The climate regionalization in China for 1981–2010. *Chin. Sci. Bull* **2013**, *58*, 3088–3099.
57. Gorelick, N.; Hancher, M.; Dixon, M.; Ilyushchenko, S.; Thau, D.; Moore, R. Google Earth Engine: Planetary-scale geospatial analysis for everyone. *Remote Sens. Environ.* **2017**, *202*, 18–27. [CrossRef]
58. Deng, M.; Chen, J.; Tao, F.; Zhu, J.; Wang, M. On the coupling and coordination development between environment and economy: A case study in the Yangtze River Delta of China. *Int. J. Environ. Res. Public Health* **2022**, *19*, 586. [CrossRef]
59. Li, W.; Yi, P.; Zhang, D.; Zhou, Y. Assessment of coordinated development between social economy and ecological environment: Case study of resource-based cities in Northeastern China. *Sustain. Cities Soc.* **2020**, *59*, 102208. [CrossRef]
60. Fan, W.; Wang, H.; Liu, Y.; Liu, H. Spatio-temporal variation of the coupling relationship between urbanization and air quality: A case study of Shandong Province. *J. Clean. Prod.* **2020**, *272*, 122812. [CrossRef]
61. Chen, J.; Li, Z.; Dong, Y.; Song, M.; Shahbaz, M.; Xie, Q. Coupling coordination between carbon emissions and the eco-environment in China. *J. Clean. Prod.* **2020**, *276*, 123848. [CrossRef]
62. Cheng, X.; Long, R.; Chen, H.; Li, Q. Coupling coordination degree and spatial dynamic evolution of a regional green competitiveness system–A case study from China. *Ecol. Indic.* **2019**, *104*, 489–500. [CrossRef]
63. Liao, S.; Wu, Y.; Wong, S.W.; Shen, L. Provincial perspective analysis on the coordination between urbanization growth and resource environment carrying capacity (RECC) in China. *Sci. Total Environ.* **2020**, *730*, 138964. [CrossRef] [PubMed]
64. Li, Y.F.; Li, Y.; Zhou, Y.; Shi, Y.L.; Zhu, X.D. Investigation of a coupling model of coordination between urbanization and the environment. *J. Environ. Manag.* **2012**, *98*, 127–133. [CrossRef]

65. Xu, H. A remote sensing urban ecological index and its application. *Acta Ecol. Sin.* **2013**, *33*, 7853–7862.
66. Xu, H.; Wang, Y.; Guan, H.; Shi, T.; Hu, X. Detecting Ecological Changes with a Remote Sensing Based Ecological Index (RSEI) Produced Time Series and Change Vector Analysis. *Remote Sens.* **2019**, *11*, 2345. [CrossRef]
67. Zhou, J.X.; Chen, J.; Chen, X.H.; Zhu, X.L.; Qiu, Y.A.; Song, H.H.; Rao, Y.H.; Zhang, C.S.; Cao, X.; Cui, X.H. Sensitivity of six typical spatiotemporal fusion methods to different influential factors: A comparative study for a normalized difference vegetation index time series reconstruction. *Remote Sens. Environ.* **2021**, *252*, 112130. [CrossRef]
68. Solomos, S.; Abuelgasim, A.; Spyrou, C.; Binietoglou, I.; Nickovic, S. Development of a dynamic dust source map for NMME-DREAM v1.0 model based on MODIS Normalized Difference Vegetation Index (NDVI) over the Arabian Peninsula. *Geosci. Model Dev.* **2019**, *12*, 979–988. [CrossRef]
69. Nguyen, C.T.; Chidthaisong, A.; Diem, P.K.; Huo, L.Z. A Modified Bare Soil Index to Identify Bare Land Features during Agricultural Fallow-Period in Southeast Asia Using Landsat 8. *Land* **2021**, *10*, 231. [CrossRef]
70. Suharyadi, R.; Umarhadi, D.A.; Awanda, D.; Widyatmanti, W. Exploring Built-Up Indices and Machine Learning Regressions for Multi-Temporal Building Density Monitoring Based on Landsat Series. *Sensors* **2022**, *22*, 4716. [CrossRef]
71. Nanjam, R.; Ahmadi, F.F. A New Spatiotemporal Model for Analyzing the Variations of Urban Heat Islands Using Remotely Sensed Multi Spectral Images: The Case of Mashhad City, Iran. *J. Indian. Soc. Remote* **2021**, *49*, 2489–2502. [CrossRef]
72. Xie, F.; Fan, H. Deriving drought indices from MODIS vegetation indices (NDVI/EVI) and Land Surface Temperature (LST): Is data reconstruction necessary? *Int. J. Appl. Earth Obs.* **2021**, *101*, 102352. [CrossRef]
73. Jolliffe, I.T.; Cadima, J. Principal component analysis: A review and recent developments. *Philos. Trans. R. Soc. A* **2016**, *374*, 20150202. [CrossRef] [PubMed]
74. Dos Santos, P.H.; Neves, S.M.; Sant'Anna, D.O.; de Oliveira, C.H.; Carvalho, H.D. The analytic hierarchy process supporting decision making for sustainable development: An overview of applications. *J. Clean. Prod.* **2019**, *212*, 119–138. [CrossRef]
75. Zhu, Y.X.; Tian, D.Z.; Yan, F. Effectiveness of Entropy Weight Method in Decision-Making. *Math. Probl. Eng.* **2020**, *2020*, 3564835. [CrossRef]
76. Dou, S.; Zhu, Y.; Xu, D.; Amuakwa-Mensah, F. Ecological challenges in the economic recovery of resource-depleted cities in China. *J. Environ. Manag.* **2023**, *333*, 117406. [CrossRef]
77. Xu, Y.; Dai, Q.-Y.; Lu, Y.-G.; Zhao, C.; Huang, W.-T.; Xu, M.; Feng, Y.-X. Identification of ecologically sensitive zones affected by climate change and anthropogenic activities in Southwest China through a NDVI-based spatial-temporal model. *Ecol. Indic.* **2024**, *158*, 111482. [CrossRef]
78. Praveen, B.; Talukdar, S.; Shahfahad; Mahato, S.; Mondal, J.; Sharma, P.; Islam, A.M.T.; Rahman, A. Analyzing trend and forecasting of rainfall changes in India using non-parametrical and machine learning approaches. *Sci. Rep.* **2020**, *10*, 10342. [CrossRef]
79. Zhang, M.M.; Kafy, A.A.; Ren, B.; Zhang, Y.W.; Tan, S.K.; Li, J.X. Application of the Optimal Parameter Geographic Detector Model in the Identification of Influencing Factors of Ecological Quality in Guangzhou, China. *Land* **2022**, *11*, 1303. [CrossRef]
80. Wei, W.; Jin, C.G.; Han, Y.; Huang, Z.H.; Niu, T.; Li, J.K. The Coordinated Development and Regulation Research on Public Health, Ecological Environment and Economic Development: Evidence from the Yellow River Basin of China. *Int. J. Environ. Res. Public Health* **2022**, *19*, 6927. [CrossRef]
81. Park, Y.S.; Lek, S. Chapter 7—Artificial Neural Networks: Multilayer Perceptron for Ecological Modeling. In *Developments in Environmental Modelling*; Jørgensen, S.E., Ed.; Elsevier: Amsterdam, The Netherlands, 2016; Volume 28, pp. 123–140.
82. Chen, Q.; Chen, H.; Zhang, J.; Hou, Y.; Shen, M.; Chen, J.; Xu, C. Impacts of climate change and LULC change on runoff in the Jinsha River Basin. *J. Geogr. Sci.* **2020**, *30*, 85–102. [CrossRef]
83. Zhou, Y.; Yi, Y.-J.; Liu, H.-X.; Tang, C.-H.; Zhu, Y.-l.; Zhang, S.-H. Effect of geomorphologic features and climate change on vegetation distribution in the arid hot valleys of Jinsha River, Southwest China. *J. Mt. Sci.* **2022**, *19*, 2874–2885. [CrossRef]
84. Zhou, J.; Lu, T. Long-Term Spatial and Temporal Variation of Near Surface Air Temperature in Southwest China During 1969–2018. *Front. Earth Sci.* **2021**, *9*, 753757. [CrossRef]
85. Yetemen, O.; Istanbulluoglu, E.; Flores-Cervantes, J.H.; Vivoni, E.R.; Bras, R.L. Ecohydrologic role of solar radiation on landscape evolution. *Water Resour. Res.* **2015**, *51*, 1127–1157. [CrossRef]
86. Yang, J.; El-Kassaby, Y.A.; Guan, W. The effect of slope aspect on vegetation attributes in a mountainous dry valley, Southwest China. *Sci. Rep.* **2020**, *10*, 16465. [CrossRef]
87. Li, J.; Dong, S.; Yang, Z.; Peng, M.; Liu, S.; Li, X. Effects of cascade hydropower dams on the structure and distribution of riparian and upland vegetation along the middle-lower Lancang-Mekong River. *For. Ecol. Manag.* **2012**, *284*, 251–259. [CrossRef]
88. Zhou, Y.; Yi, Y.; Liu, H.; Tang, C.; Zhang, S. Spatiotemporal dynamic of soil erosion and the key factors impact processes over semi-arid catchments in Southwest China. *Ecol. Eng.* **2024**, *201*, 107217. [CrossRef]
89. Liu, S.; Wang, C.; Wang, P.; Chen, J.; Wang, X.; Yuan, Q. Anthropogenic disturbances on distribution and sources of pharmaceuticals and personal care products throughout the Jinsha River Basin, China. *Environ. Res.* **2021**, *198*, 110449. [CrossRef]
90. Schmidt, A.H.; Denn, A.R.; Hidy, A.J.; Bierman, P.R.; Tang, Y. Human and natural controls on erosion in the Lower Jinsha River, China. *J. Asian Earth Sci.* **2019**, *170*, 351–359. [CrossRef]
91. Liu, F.; Yang, L.; Wang, S. Spatial and temporal evolution and correlation analysis of landscape ecological risks and ecosystem service values in the Jinsha River Basin. *J. Resour. Ecol.* **2023**, *14*, 914–927. [CrossRef]

92. Zhang, Y.; Shang, K.; Shi, Z.; Wang, H.; Li, X. Spatial pattern of the vitality of Chinese characteristic towns: A perspective from nighttime lights. *Land* **2022**, *11*, 85. [CrossRef]
93. Yuan, J.; Li, R.; Huang, K. Driving factors of the variation of ecosystem service and the trade-off and synergistic relationships in typical karst basin. *Ecol. Indic.* **2022**, *142*, 109253. [CrossRef]
94. Lian, Y.; You, G.J.-Y.; Lin, K.; Jiang, Z.; Zhang, C.; Qin, X. Characteristics of climate change in southwest China karst region and their potential environmental impacts. *Environ. Earth Sci.* **2015**, *74*, 937–944. [CrossRef]
95. Soga, M.; Gaston, K.J. The ecology of human–nature interactions. *Proc. R. Soc. B* **2020**, *287*, 20191882. [CrossRef] [PubMed]
96. Liu, Q.; Hao, Y.; Stebler, E.; Tanaka, N.; Zou, C.B. Impact of plant functional types on coherence between precipitation and soil moisture: A wavelet analysis. *Geophys. Res. Lett.* **2017**, *44*, 12197–112207. [CrossRef]
97. Ding, X.; Shao, X.; Wang, J.; Peng, S.; Shi, J. Research on the Spatial-Temporal Pattern Evolution and Driving Force of Ecological Environment Quality in Kunming City Based on Remote Sensing Ecological Environment Index in the Past 25 Years. *Pol. J. Environ. Stud.* **2023**, *33*, 1073–1089. [CrossRef]
98. Li, W.; Xie, S.; Wang, Y.; Huang, J.; Cheng, X. Effects of urban expansion on ecosystem health in Southwest China from a multi-perspective analysis. *J. Clean. Prod.* **2021**, *294*, 126341. [CrossRef]
99. Yao, S.; Jiang, Y.; Su, X.; Wang, R. Evaluation of Regional Eco-Efficiency and Improvement Paths in the Yangtze River Economic Belt under the "Double Carbon" Target. *Pol. J. Environ. Stud.* **2024**, *33*, 4879–4892. [CrossRef]
100. Dai, X.; Li, W.; Liu, Z.; Tong, C.; Li, C.; Zeng, J.; Ye, Y.; Li, W.; Shan, Y.; Zhou, J.; et al. Effects of Mining on Urban Environmental Change: A Case Study of Panzhihua. *Remote Sens.* **2022**, *14*, 6004. [CrossRef]
101. Yang, R.Y.; Zhong, C.B. Land Suitability Evaluation of Sorghum Planting in Luquan County of Jinsha River Dry and Hot Valley Based on the Perspective of Sustainable Development of Characteristic Poverty Alleviation Industry. *Agriculture* **2022**, *12*, 1852. [CrossRef]
102. Ma, L.; Chung, J.; Thorson, S. E-government in China: Bringing economic development through administrative reform. *Gov. Inf. Q.* **2005**, *22*, 20–37. [CrossRef]
103. Qi, Y.; Yang, Y.; Jin, F. China's economic development stage and its spatio-temporal evolution: A prefectural-level analysis. *J. Geogr. Sci.* **2013**, *23*, 297–314. [CrossRef]
104. Sun, M.; Yang, R.; Li, X.; Zhang, L.; Liu, Q. Designing a path for the sustainable development of key ecological function zones: A case study of southwest China. *Glob. Ecol. Conserv.* **2021**, *31*, e01840. [CrossRef]
105. Zhao, L. Spatial Pattern Evolution and Spatial Transformation Trend of Industry in China. *Sci. Geogr. Sin.* **2021**, *41*, 387–396. [CrossRef]
106. Sang, Y.-F.; Wang, Z.; Liu, C. Spatial and temporal variability of daily temperature during 1961–2010 in the Yangtze River Basin, China. *Quat. Int.* **2013**, *304*, 33–42. [CrossRef]
107. Ning, Y.; Liu, S.G.; Zhao, S.Q.; Liu, M.C.; Gao, H.Q.; Gong, P. Urban growth rates, trajectories, and multi-dimensional disparities in China. *Cities* **2022**, *126*, 103717. [CrossRef]
108. Ge, Y.; Yuan, Y.; Hu, S.; Ren, Z.; Wu, Y. Space–time variability analysis of poverty alleviation performance in China's poverty-stricken areas. *Spat. Stat.* **2017**, *21*, 460–474. [CrossRef]
109. Teng, Y.; Yang, J.; Zuo, R.; Wang, J. Impact of urbanization and industrialization upon surface water quality: A pilot study of Panzhihua mining town. *J. Earth Sci.* **2011**, *22*, 658–668. [CrossRef]
110. Liu, M.; Zhang, Z.; Zhang, H.; Yang, M.; Song, D.; Ou, X. Spatial-Temporal Monitoring of Urban Growth: A Case in Kunming, Southwest China. In Proceedings of the Geo-Informatics in Resource Management and Sustainable Ecosystem, Hong Kong, China, 18–20 November 2016; pp. 116–127.
111. Liu, T.; Yu, L.; Chen, X.; Li, X.; Du, Z.; Yan, Y.; Peng, D.; Gong, P. Utilizing nighttime light datasets to uncover the spatial patterns of county-level relative poverty-returning risk in China and its alleviating factors. *J. Clean. Prod.* **2024**, *448*, 141682. [CrossRef]
112. Li, H.B.; Jin, X.B.; McCormick, B.P.; Tittonell, P.; Liu, J.; Han, B.; Sun, R.; Zhou, Y.K. Analysis of the contribution of land consolidation to sustainable poverty alleviation under various natural conditions. *Land Use Policy* **2023**, *133*, 106871. [CrossRef]
113. Deng, Y.; Wang, S.; Bai, X.; Tian, Y.; Wu, L.; Xiao, J.; Chen, F.; Qian, Q. Relationship among land surface temperature and LUCC, NDVI in typical karst area. *Sci. Rep.* **2018**, *8*, 641. [CrossRef] [PubMed]
114. Chen, X.; Li, X. The Imbalanced Pattern of Population and Economy and Its Influencing Factors in the Wumeng Mountain Area. *J. Resour. Ecol.* **2024**, *15*, 937–950. [CrossRef]
115. Yang, Y.; Fang, Y.-p.; Xu, Y.; Zhang, Y. Assessment of urban resilience based on the transformation of resource-based cities: A case study of Panzhihua, China. *Ecol. Soc.* **2021**, *26*, 20. [CrossRef]
116. Emran, M.S.; Hou, Z. Access to markets and rural poverty: Evidence from household consumption in China. *Rev. Econ. Stat.* **2013**, *95*, 682–697. [CrossRef]
117. Su, J.; Shen, T.; Jin, S. Ecological efficiency evaluation and driving factor analysis of the coupling coordination of the logistics industry and manufacturing industry. *Environ. Sci. Pollut. Res.* **2022**, *29*, 62458–62474. [CrossRef]
118. Wang, J.; Han, Q.; Wu, K.; Xu, Z.; Liu, P. Spatial-temporal patterns and evolution characteristics of the coordinated development of industrial economy, natural resources and environment in China. *Resour. Policy* **2022**, *75*, 102463. [CrossRef]

119. Jiang, S.; Chen, X.; Smettem, K.; Wang, T. Climate and land use influences on changing spatiotemporal patterns of mountain vegetation cover in southwest China. *Ecol. Indic.* **2021**, *121*, 107193. [CrossRef]
120. Li, W.; Wang, Y.; Xie, S.; Cheng, X. Spatiotemporal evolution scenarios and the coupling analysis of ecosystem health with land use change in Southwest China. *Ecol. Eng.* **2022**, *179*, 106607. [CrossRef]

remote sensing

MDPI

*Article*

# OMAD-6: Advancing Offshore Mariculture Monitoring with a Comprehensive Six-Type Dataset and Performance Benchmark

**Zewen Mo [1], Yinyu Liang [1], Yulin Chen [1], Yanyun Shen [2], Minduan Xu [1], Zhipan Wang [1] and Qingling Zhang [1,*]**

[1] School of Aeronautics and Astronautics, Sun Yat-sen University, Shenzhen Campus, 66 Gongchang Road, Guangming District, Shenzhen 518107, China; mozw6@mail2.sysu.edu.cn (Z.M.); liangyy255@mail2.sysu.edu.cn (Y.L.); chenylin223@mail2.sysu.edu.cn (Y.C.); xumd6@mail2.sysu.edu.cn (M.X.); wangzhp25@mail2.sysu.edu.cn (Z.W.)

[2] Xi'an Modern Control Technology Research Institute of China North Industries Group, Xi'an 710065, China; shenyanyun_203@126.com

* Correspondence: zhangqling@mail.sysu.edu.cn; Tel.: +86-181-2709-5863

**Abstract:** Offshore mariculture is critical for global food security and economic development. Advances in deep learning and data-driven approaches, enable the rapid and effective monitoring of offshore mariculture distribution and changes. However, detector performance depends heavily on training data quality. The lack of standardized classifications and public datasets for offshore mariculture facilities currently hampers effective monitoring. Here, we propose to categorize offshore mariculture facilities into six types: TCC, DWCC, FRC, LC, RC, and BC. Based on these categories, we introduce a benchmark dataset called OMAD-6. This dataset includes over 130,000 instances and more than 16,000 high-resolution remote sensing images. The images with a spatial resolution of 0.6 m were sourced from key regions in China, Chile, Norway, and Egypt, from the Google Earth platform. All instances in OMAD-6 were meticulously annotated manually with horizontal bounding boxes and polygons. Compared to existing remote sensing datasets, OMAD-6 has three notable characteristics: (1) it is comparable to large, published datasets in instances per category, image quantity, and sample coverage; (2) it exhibits high inter-class similarity; (3) it shows significant intra-class diversity in facility sizes and arrangements. Based on the OMAD-6 dataset, we evaluated eight state-of-the-art methods to establish baselines for future research. The experimental results demonstrate that the OMAD-6 dataset effectively represents various real-world scenarios, which have posed considerable challenges for current instance segmentation algorithms. Our evaluation confirms that the OMAD-6 dataset has the potential to improve offshore mariculture identification. Notably, the QueryInst and PointRend algorithms have distinguished themselves as top performers on the OMAD-6 dataset, robustly identifying offshore mariculture facilities even with complex environmental backgrounds. Its ongoing development and application will play a pivotal role in future offshore mariculture identification and management.

**Keywords:** offshore mariculture; deep learning; food security; marine economy; benchmark dataset

## 1. Introduction

Aquaculture is central to global food security and economic prosperity. The Food and Agriculture Organization of the United Nations has reported a pivotal milestone: in 2022, aquaculture production exceeded traditional fishing for the first time, with mariculture alone contributing 54.79% to total aquaculture output [1]. This sector bolsters the sustainability of seafood resources and concurrently mitigates the pressure of overfishing on marine ecosystems [2].

Amid the stagnation of the global fishing industry and the concurrent rise in seafood demand, the imperative for efficient and sustainable mariculture management has become increasingly critical. Conventional approaches to monitoring offshore mariculture are often labor-intensive and prone to human error, making them costly and inefficient [3]. Advances in deep learning have enabled the development of semantic segmentation algorithms capable of accurately identifying and classifying offshore mariculture areas [2]. However, their effectiveness heavily depends on access to high-quality, diverse datasets for training and validation [4].

Upon thorough examination of the existing body of literature, several pivotal concerns have been identified in the realm of offshore mariculture identification:

a. Classification Inconsistencies: A predominant trend in the literature is the narrow thematic scope of studies, which tend to concentrate on a limited range of mariculture facilities. For instance, research often zeroes in on floating raft aquaculture [5] and raft laver aquaculture [6], thereby restricting the broader applicability of the insights gleaned. This selective focus suggests a pressing need for a more inclusive classification framework that accounts for a wider array of mariculture systems to bolster the universality and practicality of research findings.
b. Geographical Limitations: The geographical scope of research is frequently restricted to particular locales, such as the northern Yellow Sea and Liaodong Bay area [7], the surrounding Qinshan Island [8], and the Bohai Rim area [9]. This regional confinement impedes the formulation of algorithms with enhanced generalizability and adaptability across diverse marine environments.
c. Fragmented and Private Datasets: The datasets utilized in algorithmic studies are often not only limited in size but also privately owned and inaccessible to the public. A case in point is the disclosure of the regional distribution of offshore rafts and cage aquaculture along China's coastline in 2018 [10]. Despite this revelation, the absence of alignment between remote sensing imagery and vector maps has precluded the formation of a holistic dataset, essential for robust algorithmic training and validation.

We present a Six-Type Offshore Mariculture Areas Dataset (OMAD-6), crafted to tackle the aforementioned challenges. Our classification schema is the result of an exhaustive literature review coupled with meticulous manual visual analysis of global offshore mariculture areas captured in remote sensing imagery. This rigorous process has enabled us to categorize offshore mariculture into six distinct types: traditional cage culture, deep-water cage culture, floating raft culture, longline culture, raft culture, and bouchot culture. The OMAD-6 dataset leverages high-resolution (0.6 m) Google Earth remote sensing imagery, which has been collected from 2021 to 2023, encompassing key mariculture regions worldwide, such as China, Chile, Norway, and Egypt. By providing a comprehensive and unified dataset for instance segmentation research, OMAD-6 is poised to enhance the automation and efficiency of mariculture management practices. This dataset not only addresses the inconsistencies and limitations previously identified in the literature but also paves the way for more generalized and diverse algorithm development, ultimately contributing to the advancement of sustainable mariculture practices worldwide.

The main contributions of this article are manifold:

a. A Unified Classification: We have consolidated existing literature and FAO reports to categorize offshore mariculture facilities into six types: traditional cage, deep-water cage, floating raft, longline, raft, and bouchot culture.
b. A Publicly Available Dataset: We have constructed a publicly available offshore mariculture dataset, OMAD-6, which covers multiple countries and includes detailed annotations.
c. A Performance Benchmark Testing: We conducted performance benchmark tests on eight prominent instance segmentation algorithms, including Mask R-CNN [11], Cascade R-CNN [12], YOLACT [13], PointRend [14], QueryInst [15], CondInst [16], BoxInst [17], and ConvNeXt [18]. Among these, QueryInst and PointRend demonstrated superior performance on the OMAD-6 dataset, achieving mask mAP values of 80.9 and 80.5, respectively. These results underscore their effectiveness in addressing the challenges of offshore mariculture area segmentation, providing a robust foundation for improving the precision and efficiency of mariculture management practices.

The remainder of this paper is organized as follows: Section 2 reviews related works, with a focus on remote sensing datasets and advances in offshore mariculture algorithms. Section 3 details the construction of the OMAD-6 dataset, including data acquisition, annotation, and dataset characteristics. In Section 4, we evaluate the performance of eight state-of-the-art instance segmentation algorithms applied to the OMAD-6 dataset, including benchmark testing and results analysis. Finally, Section 5 summarizes the study's key findings and proposes future research directions, such as dataset expansion and algorithmic advancements.

## 2. Related Works

### 2.1. Datasets for Object Recognition in Remote Sensing Images

The rapid expansion of remote sensing technology in recent years has heralded a new era characterized by a wealth of diverse and comprehensive datasets. These datasets have emerged as the foundational pillars for geospatial object detection and classification, markedly advancing the cutting edge of remote sensing image analysis. As detailed in Table 1, the array of available datasets provides indispensable data support, which is crucial for the evolution of research methodologies and outcomes within this field. Well-crafted and meticulously annotated datasets are indispensable for the training and validation of algorithms designed to detect intricate patterns and structures.

**Table 1.** Comparison between the OMAD-6 dataset and other datasets containing remote sensing images.

| Dataset | Source | Categories | Instances | Instances PC | Images | Image Width | Annotation | Year |
|---|---|---|---|---|---|---|---|---|
| NWPU VHR-10 [19] | MS | 10 | 3651 | 365.1 | 800 | ~1000 | HBB | 2014 |
| VEDAI [20] | satellite | 9 | 2950 | 327.8 | 1268 | 512, 1024 | OBB | 2015 |
| DLR 3k [21] | aerial | 8 | 14,235 | 1779.4 | 20 | 5616 | OBB | 2015 |
| HRSC2016 [22] | GE | 26 | 2976 | 114.5 | 1061 | ~1100 | OBB | 2016 |
| RSOD [23] | GE | 4 | 6950 | 1737.5 | 976 | ~1000 | HBB | 2017 |
| DIOR [24] | GE | 20 | 190,288 | 9514.4 | 23,463 | 800 | HBB | 2019 |
| SIMD [25] | GE | 15 | 45,096 | 3006.4 | 5000 | 1024 | HBB | 2020 |
| DOTA-v2.0 [26] | MS | 18 | 1,793,658 | 99,647.7 | 11,268 | 800–20,000 | OBB | 2021 |
| FAIR1M [27] | MS | 37 | 1,020,579 | 27,583.2 | 42,796 | 600–10,000 | OBB, polygon | 2021 |
| CORS-ADD [28] | MS | 7 | 32,285 | 4612.1 | 7339 | 640 | HBB, OBB | 2023 |
| HRSI [29] | MS | 3 | 25,256 | 8418.7 | 796 | 512–4000 | OBB | 2024 |
| OMAD-6 (Ours) | GE | 6 | 132,144 | 22,024 | 16,270 | 512 | HBB, polygon | 2024 |

PC: per categories; MS: multi-source; GE: Google Earth; HBB: horizontal bounding box; OBB: oriented bounding box.

For instance, The NWPU VHR-10 [19] dataset, created by Northwestern Polytechnical University, is sourced from Google Earth and the Vaihingen datasets. It contains 800 images and 3651 instances, playing a significant role in object detection and geographic image classification in remote sensing images. Similarly, the VEDAI [20] dataset focuses on vehicle detection, containing 1268 images and 2950 instances, and supports both HBB and OBB annotations. Additionally, the DLR 3k [21] dataset, created by the German Aerospace Center, includes 20 aerial images with 14,235 instances and is widely used for object detection and image segmentation tasks. Furthermore, the HRSC2016 [22] dataset, developed by Wuhan University, contains 1061 high-resolution images and 2976 instances, supports OBB annotations, and is extensively used in ship detection research. In addition, Wuhan University also provided the RSOD [23] dataset, which includes four categories and a total of 6950 instances, primarily used for detecting airplanes, playgrounds, overpasses, and oil drums.

Moreover, the DIOR [24] dataset comprises 23,463 images and 190,288 instances, covering 20 target types such as airplanes, ships, and vehicles, making it one of the largest remote sensing image object detection datasets available. The SIMD [25] dataset contains 5000 images and 45,096 instances annotated using HBB for maritime target detection. In contrast, the DOTA [26] dataset is a large-scale remote sensing image dataset covering 15 categories with 1793,658 instances, widely used in object detection and classification research. Furthermore, the FAIR1M [27] dataset, created by the Institute of Aeronautical

Information of the Chinese Academy of Sciences, contains 42,796 images and 1,020,579 instances, covering 37 categories and is one of the most fine-grained remote sensing image classification datasets available. Lastly, the CORS-ADD [28] dataset, created by the Harbin Institute of Technology, contains 7339 images and 32,285 instances, supporting both HBB and OBB annotations, and is mainly used for aircraft detection research. The HRSI [29] dataset, developed by Anhui University, contains 796 hazy weather images with 25,256 instances, focusing on large vehicles, ships, and airplanes. Supporting OBB annotations, it addresses the challenge of detecting objects in degraded imagery, providing a crucial resource for adverse weather studies in remote sensing.

Although these datasets have achieved significant success in their respective fields, they lack a specific focus on offshore mariculture areas. To bridge this gap, we introduce the OMAD-6 dataset, specifically designed for the classification and identification of offshore mariculture areas. It is derived from high-resolution images from Google Earth, encompassing multiple major marine mariculture regions. The dataset is annotated using horizontal bounding box (HBB) and polygon methods and includes six categories with a total of 132,144 instances. Compared to existing datasets (Table 1), OMAD-6 exhibits comparable characteristics regarding instance quantity, category, annotation type, and image quantity. The average number of instances per category in OMAD-6 is 22,024, second only to DOTA-v2.0 and FAIR1M, providing a solid data foundation for researching and extracting offshore mariculture areas.

### 2.2. *Advancements in Offshore Mariculture Algorithms*

Detecting and extracting offshore mariculture areas have long been critical tasks in the realm of remote sensing applications. With the continuous development of remote sensing technology and image processing algorithms, researchers have proposed various methods to improve the accuracy of mariculture area identification. These methods can be broadly categorized into traditional spectral feature-based methods, object-oriented methods, and deep learning approaches.

#### 2.2.1. Traditional Spectral Feature-Based Methods

Initial studies predominantly relied on the spectral characteristics of remote sensing images to differentiate mariculture areas from adjacent water bodies. Commonly used water indices, such as the Normalized Difference Water Index (NDWI) and its variants, were employed for water extraction to identify mariculture regions [30]. For instance, utilizing combinations of spectral bands from high-resolution remote sensing images significantly enhanced the accuracy of differentiating floating raft aquaculture areas from surrounding seawater [5]. However, these pixel-based spectral methods often faced challenges in complex environments, such as turbid waters or regions with intricate backgrounds, leading to decreased identification accuracy.

#### 2.2.2. Object-Oriented Methods

To address the limitations of traditional methods in complex environments, researchers introduced object-oriented image analysis techniques. This approach segments images into meaningful objects rather than individual pixels, leveraging the spatial relationships between objects to improve identification accuracy. A method combining object-based neighborhood features was employed to accurately identify and delineate mariculture facilities from high-resolution images [31]. By considering the spatial relationships among image objects, this method enhanced the detection capability for individual mariculture structures. Additionally, the Case-Based Reasoning (CBR) method achieved successful multi-scale image segmentation and classification, demonstrating excellent performance in turbid waters and complex backgrounds [32]. This object-oriented approach has proven to be more robust in environments where traditional spectral methods fall short, thereby offering a more reliable means of mariculture area identification.

#### 2.2.3. Development of Deep Learning Methods

The advent of deep learning has heralded a new wave of research in the extraction of mariculture areas, with an increasing number of studies leveraging these techniques to achieve remarkable results. Deep learning models, particularly Convolutional Neural Networks (CNNs), have become research hotspots in this field due to their powerful feature learning and pattern recognition capabilities.

The U-EFF-NET model utilizes EfficientNet as the feature extraction network and incorporates the Convolutional Block Attention Module (CBAM) in the decoder to enhance extraction accuracy [33]. RaftNet, a deep neural network based on the UNet architecture, incorporates enhanced dual-channel and residual mixed dilated convolutional blocks to effectively handle turbid water environments and floating raft culture areas of different scales [34]. Additionally, UPS-Net, an improved Unet model, integrates pyramid upsampling and compressed excitation (PSE) structures to capture detailed boundary and contextual information, addressing the "adhesion" phenomenon in raft aquaculture area extraction, and performs excellently well in high-resolution images [35].

Other models, such as HCNet [36], HCHNet [37], IDUDL [38], and RADNet [6], have also demonstrated outstanding performance. HCNet and HCHNet enhance traditional convolutional methods by improving their multi-resolution analysis capabilities; IDUDL leverages the cloud and fog penetration ability of SAR data to segment mariculture areas; RADNet enhances the spectral response and boundary detection of raft laver aquaculture areas through reverse attention and boundary guidance modules.

Despite significant advancements in extraction accuracy and efficiency through deep learning techniques, several challenges persist. Deep learning models achieve high detection accuracy by precisely identifying and delineating mariculture areas, effectively adapting to complex environments with intricate backgrounds and varying mariculture structures. They exhibit strong feature learning capabilities, automatically deriving high-level features and minimizing dependence on manual feature design. However, these models are limited by the absence of unified classification systems and datasets, resulting in inconsistencies in data and classification standards across studies, hindering algorithm comparisons. Furthermore, their generalization ability is constrained; models may perform poorly in different regions or conditions due to limited training data. In summary, mariculture area extraction algorithms have progressed from traditional methods to deep learning approaches, each offering unique contributions and facing specific limitations. To advance this field, establishing unified classification standards and shared datasets is essential for effective comparison and application of various algorithms. This study aims to address these issues, furthering the detection and analysis of offshore mariculture areas.

## 3. Dataset Construction

### *3.1. Selection of Study Areas*

The Food and Agriculture Organization of the United Nations, reports that Asia accounted for 91.61% of global aquaculture production in 2020, with China contributing 62.77% of Asia's output. In Africa, Americas and Europe, the countries with the highest proportions of aquaculture production are Egypt, Chile, and Norway, respectively, at 67.62%, 34.21%, and 45.28% [39]. Thus, selecting China, Egypt, Chile, and Norway as research areas effectively represents typical conditions of offshore mariculture worldwide.

### *3.2. Classification of Offshore Mariculture Areas*

Through an extensive literature review and detailed manual analysis of remote sensing images, we have systematically classified offshore mariculture facilities into six distinct categories (as illustrated in Figure 1). This classification is based on a combination of structural characteristics, operational practices, and spatial distribution patterns observed in the imagery. Each facility type represents a unique approach to mariculture, influenced by both environmental conditions and regional technological preferences. The six categories encompass a range of designs and scales, reflecting the diversity and complexity of modern offshore mariculture systems.

**Figure 1.** Samples of each category in the OMAD-6 dataset. (**a–f**) are in the order of TCC, DWCC, FRC, LC, RC, and BC.

Traditional cage culture (TCC) [31,40,41] is primarily located in nearshore waters. The net cage, typically constructed from materials like wood, bamboo, synthetic fiber mesh, or metal mesh, has a simple structure comprising a floating frame and netting, secured by piles or anchors on both sides. This method offers advantages such as low cost, ease of operation, and high flexibility, making it suitable for cultivating various fish and shellfish.

Deep-water cage culture (DWCC) [42–45] typically occurs in open waters far from the coast. The mariculture facilities consist of large circular cages made of steel or high-strength plastic, which feature a robust structure providing ample space and higher carrying capacity.

This method minimizes interference from human activities with the mariculture fish, better protects the benthic environment, reduces organic pollution, and is suitable for large-scale commercial mariculture.

Floating raft culture (FRC) [46–48] primarily cultivates shellfish and algae. Floating rafts consist of floating barrels or bamboo rafts and fixed ropes, floating on the water's surface. This type of facility features a simple structure, low cost, and convenient management, making it widely used worldwide.

Longline culture (LC) [49,50] is also utilized for cultivating shellfish and algae. It involves pulling a long rope on the water surface to suspend mariculture products in the water. This method makes efficient use of three-dimensional water space, is suitable for large-scale production, is cost-effective, and is environment friendly.

Raft culture (RC) [4,35] is mainly employed for cultivating seaweed. Bamboo poles, ropes, and other equipment are used in shallow coastal waters to create mariculture racks fixed to the seabed. This method offers a stable structure, wide adaptability, and ease of management.

Bouchot culture (BC) [51,52] is primarily used for shellfish farming. It typically consists of wooden stakes and rafts. The stakes are driven into the seabed, and the rafts are fixed to the stakes at a height of about 2 m above the seabed. The structure remains stable, even as the raft may be exposed during low tide and submerged during high tide. This method effectively utilizes vertical space, increases yield per unit area, and has minimal environmental impact.

By providing a detailed classification of offshore mariculture areas, we have established a solid foundation for the construction of a comprehensive dataset of offshore aquaculture areas. This classification system offers valuable references for future research, enabling more accurate and efficient identification and analysis of mariculture practices worldwide.

### 3.3. Data Acquisition and Annotation

#### 3.3.1. Data Acquisition

Offshore mariculture areas exhibit significant variations in imagery due to factors such as lighting, humidity, growth cycles of cultivated organisms and tidal fluctuations (Figure 1). To ensure data integrity and accurate representation, we obtained 552 optical remote sensing images with dimensions of 28,416 × 28,416 pixels, and 196 images of varying sizes, from Google Earth. These images, captured between 2021 and 2023, possess a spatial resolution of 0.6 m, providing high-detailed imagery for analysis.

#### 3.3.2. Image Cropping and Preprocessing

From the collected imagery, we selected 100 high-quality images of mariculture areas for cropping. To enhance dataset robustness, diversity and reduce edge effects [53], each cropped block incorporates a 10% horizontal and vertical overlap with neighboring blocks. The selection process was guided by three key criteria: a substantial total image count, balanced category distribution and sufficient instances per category. Based on these criteria, we cropped and manually selected 16,270 image blocks with a size of 512 × 512 pixels, to construct a comprehensive dataset (OMAD-6) optimized for fine-grained classification of offshore mariculture areas, providing a rich resource for the analysis and understanding of mariculture practices worldwide.

#### 3.3.3. Image Annotation

Following the classification outlined in Section 3.1, we manually annotated 132,144 instances with the SAM-Tool (https://github.com/zhouayi/SAM-Tool, accessed on 2 December 2023) and Labelme (version 4.5.7). Annotations included both HBB (horizontal bounding box) and Polygon formats, with examples presented in Figure 2. The annotation process involved meticulously annotating each instance via the dot prompt operation within the Segment Anything Model (SAM) [54] and exporting the annotations in COCO format using the SAM-Tool. When annotations were missing or infeasible using SAM, we converted the COCO format annotations to a Labelme-compatible format. We then manually used Labelme to perform Polygon annotations on these problematic instances, followed by conversion back to COCO format.

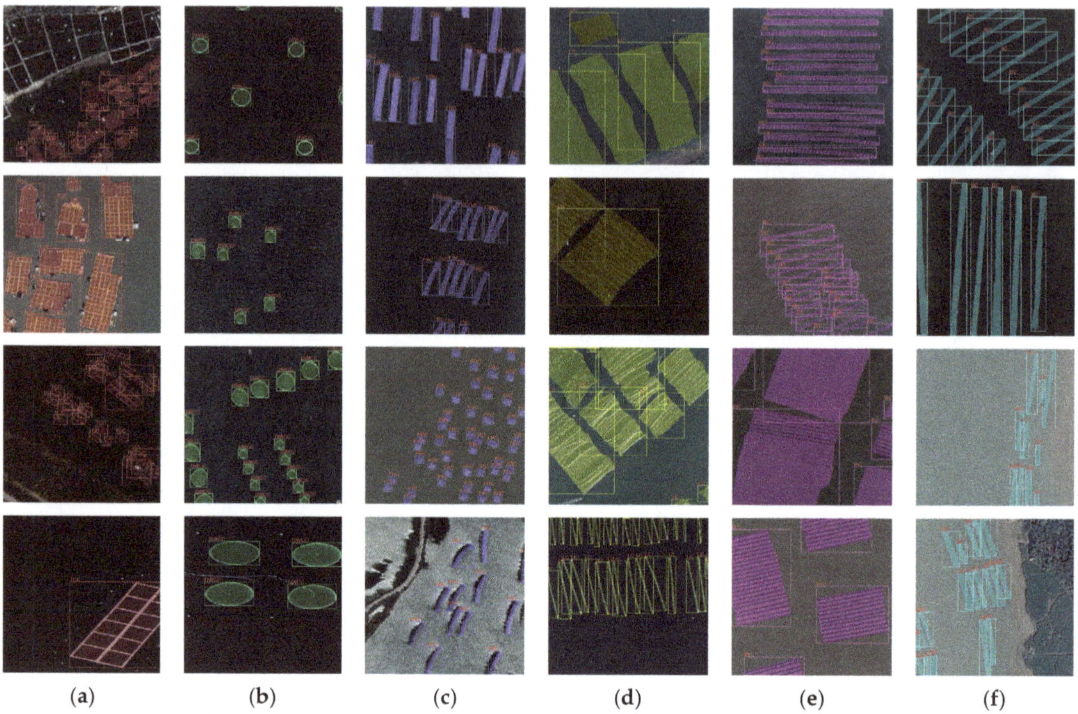

**Figure 2.** Visualization of representative OMAD-6 dataset annotations; (**a**–**f**) are in the order of TCC, DWCC, FRC, LC, RC, and BC.

3.3.4. Dataset Split

To reduce the impact of subjectivity on the dataset, we randomly divided it into training, validation, and test sets at a 6:1:3 ratio stratified by image category. This resulted in 9761 images for training, 1624 for validation, and 4885 for testing. Table 2 details the instance distribution per category across each dataset split. The six-type offshore mariculture areas recognition dataset (OMAD-6) is available at https://github.com/Ainult/OMAD-6 (accessed on 30 September 2024).

**Table 2.** The number of instances for each category and dataset split.

| Category | Train | Validation | Test |
|---|---|---|---|
| TCC | 14,649 | 2389 | 7566 |
| DWCC | 9151 | 1525 | 4531 |
| FRC | 26,199 | 4206 | 12,335 |
| LC | 7213 | 1193 | 3469 |
| RC | 7598 | 1162 | 3853 |
| BC | 15,131 | 2408 | 7566 |
| Total | 79,941 | 12,883 | 39,320 |

### 3.4. OMAD-6 Dataset Characteristics

#### 3.4.1. Wide Sample Coverage

The OMAD-6 dataset is geographically and typologically diverse, enhancing its applicability and representativeness for fine-grained classification of offshore mariculture areas (Figure 3). High-resolution imagery (0.6 m) was collected from the following key regions:

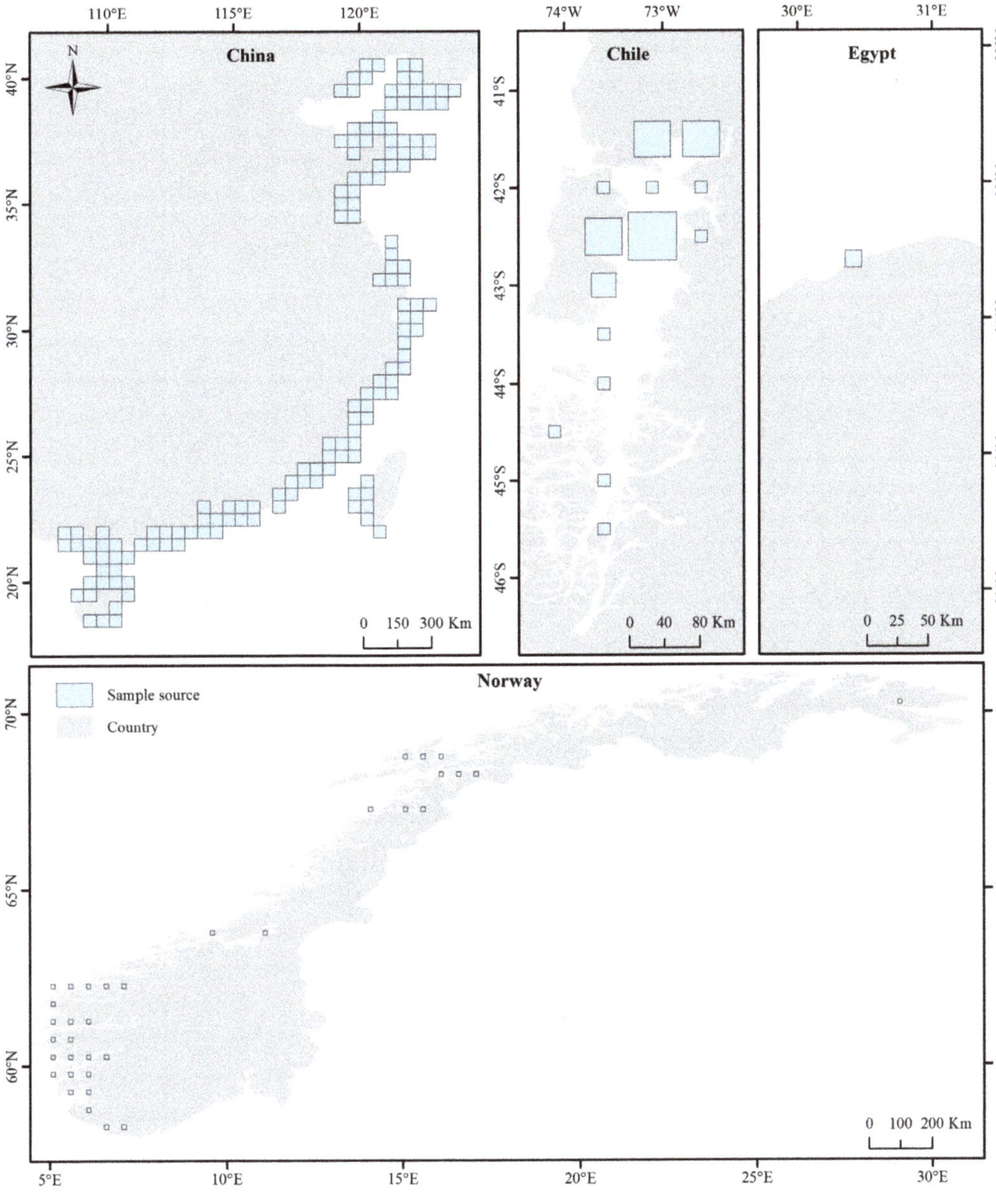


**Figure 3.** Geographic sources of OMAD-6 dataset: coastal China, Chile, Norway, and Nile Basin, Egypt.

Coastal areas of China: Samples were selected from 552 optical remote sensing images, each measuring 28,416 × 28,416 pixels. These images encompass all six types of offshore mariculture.

Coastal areas of Chile: Samples were selected from 51 optical remote sensing images, with pixel dimensions ranging from 3108 × 2080 to 39,936 × 23,808. These images predominantly represent three types of mariculture: deep-water cage culture, traditional cage culture, and longline culture.

Nile River Basin in Egypt: Samples were selected from 4 optical remote sensing images, with pixel dimensions ranging from 5343 × 6359 to 16,128 × 12,288. These scenes solely depict traditional cage culture.

Coastal areas of Norway: Samples were selected from 141 optical remote sensing images, with pixel dimensions ranging from 668 × 642 to 13,503 × 9392. These views primarily include two types of aquaculture: deep-water cage culture and traditional cage culture.

By amalgamating these diverse datasets, we have ensured that the OMAD-6 dataset is not only geographically expansive but also typologically varied, thereby enhancing its practical utility and representativeness for fine-grained classification research in offshore mariculture areas.

#### 3.4.2. High Inter-Class Similarity

A notable challenge presented by OMAD-6 is the high degree of similarity between certain categories in optical remote sensing imagery (Figure 4). Distinguishing categories like BC from FRC, RC from LC, and TCC from FRC can be particularly difficult. This similarity is further accentuated by the overlapping size distributions evident in the example pixel area distribution map (Figure 5).

Figure 4a, shows that BC and FRC appear similar in the RGB image, but exhibit noticeable differences. Firstly, BC is longer than FRC. Secondly, most of the BC structures are submerged below the sea surface, resulting in a darker coloration due to limited light reflection. In contrast, the majority of FRC structures float on the sea surface, presenting a pronounced three-dimensional appearance that protrudes above the water.

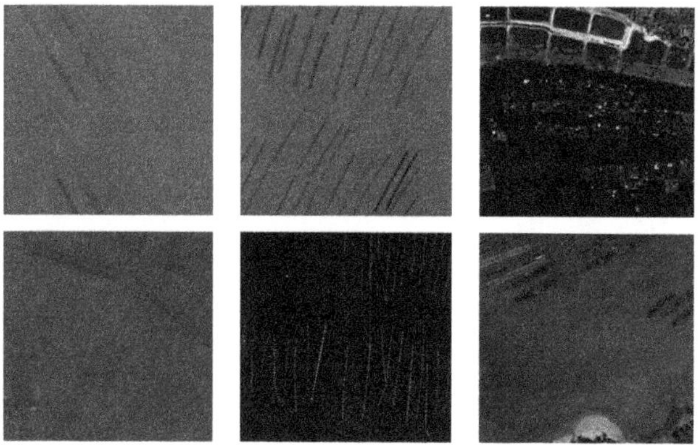

(a) (b) (c)

**Figure 4.** High inter-class similarity. (**a**–**c**) are in the order of BC vs. FRC, RC vs. LC, and TCC vs. FRC.

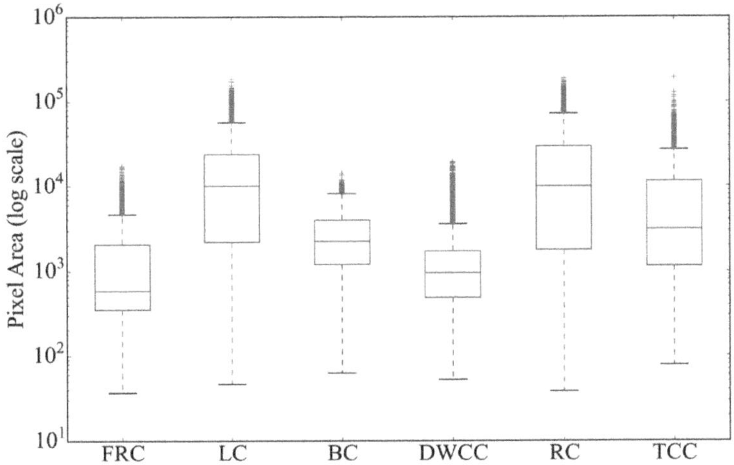


**Figure 5.** Object size distribution of each class.

Figure 4b, shows that RC and LC appear similar in the RGB image, but exhibit significant differences. Firstly, the external arrangement of RC displays more pronounced rectangular characteristics. Secondly, RC contains internal dividing lines that partition its interior into smaller rectangles.

Figure 4c, shows that TCC and FRC share some similarities in certain RGB images, both exhibiting portions that float on the sea surface, but display significant differences. Firstly, TCC is composed internally of distinct small rectangles. Secondly, while FRC also has internal small rectangular features, at the current resolution of 0.6 m, these features are essentially indistinguishable.

To effectively address these challenges, classification algorithms with enhanced resolution and superior feature extraction capabilities are essential for accurately differentiating these similar categories. Future research on algorithms should focus on refining fine-grained recognition abilities and improving feature extraction modules, thereby facilitating accurate and efficient identification and classification of diverse offshore mariculture areas in real-world applications. This advancement will be crucial for enhancing the effectiveness of monitoring and managing offshore mariculture practices.

### 3.4.3. High Intra-Class Diversity

The dataset exhibits high intra-class diversity, primarily reflected in the size and arrangement of mariculture facilities (Figure 6). Factors like environmental conditions, mariculture technology and equipment, economic and market demands, species-specific requirements, and remote sensing image influences (lighting, humidity, growth cycles, and tides) all contribute to variations in design and usage of facilities for the same mariculture type across different regions. As shown in Figures 5 and 6, the pixel area of these facilities ranges from tens to nearly 200,000, and their arrangements exhibit substantial diversity. This diversity places higher demands on the extraction algorithm, requiring stronger adaptability and generalization capabilities to effectively handle this highly varied mariculture facility data.

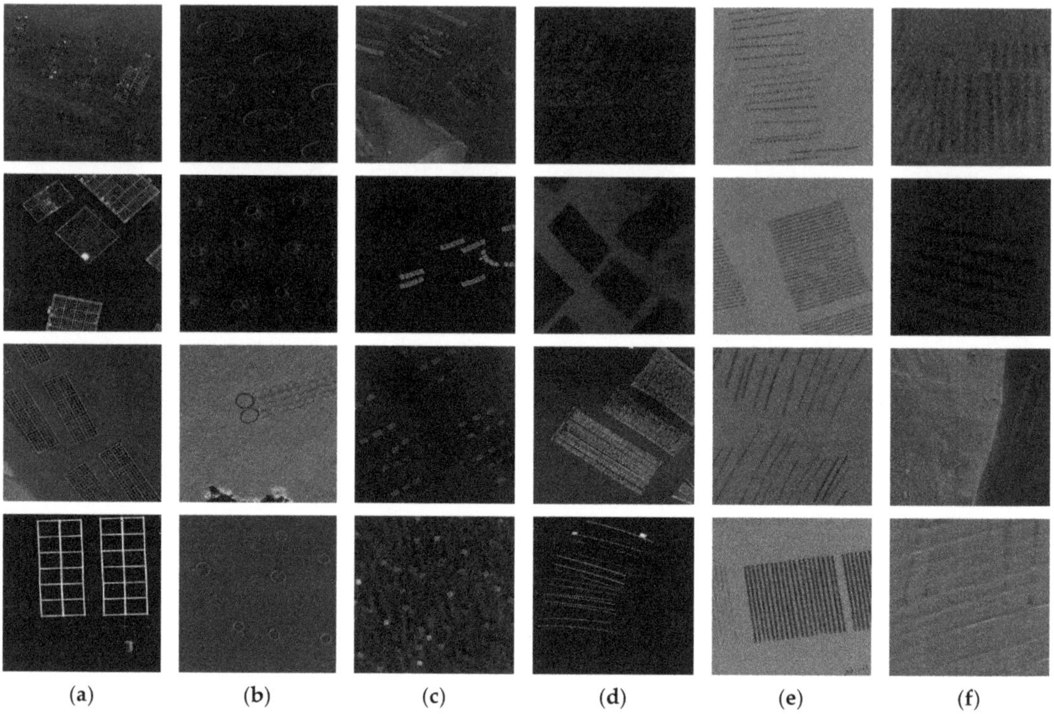

Figure 6. High intra-class diversity in the dataset. (a–f) are in the order of TCC, DWCC, FRC, LC, RC and BC.

## 4. BENCHMARK

### 4.1. Evaluation Metrics

To ensure the reliability of the experiment, we organized the OMAD-6 dataset into the standard MS-COCO [55] format, which includes three parts: train-2017, val-2017, and test-2017, containing 9761, 1624, and 4885 images, respectively. We use the standard evaluation metrics of COCO-2017: mAP, $mAP_{50}$, $mAP_{75}$, $mAP_S$, $mAP_M$ and $mAP_L$, with the last three metrics used to evaluate the algorithm's performance on different target sizes.

Mean Average Precision (mAP): Mean Average Precision (mAP) is a metric that measures the overall performance of instance segmentation algorithms across all categories. It calculates the average precision (AP) for each category and then takes the average value. The definitions of Precision (P) and Recall I are as follows:

Precision (P): Precision is the ratio of correctly predicted positive observations to the total predicted positives. It indicates the proportion of predicted instances that are correct.

Recall I: Recall is the ratio of correctly predicted positive observations to all observations in the actual class. It indicates the proportion of actual positive instances captured by the predictions.

$$P = \frac{TP}{TP + FP} \quad (1)$$

$$R = \frac{TP}{TP + FN} \quad (2)$$

where TP is the number of true positives, FP is the number of false positives, and FN is the number of false negatives. TP is determined by IoU, an indicator used to evaluate the degree of overlap between predicted and actual bounding boxes. It is defined as follows:

$$\mathrm{IoU} = \frac{\text{Area of Overlap}}{\text{Area of Union}} \tag{3}$$

Among them, the Area of Overlap is the intersection area between the predicted bounding box and the actual bounding box, and the Area of Union is the union area of the two. The predicted bounding box is considered correct when the IoU exceeds a predetermined threshold (e.g., 0.75).

The average precision (AP) is calculated based on the area under the precision-recall curve ($\mathrm{P-R}$ Curve). The specific formula is as follows:

$$\mathrm{AP} = \int_0^1 \mathrm{PIdR} \tag{4}$$

The calculation formula for mAP in multi-category situations is:

$$\mathrm{mAP} = \frac{1}{\mathrm{N}} \sum_{\mathrm{i}=1}^{\mathrm{N}} \mathrm{AP_i} \tag{5}$$

Among them, N is the total number of categories, and $\mathrm{AP_I}$ is the average precision of the i-th category. The mAP is defined as the average precision across all categories and IoU thresholds:

mAP ($\mathrm{IoU} = 0.50{:}0.95$): This metric calculates average precision at IoU thresholds ranging from 0.50 to 0.95, with a step size of 0.05. It provides a comprehensive evaluation of the algorithm's performance across various levels of overlap.

$\mathrm{mAP_{50}}$: This metric calculates average precision at an IoU threshold of 0.50. It is less stringent, allowing for more lenient matching between predicted and actual bounding boxes.

$\mathrm{mAP_{75}}$: This metric calculates average precision at an IoU threshold of 0.75. It is more stringent, requiring greater overlap between predicted and actual bounding boxes.

To evaluate the performance of the algorithm on targets of different sizes, $\mathrm{mAP_S}$, $\mathrm{mAP_M}$ and $\mathrm{mAP_L}$ represent the mean Average Precision for small, medium, and large objects, respectively. The definitions are as follows:

Small objects: Objects with an area of less than $32^2$ pixels.

Medium objects: Objects with an area between $32^2$ and $96^2$ pixels.

Large objects: Objects with an area greater than $96^2$ pixels.

These metrics help understand how well the algorithm performs across varying object sizes, ensuring comprehensive evaluation and highlighting any strengths or weaknesses in detecting objects of different scales.

### *4.2. Experimental Setup and Results Analysis*

The lack of publicly available code for existing offshore aquaculture extraction algorithms motivated us to test the performance of eight typical instance segmentation algorithms on the OMAD-6 dataset. Each algorithm exhibits unique advantages in instance segmentation tasks. Mask R-CNN [11], built on Faster R-CNN [56], adds a branch for predicting segmentation masks and uses RoIAlign technology to improve mask accuracy, being widely regarded as the gold standard for instance segmentation. Cascade R-CNN [12] employs multi-stage detection, increasing the IoU threshold at each stage to enhance detection accuracy, especially for objects of different scales and occlusions. YOLACT [13] divides the instance segmentation task into two parallel subtasks—generating prototype masks and predicting mask coefficients and combines these to generate the final mask, known for its real-time performance and fast inference. PointRend [14] iteratively refines the segmentation mask, focusing on uncertain areas to significantly improve mask quality at object boundaries while maintaining computational efficiency. QueryInst [15] integrates instance queries into detection pipelines and directly predicts masks from queries, simplifying the instance segmentation process. CondInst [16] uses conditional convolution to filter instance-specific features dynamically, reducing computational overhead while maintaining high accuracy. BoxInst [17] improves the mask loss function and uses only bounding box annotations to achieve high-quality instance segmentation, narrowing the performance gap between weakly supervised and fully supervised methods. ConvNeXt [18], an efficient convolutional neural network achieves performance competitive with Transformer models across multiple visual recognition tasks through modern design enhancements. The performance evaluation of these algorithms on OMAD-6 will provide important references for future research and applications.

Based on the constructed dataset, we trained and evaluated the above eight state-of-the-art object detection and instance segmentation algorithms. We utilized the MMDetection toolbox provided by OpenMMLab for implementation. All experiments were conducted on a workstation equipped with an Intel® Core™ i9-10900K CPU and an NVIDIA GeForce RTX 3090 GPU.

For training, we initialized all models with pre-trained weights on the COCO dataset to accelerate convergence. The initial learning rates were set to 0.00125 for all models, following standard practice for fine-tuning on custom datasets. Other hyperparameters, such as batch size, weight decay, and momentum, were kept at the default values specified in the MMDetection toolbox for each respective algorithm.

Each model was trained for 12 epochs using the stochastic gradient descent (SGD) optimizer. Data augmentation techniques, including random horizontal flipping and multi-scale resizing, were employed to enhance model generalization. The dataset was split into 60% for training, 10% for validation and 30% for testing.

Evaluation involved using standard metrics such as Average Precision (AP) for both bounding box (Box) and mask (Mask) predictions at various Intersection over Union (IoU) thresholds (e.g., 0.5, 0.75) and across object scales (small, medium, large). By comparing these indicators, we analyzed performance characteristics of each method under different target scales and IoU thresholds, providing insights into their strengths and weaknesses in various scenarios.

#### 4.2.1. Mask mAP Performance

Table 3 highlights the performance of evaluated algorithms in terms of Mask mAP. QueryInst and PointRend emerge as frontrunners, achieving mAP values of 80.9 and 80.5, respectively. These two methods perform particularly well at high IoU thresholds (0.5 and 0.75), signifying powerful capability in detailed segmentation tasks. Specifically, QueryInst achieved $mAP_{50}$ and $mAP_{75}$ values of 93.4 and 88.2, respectively, while PointRend achieved 94.0 and 87.7, demonstrating robust performance at different thresholds. CondInst followed closely with an mAP of 80.2, reaching the highest value of 94.5 on $mAP_{50}$, indicating excellent detection performance at lower thresholds.

**Table 3.** Results of Eight Instance Segmentation Methods on the OMAD-6 test subset in terms of Mask mAP.

| Method | mAP | $mAP_{50}$ | $mAP_{75}$ | $mAP_S$ | $mAP_M$ | $mAP_L$ |
|---|---|---|---|---|---|---|
| Mask R-CNN [11] | 78.2 | 92.8 | 85.7 | 66.9 | 76.3 | 71.2 |
| Cascade R-CNN [12] | 77.2 | 91.8 | 83.6 | 64.6 | 75.3 | 69.2 |
| YOLACT [13] | 62.3 | 82.1 | 69.1 | 45.1 | 59.1 | 60.6 |
| PointRend [14] | 80.5 | 94.0 | 87.7 | **<u>69.1</u>** | 79.4 | 72.9 |
| QueryInst [15] | **<u>80.9</u>** [1] | 93.4 | 88.2 | 67.6 | **<u>80.1</u>** | **<u>75.8</u>** |
| CondInst [16] | 80.2 | **<u>94.5</u>** | **<u>88.4</u>** | 68.4 | 78.0 | 72.7 |
| BoxInst [17] | 68.7 | 92.5 | 78.4 | 49.6 | 65.9 | 70.0 |
| ConvNeXt [18] | 78.7 | 93.9 | 87.2 | 65.7 | 76.9 | 75.1 |

[1] Bold and underline indicate the best-performing values. The following highlights are the same.

#### 4.2.2. Box mAP Performance

Table 4 shows that Box mAP, QueryInst, and Cascade R-CNN lead the pack, achieving $mAP^{bb}$ scores of 82.5 and 82.0, respectively. QueryInst performs consistently across various metrics, especially excelling in high IoU threshold (0.75) and small target ($mAP^{bb}_{S}$) detection, achieving 89.9 and 71.3, respectively. Cascade R-CNN also thrives at high IoU thresholds, reaching an $mAP^{bb}_{75}$ of 90.3, indicating its advantage in precise target localization. The overall performance of CondInst is very stable, with an $mAP^{bb}$ value of 81.4, and it stands out particularly in $mAP^{bb}_{50}$, reaching 95.4. ConvNeXt shows balanced performance across all target scales, notably achieving 71.6 in large object detection ($mAP^{bb}_{L}$), demonstrating its strong capability in handling large target tasks.

**Table 4.** Results of Eight Instance Segmentation Methods on the OMAD-6 test subset in terms of Box mAP.

| Method | $mAP^{bb}$ | $mAP^{bb}_{50}$ | $mAP^{bb}_{75}$ | $mAP^{bb}_{S}$ | $mAP^{bb}_{M}$ | $mAP^{bb}_{L}$ |
|---|---|---|---|---|---|---|
| Mask R-CNN [11] | 78.8 | 94.2 | 88.5 | 70.1 | 79.7 | 66.1 |
| Cascade R-CNN [12] | 82.0 | 94.5 | 90.3 | 72.5 | 82.7 | 69.0 |
| YOLACT [13] | 65.5 | 89.5 | 74.6 | 59.3 | 62.9 | 58.3 |
| PointRend [14] | 78.0 | 94.8 | 88.5 | 70.5 | 78.4 | 63.6 |
| Querylnst [15] | **82.5** | 93.9 | 89.9 | 71.3 | **83.7** | 69.3 |
| CondInst [16] | 81.4 | 95.4 | 89.9 | **74.1** | 80.5 | 69.1 |
| Boxlnst [17] | 80.9 | **95.6** | 89.9 | 73.2 | 80.2 | 69.0 |
| ConvNeXt [18] | **82.5** | 95.0 | **90.6** | 73.5 | 83.2 | **71.6** |

#### 4.2.3. Category Mask Performance

Precise pixel-level segmentation is essential for accurately identifying and managing mariculture facilities and organisms within complex offshore environments. Therefore, we conducted a more granular analysis of the three top-performing Mask mAP algorithms: QueryInst, PointRend, and CondInst. This analysis categorized their performance based on specific mariculture types (Table 5).

**Table 5.** Results of the three top-performing Instance Segmentation Methods on the OMAD-6 Test Subset by Category in Terms of Mask mAP.

| Category | Querylnst [15] | PointRend [14] | CondInst [16] |
|---|---|---|---|
| TCC | 84.5 | 85.4 | **86.2** |
| DWCC | **93.3** | 93.1 | 91.2 |
| FRC | 78.0 | **80.2** | 79.2 |
| LC | **73.5** | 69.6 | 70.3 |
| RC | **86.4** | 85.0 | 83.5 |
| BC | 69.9 | 69.6 | **70.6** |

In the TCC category, CondInst performed the best with an mAP of 86.2, demonstrating excellent performance in identifying TCC. In the DWCC category, QueryInst leads with an mAP of 93.3, but the other two methods also perform well at 93.1 (PointRend) and 91.2 (CondInst). In the FRC category, PointRend performs best with an mAP of 80.2, indicating high accuracy in FRC recognition. In the LC category, CondInst performs slightly better with an mAP of 70.3. In the RC category, QueryInst leads with an mAP of 86.4, reflecting an advantage in RC recognition. In the BC category, CondInst has the best mAP of 70.6. These results emphasize that different algorithms have varying strengths across mariculture categories. Selecting the most suitable algorithm based on the target mariculture type can optimize segmentation performance.

In addition, we analyzed the confusion matrices of the three instance segmentation methods on the OMAD-6 dataset (as shown in Figure 7). The confusion matrix reveals classification confusion problems in certain categories for various methods, providing insights for future improvement and optimization. It can be seen that QueryInst has significant advantages in detecting LC, RC, and BC, while PointRend performs excellently in detecting TCC, DWCC, and FRC. CondInst performs relatively well in detecting other categories except for LC. By analyzing these misclassification patterns, we can identify specific areas for improvement in future algorithm development. For example, future iterations of CondInst could focus on enhancing its ability to detect LC accurately.

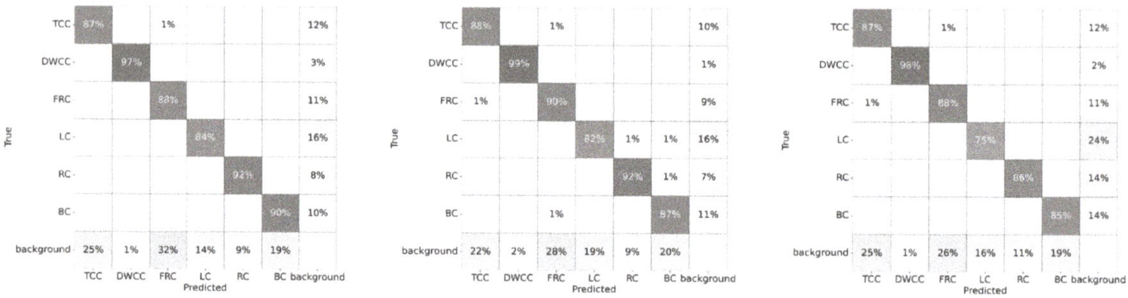


(**a**) (**b**) (**c**)

**Figure 7.** The mask (IoU = 0.75) confusion matrices of the three instance segmentation methods on the OMAD-6 datasets. (**a–c**) are in order Querylnst, PointRend, and CondInst.

Figure 8 visually illustrates the detection results of the top-three performing algorithms (QueryInst, PointRend, and CondInst) on the OMAD-6 dataset. Comparing the ground truth labels with the predicted results reveals that each method exhibits superior segmentation performance for specific categories.

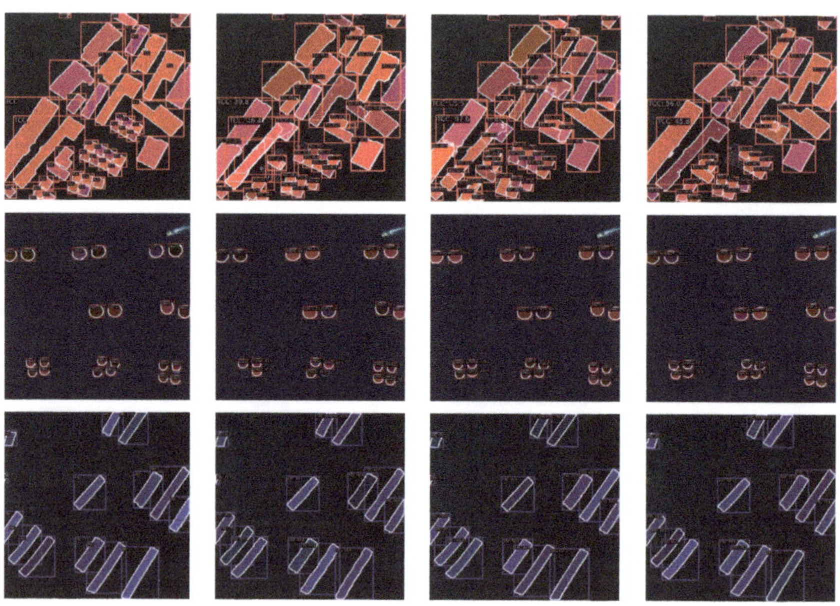

**Figure 8.** *Cont.*

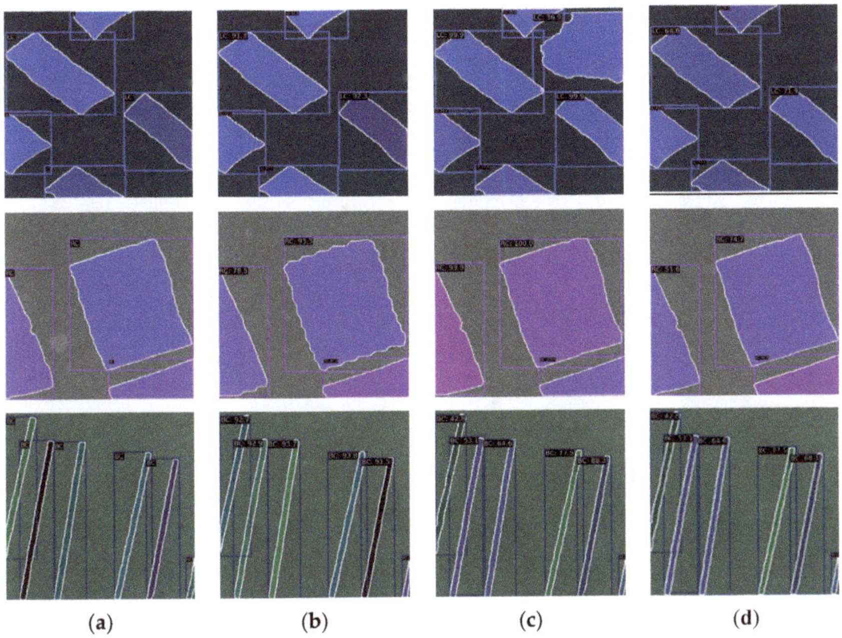

**Figure 8.** Detection results from the three methods: (**a–d**) are in the order of Ground Truth, QueryInst, PointRend, and CondInst.

Overall, QueryInst and PointRend led in overall segmentation accuracy on the OMAD-6 dataset. Their ability to effectively identify and segment offshore mariculture facilities in complex backgrounds makes them valuable tools for supporting operations.

*4.3. Discussion*

Although both QueryInst and PointRend exhibited robust performance on the OMAD-6 dataset, there is still room for enhancement. Figure 9 presents a detailed analysis of the errors in their test set predictions, juxtaposed with the ground truth annotations for comparative assessment.

QueryInst's performance in detecting small and medium-sized objects is relatively subpar. This could be attributed to the inherent limitations of the instance query process, which might not be as sensitive to these sizes. As illustrated in Figure 9, target adhesion and missed detection predominantly affect small-sized targets, particularly those of TCC, FRC, and BC. Here are some improvement suggestions:

a. Multi-Scale Feature Fusion: The query system can better handle objects of different scales by integrating a Feature Pyramid Network (FPN) module [57]. FPN enables the model to extract and combine features from multiple levels, enhancing the detection of small and medium-sized objects.
b. Attention Mechanisms: Implement attention mechanisms such as Squeeze-and-Excitation Network (SENet) [58] or Convolutional Block Attention Module (CBAM) [59]. These mechanisms help the model focus more on relevant parts of the image, improving detection accuracy for small and medium-sized objects.
c. Improved Sampling Strategies: Utilizing more sophisticated sampling strategies, such as balanced positive and negative sampling or focal loss, can address class imbalance issues, ensuring that small and medium-sized objects are adequately represented during training.

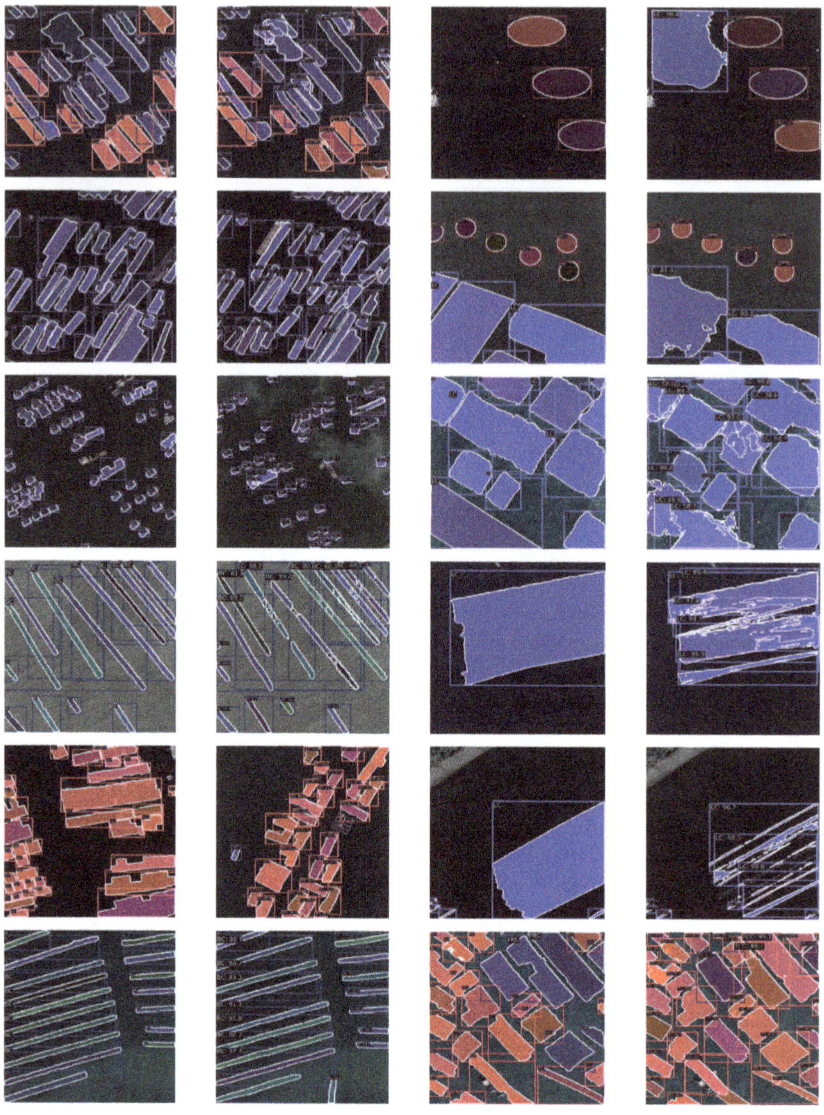

**Figure 9.** Comparison of misdetections: (**b**) QueryInst and (**d**) PointRend vs. (**a**,**c**) Ground Truth.

PointRend exhibits relatively subpar performance in detecting medium and larger targets. This might be due to insufficient feature extraction and refinement processing for these sizes. As shown in Figure 9, during detection, issues such as target adhesion, missed detection, and false detection are particularly prominent when detecting medium and larger targets such as LC, RC, and BC. Here are some improvement suggestions:

a. Enhanced Feature Extraction: Utilizing deeper or more advanced backbone networks, such as ResNeXt [60] or EfficientNet [61], can improve the quality of feature extraction, especially for medium and large objects. These advanced backbone networks provide richer and more detailed feature representations, which help to detect and segment medium and larger targets better.

b. Refinement Network: Adding a refinement network specifically focused on improving mask quality for medium to large objects can be beneficial. This network would leverage the backbone network's features to iteratively refine predicted masks, enhancing segmentation accuracy. Multiple refinement steps can better capture target boundaries and details, reducing adhesion and missed detections.
c. Error Correction Module: Integrating DenseCRF [62] or post-processing techniques like Mask Scoring R-CNN [63] can help rectify detection errors and improve overall segmentation quality. DenseCRF optimizes segmentation results by considering the relationships between pixels, while Mask Scoring R-CNN improves accuracy by scoring and adjusting predicted masks. These technologies can effectively reduce false positives and enhance detection performance for medium and larger targets.

By integrating the aforementioned strategies, we can effectively mitigate the specific weaknesses of QueryInst and PointRend, thereby significantly bolstering the robustness and accuracy of these models. This comprehensive approach ensures that the algorithms are better equipped to handle the complexities and challenges inherent in the detection and segmentation of offshore mariculture facilities. Through continuous refinement and optimization, these models can be further enhanced to provide more reliable and precise results, which are crucial for advancing the field of remote sensing and maritime resource management.

## 5. Conclusions

This study addresses the pressing need for automation and efficiency enhancement in mariculture management, given aquaculture's significant role in global food security and economic development. Here, we introduce the OMAD-6 dataset, a meticulously curated and comprehensive resource specifically designed for identifying offshore mariculture facilities. The dataset leverages high-resolution imagery from various global regions, applying a unified classification to ensure accurate and efficient recognition of offshore mariculture facilities. Our benchmark testing on the OMAD-6 dataset has underscored the effectiveness of state-of-the-art instance segmentation algorithms. This evaluation demonstrates the dataset's utility and identifies challenges and opportunities for improving offshore mariculture identification. Notably, the QueryInst and PointRend algorithms have distinguished themselves as top performers on the OMAD-6 dataset, robustly identifying offshore mariculture facilities amidst complex environmental backgrounds.

While OMAD-6 provides a comprehensive dataset for surface-visible offshore mariculture, it lacks underwater farming scenes, such as crab farming, where no visible aquaculture facilities are present on the water surface. Underwater detection introduces additional complexities, including limited visibility, variations in light penetration, and dynamic underwater environments requiring specialized imaging techniques and algorithms. Future work should explore the inclusion of underwater aquaculture datasets to address these challenges, especially in scenarios where surface-based observations are insufficient, thereby further expanding the scope of offshore mariculture monitoring. Our benchmark testing provides a solid foundation for future research directions. We plan to expand the OMAD-6 dataset by incorporating a more extensive imagery sample, enriching its diversity and representativeness. Concurrently, efforts will be made to enhance the robustness of existing algorithms to handle the intricacies and variability present in real-world offshore mariculture environments.

In anticipation of future developments, the incorporation of advanced artificial intelligence technologies, including machine learning, deep learning, and reinforcement learning, presents notable prospects for substantial advancements within this domain. These methodologies possess the capability to significantly augment detection and monitoring systems for offshore mariculture by enhancing precision, efficiency, and adaptability. Illustratively, deep learning models may be trained to discern intricate patterns inherent within marine environments, whereas reinforcement learning can facilitate systems that accommodate dynamic alterations in such contexts. However, challenges, such as the prerequisite for extensive high-quality training data and ensuring the interpretability of AI models in pivotal applications, must be meticulously tackled. Consequently, future research endeavors should concentrate on devising innovative solutions that harness these technologies while preserving reliability and transparency.

In conclusion, the introduction of the OMAD-6 dataset represents a significant advancement in remote sensing image analysis for offshore mariculture. Its comprehensive nature, coupled with successful benchmark testing of leading instance segmentation algorithms, positions OMAD-6 as a crucial asset for researchers and practitioners. Expanding future datasets to include underwater aquaculture scenes, particularly in scenarios where surface-visible facilities are absent, will address current limitations and enable a more holistic understanding and monitoring of offshore mariculture practices. This addition will also facilitate the development of specialized algorithms capable of handling the unique challenges posed by underwater environments, thereby enhancing the dataset's applicability to real-world conditions. The ongoing development and application of this dataset will undoubtedly play a pivotal role in shaping the future of identifying and managing offshore mariculture areas.

**Author Contributions:** Conceptualization, Z.M. and Q.Z.; data curation, Y.L. and Y.C.; formal analysis, Z.M.; funding acquisition, Q.Z.; investigation, Y.L.; methodology, Z.M.; project administration, Q.Z.; resources, Z.M.; software, Z.M.; supervision, Z.W. and Q.Z.; validation, Y.L., Y.C., Y.S. and Z.W.; writing—original draft, Z.M.; writing—review and editing, M.X. and Q.Z. All authors have read and agreed to the published version of the manuscript.

**Funding:** This research was funded by the National Key Research and Development Program of China, grant number 2022YFE0209300. The funder is the corresponding author, Qingling Zhang. The funding period for this project is from 1 November 2022, to 31 October 2025.

**Data Availability Statement:** The OMAD-6 dataset is available at https://github.com/Ainult/OMAD-6 (accessed on 30 September 2024).

**Conflicts of Interest:** Author Yanyun Shen was employed by the company Xi'an Modern Control Technology Research Institute of China North Industries Group. The remaining authors declare that the research was conducted in the absence of any commercial or financial relationships that could be construed as a potential conflict of interest.

## Abbreviations

The following abbreviations are used in this manuscript:

| | |
|---|---|
| OMAD-6 | A Six-Type Offshore Mariculture Areas Recognition Dataset |
| TCC | Traditional Cage Culture |
| DWCC | Deep-Water Cage Culture |
| FRC | Floating Raft Culture |
| LC | Longline Culture |
| RC | Raft Culture |
| BC | Bouchot Culture |
| HBB | Horizontal Bounding Box |
| OBB | Oriented Bounding Box |
| PC | Per Categories |
| MS | Multi-source |
| GE | Google Earth |
| COCO | Common Objects in Context |
| AP | Average Precision |
| mAP | Mean Average Precision |
| IoU | Intersection over Union |
| FPN | Feature Pyramid Network |
| CBAM | Convolutional Block Attention Module |
| SENet | Squeeze-and-Excitation Network |

## References


1. FAO. *The State of World Fisheries and Aquaculture 2024. Blue Transformation in Action*; FAO: Rome, Italy, 2024; p. 264.
2. Zhang, W.; Belton, B.; Edwards, P.; Henriksson, P.J.G.; Little, D.C.; Newton, R.; Troell, M. Aquaculture will continue to depend more on land than sea. *Nature* **2022**, *603*, E2–E4. [CrossRef] [PubMed]
3. DeWeerdt, S. Can aquaculture overcome its sustainability challenges? *Nature* **2020**, *588*, S60–S62. [CrossRef]
4. Liu, J.; Lu, Y.; Guo, X.; Ke, W. A Deep Learning Method for Offshore Raft Aquaculture Extraction Based on Medium Resolution Remote Sensing Images. *IEEE J. Sel. Top. Appl. Earth Obs. Remote Sens.* **2023**, *16*, 6296–6309. [CrossRef]
5. Hou, T.; Sun, W.; Chen, C.; Yang, G.; Meng, X.; Peng, J. Marine floating raft aquaculture extraction of hyperspectral remote sensing images based decision tree algorithm. *Int. J. Appl. Earth Obs. Geoinf.* **2022**, *111*, 102846. [CrossRef]
6. Cui, B.; Zhao, Y.; Yang, M.; Huang, L.; Lu, Y. Reverse attention dual-stream network for extracting laver aquaculture areas from GF-1 remote sensing images. *IEEE J. Sel. Top. Appl. Earth Obs. Remote Sens.* **2023**, *16*, 5271–5283. [CrossRef]
7. Kang, J.; Sui, L.; Yang, X.; Liu, Y.; Wang, Z.; Wang, J.; Yang, F.; Liu, B.; Ma, Y. Sea surface-visible aquaculture spatial-temporal distribution remote sensing: A case study in Liaoning province, China from 2000 to 2018. *Sustainability* **2019**, *11*, 7186. [CrossRef]
8. Jiang, Z.; Ma, Y. Accurate extraction of offshore raft aquaculture areas based on a 3D-CNN model. *Int. J. Remote Sens.* **2020**, *41*, 5457–5481. [CrossRef]
9. Wang, S.; Huang, C.; Li, H.; Liu, Q. Synergistic Integration of Time Series Optical and SAR Satellite Data for Mariculture Extraction. *Remote Sens.* **2023**, *15*, 2243. [CrossRef]
10. Liu, Y.; Wang, Z.; Yang, X.; Zhang, Y.; Yang, F.; Liu, B.; Cai, P. Satellite-based monitoring and statistics for raft and cage aquaculture in China's offshore waters. *Int. J. Appl. Earth Obs. Geoinf.* **2020**, *91*, 102118. [CrossRef]
11. He, K.; Gkioxari, G.; Dollár, P.; Girshick, R. Mask r-cnn. In Proceedings of the IEEE International Conference on Computer Vision, Venice, Italy, 22–29 October 2017; pp. 2961–2969.
12. Cai, Z.; Vasconcelos, N. Cascade R-CNN: High quality object detection and instance segmentation. *IEEE Trans. Pattern Anal. Mach. Intell.* **2019**, *43*, 1483–1498. [CrossRef]
13. Bolya, D.; Zhou, C.; Xiao, F.; Lee, Y.J. Yolact: Real-time instance segmentation. In Proceedings of the IEEE/CVF International Conference on Computer Vision, Seoul, Republic of Korea, 27 October–2 November 2019; pp. 9157–9166.
14. Kirillov, A.; Wu, Y.; He, K.; Girshick, R. Pointrend: Image segmentation as rendering. In Proceedings of the IEEE/CVF Conference on Computer Vision and Pattern Recognition, Seattle, WA, USA, 13–19 June 2020; pp. 9799–9808.
15. Fang, Y.; Yang, S.; Wang, X.; Li, Y.; Fang, C.; Shan, Y.; Feng, B.; Liu, W. Instances as queries. In Proceedings of the IEEE/CVF International Conference on Computer Vision, Montreal, BC, Canada, 11–17 October 2021; pp. 6910–6919.
16. Tian, Z.; Shen, C.; Chen, H. Conditional convolutions for instance segmentation. In Proceedings of the Computer Vision–ECCV 2020: 16th European Conference, Glasgow, UK, 23–28 August 2020; Proceedings, Part I 16, 2020. pp. 282–298.
17. Tian, Z.; Shen, C.; Wang, X.; Chen, H. Boxinst: High-performance instance segmentation with box annotations. In Proceedings of the IEEE/CVF Conference on Computer Vision and Pattern Recognition, Nashville, TN, USA, 20–25 June 2021; pp. 5443–5452.
18. Todi, A.; Narula, N.; Sharma, M.; Gupta, U. ConvNext: A Contemporary Architecture for Convolutional Neural Networks for Image Classification. In Proceedings of the 2023 3rd International Conference on Innovative Sustainable Computational Technologies (CISCT), Dehradun, India, 8–9 September 2023; pp. 1–6.
19. Cheng, G.; Han, J.; Zhou, P.; Guo, L. Multi-class geospatial object detection and geographic image classification based on collection of part detectors. *ISPRS J. Photogramm. Remote Sens.* **2014**, *98*, 119–132. [CrossRef]
20. Razakarivony, S.; Jurie, F. Vehicle detection in aerial imagery: A small target detection benchmark. *J. Vis. Commun. Image Represent.* **2016**, *34*, 187–203. [CrossRef]
21. Liu, K.; Mattyus, G. Fast multiclass vehicle detection on aerial images. *IEEE Geosci. Remote Sens. Lett.* **2015**, *12*, 1938–1942. [CrossRef]
22. Liu, Z.; Wang, H.; Weng, L.; Yang, Y. Ship rotated bounding box space for ship extraction from high-resolution optical satellite images with complex backgrounds. *IEEE Geosci. Remote Sens. Lett.* **2016**, *13*, 1074–1078. [CrossRef]
23. Long, Y.; Gong, Y.; Xiao, Z.; Liu, Q. Accurate object localization in remote sensing images based on convolutional neural networks. *IEEE Trans. Geosci. Remote Sens.* **2017**, *55*, 2486–2498. [CrossRef]
24. Li, K.; Wan, G.; Cheng, G.; Meng, L.; Han, J. Object detection in optical remote sensing images: A survey and a new benchmark. *ISPRS J. Photogramm. Remote Sens.* **2020**, *159*, 296–307. [CrossRef]
25. Haroon, M.; Shahzad, M.; Fraz, M.M. Multisized object detection using spaceborne optical imagery. *IEEE J. Sel. Top. Appl. Earth Obs. Remote Sens.* **2020**, *13*, 3032–3046. [CrossRef]
26. Ding, J.; Xue, N.; Xia, G.-S.; Bai, X.; Yang, W.; Yang, M.Y.; Belongie, S.; Luo, J.; Datcu, M.; Pelillo, M. Object detection in aerial images: A large-scale benchmark and challenges. *IEEE Trans. Pattern Anal. Mach. Intell.* **2021**, *44*, 7778–7796. [CrossRef] [PubMed]
27. Sun, X.; Wang, P.; Yan, Z.; Xu, F.; Wang, R.; Diao, W.; Chen, J.; Li, J.; Feng, Y.; Xu, T. FAIR1M: A benchmark dataset for fine-grained object recognition in high-resolution remote sensing imagery. *ISPRS J. Photogramm. Remote Sens.* **2022**, *184*, 116–130. [CrossRef]
28. Shi, T.; Gong, J.; Jiang, S.; Zhi, X.; Bao, G.; Sun, Y.; Zhang, W. Complex optical remote-sensing aircraft detection dataset and benchmark. *IEEE Trans. Geosci. Remote Sens.* **2023**, *61*, 5612309. [CrossRef]
29. Liu, B.; Chen, S.-B.; Wang, J.-X.; Tang, J.; Luo, B. An Oriented Object Detector for Hazy Remote Sensing Images. *IEEE Trans. Geosci. Remote Sens.* **2024**, *62*, 1001711. [CrossRef]

30. Zhou, C.; Wong, K.; Tsou, J.Y.; Zhang, Y. Detection and Statistics of Offshore Aquaculture Rafts in Coastal Waters. *J. Mar. Sci. Eng.* **2022**, *10*, 781. [CrossRef]
31. Fu, Y.; Deng, J.; Ye, Z.; Gan, M.; Wang, K.; Wu, J.; Yang, W.; Xiao, G. Coastal aquaculture mapping from very high spatial resolution imagery by combining object-based neighbor features. *Sustainability* **2019**, *11*, 637. [CrossRef]
32. Xu, Y.; Lu, L. An Attention-Fused Deep Learning Model for Accurately Monitoring Cage and Raft Aquaculture at Large-Scale Using Sentinel-2 Data. *IEEE J. Sel. Top. Appl. Earth Obs. Remote Sens.* **2024**, *17*, 9099–9109. [CrossRef]
33. Qin, G.; Wang, S.; Wang, F.; Zhou, Y.; Wang, Z.; Zou, W. U_EFF_NET: High-precision segmentation of offshore farms from high-resolution SAR remote sensing images. *IEEE J. Sel. Top. Appl. Earth Obs. Remote Sens.* **2022**, *15*, 8519–8528. [CrossRef]
34. Su, H.; Wei, S.; Qiu, J.; Wu, W. RaftNet: A new deep neural network for coastal raft aquaculture extraction from Landsat 8 OLI data. *Remote Sens.* **2022**, *14*, 4587. [CrossRef]
35. Cui, B.; Fei, D.; Shao, G.; Lu, Y.; Chu, J. Extracting raft aquaculture areas from remote sensing images via an improved U-net with a PSE structure. *Remote Sens.* **2019**, *11*, 2053. [CrossRef]
36. Fu, Y.; Ye, Z.; Deng, J.; Zheng, X.; Huang, Y.; Yang, W.; Wang, Y.; Wang, K. Finer resolution mapping of marine aquaculture areas using worldView-2 imagery and a hierarchical cascade convolutional neural network. *Remote Sens.* **2019**, *11*, 1678. [CrossRef]
37. Fu, Y.; Deng, J.; Wang, H.; Comber, A.; Yang, W.; Wu, W.; You, S.; Lin, Y.; Wang, K. A new satellite-derived dataset for marine aquaculture areas in China's coastal region. *Earth Syst. Sci. Data* **2021**, *13*, 1829–1842. [CrossRef]
38. Wang, X.; Zhou, J.; Fan, J. IDUDL: Incremental double unsupervised deep learning model for marine aquaculture SAR images segmentation. *IEEE Trans. Geosci. Remote Sens.* **2022**, *60*, 4209412. [CrossRef]
39. FAO. *The State of World Fisheries and Aquaculture 2022. Towards Blue Transformation*; Food and Agriculture Organization of the United Nations: Rome, Italy, 2022; p. 266.
40. Espinasse, M.; Mikkelsen, E.; Sørbye, S.H.; Skern-Mauritzen, M.; Falk-Andersson, J.; Fauchald, P. Seafood production in Northern Norway: Analyzing variation and co-development in aquaculture and coastal fisheries. *Mar. Policy* **2023**, *155*, 105777. [CrossRef]
41. Han, Q.; Zhang, X.; Xu, X.; Wang, X.; Yuan, X.; Ding, Z.; Zhao, S.; Wang, S. Antibiotics in marine aquaculture farms surrounding Laizhou Bay, Bohai Sea: Distribution characteristics considering various culture modes and organism species. *Sci. Total Environ.* **2021**, *760*, 143863. [CrossRef]
42. Urbina, M.; Cumillaf, J.; Paschke, K.; Gebauer, P. Effects of pharmaceuticals used to treat salmon lice on non-target species: Evidence from a systematic review. *Sci. Total Environ.* **2019**, *649*, 1124–1136. [CrossRef] [PubMed]
43. Wang, M.; Cui, Q.; Wang, J.; Ming, D.; Lv, G. Raft cultivation area extraction from high resolution remote sensing imagery by fusing multi-scale region-line primitive association features. *ISPRS J. Photogramm. Remote Sens.* **2017**, *123*, 104–113. [CrossRef]
44. Chatziantoniou, A.; Spondylidis, S.C.; Stavrakidis-Zachou, O.; Papandroulakis, N.; Topouzelis, K. Dissolved oxygen estimation in aquaculture sites using remote sensing and machine learning. *Remote Sens. Appl. Soc. Environ.* **2022**, *28*, 100865. [CrossRef]
45. Wang, F.; Gao, Y.; Guan, C.; Jia, Y. Growth performance, hematology, antioxidant capacity, immunity, and intestinal microbiota of spotted knifejaw (*Oplegnathus punctatus*) reared in recirculating aquaculture system and offshore aquaculture net pen. *Aquaculture* **2023**, *562*, 738816. [CrossRef]
46. Liang, C.; Cheng, B.; Xiao, B.; He, C.; Liu, X.; Jia, N.; Chen, J. Semi-/weakly-supervised semantic segmentation method and its application for coastal aquaculture areas based on multi-source remote sensing images—Taking the Fujian coastal area (mainly Sanduo) as an example. *Remote Sens.* **2021**, *13*, 1083. [CrossRef]
47. Sui, B.; Jiang, T.; Zhang, Z.; Pan, X.; Liu, C. A modeling method for automatic extraction of offshore aquaculture zones based on semantic segmentation. *ISPRS Int. J. Geo-Inf.* **2020**, *9*, 145. [CrossRef]
48. Cheng, Z.; Hong, G.; Li, Q.; Liu, S.; Wang, S.; Ma, Y. Seasonal dynamics of coastal pollution migration in open waters with intensive marine ranching. *Mar. Environ. Res.* **2023**, *190*, 106101. [CrossRef]
49. Zhang, X.; Ma, S.; Su, C.; Shang, Y.; Wang, T.; Yin, J. Coastal oyster aquaculture area extraction and nutrient loading estimation using a GF-2 satellite image. *IEEE J. Sel. Top. Appl. Earth Obs. Remote Sens.* **2020**, *13*, 4934–4946. [CrossRef]
50. Zhang, Y.; Wang, C.; Chen, J.; Wang, F. Shape-constrained method of remote sensing monitoring of marine raft aquaculture areas on multitemporal synthetic sentinel-1 imagery. *Remote Sens.* **2022**, *14*, 1249. [CrossRef]
51. FAO. Fishery and Aquaculture Country Profiles. China, 2017. Country Profile Fact Sheets. In Fisheries and Aquaculture. Available online: https://www.fao.org/fishery/en/facp/chn (accessed on 1 July 2024).
52. Toupoint, N.; Barbier, P.; Tremblay, R.; Archambault, P.; McKindsey, C.W.; Winkler, G.; Olivier, F. Influence of intertidal recreational fisheries and 'bouchot'mussel culture on bivalve recruitment. *Mar. Environ. Res.* **2016**, *117*, 1–12. [CrossRef] [PubMed]
53. Vuttipittayamongkol, P.; Elyan, E. Improved overlap-based undersampling for imbalanced dataset classification with application to epilepsy and parkinson's disease. *Int. J. Neural Syst.* **2020**, *30*, 2050043. [CrossRef] [PubMed]
54. Kirillov, A.; Mintun, E.; Ravi, N.; Mao, H.; Rolland, C.; Gustafson, L.; Xiao, T.; Whitehead, S.; Berg, A.C.; Lo, W.-Y. Segment anything. In Proceedings of the IEEE/CVF International Conference on Computer Vision, Paris, France, 4–6 October 2023; pp. 4015–4026.
55. Lin, T.-Y.; Maire, M.; Belongie, S.; Hays, J.; Perona, P.; Ramanan, D.; Dollár, P.; Zitnick, C.L. Microsoft coco: Common objects in context. In Proceedings of the Computer Vision–ECCV 2014: 13th European Conference, Zurich, Switzerland, 6–12 September 2014; Proceedings, Part V 13, 2014. pp. 740–755.
56. Ren, S.; He, K.; Girshick, R.; Sun, J. Faster r-cnn: Towards real-time object detection with region proposal networks. *Adv. Neural Inf. Process. Syst.* **2015**, *39*, 1137–1149. [CrossRef]

57. Gong, Y.; Yu, X.; Ding, Y.; Peng, X.; Zhao, J.; Han, Z. Effective fusion factor in FPN for tiny object detection. In Proceedings of the IEEE/CVF Winter Conference on Applications of Computer Vision, Virtual, 5–9 January 2021; pp. 1160–1168.
58. Cheng, D.; Meng, G.; Cheng, G.; Pan, C. SeNet: Structured edge network for sea–land segmentation. *IEEE Geosci. Remote Sens. Lett.* **2016**, *14*, 247–251. [CrossRef]
59. Woo, S.; Park, J.; Lee, J.-Y.; Kweon, I.S. Cbam: Convolutional block attention module. In Proceedings of the European Conference on Computer Vision (ECCV), Munich, Germany, 8–14 September 2018; pp. 3–19.
60. Xie, S.; Girshick, R.; Dollár, P.; Tu, Z.; He, K. Aggregated residual transformations for deep neural networks. In Proceedings of the IEEE Conference on Computer Vision and Pattern Recognition, Honolulu, HI, USA, 21–26 July 2017; pp. 1492–1500.
61. Koonce, B.; Koonce, B. *EfficientNet. Convolutional Neural Networks with Swift for Tensorflow: Image Recognition and Dataset Categorization*; Springer Nature: Dordrecht, The Netherlands, 2021; pp. 109–123. [CrossRef]
62. Cheng, X.; Lei, H. Semantic segmentation of remote sensing imagery based on multiscale deformable CNN and DenseCRF. *Remote Sens.* **2023**, *15*, 1229. [CrossRef]
63. Huang, Z.; Huang, L.; Gong, Y.; Huang, C.; Wang, X. Mask scoring r-cnn. In Proceedings of the IEEE/CVF Conference on Computer Vision and Pattern Recognition, Long Beach, CA, USA, 15–20 June 2019; pp. 6409–6418.

land

MDPI

Article

# Construction and Evaluation of Urban Green Infrastructure in Ecologically Vulnerable Areas Based on Multi-Scale and Multi-Objective Approaches: Taking the Four-Lake Hydrographic Network in the Upper Yellow River Basin as an Example

**Wenhao Cheng [1,2] and Caihong Ma [2,*]**

1 College of Urban and Environment Sciences, Northwest University, Xi'an 710127, China; chengwh@nxu.edu.cn
2 College of Geography and Planning, Ningxia University, Yinchuan 750021, China
* Correspondence: mchyanni@nxu.edu.cn

**Abstract:** The construction and optimization of urban green infrastructure (UGI) are regarded as effective strategies for harmonizing the natural landscape with human activities, particularly in ecologically vulnerable areas in the Upper Yellow River Basin, China. However, there is little attention paid to the scale effects and object effects as well as an absence of comprehensive assessments regarding landscape stability. Taking the Four-Lake Hydrographic Network (FLHN) in Shizuishan, a prefecture-level city, as an example, this study focuses on identifying the important sources of UGI by integrating both regional and interregional perspectives utilizing morphological spatial pattern analysis (MSPA). UGI networks were constructed and optimized based on trade-offs and synergizing relationships between individual objects using the minimum cumulative resistance (MCR) model, and the UGI network's stability combined centrality and connectivity aspects, which were subsequently assessed. The results showed that a total of 19 important sources covering an area of 105.07 km$^2$ were identified in the FLHN, integrating both regional and interregional levels. It was deemed unnecessary to maintain lengths of 7.79 km key corridors, 9.42 km general corridors, and 29.89 km fragile corridors; furthermore, there was no longer a requirement to upgrade an additional 5.51 km of general corridors and 25.78 km of any corridor, as UGI corridors were extracted based on a trade-off and synthesized objective methodology. The overall connectivity index value ($O_G$) of UGI stability with respect to the multi-objective model demonstrated superior performance compared to the same index in scenarios involving the use of a single-objective approach and the straightforward overlay of each object. This study reveals the multifaceted requirements of urban landscape security and sustainability, indicating that multi-scale and multi-objective approaches in territorial space planning not only ensure the integrity of the landscape patterns but also reduce the costs associated with landscape construction. This model can be utilized to implement urban landscape entity protection and restoration for landscapes with various geographical characteristics, and it can provide valuable guidance for similar areas.

**Keywords:** regional and interregional; trade-off and synergy objectives; urban green infrastructure; Four-Lake Hydrographic Network; ecological restoration

## 1. Introduction

In the context of excessive resource extraction and extensive deforestation, the Upper Yellow River Basin of China is regarded as an ecologically vulnerable zone [1]. It presents typical features, such as ecological environment fragility and landscape pattern fragmentation [2]. Additionally, the accelerating urbanization process has either deliberately or inadvertently disrupted the equilibrium between ecological conservation and sustainable development for urban agglomeration, resulting in biodiversity loss, air pollution, and heat islands [3,4]. Urban green infrastructure (UGI) is defined as a strategically planned, high-quality network situated in natural and semi-natural areas that is intended to reverse ecosystem degradation and enhance landscape structure in urban centers, suburbs, and metropolitan regions [5], including green spaces such as forests and blue spaces such as lakes, wetlands, and other water systems [6,7]. It not only provides recreational services, manages regional biodiversity, and guarantees water supply but also plays a crucial role in enhancing urban transformation dynamics and promoting human well-being [8,9].

A significant number of studies on the construction of UGI have been executed across various regions [10]. Nevertheless, most of these studies were carried out from single- or partial-scale and single-objective perspectives, lacking comprehensive assessments of landscape stability. Neglecting scale variations and objective transformations during the UGI construction process can lead to significant oversights [11]. Firstly, the development of UGI represents a process characterized by landscape heterogeneity, with key factors varying according to scale transformation [12]. The essential aspect is the comprehensive consideration of landscape factors at the micro-scale, with an emphasis on landscape restoration. Additionally, it is crucial to optimize both the quantity and quality of UGI corridors at a mesoscopic scale, thereby facilitating the integration of landscape units. Finally, the establishment of a landscape security pattern at the macro-scale is vital for balancing natural ecosystems and human needs [13]. Secondly, the study areas consistently align with administrative boundaries at a singular scale, i.e., provincial-, prefecture city-, or county-level boundaries. It is essential to transcend these boundary limitations by adopting an interregional perspective for the integration of landscape systems [14]. Furthermore, it is crucial to harmonize the relationships between administrative districts to enhance human welfare, reflecting contemporary trends in urban development [15]. Thirdly, strategies that integrate diverse objectives should be developed to address the complexities inherent to regional development. Additionally, the stability of UGI networks often exhibits deviations due to fluctuations in evaluation factors and varying values of individual indices across different scales and objectives [16,17]. Therefore, it is necessary to explore methods for constructing UGI networks based on multi-scale and multi-objective approaches within the framework of territorial spatial planning. Meanwhile, there is also a need to explore methods for assessing UGI network stability more comprehensively.

The construction and optimization of UGI have evolved into a fundamental paradigm of "identifying the source areas, determining the resistance surface, and extracting the corridors, evaluating the network stability" [18]. Currently, the primary methods used to identify UGI sources involve calculating ecological service importance in conjunction with ecological sensitivity [19], performing morphological spatial pattern analysis (MSPA) alongside landscape connectivity analysis (LCA) [20,21], and employing spatially continuous wavelet transforms [22]. The MSPA approach is extensively utilized, as it can automatically categorize the pixel data of focal feature classes into new structural connectivity classes. It has also been demonstrated that it can effectively establish network connectivity regardless of scale variations [16]. In this context, a comprehensive resistance surface map is constructed using the analytic hierarchy process (AHP) to determine the landscape resistance surface. Furthermore, researchers have modified the resistance surface by incorporating night-time light data, which effectively capture landscape variability in the study area while mitigating the biases introduced by anthropogenic factors [23]. There are two predominant methodologies used for extracting UGI corridors nowadays: circuit theory and minimum cumulative resistance (MCR). MCR is diffusely applied due to its consideration of landscape heterogeneity and the horizontal connectivity among landscape units [24,25]. Drawing upon graph theory, the metrics employed to assess UGI patterns can be categorized into two primary aspects: the centrality aspect and the connectivity aspect. The centrality metrics represent the concentration degree and comprehensive influence on the overall network, whereas the connectivity metrics describe the landscape process and interconnective performance [26]. Shen and Wu managed to create a quantifiable system to assess the dimensions of centrality and connectivity, determined by the indices of the average node degree ($k$), average shortest path ($l$), global efficiency ($E_G$), and clustering coefficient ($cc$). Moreover, an overall value ($O_G$) was introduced to synthetically evaluate landscape structure through an equally weighted assessment of $k$, $l$, $E_G$, and $cc$ [16].

The Four-Lake Hydrographic Network (FLHN) of the prefecture-level city of Shizuishan in the Ningxia Hui Autonomous Region is a typical vulnerable area in the Upper Yellow River Basin of China. Urban sprawl has led to many serious problems in the corresponding ecosystem, characterized by pronounced landscape heterogeneity [27]. Aiming to address the integration of sources across various scales and the coordination of corridors with nodes based on diverse objectives within the landscape network, a collaborative framework of UGI network construction and evaluation was proposed based on multi-scale and multi-objective approaches. The main aims of this study were ① to assess the advantage of integrating regional and interregional perspectives, comparing the numbers and areas of UGI sources between multi-scale and single-scale approaches, and ② to explore the characteristics of this network by making trade-offs in regard to and synergizing the complex relationships between individual objects, evaluating the superiority of the UGI network in terms of both stability and final cost.

## 2. Study Area and Data

### *2.1. Study Area*

The Four-Lake Hydrographic Network (FLHN) of the prefecture-level city of Shizuishan in the Ningxia Hui Autonomous Region, China, is composed of Xinghaihu Lake, Zhenshuohu Lake, Shahu Lake, and Hanquanhai Lake (Figure 1), which constitute a lake group, a wetland system whose members are spatially distinct yet interconnected through dispersal and energy flow. Both Xinghaihu Lake and Zhenshuohu Lake are recognized as national wetland parks, while Shahu Lake is regarded as a provincial natural reserve. The FLHN is situated at the border junction of Dawukou District and Pingluo County in the prefectural city of Shizuishan, constituting an essential component of the "Dawukou-Pingluo-Shahu" urban cluster. It plays an important role in safeguarding the integrity of the natural landscape system in the northern Ningxia Plain and serves as a vital link between the Helan Mountain National Nature Reserve and the Yellow River Ecological Corridor. The boundaries of the study area are defined by the influence of lakes and other water systems, extending from Huinongqu Canal in the east to Xiganqu Canal in the west, from the border with Huinong District in the prefecture-level city of Shizuishan in the north to the border with Helan County administrated by the prefecture-level city of Yinchuan in the south.

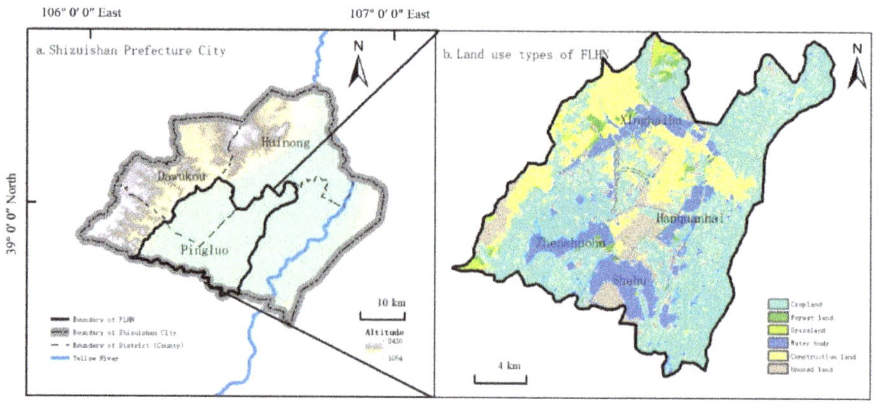


**Figure 1.** Location of the study area.

*2.2. Data Processing*

The main data and their sources and formats are presented in Table 1. In this study, the land-use types were resampled into six categories: cropland, forestland, grassland, water body, construction land, and unused land. The raster grids were reclassified into 30 m × 30 m for further calculation.

**Table 1.** The main data used in this study.

| Data | Format | Source |
|---|---|---|
| Land-use type | Vector data in 2020 | Ningxia Nature Resource Department |
| Administrative boundary | Polygons | http://www.hydrosheds.org/ |
| Digital elevation model (DEM) | Grid at 30 m | http://www.gscloud.cn/ |
| Slope | Grid at 90 m | http://www.gscloud.cn/ |
| Normalized difference vegetation Index (NDVI) | Grid at 1000 m in 2020 | http://www.resdc.cn |
| Road and railway | Polyline in 2020 | http://www.openstreetmap.org/ |
| Nighttime light | Grid at 500 m in 2020 | https://payneinstitute.mines.edu// |

## 3. Methods

A flowchart of the stages of this study is shown in Figure 2. It consists of three major steps: identifying and integrating UGI sources based on multi-scale data (see Section 3.1), establishing landscape resistance surfaces using similar factors and weights (see Section 3.2), and extracting corridors and performing a structural evaluation of UGI based on a multi-objective approach (see Section 3.3).

*3.1. Identifying and Integrating the UGI Sources Based on Multi-Scale Data*

3.1.1. Identification of UGI Sources

The UGI sources are of great importance in accurately determining the total number and area of landscape elements in a region, protecting biodiversity, and maintaining the structural stability of a landscape [28]. The MSPA method is utilized to determine the spatial patterns of raster images [29]. Using this method, important patches can be selected from the perspective of spatial morphology and connectivity, providing a scientific basis for the identification of UGI sources. Considering the integrity of green and blue space systems, forestland and waterbodies were designated as foreground elements, while other land-use types were set as background elements. The binary raster data were processed via the eight-neighborhood analysis method using Guidos Toolbox 3.0 software. The default values of edge width, transition, and intext were set as 1, off, and on, respectively. Subsequently, the land-use types were reclassified into 7 basic spatial elements: core, islet, perforation, edge, loop, bridge, and branch. Consequently, the landscape patches, with a value of "core" and an area exceeding 2 km$^2$, were selected as alternative ecological patches of UGI. Furthermore, based on island bio-geography theory, the area of patches and their connectivity respond to landscape spatial structure continuity and species carrying capacity [30]. Building upon the MSPA result, we selected LCA to calculate the structural connectivity and dispersal potential of the habitat patches using Confer 2.6. The patch connectivity index (*dIIC*) and patch importance index (*dPC*) were selected to rank the patches' importance as LCA indicators [31]; the higher the *dPC* index, the greater the patch importance, and, accordingly, numerous patches can be identified as UGI sources. The corresponding equations are as follows [20]:

$$dIIC = \frac{\sum_{i=1}^{n}\sum_{j=1}^{n}\frac{a_i a_j}{1+nl_{ij}}}{A_L^2} \qquad dPC = \frac{PC - PC_{remove}}{PC} * 100\%$$

Here, *dIIC* is the connectivity probability of the UGI network, $n$ is the number of total patches, and $a_i$ and $a_j$ denote the areas of patches $i$ and $j$, respectively. $l_{ij}$ is the probability of maximum diffusion between patch $i$ and patch $j$, and $A_L$ is the total area of the study region. $PC_{remove}$ is the connectivity probability of the remaining patches after removing a patch, and $dPC$ is the importance degree of a patch. Finally, the patches for which the $dPC$ index was larger than 2 were identified as the UGI sources.

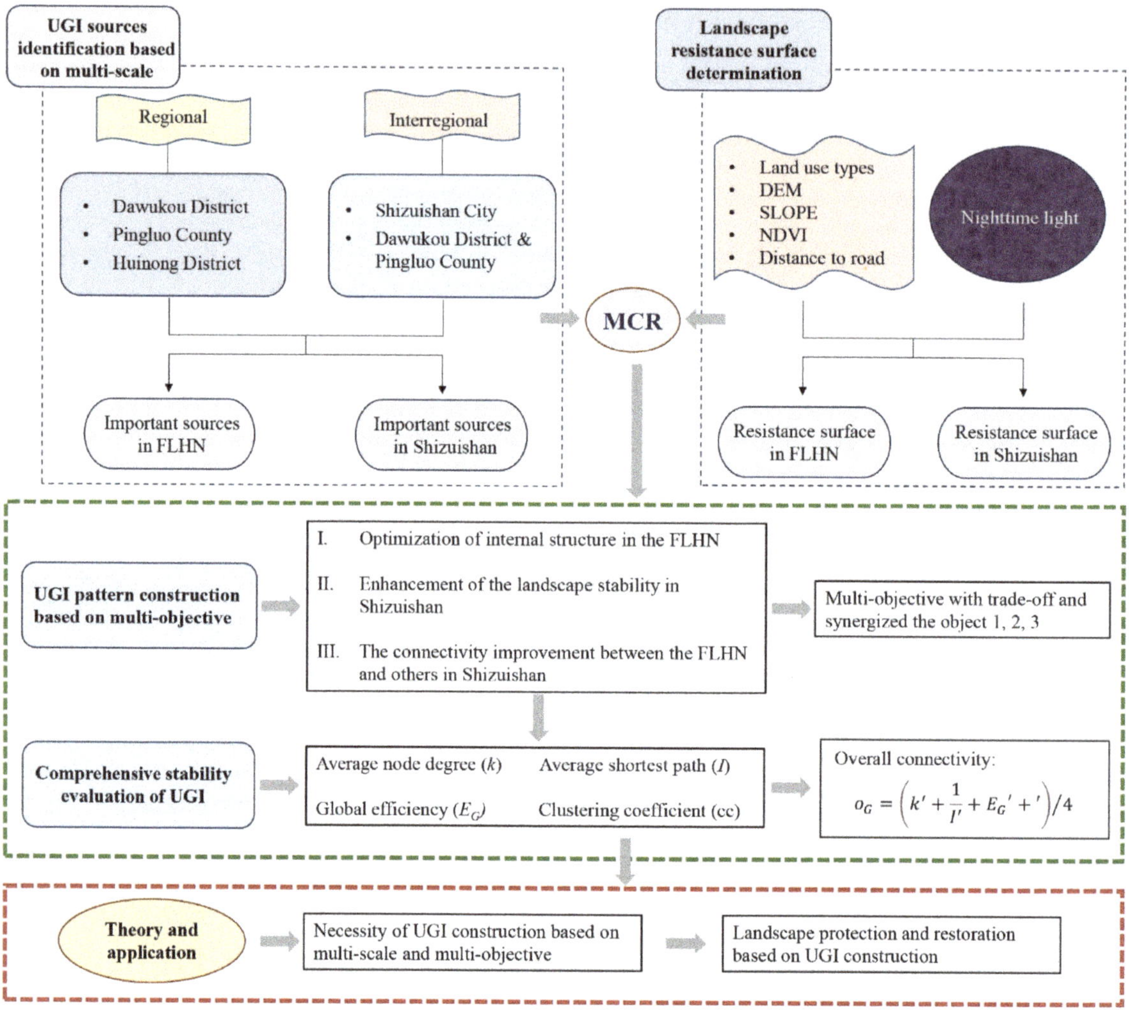


**Figure 2.** The research framework of urban green infrastructure (UGI) pattern construction and optimization.

### 3.1.2. Integration and Evaluation of UGI Sources Based on Multi-Scale Data

Combined with adhering to the spatial principle of importance in itself and strong connectivity among UGI sources [32], it was necessary to integrate the UGI sources for the levels of the FLHN and the prefecture-level city of Shizuishan to a different extent, i.e., from regional and interregional perspectives, respectively. Based on the result, we evaluated the spatial effects of extent changes and calculated the number and areas of identifying sources from a multi-scale perspective.

### 3.2. Establishing the Landscape Resistance Surface

A landscape resistance surface reveals the status of elements' movement in a whole region [33]. Based on landscape-scale interactions, different spatial extents can transmit information between each other [34]. In order to replicate the scale effect, landscape-resistant surfaces were established using similar factors and weights at the levels of the FLHN and the prefecture-level city of Shizuishan, respectively. Based on previous studies [35,36] and considering the study area's characteristics and the data obtainable, we derived five resistance indicators, including land-use types, the DEM, slope, distance to roads, and NDVI, at the two levels. The resistance factors were reclassified into five values, denoted as 1, 2, 3, 4, and 5, with equal weights applied across the factors, and determined by experts (Table 2). Finally, the comprehensive resistance surfaces were set to five grades, ranging from lower, low, medium, high, and higher; the lower the resistance grade, the weaker the impact of external disturbance on the landscape. Using Raster Calculator tool available with ArcGIS 10.8, the landscape resistance surfaces were calculated.

**Table 2.** Landscape resistance evaluation index system.

| Resistance Value | Land Use Type | DEM (m) | SLOPE (°) | Distance to Road (m) | NDVI |
|---|---|---|---|---|---|
| 1 | Forestland | 1054–1500 | 0–4.45 | >2051 | 0.70–0.99 |
| 2 | Cropland, Grassland | 1501–2000 | 4.46–11.93 | 1051–2050 | 0.51–0.69 |
| 3 | Waterbody | 2001–2500 | 11.94–19.63 | 551–1050 | 0.35–0.50 |
| 4 | Unused land | 2501–3000 | 19.64–28.27 | 51–550 | 0.20–0.34 |
| 5 | Construction land | 3001–3430 | 28.28–59.56 | Itself (50) | 0.00–0.19 |
| Weight | 0.2 | 0.2 | 0.2 | 0.2 | 0.2 |

In addition, as the nighttime light index can be applied as a proxy for characterizing human activities, it was utilized to modify the basic resistance surfaces, improving the accuracy of corridor simulation. The corresponding formula is as follows [37]:

$$R' = \frac{NL_i}{NL_a} R$$

where $R'$ denotes the modified landscape resistance of grid $i$, $R$ denotes the basic resistance coefficient corresponding to the land-use type of grid $i$, $NL_i$ denotes the nighttime light intensity of grid $i$; and $NL_a$ denotes the average nighttime light intensity corresponding to the land-use type of grid $i$. The landscape resistance surfaces revised for nighttime light were calculated using the Raster Calculator available with ArcGIS 10.8.

### 3.3. Developing and Evaluating the UGI Pattern

#### 3.3.1. Construction and Optimization of UGI Network Based on Multi-Objective Approach

The objectives of this study were to ① optimize the internal structure of the FLHN, ② enhance landscape stability in the prefecture-level city of Shizuishan, ③ improve the connectivity between the FLHN and other groups of the prefecture-level city of Shizuishan, and ④ achieve the multi-objective goal of trade-offs and synergy between objects 1, 2, and 3. Based on the differences for a single object, we nested the patterns of the sources, corridors, and key nodes of UGI at a corresponding object and devised strategies for optimizing the landscape pattern.

As critical paths and points are utilized to facilitate a flow of ecological processes between important patches, the sources, corridors, and key nodes of UGI can provide shelters for migrating animals and even promote species diffusion [38]. The MCR model is widely applied to extract corridors and nodes since it considers the cost of species migration from a source to a target. The corresponding formula for calculation is as follows [39]:

$$MCR = f_{min}\sum_{i=1}^{m}\sum_{j=1}^{n} D_{ij}R_i,$$

where $MCR$ denotes the minimum cumulative resistance; $f$ denotes the positive correlation function between the minimum cumulative resistance and an ecological process; $D_{ij}$ is the migration distance traversed by source $i$ to target $j$; and $R_i$ is the resistance unit coefficient of species diffusion. Corridors were obtained using the Linkage Mapper Toolbox of ArcGIS 10.8.

The gravity model was applied to rank corridors, with the corridors' significance being measured by the mutual gravitational force of different patches. The greater the value of the gravitational force between each pair of patches, the greater the importance of corridors. The corresponding formula is as follows [40]:

$$G_{ab} = \frac{N_a N_b}{D_{ab}^2} = \left(\frac{1}{p_a} \times lnS_a\right)\left(\frac{1}{p_b} \times lnS_b\right) \Big/ \left(\frac{L_{ab}}{L_{max}}\right)^2$$

where $G_{ab}$ denotes the mutual gravitational force between patches $a$ and $b$; $D_{ab}$ is the normalized value of the resistance of UGI corridors between two important sources; $N_a(N_b)$, $p_a(p_b)$, and $S_a(S_b)$ denote the weight, resistance value, and area of important source $a(b)$, respectively; $L_{ab}$ denotes the cumulative resistance value of sources $a$ and $b$; and $L_{max}$ denotes the maximum resistance value. The gravity model was run using Excel.

In this study, the corridors with mutual forces greater than 3000 between patches were identified as key corridors, those with mutual forces ranging from 2000 to 3000 between patches were identified as general corridors, and those with mutual forces ranging from 1000 to 2000 between patches were identified as fragile corridors. The points at the intersection of key corridors, and of key corridors and general corridors, were termed key nodes; those at the intersection of general corridors, and key corridors and fragile corridors, were defined as general nodes; and those at the intersection of fragile corridors, and general corridors and fragile corridors, were defined as fragile nodes. The key nodes were the narrow areas in the corridors with the highest current values, which should be prioritized in conservation strategies [41]. The thresholds of corridors' grades and points' interactions were determined by the experts according to the actual conditions of the study area.

### 3.3.2. Comparison of Landscape Stability Evaluations

The evaluation of UGI patterns is a prerequisite for enhancing landscape pattern stability [42]. The tools Ucinet and Pajek are utilized to abstract landscape entities into complex networks and explore the structural characteristics of UGI based on various metrics. The metrics consist of nodes and lines representing landscape structures (sources and corridors). The system of metrics is summarized in Table 3 via the integration of previous studies by Shen and Wu [16].

**Table 3.** Metrics for quantifying landscape stability.

| Metric | Equation | Description | Type | References |
|---|---|---|---|---|
| Average node degree ($k$) | $k = \frac{1}{N} \sum_{i \neq j \in G} a_{ij}$ | $K$ denotes the average node degree of points in network $G$; $N$ is the number of total points; $a_{ij}$ is the number of nodes $i$ directly connected to nodes $j$; and a higher $k$ value indicates better convenience between nodes. | Centrality | [43,44] |
| Average shortest path ($I$) | $I = \frac{2}{N(N-1)} \sum_{i \neq j \in G} d_{ij}$ | $I$ denotes the cost of migration between different nodes, which is opposite to node connectivity, and $d_{ij}$ is the distance of the shortest path connecting nodes. | Connectivity | [45] |

Table 3. *Cont.*

| Metric | Equation | Description | Type | References |
|---|---|---|---|---|
| Global efficiency ($E_G$) | $E_G = \frac{1}{N(N-1)} \sum_{i \neq j \in G} \frac{1}{d_{ij}}$ | $E_G$ describes the efficiency of movement in the whole network, which is applied to avoid the divergence of the $I$ index; the greater the $E_G$ index, the less energy consumed for moving between nodes. | Centrality | [46,47] |
| Clustering coefficient ($cc$) | $cc = \frac{1}{N} \sum_{i=1}^{N} \frac{2E_i}{k_i(k_{i-1})}$ | $cc$ denotes the agglomeration degree of points in the UGI network $G$, $E_i$ is the actual number of connecting lines between the node $i$ and its neighboring nodes, and $k_i$ is the degree of node $i$; the greater the value of $cc$, the greater the agglomeration degree of network $G$. | Connectivity | [26,48,49] |
| Overall connectivity ($O_G$) | $o_G = \left(k' + \frac{1}{I'} + E_G' + cc'\right)/4$ | $O_G$ denotes the overall connectivity of UGI network $G$, the weights for which correspond to the equality of $k$, $I$, $E_G$, and $cc$, and $k'$, $I'$, $E_G{}'$, $cc'$ are the normalized values of each metric; a greater $O_G$ value indicates greater comprehensive stability | Comprehensiveness | [16] |

## 4. Results

### 4.1. Identification of UGI-Important Sources Based on Multi-Scale Information

#### 4.1.1. Spatial Patterns of Important Sources in Each Scale

Upon analyzing the spatial elements based on the extent of the FLHN, 79.6% of the total areas in the FLHN were identified as core elements, and 8.6% of the total areas were identified as edge elements, whereas the proportions of other elements were relatively low. This result not only showed that the FLHN was the most significant natural landscape in the region but also indicated that landscape connectivity across the four lakes should be improved. Based on the spatial distribution of the sources obtained at each scale, it was evident that the majority of the FLHN constituted the important blue and green space, serving as critical hubs for species movement. There were four groups of UGI sources in the prefecture-level city of Shizuishan, including the FLHN, the Helanshan Mountain Nature Reserve, the main stream of the Yellow River, and the wind-sheltering and sand-fixing forest belt on the edge of the Mu Us Desert. The overlap rates of the important sources in the FLHN were approximately equivalent from both regional and interregional perspectives, ranging from 78.1% to 95.4%. Furthermore, we found that the areas of these important sources throughout the FLHN were significantly higher than those at the other scales; nevertheless, this was the case from regional or interregional perspectives, indicating that it was necessary to explore the sources in accordance with the principle of landscape stability and sustainability at multiple scales. The spatial distribution of UGI sources from a regional perspective, such as the scales of Dawukou District, Pingluo County, and Huinong District, was evident; however, it failed to adequately reflect landscape connectivity across different scales. In contrast to the regional perspective, the connectivity of important sources exhibited significant transmission across scales from the interregional perspective, such as the extent of the prefecture-level city of Shizuishan, Dawukou District, and Pingluo

County. However, despite their critical role in administrative jurisdiction (and yet low connectivity between pairs of patches), these sources were not thoroughly explored.

4.1.2. Integration and Assessment of Important Sources

Considering both their intrinsic significance and mutual landscape connectivity, the important sources of the FLHN and the prefecture-level city of Shizuishan were obtained by integrating the sources from each scale (Figure 3). There were 19 important sources in the FLHN, encompassing an area of 105.07 km$^2$, which accounts for 15.0% of the four lakes and their tributaries. At the same time, there were 41 important sources across the entire extent of the prefecture-level city of Shizuishan, covering an area of 433.10 km$^2$, which accounts for 8.9% of the entire region. It contained four source groups: the FLHN, the Helanshan Mountain Nature Reserve, the main stream of the Yellow River, and the wind-sheltering and sand-fixing forest belt on the edge of the Mu Us Desert. The FLHN was located in the intersection of the other three source groups.

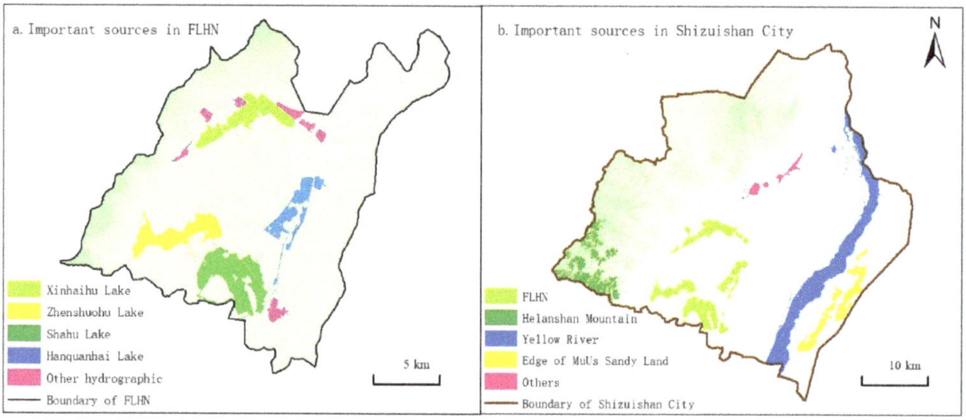


**Figure 3.** Important sources of urban green infrastructure (UGI) in the study area.

The number and areas of important sources in the FLHN and the prefecture-level city of Shizuishan are shown in Tables 4 and 5, respectively. As shown, the number and areas of important sources at both levels increased, when combined by each scale. At the level of the FLHN, the important sources in the main parts of Shahu Lake and Hanquanhai Lake were all identified as a cohesive whole, underscoring the necessity of optimizing the natural landscape pattern within each lake. Parts of the important sources were identified at multiple scales, while there was failure from a regional perspective, which could better clarify the pathway for species dispersal in the FLHN. At the Shizuishan level, the main stream of the Yellow River was regarded as an essential source in its entirety, mitigating deficiencies in identifying important sources created by administrative boundary limitations. It also revealed the importance of enhancing landscape restoration efforts since smaller patches, such as the Jianquanhu Wetland and the south of the Mu Us Desert, could be identified through integration across scales.

**Table 4.** Number and areas of important sources in the Four-Lake Hydrographic Network (FLHN) at different scales.

| Scale | FLHN | Dawujou District | Pingluo County | Dawukou District and Pingluo County | Prefecture-Level City of Shizuishan | Multi-Scale |
|---|---|---|---|---|---|---|
| Number | 19 | 9 | 9 | 16 | 16 | 19 |
| Area (km$^2$) | 105.07 | 24.38 | 75.81 | 97.81 | 97.81 | 105.07 |

**Table 5.** Number and areas of important sources in the prefecture-level city of Shizuishan at different scales.

| Scale | FLHN | Dawujou District | Pingluo County | Huinong County | Dawukou District and Pingluo County | Shizuishan Prefecture City | Multi-Scale |
|---|---|---|---|---|---|---|---|
| Number | 19 | 9 | 26 | 7 | 30 | 31 | 41 |
| Area ($km^2$) | 105.07 | 24.38 | 343.67 | 59.74 | 342.9 | 389.06 | 433.14 |

### 4.2. Landscape Resistance Surfaces from Two Levels

The landscape resistance surfaces at the level of the FLHN and the prefecture-level city of Shizuishan were obtained using identical factors and weights, respectively (Figure 4). At the FLHN level, the resistance values for most of factors were lower or low, except for the NDVI, which exhibited a significant percentage for high and higher. The comprehensive resistance surface was combined with the factors in equal weight and divided into five ranks based on natural breaks. The ratios of zones with lower, low, medium, and high ranks were 30.6%, 36.1%, 21.6%, and 11.8%, while zones with higher ranking were nearly absent. The result showed that it was feasible to construct and optimize the UGI network. In addition, we should be more focused on reflecting the contradiction between human activities and landscape restoration. The zones with high and higher resistance values were mainly concentrated in central urban and surrounding areas, which were affected by impermeable ground and traffic. At the level of the prefectural city of Shizuishan, zones with higher resistance values were primarily located in high-elevation areas in Helanshan Mountain Natural Reserve and distributed in the regions with little vegetation in the Mu Us Desert. The zones with higher resistance values were found to be adjacent to urban built-up areas and traffic routes as determined by incorporating the nighttime light index, also indicating that human activities have had a significant impact on landscape stability.

### 4.3. Construction and Evaluation of UGI Network Based on a Multi-Objective Approach

#### 4.3.1. Construction and Optimization of a UGI Network Based on a Multi-Objective Approach

Objective 1: optimizing the internal structure of the FLHN. We extracted corridors and nodes at the FLHN level (Figure 5a). Six key corridors with a length of 83.18 km, six general corridors with a length of 121.90 km, and two fragile corridors with a length of 32.92 km were extracted, and eight key nodes, five general nodes, and four fragile nodes were also extracted. The key corridors were located in the east of the FLHN, establishing a sequential connection between Xinghaihu Lake, Hanquanhai Lake, and Shahu Lake. At the same time, the general corridors connected Xinghaihu Lake and Zhenshuo Lake via irrigational canals and connected Xinghaihu Lake and Shahu Lake via green belts along roads. The key nodes were associated with the key corridors, making the Hanquanhai Lake the axis in the FLHN. In order to optimize the internal structure of the FLHN, we needed to upgrade four general corridors, turn two fragile corridors into key corridors corresponding to a length of 86.91 km, and incorporate four key nodes from the existing general nodes. There were 10 important sources that were connected by key corridors, an increase from the value of only 7 before optimization. The new key nodes were situated at the intersection of key corridors between the southwest of Xinghaihu Lake and the southeast of Shahu Lake, serving as the stepping stones for species diffusion. The optimization of the UGI network not only established direct connections between Zhenshuo Lake and other lakes but also enhanced connectivity between Xinghaihu Lake and Shahu Lake. Furthermore, this approach was employed to improve overall efficiency in the western region of the FLHN.

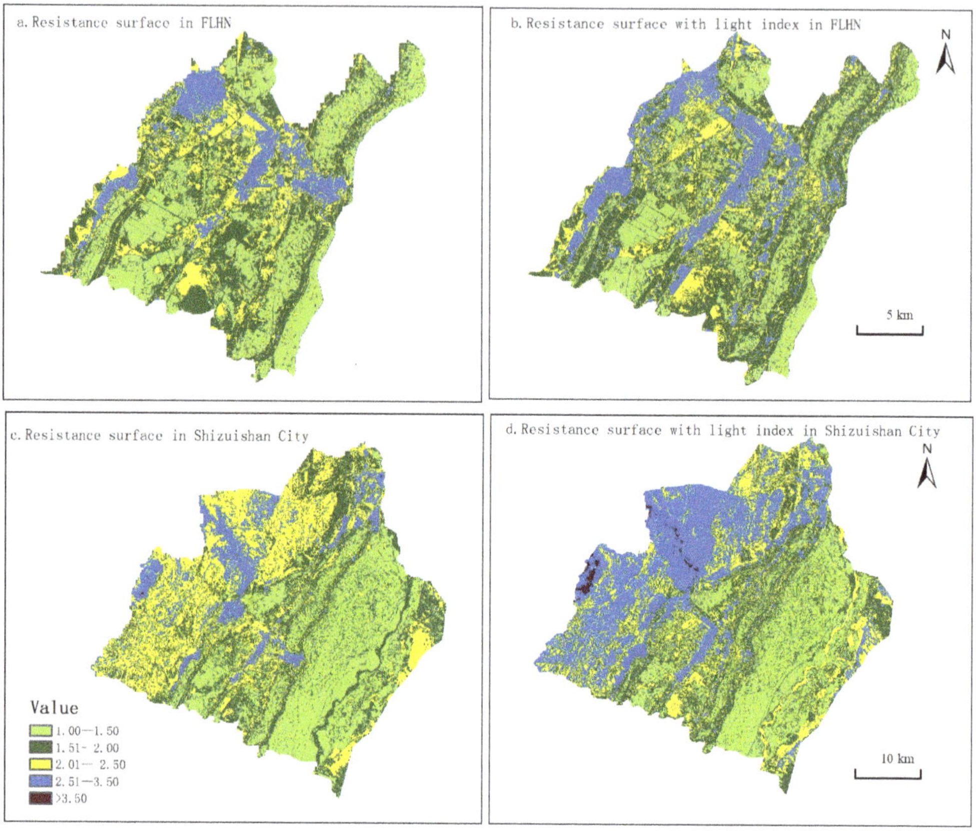


**Figure 4.** Resistance surfaces of the study area.

Objective 2: enhancing landscape stability in the prefecture-level city of Shizuishan (Figure 5b). We extracted corridors and nodes at the level of the prefectural city of Shizuishan. There were seven key corridors with a total length of 86.34 km, four general corridors measuring 60.51 km, and four fragile corridors spanning 62.13 km, and eight key nodes, five general nodes, and four fragile nodes were extracted. The spatial characteristics of the key corridors and nodes are similar to those relating to objective 1. Aiming at enhancing landscape stability in the prefecture-level city of Shizuishan, it was necessary to add a new key corridor, with a length of 14.16 km, and turn two general corridors spanning 86.91 km into corresponding key corridors. In addition, two fragile corridors with a length of 20.79 km needed to be upgraded to key corridors, while another two fragile corridors measuring 30.79 km needed to be upgraded to general corridors. Furthermore, it was also necessary to add three key nodes from existing general and fragile nodes. There were 12 important sources that were connected by key corridors, an increase from that of only 10 prior to optimization. The distances of short paths between Xinghaihu Lake and Shahu Lake were shortened by the new key corridor. Simultaneously, it established direct connections between Zhenshuohu Lake and both Shahu Lake and Hanquanhai Lake, as the fragile corridors were optimized. Moreover, the newly identified key nodes provided stepping stones for animal migration along pathways originating from the north, center, and south.

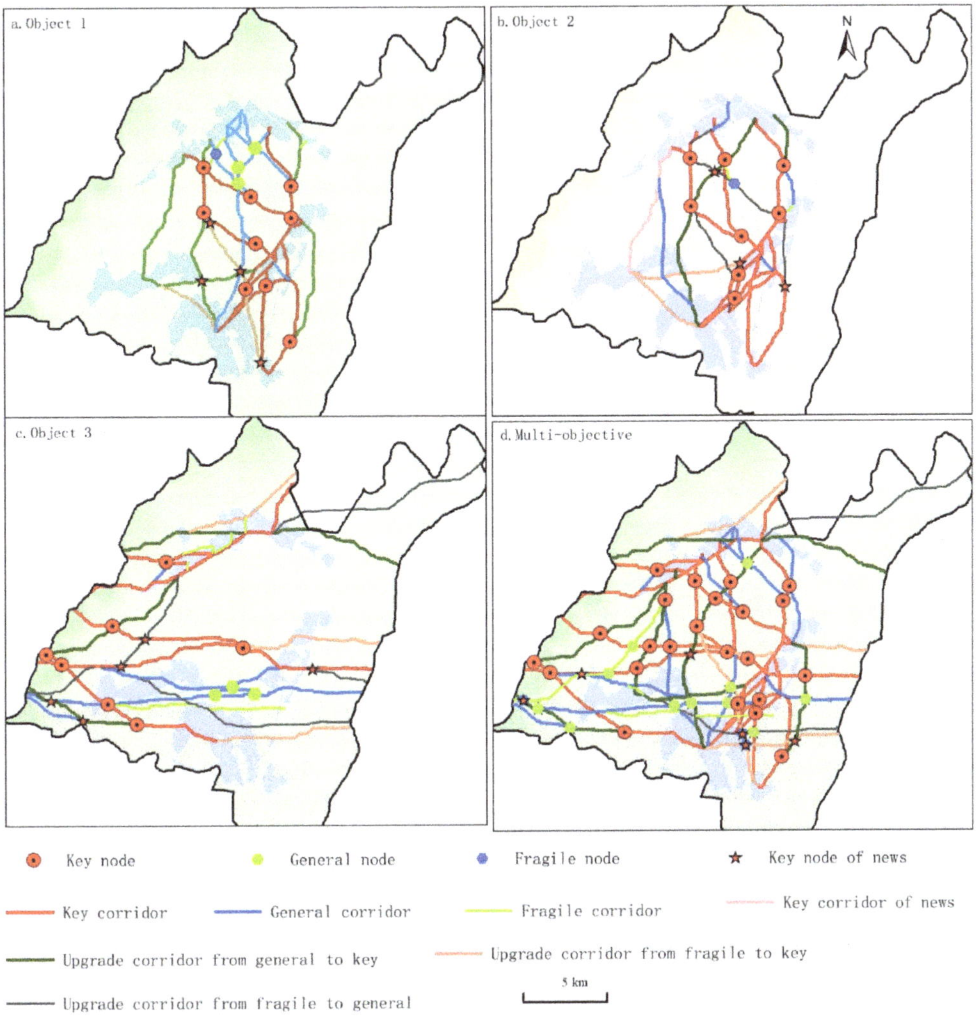


**Figure 5.** Construction and optimization of a UGI network in the Four-Lake Hydrographic Network (FLHN) based on different objectives.

Objective 3: improving the connectivity between the FLHN and other areas in the prefecture-level city of Shizuishan (Figure 5c). Based on the connectivity between the FLHN and the other three groups, a total of five key corridors spanning 111.72 km, five general corridors measuring 122.16 km, and eight fragile corridors with a length of 142.93 km were extracted. Additionally, three key nodes, eight general nodes, and one fragile node were also explored. In this case, the key corridors, which were used to link the Helanshan Mountain with the FLHN, were relatively effective. However, they did not adequately connect the other two groups, such as the main stream of the Yellow River and the wind-sheltering and sand-fixing forest belt on the edge of the Mu Us Desert. In order to enhance the axis of the FLHN to allow the interaction of three landscape groups, it was essential to upgrade two general corridors with a length of 55.30 km to corresponding key corridors. Additionally, two fragile corridors with a length of 44.84 km were upgraded to corresponding key corridors, while another three fragile corridors with a length of 71.85 km were upgraded to general corridors. Furthermore, it was necessary to incorporate five key nodes from existing general and fragile nodes. When the UGI network was optimized, the number of important sources, which were connected by key corridors, increased from 5 to 7. The species could thus consume less energy, as the pathway to Helanshan Mountain was optimized. Furthermore, it improved the degree of connectivity to the Yellow River and the Mu Us Desert, as the fragile corridors were upgraded. The new key nodes provided a supplement for enhancing pathways facilitating biological migration from Hanquanhai Lake to both the Yellow River and Helanshan Mountain.

Multi-objective goal: achieving a trade-off and synergy between objectives 1, 2, and 3 not only to optimize the internal structure of the FLHN but also enhance the importance of the FLHN for the construction of UGI in the prefecture-level city of Shizuishan (Figure 5d). Aiming at maximizing overall efficiency, the UGI network was optimized by considering trade-offs and synergy in the relationship between objects 1, 2, and 3. In order to achieve the multi-objective goal, the 10 existing key corridors measuring 120.50 km in the FLHN were maintained, and it was necessary to preserve 7 key corridors spanning 105.68 km between the FLHN and the other three groups. At the same time, eight general corridors totaling 117.42 km and seven fragile corridors exceeding 56.21 km were upgraded to corresponding key corridors. Additionally, the existing nine general corridors with a length of 113.86 km were retained, while three fragile corridors with a length of 41.99 km were upgraded to general corridors. Meanwhile, it was essential to incorporate three key nodes from both general and fragile nodes. There were 12 important sources that were connected by key corridors, an increase from that of only 9 prior to optimization. In comparison to the overlap of the individual objectives 1, 2, and 3, the maintenance of key, general, and fragile corridors was reduced by lengths of 7.79 km, 9.42 km, and 29.89 km, respectively, based on the multi-objective goal. Additionally, the upgrade requirements for general and fragile corridors decreased, with lengths of 5.51 km and 32.68 km, respectively.

#### 4.3.2. Comparison of Landscape Stability

Table 6 shows the stability of the landscape networks in the FLHN in accordance with different objectives. Based on the landscape networks obtained from applying the multiple-objective approach, we attempted to upgrade the general corridors to important corridors for ecological security and subsequently compute the metrics presented in Table 3. The centrality metrics of the UGI network, including the average node degree ($k$) and global efficiency ($E_G$) derived from the multi-objective approach, were found to be greater than those corresponding to the single-objective frameworks. This result confirmed that the overall structural stability of the UGI network could be improved by effectively balancing and coordinating the interrelationships among various objectives. The connectivity metrics of the UGI network, including the average shortest path ($l$) and clustering coefficient ($cc$), were less than those of similar indexes for objectives 1 and 2 and yet higher than those for objective 3, demonstrating that the landscape connectivity in the internal region of the FLHN was relatively high, while the situation between the FLHN and other three groups needed to be strengthened. The evaluation indices based on the multi-objective goal, with the exception of the average shortest path ($l$), were less than those after the simple addition of each single objective. The multi-objective approach led to the highest overall connectivity ($O_G$) value; however, it exhibited lower values in comparison to both the single objective and the simple superposition of each objective. The results show that a single index exhibited deviations in the landscape network stability evaluation because of variations in evaluation factors and the varying values of this index in different scenarios. It also showed that the comprehensive approach avoided this deficiency through trade-offs and synergizing the metrics.

**Table 6.** Comparison of urban green infrastructure (UGI) stability based on different objectives.

| | $k$ | $I$ | $E_G$ | $cc$ | $O_G$ |
|---|---|---|---|---|---|
| Objective 1 | 1.2587 | 0.1472 | 0.5280 | 0.9143 | 0.2946 |
| Objective 2 | 1.2587 | 0.1506 | 0.1506 | 1.4358 | 0.3422 |
| Objective 3 | 1.4286 | 0.3155 | 0.4989 | 0.4762 | 0.3405 |
| Multi-objective | 2.8571 | 0.2750 | 0.7748 | 0.4508 | 0.6146 |
| Simple addition of objectives 1, 2, and 3 | 4.5000 | 0.2953 | 0.7954 | 0.3809 | 0.5023 |

## 5. Discussion

### 5.1. Necessity of UGI Construction Based on Multi-Scale and Multi-Objective Goals

The essence of UGI pattern construction is delineating the spatial range and distribution of ecological elements, providing a target for landscape protection [50]. However, such a target could hardly be found for the optimal scale, within which the UGI sources were consistent from both a regional and interregional perspective. Firstly, the restriction of administrative boundaries may disrupt the continuity of spatial structures and the sustainability of natural ecosystems. A multi-scale approach can maintain landscape integrity by addressing the extraterritorial effects of landscape security [11]. On the other hand, spatial extent changes can lead to variations in the identification of important sources. If the extent of a study area is large enough that it includes multiple administrative areas, it may be difficult to address requirements concerning the ecological conservation and landscape restoration of the target administrative region. In contrast, if the study area extent is small, i.e., a county, a township, etc., the priority tends to be maintaining the significance of UGI sources within the administrative division while neglecting functional and structural connections with habitat patches beyond its boundaries [51]. The multi-scale approach not only accounts for landscape connectivity between UGI sources but also addresses areas with poor connectivity that are significant within an administrative region [22].

Previous studies revealed that an increment in the UGI evaluation index is of significance for different objectives and different species [52]. Furthermore, a change in one index can have an impact on multiple ecosystem services across transitional areas [53]. In terms of the spatial distribution of ecological elements, when planning to protect and restore urban landscape patterns, there can be trade-offs or a synergy of mutual gains for achieving different objectives, resulting in multiple goals, so patterns should not be optimized via the simple superposition of a single goal [36]. Spatial stability is regarded as the most essential issue in ecosystem restoration involving territorial space planning, where both corridors and nodes play important and complementary roles. The multi-objective approach can be used to build UGI networks with different objectives, thereby supporting the allocation of spatial elements and the formulation of landscape restoration strategies [54,55].

This study provides approaches to UGI planning based on multiple scales and objectives, revealing the diverse requirements of urban landscape security and sustainability. It can be applied to implement protection and restoration programs for urban landscapes according to different geographical conditions. It also provides a valuable example of dealing with similar areas, such as Yinchuan Wetland Park, Wuzhong Wetland Park in Ningxia, and Longyangxia Gorge in Qinghai in the upper Yellow River Basin. They were classified as arid or semiarid regions and determined to be ecologically vulnerable areas. Additionally, they were both located at the junction of different counties (district). Based on the advantages of approaches using multiple scales and objectives, researchers and governors can determine landscape security patterns to reflect the real characteristics of a region.

### 5.2. Landscape Protection and Restoration Based on UGI Construction

UGI construction is the basis for territorial spatial planning, reflecting the multiple functionalities and comprehensive significance of landscapes, thereby allowing the establishment of appropriate management suggestions [36]. Determining landscape patterns based on multiple scales and objectives indicates the trade-offs and synergy between green infrastructure and a multifunctional landscape, serving as the priority for prevention, conservation, restoration, and reconstruction [56,57].

At prefectural city level, the landscape pattern primarily involved trade-offs and synergizing for multiple objectives, and the overall efficiency of the UGI network was emphasized. At the county (district) scale, the focus was on the connection between the FLHN and UGI groups in corresponding areas. For example, the major engineering project was to improve the connective stability between Xinghaihu Lake and forest ecosystems, such as North Wudang Forest Park in Dawukou District. It was necessary to dredge ditches between the FLHN and the main stream of the Yellow River in Pingluo County. Huinong District played a major role in optimizing the pathway between Xinghaihu Lake and the main stream of the Yellow River nearby Jianquan Wetland and the Yanwochi Reservoir. Attention should be paid to the main engineering projects, including the optimization of the wetland system along the corridors between the Shahu and Xinghaihu Lakes, the silt displacement of the Sanerzhigou Canal, the management of the water environment of the Diannong River, and the comprehensive improvement project consisting of returning fishponds to natural lakes. Simultaneously, focus should be placed on ecological restoration and water quality improvement in the north and silt displacement in the center and south of the region, which resulted in the mitigation of eutrophication at Xinghaihu Lake. The coordination of water use with Pingluo Fine Chemical Industry Park was the major issue at Zhenshuohu Lake. It was imperative to desilt the bottom of the lake and purify the inner circulation in Shahu Lake. Regarding Hanquanhai Lake, the key issue was improving the internal structure.

To achieve objective 1, we should improve the grades of corridors along the west of Xinghaihu Lake and the south of Shahu Lake. In order to achieve objective 2, we should concentrate on the movement efficiency and agglomeration degree along eastern Xinghaihu Lake and the main body of Shahu Lake. To achieve objective 3, we should focus on improving the convergence of the important sources and key nodes in various corridors while also coordinating the internal structure with external links within the FLHN. Additionally, it is vital to integrate academic theory with practical application in planning a UGI network. Theoretically, the more corridors of the UGI, the greater the possibility of a stable network. However, the security of landscape patterns and economic construction costs should be considered comprehensively in practice.

### *5.3. Limitations and Future Research*

The framework presented in the study could be utilized as a decision-making tool for landscape conservation. However, this study has certain limitations. The MSPA and MCR models were applied to identify landscape patches based on structure-oriented analysis, but no attention was paid to landscape-function-oriented aspects, such as ecosystem service value. Additionally, the ecological sensitivity of UGI sources was overlooked, and variations in corridor widths for different species were not considered [20,58]. The method of evaluating UGI network stability mainly focuses on the centrality and connectivity of corridors and nodes, resulting in a lack of comprehensive comparative assessments of UGI source patterns. In the future, we will explore the relationship between the UGI and mental health, economic development, urban metabolism digital twins, etc. [59,60].

## 6. Conclusions

Using the Four-Lake Hydrographic Network (FLHN) in Shizuishan as the research area, a framework based on multi-scale and multi-objective approaches was proposed, UGI sources were identified by integrating regional and interregional perspectives, and UGI networks were constructed and optimized via trade-offs and synergy between the complex relationships of each objective. In this study, a total of 19 important sources with an area of 105.07 km$^2$ were recognized in the FLHN, while 41 key source areas covering an area of 433.10 km$^2$ were identified in the prefecture-level city of Shizuishan. Both the intrinsic importance of each source and the connectivity between them were considered when integrating UGI sources, avoiding discontinuity of UGI sources among different administrative areas. Based on the multi-objective goal, the UGI pattern did not have to maintain a length of 7.79 km for key corridors, 9.42 km for general corridors and 29.89 km for fragile corridors. Meanwhile, it was not necessary to upgrade 5.51 km of general corridors and 25.78 km of fragile corridors.

By comparing land stability across various scenarios, we determined that the centrality metrics of UGI stability were the greatest under the multi-objective approach, whereas the connectivity metric of UGI stability wavered for each objective. However, the multi-objective approach obtained the highest overall connectivity ($O_G$) value. It not only addressed the diverse landscape pattern requirements in an uncertain situation but also decreased the landscape construction costs associated with complex characteristics.

In this study, we determined landscape patterns through scale nesting and objective transmission and made up for the deficiency of UGI network construction and optimization based on one scale and objective. It is important to pay more attention to the relationship between UGI construction and the integration of urban and rural development, as UGI has emerged as a key asset for attracting spatial agglomeration under the new quality productivity forces.

**Author Contributions:** W.C.: conceptualization, formal analysis, and writing—original draft. C.M.: methodology, supervision, writing—review, and funding acquisition. All authors have read and agreed to the published version of the manuscript.

**Funding:** This research was funded by the National Science Research Project Foundation of China (2023YFF1304705), the National Natural Science Foundation of China (2024AAC02022), and the Ningxia Education Science Foundation of China (NYG2024007).

**Data Availability Statement:** Data will be made available on request.

**Acknowledgments:** We thank the reviewers for providing constructive comments on this manuscript. We also thank the editors for their careful work.

**Conflicts of Interest:** The authors declare that they have no known competing financial interests or personal relationships that could have appeared to influence the work reported in this paper.

## References

1. Yang, X.; Bai, Y.; Che, L.; Qiao, F.; Xie, L. Incorporating ecological constraints into urban growth boundaries: A case study of ecologically fragile areas in the Upper Yellow River. *Ecol. Indic.* **2021**, *124*, 107436. [CrossRef]
2. Zhou, Z.; Sun, X.; Zhang, X.; Wang, Y. Inter-regional ecological compensation in the Yellow River Basin based on the value of ecosystem services. *J. Environ. Manag.* **2022**, *322*, 116073. [CrossRef]
3. Guan, H.; Bai, Y.; Tang, y.; Zhang, C.; Zou, J. Spatial identification and optimization of ecological network in desert-oasis area of Yellow River Basin, China. *Ecol. Indic.* **2023**, *147*, 109999. [CrossRef]
4. He, Y.; Liang, Y.; Liu, L.; Yin, Z.; Huang, J. Loss of green landscapes due to urban expansion in China. *Resour. Conserv. Recycl.* **2023**, *199*, 107228. [CrossRef]
5. Liquete, C.; Kleeschulte, S.; Dige, G.; Maes, J.; Grizzetti, B.; Olah, B.; Zulian, G. Mapping green infrastructure based on ecosystem services and ecological networks: A Pan-European case study. *Environ. Sci. Policy* **2015**, *54*, 268–280. [CrossRef]
6. Berglihn, E.C.; Gómez-Baggethun, E. Ecosystem services from urban forests: The case of Oslomarka, Norway. *Ecosyst. Serv.* **2021**, *51*, 101358. [CrossRef]
7. Zhang, B.; MacKenzie, A. Trade-offs and synergies in urban green infrastructure: A systematic review. *Urban For. Urban Green.* **2024**, *94*, 128262. [CrossRef]
8. Jakstis, K.; Dubovik, M.; Laikari, A.; Mustajärvi, K.; Wendling, L.; Fischer, L.K. Informing the design of urban green and blue spaces through an understanding of Europeans' usage and preferences. *People Nat.* **2022**, *5*, 162–182. [CrossRef]
9. Sousa-Silva, R.; Zanocco, C. Assessing public attitudes towards urban green spaces as a heat adaptation strategy: Insights from Germany. *Landsc. Urban Plan.* **2024**, *245*, 105013. [CrossRef]
10. Gantchoff, M.G.; Erb, J.D.; MacFarland, D.M.; Norton, D.C.; Price Tack, J.L.; Roell, B.J.; Belant, J.L. Potential distribution and connectivity for recolonizing cougars in the Great Lakes region, USA. *Biol. Conserv.* **2021**, *257*, 109144. [CrossRef]
11. Lu, J.; Jiao, S.; Han, Z.; Yin, J. Promoting ecological restoration of deeply urbanized hilly areas: A multi-scale ecological networks approach. *Ecol. Indic.* **2023**, *154*, 110655. [CrossRef]

12. Scolozzi, R.; Geneletti, D. A multi-scale qualitative approach to assess the impact of urbanization on natural habitats and their connectivity. *Environ. Impact Assess. Rev.* **2012**, *36*, 9–22. [CrossRef]
13. Sun, X.; Wu, J.; Tang, H.; Yang, P. An urban hierarchy-based approach integrating ecosystem services into multiscale sustainable land use planning: The case of China. *Resour. Conserv. Recycl.* **2022**, *178*, 106097. [CrossRef]
14. Dong, J.; Peng, J.; Xu, Z.; Liu, Y.; Wang, X.; Li, B. Integrating regional and interregional approaches to identify ecological security patterns. *Landsc. Ecol.* **2021**, *36*, 2151–2164. [CrossRef]
15. Xu, T.; Nie, W.; Wang, Z.; Yang, G.; Zhang, Y.; Xu, B. Research on the coupled mechanism of landscape connectivity simulation by integrating multi-level ecological security patterns and multi-scenario simulation: A case study of the main urban area of Hangzhou. *Land Degrad. Dev.* **2024**, *35*, 2716–2737. [CrossRef]
16. Shen, Z.; Wu, W.; Tian, S.; Wang, J. A multi-scale analysis framework of different methods used in establishing ecological networks. *Landsc. Urban Plan.* **2022**, *228*, 104579. [CrossRef]
17. Men, D.; Pan, J. Ecological network identification and connectivity robustness evaluation in the Yellow River Basin under a multi-scenario simulation. *Ecol. Model.* **2023**, *482*, 110384. [CrossRef]
18. Lou, Y.; Yang, D.; Zhang, P.; Zhang, Y.; Song, M.; Huang, Y.; Jing, W. Multi-Scenario Simulation of Land Use Changes with Ecosystem Service Value in the Yellow River Basin. *Land* **2022**, *11*, 992. [CrossRef]
19. Zhong, B.; Wu, S.; Wu, N.; Sun, G.; He, C.; Liu, L.; Chen, H. Constructing ecological security pattern based on spatio-temporal evaluation of ecosystem services and ecological health in Chengdu, Southwest China. *Front. Environ. Sci.* **2024**, *12*, 992. [CrossRef]
20. Qian, W.; Zhao, Y.; Li, X. Construction of ecological security pattern in coastal urban areas: A case study in Qingdao, China. *Ecol. Indic.* **2023**, *154*, 110754. [CrossRef]
21. Chen, J.; Kinoshita, T.; Li, H.; Luo, S.; Su, D. Which green is more equitable? A study of urban green space equity based on morphological spatial patterns. *Urban For. Urban Green.* **2024**, *91*, 128178. [CrossRef]
22. Dong, J.; Peng, J.; Liu, Y.; Qiu, S.; Han, Y. Integrating spatial continuous wavelet transform and kernel density estimation to identify ecological corridors in megacities. *Landsc. Urban Plan.* **2020**, *199*, 103815. [CrossRef]
23. Tang, H.; Peng, J.; Jiang, H.; Lin, Y.; Dong, J.; Liu, M.; Meersmans, J. Spatial analysis enables priority selection in conservation practices for landscapes that need ecological security. *J. Environ. Manag.* **2023**, *345*, 118888. [CrossRef]
24. Dai, L.; Liu, Y.; Luo, X. Integrating the MCR and DOI models to construct an ecological security network for the urban agglomeration around Poyang Lake, China. *Sci. Total Environ.* **2021**, *754*, 141868. [CrossRef] [PubMed]
25. Wang, Y.; Qu, Z.; Zhong, Q.; Zhang, Q.; Zhang, L.; Zhang, R.; Yi, Y.; Zhang, G.; Li, X.; Liu, J. Delimitation of ecological corridors in a highly urbanizing region based on circuit theory and MSPA. *Ecol. Indic.* **2022**, *142*, 109258. [CrossRef]
26. Mao, H.; Zhou, L.; Ying, R.; Pan, D. Time Preferences and green agricultural technology adoption: Field evidence from rice farmers in China. *Land Use Policy* **2021**, *109*, 105627. [CrossRef]
27. Cheng, W.; Ma, C.; Li, T.; Liu, Y. Construction of Ecological Security Patterns and Evaluation of Ecological Network Stability under Multi-Scenario Simulation: A Case Study in Desert–Oasis Area of the Yellow River Basin, China. *Land* **2024**, *13*, 1037. [CrossRef]
28. Wu, J.G. *Landscape Ecology: Pattern, Process, Scale and Hierarchy*; Higher Education Press: Beijing, China, 2007.
29. Vogt, P.; Riitters, K. GuidosToolbox: Universal digital image object analysis. *Eur. J. Remote Sens.* **2017**, *50*, 352–361. [CrossRef]
30. MacArthur, R.H.; Wilson, E.O. *The Theory of Island Biogeography*; Princeton University Press: Princeton, NJ, USA, 2001.
31. Saura, S.; Torné, J. Conefor Sensinode 2.2: A software package for quantifying the importance of habitat patches for landscape connectivity. *Environ. Model. Softw.* **2009**, *24*, 135–139. [CrossRef]
32. Molne, F.; Donati, G.F.A.; Bolliger, J.; Fischer, M.; Maurer, M.; Bach, P.M. Supporting the planning of urban blue-green infrastructure for biodiversity: A multi-scale prioritisation framework. *J. Environ. Manag.* **2023**, *342*, 118069. [CrossRef]
33. Koen, E.L.; Bowman, J.; Walpole, A.A. The effect of cost surface parameterization on landscape resistance estimates. *Mol. Ecol. Resour.* **2012**, *12*, 686–696. [CrossRef] [PubMed]
34. Dhanjal-Adams, K.L.; Klaassen, M.; Nicol, S.; Possingham, H.P.; Chades, I.; Fuller, R.A. Setting conservation priorities for migratory networks under uncertainty. *Conserv. Biol.* **2017**, *31*, 646–656. [CrossRef] [PubMed]
35. Saura, S.; Pascual-Hortal, L. A new habitat availability index to integrate connectivity in landscape conservation planning: Comparison with existing indices and application to a case study. *Landsc. Urban Plan.* **2007**, *83*, 91–103. [CrossRef]
36. Jiang, H.; Peng, J.; Xu, D.; Tang, H. Constructing ecological security patterns with differentiated management intensity based on multifunctional landscape identification and multi-criteria decision-making. *Glob. Ecol. Conserv.* **2024**, *50*, e02862. [CrossRef]
37. Liu, M.; Peng, J.; Dong, J.; Jiang, H.; Xu, D.; Meersmans, J. Trade-offs of landscape connectivity between regional and interregional ecological security patterns in a junction area of Beijing-Tianjin-Hebei region. *Appl. Geogr.* **2024**, *167*, 103272. [CrossRef]
38. Labib, S.M. Investigation of the likelihood of green infrastructure (GI) enhancement along linear waterways or on derelict sites (DS) using machine learning. *Environ. Model. Softw.* **2019**, *118*, 146–165. [CrossRef]
39. Santos, J.S.; Leite, C.C.C.; Viana, J.C.C.; dos Santos, A.R.; Fernandes, M.M.; de Souza Abreu, V.; do Nascimento, T.P.; dos Santos, L.S.; de Moura Fernandes, M.R.; da Silva, G.F.; et al. Delimitation of ecological corridors in the Brazilian Atlantic Forest. *Ecol. Indic.* **2018**, *88*, 414–424. [CrossRef]
40. Nie, W.; Shi, Y.; Siaw, M.J.; Yang, F.; Wu, R.; Wu, X.; Zheng, X.; Bao, Z. Constructing and optimizing ecological network at county and town Scale: The case of Anji County, China. *Ecol. Indic.* **2021**, *132*, 108294. [CrossRef]

41. Wang, Y.; Zhang, F.; Li, X.; Johnson, V.C.; Tan, M.L.; Kung, H.-T.; Shi, J.; Bahtebay, J.; He, X. Methodology for Mapping the Ecological Security Pattern and Ecological Network in the Arid Region of Xinjiang, China. *Remote Sens.* **2023**, *15*, 2836. [CrossRef]
42. Sahraoui, Y.; De Godoy Leski, C.; Benot, M.-L.; Revers, F.; Salles, D.; van Halder, I.; Barneix, M.; Carassou, L. Integrating ecological networks modelling in a participatory approach for assessing impacts of planning scenarios on landscape connectivity. *Landsc. Urban Plan.* **2021**, *209*, 104039. [CrossRef]
43. Rayfield, B.; Fortin, M.J.; Fall, A. Connectivity for conservation: A framework to classify network measures. *Ecology* **2011**, *92*, 847–858. [CrossRef] [PubMed]
44. Liccari, F.; Boscutti, F.; Bacaro, G.; Sigura, M. Connectivity, landscape structure, and plant diversity across agricultural landscapes: Novel insight into effective ecological network planning. *J. Environ. Manag.* **2022**, *317*, 115358. [CrossRef] [PubMed]
45. Upadhyay, S.; Roy, A.; Ramprakash, M.; Idiculla, J.; Kumar, A.S.; Bhattacharya, S. A network theoretic study of ecological connectivity in Western Himalayas. *Ecol. Model.* **2017**, *359*, 246–257. [CrossRef]
46. Luo, Y.; Wu, J.; Wang, X.; Wang, Z.; Zhao, Y. Can policy maintain habitat connectivity under landscape fragmentation? A case study of Shenzhen, China. *Sci. Total Environ.* **2020**, *715*, 136829. [CrossRef] [PubMed]
47. Kim, B.; Park, J. Random ecological networks that depend on ephemeral wetland complexes. *Ecol. Eng.* **2020**, *156*, 105972. [CrossRef]
48. Huang, X.; Ye, Y.; Zhao, X.; Guo, X.; Ding, H. Identification and stability analysis of critical ecological land: Case study of a hilly county in southern China. *Ecol. Indic.* **2022**, *141*, 109091. [CrossRef]
49. De Montis, A.; Caschili, S.; Mulas, M.; Modica, G.; Ganciu, A.; Bardi, A.; Ledda, A.; Dessena, L.; Laudari, L.; Fichera, C.R. Urban–rural ecological networks for landscape planning. *Land Use Policy* **2016**, *50*, 312–327. [CrossRef]
50. Hou, W.; Liu, J.; Walz, U. Optimization of green infrastructure networks in the perspectives of enhancing structural connectivity and multifunctionality in an urban megaregion. *J. Environ. Manag.* **2023**, *348*, 119084. [CrossRef]
51. Peng, J.; Zhao, S.; Dong, J.; Liu, Y.; Meersmans, J.; Li, H.; Wu, J. Applying ant colony algorithm to identify ecological security patterns in megacities. *Environ. Model. Softw.* **2019**, *117*, 214–222. [CrossRef]
52. Ortega, U.; Ametzaga-Arregi, I.; Sertutxa, U.; Peña, L. Identifying a green infrastructure to prioritise areas for restoration to enhance the landscape connectivity and the provision of ecosystem services. *Landsc. Ecol.* **2023**, *38*, 3751–3765. [CrossRef]
53. Chen, L.; Pei, S.; Liu, X.; Qiao, Q.; Liu, C. Mapping and analysing tradeoffs, synergies and losses among multiple ecosystem services across a transitional area in Beijing, China. *Ecol. Indic.* **2021**, *123*, 107329. [CrossRef]
54. Barron, S.; Rugel, E.J. Tolerant greenspaces: Designing urban nature-based solutions that foster social ties and support mental health among young adults. *Environ. Sci. Policy* **2023**, *139*, 1–10. [CrossRef]
55. Xu, D.; Peng, J.; Jiang, H.; Dong, J.; Liu, M.; Chen, Y.; Wu, J.; Meersmans, J. Incorporating barriers restoration and stepping stones establishment to enhance the connectivity of watershed ecological security patterns. *Appl. Geogr.* **2024**, *170*, 103347. [CrossRef]
56. Peng, J.; Hu, X.; Qiu, S.; Hu, Y.; Meersmans, J.; Liu, Y. Multifunctional landscapes identification and associated development zoning in mountainous area. *Sci. Total Environ.* **2019**, *660*, 765–775. [CrossRef]
57. Thaweepworadej, P.; Evans, K.L. Urbanisation of a growing tropical mega-city during the 21st century—Landscape transformation and vegetation dynamics. *Landsc. Urban Plan.* **2023**, *238*, 104812. [CrossRef]
58. Huang, L.; Wang, J.; Cheng, H. Spatiotemporal changes in ecological network resilience in the Shandong Peninsula urban agglomeration. *J. Clean. Prod.* **2022**, *339*, 130681. [CrossRef]
59. Nigg, C.; Fiedler, J.; Burchartz, A.; Reichert, M.; Niessner, C.; Woll, A.; Schipperijn, J. Associations between green space availability and youth's physical activity in urban and rural areas across Germany. *Landsc. Urban Plan.* **2024**, *247*, 105068. [CrossRef]
60. Geremicca, F.; Bilec, M.M. Searching for new Urban Metabolism techniques: A review towards future development for a city-scale Urban Metabolism Digital Twin. *Sustain. Cities Soc.* **2024**, *107*, 105445. [CrossRef]

International Journal of
*Geo-Information*

MDPI

*Article*

# Impact of Land Use Change on Carbon Storage in Complex Terrains: A Case Study of Sichuan–Chongqing, China

**Tiankuo Chu [1,2], Jun Li [1,*], Chengye Zhang [1], Xiaoai Dai [2], Yikun Qing [3], Huan Huang [4,5] and Manling Peng [1]**

1 College of Geoscience and Surveying Engineering, China University of Mining and Technology, Beijing 100083, China; tiankuochu@student.cumtb.edu.cn (T.C.); czhang@cumtb.edu.cn (C.Z.); pml@student.cumtb.edu.cn (M.P.)
2 College of Geography and Planning, Chengdu University of Technology, Chengdu 610059, China; daixiaoa@cdut.edu.cn
3 College of Social Science, Duke Kunshan University, Kunshan 215300, China; yq92@duke.edu
4 College of Business, Chengdu University of Technology, Chengdu 610059, China; huan@cdut.edu.cn
5 Digital Hu Huanyong Line Research Institute, Chengdu University of Technology, Chengdu 610059, China
* Correspondence: junli@cumtb.edu.cn

**Abstract:** Understanding human activity and global climate change requires an understanding of the impact of land-use change on carbon storage. Nevertheless, the number of studies examining carbon storage in complex terrain areas remains relatively limited. This study quantified land use shifts and carbon storage variations from 2010 to 2030 in the Sichuan–Chongqing region in China by combining InVEST and PLUS models. The results show that (1) the Sichuan–Chongqing region exhibits a predominantly concentrated pattern of land utilization, with forests occupying a dominant position among various land types in the locality. (2) The total carbon storage loss between 2010 and 2020 is 24.19 Tg. Except for the ecological development scenario, compared to 2020, all multi-scenario projections show less carbon storage in 2030. (3) The primary cause of carbon loss is the sprawl of built-up areas into farmlands, forests, and grasslands. (4) The geomorphology of complex terrain greatly affects carbon storage; flat terrain tends to be less carbon-intensive than more steeply sloping terrain. This study offers worthwhile insights for sustainable land management and carbon mitigation strategies in complex terrains worldwide.

**Keywords:** land-use change; carbon storage; complex terrain; multiscenario simulation; Sichuan–Chongqing region

## 1. Introduction

Carbon storage in terrestrial ecosystems is influenced both by non-anthropogenic factors, like climate and atmospheric composition, and human activities, for example, land cover and management measures [1]. Understanding and monitoring carbon storage on terrestrial ecosystems could prove crucial in elucidating the linkages among carbon storage, climate change, and biodiversity for sustainable development [2,3]. Land use change is the second-most influential factor in terms of its impact on increasing atmospheric carbon dioxide concentrations, and the variations in carbon storage within terrestrial ecosystems correlate strongly with land use change [4].

Significant variations in carbon sequestration capacities are observed across different land use [5], and grassland contains more carbon storage than land does, such as construction land and farmland. Land use dynamics significantly affect the overall carbon stock of regional terrestrial ecosystems [6]. Therefore, assessing ecosystem carbon storage under current land use and modelling changes in it subject to future changes in land use are conducive to the efficient and rational use the limited natural resources. This is highly important for achieving sustainable development and carbon neutrality.

In recent years, studies on regional carbon storage estimation can be split into two categories. The first category is grounded in measurements at sample sites, with the results extended to the regional scale via various methods, such as calculating the change in regional carbon storage using the biomass conversion method [7,8] or analysing the exchange relationship between terrestrial ecosystems and atmospheric carbon dioxide to estimate carbon storage from the viewpoint of micrometeorology [9]. However, conventional carbon storage estimation methodologies are time-consuming, labour-intensive, constrained by observational outcomes, and prone to data gaps [10]. Moreover, these approaches do not adequately consider spatial coherence of terrestrial ecosystems, and there is uncertainty and error in the transformation process from the sampling point to the whole study area, which makes it impossible to calculate regional carbon stocks for a larger area [11]. The second category involves estimating regional carbon storage using geographic information science (GIS) and remote sensing models. An illustrative example is the Geographically Weighted Regression (GWR) model [12], which can be integrated with sample data and other remotely sensed metrics to model the spatial heterogeneity of ecosystem carbon density. Furthermore, the Terrestrial Ecosystem Dynamic Model (TEM) [13] and the Ecological Process Model (Biome-BGC) [14] are also widely employed, these models elucidate the carbon cycling patterns of ecosystems by calculating the relevant physiological and ecological parameters for different ecosystem types. Another notable model is the CASA model [15], which estimates carbon storage based on factors such as effective solar radiation, temperature, and water availability. Moreover, researchers have increasingly combined deep learning techniques [16] with other remote sensing models to estimate carbon storage [17]. While these models contribute to a comprehensive understanding of the dynamics of regional carbon stocks, they often overlook certain environmental and ecological factors, may lack accuracy, and generally lack the flexibility for adjustments and refinements based on specific conditions. The Integrated Valuation of Ecosystem Services and Tradeoffs (InVEST) model is commonly used to assess and estimate the carbon storage and sequestration capacity of ecosystems. Compared to other traditional estimation models, this module is mainly based on the carbon pools of forest ecosystems but also takes into account other ecosystems like grass-lands, wetlands and farmlands [18]. The model requires less data and is easier to obtain, so it achieves good results in calculating carbon storage [19–21].

Given the accelerated rate of urbanisation and global warming, it is crucial to allocate land resources correctly, maintain land use patterns, and ensure a stable level of carbon storage [22]. A single carbon storage estimation model can only determine the total amount of carbon storage in the current region, but it is unable to predict and estimate the dynamic pattern of regional carbon storage changes in future time scales, nor can the drivers of regional carbon storage changes be explored through modelling approaches. To tackle this challenge, researchers are assessing regional carbon storage while considering the multidimensional socioeconomic and ecological benefits. They have estimated future carbon storage in regional terrestrial ecosystems and the factors driving carbon storage changes by combining traditional carbon storage estimation models with land type dynamics prediction models. In these land use type prediction models, the Conversion of Land Use and its Effects at Small Regional Extent (CLUE-S) model evaluates the suitability of regional unused land and simulates the optimisation of land use in accordance with the local ecological environment [23]. Moreover, the Future Land Use Simulation (FLUS) model combines system dynamics and meta-cellular automata models to simulate multiple scenarios considering various socioeconomic and natural climate factors [24]. However, the Patch-Generating Land Use Simulation (PLUS) model has benefits in predicting the status of carbon storage under future development scenarios. It is more effective than models such as FLUS at determining the drivers behind land expansion and predicting the distribution of land-use landscapes at the pixel scale [25].

Combining the strengths of the InVEST model and the PLUS model can address the limitations of a single model for carbon storage estimation and deepen the understanding of the dynamic relationship between land use change and carbon storage, this research examined patterns in land use and regional carbon storage changes over time and space in a complex terrain region, with the Sichuan–Chongqing area being the focal point. Multiple scenarios of 2030 land use were performed using the PLUS model, while changes in drivers were simultaneously analysed and subsequently, integrated with the InVEST model, we delved into the space-time-evolving patterns of carbon storage and intended to delve into the following key inquiries: (1) What are the distinctions in land use between the Sichuan–Chongqing region over the past decade? (2) What are the spatial and temporal patterns for carbon stocks estimated from the land use distribution results from multiple scenario simulations in this region? (3) What are the main drivers of land change carbon change in Sichuan–Chongqing? The findings of this study will aid in the optimisation of land use structure in complex terrain regions such as the Sichuan–Chongqing region. This approach will facilitate the rationalization of spatial resource planning, sustain carbon sequestration at optimal levels, and expedite the achievement of carbon neutrality.

## 2. Materials and Methods

### 2.1. Study Area

The Sichuan–Chongqing area is situated in the southwest of China. It comprises the province of Sichuan and the municipality of Chongqing (Figure 1). The region spans approximately 568,000 square kilometres and features a complex topography, including mountainous, plateau, and basin landscapes, as well as abundant natural resources. The climate is typical of a subtropical monsoon, with high temperatures and precipitation during the summer. The study areas of Sichuan and Chongqing are experiencing rapid economic development, with the city cluster centred around Chongqing becoming a strategic hub for China. The local government has ordered the full implementation of the new development concept and promote the early formation of a carbon peak and a carbon-neutral pattern.

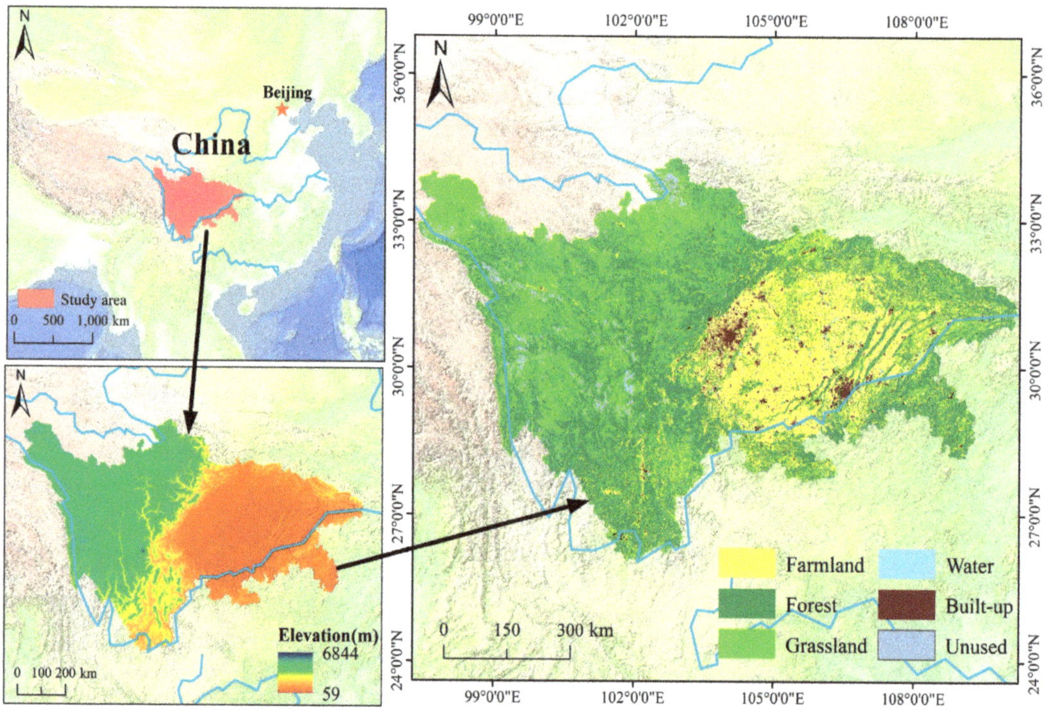


**Figure 1.** Location of Sichuan–Chongqing region in China.

### 2.2. Data Resources

In order to investigate the spatial and temporal patterns and drivers of carbon storage under land use change, this study collected multi-source datasets, as shown in Table 1. These datasets include land use data for carbon storage estimation and prediction, as well as natural factors (DEM, slope, rainfall, etc.) and anthropogenic factors (GDP, population, etc.) for the purpose of analysing the drivers of carbon storage change. The land use data was classified into six categories: farmland, grassland, forest, water, built-up, and unused. Additionally, the study used population and GDP data from 2019 because 2020 data were unavailable. The city-level road data and motorways were processed via the ArcGIS Euclidean distance function. All the data were uniformly adjusted to a 1-kilometre resolution via the resampling method.

**Table 1.** List of data base and its sources used for this study.

| Data | Year | Resolution | Data Resource |
|---|---|---|---|
| Land use | 2010–2020 | 1000 m | https://www.resdc.cn/ (accessed on 5 May 2024) |
| Land use constraints | 2020 | 1000 m | Obtained through processing of land use data (accessed on 5 May 2024) |
| Soil type | 1995 | 1000 m | https://www.resdc.cn/ (accessed on 5 May 2024) |
| Average annual temperature | 2020 | 1000 m | https://www.resdc.cn/ (accessed on 5 May 2024) |
| Average annual precipitation | 2020 | 500 m | https://cds.climate.copernicus.eu/ (accessed on 5 May 2024) |
| DEM | 2020 | 90 m | http://www.gscloud.cn/home (accessed on 5 May 2024) |
| Slope | 2020 | 90 m | Obtained through processing of DEM data |
| GDP | 2019 | 1000 m | https://www.resdc.cn/ (accessed on 5 May 2024) |
| Population | 2019 | 1000 m | https://www.resdc.cn/ (accessed on 5 May 2024) |
| Urban primary roads | 2020 | Vector data | https://www.openhistoricalmap.org/ (accessed on 5 April 2024) |
| Urban secondary roads | 2020 | Vector data | https://www.openhistoricalmap.org/ (accessed on 5 April 2024) |
| Urban tertiary roads | 2020 | Vector data | https://www.openhistoricalmap.org/ (accessed on 29 April 2024) |
| Highways | 2020 | Vector data | https://www.openhistoricalmap.org/ (accessed on 29 April 2024) |
| Rivers | 2020 | Vector data | Obtained through processing of DEM data |
| Government locations | 2020 | Vector data | https://www.webmap.cn/ (accessed on 5 May 2024) |

### 2.3. Methods

This study aimed to examine the spatial and temporal variations of land use and carbon storage in complex terrain areas. To achieve this, land use data, carbon intensity data and related driver information were used in conjunction with the InVEST and PLUS models. Furthermore, a multi-scenario emulation on land use in the Sichuan–Chongqing region was conducted along with an investigation into the driving factors and distributional differences of carbon density in complex terrain areas. Depicted in Figure 2 is the study's flowchart.

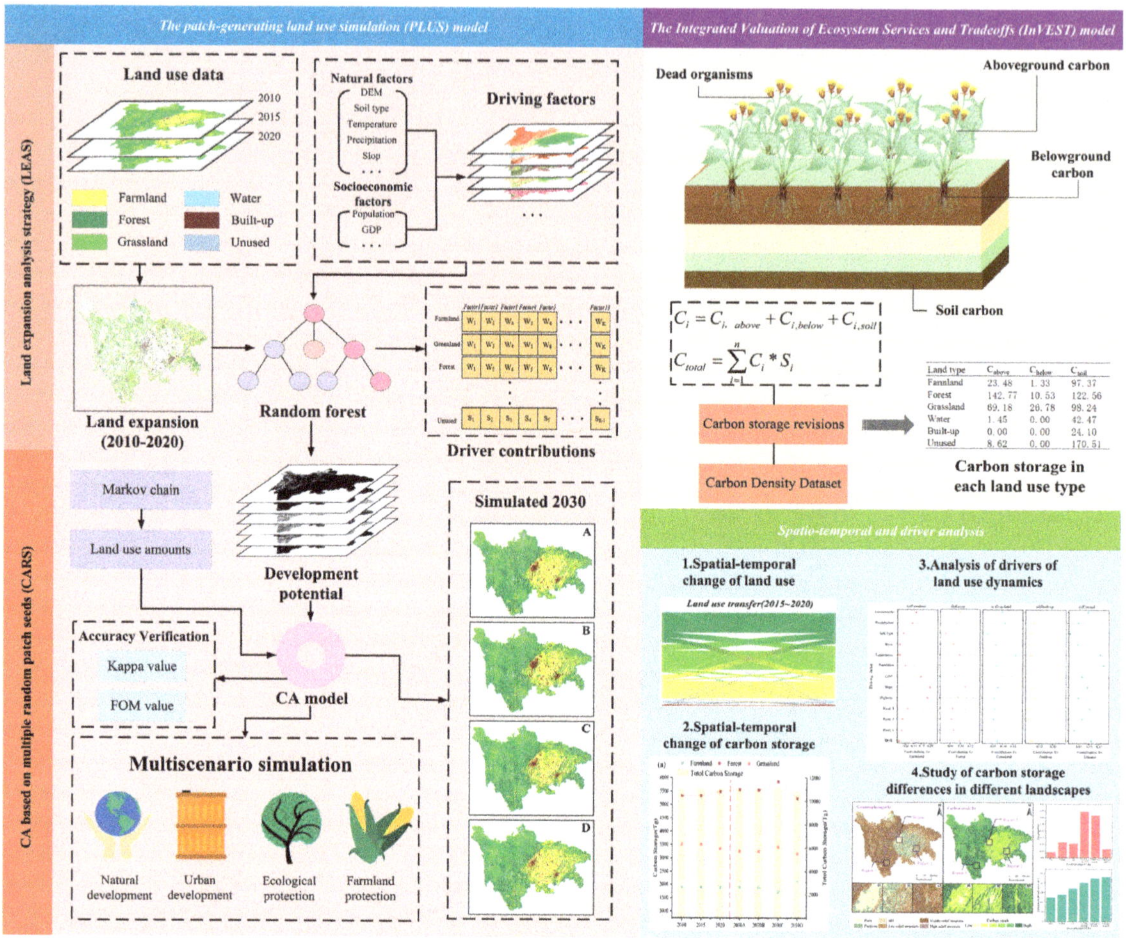


**Figure 2.** Flowchart.

### 2.3.1. Carbon Storage Estimation Modelling and Correction

To estimate regional carbon storage, the carbon assessment feature of the InVEST framework was utilized, which distinguishes carbon stock in terrestrial ecosystems into four elements: (1) carbon storage in aboveground biomass, (2) carbon storage in below-ground biomass, (3) carbon storage in organic carbon in soils, and (4) carbon storage in dead organisms [26]. The model excludes highly unstable carbon that can easily oxidise, decompose and be released into the atmosphere. In addition, only three carbon components were considered in this study because of limitations in data acquisition for the fourth component [27].

$$C_t = C_{t,above} + C_{t,below} + C_{t,soil} \tag{1}$$

$$C_{sum} = \sum_{t=1}^{n} C_t * S_t \tag{2}$$

where $C_t$ stands for the carbon density for land-use type $t$; the variables $C_{t,above}$, $C_{t,below}$, and $C_{t,soil}$ represent the three components contributing to carbon density within this land use classification: carbon in aboveground biomass, carbon in belowground biomass, and soil organic carbon at a depth of 0–100 cm; and $C_{sum}$ is the sum carbon storage in this region.

$S_t$ refers to area of tth type; variable n denotes quantity of defined land types in the study area, which is set to 6 for this analysis.

The above-ground biomass carbon storage and 0–100 cm soil carbon storage data were acquired from the 2010 China Terrestrial Ecosystem Carbon Density Dataset [28]. For the below-ground biomass component, the carbon storage was calculated using the biomass factor conversion method as follows:

$$C_{t,below} = a * b * W_{D,t} \tag{3}$$

where $W_{D,t}$ is above-ground biomass for land type *t*; *a* is the transformation factor; *b* is the proportion of below-ground to above-ground biomass; and above-ground biomass refers to the relevant research results [29,30], with a values of 5 and b values of 0.19, 0.25, and 5.2 for farmland, forest, and grassland, respectively.

In terms of the sources of carbon for these two components of aboveground biomass and soil, due to data limitations, it is difficult to directly obtain the relevant carbon storage parameters in our study region. To increase the reliability of carbon stock calculations, this paper uses a connection model between carbon density and precipitation to calibrate the results [31]. The relation model is as follows: (The final result calculated by InVEST carbon model is shown in Table A1).

$$C_{BP} = 6.789 * e^{0.0054 * P_{MA}} \tag{4}$$

$$C_{SP} = 3.3968 * P_{MA} + 3996.1 \tag{5}$$

$$K_{BP} = C_{BP1} / C_{BP2} \tag{6}$$

$$K_{SP} = C_{SP1} / C_{SP2} \tag{7}$$

where $C_{BP}$ is the carbon density determined from annual precipitation, $P_{MA}$ is the annual precipitation, $K_{BP}$ is the calibration factor for the precipitation coefficient of aboveground biomass carbon density, $K_{SP}$ is the calibration factor for the precipitation coefficient of soil carbon density, and $C_{BP1}$ and $C_{BP2}$ are the aboveground carbon density correction coefficients, which are taken from the mean annual precipitation of Sichuan–Chongqing region and the nationwide average annual precipitation separately in Formula (4). $C_{SP1}$ and $C_{SP2}$ are brought into Formula (5) according to the average annual precipitation of the Sichuan–Chongqing region as well as the national average annual precipitation.

### 2.3.2. Land Expansion Analysis Strategy (LEAS) Modelling

LEAS module of the PLUS model assesses the feasibility and suitability of land use by analysing the current status and trends of land use in a city or region. The model uses a random forest algorithm to examine land use expansion and its drivers on a case-by-case basis. It calculates the probability of development for different sites and assesses the contribution of each driver to land expansion over a specified time period, which is calculated using the formula shown below [25]:

$$P^{d}_{i,k(X)} = \sum_{n=1}^{M} I[h_n(X) = d] / M \tag{8}$$

where *X* represents a vector of various driving factors and *M* represents the decision tree number employed. *d* is a binary variable that takes values of 0 and 1, where 0 means that land type *k* does not allow the transfer of other land and 1 indicates that they can be transformed; $h_n(X)$ is its forecasted land type obtained by evaluating the decision tree *n* in the set. $I[h_n(X) = d]$ is a function that represents the probability that a land type *k* undergoes expansion in image *i* conditioned on the value of *d* and given the input X.

### 2.3.3. CA Based on Multiple Random Patch Seeds (CARS) Modelling

The CARS modelling within the PLUS framework, building upon the traditional binary state space, introduces multiple "types" or "species" that can interact with each other, enabling the simulation of intricate phenomena. This model is capable of simulating future land use across vast regions. In this study, the CARS model was employed to calculate the spatial distribution and development of future carbon storage by obtaining the development probability of different land to emulate and predict future land change. The total conversion probability $P_{O,i,k}^{d=1,t}$ of land type $k$ is given by [25]:

$$P_{O,i,k}^{d=1,t} = P_{i,k}^{d=1} * \Omega_{i,k}^{t} * D_k^t \tag{9}$$

$$\Omega_{i,k}^{t} = \frac{\mathrm{con}(c_i^{t-1} = k)}{n * n - 1} * w_k \tag{10}$$

$$D_k^t = \begin{cases} D_k^{t-1} & (\left|G_k^{t-1}\right| \leq \left|G_k^{t-2}\right|) \\ D_k^{t-1} * \frac{G_k^{t-2}}{G_k^{t-1}} & (G_k^{t-1} < G_k^{t-2} < 0) \\ D_k^{t-1} * \frac{G_k^{t-1}}{G_k^{t-2}} & (0 < G_k^{t-2} < G_k^{t-1}) \end{cases} \tag{11}$$

where $P_{O,i,k}^{d=1,t}$ is the likelihood that pixel $i$ will be a land of class $k$ in the future; $\Omega_{i,k}^{t}$ is the perturbation of this pixel to the surrounding neighbourhood; and $D_k^t$ represents the level of demand for land $k$. In Equations (10) and (11), con is the number of pixels of $k$th class in the $n * n$ window of the results of the final iteration; $w_k$ is the value of the weights among the land types, defaulting to 1; and $G_k^{t-1}$ and $G_k^{t-2}$ are the variance in requirements for land type $k$ between the present demand at iteration $t-1$ and the future demand at iteration $t-2$.

### 2.3.4. Setting of Land Development Scenarios

In order to clarify the differences in land use patterns and carbon storage under different development plans in topographically complex regions, and to capture the spatial and temporal variation patterns of carbon storage, this study sets up four different development scenarios applicable to the Sichuan–Chongqing region, taking into account the current development status and policies of the study region, as well as the land use conditions, and referring to the existing multi-scenario land use simulation framework [32,33]. The scenario setting process takes into account the differences in development policies and the rationality of scenario setting, and determines the differences in land use transfer under different scenarios [34]:

(1) Natural development scenario: Without any other modifications to the land use transfer matrix, the previous land transfer probabilities continued, but owing to the relative stability of water, the transfer of water is constrained in the forecasting process of this study.

(2) Urban development scenario: In this context, economic development is identified as the overarching objective, especially the development of built-up land. In the forecast analysis, the likelihood of conversion of built-up land to alternative land types decreases by 20 percent, while the likelihood of other land being converted to built-up areas increases by 20 percent.

(3) Ecological protection scenario: Green development is set as the primary objective in this scenario, with a 20 percent reduction in the odds of converting forests and grasslands to different types. The probability of converting land types other than building land to forest and grassland increases by 20 percent.

(4) Farmland protection scenario: The probability of transferring farmland to other land types decreases by 20 percent, whereas the probability of transferring other land types to farmland increases by 20 percent under this scenario, compared to the transfer probability under the natural development scenario.

## 3. Results

### *3.1. Spatial–Temporal Change of Land Use in the Sichuan–Chongqing Region from 2010 to 2020*

#### 3.1.1. Land Use Distribution in the Sichuan–Chongqing Region from 2010 to 2020

The spatial distribution of land use patterns in Sichuan–Chongqing region was relatively concentrated (Table 2, Figure 3). Forests, encompassing woodlands, open forests, and scrublands, dominated the region, occupying between 34.59% to 35.20% of the total area. They were primarily distributed in the western Sichuan plateau region, one of the most topographically complex regions in China, located somewhere in the middle of the Qinghai–Tibet plateau and the Sichuan Basin. Grassland, farmland, and unused land accounted for 30.63% to 32.52%, 27.27% to 27.89%, and 3.07% to 3.27% of the total area, respectively. Water and built-up land occupy a smaller proportion of the whole region, accounting for 0.86% to 1.04% and 0.83% to 1.52%, respectively. The majority of built-up was concentrated in Chengdu and its adjacent regions, as well as within the Chongqing Metropolitan Area and the Sichuan–Chongqing economic region. Notably, the proportion of farmland in this region surpassed the national average of 14.3%, whereas the proportion of built-up land remained below the national average of 4%.

**Table 2.** Distribution of land types by year in Sichuan and Chongqing (km$^2$).

| Land Type | 2010 | | 2015 | | 2020 | |
|---|---|---|---|---|---|---|
| | Area | Proportion | Area | Proportion | Area | Proportion |
| Farmland | 155,858 | 27.89% | 155,236 | 27.78% | 154,376 | 27.27% |
| Forest | 193,484 | 34.62% | 193,263 | 34.59% | 199,213 | 35.20% |
| Grassland | 181,707 | 32.52% | 181,516 | 32.48% | 173,366 | 30.63% |
| Water | 4831 | 0.86% | 5137 | 0.92% | 5892 | 1.04% |
| Built-up | 4661 | 0.83% | 6348 | 1.14% | 8604 | 1.52% |
| Unused | 18,262 | 3.27% | 17,303 | 3.10% | 17,352 | 3.07% |

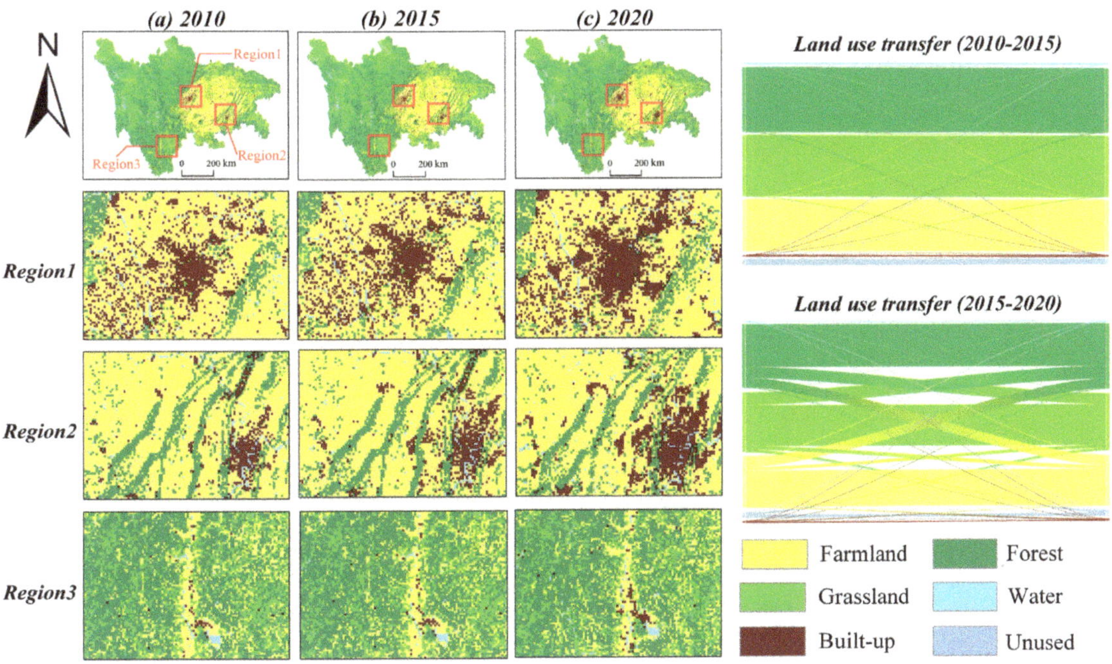


**Figure 3.** Land use distribution and land transfer in the Sichuan–Chongqing area.

From a temporal perspective, area changes between 2010 and 2020 varies considerably among land types, with the most notable change occurring in built-up, which increased from 0.83% to 1.52%, almost doubling in just a decade, expanding by 3943 km$^2$. Additionally, the area of forests (+5729 km$^2$, +0.58%) and watersheds (+1061 km$^2$, +0.18%) also showed increasing trends. In contrast, the other three land types in the region exhibited a downward trend during this period. In terms of area change, grassland decreased the most (−8341 km$^2$, −1.89%), followed by farmland (−1482 km$^2$, −0.62%) and unused land (−910 km$^2$, −0.20%).

#### 3.1.2. Land Use Transfer in the Sichuan–Chongqing Region from 2010 to 2020

To gain insight into the diversity of land use changes in the Sichuan–Chongqing region, as well as the specific direction of flow, a land use transfer study was conducted using a land use matrix (Table A2). The findings indicated that a total of 5,879,400 hm$^2$ of grassland was converted from 2010 to 2020. The majority of these conversions were into forest, accounting for 3,852,300 hm$^2$ or 65.52% of the total converted area. In addition, approximately 19.58% of the grassland area transferred to farmland, and approximately 1.63% was converted into built-up land. A total of 4,665,700 hm$^2$ of farmland was transferred out, of which forest was the largest source of transfer, with an area of 2,985,100 hm$^2$, accounting for 63.98%. Another 20.55% of the area was converted to grassland, with an area of 958,700 hm$^2$. A total of 646,600 hm$^2$ of built-up land was transferred, with 491,700 hm$^2$ (76.04%) sourced from farmland, marking it as a primary reason for the enlargement of built-up land over the past decade. Moreover, a discernible interplay existed between farmland, built-up land, and water, with 22.59 hm$^2$ of farmland transformed to water, while 18.81 hm$^2$ of water was converted back to farmland. Additionally, 49.17 hm$^2$ of farmland was converted to built-up land, and 18.08 hm$^2$ of built-up land was returned to farmland.

### *3.2. Multi-Scenario Simulation of Spatial and Temporal Carbon Storage Changes in the Sichuan–Chongqing Region from 2010 in 2030*

#### 3.2.1. Spatiotemporal Changes in Land Use in the Sichuan–Chongqing Region Under Multiple Scenarios

Taking the land use status of Sichuan–Chongqing region in 2020 as the initial state of simulation, we applied the CARS model combined with the Markov chain model to derive land use distribution situation of each scenario in 2030 (Table 3, Figure 4). The CARS model parameters are shown in Table A3. In this study, the 2010 land use data was employed to simulate and validate the 2020 land use data, resulting in a kappa coefficient of 0.892 and an FOM value of 0.191. These findings substantiate the feasibility of multi-scenario simulation in this study area. The research findings show that the spatial distribution pattern of land use will not undergo significant changes by 2030 under various development scenarios. In terms of development scenarios, the expansion directions of built-up in all four development scenarios are essentially the same, extending outwards from the areas in 2020, albeit at varying speeds. The urban development scenario has the fastest expansion rate, while the farmland protection scenario has the slowest expansion rate. In terms of regional development, the original urban clusters of Chengdu and Chongqing are undergoing the fastest expansion. Furthermore, regions such as Yibin, Luzhou, and Nanchong are developing increasingly prominent construction land clusters, especially under the urban development scenario. Changes in forest and grassland are primarily concentrated around the Sichuan Basin, expanding towards the centre of the basin, with particularly evident ecological protection scenarios. The protection of farmland slows down the expansion rate of forest and grassland towards the centre in the farmland protection scenario.

**Table 3.** The area of each land use ($km^2$) (A, B, C, and D represent the scenarios of natural development, urban development, ecological protection, and farmland protection).

| Land Type | Scenario | | | |
|---|---|---|---|---|
| | A | B | C | D |
| Farmland | 155,608 | 154,570 | 139,813 | 174,016 |
| Forest | 201,578 | 200,852 | 212,630 | 189,953 |
| Grassland | 168,498 | 168,298 | 176,160 | 163,638 |
| Water | 5892 | 5892 | 5892 | 5892 |
| Built-up | 10,444 | 12,588 | 9854 | 8882 |
| Unused | 16,783 | 16,603 | 14,454 | 16,422 |

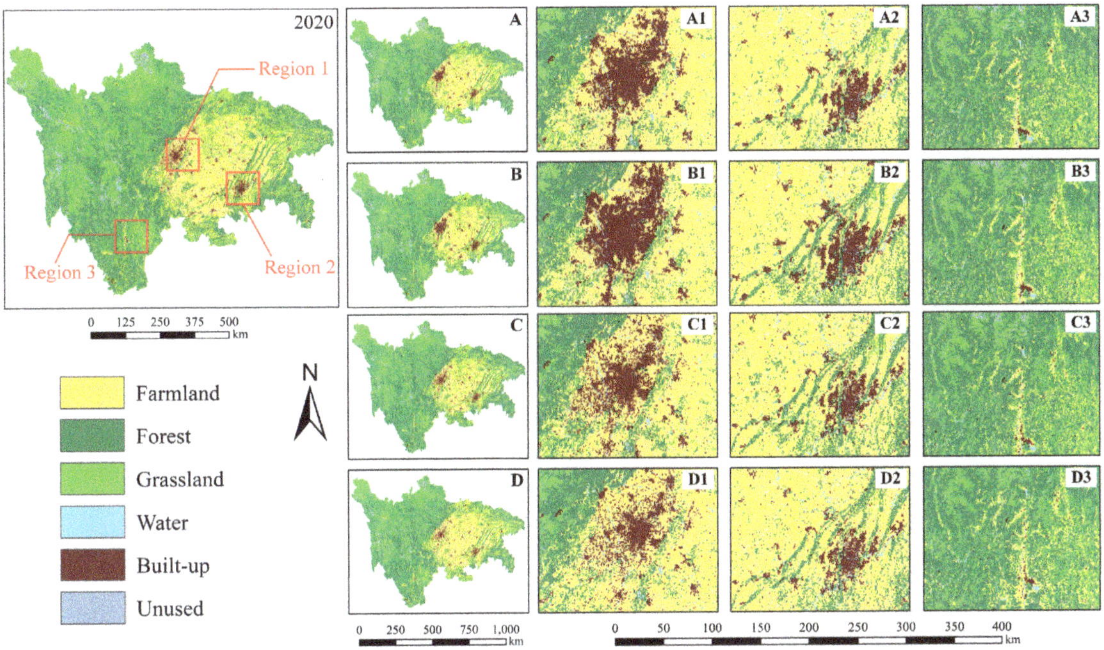


**Figure 4.** Land use under each scenario (**A–D** represent the scenarios of natural development, urban development, ecological protection, and farmland protection in the Sichuan–Chongqing region in 2030, respectively. Regions 1, 2, and 3 are enlarged example regions).

#### 3.2.2. Estimation and Prediction of the Spatial and Temporal Evolution of Carbon Storage in Sichuan–Chongqing Region

The InVEST model was employed to quantify carbon storage for individual land type as well as the entire Sichuan–Chongqing region's total carbon storage from 2010 to 2030 (Table 4, Figure 5). The findings showed that the total carbon storage decreased from 11,130.07 Tg to 11,105.88 Tg between 2010 and 2020, resulting in a decrease of 24.19 Tg in total carbon storage. It also shows a downward trend from 2020 to 2030 across various development scenarios, with the exception of the eco-development scenario. The total carbon stock in the natural development scenario is projected to be reduced by 20.00 Tg compared to 2020, with grassland being the type with the largest change in overall carbon storage under the scenario, with a projected decrease of 94.54 Tg, an increase of 4.43 Tg in carbon storage in built-up, and an increase of 65.24 Tg in carbon storage in forests. In the urban development scenario, total carbon storage decreases by 54.65 Tg, mainly attributable to lower grassland area, which reduces carbon storage in the grassland area by

98.42 Tg, and extensive land is transformed to built-up, this would result in an increase of 9.60 Tg of carbon storage on building sites. In the context of ecological protection scenario, the total carbon storage is projected to increase by 197.55 Tg, comprising significant gains. Specifically, the carbon storage of forest is anticipated to rise by 370.12 Tg, and grassland by 544.26 Tg. Conversely, farmland carbon storage is expected to decrease by 177.93 Tg, while that of unutilised land is projected to decline by 51.91 Tg. Despite maintaining a trend of expansion, the rate of growth in built-up land has slowed down, contributing an additional 3.01 Tg to carbon storage over a decade. The farmland protection scenario is anticipated to result in the most significant reduction in carbon stocks, with total carbon storage reduced by 220.39 Tg. This primary loss is attributable to greenfield, with the carbon of grassland projected to decline by 188.92 Tg and forest anticipated to reduce by 255.45 Tg.

**Table 4.** Estimated and projected carbon storage (Tg) of different land type and total carbon storage in the Sichuan–Chongqing region (A, B, C, and D represent the scenarios of natural development, urban development, ecological protection, and farmland protection).

| Land Type | Year | | | Scenario | | | |
|---|---|---|---|---|---|---|---|
| | 2010 | 2015 | 2020 | A | B | C | D |
| Farmland | 1904.27 | 1896.67 | 1886.17 | 1901.22 | 1888.54 | 1708.24 | 2126.13 |
| Forest | 5337.45 | 5331.35 | 5495.49 | 5560.73 | 5540.70 | 5865.61 | 5240.04 |
| Grassland | 3528.75 | 3525.04 | 3366.77 | 3272.23 | 3268.35 | 3421.03 | 3177.85 |
| Water | 21.22 | 22.56 | 25.88 | 25.88 | 25.88 | 25.88 | 25.88 |
| Built-up | 11.23 | 15.30 | 20.74 | 25.17 | 30.34 | 23.75 | 21.41 |
| Unused | 327.15 | 309.97 | 310.84 | 300.65 | 297.43 | 258.93 | 294.18 |
| Total | 11,130.07 | 11,100.89 | 11,105.88 | 11,085.88 | 11,051.23 | 11,303.43 | 10,885.49 |

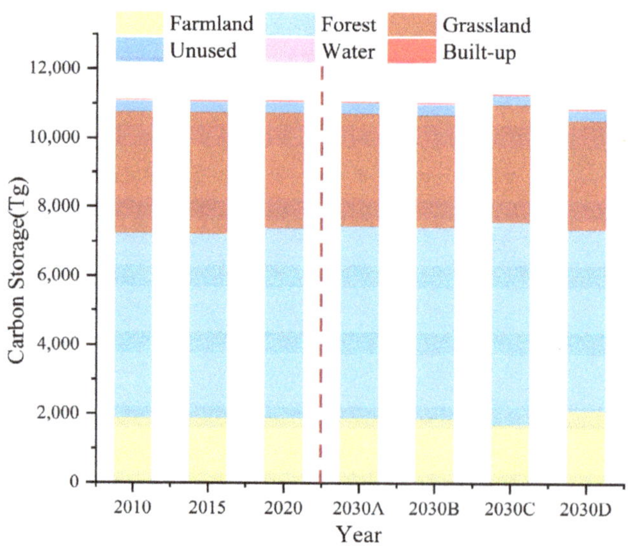


**Figure 5.** Estimated and projected carbon stocks from 2010 to 2030 (the left side of the red dotted line represents estimated carbon stocks from 2010 to 2020, while the right side displays projected carbon stock values in 2030 under various scenarios. A, B, C, and D represent the scenarios of natural development, urban development, ecological protection, and farmland protection in the Sichuan–Chongqing region in 2030, respectively).

### 3.3. The Drivers of Land Types and Carbon Storage Dynamics in the Sichuan–Chongqing Area

For clarification of the causes of land change in the Sichuan–Chongqing region, the LEAS module was applied to find the primary drivers of dynamic changes in each land type. The parameterisation of the LEAS module were detailed in Table A4. As illustrated in Figure 6, there are notable differences in the primary drivers of land type and carbon storage dynamics. For farmland, the primary driver was slope, with a contribution rate of 0.203, while economic factors such as GDP also had a substantial influence, contributing 0.147. Forests were primarily driven by slope (0.124) and DEM (0.098). Grassland dynamics were more closely tied to the climatic conditions of the Sichuan–Chongqing region, with temperature contributing 0.164. The drivers of built-up land dynamics differed significantly from those of farmland, forest, and grassland, as the most significant influences were no longer natural factors like slope and temperature, but rather economic factors such as GDP and population, contributing 0.264 and 0.248 respectively, far exceeding the contributions of other drivers. The drivers of unused land were diverse, with both natural factors like DEM and temperature and anthropogenic factors such as population size playing significant roles.

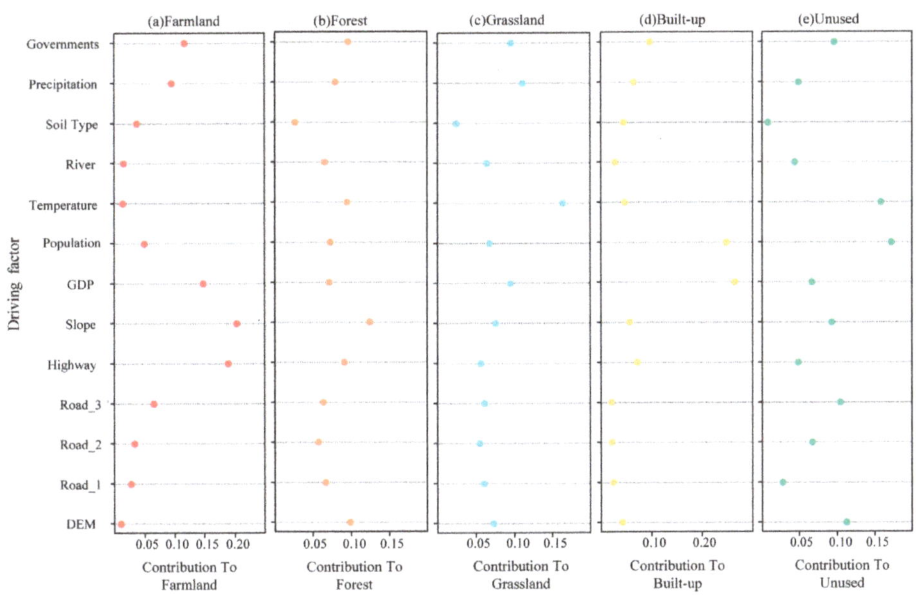


**Figure 6.** Contribution of land use and carbon storage change drivers to individual land types in the Sichuan–Chongqing region.

## 4. Discussion

### 4.1. Impacts of Natural Conditions and Anthropogenic Activities upon the Dynamics of Land Types

The distribution of land use in the Sichuan–Chongqing region is characterised by a high degree of complexity. Farmland, forest and grassland are found in patterns that are adapted to the natural conditions of the region. The Sichuan–Chongqing area is distinguished by strong topographical differences, an uneven temporal and spatial distribution of precipitation and significant regional heterogeneity [35]. In the western and southern regions of Sichuan, mountain plains and valleys alternating, and owing to the complex topography, higher average altitude and lower average annual temperature, these conditions are not conducive to the development of farmland. This has resulted in a relatively slower pace of agricultural development and a cultivated land area that is smaller than that of the eastern region. On the other hand, the Sichuan Basin in the eastern part of the

Sichuan–Chongqing region lies on a flat terrain and is abundant in water resources, providing unique natural conditions for farmland development, and the interactions and changes in these natural factors affect farmers' decisions concerning farmland use. In addition, the dynamics of forests and grasslands are closely related to natural factors. Western Sichuan has 60% of the forest resources and more than 90% of the grasslands in Sichuan Province, with forests consisting mainly of alpine coniferous forests with high resistance to cold and grasslands consisting mainly of thickets and grasslands. The spatial heterogeneity of these natural conditions has important implications for differences in the spatial distribution of carbon density [36].

In addition to natural conditions, human activities also have a major impact on land use dynamics, especially built-up land in the Sichuan–Chongqing region, where the driving force for built-up land is mainly population and GDP. Regions with large populations tend to generate new built-up land, and population growth drives urban development and expansion [37], which causes cities to continue to extend outwards, occupying increasingly undeveloped land. On the other hand, the area of Sichuan is unbalanced in its development, with economic and social development more rapid than agricultural development [38], and urban economic development requires more land to improve infrastructure such as housing, factories and commercial facilities. As industrialisation and urbanisation progress, some forests and grasslands may be used for built-up infrastructure. It is thus imperative to develop and utilise finite resources in a rational manner, formulate scientific planning and management measures, and give due consideration to factors such as topography, natural conditions and the ecological environment in order to achieve sustainable development.

### *4.2. The Impact of Urbanisation on the Change in Carbon Storage in the Sichuan–Chongqing Region*

Over the past decade, there has been intense urban expansion in the Sichuan–Chongqing region [39], resulting in an increasing demand for construction and industrial land and a decreasing area of collective rural land and farmland. Specifically, in the context of urbanisation, activities such as the logging of forests and the reclamation of wasteland have been observed, and the expansion of land development has become increasingly frequent, causing severe damage to the ecosystem. Areas of land with high carbon density, such as forests and grasslands, has been decreasing, while the size of built-up land with relatively low carbon density has been increasing dramatically, which has led to a decrease in the overall carbon sequestration capacity in the Sichuan–Chongqing region. Furthermore, excessive grazing and exploitation have also harmed grasslands and soils, leading to a decline in soil quality and a reduction in plant cover, which has significantly affected the ability to store and fix carbon. Between 2015 and 2020, the total carbon storage in this region slightly rebounded. This suggests that the region has accelerated its ecological development and environmental protection efforts in recent years. Effective measures, such as scientific planting, regular cleaning, and fire prevention, have been taken to improve the ecological level during the urbanisation process. Policies promoting forest growth and protection have increased carbon storage. The "double carbon" policy has also facilitated the conversion of farmland to more carbon-intensive forests and grasslands, resulting in significant progress.

### *4.3. Differences in Land Types and Carbon Storage Under the Complex Terrain of the Sichuan–Chongqing Region*

The diverse and intricate terrain units in the Sichuan–Chongqing region, including mountains, plateaus, hills, basins, and plains, result in rugged and uneven land surfaces, as well as heterogeneous vegetation distributions [40]. Topographic factors such as slope and DEM play key roles in land distribution patterns and carbon storage distributions in complex terrain areas. In this study, the degree of surface undulation was calculated from DEM data, and the landforms of the Sichuan–Chongqing region were classified into six categories on the basis of the degree of surface undulation: plains, plateaus, hills, low-relief mountains, middle-relief mountains, and high-relief mountains [41]. As shown in Figures 7 and 8, farmland and built-up land are concentrated in relatively flat plains and tableland areas that exhibit low carbon storage, such as Chengdu, Chongqing Municipality and the surrounding areas. In contrast, the land types in medium and high rolling hills are dominated by those with higher carbon storage per unit area, such as forests and grasslands. Examples of these landscapes include the western Sichuan plateau region. Overall, the distributions of land use and carbon storage in the Sichuan–Chongqing region are closely related to topographic relief. Carbon storage is lower in flat areas and higher in steep mountainous areas. This study's results provide an essential reference for land resource management and carbon storage assessment in the region. This study also offers a new perspective on the influence of complex topography on the carbon cycle. In the future, further investigations can be conducted on the interactions between topographic factors and other factors, such as vegetation type and soil properties, to comprehensively explore the mechanism of topographic influences on carbon storage distribution.

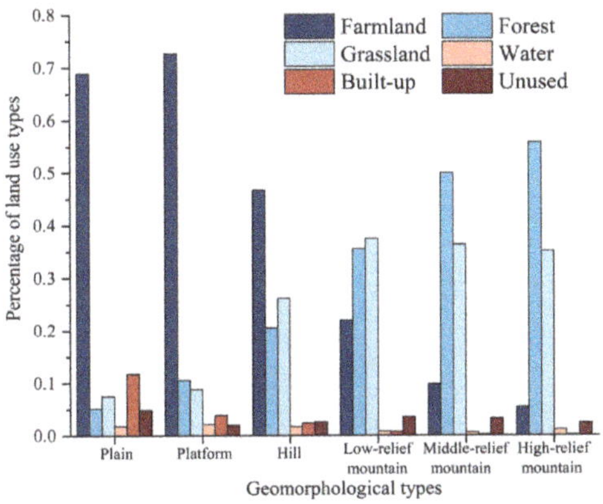


**Figure 7.** Land use distribution in the Sichuan–Chongqing region under various landforms.

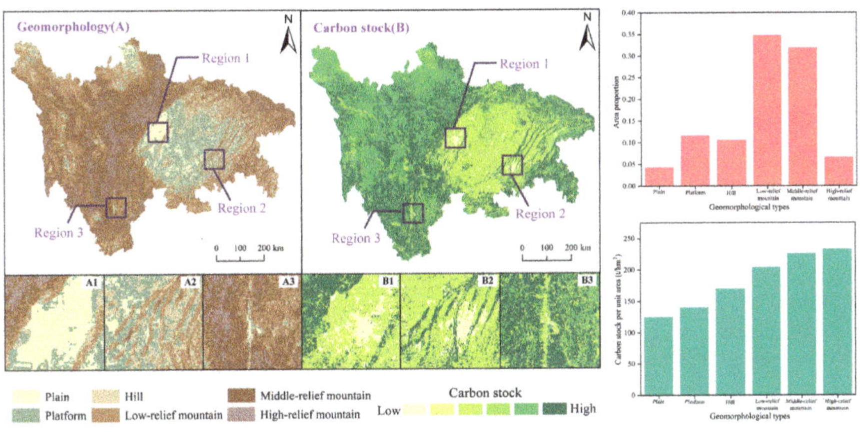


**Figure 8.** Carbon stock distribution and geomorphological distribution in the Sichuan–Chongqing region ((**A**,**B**) represent the geomorphological distribution and carbon stock distribution in the Sichuan–Chongqing region, respectively; Regions 1, 2 and 3 are enlarged example regions; and the statistical graphs on the right side represent the proportion of geomorphological landforms and the average carbon stock of geomorphological landforms, respectively).

### 4.4. The Uncertainty of the Research Study

In this study, the InVEST model was employed to calculate the carbon density of land types per unit area in the Sichuan–Chongqing region as a basis for carbon storage estimation. Given the disparate sources of carbon density data and references, as well as the discrepancies in the natural conditions of the specific study area and the human activity conditions, there are some discrepancies between the estimated regional carbon storage in this study and that of some previous relevant studies [42,43]. The InVEST model, however, is a simplified carbon density estimation model that divides the carbon pools into four categories and assumes that the carbon density per unit area of each land type remains constant. Nevertheless, it is important to note that carbon storage may change over time due to a number of factors, including natural conditions, human activities or other factors. In this study, to account for the differences in climatic conditions between the Sichuan–Chongqing region and other regions, precipitation was introduced to correct the InVEST model, with the aim of ensuring that the carbon storage estimation results are as close to the real values as possible. However, it should be noted that the final results inevitably contain some degree of error.

To explore the main drivers of land use change and carbon storage change in the Sichuan–Chongqing region, the PLUS model was used to select natural drivers and socioeconomic factors, such as DEM, slope, GDP, and population, that could be used to assess the drivers of change in land accurately and further analyse the drivers of carbon storage [44–46]. Moreover, the CA module was also used to model future land use under different policy priorities, with a kappa coefficient of 0.892 and a FOM value of 0.191 for simulating the 2020 results against the 2010 land use results prior to the formal simulation. However, although such accuracy indicates that the PLUS model has high precision and reliability and can be used to predict the future distribution of land use in the Sichuan–Chongqing region, when it is applied to project future land use scenarios, deviations from actual trends may occur and could be exacerbated by the lack of relevant data post-2020 for more precise estimations.

In light of the above, future research should more fully consider the conditions of natural and anthropogenic factors, utilise more accurate data to estimate carbon storage in combination with field surveys, and consider the spatial inhibition of land types and the age structure of vegetation among other factors which have an impact on carbon density. Furthermore, a greater number of factors should be considered during the process of multi-scenario modelling, and more flexible and accurate scenario rules should be formulated in conjunction with different policies with the aim of obtaining more accurate land use modelling results and calculating carbon storage with greater precision.

## 5. Conclusions

This study focuses on the Sichuan–Chongqing region. Land use data and carbon density data were used to calculate the carbon storage per unit area for each land type according to the InVEST model. The model was refined via annual precipitation data. The analysis of changes in carbon storage linked to land use transformations and the driving forces behind such changes from 2010 to 2020 was enabled by employing the PLUS model alongside DEM, GDP, population, and other relevant data. Additionally, the study examined differences in the distributions of land types across diverse landforms within the intricate topography of the Sichuan–Chongqing region and their associated carbon storage implications. A multi-scenario simulation for the land use projection in 2030 was conducted to assess carbon storage under various scenarios and explore temporal and spatial shifts in carbon storage trends. The following conclusions were obtained:

(1) The land types in the Sichuan–Chongqing region are concentrated, with forests being the dominant type. The proportion of built-up land exceeds China's average farmland share (14.3%) but is below the national average for built-up land (4%). Built-up land exhibited the most significant temporal fluctuations—increasing from 0.83% in 2010 to 1.52% in 2020—whereas areas of grassland, farmland, and unused land decreased.

(2) The total carbon storage for each land type was calculated using the modified InVEST model. Forestland had the highest carbon storage per unit area at 275.86 t/hm$^2$. From 2010 to 2020, the Sichuan–Chongqing region experienced a decline in total carbon storage

from 111,130.07 Tg to 11,105.88 Tg, indicating a reduction of 24.19 Tg. The PLUS model demonstrated a kappa value of 0.892 and an FOM value of 0.191. These results suggest that the PLUS model is appropriate for simulating the expected land use distribution in the Sichuan–Chongqing region. The results of our multi-scenario prediction show that except for the ecological development scenario, the total carbon storage in different development scenarios will decline. The farmland protection scenario shows the most significant decline in total carbon storage, with a projected reduction of 220.39 Tg.

(3) The decline in carbon storage is attributed primarily to the encroachment of urban development into agricultural land, forests, and grasslands, leading to a reduction in high-carbon-density regions and an increase in low-carbon-density areas. Natural factors, including slope and temperature, significantly influence agricultural land, forests, and grasslands, whereas economic factors such as GDP and population growth are driving forces behind the continuous expansion of urbanised areas. Moreover, the distribution of land use and carbon storage in the Sichuan–Chongqing region is intricately linked to its diverse topography, where flat regions present lower carbon storage levels and steep mountainous areas present higher carbon storage capacities.

**Author Contributions:** Conceptualization, Tiankuo Chu and Xiaoai Dai; methodology, Tiankuo Chu; validation, Tiankuo Chu, Jun Li, and Chengye Zhang; formal analysis, Tiankuo Chu; investigation, Tiankuo Chu; resources, Tiankuo Chu; data curation, Tiankuo Chu; writing—original draft preparation, Tiankuo Chu, Huan Huang, and Manling Peng; writing—review and editing, Jun Li, Chengye Zhang, Xiaoai Dai, and Yikun Qing; supervision, Xiaoai Dai; funding acquisition, Jun Li and Chengye Zhang All authors have read and agreed to the published version of the manuscript.

**Funding:** This work was founded by the National Natural Science Foundation of China Grant (42271480, 42371347) and The Science and Technology Development Plan Project of the Silk Road Economic Belt Innovation-Driven Development Pilot Zone and the Urumqi-Changji-Shihezi National Innovation Demonstration Zone Grant (2023LQY02).

**Institutional Review Board Statement:** Not applicable.

**Informed Consent Statement:** Not applicable.

**Data Availability Statement:** Data are contained within the article.

**Conflicts of Interest:** The authors declare no conflicts of interest.

## Appendix A

**Table A1.** Carbon density of each land use in Sichuan–Chongqing region (t/hm$^2$).

| Land Type | Above-Ground Carbon Storage | Underground Carbon Storage | Soil Carbon Storage |
|---|---|---|---|
| Farmland | 23.48 | 1.33 | 97.37 |
| Forest | 142.77 | 10.53 | 122.56 |
| Grassland | 69.18 | 26.78 | 98.24 |
| Water | 1.45 | 0.00 | 42.47 |
| Built-up | 0.00 | 0.00 | 24.10 |
| Unused | 8.62 | 0.00 | 170.51 |

**Table A2.** Land use transfer matrix for the Sichuan–Chongqing region, 2010–2020 (million hm$^2$).

| Land Type | | 2020 | | | | | | | |
|---|---|---|---|---|---|---|---|---|---|
| | | Grassland | Farmland | Built-up | Forest | Water | Unused | Total | Transfer out |
| 2010 | Grassland | 1229.13 | 115.13 | 4.39 | 385.23 | 9.56 | 73.63 | 1817.07 | 587.94 |
| | Farmland | 95.87 | 1092.01 | 49.17 | 298.51 | 22.59 | 0.43 | 1558.58 | 466.57 |
| | Built-up | 1.5 | 18.08 | 21.38 | 3.71 | 1.9 | 0.04 | 46.61 | 25.23 |
| | Forest | 330.98 | 295.81 | 7.47 | 1285.43 | 7.86 | 7.29 | 1934.84 | 649.41 |

**Table A2.** *Cont.*

| Land Type | 2020 | | | | | | | |
|---|---|---|---|---|---|---|---|---|
| | Grassland | Farmland | Built-up | Forest | Water | Unused | Total | Transfer out |
| Water | 5.13 | 18.81 | 3.51 | 4.34 | 15.45 | 1.07 | 48.31 | 32.86 |
| Unused | 71.05 | 3.92 | 0.12 | 14.91 | 1.56 | 91.06 | 182.62 | 91.56 |
| Total | 1733.66 | 1543.76 | 86.04 | 1992.13 | 58.92 | 173.52 | 5588.03 | 1853.57 |
| Transfer to | 504.53 | 451.75 | 64.66 | 706.7 | 43.47 | 82.46 | 1853.57 | |

**Table A3.** CA module parameter setting.

| Parameters | Neighbourhood Size: 5 | Patch Generate: 0.5 | Expansion Coefficient: 0.1 | | Percentage of Seeds: 0.0001 |
|---|---|---|---|---|---|
| Domain weight | Farmland 0.196 | Forest 0.130 | Grassland 0.427 | Built-up 0.140 | Unused 0.073 |

**Table A4.** LEAS module parameter setting.

| Parameters | Sampling Rate | Number of Regression Tree | Sample Module | mTry | Thread |
|---|---|---|---|---|---|
| | 0.02 | 20 | Random sampling | 13 | 4 |

## References


1. Tian, H.; Melillo, J.; Lu, C.; Kicklighter, D.; Liu, M.; Ren, W.; Xu, X.; Chen, G.; Zhang, C.; Pan, S.; et al. China's terrestrial carbon balance: Contributions from multiple global change factors. *Glob. Biogeochem. Cycles* **2011**, *25*, GB1007. [CrossRef]
2. Yang, H.; Huang, J.; Liu, D. Linking climate change and socioeconomic development to urban land use simulation: Analysis of their concurrent effects on carbon storage. *Appl. Geogr.* **2020**, *115*, 102135. [CrossRef]
3. Midgley, G.F.; Bond, W.J.; Kapos, V.; Ravilious, C.; Scharlemann, J.P.W.; Woodward, F.I. Terrestrial carbon stocks and biodiversity: Key knowledge gaps and some policy implications. *Curr. Opin. Environ. Sustain.* **2010**, *2*, 264–270. [CrossRef]
4. Quay, P.D.; Tilbrook, B.; Wong, C.S. Oceanic Uptake of Fossil Fuel $CO_2$: Carbon-13 Evidence. *Science* **1992**, *256*, 74–79. [CrossRef]
5. Abu-Hashim, M.; Elsayed, M.; Belal, A.-E. Effect of land-use changes and site variables on surface soil organic carbon pool at Mediterranean Region. *J. Afr. Earth Sci.* **2016**, *114*, 78–84. [CrossRef]
6. Houghton, R.; Goodale, C. Effects of land-use change on the carbon balance of terrestrial ecosystems. *Ecosyst. Land Use Change* **2004**, *9*, 85–98. [CrossRef]
7. Fang, J.; Chen, A.; Peng, C.; Zhao, S.; Ci, L. Changes in forest biomass carbon storage in China between 1949 and 1998. *Science* **2001**, *292*, 2320–2322. [CrossRef]
8. Guo, J.; Zhang, R.P.; Wang, R.H.; Yusupujiang, A.; Asiyemu, T.; Zhang, D.Y. Estimation of carbon storage and carbon density of forest vegetation in Ili River Valley, Xinjiang. In Proceedings of the IOP Conference Series: Earth and Environmental Science, Beijing, China, 7–8 July 2016; pp. 12–27. [CrossRef]
9. Jung, M.; Reichstein, M.; Margolis, H.A.; Cescatti, A.; Richardson, A.D.; Arain, M.A.; Arneth, A.; Bernhofer, C.; Bonal, D.; Chen, J.Q.; et al. Global patterns of land-atmosphere fluxes of carbon dioxide, latent heat, and sensible heat derived from eddy covariance, satellite, and meteorological observations. *J. Geophys. Res. Biogeosci.* **2011**, *116*, G00J07. [CrossRef]
10. Pan, Y.D.; Birdsey, R.A.; Fang, J.Y.; Houghton, R.; Kauppi, P.E.; Kurz, W.A.; Phillips, O.L.; Shvidenko, A.; Lewis, S.L.; Canadell, J.G.; et al. A Large and Persistent Carbon Sink in the World's Forests. *Science* **2011**, *333*, 988–993. [CrossRef]
11. Piao, S.; He, Y.; Wang, X.; Chen, F. Estimation of China's terrestrial ecosystem carbon sink: Methods, progress and prospects. *Sci. China Earth Sci.* **2022**, *65*, 641–651. [CrossRef]
12. Sun, B.Q.; Du, J.Q.; Chong, F.F.; Li, L.J.; Zhu, X.Q.; Zhai, G.Q.; Song, Z.B.; Mao, J.L. Spatio-Temporal Variation and Prediction of Carbon Storage in Terrestrial Ecosystems in the Yellow River Basin. *Remote Sens.* **2023**, *15*, 3866. [CrossRef]
13. Melillo, J.M.; McGuire, A.D.; Kicklighter, D.W.; Moore, B.; Vorosmarty, C.J.; Schloss, A.L. Global climate-change and terrestrial net primary production. *Nature* **1993**, *363*, 234–240. [CrossRef]
14. Sun, Q.L.; Li, B.L.; Zhang, T.; Yuan, Y.C.; Gao, X.Z.; Ge, J.S.; Li, F.; Zhang, Z.J. An improved Biome-BGC model for estimating net primary productivity of alpine meadow on the Qinghai-Tibet Plateau. *Ecol. Model.* **2017**, *350*, 55–68. [CrossRef]
15. Potter, C.S.; Randerson, J.T.; Field, C.B.; Matson, P.A.; Vitousek, P.M.; Mooney, H.A.; Klooster, S.A. Terrestrial ecosystem production—A process model-based on global satellite and surface data. *Glob. Biogeochem. Cycles* **1993**, *7*, 811–841. [CrossRef]

16. Zhang, C.; Li, F.; Li, J.; Xing, J.; Yang, J.; Guo, J.; Du, S. Recognition of land use on open-pit coal mining area based on DeepLab v3+ and GF-2 high-resolution images. *Coal Geol. Explor.* **2022**, *50*, 94–103. [CrossRef]
17. Yan, M.; Li, Z.Y.; Tian, X.; Chen, E.X.; Zhang, W.F.; Guo, Y.; Li, C.M. Simulation of carbon flux of forest ecosystem by Biome-BGC and MODIS-PSN models. In Proceedings of the IEEE Geoscience and Remote Sensing Symposium, Quebec City, QC, Canada, 13–18 July 2014; pp. 1065–1068. [CrossRef]
18. Guo, L.B.; Gifford, R.M. Soil carbon stocks and land use change: A meta analysis. *Glob. Change Biol.* **2022**, *8*, 345–360. [CrossRef]
19. Zhao, Z.; Liu, G.; Mou, N.; Xie, Y.; Xu, Z.; Li, Y. Assessment of carbon storage and its influencing factors in Qinghai-Tibet Plateau. *Sustainability* **2018**, *10*, 1864. [CrossRef]
20. Zhao, M.; He, Z.; Du, J.; Chen, L.; Lin, P.; Fang, S. Assessing the effects of ecological engineering on carbon storage by linking the CA-Markov and InVEST models. *Ecol. Indic.* **2019**, *98*, 29–38. [CrossRef]
21. Babbar, D.; Areendran, G.; Sahana, M.; Sarma, K.; Raj, K.; Sivadas, A. Assessment and prediction of carbon sequestration using Markov chain and InVEST model in Sariska Tiger Reserve, India. *J. Clean Prod.* **2021**, *278*, 123333. [CrossRef]
22. Krcmar, E.; Stennes, B.; van Kooten, G.C.; Vertinsky, I. Carbon sequestration and land management under uncertainty. *Eur. J. Oper. Res.* **2001**, *135*, 616–629. [CrossRef]
23. Chen, Y.; Cui, J.; He, L.; Men, M. Optimal Simulation of the Land-Use Structure Based on the Local Ecological Environment: The Case of Huanghua City in Hebei Province, China. *Rev. Bras. Cienc. Solo* **2019**, *43*, e0180063. [CrossRef]
24. Liu, X.; Liang, X.; Li, X.; Xu, X.; Ou, J.; Chen, Y.; Li, S.; Wang, S.; Pei, F. A future land use simulation model (FLUS) for simulating multiple land use scenarios by coupling human and natural effects. *Landsc. Urban Plan.* **2017**, *168*, 94–116. [CrossRef]
25. Liang, X.; Guan, Q.; Clarke, K.C.; Liu, S.; Wang, B.; Yao, Y. Understanding the drivers of sustainable land expansion using a patch-generating land use simulation (PLUS) model: A case study in Wuhan, China. *Comput. Environ. Urban Syst.* **2021**, *85*, 101569. [CrossRef]
26. Sharp, R.; Tallis, H.; Ricketts, T.; Guerry, A.; Wood, S.; Chaplin-Kramer, R.; Nelson, E.; Ennaanay, D.; Wolny, S.; Olwero, N. *InVEST 3.10 User's Guide*; Collaborative Publication by the Natural Capital Project; Stanford University: Stanford, CA, USA; The University of Minnesota: Minneapolis, MN, USA; The Nature Conservancyy: Arlington, VA, USA; World Wildlife Fund: Morges, Switzerland, 2018.
27. Zhou, J.J.; Zhao, Y.R.; Huang, P.; Zhao, X.; Feng, W.; Li, Q.Q.; Xue, D.X.; Dou, J.; Shi, W.; Wei, W.; et al. Impacts of ecological restoration projects on the ecosystem carbon storage of inland river basin in arid area, China. *Ecol. Indic.* **2020**, *118*, 106803. [CrossRef]
28. Xu, L.; He, N.; Yu, G. A dataset of carbon density in Chinese terrestrial ecosystems (2010s). *China Sci. Data* **2019**, *4*, 90–96. [CrossRef]
29. Fang, J.-Y.; Wang, Z.M. Forest biomass estimation at regional and global levels, with special reference to China's forest biomass. *Ecol. Res.* **2001**, *16*, 587–592. [CrossRef]
30. Huang, M.; Ji, J.; Cao, M.; Li, K. Modeling study of vegetation shoot and root biomass in China. *Acta Ecol. Sin.* **2006**, *26*, 4156–4163. [CrossRef]
31. Alam, S.A.; Starr, M.; Clark, B.J. Tree biomass and soil organic carbon densities across the Sudanese woodland savannah: A regional carbon sequestration study. *J. Arid. Environ.* **2013**, *89*, 67–76. [CrossRef]
32. Gao, L.; Tao, F.; Liu, R.; Wang, Z.; Leng, H.; Zhou, T. Multi-scenario simulation and ecological risk analysis of land use based on the PLUS model: A case study of Nanjing. *Sustain. Cities Soc.* **2022**, *85*, 104055. [CrossRef]
33. Lou, Y.; Yang, D.; Zhang, P.; Zhang, Y.; Song, M.; Huang, Y.; Jing, W. Multi-Scenario Simulation of Land Use Changes with Ecosystem Service Value in the Yellow River Basin. *Land* **2022**, *11*, 992. [CrossRef]
34. Wu, Y.Z.; Zhang, X.L.; Shen, L.Y. The impact of urbanization policy on land use change: A scenario analysis. *Cities* **2011**, *28*, 147–159. [CrossRef]
35. Ji, T.; Li, G.; Yang, H.; Liu, R.; He, T. Comprehensive drought index as an indicator for use in drought monitoring integrating multi-source remote sensing data: A case study covering the Sichuan-Chongqing region. *Int. J. Remote Sens.* **2018**, *39*, 786–809. [CrossRef]
36. Gao, X.; Gray, J.; Cohrs, C.W.; Cook, R.; Albaugh, T.J. Longer greenup periods associated with greater wood volume growth in managed pine stands. *Agric. For. Meteorol.* **2021**, *297*, 108237. [CrossRef]
37. Zou, D.; Cong, H. Data driven based coordinated development pattern and influencing factor analysis using spatial and temporal data. *Clust. Comput.* **2019**, *22*, S9991–S10007. [CrossRef]
38. Xiao, M.; Zhang, Q.; Qu, L.; Hussain, H.A.; Dong, Y.; Zheng, L. Spatiotemporal Changes and the Driving Forces of Sloping Farmland Areas in the Sichuan Region. *Sustainability* **2019**, *11*, 906. [CrossRef]
39. Song, Y.; Jiao, L.; Liu, J.; Xu, G. Analysis of the factors affecting the degree of urban expansion: Taking Wuhan city as an example. *Geomat. Inf. Sci. Wuhan Univ.* **2021**, *46*, 417–426. [CrossRef]
40. Shao, J.A.; Li, Y.; Ni, J. The characteristics of temperature variability with terrain, latitude and longitude in Sichuan-Chongqing Region. *J. Geogr. Sci.* **2012**, *22*, 223–244. [CrossRef]
41. Cheng, W.; Zhou, C.; Chai, H.; Zhao, S.; Liu, H.; Zhou, Z. Research and compilation of the Geomorphologic Atlas of the People's Republic of China (1:1,000,000). *J. Geogr. Sci.* **2011**, *21*, 89–100. [CrossRef]
42. Wu, D.; Zhu, K.; Zhang, S.; Huang, C.; Li, J. Evolution Analysis of Carbon Storage in Chengdu-Chongqing Economic Zone Based on PLUS Model and InVEST Model. *Ecol. Environ. Monit. Three Gorges* **2022**, *7*, 1–18. [CrossRef]

43. Xiang, M.; Wang, C.; Tan, Y.; Yang, J.; Duan, L.; Fang, Y.; Li, W.; Shu, Y.; Liu, M. Spatio-temporal evolution and driving factors of carbon storage in the Western Sichuan Plateau. *Sci. Rep.* **2022**, *12*, 8114. [CrossRef]
44. Li, P.; Chen, J.; Li, Y.; Wu, W. Using the InVEST-PLUS Model to Predict and Analyze the Pattern of Ecosystem Carbon storage in Liaoning Province, China. *Remote Sens.* **2023**, *15*, 4050. [CrossRef]
45. Wang, Z.; Zeng, J.; Chen, W. Impact of urban expansion on carbon storage under multi-scenario simulations in Wuhan, China. *Environ. Sci. Pollut. Res.* **2022**, *29*, 45507–45526. [CrossRef] [PubMed]
46. Zhu, L.; Song, R.; Sun, S.; Li, Y.; Hu, K. Land use/land cover change and its impact on ecosystem carbon storage in coastal areas of China from 1980 to 2050. *Ecol. Indic.* **2022**, *142*, 109178. [CrossRef]

remote sensing

MDPI

Article

# The Link between Surface Visible Light Spectral Features and Water–Salt Transfer in Saline Soils—Investigation Based on Soil Column Laboratory Experiments

Shaofeng Qin [1,2,†], Yong Zhang [3,4,†], Jianli Ding [1,2,5,*], Jinjie Wang [1,2], Lijing Han [6], Shuang Zhao [1,2] and Chuanmei Zhu [1,2]

1 College of Geography and Remote Sensing Sciences, Xinjiang University, Urumqi 830017, China; 107552101118@stu.xju.edu.cn (S.Q.); wangjj@xju.edu.cn (J.W.); zhaoshuang@stu.xju.edu.cn (S.Z.); zcm_rs@xju.edu.cn (C.Z.)
2 Xinjiang Key Laboratory of Oasis Ecology, Xinjiang University, Urumqi 830017, China
3 Visiontek Inc. Wuhan, 6 Phoenix Avenue, Wuhan 430205, China; zyong@whu.edu.cn
4 School of Electronics and Information Engineering, Wuzhou University, Wuzhou 543003, China
5 Xinjiang Institute of Technology, Aksu 843100, China
6 College of Geodesy and Geomatics, Shandong University of Science and Technology, Qingdao 266590, China; hanlijing@sdust.edu.cn
* Correspondence: dingjl@xju.edu.cn
† These authors contributed equally to this work.

**Abstract:** Monitoring soil salinity with remote sensing is difficult, but knowing the link between saline soil surface spectra, soil water, and salt transport processes might help in modeling for soil salinity monitoring. In this study, we used an indoor soil column experiment, an unmanned aerial vehicle multispectral sensor camera, and a soil moisture sensor to study the water and salt transport process in the soil column under different water addition conditions and investigate the relationship between the soil water and salt transport process and the spectral reflectance of the image on the soil surface. The observation results of the soil column show that the soil water and salt transportation process conforms to the basic transportation law of "salt moves together with water, and when water evaporates, salt is retained in the soil weight". The salt accumulation phenomenon increases the image spectral reflectance of the surface layer of the soil column, while soil temperature has no effect on the reflectance. As the water percolates down, water and salt accumulate at the bottom of the soil column. The salinity index decreases instantly after the addition of brine and then tends to increase slowly. The experimental results indicate that this work can capture the relationship between the water and salt transport process and remote sensing spectra, which can provide theoretical basis and reference for soil water salinity monitoring.

**Keywords:** soil salinization; visible spectrum; small sample experiment; effects of boundary layer

## 1. Introduction

Soil salinization refers to the phenomenon of salt ions in the soil that accumulate in the soil mass as a result of water movement, and is one of the major forms of land degradation in the world [1–3]. Crops growing in high-saline soils are subjected to salt stress due to low osmotic potential [4–6]. High soil salinity reduces vegetation productivity and decomposition capacity, posing a growing threat to food security and ecosystem services [7,8]. Therefore, monitoring the spatial and temporal distribution of salinization in surface soils is essential for the implementation of appropriate measures for land resource allocation, soil improvement, and ecological protection [9].

Traditional soil salinity surveys require the collection of dense soil sample points utilizing spatial interpolation, which is time-consuming and laborious [10]. At present, remote sensing technology has become the mainstream means of soil salinity investigation, with its advantages of easy access to image information and rich wavebands [11–13]. Remote sensing technology is primarily used to extract image information at sampling points and calculate the salinity index, which is highly correlated with soil salinity [14,15]. This process is supplemented with environmental variables related to the soil-forming environment to construct models, such as multiple linear regression, random forests, XGBoost, and other machine learning models, to accurately invert surface salinity [15–17]. However, satellite optical sensors only scan the soil surface and cannot penetrate the soil, but soil surveys should involve the entire soil profile [18,19]. According to Farifteh et al. [20], a means to break through this limitation is to use physical models of solute transport capable of monitoring the entire soil layer in conjunction with remote sensing modeling approaches. Many studies have confirmed the applicability of physical models in assessing water and salt changes in soil layers at different burial depths [21–23]. General physical models of soil water and salt transport are based on Darcy's law and are constrained by realistic physical boundary conditions to simulate soil water and salt movement processes.

Clarifying the relationship between the physical model and the remote sensing model is essential to solve the problem posed by Farifteh et al. [20]. The use of traditional in situ field observations to obtain soil water and salt information and explore its relationship with satellite imagery is frequently affected by weather conditions, leading to missing data [24]. An alternative approach is to use indoor soil column experiments, where water and salt changes in the soil column are controlled by adjusting environmental conditions similar to those in the field, and the surface salinity information characteristics can be observed using cameras, spectrometers, and other instruments [25]. The effects of relative humidity, ambient temperature, and salt type on the salt precipitation dynamics on the soil surface were investigated by designing soil column experiments. The effects of different measurement heights on the spectral measurements of saline soils were investigated by using a soil column experiment. In view of the previous studies, the relationship between water and salt transport processes and remote sensing spectra can be explored by designing soil column experiments using equipment such as unmanned aircraft sensors [26].

The purpose of this study is to explore the relationship between water and salt transport processes and remote sensing spectra to provide theoretical support for the remote sensing monitoring of soil salinization, and to provide a reference for coupling the physical model of soil water and salt transport with a remote sensing model. The main objectives of this study are as follows: (1) To apply the law of the soil water–salt transport process under the condition of adding water with different salinity to the soil column. (2) To investigate the relationship between soil water and salt transport processes and the spectral reflectance of soil surface images. (3) To determine the relationship between the water–salt transport process and the salinity index of the soil surface.

## 2. Materials and Methods

### 2.1. Design of the Experimental Setup for the Indoor Soil Column Experiment

According to the soil column experiment of Shokri-Kuehni et al. [25], as shown in Figure 1a, a plexiglas tube with a height of 60 cm and an outer diameter of 20 cm was used for this experiment. Figure 1b shows the precipitation deposition of soil salts on the surface over time. Small holes were made in the bottom of the soil column for drainage, and for installation of the soil water and salt monitoring instrument, 5 cm diameter holes were drilled laterally at 7 cm, 22 cm, 37 cm, and 52 cm from the upper edge of the plexiglas tube, named Soil layer-1, Soil layer-2, Soil layer-3, and Soil layer-4 in order of depth. Soil volumetric water content, bulk soil electrical conductivity ($EC_b$, us/cm), and temperature (°C) were obtained using a soil moisture monitoring instrument (FT-TS400, Shandong Fengtu IOT Technology Co., Ltd., Weifang, China). The soil moisture monitoring instrument was factory-calibrated and validated. Data were automatically collected every 5 min. Soil volumetric water content: range: 0–100%; accuracy: ±3%. Bulk soil electrical conductivity: range: 0–20,000 μS/cm; accuracy: ±3% (0–10,000 μS/cm), ±5% (10,000–20,000 μS/cm); resolution: 10 μS/cm (0–10,000 μS/cm), 50 μS/cm (10,000–20,000 μS/cm). Temperature: range: −40 to +125 °C; accuracy: ±0.5 °C; resolution: 0.1 °C. The experiment was designed with three soil columns, each filled and compacted with sandy soil of different water content and salinity gradients (see Table 1 for initial water content and salinity), with the upper surface 2 cm from the upper edge of the soil column. The salt we used is a mixture of sodium chloride and sodium sulfate. The milligram equivalent ratio of chloride to sulfate ions is 0.676056. The weight of the soil column was weighed below the column using an electronic scale (accuracy ±0.01 kg) for measuring evaporation, and the weight was weighed twice a day at regular intervals in the morning and evening. The laboratory was supplied with radiators, so there was little difference in temperature and humidity between day and night, with an average room air temperature of about 25.33 ± 0.83 °C and an average humidity of 22.24 ± 3.42% throughout the experiment. Soil texture was measured by a Malvern laser particle size analyzer (Mastersizer 3000; Malvern Panalytical, Malvern, UK) after pretreatment of soil column soil. The measured soil texture contained 0.47%, 4.38%, and 95.15% clay, silt, and sand, respectively, and the ring knife method measured a field capacity of 1.537 g/cm$^3$, making the soil texture type homogeneous sandy (Table 2). Soil hydraulic parameters were measured using the centrifugal method (H-1400 pF centrifuge for soil, Kokusan Corporation, Kyoto, Japan); the soil moisture characteristic curve was fitted using the Pyswr module in Python 3.9 and the fitted curve is shown in Figure 2.

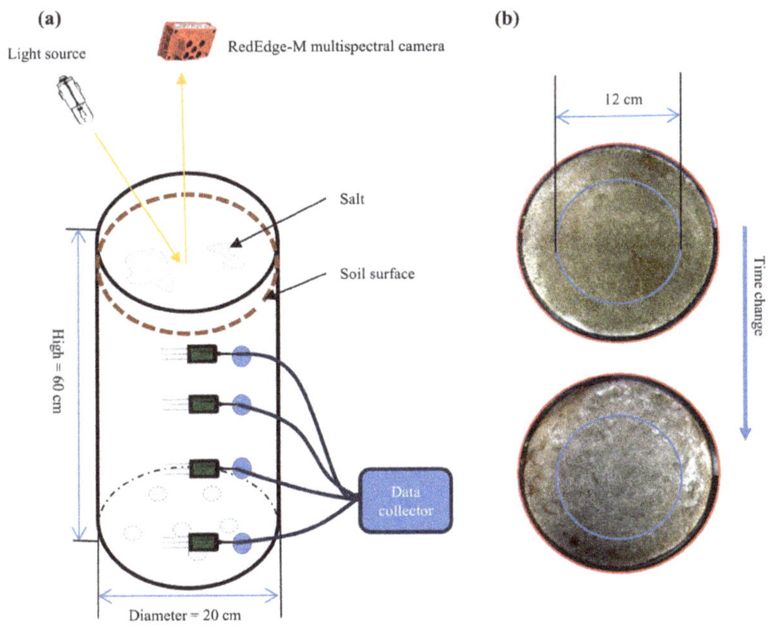


**Figure 1.** Schematic diagram of the indoor soil column experimental setup. (**a**) shows a schematic diagram of the installation setup of the soil column dimensions, light source, camera, and soil sensor. (**b**) shows a true-color image of the soil surface as it changes over time and soil salts crystallize and precipitate.

**Table 1.** Initial water content and salinity of soil columns.

| Soil Column | Layer Name | Initial Water Content (%) | Initial Salinity (μS/cm) |
|---|---|---|---|
| Soil column A | Soil layer-1 | 1.704 | 0.000 |
| | Soil layer-2 | 1.280 | 0.000 |
| | Soil layer-3 | 2.093 | 0.000 |
| | Soil layer-4 | 1.160 | 0.000 |
| Soil column B | Soil layer-1 | 9.020 | 260.000 |
| | Soil layer-2 | 6.660 | 268.150 |
| | Soil layer-3 | 2.800 | 143.000 |
| | Soil layer-4 | 1.704 | 1.000 |
| Soil column C | Soil layer-1 | 16.675 | 647.250 |
| | Soil layer-2 | 11.255 | 390.750 |
| | Soil layer-3 | 11.730 | 486.250 |
| | Soil layer-4 | 15.490 | 561.500 |

**Table 2.** Soil physical properties.

| Soil Type | Upper Cumulative Results (2 μm) | Results in the Interval (2, 50 μm) | Results in the Interval (50, 2000 μm) | Bottom Cumulative Result (2000 μm) | Specific Surface Area Result |
|---|---|---|---|---|---|
| | 0.47 | 4.38 | 95.15 | $1 \times 10^{-11}$ | 37.67 |

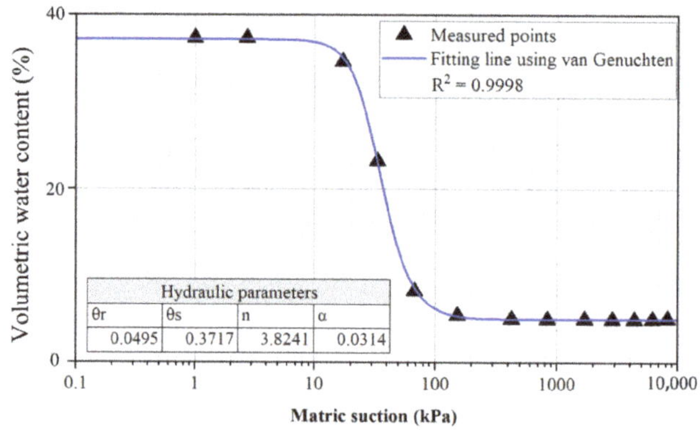


**Figure 2.** Characteristic soil moisture profiles of soil samples from soil columns measured by centrifugation.

In order to obtain continuous soil spectral data during the experimental period, a Micasense RedEdge-M multispectral camera (RedEdge-M, MicaSense Inc., Seattle, WA, USA) was used to capture multispectral images of the soil surface with detailed spectral information shown in Table 3, referring to the study of Shi et al. [27]. The camera was fixed at a height of 50 cm from the soil surface using a bracket and the camera was used in the manual mode, with the angle of capture being perpendicular to the soil surface. A halogen lamp was used to provide a light source to simulate sunlight; the angle of the light source was 30° from the shooting angle, and the light source was 45 cm from the center of the soil surface. At the beginning of the experiment, the soil surface was photographed under darkroom conditions, and the initial image was taken once before the addition of saline water, and then hourly due to the rapid changes in the soil surface in the early part of the experiment. Because the evaporation process slowed down significantly after about 2 days, the soil surface changed more slowly and was photographed about once every 2 or 4 h, and was not photographed at night. The saline water addition experiments were conducted three times, with the first experiment lasting 60 h and the second and third experiments totaling 80 h. In the first experiment, the water addition was set at 1 mm of precipitation or irrigation, and the salinity was 1 g/L. In the second and third experiments, the water addition was set at 2 mm and 4 mm, and the salinity was 2 g/L. The soil surface was not changed at night.

**Table 3.** Micasense RedEdge-M camera band spectral information.

| Band Number | Band Name | Center Wavelength (nm) | Bandwidth (nm) |
|---|---|---|---|
| Band 1 | Blue band | 475 | 20 |
| Band 2 | Green band | 560 | 20 |
| Band 3 | Red band | 668 | 10 |
| Band 4 | Near-infrared band | 840 | 40 |
| Band 5 | RedEdge band | 717 | 10 |

### 2.2. Multispectral Camera Data Processing

The RedEdge-M multispectral camera utilizes a multi-lens imaging mechanism in a multi-camera imaging system. The RedEdge-M multispectral camera contains five independent imaging lenses, each of which performs imaging independently and stores the results as separate image files. Therefore, when combining the images of different bands, there will be an obvious misalignment offset phenomenon, which requires image alignment work.

Virtual imaging is a method of re-projecting each lens onto an ideal virtual image by utilizing the positional relationship between them [28,29]. When processing RedEdge-M images, the use of a virtual imaging technique can easily realize the band alignment without the need to know the outer azimuthal elements of the image and the Digital Elevation Model (DEM) of the survey area in advance, but only the average relative speed at the time of aerial photography. Therefore, in this study, we adopt the virtual imaging technique as an alignment method for the images between different bands of the RedEdge-M camera. The general adoption of the virtual imaging method realizes the waveband alignment with sub-image-element accuracy, and the alignment accuracy is better than 0.3 image pixels.

### 2.3. Calculation of the Salinity Spectral Index

In order to assess the relationship between soil water content and salinity on the surface of the soil column and the changes in remotely sensed salinity indices, one vegetation index and ten salinity indices were calculated (Table 4), and all of these remotely sensed indices were shown to be significantly correlated with soil surface salinity. In general, the vegetation index showed a moderately significant negative correlation with soil salinity, while the salinity index related to soil background showed a moderately significant positive correlation with soil salinity.

**Table 4.** Spectral indices used in the experiment and calculation formulae.

| Index Names | Calculation Equation | Reference |
|---|---|---|
| NDVI | $(NIR - Red)/(NIR + Red)$ | [30] |
| SI | $(Blue + Red)^{0.5}$ | [31] |
| SI1 | $\left(Green^2 \times Red^2\right)^{0.5}$ | |
| SI2 | $\left(Green^2 + Red^2 + NIR^2\right)^{0.5}$ | [32] |
| SI3 | $\left(Green^2 + Red^2\right)^{0.5}$ | |
| S1 | $Blue/Red$ | |
| S2 | $(Blue - Red)/(Blue + Red)$ | [33] |
| S3 | $Green \times Red/Blue$ | |
| S4 | $(Blue \times Red)^{0.5}$ | |
| S5 | $Blue \times Red/Gree$ | [34] |
| S6 | $Red \times NIR/Green$ | |

## 3. Results

### *3.1. Variation in Band Reflectance on the Surface of Soil Column*

To compare the change in reflectance of different bands on the surface of the soil column with time after water addition, the first multiband image acquired after water addition and the last multiband image at the end period of the experiment were subjected to pixel counting, and the counting results are shown in box plots (Figure 3). The overall results show an increasing trend in band reflectance with increasing band wavelength. A comparison of the statistical characteristics of the first and the last images after water addition reveals that the mean, median, and quartile of the image element statistics of the last image have increased reflectance in all bands compared to the first image. Among them, the exception is soil column A of the first set of experiments, which may be due to too little precipitated salts. Within the same set of experiments, the reflectance increase for the three soil columns will be arranged in accordance with the gradient of initial salinity conductivity, i.e., it satisfies soil column C > soil column B > soil column A, which suggests that the initial soil salinity will affect the process of the increase in soil surface salinity reflectance.

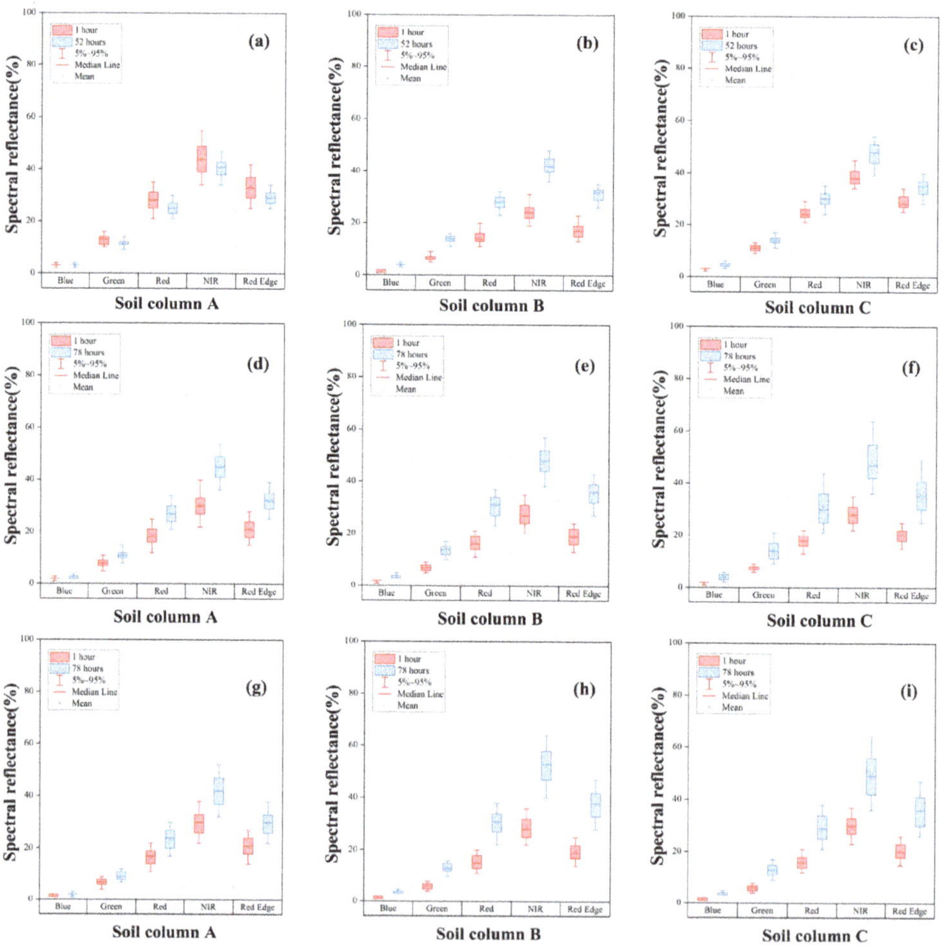


**Figure 3.** Comparison results of initial and end surface reflectance of three soil column experiments in three groups of experiments, where (**a**–**c**) are the first set of experiments, (**d**–**f**) are the second set of experiments, and (**g**–**i**) are the third set of experiments.

### 3.2. Characteristics of Water Content Change in Soil Column Layers

Figure 4 shows the process of soil water content change in the stratified layers of the three soil columns during the three sets of experiments. The experimental results show that in the first set of experiments, the soil water content of the first soil layer in the three soil columns changed the most, in which the range of change was sorted as soil column C > soil column B > soil column A. This result was mainly controlled by the initial water content. Soil column A was in a dry state and the initial water content was very low; the first set of experiments added water to infiltrate the surface soil and for water infiltration into the second layer. The water content of the surface layer of soil column B showed a peak after adding water, but the water content did not exceed 20%, and the water content of the second layer gradually increased, which was controlled by water infiltration. The water content in the surface layer of soil column C also showed a peak, but the variation was larger than that in soil columns A and B, with a value of about 65%, and the water content in its second layer also showed a slow increase in the phenomenon, but it did not exceed 20% and then leveled off. In the second set of experiments, the overall change in the three soil columns also satisfies the order of soil column C > soil column B > soil column A. In the three soil columns, the water content in the surface layer and the second layer increased rapidly at first and then decreased slowly, which was mainly controlled by water infiltration. Among them, the peak of the second layer in soil column B was larger than the peak of the first layer, which might be caused by the infiltration of water from the upper layer together with the added water. The third layer of soil column C experienced a gradual increase in water content after a period of plateauing, which was also caused by the infiltration of water from the upper layer. In the third set of experiments, the surface layer and the third layer satisfied the same pattern. However, in soil columns B and C, the water content of the fourth soil layer increased and then leveled off, with soil column B leveling off at about 60% of the water content, while the fourth layer in soil column C was saturated.

Overall, the water movement process in the three sets of experiments was mainly controlled by the initial water content and water infiltration process. When the soil water content is not saturated, there will be a peak and then a slow decline phenomenon, and the soil water will eventually accumulate in a certain layer.

### 3.3. Characteristics of Conductivity Change in Soil Column Layers

The $EC_b$ change process of stratified soil in the different soil columns of the three experimental groups in this study is shown in Figure 5. The results showed that the $EC_b$ change rule of stratified soil was similar to that of soil moisture transport. In the first group of experiments, the $EC_b$ of the surface layer of soil column A showed the change rule of increasing and then slowly decreasing, and the surface layer of soil column B and soil column C also showed a peak and then slowly decreased, but the second layer of soil column C showed a slow increase and then leveled off, indicating that the second layer of soil column C showed the phenomenon of salt accumulation. The peak fluctuations for soil column C were in the order of amplitude of soil column C > soil column B > soil column A, which was mainly controlled by the initial water content and salinity of the soil. In the second set of experiments, the surface layer also showed the phenomenon of peak increase and then decrease, and the peak fluctuations were also in the order of magnitude of soil column C > soil column B > soil column A. The $EC_b$ in the third layer of soil column B and soil column C showed the phenomenon of slow increase, which was also controlled by the initial water content and salinity. In soil column B, the $EC_b$ in the second layer was larger than that in the first layer after the $EC_b$ leveled off, which was mainly caused by the water washing the surface soil and bringing salts into the second layer. In the third set of experiments, the surface layer, the second layer, and the third layer all showed the phenomenon of increasing peaks and then decreasing peaks, and the peak fluctuations were in the order of magnitude of soil column C > soil column B > soil column A. Among them, the values of the third layer of $EC_b$ were larger than those of the first layer and the second layer after leveling off, which was also caused by the water washing of the surface soil and bringing salts into the second layer. In the fourth layer of soil columns B and C, the $EC_b$ increased slowly and then leveled off, suggesting that soil salts accumulated at this location after water washing.

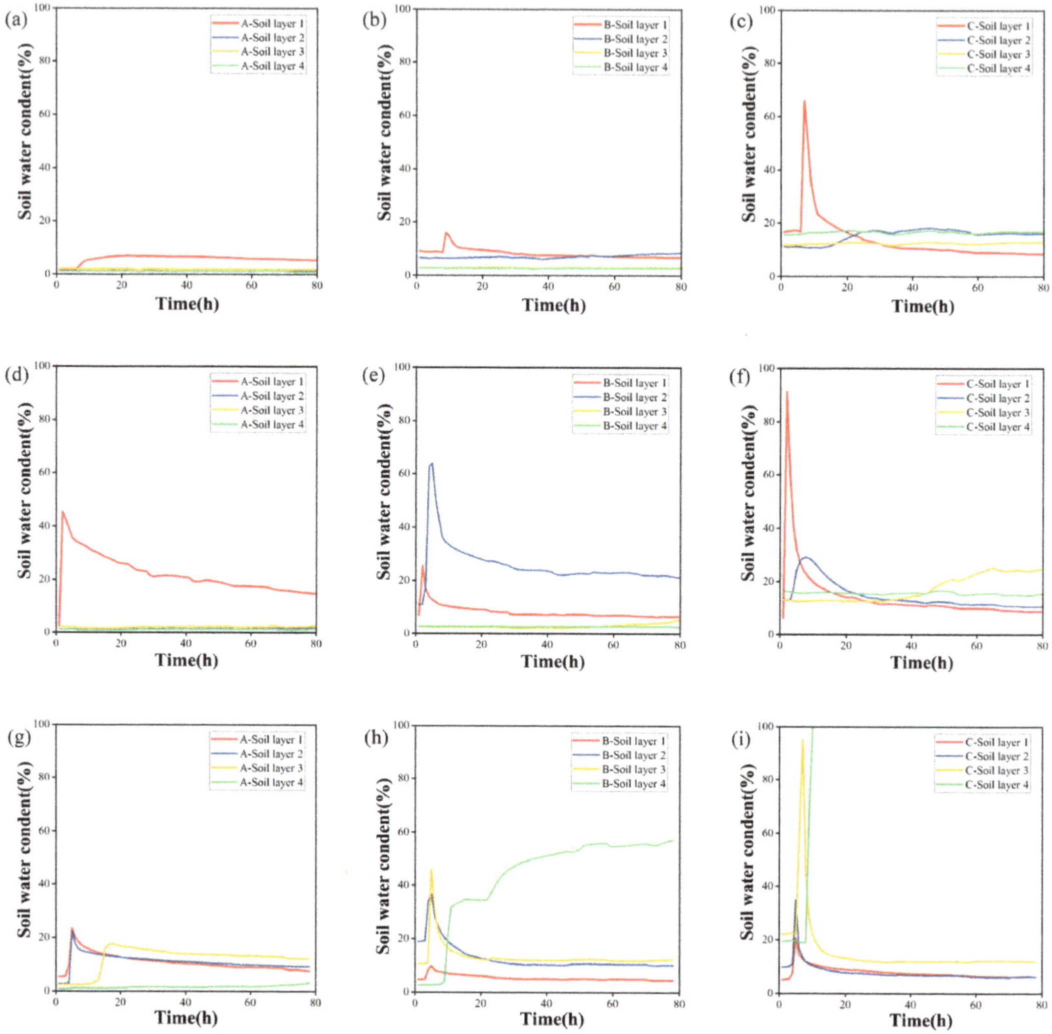


**Figure 4.** Results of soil moisture content of different soil layers during the experiments of three soil columns in three groups of experiments, where (**a**–**c**) is the first set of experiments, (**d**–**f**) is the second set of experiments, and (**g**–**i**) is the third set of experiments.

Overall, the salt movement process in the three sets of experiments was mainly controlled by the combination of initial water content and salt content and the water infiltration process. After each water addition experiment, there was a peak and then a slow decline phenomenon, indicating the washing effect of water on the soil salts, and the soil salts eventually accumulated in a layer due to the water movement. This process fully demonstrates the water–salt transport law of "salt moves together with water, and when water evaporates, salt is retained in the soil weight".

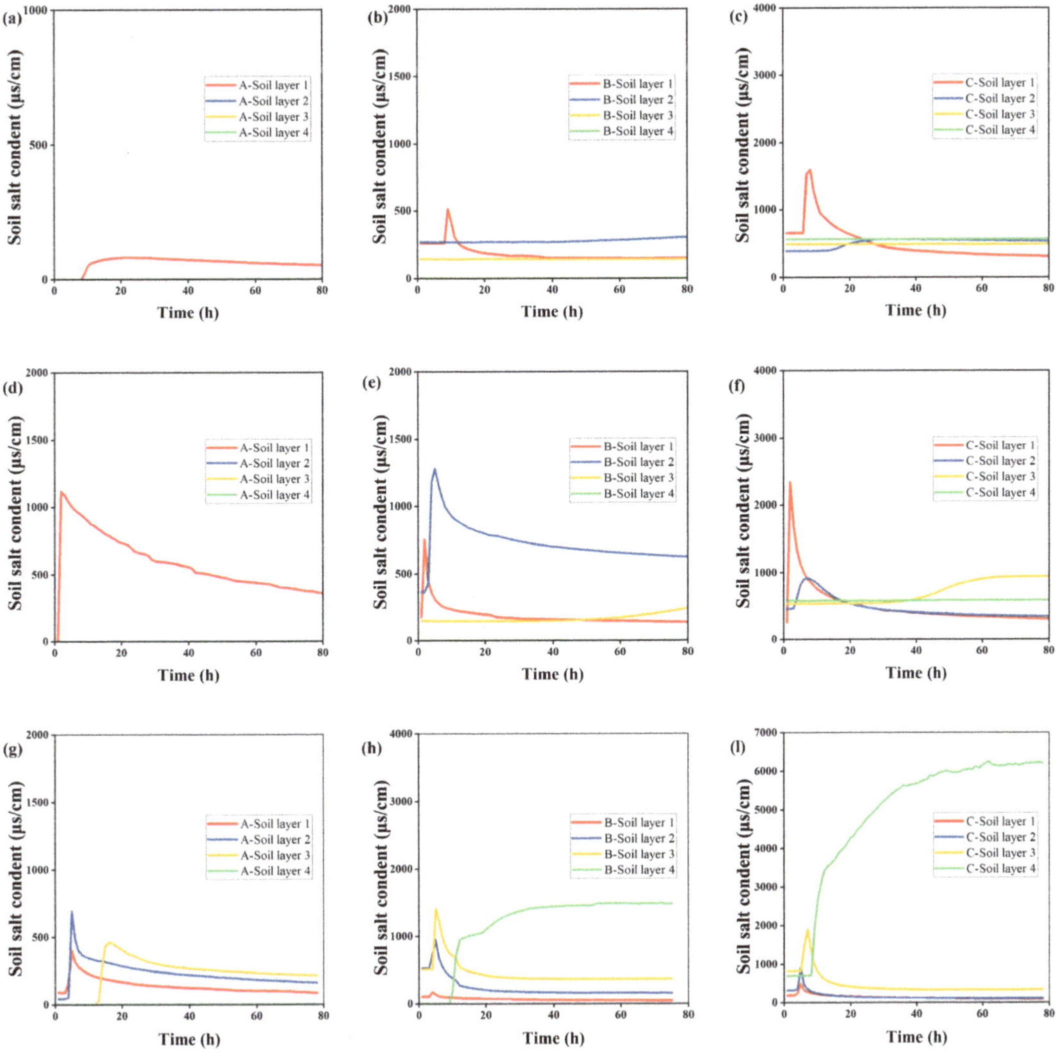


**Figure 5.** Soil conductivity results for different soil layers during the three soil column experiments in the three groups of experiments, where (**a**–**c**) is the first set of experiments, (**d**–**f**) is the second set of experiments, and (**g**–**i**) is the third set of experiments.

### *3.4. Temperature Changes in Soil Column Layers*

The soil sensors simultaneously acquired the temperature changes in the soil column layer, and the results are shown in Figure 6. Overall, the temperature results of the soil column layering in different groups of experiments did not change much during the experimental period, showing fluctuation, and the process of change was consistent in each layer. Among them, the largest fluctuation was in the topsoil layer, and the temperature was lower than the temperature at the bottom of the soil column overall. This is different from the natural state of the soil profile temperature from top to bottom according to the decreasing law; the reason may be that the natural state of the deep soil cannot be directly exposed to solar radiation, and the experimental environment in which the soil column is located cannot completely simulate field conditions. However, it is not difficult to see that the temperature in different soil layers has little effect on the water and salt transport process in the soil.

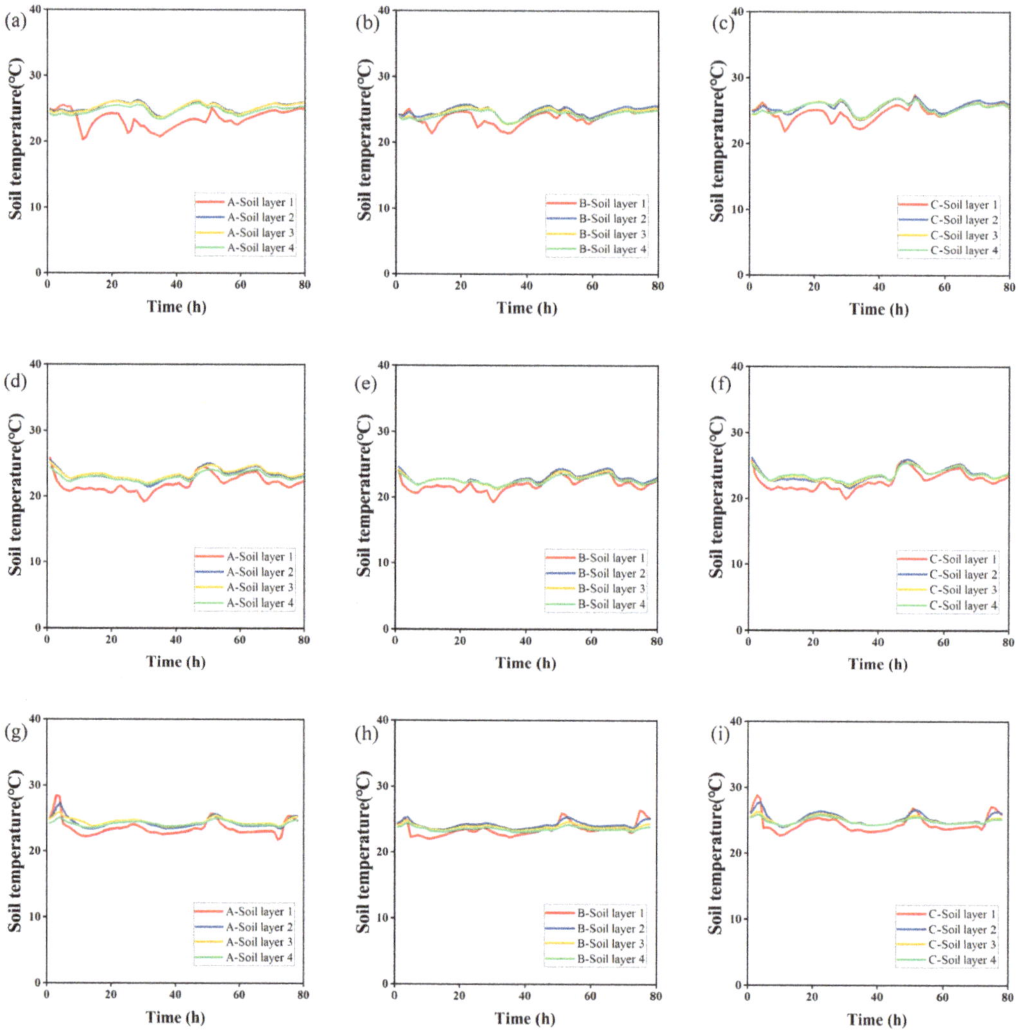

**Figure 6.** Soil temperature results for different soil layers during the three soil column experiments in the three groups of experiments, where (**a–c**) is the first set of experiments, (**d–f**) is the second set of experiments, and (**g–i**) is the third set of experiments.

### 3.5. Characteristics of Changes in the Surface Salinity Index of Soil Columns

In order to investigate the characteristics of the variation in the surface salinity index of the soil columns, the mean value of the soil surface salinity index of each group of experiments was nonlinearly fitted to time. The fitting results showed that most of the soil surface salinity index mean values versus time satisfy the Freundlich equation:

$$Salinity\ index = aT^{bT^{-c}}$$

where *Salinity index* is various salinity indices; *T* is time; and *a*, *b*, and *c* are constant covariates.

The fitting results that were not significant and not converged were excluded, and all the results were averaged according to different remote sensing indices; the results are shown in Table 5. The results showed that the vegetation index NDVI had the lowest fitting accuracy, with an $R^2$ of 0.274 and an RMSE of 0.019. This was because there was no vegetation information on the surface of the bare soil of the soil column, and it would be less accurate to use the vegetation index to indicate the change in salinity on the bare soil. The highest fitting accuracy was found in salinity index S5 with an $R^2$ of 0.734 and RMSE of 0.009. Overall, most of the salinity indices had high fitting accuracy and significant fitting models. The salinity indices with lower fitting accuracies were salinity indices S1 and S2, which may indicate that salinity indices S1 and S2 are less applicable in monitoring soil salinity changes. All other fitting results are presented in the Supplementary Materials.

**Table 5.** Mean accuracy of the model fitted to the salinity index vs. time.

| Salinity Index | $R^2$ | RMSE |
|---|---|---|
| NDVI | 0.274 | 0.019 |
| S1 | 0.451 | 0.010 |
| S2 | 0.469 | 0.017 |
| S3 | 0.624 | 0.123 |
| S4 | 0.726 | 0.011 |
| S5 | 0.734 | 0.009 |
| S6 | 0.704 | 0.115 |
| SI | 0.719 | 0.020 |
| SI1 | 0.730 | 0.057 |
| SI2 | 0.729 | 0.032 |
| SI3 | 0.702 | 0.038 |

In order to clearly demonstrate the characteristics of the salinity index with time, salinity index S5 with the highest fitting accuracy is displayed in Figure 7, and the other various indices are shown in Figure 7 in relation to time; the fitting accuracies are shown in the Supplementary Materials. The results showed that salinity index S5 had a high starting value at the beginning of each group of experiments, decreased instantly after water addition, and then increased slowly with time. This process indicates that the effect of water addition on soil surface salinity is in accordance with the basic water–salt transport law of "salt moves together with water, and when water evaporates, salt is retained in the soil weight". As for the other spectral indices in the annex, except for the NDVI, which has no obvious change characteristics, all the other salinity indices conform to the rule of change.

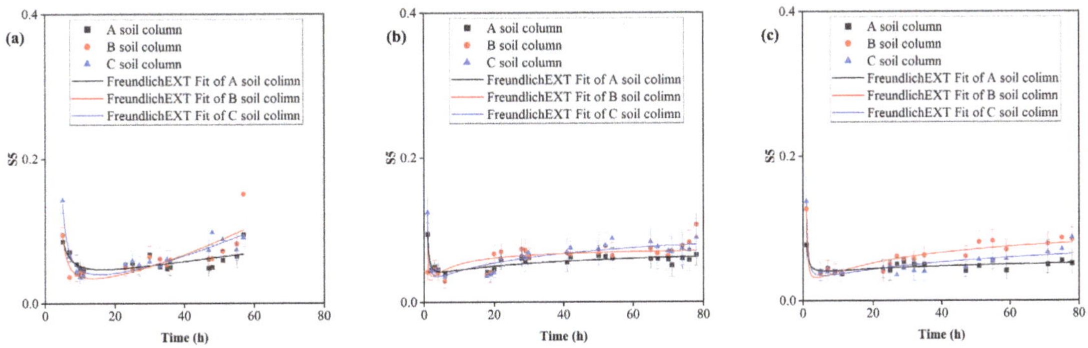


**Figure 7.** Results of salinity index S5 with time for each soil column in the three sets of experiments. (**a**–**c**) are for the first, second, and third experiments, respectively.

In order to clearly show the change process of salinity index on the soil surface, the change process of S5 of soil column A in the first group of experiments is shown in Figure 8, and the change process of S5 on the surface of soil columns of other experiments is similar to that of Figure 8, and therefore it is put in the annex. The results show that the process of salinity index S5's changes with time for soil column A in the first set of experiments is consistent with the above process. The explanation of the process of salinity index S5's change on the surface of the soil column can be that in the initial state (original analyzed without salt or analyzed with salt), S5 decreased instantly after the addition of brine, and after that, under evaporation, salt precipitated to form salt crystals on the surface of the soil, which led to the process of S5 showing a decrease and then a slow increase. The results also showed that there was heterogeneity in the salinity changes in the surface space of the soil columns, presenting inhomogeneous variations, especially in the later stages of the experiment, when the values of salinity index S5 on the surface of the soil columns were in the range of 0.069 to 0.229. This agrees with the manifestation of a strong heterogeneity of the soil salinity in the space in the natural state.

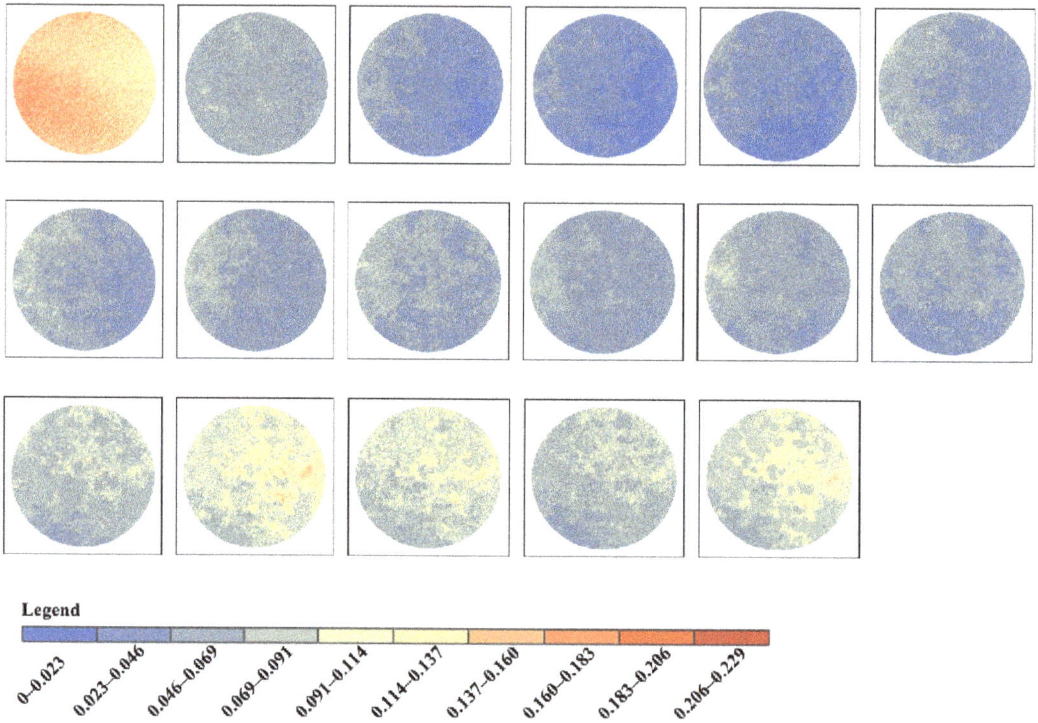


**Figure 8.** The change process of salinity index S5 of soil column A in the first experiment.

### 3.6. Accumulation of Mass Loss in the Soil Column during the Experiment

The process of salt accumulation on the soil surface is related to the evaporation process. The relationship between the evaporation process and salt accumulation was investigated by weighing the mass of each soil column in the three groups of experiments during the experimental period to obtain the cumulative mass loss of the soil columns, and the results are shown in Figure 9. The results show that the process of mass loss accumulation for each soil column in each group of experiments was similar, and the overall process of mass loss accumulation increased first and then the process of increase slowed down, i.e., the process of mass loss accumulation showed a logarithmic pattern. A comparison of the different groups of experiments shows that soil column C has the smallest cumulative magnitude in mass loss accumulation, while soil column A has the largest cumulative magnitude in mass loss accumulation. The reasons may consist of two parts, firstly, the initial water content of each soil column is different, which leads to a difference in mass loss accumulation in the evaporation process; secondly, the salt content of soil column C is the highest among the soil columns, and after the salt precipitates out and crystallizes on the surface of the soil, the salt crust or the soil crust has an inhibitory effect on the evaporation process of the soil columns.

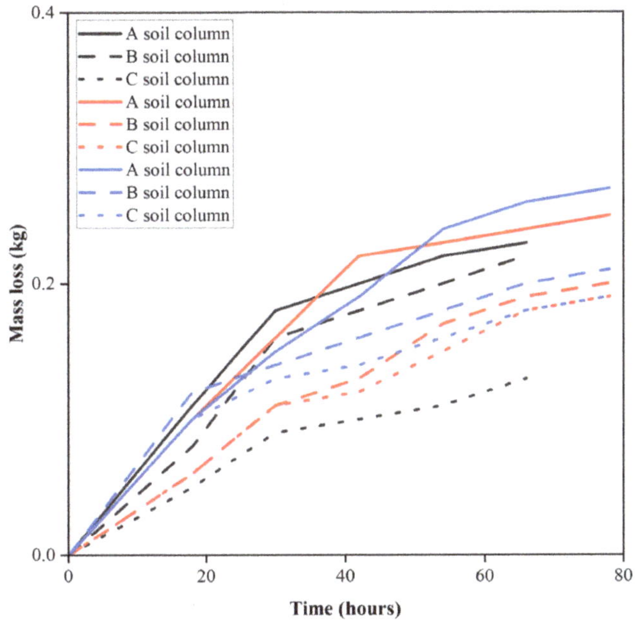


**Figure 9.** Cumulative mass loss of each soil column in three groups of experiments. The line color black is for the first group of experiments, red is for the second group of experiments, and blue is for the third group of experiments.

## 4. Discussion

### 4.1. Changes in Soil Spectral Reflectance

In this study, a multispectral camera was used to photograph the soil surface, which allowed us to investigate the characteristics of the change in spectral reflectance of the soil surface, and the results showed that the spectral reflectance of each band of the image at the end of the experiment was higher than that of the band at the beginning of the experiment after the addition of water during the test period. Many studies using remote sensing to monitor surface salinity have also pointed out that the spectral reflectance of images of high-salinity soils is higher than that of low-salinity soils [35–37]. Previous studies, however, mostly focused on the same scene image and explored the spatial variation in salinity, while it is difficult to explain the difference in spectral reflectance of the same site over time [19]. The reason for this is mainly water–salt transport, in which water accumulates and precipitates salts on the surface by capillary forces and evaporation, resulting in an increase in spectral reflectance in the band [38]. Similarly, it is caused by changes in the matric potential difference of moisture, which cause moisture to move from the wet to the dry zone, leading to salt accumulation and precipitation on the surface and resulting in an increase in the spectral reflectance in the band [39]. Different initial salt states of the soil column caused different results, so it showed that the initial state of the soil affects the salt accumulation process. Interpreting the addition of water in the experiment as precipitation or irrigation, the results of this study also showed that the spectral reflectance of saline soils at the same site was affected by precipitation or irrigation. Over-irrigation can cause soil salinization and reduce water use efficiency [40]. Soil surface spectral reflectance is affected by many factors, such as salt type, soil type, light conditions, and salt crust conditions. The available indoor spectral measurements are mostly hyperspectral, and in general the spectral reflectance increases with increasing salinity and is strongly influenced by the type of salinity [41,42]. The effect of salt type is greater than that of soil texture, especially for some hygroscopic salts ($MgCl_2$ and $CaCl_2$). Spectral measurements of saline soils under natural light sources can distinguish between crusted and non-crusted surfaces and help assess the extent of crust development [43]. Meanwhile, some scholars pointed out that the height of the soil column affects the moisture and salt content in the soil column, which indirectly affects the measurement of laboratory soil spectra [26]. Soil temperature is regarded by many scholars as an important environmental indicator for soil salinity monitoring, but the results of this study show that soil temperature has little effect on the process of spectral changes on the surface of soil columns.

### 4.2. Relationship between Soil Column Water and Salt Transport Laws and Surface Soil Spectral Indices

The results of water and salt transport in this study fully reflected the basic water and salt transport law of "salt moves together with water, and when water evaporates, salt is retained in the soil weight", but it is not clear whether some commonly used spectral indices are consistent with this law and its mechanism. When remote sensing is used to monitor soil salinity, the purpose of constructing the soil salinity index is to make the soil surface salinity information easier to distinguish, differentiate the surface salinity from other features, and to make the salinity gradient easier to represent. Some scholars analyzed the relationship between surface spectral reflectance data and the salt content of subsurface soils (5 cm and 15 cm) by using the structural equation modeling (SEM) method, and the results pointed out that there were significant correlations between soil surface spectral reflectance and subsurface soil salinity, as well as between different types of multispectral data (e.g., Landsat data) and subsurface soil salinity [44].

Considering the complexity of the salinization process, it is very difficult to achieve accurate detection in the early stages of occurrence. Often, high concentrations of salt affect some physical and chemical properties of the soil, and these properties may be detected even if there is no salt weathering on the soil surface. This is especially the case when no salt crust is formed on the surface; however, it is difficult to meet the application requirements for the salinization process and the deep salt accumulation phenomenon using only remote sensing techniques. One idea to solve this problem is to couple numerical simulation with remote sensing techniques, but first the mapping relationship between soil surface spectra and water–salt transport needs to be understood. As the results of this study show, the soil salinity index on the surface of the soil column decreases rapidly after the addition of water and shows a slow increase process with the evaporation process. In the natural state, saline soils affected by precipitation or irrigation will produce changes in the soil salinity index, which may affect the accuracy of large-scale remote sensing to monitor soil salinization [45].

### 4.3. Limitations and Future

Analyzing the experimental process and results, the limitations of this work are mainly reflected in three points: The first is that the experimental process cannot accurately obtain the salinity of the pore water, which has been difficult to accurately obtain with the current technical means. The second point is that the work did not achieve the process of coupling remote sensing and physical modeling to model the monitoring of soil water and salt changes. The third point is the difference between indoor and outdoor environments, for example, the results of indoor experiments cannot be matched with outdoor results due to the fact that indoor heating leads to a very weak effect of temperature changes, and similarly indoors is very weakly affected by wind. Despite the limitations of this study, these limitations did not affect the conclusions of this work. In future work, it is necessary to construct a model and framework method for monitoring soil water salinity changes by coupling remote sensing and physical modeling, and this study has provided some theoretical basis for this direction. Our experiments mainly used sandy soils, and in the next research work we need to investigate the use of other types of soils such as clay and loam and compare their saline properties and behavioral results with the current results.

## 5. Conclusions

In this study, we used an indoor soil column experiment, an unmanned aerial vehicle multispectral sensor camera, and a soil moisture sensor to study the water and salt transport process in the soil column under different water addition conditions, investigate the relationship between the soil water and salt transport process and the spectral reflectance of the image on the soil surface, and analyze the relationship between the water and salt transport process and the salinity index of the soil surface. The observation results of the soil column showed that the soil water and salt transport process conformed to the basic transport law of "salt moves together with water, and when water evaporates, salt is retained in the soil weight", and water and salt would accumulate at the bottom of the soil column in the late stage of the experiment. The phenomenon of salt accumulation affects the image reflectance results. With the precipitation and deposition of salt, the reflectance of the soil surface increases, and the soil temperature does not show any effect on the spectra, while the salinity index decreases instantly after the addition of water and then shows a slow increasing trend. The innovation of this experiment is the use of an unmanned aerial vehicle multispectral sensor camera to calculate different salinity indices, which is able to obtain finer soil salinity information and improve the soil water and salt surface observation technique during the experiment. The results of our work are more applicable to the process of water and salt transport changes in an arid zone. This work can provide a theoretical basis and reference for constructing a coupled model of remote sensing observation technology and a soil physical model for soil water salinity monitoring.

**Supplementary Materials:** The following supporting information can be downloaded at: https://www.mdpi.com/article/10.3390/rs16183421/s1. Figure S1: Results of salinity index NDVI with time for each soil column in the three sets of experiments. (a), (b), and (c) are for the first, second, and third experiments, respectively; Figure S2: Results of salinity index S1 with time for each soil column in the three sets of experiments. (a), (b), and (c) are for the first, second, and third experiments, respectively; Figure S3: Results of salinity index S2 with time for each soil column in the three sets of experiments. (a), (b), and (c) are for the first, second, and third experiments, respectively; Figure S4: Results of salinity index S2 with time for each soil column in the three sets of experiments. (a), (b), and (c) are for the first, second, and third experiments, respectively; Figure S5: Results of salinity index S3 with time for each soil column in the three sets of experiments. (a), (b), and (c) are for the first, second, and third experiments, respectively; Figure S6: Results of salinity index S4 with time for each soil column in the three sets of experiments. (a), (b), and (c) are for the first, second, and third experiments, respectively; Figure S7: Results of salinity index S6 with time for each soil column in the three sets of experiments. (a), (b), and (c) are for the first, second, and third experiments, respectively; Figure S8: Results of salinity index SI with time for each soil column in the three sets of experiments. (a), (b), and (c) are for the first, second, and third experiments, respectively; Figure S9: Results of salinity index SI1 with time for each soil column in the three sets of experiments. (a), (b), and (c) are for the first, second, and third experiments, respectively; Figure S10: Results of salinity index SI2 with time for each soil column in the three sets of experiments. (a), (b), and (c) are for the first, second, and third experiments, respectively; Figure S11: Results of salinity index SI3 with time for each soil column in the three sets of experiments. (a), (b), and (c) are for the first, second, and third experiments, respectively; Figure S12: The change process of salinity index S5 of soil column B in the first experiment; Figure S13: The change process of salinity index S5 of soil column C in the first experiment; Figure S14: The change process of salinity index S5 of soil column A in the second experiment; Figure S15: The change process of salinity index S5 of soil column B in the second experiment; Figure S16: The change process of salinity index S5 of soil column C in the second experiment; Figure S17: The change process of salinity index S5 of soil column A in the third experiment; Figure S18: The change process of salinity index S5 of soil column B in the third experiment; Figure S19: The change process of salinity index S5 of soil column C in the third experiment.

**Author Contributions:** Conceptualization, S.Q. and Y.Z.; methodology, L.H.; software, C.Z.; validation, J.D., J.W. and S.Z.; formal analysis, Y.Z.; investigation, S.Q.; resources, J.D.; data curation, L.H.; writing—original draft preparation, S.Q. and Y.Z.; writing—review and editing, J.D. and J.W.; visualization, L.H.; supervision, J.W.; project administration, J.D.; funding acquisition, J.D. and J.W. All authors have read and agreed to the published version of the manuscript.

**Funding:** This research was funded by The Technology Innovation Team (Tianshan Innovation Team), the Innovative Team for Efficient Utilization of Water Resources in Arid Regions (No. 2022TSYCTD0001), the Key Project of Natural Science Foundation of Xinjiang Uygur Autonomous Region (No. 2021D01D06), and the National Natural Science Foundation of China (No. 41961059). This research was also funded by the Graduate Student Innovation Project of Xinjiang Uygur Autonomous Region Postgraduate Education Innovation Project (No. XJ2021G042), the Excellent Doctoral Innovation Project of Xinjiang University (No. XJU2024BS075), and the Research Project on Spatial and Temporal Evolution of Soil Salinization in the Aksu River Basin (No. 11N457603776202312202).

**Data Availability Statement:** Data available on request due to restrictions. The data presented in this study are available on request from the corresponding author.

**Acknowledgments:** We are sincerely grateful to the reviewers and editors for their constructive comments towards the improvement of the manuscript.

**Conflicts of Interest:** Author Yong Zhang was employed by the company Visiontek Inc. The remaining authors declare that the research was conducted in the absence of any commercial or financial relationships that could be construed as a potential conflict of interest.

## References

1. Wu, H.; Guo, B.; Fan, J.; Yang, F.; Han, B.; Wei, C.; Lu, Y.; Zang, W.; Zhen, X.; Meng, C. A novel remote sensing ecological vulnerability index on large scale: A case study of the China-Pakistan Economic Corridor region. *Ecol. Indic.* **2021**, *129*, 107955. [CrossRef]
2. Vineis, P.; Khan, A. Climate Change–Induced Salinity Threatens Health. *Science* **2012**, *338*, 1028–1029. [CrossRef] [PubMed]
3. Prăvălie, R. Exploring the multiple land degradation pathways across the planet. *Earth-Sci. Rev.* **2021**, *220*, 103689. [CrossRef]
4. Sahab, S.; Suhani, I.; Srivastava, V.; Chauhan, P.S.; Singh, R.P.; Prasad, V. Potential risk assessment of soil salinity to agroecosystem sustainability: Current status and management strategies. *Sci. Total Environ.* **2021**, *764*, 144164. [CrossRef]
5. McNeill, J.R.; Winiwarter, V. Breaking the Sod: Humankind, History, and Soil. *Science* **2004**, *304*, 1627–1629. [CrossRef] [PubMed]
6. Godfray, H.C.J.; Beddington, J.R.; Crute, I.R.; Haddad, L.; Lawrence, D.; Muir, J.F.; Pretty, J.; Robinson, S.; Thomas, S.M.; Toulmin, C. Food Security: The Challenge of Feeding 9 Billion People. *Science* **2010**, *327*, 812–818. [CrossRef]
7. Zhao, Q.; Bai, J.; Zhang, G.; Jia, J.; Wang, W.; Wang, X. Effects of water and salinity regulation measures on soil carbon sequestration in coastal wetlands of the Yellow River Delta. *Geoderma* **2018**, *319*, 219–229. [CrossRef]
8. Aimen, A.; Kuandykova, G.; Suleimenova, I.; Tashmukhamedov, F.; Demeuova, G.; Anarova, G.; Aimenova, S.; Moldasheva, A.J.I.J.O.E.; Sciences, E. Environmental evaluation of water salt exchange process in soil degradation in the arid zone. *Int. J. Ecosyst. Ecol. Sci.* **2022**, *12*, 33. [CrossRef]
9. Daliakopoulos, I.N.; Tsanis, I.K.; Koutroulis, A.; Kourgialas, N.N.; Varouchakis, A.E.; Karatzas, G.P.; Ritsema, C.J. The threat of soil salinity: A European scale review. *Sci. Total Environ.* **2016**, *573*, 727–739. [CrossRef]
10. Abrol, I.; Yadav, J.S.P.; Massoud, F. *Salt-Affected Soils and Their Management*; Food & Agriculture Organization: Rome, Italy, 1988; Volume 39.
11. Wu, W.; Al-Shafie, W.M.; Mhaimeed, A.S.; Ziadat, F.; Nangia, V.; Payne, W.B. Soil Salinity Mapping by Multiscale Remote Sensing in Mesopotamia, Iraq. *IEEE J. Sel. Top. Appl. Earth Obs. Remote Sens.* **2014**, *7*, 4442–4452. [CrossRef]
12. Hassani, A.; Azapagic, A.; Shokri, N. Global predictions of primary soil salinization under changing climate in the 21st century. *Nat. Commun.* **2021**, *12*, 6663. [CrossRef] [PubMed]
13. Lhissoui, R.; Harti, A.E.; Chokmani, K. Mapping soil salinity in irrigated land using optical remote sensing data. *Eurasian J. Soil Sci.* **2014**, *3*, 82–88. [CrossRef]
14. Chaaou, A.; Chikhaoui, M.; Naimi, M.; Miad, A.K.E.; Bokoye, A.I.; Ennasr, M.S.; Harche, S.E. Potential of land degradation index for soil salinity mapping in irrigated agricultural land in a semi-arid region using Landsat-OLI and Sentinel-MSI data. *Environ. Monit. Assess.* **2024**, *196*, 843. [CrossRef] [PubMed]
15. Metternicht, G.I.; Zinck, J.A. Remote sensing of soil salinity: Potentials and constraints. *Remote Sens. Environ.* **2003**, *85*, 1–20. [CrossRef]

16. Sahbeni, G.; Ngabire, M.; Musyimi, P.K.; Székely, B. Challenges and Opportunities in Remote Sensing for Soil Salinization Mapping and Monitoring: A Review. *Remote Sens.* **2023**, *15*, 2540. [CrossRef]
17. Guo, B.; Zang, W.; Luo, W.; Wen, Y.; Yang, F.; Han, B.; Fan, Y.; Chen, X.; Qi, Z.; Wang, Z.; et al. Detection model of soil salinization information in the Yellow River Delta based on feature space models with typical surface parameters derived from Landsat8 OLI image. *Geomat. Nat. Hazards Risk* **2020**, *11*, 288–300. [CrossRef]
18. Miller, B.A.; Lee Burras, C. Comparison of Surficial Geology Maps Based on Soil Survey and in Depth Geological Survey. *Soil Horiz.* **2015**, *56*, 1–12. [CrossRef]
19. Ben-Dor, E.; Chabrillat, S.; Demattê, J.A.M.; Taylor, G.R.; Hill, J.; Whiting, M.L.; Sommer, S. Using Imaging Spectroscopy to study soil properties. *Remote Sens. Environ.* **2009**, *113*, S38–S55. [CrossRef]
20. Farifteh, J.; Farshad, A.; George, R.J. Assessing salt-affected soils using remote sensing, solute modelling, and geophysics. *Geoderma* **2006**, *130*, 191–206. [CrossRef]
21. Han, M.; Zhao, C.; Šimůnek, J.; Feng, G. Evaluating the impact of groundwater on cotton growth and root zone water balance using Hydrus-1D coupled with a crop growth model. *Agric. Water Manag.* **2015**, *160*, 64–75. [CrossRef]
22. Yang, T.; Šimůnek, J.; Mo, M.; McCullough-Sanden, B.; Shahrokhnia, H.; Cherchian, S.; Wu, L. Assessing salinity leaching efficiency in three soils by the HYDRUS-1D and -2D simulations. *Soil Tillage Res.* **2019**, *194*, 104342. [CrossRef]
23. Ramos, T.B.; Darouich, H.; Oliveira, A.R.; Farzamian, M.; Monteiro, T.; Castanheira, N.; Paz, A.; Alexandre, C.; Gonçalves, M.C.; Pereira, L.S. Water use, soil water balance and soil salinization risks of Mediterranean tree orchards in southern Portugal under current climate variability: Issues for salinity control and irrigation management. *Agric. Water Manag.* **2023**, *283*, 108319. [CrossRef]
24. Han, L.; Ding, J.; Ge, X.; He, B.; Wang, J.; Xie, B.; Zhang, Z. Using spatiotemporal fusion algorithms to fill in potentially absent satellite images for calculating soil salinity: A feasibility study. *Int. J. Appl. Earth Obs. Geoinf.* **2022**, *111*, 102839. [CrossRef]
25. Shokri-Kuehni, S.M.S.; Norouzi Rad, M.; Webb, C.; Shokri, N. Impact of type of salt and ambient conditions on saline water evaporation from porous media. *Adv. Water Resour.* **2017**, *105*, 154–161. [CrossRef]
26. Xu, L.; Wang, Z.-c.; Nyongesah, J.M. Soil Column Sample Height Influences Soil Spectral Reflectance in Laboratory Experiment. *J. Indian Soc. Remote Sens.* **2019**, *47*, 1187–1196. [CrossRef]
27. Shi, T.; Cui, L.; Wang, J.; Fei, T.; Chen, Y.; Wu, G. Comparison of multivariate methods for estimating soil total nitrogen with visible/near-infrared spectroscopy. *Plant Soil* **2013**, *366*, 363–375. [CrossRef]
28. Jhan, J.-P.; Rau, J.-Y.; Huang, C.-Y. Band-to-band registration and ortho-rectification of multilens/multispectral imagery: A case study of MiniMCA-12 acquired by a fixed-wing UAS. *ISPRS J. Photogramm. Remote Sens.* **2016**, *114*, 66–77. [CrossRef]
29. Jhan, J.-P.; Rau, J.-Y.; Haala, N. Robust and adaptive band-to-band image transform of UAS miniature multi-lens multispectral camera. *ISPRS J. Photogramm. Remote Sens.* **2018**, *137*, 47–60. [CrossRef]
30. Deering, D.; Haas, R. Using Landsat digital data for estimating green biomass. In Proceedings of the Ann. Meeting of the Soc. for Range Management Symp. on Rangeland Remote Sensing, Portland, OR, USA, 17 February 1980; Available online: http://ntrs.nasa.gov/archive/nasa/casi.ntrs.nasa.gov/19800024311.pdf (accessed on 25 January 2015).
31. Khan, N.M.; Rastoskuev, V.V.; Sato, Y.; Shiozawa, S. Assessment of hydrosaline land degradation by using a simple approach of remote sensing indicators. *Agric. Water Manag.* **2005**, *77*, 96–109. [CrossRef]
32. Douaoui, A.E.K.; Nicolas, H.; Walter, C. Detecting salinity hazards within a semiarid context by means of combining soil and remote-sensing data. *Geoderma* **2006**, *134*, 217–230. [CrossRef]
33. Abbas, A.; Khan, S. Using remote sensing techniques for appraisal of irrigated soil salinity. In Proceedings of the International Congress on Modelling and Simulation (MODSIM), Christchurch, New Zealand, 10–13 December 2007; Modelling and Simulation Society of Australia and New Zealand: Canberra, Australia, 2007; pp. 2632–2638.
34. Allbed, A.; Kumar, L.; Aldakheel, Y.Y. Assessing soil salinity using soil salinity and vegetation indices derived from IKONOS high-spatial resolution imageries: Applications in a date palm dominated region. *Geoderma* **2014**, *230–231*, 1–8. [CrossRef]
35. Fan, X.; Liu, Y.; Tao, J.; Weng, Y. Soil Salinity Retrieval from Advanced Multi-Spectral Sensor with Partial Least Square Regression. *Remote Sens.* **2015**, *7*, 488–511. [CrossRef]
36. Abuelgasim, A.; Ammad, R. Mapping soil salinity in arid and semi-arid regions using Landsat 8 OLI satellite data. *Remote Sens. Appl. Soc. Environ.* **2019**, *13*, 415–425. [CrossRef]
37. Masoud, A.A. Predicting salt abundance in slightly saline soils from Landsat ETM+ imagery using Spectral Mixture Analysis and soil spectrometry. *Geoderma* **2014**, *217–218*, 45–56. [CrossRef]
38. Xing, X.; Li, X.; Ma, X. Capillary rise and saliferous groundwater evaporation: Effects of various solutes and concentrations. *Hydrol. Res.* **2019**, *50*, 517–525. [CrossRef]
39. Hu, S.; Zhang, C.; Lu, N. Quantifying Coupling Effects between Soil Matric Potential and Osmotic Potential. *Water Resour. Res.* **2023**, *59*, e2022WR033779. [CrossRef]
40. Qin, S.; Ding, J.; Ge, X.; Wang, J.; Wang, R.; Zou, J.; Tan, J.; Han, L. Spatio-Temporal Changes in Water Use Efficiency and Its Driving Factors in Central Asia (2001–2021). *Remote Sens.* **2023**, *15*, 767. [CrossRef]
41. Farifteh, J.; van der Meer, F.; van der Meijde, M.; Atzberger, C. Spectral characteristics of salt-affected soils: A laboratory experiment. *Geoderma* **2008**, *145*, 196–206. [CrossRef]
42. Farifteh, J.; van der Meer, F.; Carranza, E.J.M. Similarity measures for spectral discrimination of salt-affected soils. *Int. J. Remote Sens.* **2007**, *28*, 5273–5293. [CrossRef]

43. Eshel, G.; Levy, G.J.; Singer, M.J. Spectral Reflectance Properties of Crusted Soils under Solar Illumination. *Soil Sci. Soc. Am. J.* **2004**, *68*, 1982–1991. [CrossRef]
44. Liu, Y.; Pan, X.; Wang, C.; Li, Y.; Shi, R. Can subsurface soil salinity be predicted from surface spectral information?—From the perspective of structural equation modelling. *Biosyst. Eng.* **2016**, *152*, 138–147. [CrossRef]
45. Han, L.; Ding, J.; Zhang, J.; Chen, P.; Wang, J.; Wang, Y.; Wang, J.; Ge, X.; Zhang, Z. Precipitation events determine the spatiotemporal distribution of playa surface salinity in arid regions: Evidence from satellite data fused via the enhanced spatial and temporal adaptive reflectance fusion model. *Catena* **2021**, *206*, 105546. [CrossRef]

land

Article

# Twenty Years of Land Accounts in Europe

Eva Ivits [1,*], Erika Orlitova [2], Roger Milego [3], Gergely Maucha [4], Barbara Kosztra [4], Emanuele Mancosu [5], Jaume Fons [3], Mirko Gregor [6], Manuel Löhnertz [6] and Gerard Hazeu [7]

1 European Environment Agency, Kongens Nytorv 6, 1050 Copenhagen, Denmark
2 Gisat s.r.o., Milady Horákové 57a, 170 00 Praha, Czech Republic; erika.orlitova@gisat.cz
3 Departament de Geografia, Universitat Autònoma de Barcelona, Edifici B, Carrer de la Fortuna, Campus UAB, 08193 Bellaterra, Spain; roger.milego@uab.cat (R.M.); jaume.fons@uab.cat (J.F.)
4 Land Monitoring Unit, Lechner Non-Profit Ltd., Budafoki út 59, 1111 Budapest, Hungary; gergely.maucha@lechnerkozpont.hu (G.M.); barbara.kosztra@lechnerkozpont.hu (B.K.)
5 European Topic Centre, University of Malaga, Ada Byron Research Building, 29010 Málaga, Spain
6 Space4environment, 48, rue Gabriel Lippmann, L-6947 Niederanven, Luxembourg; mirko.gregor@gmail.com (M.G.); loehnertz@space4environment.com (M.L.)
7 Wageningen Environmental Research, Wageningen University & Research, P.O. Box 47, 6700 AA Wageningen, The Netherlands; gerard.hazeu@wur.nl
* Correspondence: eva.ivits@eea.europa.eu

**Abstract:** Land use and its change impact food security, carbon cycling, biodiversity, and, hence, the condition of ecosystems to mitigate and adapt to climate change, support economic prosperity, and human well-being. To support and guide policy actions between the economy and the environment, harmonized time series datasets, transparent methodologies, and easily interpretable statistics are needed. Therefore, monitoring of the function and condition of lands and their change, along with properly agreed methodologies and freely accessible data, are essential. The Copernicus Land Monitoring Service has produced over 20 years of Corine Land Cover datasets for 39 countries in Europe, which allows continental-wide harmonized and comparable monitoring and accounting of land cover and land use change at a high thematic resolution and in a long time series (2000–2018). With the upcoming 2024 update, the time series will reach a unique product worldwide in terms of time series length, spatial resolution, extent, and thematic detail, enabling policymakers and the scientific community to address the main anthropogenic drivers of land and ecosystem degradation. This paper describes a unified approach for producing continental-wide land accounts that aligns with internationally agreed-upon standards for measuring the environment and its relationship with the economy. Furthermore, the study provides a harmonized time series of geospatial data for deriving land accounts and provides statistics of land cover and land use status and changes for a twenty-year period. All geospatial data and statistics presented in this paper are freely accessible and downloadable to serve other studies.

**Keywords:** land accounts; Copernicus Land Monitoring Service; Corine Land Cover; geospatial database; policy support

## 1. Introduction

Land is a finite resource, and its use is one of the principal drivers of environmental change, including land use and its change impact food security; carbon cycling; and water cycles, which in turn influence the integrity of ecosystems and our natural capital [1]. These changes are directly associated with a healthy environment and human well-being. In Europe, the proportion of total land use occupied by production (agriculture, forestry, etc.) is one of the highest on the planet [2], and conflicting land use demands require decisions that involve hard trade-offs. Factors driving land use in Europe include increasing demand for living space per person, the need for food, feed, and fiber, and the link between economic activity, increased mobility, and the growth of transport infrastructure. Effective land and soil management is fundamental for sustainable resource use and the delivery of ecosystem services [3]. For policy to counteract negative impacts and to plan incentives for better land management it is essential to monitor land use and its changes using transparent, repeatable, and harmonized methods.

Land cover and land use status and change are critical to inform the development and implementation of policies, such as agriculture, forestry, biodiversity, climate policies, or regional planning. Although a binding legislative mechanism for the sustainable management of land at the level of the European Union (EU) is lacking [4], the sustainable management of land is crucial to ensure that land continues to provide its functions now and in the future. Some EU policies already frame conditions for land use, e.g., the Common Agricultural Policy or the EU regulation for land-based carbon accounting (land use, land use change, and forestry—LULUCF), as well as the EU renewable energy and climate goals. At the same time a wide range of EU environmental policies have a major stake in land management, for example the EU Biodiversity Strategy with its target to 'maintain and restore ecosystems and their services' [5]. With the adoption of the UN Sustainable Development Goals in 2016, European countries, which are party to the UNCCD, and the EU have committed themselves to implement 'Land Degradation Neutrality' (LDN) in their mandate areas in the period up to 2030. The corresponding SDG target is target 15 on land to "protect, restore and promote sustainable use of terrestrial ecosystems, sustainably manage forests, combat desertification, and halt and reverse land degradation and halt biodiversity loss".

The international System of Environmental-Economic Accounting [6] is a key reporting framework for better ecosystem management in general. It comprises, inter alia, standards on land and water accounting, published in the 'Central Framework' (SEEA CF), and guidance on ecosystem accounting. In 2021, the United Nations Statistical Commission adopted chapters 1–7 of the System of Environmental-Economic Accounting—Ecosystem Accounting (SEEA EA) as an international statistical standard [7]. These methods provide a systematic way to measure and report stocks and flows of natural capital and anthropogenic impacts on them [8–12]. The complex interaction between elements of natural capital calls for a coherent monitoring approach to facilitate the understanding of the coupled human–environment system [13–15]. This monitoring must be supported by quantitative, robust, reliable, and comparable methods to map the condition and degradation of ecosystems and their services, and thus supplying a standardized framework for ecosystem assessment and accounting. Therefore, land accounting based on harmonized geospatial time series is a key element of the SEEA CF.

Besides accounting for the status of land cover, another key issue is the land use transforms land cover over time, as well as the impacts these changes cause. Land use change has manifold impacts on the carbon cycle [16–22]: it accelerates climate change [23,24]; it drives soil erosion [25]; and it strongly affects biodiversity [26–30]. Land accounts support the monitoring of land use change impacts as they describe how land resources change over time in a consistent and systematic way so that the implications of change can be better understood. Land use changes and the resulting change in land cover can lead to the total elimination of ecosystem types. However, for land, it is mostly their functions that are vulnerable to depletion through human actions, and such transformations are either permanent (e.g., soil sealing) or cannot be recreated over periods relevant to most policies.

The need to manage land and ecosystem resources effectively is pressing, especially given the increasing pressures from climate change, biodiversity loss and socio-economic trends. There is a growing research effort on ecosystem and natural capital accounting both in terms of conceptual developments [31–36] and applications [37–39]. Many conceptual frameworks and methodologies have emerged on ecosystems condition accounting [40,41]. Overviews on ecosystem accounting [42] and syntheses of particular topics of ecosystem accounting [43] have been produced in the literature and in the context of the United Nations with the development of the [7,44–46]. Advances in Earth observation, particularly through the EU Copernicus Programme and its Land Monitoring Service (CLMS) enhance our capacity to monitor natural capital trends. With the production and publication of the Corine Land Cover of the CLMS an efficient land and ecosystem accounting system could be developed and implemented for the presented work to connect policy information needs with adequate analytical tools.

While the land accounting concept is well documented in the scientific literature, according to our review continental-wide quantification of land cover change statistics with geospatial input to reproduce results do not yet exist. In our view the present paper is a first of its kind that brings several types of stakeholders close to land accounting. Accordingly, the purpose of this paper is to support the scientific and policy making community with quantitative information on land cover as well as land use status and change for a twenty-year period (2000–2018). The paper provides a sound methodology, a land accounting geospatial time series and derived statistical information using CLMS land cover data. Land accounting statistics are complemented by illustrative examples of how land accounting methods can support the analysis of environmental pressures arising from the land use sector. Besides the statistics presented in the paper to illustrate how land accounts can support various assessments, a vast amount of information is made available in interactive online dashboards. The geospatial and statistical data presented here and in interactive dashboards can be downloaded for further analytics that bring the user closer to monitoring the impact of land cover and land use change and may support further scientific and policy impact studies.

## 2. Materials and Methods

### 2.1. The Corine Land Cover Accounting Layers

The first set of Corine Land Cover (CLC) datasets of the Copernicus Land Monitoring Service was produced between 1986 and 1989, and since 2000, the CLC mapping has been performed in a harmonized manner, focusing on a single reference year for all participating European countries. Altogether, five mapping inventories have been implemented in this period, based predominantly on the visual interpretation of high-resolution (10–30 m) satellite imagery and ancillary, in situ datasets. These inventories have produced five status layers (referred to as CLC1990, CLC2000, CLC2006, CLC2012, and CLC2018, based on their nominal reference years) and four CLC-change (CLCC) layers for the corresponding periods (1990–2000, 2000–2006, 2006–2012, and 2012–2018).

The minimum mapping unit (MMU) of 25 ha was specified for CLC status products as a compromise between spatial details and feasibility, considering the intensive manual workload required to map large areas. Mapping changes with a higher spatial resolution was a user requirement, as most relevant land cover/land use changes happen on a larger scale than the 25 ha area limit. The MMU for CLCC layers was therefore defined at 5 ha. The original mapping instructions created for the CLC2000 inventory included the recommendation to create CLC1990–2000 changes as the difference in CLC1990 and CLC2000 status layers. However, due to the scale difference between CLC and CLCC data determined by varying MMUs, the simple difference between the datasets resulted in the issue that (1) the change layer included lots of noise and false changes, and (2) a significant portion of valid CLC changes between 5 and 25 ha were missing from the CLCC database. As the drawbacks of this strategy were clarified only after the CLC2000 mapping started, some countries created CLCC1990–2000 data by intersecting two status layers, while others applied the newly developed "change mapping first" approach. The consequence is that the CLCC1990–2000 data and statistics are not fully comparable between countries.

From the CLC2006 inventory onward, the only method that has been used for the derivation of CLC-Change database has been the "change mapping first" approach, producing a change database directly, by means of computer-aided visual image interpretation, comparing satellite images for the two reference years. The key steps of CLC update and CLCC mapping, according to the "change mapping first" approach [47], are as follows:

1. Revision of a previous CLC status layer (photo-interpretation);
2. Direct delineation of CLC changes (photo-interpretation);

3. Creation of a "new" CLC status layer (GIS operation automated):

$$CLC_{new} = CLC_{old/revised} + CLC_{changes} \quad (1)$$

4. Generalization (eliminating polygons smaller than 25 ha by semi-automatic GIS operations).

The step of revision of the previous CLC status layer was deemed necessary because of better reference data (higher resolution satellite images) from one inventory to the next, allowing the detection of formerly unnoticed features. The key advantage of a workflow based on change mapping first is the direct, visually controlled delineation of CLCC features, which ensures a significantly higher reliability of CLCC data than any other change derivation method could provide.

Europe-wide CLC and CLCC data are available as vector and raster products for the EEA-38 region[1] and the UK. The European vector and raster mosaics are distributed via the Copernicus Land portal[2]. Most of the modeling and statistical applications are based on the 100-m raster version of CLC data, which is also used in this study. The CORINE Land Cover Product User Manual [47] gives a detailed overview of CLC characteristics, product methodology and workflows, user requirements, and potential use cases.

Due to the technical characteristics of CLC and CLCC data, evolution of CLC update methodology and refinements in the understanding of thematic content [48], the statistics derived directly from historical CLC time series include several inconsistencies. To create a solid basis for CLC based statistical time series analysis a harmonization methodology was elaborated, resulting the so-called 'CLC accounting layers', which are the data sets recommended to be used for land or ecosystem accounting purposes and which are used in this study. The solution applied for the harmonization of CLC time series is applicable for the European CLC mosaics from 2000 onwards. It is based on combining the lower resolution CLC status and higher resolution CLC change layers to create a homogenous quality time series fulfilling the relation:

$$CLC_{change} = CLC_{accounting\ new\ status} - CLC_{accounting\ old\ status}. \quad (2)$$

Additional criteria are:

- Add more detail to the 25 ha resolution CLC status layer (CLC2018) from the higher resolution (5 ha) CLCC layers and use this "adjusted" layer as a reference.
- Create new CLC status layers by "backdating" the reference layer, realized as subtracting CLCC based information for CLC2018.

The processing steps for creating the CLC accounting layers are shown in Figure 1 and are described in detail in Appendix A with a public link to download the datasets. Harmonization leads to increased comparability in CLC time series statistics as many effects causing status layer differences are falsely evaluated as real changes are filtered out. Furthermore, the combination of the CLC change layers with the status layers also increased the spatial detail of the accounting layers. Due to this combination, the accounting layers cannot be described with an overall MMU as the status information remains on 25 ha and the land cover change information is stored on 5 ha. We consider, however, that an improved monitoring of land cover change has a much higher importance than fully confirming with cartographic conventions. The resulting data layers are known as the Corine Land Cover accounting data layers and are used by many land and ecosystem accounting systems and applications [49–53]. The reader is invited to consult Appendix A to explore the datasets as web map services.

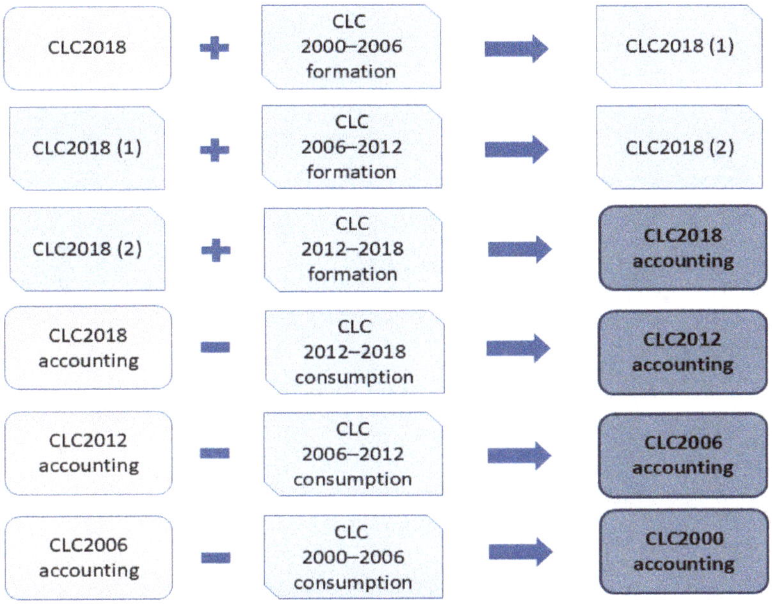


**Figure 1.** Processing steps for creating the CLC accounting layers.

### 2.2. Administrative Reference Layers

A detailed understanding of European land cover and use dynamics is facilitated by the inclusion of thematic or geographical dimensions, such as the political subdivisions of countries, regions, and municipalities. Within the land accounting system, a standard administrative boundaries layer that covers the whole EEA-38 region is key for identifying regional trends to support national policy objectives.

The administrative boundaries layer is a harmonized dataset that combines boundaries represented by the Nomenclature of Territorial Units for Statistics (NUTS) relative to the EU-28 Member States (in 2019), with the equivalent administrative units relative to the five non-EU EEA member countries (Iceland, Liechtenstein, Norway, Switzerland, and Turkey) and the six cooperating countries (Albania, Bosnia and Herzegovina, Kosovo [12], Montenegro, North Macedonia, and Serbia). In addition to the official release of Eurostat GISCO [13], the administrative boundary layer has been combined with the Economic Exclusive Zone dataset [14] to assign a country code to the coastal area of the CLC accounting layers not covered by the NUTS. This way, a perfect match between the CLC coverage and administrative boundaries was achieved[3].

The dataset is produced in raster format (GeoTiff) at the same resolution as the accounting layers (100 m) and contains the aggregation level by Country (NUTS0 + coastal areas. It is freely accessible from the datahub of the European Environment Agency[4]. Although the present study only explores regional land cover and use change patterns, note that analytics of any other spatial aggregations are possible as long as the scale, spatial resolution, and grid size of the various new layers match those of the CLC accounting layers. Such an application was produced, for example, for the Natura2000 areas [54].

### 2.3. Land and Ecosystem Accounting Nomenclature

The hierarchical nomenclature of the CLC datasets is shown in [48]. CLC level 3 has the most detailed nomenclature with 44 classes, which are aggregated to fifteen level 2 classes and five broad classes at level 1. Accounting for land cover change can in principle be implemented for individual level 3 land cover classes, but for ease of interpretation due to the numerous possible combinations of the 44 classes, a more concise class combination was developed. CLC level 1 and 2 classes are combined into 8 new groups, the so-called LEAC (Land and Ecosystems Accounting) categories. The combinations and cross-walks between the LEAC and CLC classes are shown in Table 1.

**Table 1.** Aggregation used for land and ecosystem accounting (LEAC classes).

| LEAC Groups | | CLC Classes |
|---|---|---|
| 1 | Artificial surfaces | 1 |
| 2A | Arable land and permanent crops | 2.1 + 2.2 + 2.4.1 |
| 2B | Pastures and mosaic farmland | 2.3 + 2.4.2 + 2.4.3 + 2.4.4 |
| 2B1 | Pastures | 2.3 |
| 2B2 | Mosaic farmland | 2.4.2 + 2.4.3 + 2.4.4 |
| 3A | Forests and transitional woodland shrub | 3.1 + 3.2.4 |
| 3A1 | Standing forests | 3.1 |
| 3A2 | Transitional woodland and shrub | 3.2.4 |
| 3B | Natural grassland, heathland, sclerophyllous vegetation | 3.2.1 + 3.2.2 + 3.2.3 |
| 3C | Open space with little or no vegetation | 3.3 |
| 4 | Wetlands | 4 |
| 5 | Water bodies | 5 |

These LEAC categories are defined for describing land cover changes that group CLC classes with similar land use and/or environmental characteristics. The agriculture class, for example, is split into 'Arable land and permanent crops' and 'Pastures and mosaic farmland'. Forests are also split into two classes: 'Standing forests' and 'Transitional woodland and shrub'. The latter mainly maps areas that have been recently felled or new plantations. By treating them as part of a more general class of forested land, normal forest rotations are not confused with the losses or gains of woodland that come about through deforestation or afforestation [49].

### 2.4. Land Cover Flows

While the extent of various land cover types is of high importance for policies and research, it is the land cover change that is of indicative value for the condition of ecosystems. This is because the change in a given land cover to another one is indicative for various anthropogenic and environmental pressures where policies can be developed to counteract negative impacts. Land-take, for example, is the conversion of non-urban areas to artificial urban surfaces, a process which destroys habitats and hence impacts biodiversity and increases pollution. Soil sealing, which is part of land-take, is an irreversible process where land is covered with impermeable material and as a result land loses several functions, such as flood protection, carbon sequestration, or the cooling effect in case of heatwaves. Another example for land cover change of high policy importance is the drainage of wetlands for agriculture use with important biodiversity hotspots being destroyed and where the emerging aerobic condition in soils causes the oxidation of organic matter releasing $CO_2$ into the atmosphere.

When it comes to land cover changes between two observation periods, the concept of land cover flows (LCFs) has been developed to facilitate transparent accounting. The Corine Land Cover has 44 classes; hence, the number of potential unique land cover change combinations could be large ($44 \times 43 = 1892$), although many combinations are impossible or at least improbable in practice. Therefore, they are grouped into meaningful change categories, i.e., flows, as a practical solution for transparent assessments and statistical analytics. LCFs are defined using a hierarchical structure up to three levels, describing different processes of land cover or land use changes. Several LCF matrices may be derived depending on user needs. Below is an example of a possible grouping of land cover flows at level 1 hierarchy, which is used in the present paper for land accounts:

- LCF1—Urban land management: internal transformation of urban areas.
- LCF2—Urban residential expansion: land uptake by residential buildings altogether with associated services and urban infrastructure (classified in CLC111 and 112) from non-urban land (extension over sea may happen).
- LCF3—Expansion of economic sites and infrastructure: land uptake by new economic sites and infrastructure (including sport and leisure facilities) from non-urban land (extension over sea may happen).
- LCF4—Agriculture internal conversions: conversion between farming types. Rotation between annual crops is not monitored by CLC.
- LCF5—Conversion from other land cover to agriculture: expansion of agriculture land use.
- LCF6—Increase in forest land cover and other semi-natural areas: farmland abandonment and other conversions from agriculture activity or others in favor of forests or semi-natural land.
- LCF7—Forest internal land cover changes: conversions between forest classes or between transitional woodland and shrubs and forest.
- LCF8—Water body and wetland creation and management: creation of dams, reservoirs and wetlands, and possible consequences of the management of the water resource on the water surface area.
- LCF9—Changes in land cover due to natural and multiple causes: changes in land cover resulting from natural phenomena with or without any human influence, plus rare or not applicable changes.

For understanding level 2 and level 3 flows in more depth, with contextual description and the formation and consumption codes building a certain LCF, the reader may consult Appendices 1 and 2 in [52]. It is important to note that any groupings of land cover flows are possible based on the objectives of the research needs (e.g., [54,55]).

### *2.5. Reference Grid and Data Cube for Data Integration*

The use of reference grids has been long recognized as key tool for the integration of heterogeneous sources of data. The standard codification of grid cells makes the reference grids suitable for splitting the territory into several regular pieces that can be used as reference units.

The 1 km$^2$ European Reference Grid (ERG) was adopted by several European stakeholders at the First Workshop on European Reference Grids in 2003 and it confirms with the INSPIRE geographical grid systems[5]. The coordinate reference system (CRS) of the reference grid is ETRS89-LAEA Europe, also known in the EPSG geodetic parameter dataset under the identifier "EPSG:3035". The geodetic datum is the European terrestrial reference system 1989 (EPSG:6258). The Lambert azimuthal equal area (LAEA) projection is centered at 10° E, 52° N. Coordinates are based on a false easting of 4,321,000 m, and a false northing of 3,210,000 m. Being based on an equal area projection, the accounting reference grid is suitable for generalizing data, statistical mapping and analytical work whenever a true area representation is required. Recommended grid resolutions are 100 m, 1 km, and 10 km.

In the reference grid the grid cells store the land cover information (i.e., the area of land cover classes) and complementary information to be used as reporting units or ancillary datasets for the assessments. Depending on the nature of the dataset or variable, the following additional information can be distinguished:

- Geographic dimensions: define the geographical unit that each cell belongs to (NUTS region, NUTS, UMZ, biogeographical unit, etc.)
- Thematic dimensions: define a physical characteristic of a grid cell, such as land cover type.
- Measures: numeric variables which can be aggregated by any combination of the data dimensions available in the accounting model. They can be biophysical vari-

ables (e.g., vegetation productivity), climate (precipitation and temperature) but also socioeconomic figures (e.g., population, unemployment, GDP, etc.).

The EEA-39 reference grid[6] consists of 5,885,212, 1 km cells, each of which can hold a data record in the LEAC database. The geographic and reference dimensions intersect with the 1 km$^2$ EEA reference grid, in order to give each grid cell a unique feature code (e.g., a NUTS3 code, a biogeographical region code, etc.).

Considering that some thematic layers have a higher resolution than 1 km$^2$ (e.g., Corine Land Cover is available at a resolution of 100 m$^2$), the combination of such information is carried out at 100 m$^2$ in this study. This way, it is possible to store, for instance, the different land cover classes and their surfaces for each grid cell, as well as their extent within any administrative boundary.

Land accounts in this paper were produced by creating a data cube. The system uses cloud infrastructure to integrate diverse data types in near real time. It is component based in order to accommodate flexibility and change, while new user requirements are shaped over time. The system reads spatial datasets into the reference grid where every cell has a unique identifier. These cells become the geo-reference identifier for every other spatial dataset integrated in the data cube. Through this common identifier the user can relate and calculate area statistics based on the same cell identifier. After reading the geospatial layers these are converted into a tabular format, i.e., dimension. All, or any user selected dimensions, can be then aggregated within user defined spatial units. These aggregated dimensions are then added into one database (a data cube), which can be created at a 10 × 10 m, 100 × 100 m, 1 × 1 km, or 10 × 10 km grid resolution. The data cube is stored in an SQL database that can be directly accessed by business intelligence software. Year-to-year changes or area statistics of land surface processes can then be calculated, and results plotted in a user-friendly, attractive, and interactive way.

For the present land accounting work, a so-called Land and Ecosystem Accounting (LEAC) cube was created. The cube contains the following dimensions:

- Administrative boundaries for EEA-38 and the UK;
- Biogeographical regions;
- Coastal zones;
- CLC accounting layers for the years 2000, 2006, 2012, and 2018;
- LCFs for the periods 2000–2006, 2006–2012, 2012–2018, and 2000–2018.

These dimensions can be flexibly complemented with any other thematic or stratification layers thereby enriching the accounting information to be deducted from the data cube.

## 3. Results

In the following sections, twenty-year land accounts are presented for Europe using the CLC accounting layers classified into the LEAC categories (see Section 2.3), and the derived land cover flows (Section 2.4) are used to explore the change between the LEAC categories. We present land cover accounts with various measurement units (km$^2$, ha, and % of the selected region), where statistics are given in tabular formats or visualized in interactive charts and maps. Land cover change accounts using the land cover flows present both consumption and formation statistics, which are only detailed on level 1 hierarchy in the present paper but can be accessed for all levels in the interactive dashboards. Furthermore, we go into more detail addressing possible land accounting applications: we explore land cover change density during 2000–2018, indicating the % of grid cells that changed their land cover type during a given accounting period. We present the spatial pattern of the dominant land cover flows for each region in the 39 countries, exploring local land cover change processes. To illustrate the power of land accounting, we map specific land cover flows related to agriculture area change on high spatial detail for the entire territory.

The land accounting statistics and the country fact sheets are presented in corresponding online dashboards [53,56], where all results in the following sections can be reproduced by the reader. The user is suggested to further explore the dashboard, as the vast amount of information stored there cannot be presented here. The dashboards allow downloading all statistics in spreadsheet format for further analysis.

### 3.1. Land Cover Status Accounts

Figure 2 presents the 2018 land cover extent accounts for the EEA-39 region using the eight main LEAC categories. Statistics are given in % of the study area and in absolute values (km$^2$). Around 34% of all ecosystems in Europe are forested, covering approximately 2 million km$^2$. Forests have a major role in many ecosystem services, which include, among others, major carbon sinks and supply food, feed, and fiber for many economic activities; thus, their monitoring is of major importance. The second largest LEAC category in Europe are arable lands and permanent crops, ensuring Europe's food security, economic prosperity, and supporting farmland biodiversity. These agricultural areas cover almost 1.5 million km$^2$, which is around 25% of the area. Pastures and mosaic farmlands cover approximately 17% of the region, which is almost 973,000 km$^2$. Europe's natural grasslands, heathlands, and sclerophyllous vegetation, supporting biodiversity and other ecosystem services, cover only 8% of the area, approximately 495,000 km$^2$. Artificial surfaces, which affect ecosystem conditions, increase flood risk, and affect human health impacts, cover 4% of Europe.

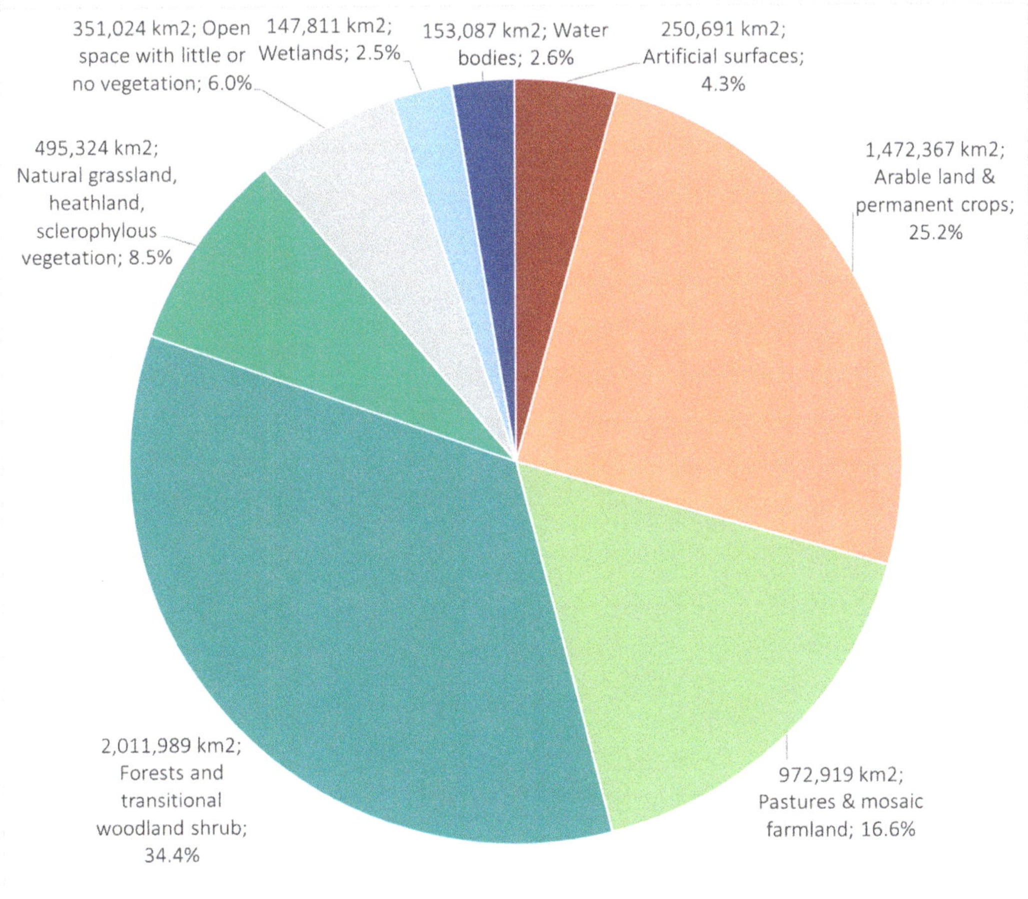


**Figure 2.** Distribution of LEAC classes in the study area for the year 2018, in km$^2$ and in % of the total study area.

Figure 3 displays the distribution of the four major LEAC categories in Europe by countries expressed in % of the country's area and Appendix B gives statistics by Europe's biogeographical regions. The Boreal region has the largest forested areas in both absolute values (612.000 km$^2$) as well as in % of the territory (69%). Forests, transitional woodlands, and shrubs mostly cover the Scandinavian region, reaching 72% of Finland and 66% of Sweden. A total of 62% of Montenegro's territory is covered by forests and transitional woodland, leading the Balkan and southern European countries. Local highly forested regions with more than 50% forest cover are, for example, the Landes forest in southwestern France, south-east Austria, Slovenia, and Northern Turkey. Most agricultural area (arable lands and permanent crops) are in the continental region, Hungary and Poland being in the lead with 52% and 43% of their area, respectively. Greece and Albania lead the countries with the largest natural grassland, heathland, and sclerophyllous vegetation areas with over 25% and 21% of their area, respectively. Ireland stands out among the European countries with its pastures and mosaic farmland area as 62% of the country has this category, and the next highest area in the proportion of the country is in Malta (49%) and the Netherlands (41%).

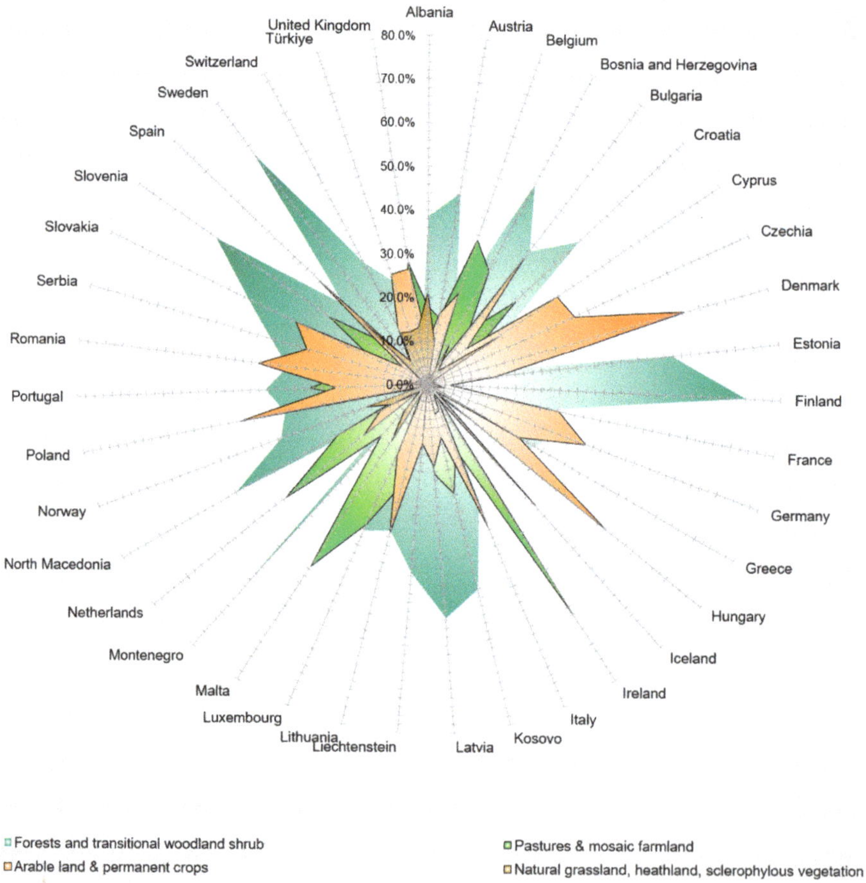


**Figure 3.** Distribution of the four main LEAC classes in the study area in % of the country's area.

*3.2. Land Cover Status Accounts*

The CLC accounting datasets, the reference layers, the reference grid and the accounting cube together allow the fast, transparent and efficient production of land accounting statistics and maps for the various accounting periods and for any spatial subset.

Figure 4 displays net change in each LEAC category in annual km$^2$ for the total period 2000–2018 (in yellow color) and within the three consecutive monitoring periods: 2000–2006, 2006–2012, and 2012–2018 (shown in different shades of blue). Positive numbers indicate that formation processes predominate, negative numbers indicate that consumption processes prevail. The main trends of net change per land cover type over the three periods are as follows: The formation within 'Artificial surfaces' is slowing down (1086 km$^2$/year, 966 km$^2$/year, and 715 km$^2$/year). The consumption of 'Arable land & permanent crops' decreases (517 km$^2$/year, 677 km$^2$/year, and 10 km$^2$/year). The net change within 'Forests and transitional woodland shrub' changed from formation during the first two periods (174 km$^2$/year and 103 km$^2$/year) to consumption during the last period (265 km$^2$/year). The consumption of grasslands and other herbaceous vegetation types fluctuated between 250 km$^2$/year and 290 km$^2$/year and the net change within 'Open space with little or no vegetation' changed from consumption during the first two periods (79.57 km$^2$/year and 42 km$^2$/year) to formation during the last period (226 km$^2$/year). The net of the land cover formations and consumptions indicate that Europe's land cover is under a dynamic change with the exception of forests and transitional woodlands where the net change is well below 0.1%. The net surface of natural grassland, heathland, and sclerophyllous vegetation and of pastures and mosaic farmland have decreased the most, with around 1% of these land covers available for use in 2018. Arable land and permanent crops has decreased by 0.5%. The largest gain was artificial surfaces with a 7% increase compared to artificial surfaces in the year 2000.

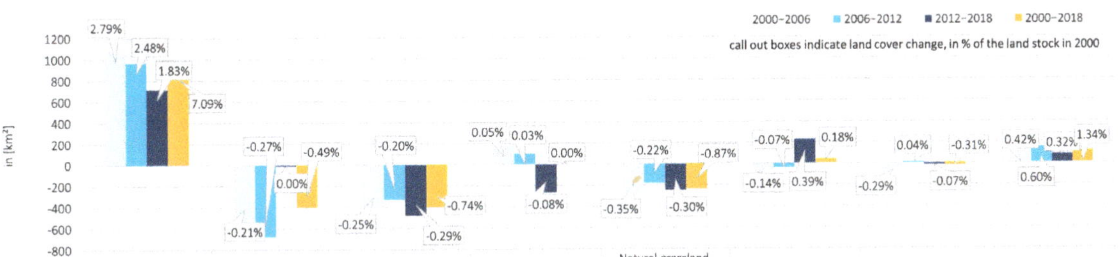


| | Artificial surfaces (4.3% of EEA39 in 2018) Change 2000-2018: 16577 km2 | Arable land & permanent crops (25.1% of EEA39 in 2018) Change 2000-2018: -7228 km2 | Pastures & mosaic farmland (16.6% of EEA39 in 2018) Change 2000-2018: -7289 km2 | Forests and transitional woodland shrub (34.4% of EEA39 in 2018) Change 2000-2018: 69 km2 | Natural grassland, heathland, sclerophylous vegetation (8.5% of EEA39 in 2018) Change 2000-2018: -467 km2 | Open space with little or no vegetation (6.0% of EEA39 in 2018) Change 2000-2018: 611 km2 | Wetlands (2.5% of EEA39 in 2018) Change 2000-2018: -4335 km2 | Water bodies (2.6% of EEA39 in 2018) Change 2000-2018: 2062 km2 |
|---|---|---|---|---|---|---|---|---|
| 2000–2006 | 1,086.7 | -517.1 | -407.6 | 173.6 | -290.3 | -79.6 | -71.7 | 105.9 |
| 2006–2012 | 966.2 | -677.1 | -328.9 | 102.6 | -181.9 | -42.0 | 10.9 | 150.3 |
| 2012–2018 | 714.8 | -10.4 | -478.4 | -265.0 | -250.5 | 226.2 | -17.0 | 80.3 |
| 2000–2018 | 922.6 | -401.5 | -405.0 | 3.7 | -240.9 | 34.9 | -25.9 | 112.2 |

**Figure 4.** Net change in the various land cover categories in the various accounting periods in the EEA-38 plus UK region, in km$^2$ and in % of the 2000 land cover extent.

Figure 5 top shows the spatial pattern of the 2000–2018 land cover change density over Europe and Figure 5 bottom displays a subset for France and Sweden selecting all three-land accounting sub-periods (2000–2006, 2006–2012, and 2012–2018). Each grid cell where any change occurred within the selected time period is colored based on the percentage of changes within the that grid cell, from light blue to red as follows: changes between 1 and 25% of the total grid area in blue, changes of 26–50% in green, changes of 51–75% in orange, and changes of 76–100% of the grid area in red. Figure 5 shows that during 2000–2018 hotspots of land cover changes in Europe were located in Portugal, France, and Sweden (top), and that both in France and Sweden the most intense period was during 2006–2012 (Figure 5 bottom). In other areas in Europe the land cover remained fairly stable during the 20-year period. Note, Figure 5 intends to show all changes that happen in a land cover, not only land use change. This is an important detail because in the case of forestry, the land use remains the same, but forest management practices are the largest in Europe. These practices do not change the land use type, i.e., forest remains forest, but the land cover changes from trees to clear cut, transitional woodland, shrubland, etc. This is what the land cover change density map reflects and, for example, as Sweden has a large proportion of forests, forest management practices show up in a land cover change density map with a higher percentage of change. The same reasoning is valid for the Foret des Landes region in France (see the subset of France in Figure 5).

Table 2 displays the comprehensive summary of land cover change accounts for the EEA-39 region for the period 2000–2018 (Appendix C displays the summary table for the period 2000–2006, 2006–2012, and 2012–2018). It gives area statistics for the main land cover types at the beginning and at the end of the period, as well as the net changes disaggregated for their formation and consumption statistics. The consumption and formation codes are reported as the area sum of the whole period ("per period [$km^2$]"), as the annual change ("per year [$km^2$]"), in proportion of all land cover changes (row "% of LC change"), and in proportion of change in the given land cover (row "% of given LC"). Note, the total net area changes are always zero because land is finite and apart from land reclamation, the process of creating new land from oceans, seas, riverbeds or lake beds, is not addressed in this study.

**Table 2.** Land cover change accounts for the period 2000–2018 with formation and consumption accounts (see Appendix C for the other periods).

| | | Artificial Surfaces | Arable Land & Permanent Crops | Pastures & Mosaic Farmland | Forests and Transitional Woodland Shrub | Natural Grassland Heathland Sclerophyllous Vegetation | Open Space with Little or No Vegetation | Wetlands | Water Bodies | Total |
|---|---|---|---|---|---|---|---|---|---|---|
| beginning of the period | LC ($km^2$) | 234,085 | 1,479,595 | 980,209 | 2,011,922 | 499,660 | 350,396 | 148,277 | 151,068 | 5,855,212 |
| consumption | per period ($km^2$) | 4286.29 | 24,388.46 | 15,265.19 | 156,795.24 | 6995.19 | 4812.87 | 1136.11 | 841.79 | 214,521.14 |
| | per year ($km^2$) | 238.13 | 1354.91 | 848.07 | 8710.85 | 388.62 | 267.38 | 63.12 | 46.77 | 11,917.84 |
| | % of LC change | 2.00% | 11.37% | 7.12% | 73.09% | 3.26% | 2.24% | 0.53% | 0.39% | 100.00% |
| | % of given LC | 1.83% | 1.65% | 1.56% | 7.79% | 1.40% | 1.37% | 0.77% | 0.56% | 3.66% |
| formation | per period ($km^2$) | 20,892.81 | 17,160.73 | 7975.50 | 156,862.50 | 2659.01 | 5440.30 | 669.56 | 2860.73 | 214,521.14 |
| | per year ($km^2$) | 1160.71 | 953.37 | 443.08 | 8714.58 | 147.72 | 302.24 | 37.20 | 158.93 | 11,917.84 |
| | % of LC change | 9.74% | 8.00% | 3.72% | 73.12% | 1.24% | 2.54% | 0.31% | 1.33% | 100.00% |
| | % of given LC | 8.33% | 1.16% | 0.81% | 7.80% | 0.53% | 1.55% | 0.45% | 1.89% | 3.66% |
| net change | per period ($km^2$) | 16,606.52 | −7227.73 | −7289.69 | 67.26 | −4336.18 | 627.43 | −466.55 | 2018.94 | 0.00 |
| | per year ($km^2$) | 922.58 | −401.54 | −404.98 | 3.74 | −240.90 | 34.86 | −25.92 | 112.16 | 0.00 |
| | % of given LC | 7.09% | −0.49% | −0.49% | 0.00% | −0.87% | 0.18% | −0.31% | 1.34% | 0.00% |
| end of period | LC (km2) | 250,691 | 1,472,367 | 972,919 | 2,011,989 | 495,324 | 351,024 | 147,811 | 153,087 | 5,855,212 |

During 2000–2018, both the largest consumption processes of a given land cover and the largest proportion of changes within all consumption processes occurred in 'Forests and transitional woodland shrub' (consumption of 8711 $km^2$/year, which represents 73% of the total consumptions in Europe). However, the largest formation processes, and also the largest proportion of changes within all formation processes, also occurred in the 'Forests and transitional woodland shrub' LC type (formation of 8715 $km^2$/year), which represents 73.1% of the total formations in Europe. Therefore, the forest land cover in Europe in fact remained fairly constant (see beginning of period and end of period rows in Table 2), and these numbers indicate that it is the forest management practices within the forest land use category that dominate European forest land use and, in fact, all land use in Europe.

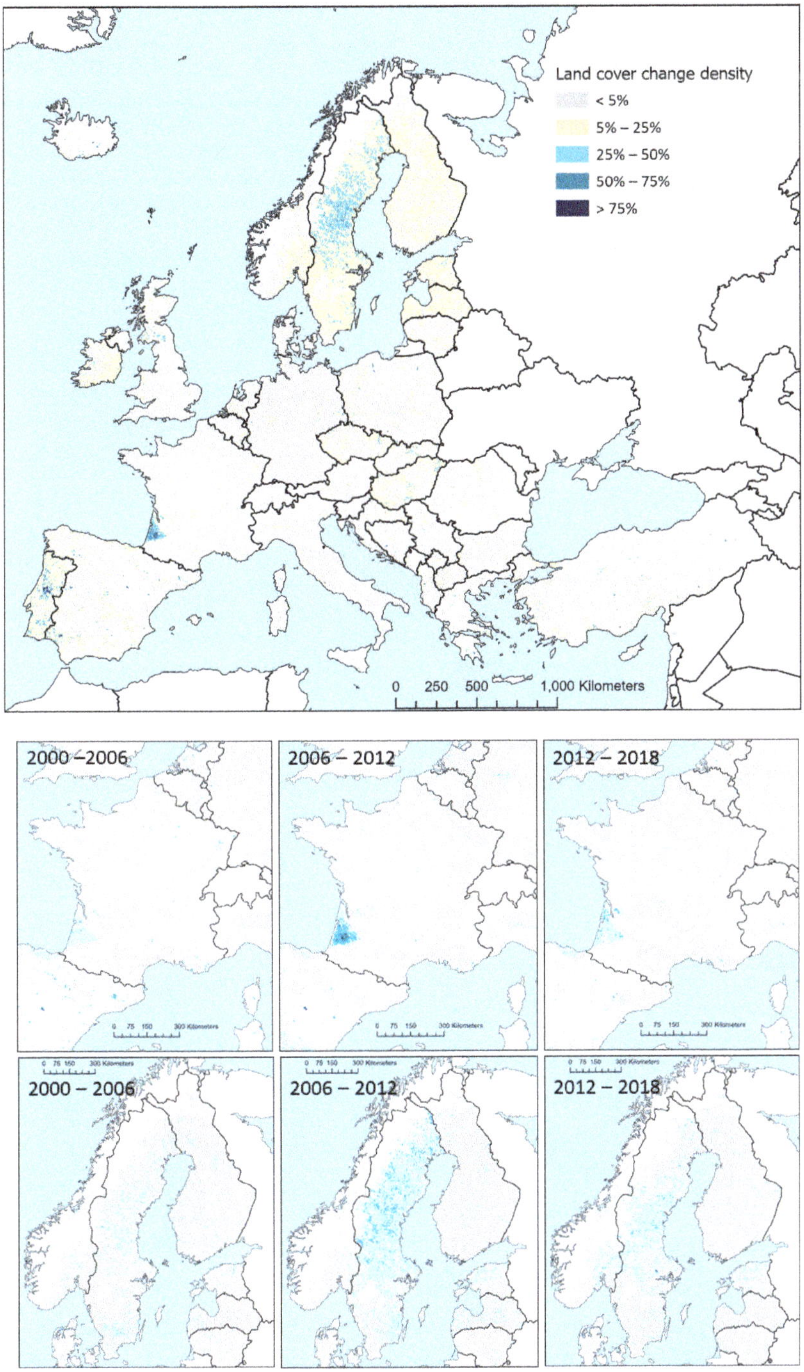


**Figure 5.** Land cover change density during 2000–2018. The 100 m accounting layers are aggregated within 5 km grids for better visibility. The % of grid cells that changed their land cover type during the accounting period is expressed in proportion of each 1 km accounting grid cell.

The largest positive change (i.e., formation) during 2000–2018 occurred within the 'Artificial surfaces' LC type (1160.7 km$^2$/year), which represents 8.3% of the total area of this land cover type and almost 10% of all land cover changes in this period. Artificial surfaces also decreased in some parts of Europe (see consumption) amounting to 238 km$^2$/year, but it was only 2% of all land cover changes. Still, the net change in artificial areas, i.e., the sum of the formations and consumptions amounted to 923 km$^2$/year increase. This means that during the 18 years period artificial areas in Europe increased from 234,085 km$^2$ to 250,691 km$^2$, thus 16,606 km$^2$ land was converted to artificial surfaces, which is roughly six times the area of Luxembourg. This increase in artificial surfaces was by far the highest in Europe, amounting to a 7% increase compared to the 2000 artificial areas, more than five times the second largest increase, which was creating more water bodies.

The largest negative net change occurred within 'Arable land & permanent crops' and 'Pastures & mosaic farmland', both decreasing roughly 400 km$^2$ annually due to consumption processes dominating formation processes. In fact, of all consumption processes, 11% happened on arable lands and permanent crops and 7% on pastures and mosaic farmlands, which together resulted in an approximately 14,500 km$^2$ net decrease in these important land cover types.

### 3.3. Land Cover Flows

Figure 6 compares the trends in the land cover flows between the accounting periods 2000–2006, 2006–2012, and 2012–2018 and the entire period 2000–2018 for the EEA38 + UK area in [km$^2$/year]. The trends in the LCFs over the three mentioned periods are relatively stable. Visible changes occur in 'LCF2: Urban residential expansion' where the values halved between the first two periods and keep decreasing (323.1 km$^2$/year, 147.2 km$^2$/year, and 106 km$^2$/year during the three periods, respectively) and in 'LCF6: Increase in forest land cover and other semi-natural areas', where the values nearly halved between the two periods 2000–2006 and 2012–2018, decreasing from 786.6 km$^2$/year to 356.8 km$^2$/year. There is an increase in 'LCF9: Changes of land cover due to natural and multiple causes' between the last two periods (435.43 km$^2$/year in 2006–2012, and 636.93 km$^2$/year in 2012–2018), which in most cases indicates the impact of wildfires. Within 'LCF7: Forest internal land cover changes' which represents changes covering the largest total area there was a sudden increase during the middle period (7502.44 km$^2$/year in 2000–2006, 11,213.66 km$^2$/year in 2006–2012, and 8428.31 km$^2$/year in 2012–2018).

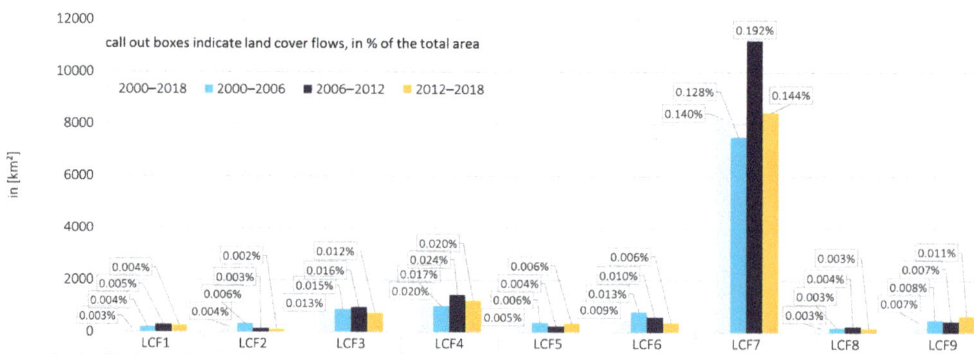


| | LCF1 | LCF2 | LCF3 | LCF4 | LCF5 | LCF6 | LCF7 | LCF8 | LCF9 |
|---|---|---|---|---|---|---|---|---|---|
| 2000–2018 | 152.2 | 225.2 | 783.3 | 1154.3 | 283.0 | 518.7 | 8223.3 | 179.8 | 398.1 |
| 2000–2006 | 206.2 | 323.1 | 884.4 | 1006.0 | 377.6 | 786.6 | 7502.4 | 188.0 | 487.9 |
| 2006–2012 | 294.5 | 147.2 | 949.0 | 1431.4 | 242.0 | 591.5 | 11213.7 | 238.2 | 435.4 |
| 2012–2018 | 250.6 | 106.0 | 731.3 | 1195.9 | 337.0 | 356.8 | 8428.3 | 165.7 | 636.9 |

**Figure 6.** Comparison of the land cover flows between the accounting periods 2000–2006, 2006–2012, 2012–2018, and the entire period 2000–2018 for the EEA39 area [in km$^2$/year].

Analyzing land cover flows on a grid cell level allows the understanding of the regional impact of land cover conversions. Figure maps the dominant land cover flows for the period 2000–2018. Each grid cell where any change occurred within the selected period is colored based on the land cover flow that dominated that grid cell, i.e., the majority of the process within that grid cell. The pie charts in Figure 7 show the distribution of the main land cover flows (LCF1–LCF9) in both km$^2$ and percentages of all land cover flows (i.e., adding up to 100%) for the 2000–2018 accounting period. During that period the total land cover change effected 214,521 km$^2$, which is also shown in Table 2, detailing the land cover change accounts (see column total formation or total consumption).

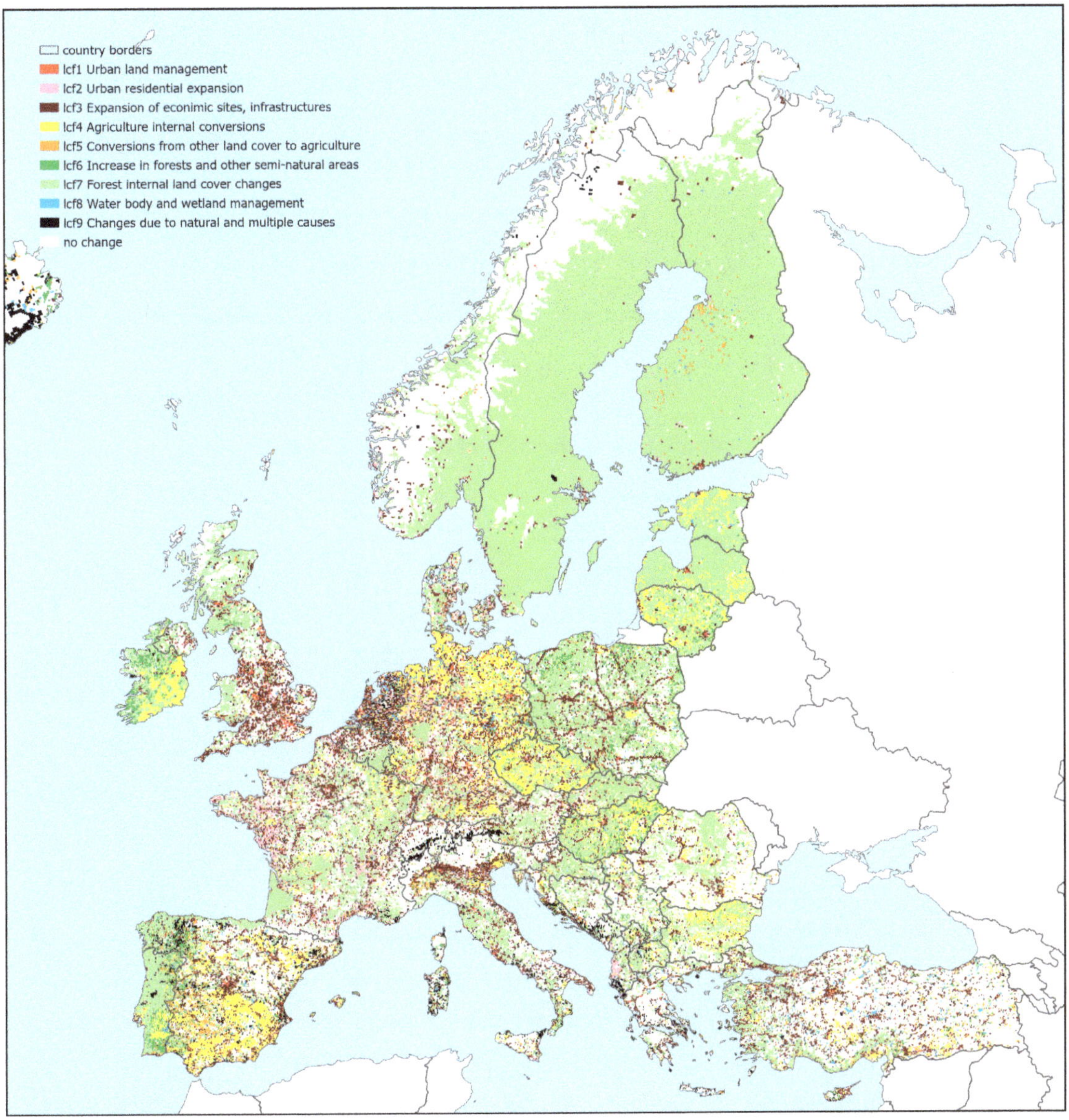


**Figure 7.** *Cont.*

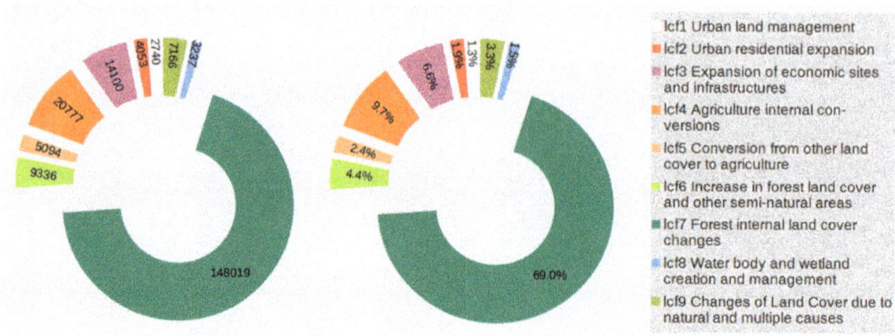


**Figure 7.** Top—map of dominant land cover flows (LCFs) for the period 2000–2018 in a 1 × 1 km grid. Each grid is colored based on the dominant land cover flow (i.e., the majority of the category). LCF1–LCF9 are explained in the legend as well as in chapter 2.4. Bottom—share of dominant land cover flows for the period 2000–2018 in km$^2$ and in % of the study area.

During 2000–2018, the land cover flow covering the largest total area was 'LCF7: Forest internal land cover changes' with 148,017 km$^2$ (69% of all land cover flows), mostly concentrating in the Nordic countries due to the large proportion of forests in these areas. This was mostly in Sweden, Norway, and Finland, where more than 90% of the land cover changes happened within their forests (Figure 8), whereas in Latvia and Slovakia, forest management accounted for more than 80% of the land cover changes. Other large land cover changes occurred due to agriculture internal conversions (LCF4, 20,777 km$^2$), and thus agriculture land management, which represented almost 10% of all changes (Figure 7). Countries where agricultural management dominated all land cover changes were Czechia (over 50% of the changes), Ireland and Germany (approximately 40% of all land cover changes).

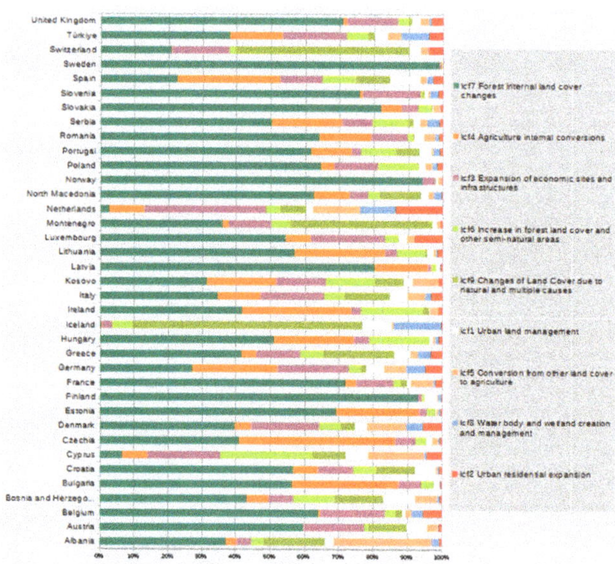


**Figure 8.** Proportion of the land cover flows (LCFs) in European countries, for the period 2000–2018 (in % of the country area).

The land cover flow "expansion of economic sites and infrastructures" (LCF3) was the third largest process in Europe and that took place over 14,100 km$^2$ (around 7% of all LCF), taking land from other land cover types (Figure 7). This land-take process mostly happened in the Netherlands (36% of all land-take processes in the country) and was around 20% of all land conversion processes in Cyprus, Denmark, Germany, and Luxembourg (Figure 8). The increase in forests and semi-natural areas (LCF6) only happened on 9336 km$^2$, which was slightly over 4% of all land cover changes. Cyprus led the way in creating these green areas, accounting for almost 27% of all land cover processes in the country, followed by Hungary and Ireland with 17%. However, all together, very few green areas were created in Europe, concerning the fact that urban residential expansion was over 4000 km$^2$, which, together with the expansion of economic sites and infrastructure, impacted over 18,000 km$^2$. This urban sprawl was also the greatest in the Netherlands (13% of land cover flows), followed by Denmark and Germany, where urban sprawl accounted for around 5% of all land use changes.

### *3.4. Land Use Drivers of Land Cover Change*

Another important element is assessing the different land use processes that have caused land cover to change and analyzing which classes have gained or lost area at the expense of other classes. In other words, what is interesting is analyzing not only the quantitative figures of changes, but also the drivers behind those changes (from which classes and/or to which classes a specific transition has occurred). Combining the land cover flows with the LEAC classes allows for the understanding of the drivers behind the change in land cover. Table 3 shows the area changes (in km$^2$) of the LEAC classes (columns) between 2000 and 2018 and the rows of the tables break down these changes according to the various land cover flow processes. The top part of Table 3 summarizes the consumption processes, whereas the bottom part details the formation processes.

**Table 3.** Consumption and formation of the LEAC classes by the various land use change processes (land cover flows) during 2000–2018. Values are presented in km$^2$.

| | Arable Land & Permanent Crops | Artificial Surfaces | Forests and Transitional Woodland Shrub | Natural Grassland, Heathland, Sclerophylous Vegetation | Open Space with Little or No Vegetation | Pastures & Mosaic Farmland | Water Bodies | Wetlands |
|---|---|---|---|---|---|---|---|---|
| | | | | Consumptions | | | | |
| lcf1 | −42 | −2652 | −5 | −9 | −1 | −30 | −2 | 0 |
| lcf2 | −1837 | 0 | −283 | −149 | −35 | −1747 | −1 | −2 |
| lcf3 | −6222 | −197 | −2587 | −1172 | −331 | −3348 | −129 | −113 |
| lcf4 | −13,181 | 0 | 0 | −663 | 0 | −6933 | 0 | 0 |
| lcf5 | 0 | −685 | −2424 | −926 | −612 | 0 | −54 | −392 |
| lcf6 | −2242 | −422 | 0 | −2264 | −1575 | −2402 | −12 | −419 |
| lcf7 | 0 | 0 | −148,019 | 0 | 0 | 0 | 0 | 0 |
| lcf8 | −834 | −242 | −600 | −306 | −387 | −721 | −75 | −72 |
| lcf9 | −32 | −88 | −2878 | −1505 | −1872 | −84 | −568 | −137 |
| Total consumption | −24,388 | −4286 | −156,795 | −6995 | −4813 | −15,265 | −842 | −1136 |
| | | | | Formations | | | | |
| lcf1 | | 2740 | | | | | | |
| lcf2 | | 4053 | | | | | | |
| lcf3 | | 14,100 | | | | | | |
| lcf4 | 14,928 | | | 196 | | 5653 | | |
| lcf5 | 2233 | | | 983 | | 1878 | | |
| lcf6 | | | 8843 | 92 | 39 | 362 | | |
| lcf7 | | | 148,019 | | | | | |
| lcf8 | | | | | 75 | | 2701 | 461 |
| lcf9 | 1 | 0 | 1 | 1388 | 5327 | 82 | 160 | 208 |
| Total formation | 17,161 | 20,893 | 156,863 | 2659 | 5440 | 7976 | 2861 | 670 |
| Net change | −7228 | 16,607 | 67 | −4336 | 627 | −7290 | 2019 | −467 |

Lcf1 = urban land management; lcf2 = urban residential expansion; lcf3 = expansion of economic sites and infrastructure; lcf4 = agriculture internal conversions; lcf5 = conversion from other land cover to agriculture; lcf6 = increase in forest land cover and other semi-natural areas; lcf7 = forest internal land cover changes; lcf8 = water body and wetland creation and management; lcf9 = changes in land cover due to natural and multiple causes.

The net change in artificial surfaces was almost 17,000 km$^2$, which can also be read from Table 2. The "Formations" section of Table 3 also explains that the major part, 14,100 km$^2$, of the increase in artificial surfaces was due to lcf3, i.e., the expansion of economic sites and infrastructure. Table 3 also explains where these processes took place, i.e., where the sprawl of economic sites and infrastructure consumed other land cover. This can be found in the "Consumptions" part of Table 3, in row lcf3, where, summing the km$^2$ consumptions of all LEAC categories, we receive the same 14,100 km$^2$ as in the formation section of artificial surfaces due to lcf3. The land accounting statistics indicate that the majority of the sprawl of economic sites and infrastructure was on arable land and permanent crops, converting 6222 km$^2$ to artificial surfaces. Another large consumption due to lcf3 took place on "Pastures and mosaic farmland" where 3348 km$^2$ land was converted to economic infrastructure. "Forests and transitional woodland, shrub" lost 2587 km$^2$ to the expansion of economic infrastructure and 1172 km$^2$ of natural grasslands, heathland, and sclerophyllous vegetation was converted to artificial surfaces through lcf3. Urban residential expansion (lcf2) happened over 4053 km$^2$ and mostly impacted "Arable land and permanent crops" and "Pastures and mosaic farmlands" where a sum of approximately 3600 km$^2$ was converted to artificial surfaces.

During the period 2000–2018, 24,388 km$^2$ of arable land were transformed, and a large portion of arable land and permanent crop changes were internal conversions (lcf4), e.g., crop rotations, converting orchards or olive plantations to arable lands, or the opposite changes. The increase in forest land cover and other semi-natural areas (lcf6) happened mostly in forests and transitional woodland, with shrubs creating 8843 km$^2$ of these green areas. From the consumption section, we can understand that the increase in forest land cover and other semi-natural areas (lcf6) mostly used arable lands and permanent crops (2242 km$^2$), natural grassland, heathland, and sclerophyllous vegetation (2264 km$^2$), open spaces with little or no vegetation (1575 km$^2$), and pastures and mosaic farmland (2402 km$^2$).

The changes in the land cover classes in Table 3 can be further analyzed regionally by mapping the differences in the formation and consumption processes for each grid cell. In this paper, we map and analyze the spatial pattern of net changes in arable land and permanent crops between 2000 and 2018 (Figure 9) together with their main contributing processes (Figures 10 and 11). Arable land and permanent crops cover 25% of Europe (Figure 3), and during the 20-year period this land cover lost most surface area (Figure 4). Hence, analyzing the net gains and losses of this large land cover through the land cover flows gives us insight into the "to" and "from" conversion processes and the connection between changes in arable land and permanent crops and other land covers.

The largest hotspots of arable land and permanent crop losses were observed around major capitals and major cities, indicating that the drivers behind the losses were the sprawl of urban areas and of economic infrastructure. While in Scandinavia the loss of arable land and permanent crops was not significant, some countries in central, western, and southern Europe show distinct agricultural area losses. Dominant patterns with decreasing agricultural areas are apparent in Poland, the Netherlands, the UK, Ireland, Hungary, the interior and southern parts of Spain and southern Portugal. In Poland and the Netherlands, around 40% of the loss of arable land and permanent crops was due to the sprawl of economic sites and infrastructure (lcf3, Figure 10), which was responsible for little more than 70% of the loss in the UK. In Spain, Portugal, and Ireland, more than 70% of the loss was due to agriculture internal conversions (lcf4), which was a major driver in Hungary as well. In Hungary, Poland, Lithuania, and Finland, the increase in forest land cover and semi-natural areas also contributed to the loss of arable lands and permanent crops with 15% of the conversions from agriculture creating these green spaces.

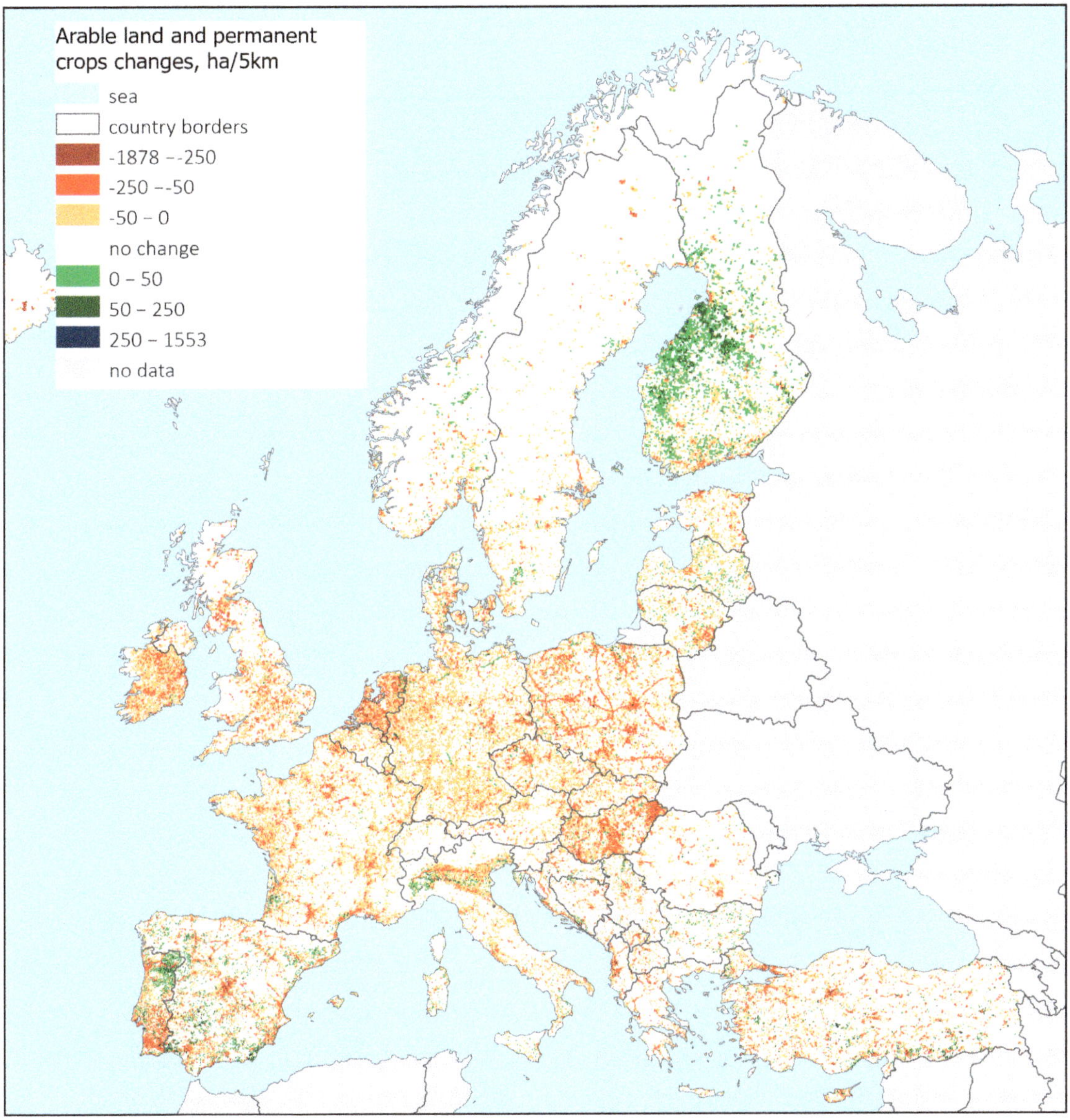


**Figure 9.** Arable land and permanent crops gain and losses during the period 2000–2018, EEA-39.

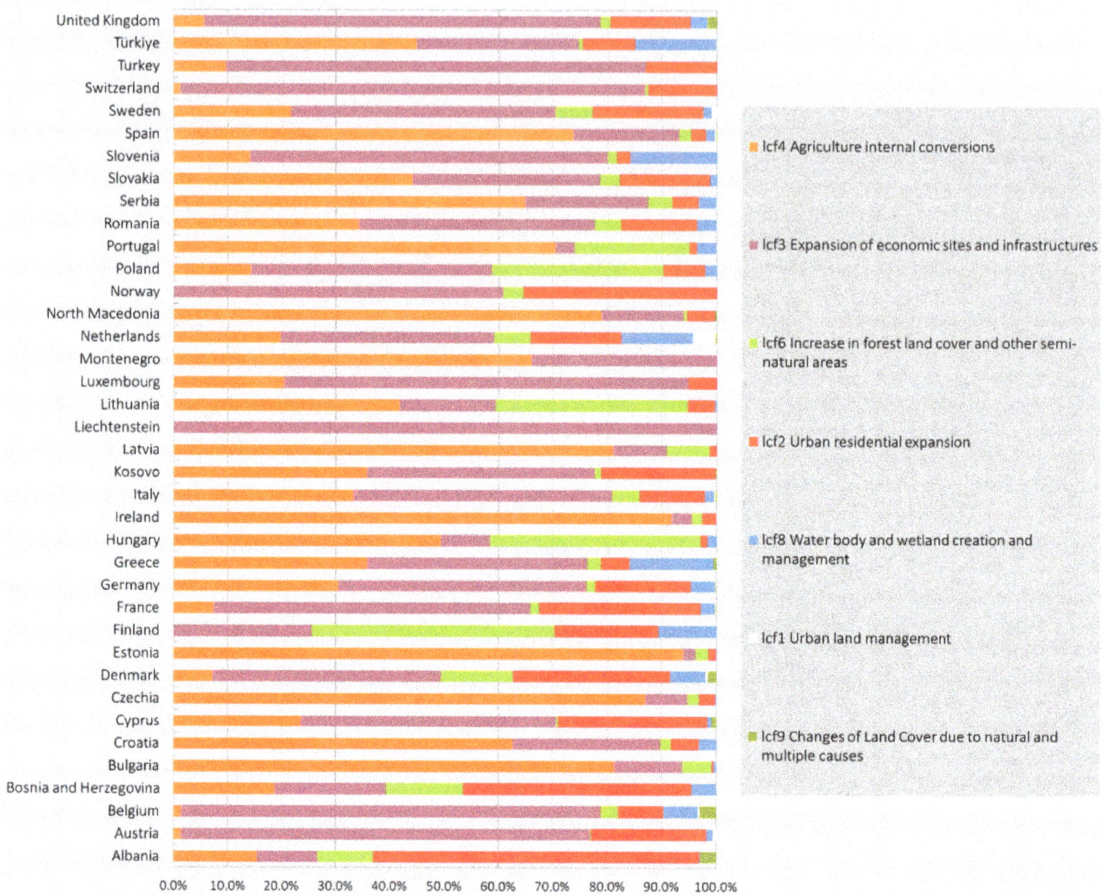


**Figure 10.** Consumption of arable land and permanent crops in the EEA-38 + UK region, during 2000–2018 by the various land cover flows.

The largest hot spots of arable land and permanent crop gains were observed in Spain, northern Portugal, southern Turkey, the Baltic countries (in particular Latvia), and from the south to central Finland (Figure 9). While in the Baltic countries of Estonia, Latvia, and Lithuania, arable land was created almost exclusively by agriculture internal conversions (converting pastures), in central Finland, all the gains were due to converting other land cover to agriculture. And a more in-depth investigation of land accounts (see tab 6 in [57]) shows that in Finland, 69% of the arable land and permanent crops-gained processes converted forested lands and 31% converted wetlands. In Portugal and Spain, the main reasons were the conversion of pastures, forest, and semi-natural land.

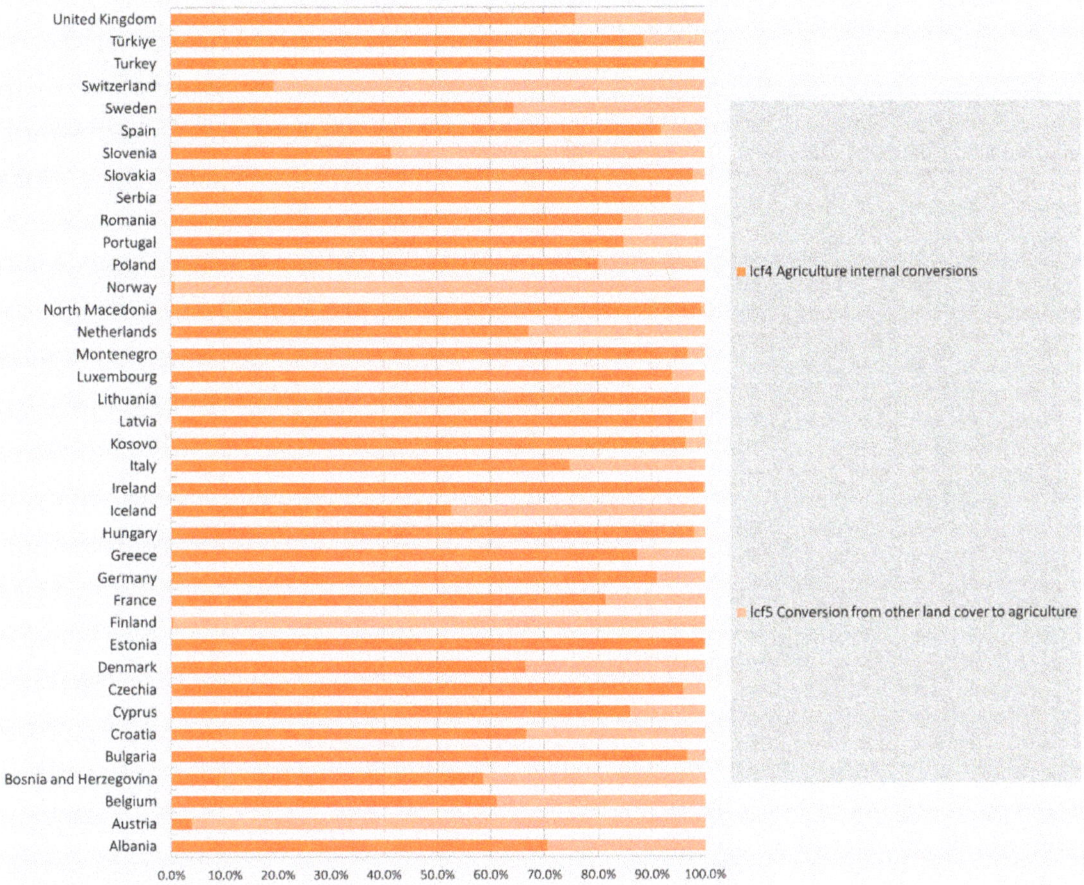


**Figure 11.** Formation of arable land and permanent crops in the EEA-38 + UK region, during 2000–2018 by the various land cover flows.

## 4. Discussion

Europe's land resources are under pressure—about 80% of Europe's land surface has been shaped by human activities, such as being covered with buildings, roads, industrial infrastructure, or used for agriculture and forestry [57]. The driving forces of change are socio-economic pressures, such as the need for land surface for industrial and economic activities, urban sprawl, and the need for food, fiber, and feed. The way we use land constitutes one of the main drivers of environmental degradation and climate change. Under sustainable conditions, land is used in a multi-functional way to satisfy economic, social, and environmental needs. When multi-functionality is interrupted by non-sustainable use, the delivery of ecosystem services is hampered. The negative impacts of intensive and non-sustainable land use may reach a level where the land starts to degrade, in some cases reaching an irreversibly degraded state.

The status and change in land and ecosystems are important elements to inform the development and implementation of policies that impact on these components of natural capital, such as water, climate, agricultural land, forest, biodiversity, and regional planning. Key EU policy documents, i.e., the Eight Environment Action Programme and the Biodiversity Strategy to 2030, set the development of natural capital accounts in the EU, with a focus on ecosystems and their services, including land, as their main objectives.

For serving the European policies especially on natural capital, climate change mitigation, adaptation, and biodiversity, the production of ecosystem and land accounts, their spatial location, and the drivers of change are important elements which inform us on the status of Europe's environment. The land and ecosystem accounting (LEAC) method presented here quantifies the land cover extent as stocks and quantifies the changes (increase and decrease in the extent) as flows between one time-step to another. The LEAC methodology and datasets align with the SEEA-CF internationally agreed-upon standards and the approach also supports the calculation of ecosystem accounts as proposed by the United Nations Statistics Department on 'Experimental Ecosystem Accounting (SEEA-EEA).

Using the described method and the geospatial time series, we have developed a vast amount of information can be derived on land cover change due to land use change. The net of the land cover formations and consumptions during 2000–2018 in Europe indicate that on the continental level, land cover extent is quite stable. The net surface of natural grassland, heathland, and sclerophyllous vegetation, of pastures and mosaic farmland, and of arable land and permanent crops have decreased the most, but only by 1% or their 2000 area. Still, this decrease together resulted in approximately 15,000 km$^2$ of net loss of this important land cover resulting from consumption processes dominating formation processes. In fact, of all consumption processes, 11% happened on arable lands and permanent crops, and 7% on pastures and mosaic farmlands. Next to the loss of the extent of these areas, large land cover changes occurred within these areas due to agriculture internal conversions; hence, agriculture land management, which represented almost 10% of all changes. Countries where agricultural management dominated all land cover changes included Czechia (over 50% of the changes), Ireland, and Germany (approx. 40% of all land cover changes).

The extent of forests and transitional woodlands net change during 2000–2018 was well below 0.1% in Europe, resulting from almost the same amount of formation and consumption processes of this land cover at the continental level. However, both the largest consumption processes and the largest formation processes in Europe occurred in this land cover type, with consumption and formation approximately 9000 km$^2$/year, representing 73% of all consumption and formation processes. Hence, while the forest land cover extent in Europe remained fairly constant (see also Table 2), the formation and consumption numbers indicate that it is the forest management practices within the forest land use category that dominates European land use. This process was predominantly in Sweden, Norway, and Finland, where more than 90% of the land cover changes happened within their forests, whereas in Latvia and Slovakia forest management, it accounted for more than 80% of the land cover changes.

The only land cover with a large, and by far the largest, change in its extent during 2000–2018 was artificial surfaces, with a 1160.7 km$^2$ annual increase, which is almost 10% of all land cover increases in this period. While we observed consumption of artificial surfaces as well, which can be attributed to land recycling, it was very little compared to the formation processes, resulting in a 7% net increase compared to artificial surfaces in the year 2000. The net increase in artificial areas, i.e., the sum of the formations and consumptions amounted to 923 km$^2$/year; hence, during the 18-year period, sealed areas in Europe increased by 16,606 km$^2$, which is roughly six times the area of Luxembourg. The processes that create artificial areas are urban sprawl (lcf2) and the sprawl of economic sites and infrastructure (lcf3), which were the third largest processes in Europe (after agriculture and forest internal conversions), converting 18,000 km$^2$ of land. Urban sprawl and the sprawl of economic sites and infrastructure mostly happened in the Netherlands (approx. 50% of all land-take processes in the country) and were around 25% of all land conversion processes in Denmark and Germany.

The major fraction of the sprawl of economic sites and infrastructure was on arable land and permanent crops converting 6222 km$^2$ to artificial surfaces and on "Pastures and mosaic farmland" where 3348 km$^2$ land was converted to economic infrastructure. Forests and transitional woodland, shrub lost 2587 km$^2$ to the expansion of economic infrastructure and 1172 km$^2$ of natural grasslands, heathland and sclerophyllous vegetation was converted to artificial surfaces through lcf3. Urban residential expansion (lcf2) happened over 4053 km$^2$ and mostly impacted "Arable land and permanent crops" and "Pastures and mosaic farmlands" where a sum of approximately 3600 km$^2$ was converted to artificial surfaces.

The core data set underpinning the presented land and ecosystem accounting system is Corine Land Cover accounting layers. The CLC accounting layers enable the consistent assessment of land accounts for the 20 years period of 2000–2018. The CLC accounting layers may be freely integrated with other datasets, such as biogeographical regions, administrative boundaries, geo-physical variables such as elevation classes, other land use information such as landscape fragmentation or socio-economic variables such as population density to understand the magnitude, the drivers and the impacts of land cover changes. The EU Copernicus Earth Observation Programme and the increasing collection of geospatial data on ecosystems and biodiversity will increase the data sets available for land and ecosystem accounting manifold. The presented land and ecosystem accounting approach can flexibly incorporate any new or updated datasets on demand.

The presented data cube approach allows the efficient, transparent, repeatable, and quality-controlled integration of geospatial datasets in accounting statistics. By converting geospatial datasets into analysis-ready formats, the workflow is faster and more effective than performing GIS analysis on raw data. If, for example, a new version of one of the input geospatial datasets becomes available, updating the accounting cubes is efficient as only the dimension in question needs to be renewed. Furthermore, as the dimensions are stored in the cloud environment, the use of the same inputs is ensured in the case of several existing versions of the same input dataset. This allows the system to deal with the increasing set of input data sets and analytical demands efficiently. With the increasing number of high-spatial and -temporal resolution Earth observation images, data are becoming bigger and bigger in size, in many cases reaching several hundreds of terabytes. The management and processing of big data is therefore a necessity in order to ensure fast, efficient, up-to-date, and state-of-the-art policy support. This is the way forward in geospatial environmental accounting systems and will allow more versatile support for a wider range of policy questions.

**Author Contributions:** Conceptualization, all authors; methodology, E.I., E.O., R.M., G.M., B.K., J.F., M.G. and G.H.; software, E.I., E.O., R.M. and E.M.; validation, R.M. and E.M.; formal analysis, E.I., E.O. and M.G.; investigation, E.I., E.O. and M.G.; data curation, R.M. and E.M.; writing—original draft preparation, E.I.; writing—review and editing, all authors; visualization, E.I. All authors have read and agreed to the published version of the manuscript.

**Funding:** This research received no external funding.

**Data Availability Statement:** The Corine Land Cover accounting layers are publicly available via the following link: https://www.eea.europa.eu/data-and-maps/data/corine-land-cover-accounting-layers/clc-accounting-layers. Accessed on 12 August 2024.

**Acknowledgments:** The authors thank the European Environment Agency for producing and making the Corine Land Cover accounting layers and the twenty years of land cover accounting statistics available.

**Conflicts of Interest:** Author Erika Orlitova was employed by the company Gisat s.r.o. Authors Gergely Maucha and Barbara Kosztra were employed by the company Lechner Non-profit Ltd. Authors Mirko Gregor and Manuel Löhnertz were employed by the company Space4environment. The remaining authors declare that the research was conducted in the absence of any commercial or financial relationships that could be construed as a potential conflict of interest.

## Appendix A. Processing Steps to Produce the CLC Accounting Layers

The accounting layers were produced with the following workflow:

Step 1: Include formation information from CLC-change layers into current CLC2018 status by creating CLC2018 accounting layer.

1. Overwrite CLC2018 with code_2006 from CLC-change 2000–2006. Intermediate result: A1_CLC2018.
2. Overwrite A1_CLC2018 with code_2012 from CLC-change 2006–2012. Intermediate result: A2_CLC2018.
3. Overwrite A2_CLC2018 with code_2018 from CLC-change 2012–2018. Result: CLC2018 accounting layer.

Step 2: Create CLC2012 accounting by including consumption information (code 2012 from CLC-change 2012–2018) into CLC2018 accounting layer. Result: CLC2012 accounting layer.

Step 3: Create CLC2006 accounting by including consumption information (code 2006 from CLC-change 2006–2012) into CLC2012 accounting layer. Result: CLC2006 accounting layer.

Step 4: Create CLC2000 accounting by including consumption information (code_2000 from CLC-change 2000–2006) into CLC2006 accounting layer. Result: CLC2000 accounting layer.

The Corine Land Cover accounting layers are publicly available via the following link: https://www.eea.europa.eu/data-and-maps/data/corine-land-cover-accounting-layers/clc-accounting-layers. Accessed on 12 August 2024.

The accounting layers can be explored as web map services under the following links: Land/CLC_accounting_layer_2000 (ImageServer) (europa.eu); Land/CLC_accounting_layer_2006 (ImageServer) (europa.eu); Land/CLC_accounting_layer_2012 (ImageServer) (europa.eu); Land/CLC_accounting_layer_2018 (ImageServer) (europa.eu). Accessed on 12 August 2024.

## Appendix B. Land Cover Extent Accounts for the Year 2018 Aggregated for Biogeographic Regions

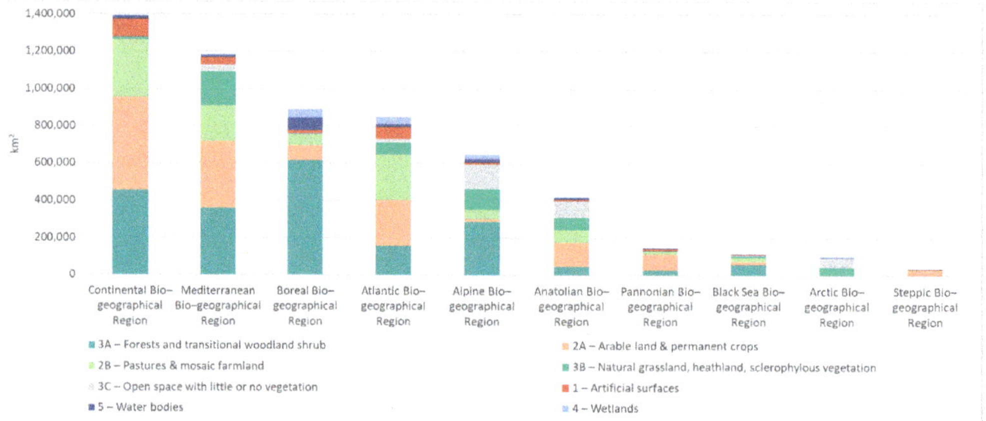


**Figure A1.** *Cont.*

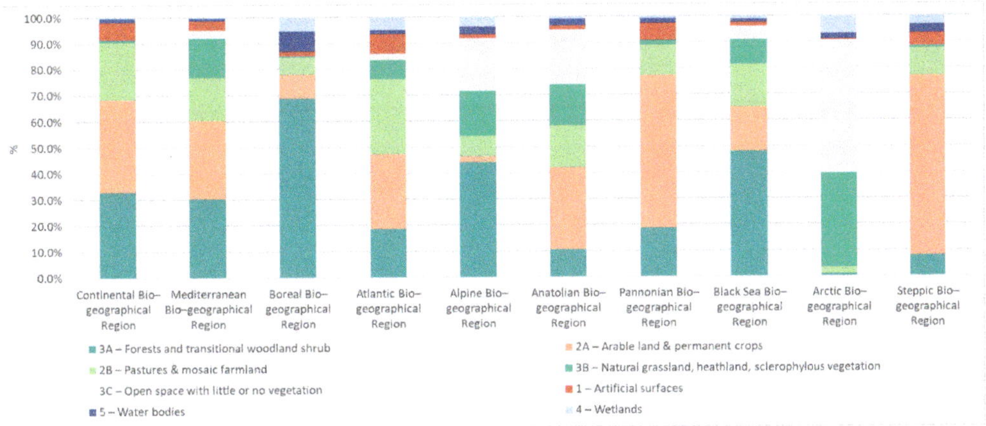


**Figure A1.** Distribution of LEAC classes in the study area (EEA-39 region) by biogeographic regions, for the year 2018. Top chart in km$^2$ of the study area, bottom chart in % of the biogeographic region.

## Appendix C. Land Cover Extent Accounts for the Periods of 2000–2006, 2006–2012, and 2012–2018

| begining of the period | | Artificial surfaces | Arable land & permanent crops | Pastures & mosaic farmland | Forests and transitional woodland shrub | Natural grassland, heathland, sclerophylou | Open space with little or no vegetation | Wetlands | Water bodies | Total |
|---|---|---|---|---|---|---|---|---|---|---|
| | LC [ km²] | 234,085 | 1,479,595 | 980,209 | 2,011,922 | 499,660 | 350,396 | 148,277 | 151,068 | 5,855,212 |
| | per period [ km²] | 1,961.60 | 8,527.69 | 5,141.29 | 48,540.55 | 2,913.89 | 2,461.75 | 655.04 | 370.64 | 70,572.45 |
| consumption | per year [ km²] | 326.93 | 1,421.28 | 856.88 | 8,090.09 | 485.65 | 410.29 | 109.17 | 61.77 | 11,762.08 |
| | % of LC change | 2.78% | 12.08% | 7.29% | 68.78% | 4.13% | 3.49% | 0.93% | 0.53% | 100.00% |
| | % of given LC | 0.84% | 0.58% | 0.52% | 2.41% | 0.58% | 0.70% | 0.44% | 0.25% | 1.21% |
| | per period [ km²] | 8,482.07 | 5,424.95 | 2,695.80 | 49,582.15 | 1,172.04 | 1,984.36 | 225.11 | 1,005.97 | 70,572.45 |
| formation | per year [ km²] | 1,413.68 | 904.16 | 449.30 | 8,263.69 | 195.34 | 330.73 | 37.52 | 167.66 | 11,762.08 |
| | % of LC change | 12.02% | 7.69% | 3.82% | 70.26% | 1.66% | 2.81% | 0.32% | 1.43% | 100.00% |
| | % of given LC | 3.53% | 0.37% | 0.28% | 2.46% | 0.23% | 0.57% | 0.15% | 0.67% | 1.21% |
| net change | per period [ km²] | 6,520.47 | -3,102.74 | -2,445.49 | 1,041.60 | -1,741.85 | -477.39 | -429.93 | 635.33 | 0.00 |
| | per year [ km²] | 1,086.75 | -517.12 | -407.58 | 173.60 | -290.31 | -79.56 | -71.66 | 105.89 | 0.00 |
| | % of given LC | 2.79% | -0.21% | -0.17% | 0.05% | -0.35% | -0.14% | -0.29% | 0.42% | 0.00% |
| end of period | LC [ km²] | 240,605 | 1,476,492 | 977,764 | 2,012,964 | 497,918 | 349,919 | 147,847 | 151,703 | 5,855,212 |

**Figure A2.** Land cover extent accounts for the periods of 2000–2006.

| begining of the period | | Artificial surfaces | Arable land & permanent crops | Pastures & mosaic farmland | Forests and transitional woodland shrub | Natural grassland, heathland, sclerophylou | Open space with little or no vegetation | Wetlands | Water bodies | Total |
|---|---|---|---|---|---|---|---|---|---|---|
| | LC [ km²] | 240,605 | 1,476,492 | 977,764 | 2,012,964 | 497,918 | 349,919 | 147,847 | 151,703 | 5,855,212 |
| | per period [ km²] | 2,546.49 | 10,037.77 | 5,744.22 | 69,987.22 | 2,200.24 | 2,181.98 | 213.51 | 345.15 | 93,256.58 |
| consumption | per year [ km²] | 424.42 | 1,672.96 | 957.37 | 11,664.54 | 366.71 | 363.66 | 35.59 | 57.53 | 15,542.76 |
| | % of LC change | 2.73% | 10.76% | 6.16% | 75.05% | 2.36% | 2.34% | 0.23% | 0.37% | 100.00% |
| | % of given LC | 1.06% | 0.68% | 0.59% | 3.48% | 0.44% | 0.44% | 0.14% | 0.23% | 1.59% |
| | per period [ km²] | 8,343.93 | 5,974.97 | 3,770.69 | 70,602.83 | 1,108.86 | 1,929.73 | 278.62 | 1,246.95 | 93,256.58 |
| formation | per year [ km²] | 1,390.66 | 995.83 | 628.45 | 11,767.14 | 184.81 | 321.62 | 46.44 | 207.83 | 15,542.76 |
| | % of LC change | 8.95% | 6.41% | 4.04% | 75.71% | 1.19% | 2.07% | 0.30% | 1.34% | 100.00% |
| | % of given LC | 3.39% | 0.40% | 0.39% | 3.51% | 0.22% | 0.39% | 0.19% | 0.82% | 1.59% |
| net change | per period [ km²] | 5,797.44 | -4,062.80 | -1,973.53 | 615.61 | -1,091.38 | -252.25 | 65.11 | 901.80 | 0.00 |
| | per year [ km²] | 966.24 | -677.13 | -328.92 | 102.60 | -181.90 | -42.04 | 10.85 | 150.30 | 0.00 |
| | % of given LC | 2.41% | -0.28% | -0.13% | 0.03% | -0.22% | -0.05% | 0.04% | 0.59% | 0.00% |
| end of period | LC [ km²] | 246,402 | 1,472,429 | 975,790 | 2,013,579 | 496,827 | 349,666 | 147,913 | 152,605 | 5,855,212 |

**Figure A3.** Land cover extent accounts for the periods of 2006–2012.

| begining of the period | | Artificial surfaces | Arable land & permanent crops | Pastures & mosaic farmland | Forests and transitional woodland shrub | Natural grassland heathland sclerophylou | Open space with little or no vegetation | Wetlands | Water bodies | Total |
|---|---|---|---|---|---|---|---|---|---|---|
| | LC [ km²] | 246.402 | 1.472.429 | 975 790 | 2.013 579 | 496.827 | 349.666 | 147.913 | 152.605 | 5.855.212 |
| consumption | per period [ km²] | 2.238 51 | 6.751.89 | 5.034.80 | 54 161.62 | 2.396.74 | 1.947.65 | 356.42 | 362.91 | 73.250.54 |
| | per year [ km²] | 373.09 | 1.125.32 | 839.13 | 9.026.94 | 399.46 | 324.61 | 59.40 | 60.49 | 12.208.42 |
| | % of LC change | 3.06% | 9.22% | 6.87% | 73.94% | 3.27% | 2.66% | 0.49% | 0.50% | 100.00% |
| | % of given LC | 0.91% | 0.46% | 0.52% | 2.69% | 0.48% | 0.56% | 0.24% | 0.24% | 1.25% |
| formation | per period [ km²] | 6.527.12 | 6.689.70 | 2.164.13 | 52.571.67 | 893.79 | 3.304.72 | 254.69 | 844.72 | 73.250.54 |
| | per year [ km²] | 1,087.85 | 1.114.95 | 360.69 | 8.761.95 | 148.97 | 550.79 | 42.45 | 140.79 | 12,208.42 |
| | % of LC change | 8.91% | 9.13% | 2.95% | 71.77% | 1.22% | 4.51% | 0.35% | 1.15% | 100.00% |
| | % of given LC | 2.60% | 0.45% | 0.22% | 2.61% | 0.18% | 0.95% | 0.17% | 0.55% | 1.25% |
| net change | per period [ km²] | 4.288.61 | -62.19 | -2,870.67 | -1.589.95 | -1.502.95 | 1.357.07 | -101.73 | 481.81 | 0.00 |
| | per year [ km²] | 714.77 | -10.36 | -478.44 | -264.99 | -250.49 | 226.18 | -16.96 | 80.30 | 0.00 |
| | % of given LC | 1.74% | 0.00% | -0.19% | -0.08% | -0.30% | 0.39% | -0.07% | 0.32% | 0.00% |
| end of period | LC [ km²] | 250.691 | 1.472.367 | 972.919 | 2.011.989 | 495.324 | 351.024 | 147.811 | 153.087 | 5.855.212 |

**Figure A4.** Land cover extent accounts for the periods of 2012−2018.

## Notes

1. https://www.eea.europa.eu/en/countries Accessed on 12 August 2024.
2. https://land.copernicus.eu/pan-european/corine-land-cover Accessed on 12 August 2024.
3. https://sdi.eea.europa.eu/catalogue/srv/api/records/8526ff78-b000-42e1-8360-a2fb3a51e4ac Accessed on 12 August 2024.
4. https://www.eea.europa.eu/en/datahub/datahubitem-view/94438969-2dd5-4ba3-b708-e4d29a8b7699 Accessed on 12 August 2024. https://sdi.eea.europa.eu/catalogue/srv/api/records/8526ff78-b000-42e1-8360-a2fb3a51e4ac Accessed on 12 August 2024.
5. https://www.eea.europa.eu/en/datahub/datahubitem-view/3c362237-daa4-45e2-8c16-aaadfb1a003b Accessed on 12 August 2024.
6. See note 5 above.

## References


1. EEA. *The European Environment—State and Outlook 2020 Knowledge for Transition to a Sustainable Europe*; EEA: Copenhagen, Denmark, 2020.
2. EEA. Net Land Take in Cities and Commuting Zones. 2023. Available online: https://www.eea.europa.eu/en/analysis/indicators/net-land-take-in-cities (accessed on 12 August 2024).
3. EEA. *Land Cover Accounts—An Approach to Geospatial Environmental Accounting*; EEA Briefing No. 12/2020; EEA: Copenhagen, Denmark, 2020; ISBN 978-92-9480-226-2. ISSN 2467-3196. [CrossRef]
4. Frelih-Larsen, A.; Bowyer, C.; Albrecht, S.; Keenleyside, C.; Kemper, M.; Nanni, S.; Naumann, S.; Mottershead, D.; Landgrebe, R.; Andersen, E.; et al. *Updated Inventory and Assessment of Soil Protection Policy Instruments in EU Member States—Final Report to DG Environment*; Ecologic Institute: Berlin, Germany, 2016; p. 462.
5. EEA. *Land System at European Level—Analytical Assessment Framework*; Briefing no. 10/2018; EEA: Copenhagen, Denmark, 2018; PDF TH-AM-18-012-EN-N; ISBN 978-92-9213-988-9. ISSN 2467-3196.
6. United Nations. System of Environmental-Economic Accounting—Ecosystem Accounting (SEEA EA). White Cover Publication, Pre-Edited Text Subject to Official Editing. 2021. Available online: https://seea.un.org/ecosystem-accounting (accessed on 12 August 2024).
7. Edens, B.; Maes, J.; Hein, L.; Obst, C.; Siikamaki, J.; Schenau, S.; Javorsek, M.; Chow, J.; Chan, J.Y.; Steurer, A.; et al. Establishing the SEEA Ecosystem Accounting as a global standard. *Ecosyst. Serv.* **2022**, *54*, 101413. [CrossRef]
8. Polce, C.; Maes, J.; Brander, L.; Cescatti, A.; Baranzelli, C.; Lavalle, C.; Zulian, G. Global change impacts on ecosystem services: A spatially explicit assessment for Europe. *One Ecosyst.* **2016**, *1*, e9990. [CrossRef]
9. Lima, J.E.F.W.; de Gois Aquino, F.; Chaves, T.A.; Lorz, C. Development of a spatially explicit approach for mapping ecosystem services in the Brazilian Savanna–MapES. *Ecol. Indic.* **2017**, *82*, 513–525. [CrossRef]
10. de Sousa, C.; Fatoyinbo, L.; Neigh, C.; Boucka, F.; Angoue, V.; Larsen, T. Cloud-computing and machine learning in support of country-level land cover and ecosystem extent mapping in Liberia and Gabon. *PLoS ONE* **2020**, *15*, e0227438. [CrossRef]
11. Vrebos, D.; Staes, J.; Broekx, S.; de Nocker, L.; Gabriels, K.; Hermy, M.; Liekens, I.; Marsboom, C.; Ottoy, S.; Van Der Biest, K.; et al. Facilitating spatially-explicit assessments of ecosystem service delivery to support land use planning. *One Ecosyst.* **2020**, *5*, e50540. [CrossRef]

12. Asante-Yeboah, E.; Koo, H.; Ros-Tonen, M.A.; Sieber, S.; Fürst, C. Participatory and Spatially Explicit Assessment to Envision the Future of Land-Use/Land-Cover Change Scenarios on Selected Ecosystem Services in Southwestern Ghana. *Environ. Manag.* **2024**, *74*, 94–113. [CrossRef]
13. Liu, J.; Dietz, T.; Carpenter, S.R.; Taylor, W.W.; Alberti, M.; Deadman, P.; Redman, C.; Pell, A.; Folke, C.; Ouyang, Z.; et al. Coupled human and natural systems: The evolution and applications of an integrated framework. *Ambio* **2021**, *50*, 1778–1783. [CrossRef]
14. Farahbakhsh, I.; Bauch, C.T.; Anand, M. Modelling coupled human–environment complexity for the future of the biosphere: Strengths, gaps and promising directions. *Philos. Trans. R. Soc. B* **2022**, *1857*, 20210382. [CrossRef] [PubMed]
15. Farahbakhsh, I.; Bauch, C.T.; Anand, M. Drivers of tipping points in coupled human-environment systems. *EGUsphere* **2023**, *preprint*. [CrossRef]
16. Houghton, R.A.; House, J.I.; Pongratz, J.; van der Werf, G.R.; DeFries, R.S.; Hansen, M.C.; Le Quéré, C.; Ramankutty, N. Carbon emissions from land use and land-cover change. *Biogeosciences* **2012**, *9*, 5125–5142. [CrossRef]
17. Perez-Quezada, J.F.; Cano, S.; Ibaceta, P.; Aguilera-Riquelme, D.; Salazar, O.; Fuentes, J.P.; Osborne, B. How do land cover changes affect carbon-nitrogen-phosphorus stocks and the greenhouse gas budget of ecosystems in southern Chile? *Agric. Ecosyst. Environ.* **2022**, *340*, 108153. [CrossRef]
18. Zhu, L.; Song, R.; Sun, S.; Li, Y.; Hu, K. Land use/land cover change and its impact on ecosystem carbon storage in coastal areas of China from 1980 to 2050. *Ecol. Indic.* **2022**, *142*, 109178. [CrossRef]
19. Li, L.; Fu, W.; Luo, M. Spatial and Temporal Variation and Prediction of Ecosystem Carbon Stocks in Yunnan Province Based on Land Use Change. *Int. J. Environ. Res. Public Health* **2022**, *19*, 16059. [CrossRef] [PubMed] [PubMed Central]
20. Tu, H.; Jiapaer, G.; Yu, T.; Zhang, L.; Chen, B.; Lin, K.; Li, X. Effects of Land Cover Change on Vegetation Carbon Source/Sink in Arid Terrestrial Ecosystems of Northwest China, 2001–2018. *Remote Sens.* **2023**, *15*, 2471. [CrossRef]
21. Legesse, F.; Degefa, S.; Soromessa, T.; Bloor, M. Carbon stock dynamics in a changing land use land cover of the Upper Awash River Basin: Implications for climate change management. *Sustain. Environ.* **2024**, *10*, 2361565. [CrossRef]
22. Muche, M.; Yemata, G.; Molla, E.; Adnew, W.; Muasya, A.M. Land use and land cover changes and their impact on ecosystem service values in the north-eastern highlands of Ethiopia. *PLoS ONE* **2023**, *18*, e0289962. [CrossRef] [PubMed] [PubMed Central]
23. Shukla, P.R.; Skea, J.; Buendia, E.C.; Masson-Delmotte, V.; Pörtner, H.-O.; Roberts, D.C.; Zhai, P.; Slade, R.; Connors, S.; van Diemen, R.; et al. (Eds.) *Climate Change and Land: An IPCC Special Report on Climate Change, Desertification, Land Degradation, Sustainable Land Management, Food Security, and Greenhouse Gas Fluxes in Terrestrial Ecosystems*; IPCC: Geneva, Switzerland, 2019, *in press*.
24. Smith, P.; Nkem, J.; Calvin, K.; Campbell, D.; Cherubini, F.; Grassi, G.; Korotkov, V.; Hoang, A.L.; Lwasa, S.; McElwee, P.; et al. Interlinkages between Desertification, Land Degradation, Food Security and Greenhouse Gas Fluxes: Synergies, Trade-offs and Integrated Response Options. In *Climate Change and Land: An IPCC Special Report on Climate Change, Desertification, Land Degradation, Sustainable Land Management, Food Security, and Greenhouse Gas Fluxes in Terrestrial Ecosystems*; Shukla, P.R., Skea, J., Buendia, E.C., Masson-Delmotte, V., Portner, H.-O., Roberts, D.C., Zhai, P., Slade, R., Connors, S., van Diemen, R., et al., Eds.; IPCC: Geneva, Switzerland, 2019; *in press*.
25. Borrelli, P.; Robinson, D.A.; Fleischer, L.R.; Lugato, E.; Ballabio, C.; Alewell, C.; Meusburger, K.; Modugno, S.; Schütt, B.; Ferro, V.; et al. An assessment of the global impact of 21st century land use change on soil erosion. *Nat. Commun.* **2017**, *8*, 2013. [CrossRef] [PubMed]
26. Tim, N. Future effects of climate and land-use change on terrestrial vertebrate community diversity under different scenarios. *Proc. R. Soc. B* **2018**, *285*, 28520180792. [CrossRef]
27. Marques, A.; Martins, I.S.; Kastner, T.; Plutzar, C.; Theurl, M.C.; Eisenmenger, N.; Huijbregts, M.A.J.; Wood, R.; Stadler, K.; Bruckner, M.; et al. Increasing impacts of land use on biodiversity and carbon sequestration driven by population and economic growth. *Nat. Ecol. Evol.* **2019**, *3*, 628–637. [CrossRef]
28. Watt, A. Land-Use Intensity and Land-Use Change: Impacts on Biodiversity. In *Life on Land. Encyclopedia of the UN Sustainable Development Goals*; Leal Filho, W., Azul, A., Brandli, L., Lange Salvia, A., Wall, T., Eds.; Springer: Cham, Switzerland, 2020. [CrossRef]
29. Abbas, S.; Yousaf, H.; Khan, S.; Rehman, M.Z.; Blueschke, D. Analysis and Projection of Transport Sector Demand for Energy and Carbon Emission: An Application of the Grey Model in Pakistan. *Mathematics* **2023**, *11*, 1443. [CrossRef]
30. Semenchuk, P.; Plutzar, C.; Kastner, T.; Matej, S.; Bidoglio, G.; Erb, K.H.; Essl, F.; Haberl, H.; Wessely, J.; Krausmann, F.; et al. Relative effects of land conversion and land-use intensity on terrestrial vertebrate diversity. *Nat. Commun.* **2022**, *13*, 615. [CrossRef] [PubMed]
31. Maes, J.; Teller, A.; Erhard, M.; Liquete, C.; Braat, L.; Berry, P.; Egoh, B.; Puydarrieux, P.; Fiorina, C.; Santos, F.; et al. Mapping and Assessment of Ecosystems and their Services. In *An Analytical Framework for Ecosystem Assessment under Action 5 of the EU Biodiversity Strategy to 2020*; Publications office of the European Union: Luxembourg, 2013.
32. Obst, C.; Hein, L.; Edens, B. National Accounting and the Valuation of Ecosystem Assets and Their Services. *Environ. Resour. Econ.* **2016**, *64*, 1–23. [CrossRef]
33. La Notte, A.; Vallecillo, S.; Maes, J. Capacity as "virtual stock" in ecosystem services accounting. *Ecol. Indic.* **2019**, *98*, 158–163. [CrossRef]
34. La Notte, A.; Vallecillo, S.; Marques, A.; Maes, J. Beyond the economic boundaries to account for ecosystem services. *Ecosyst. Serv.* **2019**, *35*, 116–129. [CrossRef] [PubMed]

35. La Notte, A.; Rhodes, C. The theoretical frameworks behind integrated environmental, ecosystem, and economic accounting systems and their classifications. *Environ. Impact Assess. Rev.* **2020**, *80*, 106317. [CrossRef]
36. Comte, A.; Campagne, C.S.; Lange, S.; Bruzón, A.G.; Hein, L.; Santos-Martín, F.; Levrel, H. Ecosystem accounting: Past scientific developments and future challenges. *Ecosyst. Serv.* **2022**, *58*, 101486. [CrossRef]
37. Schröter, M.; van der Zanden, E.H.; van Oudenhoven, A.P.E.; Remme, R.P.; Serna-Chavez, H.M.; de Groot, R.S.; Opdam, P. Ecosystem Services as a Contested Concept: A Synthesis of Critique and Counter-Arguments. *Conserv. Lett.* **2014**, *7*, 514–523. [CrossRef]
38. Barbier, E.B. The protective service of mangrove ecosystems: A review of valuation methods. *Mar. Pollut. Bull.* **2016**, *109*, 676–681. [CrossRef]
39. Caparrós, A.; Oviedo, J.; Álvarez, A.; Campos, P. Simulated exchange values and ecosystem accounting: Theory and application to free access recreation. *Ecol. Econ.* **2017**, *139*, 140–149. [CrossRef]
40. Maes, J.; Driver, A.; Czúcz, B.; Keith, H.; Jackson, B.; Nicholson, E.; Dasoo, M. A review of ecosystem condition accounts: Lessons learned and options for further development. *One Ecosyst.* **2020**, *5*, e53485. [CrossRef]
41. Maes, J.; Bruzón, A.G.; Barredo, J.I.; Vallecillo, S.; Vogt, P.; Rivero, I.M.; Santos-Martín, F. Accounting for forest condition in Europe based on an international statistical standard. *Nat. Commun.* **2023**, *14*, 3723. [CrossRef] [PubMed]
42. Edens, B.; Hein, L. Towards a consistent approach for ecosystem accounting. *Ecol. Econ.* **2013**, *90*, 41–52. [CrossRef]
43. Boyd, J.; Banzhaf, S. What Are Ecosystem Services? The Need for Standardized Environmental Accounting Units. *Ecol. Econ.* **2007**, *63*, 616–626. [CrossRef]
44. United Nations, European Commission, International Monetary Fund, Organisation for Economic Cooperation and Development, & World Bank. System of Environmental-Economic Accounting 2012—Central Framework. United Nations. 2014. Available online: https://seea.un.org/sites/seea.un.org/files/seea_cf_final_en.pdf (accessed on 12 August 2024).
45. United Nations, European Commission, International Monetary Fund, Organisation for Economic Cooperation and Development, & World Bank. System of Environmental-Economic Accounting 2012—Experimental Ecosystem Accounting. United Nations. 2014. Available online: https://seea.un.org/sites/seea.un.org/files/seea_eea_final_en_1.pdf (accessed on 12 August 2024).
46. United Nations, European Commission, International Monetary Fund, Organisation for Economic Cooperation and Development, & World Bank. System of Environmental-Economic Accounting 2012—Applications and Extensions. United Nations. 2017. Available online: https://seea.un.org/sites/seea.un.org/files/ae_final_en.pdf (accessed on 12 August 2024).
47. Corine Land Cover Product User Manual. Available online: https://land.copernicus.eu/en/technical-library/clc-product-user-manual/@@download/file (accessed on 12 August 2024).
48. Büttner, G.; Kosztra, B. CLC2018 Technical Guidelines. Internal ETC/ULS Report. 2017. Available online: https://land.copernicus.eu/user-corner/technical-library//clc2018technicalguidelines_final.pdf (accessed on 12 August 2024).
49. EEA. Land accounts for Europe 1990–2000. In *Towards Integrated Land and Ecosystem Accounting*; EEA Report, No. 11/2006; EEA: Copenhagen, Denmark, 2006.
50. EEA. 'Land Take (CSI 014/LSI 001)', European Environment Agency. 2019. Available online: https://www.eea.europa.eu/data-and-maps/indicators/land-take-3/assessment (accessed on 11 October 2019).
51. Ivits, E.; Milego, R.; Mancosu, M.; Gregor, M.; Petersen, J.E.; Buttner Gy Lohnertz, M.; Maucha, G.; Petrik, O.; Bastrup-Birk, A.; Tafi, J.; et al. Land and Ecosystem Accounts for Europe. In *Towards Geospatial Environmental Accounting*; ETC/ULS Report 02/2020; EEA: Copenhagen, Denmark, 2020; ISBN 978-3-200-07114-8.
52. EEA. *Measuring the Extent and Condition of European Ecosystems*; Briefing No. 24/2020; EEA: Copenhagen, Denmark, 2021; ISBN 978-92-9480-309-2. ISSN 2467-3196. [CrossRef]
53. EEA. Land Cover Country Fact Sheets 2000–2018. EEA Dashboard. 2021. Available online: https://www.eea.europa.eu/themes/landuse/land-cover-country-fact-sheets (accessed on 14 May 2024).
54. EEA. *Natura 2000 and Land Cover Data Viewer*; European Environment Agency: Copenhagen, Denmark, 2019. Available online: https://www.eea.europa.eu/data-and-maps/dashboards/natura-2000-data-viewer (accessed on 12 August 2024).
55. Feranec, J.; Jaffrain, G.; Soukup, T.; Hazeu, G. Determining changes and flows in European landscapes 1990-2000 using CORINE land cover data. *Appl. Geogr.* **2010**, *30*, 19–35. [CrossRef]
56. EEA. Land Cover and Change Accounts 2000–2018. EEA Dashboard. 2019. Available online: https://www.eea.europa.eu/data-and-maps/dashboards/land-cover-and-change-statistics (accessed on 14 May 2024).
57. EEA. Land Use. 2024. Available online: https://www.eea.europa.eu/en/topics/in-depth/land-use (accessed on 12 August 2024).

*remote sensing*

MDPI

*Technical Note*

# Accelerated Atmospheric to Hydrological Spread of Drought in the Yangtze River Basin under Climate

**Chengyuan Zhang [1,2], Zhiming Han [3], Shuo Wang [1,2], Jiankun Wang [1,2], Chenfeng Cui [1,2,*] and Junrong Liu [4]**

1 College of Water Resources and Architectural Engineering, Northwest A&F University, Xianyang 712100, China; summer_zcy@nwafu.edu.cn (C.Z.); shuowang@nwafu.edu.cn (S.W.); wangjiankun@nwafu.edu.cn (J.W.)

2 Key Laboratory for Agricultural Soil and Water Engineering in Arid Area of Ministry of Education, Northwest A&F University, Xianyang 712100, China

3 College of Natural Resources and Environment, Northwest A&F University, Xianyang 712100, China; hanzhiming@nwafu.edu.cn

4 China Coal Aerial Survey and Remote Sensing Group Co., Ltd., Xi'an 710100, China; mhyg_arsc@163.com

* Correspondence: cuichenfeng@nwafu.edu.cn

**Abstract:** Persistent droughts pose a threat to agricultural production, and the changing environment worsens the risk of drought exposure. Understanding the propagation of drought in changing environments and assessing possible impact factors can help in the early detection of drought, guiding agricultural production practices. The current study cannot reflect the propagation status of drought to the total terrestrial hydrological drought, so this work creatively investigated the atmospheric to hydrological drought propagation time in the Yangtze River Basin under the dynamic and static perspectives based on the Standardized Precipitation Evapotranspiration Index and the Terrestrial Water Storage Anomalous Drought Index, fine-tuned the time scale to the seasonal scale, and explored the contributing capacity of the variable interactions. The results show that: (1) under the dynamic perspective, while the propagation time is decreasing in the annual scale, the spring season shows the opposite trend; and (2) large variability exists in the timing of drought propagation at spatial scales, with elevation playing the most important influential role, and bivariate interactions contributing stronger explanations compared to single variables. This study highlights the importance of considering the impact of variable interactions and contributes to our understanding of the response of secondary droughts to upper-level droughts, providing valuable insights into the propagation of droughts to total terrestrial hydrologic drought.

**Keywords:** GRACE; drought propagation; meteorological drought; hydrological drought; water security

## 1. Introduction

Drought is an extreme weather process that is widespread, destructive, long-lasting, and easily overlooked because it is not easily detected, posing a serious hazard to agricultural production and economic growth [1]. Commonly, droughts can be summarized as meteorological, agricultural, hydrological, and socio-economic droughts [2]. Influenced by water and energy cycles, various types of droughts are connected through feedback [3]; based on this, upper-level droughts often cause secondary droughts, a phenomenon known as drought propagation [4]. Against the backdrop of global warming, the Earth's system is experiencing significant changes, such as the frequency of extreme climatic phenomena, and meteorological and hydrological systems are experiencing vulnerability, which directly alters the state of development of the hydrological cycle [5]. Studying drought propagation in a dynamic context and on smaller time scales and analyzing the properties that influence drought propagation will help reduce losses due to drought and guide drought risk management in watersheds.

Signs of meteorological drought often start with a lack of precipitation and intense evaporation, triggering a cascade of other types of droughts. As the energy cycle triggers a lack of soil moisture, triggering agricultural droughts, it further reduces the hydrological recharge of rivers, resulting in a shortage of available water resources and bringing hydrological droughts [6]. Studies show that about 454 million $km^2$ of cultivated area globally experience the threat of drought, with an accumulated loss of approximately USD 166 billion [7]. Threats posed by drought relate to ecosystem security and human security in all aspects of production, and in the context of drought conditions and the potential for further intensification of drought in the face of surges in demand for water as a result of socio-economic growth, an in-depth understanding of drought propagation would facilitate the orderly management and rational allocation of agricultural, industrial, and water supplies.

Lag, prolongation, and attenuation characterize the propagation of drought [8]. Recent studies have been concerned with drought propagation time and propagation threshold as a way to characterize the propagation of droughts. For example, based on order analysis and Copula theory, Wang et al. [9] derived drought propagation time and drought propagation threshold in central Yunnan, China. Han et al. [10] determined the risk probabilities and trigger thresholds for long-chain drought propagation in China by establishing a dynamic threshold framework and found that the drought risk probabilities and trigger thresholds had significant spatial heterogeneity. However, the analysis of drought propagation in changing environments and on smaller time scales is not clear from current studies, so it is key to explore drought propagation in altered contexts [5], particularly in the context of rapid warming humidification and accelerated hydrological cycles. Therefore, this paper studies drought propagation in changing environments, i.e., dynamic propagation, and explores it under a seasonal scale.

Various indices exist to detect and measure drought conditions and to explore drought propagation characteristics. The Standardized Precipitation Evapotranspiration Index (SPEI) [11] is a widely recognized meteorological drought index that explores multiple time scales, combining precipitation and evapotranspiration in its calculations, and that has demonstrated greater applicability in monitoring drought in changing environments than other meteorological drought indices [12]. For hydrological drought indices, rapidly evolving remote sensing technologies, such as the Gravity Recovery and Climate Experiment Satellite (GRACE) and its successor (GRACE-FO), perform well in measuring large-scale watershed storage [13,14]. Accordingly, the terrestrial water storage (TWS) settled by the satellite can reflect the changes in terrestrial water storage anomalies (TWSA), and the resulting drought index, TWSA-DSI, has been extensively utilized to explore hydrologic drought [15–17].

Previous studies have shown that meteorological and hydrological factors and teleconnection factors exert a direct consequence on drought propagation [4]. For example, a previous study [5] summarized that precipitation and the Pacific Interdecadal Oscillation have the most important roles for factors accelerating drought propagation in the Pearl River region, with precipitation having a negative correlation effect. Globally, Van Loon and Laaha [18] argued that this case is more linked to climatic scenarios and not more focused on the geographic characteristics of watersheds. However, it has been shown that the effect of the geographical status of the watershed cannot be ignored [19,20], and anthropogenic impacts such as water supply irrigation, reservoir construction, etc. have also been shown to be strong factors influencing the propagation time of droughts [21,22]. Moreover, most studies only assess the contribution of a singular variable, but the joint feedback effect of bivariate factors is not yet known, so it is crucial to explore the meteorological factors and catchment characteristics in the watershed as well as the joint effect of bivariate factors.

As China's superpower river, the Yangtze River Basin (YRB) is an important economic, cultural, and food production base, and the rational distribution of water is a guarantee of agricultural security and human life safety. Current research has concluded that the YRB is facing more severe drought pressure since 2000 [23]; therefore, the YRB is chosen as the research object of this paper, and the main objectives included in the article are to: (1) identify meteorological and terrestrial hydrological droughts in the YRB; (2) determine the timing of static propagation and dynamic propagation on annual and seasonal scales, respectively; and (3) explore the meteorological factors and the geographic characteristics of catchments that may influence the propagation time of droughts. In this paper, drought propagation time is analyzed in depth from different perspectives, and the findings will enhance the knowledge of drought propagation, which will facilitate water management and allocation in the YRB.

## 2. Study Area and Data

### 2.1. Study Area

The YRB (90°E–122°E, 24°N–35°N) (Figure 1) covers a territory of nearly 1.8 million km$^2$, about 18.8% of the country's total land area, with an average multi-year water resource volume of 996 billion m$^3$, which is close to 36% of the national total. The YRB spans three regions, and the basin is characterized by high population density and a high level of economic development, with a GDP exceeding 40%. The terrain of the basin is a three-stage ladder distribution, west high (7143 m) and east low (−142 m) (Figure 1a), the average precipitation from west to east for many years is 300 mm ~ 2400 mm, the average temperature for many years is about −4~18 °C [24], and the time and space distribution of the region's climate and hydrological characteristics is extremely uneven. The land-use types in the watershed are mainly categorized as grassland, forest, and cropland, as shown in Figure 1b; the main soil types present include Leptosol, Cambisol, Regosol, and Acrisol, as shown in Figure 1c. Under rapid global warming and accelerated urbanization, the climate and hydrology of the watersheds have been severely affected, and the region faces the threat of increasing droughts and floods [25].

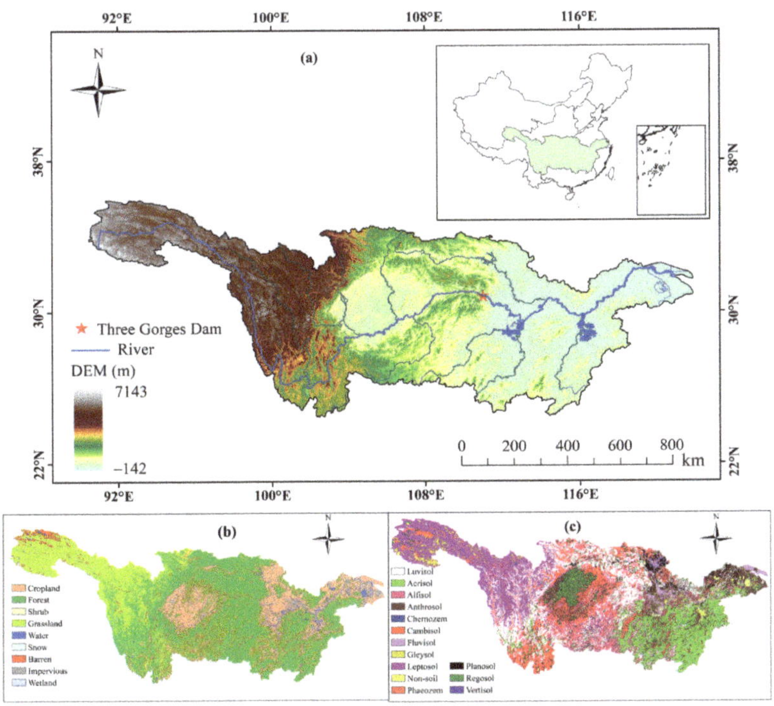


**Figure 1.** Location (**a**), land-use map (**b**), and soil map (**c**).

## 2.2. Data

### 2.2.1. GRACE Data

We used GRACE RL06 mascon data from the University of Texas Space Research Center from between April 2002 and December 2022 to derive TWSA data. This data can be obtained from https://www2.csr.utexas.edu/grace/RL06_mascons.html (accessed on 19 September 2023) with a resolution of 0.25° in space and 1 month in time. The mascon-based technique not only improves the resolution and reduces the error, but also preserves the spatial and temporal details better than classical spherical harmonic processing [26]. For technical reasons, there were 11 months of data gaps during the period from 04/2002–06/2017 (GRACE) and 06/2018–present (GRACE-FO), but the vacant data could be supplemented by a linear interpolation approach [27].

### 2.2.2. GLDAS Data

The Global Land Data Assimilation System (GLDAS) consists of four high-resolution land surface models: Noah, Mosaic, Community Land model, and Variable Infiltration Capacity [28], which contain precipitation, soil moisture, runoff, vaporization, temperature, and other land surface data. In this study, the GLDAS-2.1 Noah model data was utilized to analyze the meteorological and geographic factors influencing drought propagation, consistent with the GRACE satellite resolution (0.25°). This analysis included runoff, temperature, and soil moisture at various depths (0–10 cm, 10–40 cm, 40–100 cm, and 100–200 cm) for the period from April 2002 to December 2022. This data can be downloaded from https://disc.gsfc.nasa.gov/information?page=1&project=GRACE-DA-DM,FLDAS,GLDAS,NEWS,NLDAS,NCA-LDAS&keywords=hydrology (accessed on 25 September 2023).

### 2.2.3. Precipitation and Potential Evapotranspiration Data

This study uses climate data from the Climate Research Unit (CRU) at the University of East Anglia (UEA), which provides monthly climate data from 1901 to the present day [29]. To calculate the Meteorological Drought Index, SPEI, this study used precipitation and potential evapotranspiration data from 1971–2022, and the resolution was 0.25° in space and 1 month in time, which can be downloaded from https://crudata.uea.ac.uk/cru/data/hrg/cru_ts_4.05/cruts.2103051243.v4.05 (accessed on 6 October 2023).

### 2.2.4. Auxiliary Datasets

The 250-meter spatial resolution digital elevation maps used in this paper were obtained from Global Multi-resolution Terrain Elevation Data 2010 (GMTED2010) (GMTED 2010 | U.S. Geological Survey (usgs.gov) (accessed on 8 September 2023)), the land-use type data are from the China Land Cover Dataset (CLCD) produced by Wuhan University [30], and the soil type data from the Harmonized World Soil Database (HWSD) are from the International Organization for Food and Agriculture (IOFA) (https://www.fao.org/soils-portal (accessed on 6 August 2024)).

## 3. Methodology

The specific research steps in this paper are illustrated in Figure 2.

## 3.1. SPEI

Considering the role of temperature under global warming, SPEI [11] is computed using precipitation and potential evapotranspiration, which can capture drought characteristics on multiple time scales (1–24 months) and has been heavily applied to characterize meteorological drought [31,32]. The calculation of SPEI consists of the following three main steps: (i) calculating the cumulative difference in the water balance over the study period; (ii) standardizing the series and estimating the cumulative probability; and (iii) calculating the SPEI by converting the cumulative probability density function into a standard normally distributed sequence, which calculation process is described in more detail in [11,33].

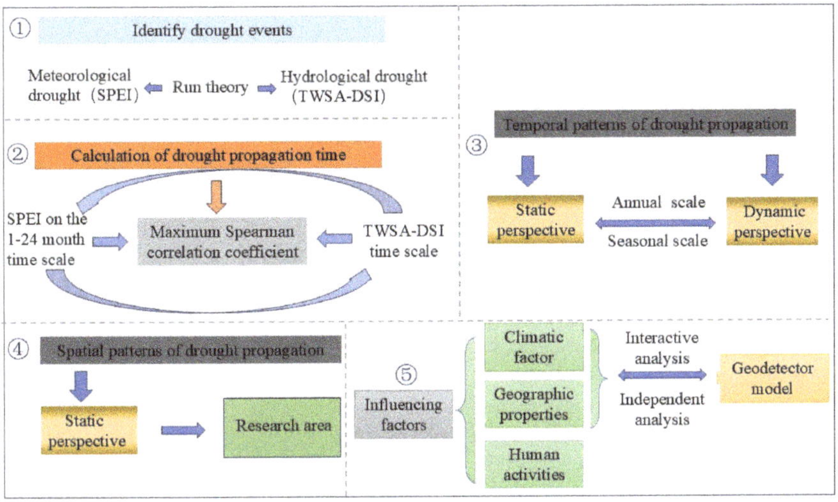


**Figure 2.** Diagram of the research process of the study.

### 3.2. TWSA-DSI

The TWS obtained from the GRACE satellite includes surface water, canopy, snow, soil, and groundwater, which is the land's vertical aquifer storage and is a key signal of hydrological drought [34]. The TWSA from GRACE is the raw data product processed for 2004–2009 distance leveling, and the resulting drought index, TWSA-DSI, is a standardized anomaly of TWS [17] and is considered a hydrological drought index [16,35].TWSA-DSI is given in in Equation (1):

$$TWSA\text{-}DSI_{i,j} = \frac{TWSA_{i,j} - \overline{TWSA_j}}{\sigma_j} \quad (1)$$

where $i$ is the year from 2002 to 2022; $j$ is the month from Jan to Dec; $\overline{TWSA_j}$ denotes the TWSA in the $j$th month, and TWSA can be obtained directly from the GRACE satellite; and $\sigma_j$ is the standard deviation. The threshold is −0.5 [35,36], where under −0.5 drought is indicated. The classification of droughts is shown in Table 1.

**Table 1.** The drought index classification.

| Category | Description | SPEI | TWSA-DSI |
|---|---|---|---|
| D0 | No Drought | (−0.5, +∞) | (−0.5, +∞) |
| D1 | Mild Drought | (−1.0, −0.5] | (−0.8, −0.5] |
| D2 | Moderate Drought | (−1.5, −1.0] | (−1.3, −0.8] |
| D3 | Severe Drought | (−2.0, −1.5] | (−1.6, −1.3] |
| D4 | Extreme Drought | (−∞, −2.0] | (−2.0, −1.6] |
| D5 | Exceptional Drought | | (−∞, −2.0] |

### 3.3. Correlation Analysis

Correlation analysis can be used to indicate the degree of correlation between two serial variables. Since drought propagation does not satisfy nonlinear propagation, the Spearman correlation coefficient, as a nonparametric rank correlation measure that satisfies the correlation measure between nonlinear sequences, has been widely used in drought propagation calculations [37,38]. The Spearman rank order is calculated as follows:

$$R = 1 - \frac{6\sum_{i=1}^{n}(y_i - x_i)^2}{n(n^2 - 1)} \quad (2)$$

where $x_i$, $y_i$ denotes the time series of the two types of drought indices, $x_i$ represents TWSA-DSI and $y_i$ denotes SPEI for the 1–24 months series, and $n$ denotes different time scales of the SPEI index, $n = 1 \ldots 24$.

### 3.4. Geodetector Model

The geodetector model, proposed by the Chinese Institute of Geographic Sciences and Resources, is a spatial analytical model for detecting dissimilarities and drivers, which is used in factor-driver analysis. Geodetector contains factor detectors, ecological detectors, interaction detectors, and risk detectors. Factor detectors can detect the explanatory power of the spatial variability of a single variable, and interaction detectors can reveal the explanatory power of a two-factor independent variable on the dependent variable [39].

(1) Factor detector

Factor detectors are calculated as follows:

$$Q = 1 - \frac{\sum_{h=1}^{L} N_h \sigma_h^2}{N\sigma^2} \quad (3)$$

where $Q$ is the ability of the factor (X) to explain the spatial heterogeneity of the variable drought propagation time (Y), with higher values of $Q$ indicating stronger explanatory power, ranging from 0–1. Y or X can be stratified into $h$ different categories; $N_h$ and $N$ are the numbers of cells in the $h$ strata and the whole region, respectively; $\sigma_h^2$ and $\sigma^2$ are the variances of Y in the $h$ strata and the whole region, respectively.

(2) Interaction detector

To identify the explanatory power of the joint action of the different factors X1 and X2 on the dependent variable Y, the explanatory power of the two factors X1 and X2 is firstly calculated separately; then the Q-value is calculated for the interaction of X1 and X2; and then $Q$ (X1) and $Q$ (X2) are compared with $Q$ (X1 ∩ X2).

Geoprobe models have been widely used for their ability to explore the contribution of two factors to the spatial variability of a variable; e.g., Lu et al. [40] explored the effect of multifactor interactions on the spatial equilibrium of water resources. This study quantifies the meteorological factors and geographic characteristics that may impact the spatial pattern of drought propagation, quantifies the extent to which each factor contributes to the propagation time of the spatial state using a factor detector, and reveals the interactions between factors on drought propagation using the interactive detector. The model can be made from http://www.geodetector.org.

## 4. Results

### 4.1. Static Drought Propagation Time

#### 4.1.1. Annual Drought Propagation Times

To find the propagation time of meteorological drought to hydrological drought propagation, we calculated multiple time scales of SPEI for the YRB at different scales for the years 1971–2022. Based on earlier studies, the propagation between them has a long time, usually ranging between 6–12 months, considering region and season [4,6]. For the convenience of subsequent calculations, the SPEI and TWSA-DSI for 2003–2022 were selected for Spearman correlation calculations. It was found that the SPEI on the 11-month cumulative time scale had the largest correlation with TWSA-DSI (r = 0.63, $p < 0.01$), which indicated that the static propagation of the two droughts in the watershed was 11 months.

This article uses the run theory to identify different types of drought events (a drought index of less than −0.5 for three consecutive months is considered a drought event), and Figure 3 illustrates the time series of SPEI11 versus TWSA-DSI. SPEI11 showed that severe meteorological droughts were mainly concentrated between 2006–2015 and that no significant severe meteorological droughts occurred after 2015 (Figure 3a); hydrological droughts were mainly concentrated between 2002–2009 as obtained from Figure 3b, TWSA and TWSA-DSI showed an upward trend, and drought durations decreased year by year, so that the hydrological drought in the YRB showed a decreasing trend, and the degree of drought was gradually alleviated. From Figure 3, it was found that severe and long-lasting meteorological droughts were also more likely to induce severe hydrological droughts, indicating that severe upper-level droughts would be transmitted to lower-level droughts.

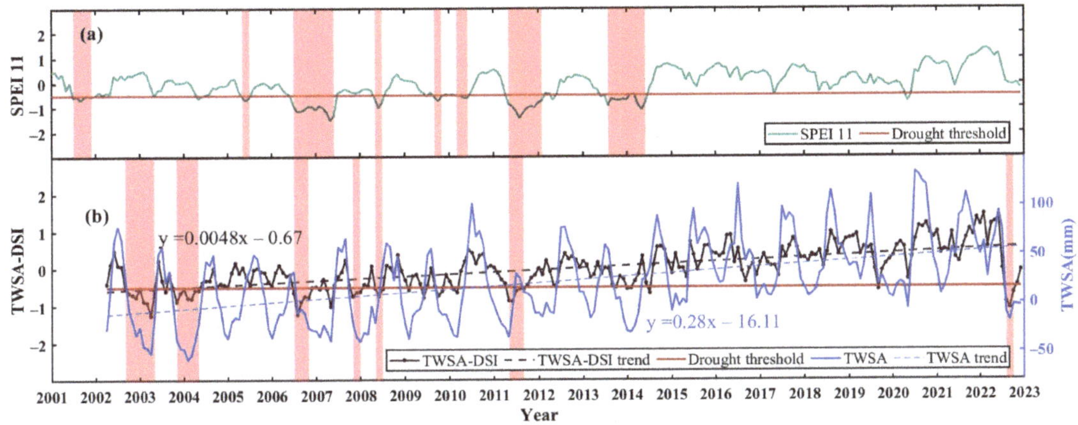


**Figure 3.** Comparison of SPEI11 (**a**) and TWSA-DSI (**b**) (red bars indicate identified drought events with durations greater than three months, and solid red lines indicate drought thresholds).

In addition, it was found from Figure 3b that abnormal severe and long-duration hydrological droughts occurred around 2003 and 2004, but SPEI11 showed that no meteorological drought occurred at this time; whereas SPEI11 showed that abnormal meteorological droughts occurred around 2014, but TWSA-DSI showed that no dramatic hydrological droughts occurred. This anomaly may be due to the different accumulation of droughts at different cumulative times of the SPEI, and therefore there was a desynchronization in the interpretation of the severity of the respective droughts by SPEI11 and TWSA-DSI. Moreover, Han et al. [41] concluded that precipitation does not influence hydrological drought only, which seems to explain this anomaly.

### 4.1.2. Seasonal Drought Propagation Times

Drought propagation not only exists on an annual scale but is also distinctly seasonal, making it necessary to conduct studies on smaller seasonal scales [41]. We analyzed the Spearman correlation between TWSA-DSI and SPEI on a 24-month time scale, and the correlation is shown in Figure 4. The SPEI cumulative scale with the largest monthly correlation is the corresponding propagation time, which is found to have a very strong seasonal cumulative effect in the basin. Table 2 shows the seasonal spreading time of the watershed, and in general, spread times were more rapid in the summer and fall than in the remaining two seasons, a result similar to the previous conclusions for other basins (e.g., [5,42]). Fall took the shortest amount of time with an average of 5 months and had the highest correlation of 0.89, while spring took the longest amount of time with an average of 8.7 months, and all passed the 99% confidence level. It can be assumed that precipitation increases rapidly in the summer, and as the TWSA contains the entire vertical portion of the aquifer, it takes some time for the precipitation to infiltrate into the soil layer and ultimately recharge the deeper groundwater; thus the correlation between TWSA-DSI and SPEI shows high values in summer and fall, while the opposite is true in winter and spring (Figure 4) [5].

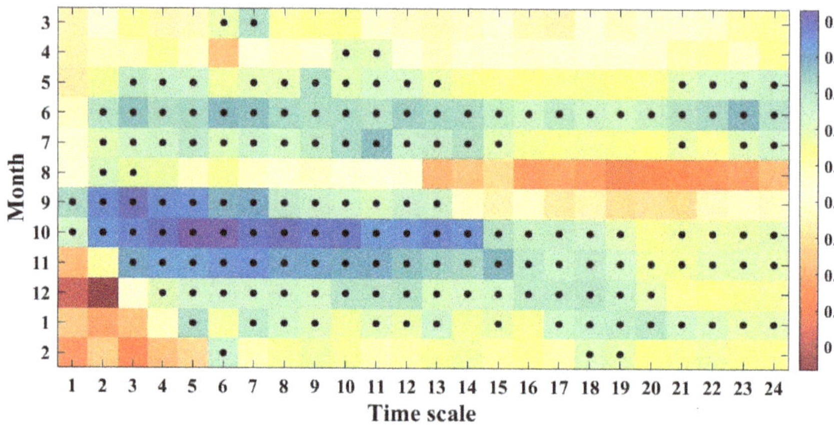


**Figure 4.** Seasonal spreading time of drought (black dots indicate 99% confidence level).

**Table 2.** Drought seasonal static propagation time.

| | Spring | Summer | Autumn | Winter |
|---|---|---|---|---|
| Propagation time | 8.7 | 6.7 | 5 | 7.7 |
| Corresponding correlation | 0.67 | 0.72 | 0.89 | 0.69 |
| Corresponding significance | $p < 0.01$ | $p < 0.01$ | $p < 0.01$ | $p < 0.01$ |

Additionally, Figure 4 reflects that TWSA-DSI in months 6 and 9–11 have the most drastic correlation with SPEI at different time scales, which suggests that severe drought events are more inclined to occur within these two time periods (summer and fall). Drought in the basin will immediately become apparent if there is insufficient precipitation in the summer and fall, and the propagation of drought will result in drought occurring more often in the fall. Therefore, when drought occurs in the summer and fall due to insufficient precipitation, this often leads to the onset of more severe next-level droughts, which require adequate attention from drought preparedness authorities.

### *4.2. Drought Dynamic Propagation Time*

#### 4.2.1. Annual Dynamic Propagation Time

As the Earth's water cycle intensifies and the hydrological environment is exposed to a dynamic context, the timing of drought propagation may also change [41]. For further analysis of the propagation time of drought under changing environments, we slid the time scales of the TWSA-DSI and the SPEI to 3 years and 5 years (to ensure that the data are mathematically statistical), respectively; carried out a correlation analysis by using Spearman; and the results are shown in Figure 5. The 3-year dynamic spreads ranged from 2–12 years with correlation coefficients clustered between 0.2 and 0.8, the 5-year dynamic spreads ranged from 2–12 years with correlation coefficients between 0.24 and 0.73, and the 3- and 5-year dynamic scales passed the test of significance except for the 17th and 18th subsequence, which failed the $p < 0.01$ test. From 2006–2015, both the 3-year sliding window and the 5-year sliding window showed that drought propagation time was concentrated in the 7–12 month period; that, except for a slight fluctuation in propagation time in 2010–2012, drought propagation time remained stable over a long period on an 11-month timescale; and that there was a significant variability in propagation time in 2004–2005 and 2016–2018, with the fastest propagation time shown to be 2 months in these two time periods. In addition, both the 3-year and 5-year sliding windows indicated that the propagation time tended to decrease and that the drought propagation time was accelerating, but this trend did not pass the MK test. However, they also suggest that anomalies are occurring in the timing of drought propagation, making it necessary to study its dynamics, as well as suggesting that the time left for drought detection and drought preparedness is decreasing.

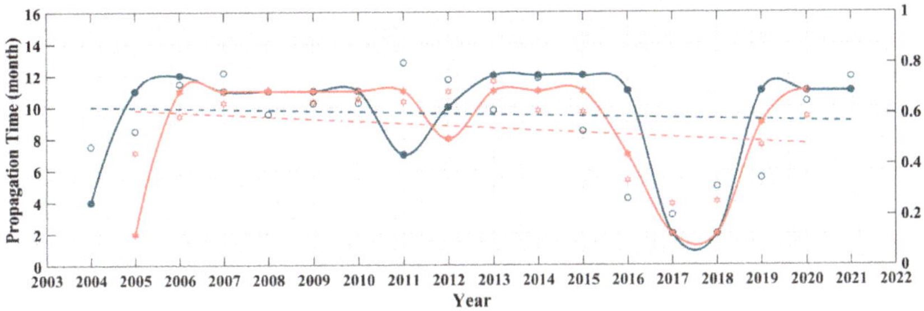


**Figure 5.** Annual dynamic propagation time (green for 3-year sliding time scales, red for 5-year sliding time scales, the dotted curve represents the drought propagation time, the dotted line represents the drought propagation time trend, and the hollow point represents the correlation coefficient of the drought propagation time (all passed the correlation test of 0.05).

#### 4.2.2. Seasonal Dynamic Propagation

Since the time of this study is 2003–2022, which is a short time series, and the sample of sliding windows should not be too small in seasonal propagation, we chose a sliding window of 10 years to obtain the propagation time of 11 subseries. The same dynamic propagation method was used to compute the seasonal propagation time based on a 10-year sliding window, and the outputs are presented in Figure 6. For the dynamic propagation, the results were subjected to the MK test as presented in Table 3. From Figure 6, it can be concluded that the seasonal propagation time of drought was also dynamic under changing environments. In this case, spring propagation time had been maintained for around 5 months propagation time until the 7th subsequence but suddenly rose to 14 months in the 7th subsequence (2009–2019), and since then it has been fluctuating between 10–16 months, with an upward trend in the spring propagation time passing the 95% confidence test that the drought propagation time is becoming slower, which is the opposite trend to the conclusion that propagation times are accelerating on an annual scale; summer propagation times show a fluctuating downward trend, although failing the 95% significance test, and drought propagation times appear to be getting shorter; fall propagation times remain stable and show no trend; and winter propagation times are more fluctuating, but also show no clear trend. In addition, Figure 6 also shows that the drought propagation time is more stable and has fewer outliers in summer and fall, with the summer drought propagation time ranging from 4–11 months; the fall propagation time remaining between 2–4 months; and the winter and spring propagation times fluctuating more and having more abnormal values, with the spring propagation time ranging between 3–16 months, and the winter propagation time ranging between 3–15 months, which are similar in both seasons. This is consistent with the findings of the static propagation scenario.

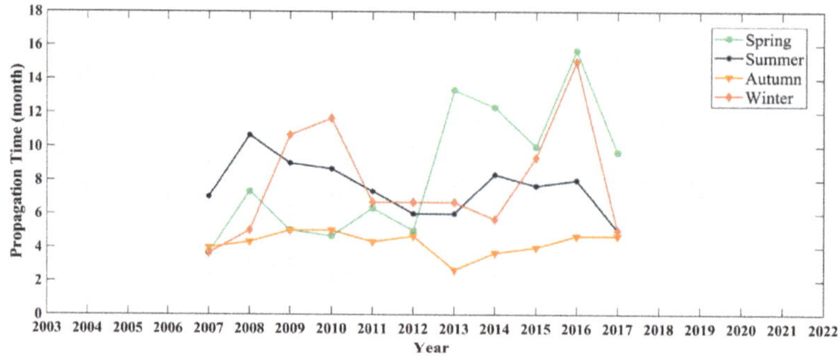


**Figure 6.** Seasonal dynamic propagation time.

**Table 3.** Trend analysis of seasonal dynamic propagation time in the YRB.

| | Spring | Summer | Autumn | Winter |
|---|---|---|---|---|
| Z value | 2.10 ** | −1.48 | 0 | 0.62 |
| trend | upward | downward | no trend | upward |

Note: ** denotes passing the test of significance; absolute values of Z greater than 1.96 indicate that a 95% confidence test was passed.

### 4.3. Spatial Propagation Time

Most previous studies only focused on drought propagation from the time perspective and seldom analyzed drought propagation from the spatial perspective; to fill this gap, we explored from a spatial perspective in the YRB (Figure 7). By studying the correlation between TWSA-DSI and SPEI, it was found that the spatial spreading time in the YRB mainly varied in a period between 4–24 months (Figure 7a), with large spatial heterogeneity, corresponding to correlation coefficients ranging from 0.23–0.81, and all the pixels passed the test of significance at 99% (Figure 7b). From Figure 7a, we found that the longest drought propagation time of 24 months mainly occurred in the headwaters of the YRB in the Jinsha River Basin (24–36°N, 90–105°E), the shortest of 4 months occurred in the Jianghan Basin (29°26′–31°37′N, 111°14′E–114°36′E), and the rest of the region was concentrated around 10–12 months.

It is noteworthy that the Spearman correlation coefficient for drought propagation in the YRB (Figure 7b) was found to be characterized by a lower level at two locations, namely, the Jinsha River region in the upper reaches as well as the Three Gorges area (31°03′ N, 109°57′E). Considering the higher elevation of the upper Jinsha River, which is highly correlated with groundwater storage, it can be interpreted that elevation is a key variable influencing the propagation of droughts; whereas the lower correlation coefficient in the middle reaches of the Three Gorges region may be linked to the storage of water in the dam [21]. Dey et al. [43] suggested that dam operation alters the seasonal timing and intensity of extreme events and spatial patterns, which will weaken the response of other types of drought to meteorological drought. Marchant and Bloomfield [44] found that the influence of watershed climate and watershed characteristics, in addition to the cumulative effect of precipitation, is critical to the timing of propagation.

### 4.4. Influence of Climatic and Geographic Characteristics on the Spatial Propagation of Drought

To investigate the impacts of spatial drought propagation, this study explored the relationship between spatial geographic characteristics (elevation and soil moisture content) and meteorological factors (precipitation, runoff, temperature, and potential evapotranspiration), and the results are shown in Figure 8. Concerning elevation, the higher the elevation, the longer the drought spread (Figure 8a), and because of the direct correlation with land use status, Yang et al. [45] found that elevation plays a key role in drought propagation. Soil moisture content had a very small effect on drought propagation time (Figure 8d). In terms of meteorological factors, higher precipitation (Figure 8b), temperature (Figure 8e), and potential evapotranspiration (Figure 8f) showed an inverse relationship with drought propagation time, while runoff did not have such a relationship, with a very small effect of runoff. Due to the strong spatial heterogeneity of the YRB, the effects of the factors need to be further analyzed.

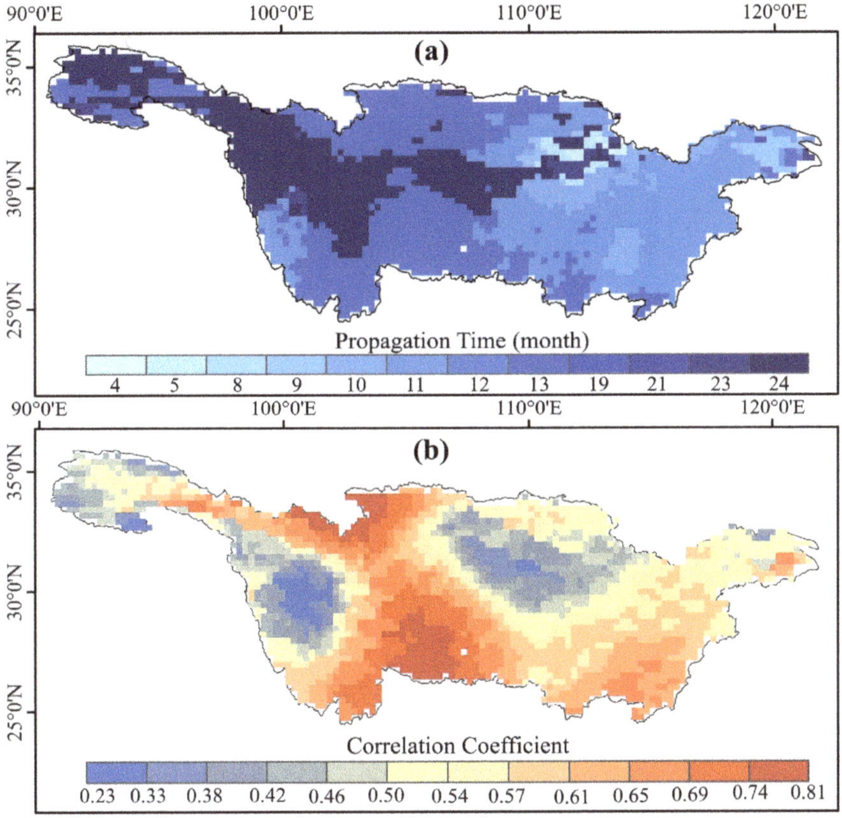


**Figure 7.** Spatial patterns of drought propagation time (**a**) and correlation (**b**).

To find out the important factors, we further analyzed the above factors based on geodetector, and the results based on single factor detection are shown in Figure 9, which shows that the influence of elevation scored the highest among the six factors and was the most important influencing factor. However, as a whole the score of single factors was not high, and the effect of each factor was not isolated but interacted with each other; further we analyzed the results based on the effect of interacting factors as shown in Figure 10. It was found that the contributions between elevation and other factors were all above 0.45, which were larger than the contributions of the remaining factor interactions, further indicating that elevation is the strongest factor influencing the spatial pattern of drought propagation in the YRB. In addition, the interactions of elevation and evapotranspiration, elevation and precipitation, and elevation and temperature were 0.71, 0.71, and 0.66, respectively (Figure 10a), which were much larger than the sum of the single drivers (e.g., E + PET = 0.59), and the effect of the interactive influences was stronger than that of the single factors.

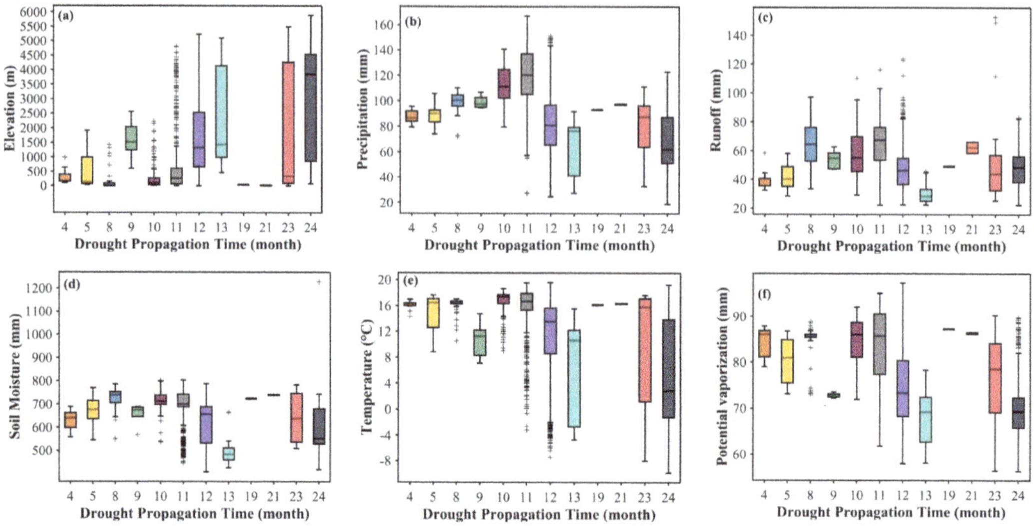


**Figure 8.** Relationship between drought propagation time and elevation (**a**), precipitation (**b**), runoff (**c**), soil moisture content (**d**), temperature (**e**), and potential evapotranspiration (**f**).

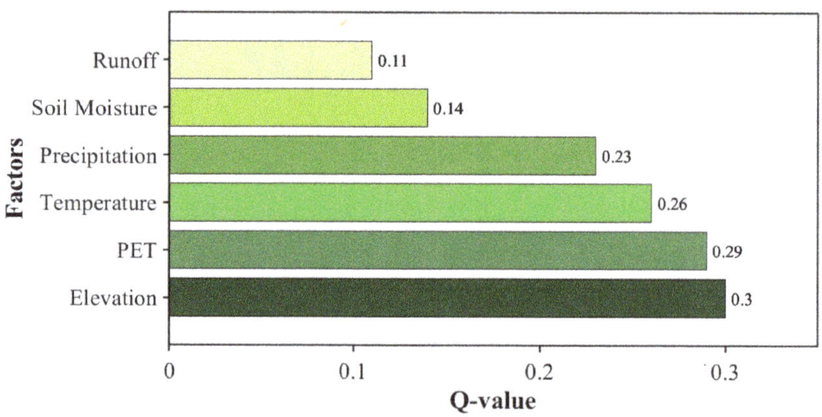


**Figure 9.** Geodetector analysis of single-factor scores affecting spatial patterns of drought propagation.

The interaction test, on the other hand, contributed a greater effect compared to the results of the single-factor test. Moreover, when the contribution of a single factor was low (soil water content and runoff), the interaction of that factor with other factors also scored lower, which is not represented in Figure 10 because the interaction between soil water content and runoff was 0.30. When the contribution of a single factor was high, the interaction of that factor with other factors also had a high contribution (e.g., elevation). Therefore, it is important to consider the interaction between factors.

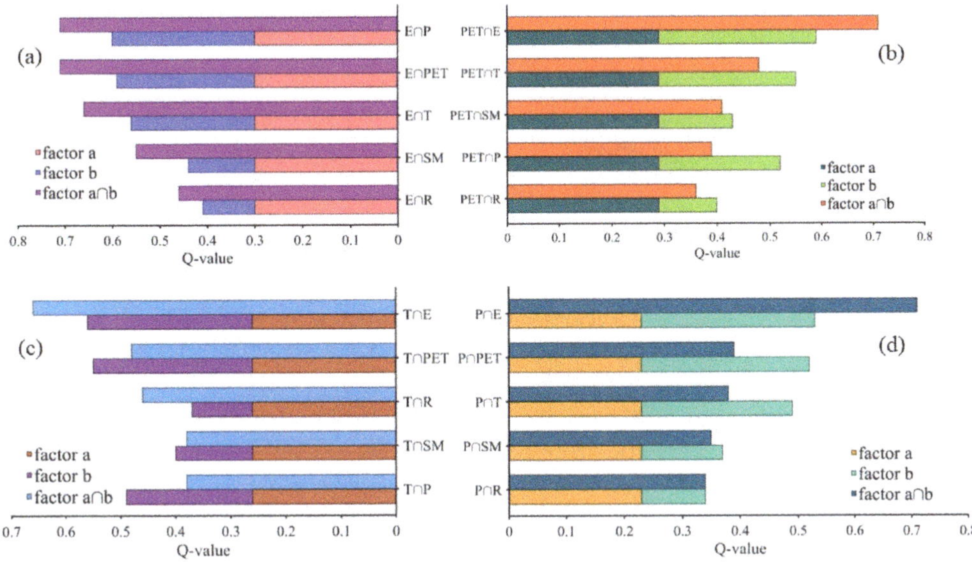


**Figure 10.** Geodetector analysis of interactive factor scores affecting the spatial distribution of drought propagation (Interactive scores of elevation (**a**), potential evapotranspiration (**b**), temperature (**c**), precipitation (**d**) with other factors).

## 5. Discussion

### 5.1. Impact of Climatic and Geographic Characteristics on the Spread of Drought

It was observed that drought propagation time was longer in alpine zones than in hot and humid regions. In Section 4.4, we analyzed the contribution of climatic and geographic characteristics on spatial drought propagation in the YRB, although the use of a geodetector model detected generally low scores on the single factors, which may be owed to the high spatial heterogeneity of the YRB [46]. The results show that potential evapotranspiration and elevation were the most critical variables in the two characteristics, respectively, but studies by previous scholars have shown that temperature and precipitation variables also play a crucial role [47,48]. In terms of climatic characteristics, on the one hand, precipitation is the source of water in the basin, and its reduction leads to a decrease in the amount of water stored in the basin, which results in a shorter drought response time in regions with less precipitation [49]; on the other hand, water vapor feedbacks are stronger in regions with higher potential evapotranspiration and temperatures, and drought propagation responds with shorter response times [50,51]. From a geographical characterization, elevation is a crucial determinant of drought propagation due to its direct correlation with snow power and groundwater [45,47]. This is similar to the conclusion of Muthuvel and Sivakumar [52] in their investigation of drought propagation in peninsular India where they found that elevation plays a crucial role in drought propagation. Additionally, Luo et al. [47] and Mao et al. [53] also found that elevation has a non-negligible role in drought propagation. Drought propagation time was relatively long at high elevations, while interactions between elevation and potential evapotranspiration, precipitation, and temperature were found to have a large contribution using interactive factor analysis. Overall, drought propagation in the YRB is strongly dependent on climatic and geographic characteristics, so it is essential to consider the interaction between factors.

Against the backdrop of global warming, the frequency of climate extremes in the YRB has risen markedly [54]. According to the conclusion of [55], the urban drought area has displayed a significant growth trend since 1981 in the region. Along with the increased risk of heat waves accompanied by an enhanced hydrological cycle, leading to changes in precipitation, temperature and potential evapotranspiration, etc. that further exacerbate the occurrence of extreme climate events [5], this will lead to a rapid spread of drought propagation, exacerbating the rate at which it spreads, as found in this study in Section 4.2.1 (Figure 3). According to Guan et al. [56], the temperature changes in the YRB are anomalously associated with climate warming, with the temperature index showing a significant rate of warming. Due to the anomalous changes in the atmospheric circulation pattern, the strength and incidence of extreme precipitation in the YRB have risen dramatically, with a more severe degree in the central and downstream areas [54]. Since the 21st century, PET in China has been growing by 3.4 mm per year, with a broadly significant increasing trend [57], and the Yarlung Tsangpo River area is expected to have a significant increase in high values of PET in the future [58]. All of these climatic factors can lead to rapid drought development and propagation [50], and in general, climate warming and humidification will further exacerbate the rate of drought propagation.

### 5.2. Limitation and Prospect

Previous studies have shown that there is a link between human activities and drought propagation [59,60]. Analyzing land-use types, grassland and cropland have longer drought propagation times; areas with higher forest cover have shorter propagation times, which may be due to high transpiration and reduced soil moisture due to forests [47], while grassland and cropland lead to reduced runoff, which slows down the drought propagation time [22]. Anthropogenically induced rapid urbanization, water abstraction, and reservoir construction activities also contribute significantly to drought propagation. Anthropogenic water abstraction leads to severe surface water loss, which prolongs the drought response time [22]. Urbanization contributes about 46.6% to meteorological drought anomalies in the YRB, and drought conditions are more pronounced in more urbanized areas [55], which will lead to a rapid response to drought and exacerbate the rate of drought propagation. Reservoir operation also alters the drought response time [61], and by altering runoff patterns [62] the Three Gorges Dam replenishes dry season water scarcity conditions thereby weakening the drought propagation processes [21]. However, this paper does not consider the impact of anthropogenic factors due to the short temporal coverage of data utilized in the present study and the difficulty in obtaining reservoir data; therefore, different datasets need to be considered to take a more holistic view of drought propagation influences in the future.

## 6. Conclusions

Based on GRACE satellite data, this paper investigates the drought propagation mechanism from the atmosphere (SPEI) to the hydrology (TWSA-DSI) in the YRB. The static propagation time from 2003 to 2022 was 11 months, and the seasonal spread times indicated that drought spread over a shorter time in summer and fall, with the shortest time in fall (on average 5 months) and the longest time in spring (on average 8.7 months). Sliding tests revealed a tendency towards shorter propagation on annual scales and significantly longer in spring on seasonal time scales. The detection of spatial drought propagation time revealed significant spatial heterogeneity in drought propagation, ranging from 4–24 months, with most areas concentrated between 11 and 12 months, and all pixels passed the 99% significance test.

The effects of climate and geography on drought were analyzed using geodetector. This study found that elevation and potential evapotranspiration played key roles in the spread of drought at the basin scale. When considering spatial propagation, the influence of a single variable should not be taken into account alone; multifactor interactions have a more important role, and therefore it is necessary to consider multifactor interactions in future research. These findings contribute to the comprehension of the static and dynamic properties of drought propagation as well as the influencing factors, which can be valuable for drought mitigation work.

**Author Contributions:** C.Z.: writing—review and editing, data curation. Z.H.: methodology, conceptualization. S.W.: software. J.W.: visualization. C.C.: methodology, conceptualization, supervision. J.L.: writing—review and editing. All authors have read and agreed to the published version of the manuscript.

**Funding:** This research was funded by Shaanxi Provincial Science and Technology Department grant number [2024QCY-KXJ-093].

**Data Availability Statement:** The GRACE satellite data can be obtained from https://www2.csr.utexas.edu/grace/RL06_mascons.html (accessed on 19 September 2023). The GLDAS satellite data can be downloaded from https://disc.gsfc.nasa.gov/information?page=1&project=GRACE-DA-DM,FLDAS,GLDAS,NEWS,NLDAS,NCA-LDAS&keywords=hydrology (accessed on 25 September 2023). Monthly precipitation and potential evapotranspiration data from CRU data can be downloaded from https://crudata.uea.ac.uk/cru/data/hrg/cru_ts_4.05/cruts.2103051243.v4.05 (accessed on 6 October 2023). The soil type data are from the International Organization for Food and Agriculture (IOFA) (https://www.fao.org/soils-portal (accessed on 6 August 2024)).

**Conflicts of Interest:** Author Junrong Liu was employed by the company China Coal Aerial Survey and Remote Sensing Group Co. The remaining authors declare that this research was conducted in the absence of any commercial or financial relationships that could be construed as potential conflicts of interest.

## References

1. Mishra, A.K.; Singh, V.P. A review of drought concepts. *J. Hydrol.* **2010**, *391*, 202–216. [CrossRef]
2. Guo, Y.; Huang, S.; Huang, Q.; Wang, H.; Wang, L.; Fang, W. Copulas-based bivariate socioeconomic drought dynamic risk assessment in a changing environment. *J. Hydrol.* **2019**, *575*, 1052–1064. [CrossRef]
3. Van Loon, A.F.; Van Lanen, H.A.J. A process-based typology of hydrological drought. *Hydrol. Earth Syst. Sci.* **2012**, *16*, 1915–1946. [CrossRef]
4. Huang, S.; Li, P.; Huang, Q.; Leng, G.; Hou, B.; Ma, L. The propagation from meteorological to hydrological drought and its potential influence factors. *J. Hydrol.* **2017**, *547*, 184–195. [CrossRef]
5. Han, Z.; Huang, S.; Huang, Q.; Leng, G.; Wang, H.; Bai, Q.; Zhao, J.; Ma, L.; Wang, L.; Du, M. Propagation dynamics from meteorological to groundwater drought and their possible influence factors. *J. Hydrol.* **2019**, *578*, 124102. [CrossRef]
6. Zhang, X.; Hao, Z.; Singh, V.P.; Zhang, Y.; Feng, S.; Xu, Y.; Hao, F. Drought propagation under global warming: Characteristics, approaches, processes, and controlling factors. *Sci. Total Environ.* **2022**, *838*, 156021. [CrossRef] [PubMed]
7. Kim, W.; Iizumi, T.; Nishimori, M. Global Patterns of Crop Production Losses Associated with Droughts from 1983 to 2009. *J. Appl. Meteorol. Climatol.* **2019**, *58*, 1233–1244. [CrossRef]
8. Van Loon, A.F. Hydrological drought explained. *Wiley Interdiscip. Rev.-Water* **2015**, *2*, 359–392. [CrossRef]
9. Wang, Z.; Chang, J.; Wang, Y.; Yang, Y.; Guo, Y.; Yang, G.; He, B. Temporal and spatial propagation characteristics of meteorological drought to hydrological drought and influencing factors. *Atmos. Res.* **2024**, *299*, 107212. [CrossRef]
10. Han, Z.; Huang, S.; Zhao, J.; Leng, G.; Huang, Q.; Zhang, H.; Li, Z. Long-chain propagation pathways from meteorological to hydrological, agricultural and groundwater drought and their dynamics in China. *J. Hydrol.* **2023**, *625*, 130131. [CrossRef]
11. Vicente-Serrano, S.M.; Begueria, S.; Lopez-Moreno, J.I. A Multiscalar Drought Index Sensitive to Global Warming: The Standardized Precipitation Evapotranspiration Index. *J. Clim.* **2010**, *23*, 1696–1718. [CrossRef]
12. Li, L.; She, D.; Zheng, H.; Lin, P.; Yang, Z.-L. Elucidating Diverse Drought Characteristics from Two Meteorological Drought Indices (SPI and SPEI) in China. *J. Hydrometeorol.* **2020**, *21*, 1513–1530. [CrossRef]
13. Chao, N.; Luo, Z.; Wang, Z.; Jin, T. Retrieving Groundwater Depletion and Drought in the Tigris-Euphrates Basin Between 2003 and 2015. *Groundwater* **2018**, *56*, 770–782. [CrossRef] [PubMed]
14. Jiang, Q.; Ferreira, V.G.; Chen, J. Monitoring groundwater changes in the Yangtze River basin using satellite and model data. *Arab. J. Geosci.* **2016**, *9*, 500. [CrossRef]
15. Jiang, Z.; Hsu, Y.-J.; Yuan, L.; Cheng, S.; Feng, W.; Tang, M.; Yang, X. Insights into hydrological drought characteristics using GNSS-inferred large-scale terrestrial water storage deficits. *Earth Planet. Sci. Lett.* **2022**, *578*, 117294. [CrossRef]
16. Liu, X.; Feng, X.; Ciais, P.; Fu, B.; Hu, B.; Sun, Z. GRACE satellite-based drought index indicating increased impact of drought over major basins in China during 2002–2017. *Agric. For. Meteorol.* **2020**, *291*, 108057. [CrossRef]
17. Zhao, M.; Geruo, A.; Velicogna, I.; Kimball, J.S. Satellite Observations of Regional Drought Severity in the Continental United States Using GRACE-Based Terrestrial Water Storage Changes. *J. Clim.* **2017**, *30*, 6297–6308. [CrossRef]
18. Van Loon, A.F.; Laaha, G. Hydrological drought severity explained by climate and catchment characteristics. *J. Hydrol.* **2015**, *526*, 3–14. [CrossRef]
19. Konapala, G.; Mishra, A. Quantifying Climate and Catchment Control on Hydrological Drought in the Continental United States. *Water Resour. Res.* **2020**, *56*, e2018WR024620. [CrossRef]

20. Wu, J.; Chen, X.; Yao, H.; Zhang, D. Multi-timescale assessment of propagation thresholds from meteorological to hydrological drought. *Sci. Total Environ.* **2021**, *765*, 144232. [CrossRef]
21. Liu, X.; Peng, J.; Liu, Y.; Yu, S.; Wang, Y. The Three Gorges Dam has weakened the drought propagation process in the Yangtze River Basin. *J. Hydrol.* **2024**, *632*, 130875. [CrossRef]
22. Wang, M.; Jiang, S.; Ren, L.; Xu, C.-Y.; Menzel, L.; Yuan, F.; Xu, Q.; Liu, Y.; Yang, X. Separating the effects of climate change and human activities on drought propagation via a natural and human-impacted catchment comparison method. *J. Hydrol.* **2021**, *603*, 126913. [CrossRef]
23. Zheng, S.; Zhang, Z.; Yan, H.; Zhao, Y.; Li, Z. Characterizing drought events occurred in the Yangtze River Basin from 1979 to 2017 by reconstructing water storage anomalies based on GRACE and meteorological data. *Sci. Total Environ.* **2023**, *868*, 161755. [CrossRef]
24. Sun, Z.; Zhu, X.; Pan, Y.; Zhang, J.; Liu, X. Drought evaluation using the GRACE terrestrial water storage deficit over the Yangtze River Basin, China. *Sci. Total Environ.* **2018**, *634*, 727–738. [CrossRef]
25. Zhai, J.Q.; Liu, B.; Hartmann, H.; Su, B.D.; Jiang, T.; Fraedrich, K. Dryness/wetness variations in ten large river basins of China during the first 50 years of the 21st century. *Quat. Int.* **2010**, *226*, 101–111. [CrossRef]
26. Scanlon, B.R.; Zhang, Z.; Save, H.; Wiese, D.N.; Landerer, F.W.; Long, D.; Longuevergne, L.; Chen, J. Global evaluation of new GRACE mascon products for hydrologic applications. *Water Resour. Res.* **2016**, *52*, 9412–9429. [CrossRef]
27. Lin, M.; Biswas, A.; Bennett, E.M. Socio-ecological determinants on spatio-temporal changes of groundwater in the Yellow River Basin, China. *Sci. Total Environ.* **2020**, *731*, 138725. [CrossRef] [PubMed]
28. Rodell, M.; Houser, P.R.; Jambor, U.; Gottschalck, J.; Mitchell, K.; Meng, C.J.; Arsenault, K.; Cosgrove, B.; Radakovich, J.; Bosilovich, M.; et al. The global land data assimilation system. *Bull. Am. Meteorol. Soc.* **2004**, *85*, 381–394. [CrossRef]
29. Elagib, N.A.; Ali, M.M.A.; Schneider, K. Evaluation and bias correction of CRU TS4.05 potential evapotranspiration across vast environments with limited data. *Atmos. Res.* **2024**, *299*, 107194. [CrossRef]
30. Yang, J.; Huang, X. The 30 m annual land cover dataset and its dynamics in China from 1990 to 2019. *Earth Syst. Sci. Data* **2021**, *13*, 3907–3925. [CrossRef]
31. Su, B.; Huang, J.; Mondal, S.K.; Zhai, J.; Wang, Y.; Wen, S.; Gao, M.; Lv, Y.; Jiang, S.; Jiang, T.; et al. Insight from CMIP6 SSP-RCP scenarios for future drought characteristics in China. *Atmos. Res.* **2021**, *250*, 105375. [CrossRef]
32. Wang, Q.; Zeng, J.; Qi, J.; Zhang, X.; Zeng, Y.; Shui, W.; Xu, Z.; Zhang, R.; Wu, X.; Cong, J. A multi-scale daily SPEI dataset for drought characterization at observation stations over mainland China from 1961 to 2018. *Earth Syst. Sci. Data* **2021**, *13*, 331–341. [CrossRef]
33. Xu, K.; Yang, D.; Yang, H.; Li, Z.; Qin, Y.; Shen, Y. Spatio-temporal variation of drought in China during 1961–2012: A climatic perspective. *J. Hydrol.* **2015**, *526*, 253–264. [CrossRef]
34. Zhu, E.; Wang, Y.; Yuan, X. Changes of terrestrial water storage during 1981–2020 over China based on dynamic-machine learning model. *J. Hydrol.* **2023**, *621*, 129576. [CrossRef]
35. Zhao, M.; Geruo, A.; Velicogna, I.; Kimball, J.S. A Global Gridded Dataset of GRACE Drought Severity Index for 2002-14: Comparison with PDSI and SPEI and a Case Study of the Australia Millennium Drought. *J. Hydrometeorol.* **2017**, *18*, 2117–2129. [CrossRef]
36. Xu, G.; Wu, Y.; Liu, S.; Cheng, S.; Zhang, Y.; Pan, Y.; Wang, L.; Dokuchits, E.Y.; Nkwazema, O.C. How 2022 extreme drought influences the spatiotemporal variations of terrestrial water storage in the Yangtze River Catchment: Insights from GRACE-based drought severity index and in-situ measurements. *J. Hydrol.* **2023**, *626*, 130245. [CrossRef]
37. Li, L.; Peng, Q.; Wang, M.; Cao, Y.; Gu, X.; Cai, H. Quantitative analysis of vegetation drought propagation process and uncertainty in the Yellow River Basin. *Agric. Water Manag.* **2024**, *295*, 108775. [CrossRef]
38. Geng, G.; Zhang, B.; Gu, Q.; He, Z.; Zheng, R. Drought propagation characteristics across China: Time, probability, and threshold. *J. Hydrol.* **2024**, *631*, 130805. [CrossRef]
39. Gu, J.; Ye, Y.; Jiang, Y.; Dong, J.; Cao, Y.; Huang, J.; Guan, H. A downscaling-calibrating framework for generating gridded daily precipitation estimates with a high spatial resolution. *J. Hydrol.* **2023**, *626*, 130371. [CrossRef]
40. Lu, Y.; Yang, X.; Bian, D.; Chen, Y.; Li, Y.; Yuan, Z.; Wang, K. A novel approach for quantifying water resource spatial equilibrium based on the regional evaluation, spatiotemporal heterogeneity and geodetector analysis integrated model. *J. Clean. Prod.* **2023**, *424*, 138791. [CrossRef]
41. Dai, M.; Huang, S.; Huang, Q.; Zheng, X.; Su, X.; Leng, G.; Li, Z.; Guo, Y.; Fang, W.; Liu, Y. Propagation characteristics and mechanism from meteorological to agricultural drought in various seasons. *J. Hydrol.* **2022**, *610*, 127897. [CrossRef]
42. Zhang, H.; Ding, J.; Wang, Y.; Zhou, D.; Zhu, Q. Investigation about the correlation and propagation among meteorological, agricultural and groundwater droughts over humid and arid/semi-arid basins in China. *J. Hydrol.* **2021**, *603*, 127007. [CrossRef]
43. Dey, P.; Swarnkar, S.; Mujumdar, P. Dam operation affects the evolution and propagation of hydrological extremes. *Hydrol. Sci. J.* **2024**, *69*, 294–308. [CrossRef]
44. Marchant, B.P.; Bloomfield, J.P. Spatio-temporal modelling of the status of groundwater droughts. *J. Hydrol.* **2018**, *564*, 397–413. [CrossRef]
45. Yang, F.; Duan, X.; Guo, Q.; Lu, S.; Hsu, K. The spatiotemporal variations and propagation of droughts in Plateau Mountains of China. *Sci. Total Environ.* **2022**, *805*, 150257. [CrossRef] [PubMed]

46. Yang, Y.; Wang, Y.; Cong, N.; Wang, N.; Yao, W. Impacts of the Three Gorges Dam on riparian vegetation in the Yangtze River Basin under climate change. *Sci. Total Environ.* **2024**, *912*, 169415. [CrossRef] [PubMed]
47. Luo, X.; Luo, X.; Ji, X.; Ming, W.; Wang, L.; Xiao, X.; Xu, J.; Liu, Y.; Li, Y. Meteorological and hydrological droughts in the Lancang-Mekong River Basin: Spatiotemporal patterns and propagation. *Atmos. Res.* **2023**, *293*, 106913. [CrossRef]
48. Valiya Veettil, A.; Mishra, A.k. Multiscale hydrological drought analysis: Role of climate, catchment and morphological variables and associated thresholds. *J. Hydrol.* **2020**, *582*, 124533. [CrossRef]
49. Zhao, W.; Wu, J.; Du, E.; Deng, X.; Sun, G.; Wang, G. Deciphering the influence of climate change and human activities on the drought propagation. *J. Hydrol. Reg. Stud.* **2024**, *51*, 101654. [CrossRef]
50. Ding, Y.; Xu, J.; Wang, X.; Cai, H.; Zhou, Z.; Sun, Y.; Shi, H. Propagation of meteorological to hydrological drought for different climate regions in China. *J. Environ. Manag.* **2021**, *283*, 111980. [CrossRef]
51. Jiang, W.; Niu, Z.; Wang, L.; Yao, R.; Gui, X.; Xiang, F.; Ji, Y. Impacts of Drought and Climatic Factors on Vegetation Dynamics in the Yellow River Basin and Yangtze River Basin, China. *Remote Sens.* **2022**, *14*, 930. [CrossRef]
52. Muthuvel, D.; Sivakumar, B. Spatial propagation of different drought types and their concurrent societal risks: A complex networks-based analysis. *J. Hydrol.* **2024**, *636*, 131247. [CrossRef]
53. Mao, R.; Li, B.; Song, J.; Li, Q.; Li, N.; Long, Y.; Feng, J. Evaluating multifaceted effects of watershed properties and human activities on drought propagation in the Wei River Basin with an integrated framework. *Sci. Total Environ.* **2024**, *926*, 171712. [CrossRef] [PubMed]
54. Li, X.; Zhang, K.; Gu, P.; Feng, H.; Yin, Y.; Chen, W.; Cheng, B. Changes in precipitation extremes in the Yangtze River Basin during 1960–2019 and the association with global warming, ENSO, and local effects. *Sci. Total Environ.* **2021**, *760*, 144244. [CrossRef] [PubMed]
55. Huang, S.; Zhang, X.; Yang, L.; Chen, N.; Nam, W.-H.; Niyogi, D. Urbanization-induced drought modification: Example over the Yangtze River Basin, China. *Urban Clim.* **2022**, *44*, 101231. [CrossRef]
56. Guan, Y.; Zhang, X.; Zheng, F.; Wang, B. Trends and variability of daily temperature extremes during 1960–2012 in the Yangtze River Basin, China. *Glob. Planet. Chang.* **2015**, *124*, 79–94. [CrossRef]
57. Yao, Y.; Zhao, S.; Zhang, Y.; Jia, K.; Liu, M. Spatial and Decadal Variations in Potential Evapotranspiration of China Based on Reanalysis Datasets during 1982-2010. *Atmosphere* **2014**, *5*, 737–754. [CrossRef]
58. Ma, D.; Wang, T.; Gao, C.; Pan, S.; Sun, Z.; Xu, Y.-P. Potential evapotranspiration changes in Lancang River Basin and Yarlung Zangbo River Basin, southwest China. *Hydrol. Sci. J.* **2018**, *63*, 1653–1668. [CrossRef]
59. Li, Z.; Huang, S.; Zhou, S.; Leng, G.; Liu, D.; Huang, Q.; Wang, H.; Han, Z.; Liang, H. Clarifying the Propagation Dynamics from Meteorological to Hydrological Drought Induced by Climate Change and Direct Human Activities. *J. Hydrometeorol.* **2021**, *22*, 2359–2378. [CrossRef]
60. Wang, Y.; Wang, S.; Chen, Y.; Wang, F.; Liu, Y.; Zhao, W. Anthropogenic drought in the Yellow River basin: Multifaceted and weakening connections between meteorological and hydrological droughts. *J. Hydrol.* **2023**, *619*, 129273. [CrossRef]
61. Xu, Y.; Zhang, X.; Wang, X.; Hao, Z.; Singh, V.P.; Hao, F. Propagation from meteorological drought to hydrological drought under the impact of human activities: A case study in northern China. *J. Hydrol.* **2019**, *579*, 124147. [CrossRef]
62. Yun, X.; Tang, Q.; Wang, J.; Liu, X.; Zhang, Y.; Lu, H.; Wang, Y.; Zhang, L.; Chen, D. Impacts of climate change and reservoir operation on streamflow and flood characteristics in the Lancang-Mekong River Basin. *J. Hydrol.* **2020**, *590*, 125472. [CrossRef]

drones

MDPI

Article

# Combining Drone LiDAR and Virtual Reality Geovisualizations towards a Cartographic Approach to Visualize Flooding Scenarios

**Ermioni Eirini Papadopoulou and Apostolos Papakonstantinou ***

Department of Civil Engineering and Geomatics, School of Engineering and Technology, Cyprus University of Technology, Lemesos 3603, Cyprus; e.papadopoulou@cut.ac.cy
* Correspondence: a.papakonstantinou@cut.ac.cy; Tel.: +357-25002184

**Abstract:** This study aims to create virtual reality (VR) geovisualizations using 3D point clouds obtained from airborne LiDAR technology. These visualizations were used to map the current state of river channels and tributaries in the Thessalian Plain, Greece, following severe flooding in the summer of 2023. The study area examined in this paper is the embankments enclosing the tributaries of the Pineios River in the Thessalian Plain region, specifically between the cities of Karditsa and Trikala in mainland Greece. This area was significantly affected in the summer of 2023 when flooding the region's rivers destroyed urban elements and crops. The extent of the impact across the entire Thessalian Plain made managing the event highly challenging to the authorities. High-resolution 3D mapping and VR geovisualization of the embarkments encasing the main rivers and the tributaries of the Thessalian Plain essentially provides information for planning the area's restoration processes and designing prevention and mitigation measures for similar disasters. The proposed methodology consists of four stages. The first and second stages of the methodology present the design of the data acquisition process with airborne LiDAR, aiming at the high-resolution 3D mapping of the sites. The third stage focuses on data processing, cloud point classification, and thematic information creation. The fourth stage is focused on developing the VR application. The VR application will allow users to immerse themselves in the study area, observe, and interact with the existing state of the embankments in high resolution. Additionally, users can interact with the 3D point cloud, where thematic information is displayed describing the classification of the 3D cloud, the altitude, and the RGB color. Additional thematic information in vector form, providing qualitative characteristics, is also illustrated in the virtual space. Furthermore, six different scenarios were visualized in the 3D space using a VR app. Visualizing these 3D scenarios using digital twins of the current antiflood infrastructure provides scenarios of floods at varying water levels. This study aims to explore the efficient visualization of thematic information in 3D virtual space. The goal is to provide an innovative VR tool for managing the impact on anthropogenic infrastructures, livestock, and the ecological capital of various scenarios of a catastrophic flood.

**Keywords:** geovisualization; LiDAR; UAS; flooding; virtual reality

## 1. Introduction

Light Detection and Ranging (LiDAR) technology in geospatial applications has seen significant advancements over the past two decades, providing precise and detailed data necessary for various domains such as environmental monitoring, urban planning, and disaster management [1]. LiDAR is a remote sensing method that uses light as a pulsed laser to measure variable distances to the Earth's surface. This technology, particularly when integrated with mobile and airborne platforms, has transformed the capabilities for 3D mapping and geovisualization [2].

Airborne LiDAR has become a critical tool for capturing high-resolution 3D representations of the Earth's surface [3]. By emitting laser pulses from an aircraft and measuring the time it takes for these pulses to return after hitting objects on the ground, LiDAR generates precise point cloud data [4]. These data are invaluable for creating digital elevation models (DEMs) and other 3D representations essential for flood modeling and risk assessment [5–8]. The ability of LiDAR to penetrate vegetation and provide accurate ground measurements makes it especially useful in densely vegetated floodplains where traditional surveying methods may be less effective [9,10].

In recent years, the integration of LiDAR with other technologies has expanded its application scope [11–13]. Combining LiDAR data with hydrological and hydraulic models enhances the accuracy of flood predictions and risk assessments. These integrations allow for the simulation of flood events under various scenarios, improving the understanding of flood behavior and aiding in developing effective mitigation strategies [14–17]. The detailed topographic information obtained from LiDAR is crucial for identifying flood-prone areas and planning infrastructure to minimize flood impacts.

Virtual reality (VR) technology represents a significant advancement in the geovisualization of geospatial data [18–21]. VR provides immersive environments where users can interact with 3D models of landscapes and infrastructure [22–24]. This interaction is particularly beneficial for visualizing flood scenarios and assessing their potential impacts. VR allows stakeholders, including scientists, policymakers, and the public, to explore complex data intuitively and engagingly [25–27]. It also enhances communication and collaborative decision-making by providing a more accessible representation of flood risks compared to traditional maps and models [28–30].

This study aims to develop VR geovisualizations utilizing 3D point clouds collected through airborne LiDAR technology. This approach is intended for the 3D mapping of the current state of river channels and tributaries within the Thessalian Plain following catastrophic flooding in the summer of 2023. This study specifically focuses on the levees of the rivers and tributaries in the Thessalian Plain, mainland Greece, an area significantly impacted by flooding that destroyed urban elements and crops. High-resolution 3D mapping and VR geovisualization of the embankments surrounding the main rivers and tributaries provide vital information for planning restoration processes and designing prevention and mitigation measures for future disasters. Integrating LiDAR data into VR platforms involves creating realistic 3D models augmented with real-time flood simulation data, allowing users to visualize potential flood scenarios dynamically. This immersive experience helps comprehend complex flood dynamics and assess the potential impact on communities and infrastructure.

This study contributes significantly to the field of cartography and 3D geovisualization. It explores the efficient transfer of thematic information into the 3D virtual space, providing an innovative tool for managing the impacts of natural disasters, specifically flooding. The high-resolution 3D mapping and VR geovisualizations could offer critical information for planning restoration processes and designing prevention and mitigation measures for similar disasters in the future. The proposed approach represents a notable advancement in integrating geospatial technologies with immersive VR applications, enhancing the ability to visualize and analyze environmental impacts in a detailed and interactive manner. This study addresses the significant challenge of flood risk management by leveraging airborne LiDAR for high-precision 3D mapping and employing VR for immersive geovisualizations of six flood scenarios. These technologies collectively enhance the ability to predict, prepare for, and mitigate the impacts of flood events, ultimately contributing to more resilient communities and better-informed decision-making processes.

## 2. Materials and Methods

This study employs a comprehensive methodology involving four main stages: (i) LiDAR flight planning, (ii) data acquisition, (iii) data processing, and (iv) geovisualization (Figure 1). The initial stage involves planning LiDAR flights. Sensor L1 was utilized, with a linear flight path and a front overlap of 85%. The flights are conducted at an altitude of 70 m., ensuring a point density exceeding 300 points per square meter. The data capture multiple returns (3 and 5), enhancing the accuracy and detail of the collected information. In the data acquisition stage, a total of 15 flights are conducted. These flights yield over 3000 photos and point clouds with a total point count of 50,462,349. The total mapped area covers 35 km in length and 100 m. in width.

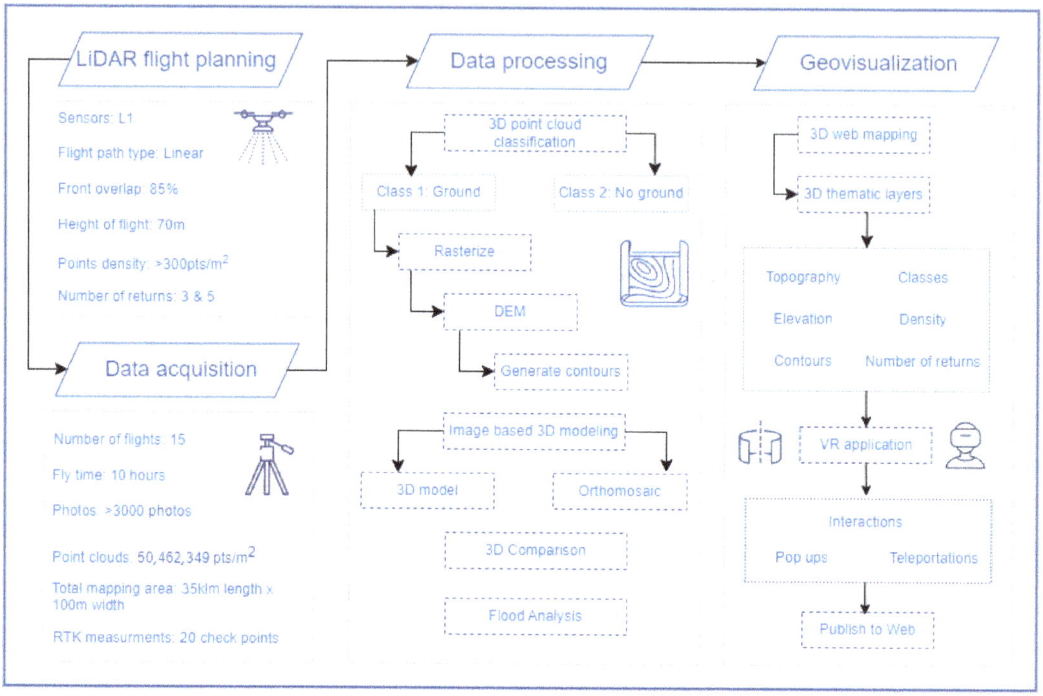


**Figure 1.** The methodological workflow followed to conduct the study.

The processing stage begins with the classification of the 3D point cloud into two primary classes: ground and no ground. The classified point cloud is then rasterized to create a digital terrain model (DTM). From the DTM, contours are generated to represent elevation changes. Concurrently, image-based 3D modeling techniques are applied to produce detailed 3D models and orthomosaics. These models are used for 3D comparisons and flood analysis, providing insights into terrain variations and potential flood zones. The final phase involves geovisualization, where the processed data are transformed into 3D web maps with thematic layers. These layers include topography, classes, elevation, density, and contours, offering a multi-faceted terrain view.

A VR application has also been developed, allowing for interactive experiences such as pop-ups and teleportation within the virtual environment. The completed visualizations and analyses are published online, making them accessible for further research and public viewing. This methodology proposes a detailed and accurate representation of the mapped area, combining advanced LiDAR technology with sophisticated data processing and visualization techniques.

### 2.1. Study Area

The Thessalian Plain, located in central Greece, is one of the country's most significant agricultural regions, known for its fertile soil and extensive farming activities (Figure 2). The plain is bordered by the Pindos Mountain range to the west and the Aegean Sea to the east, with the Pineios River crossing the area, providing essential water resources for agriculture and local communities.

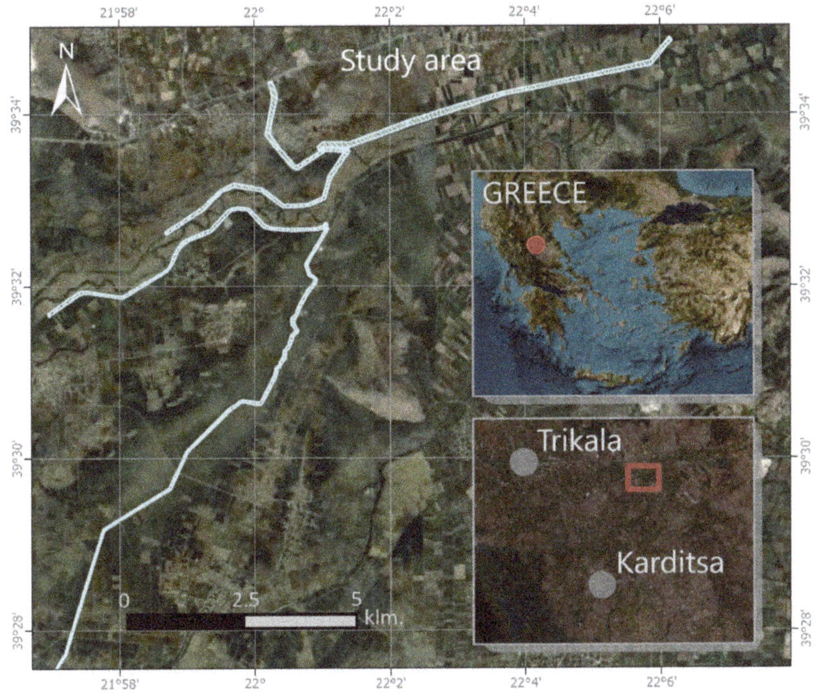


**Figure 2.** Study area. The blue color indicates the embankments enclosing the tributaries of the Pineios River between the cities of Karditsa and Trikala. The inset maps indicate the study area in red.

In the summer of 2023, the Thessalian Plain experienced an unprecedented flood due to extreme weather conditions exacerbated by climate change. Heavy and prolonged rainfall from severe thunderstorms led to the Pineios River and its tributaries overflowing their banks. The flood caused widespread devastation across the region, impacting urban and rural areas. The rainfall intensity during the event was recorded at levels significantly higher than the historical average. Meteorological data indicated that some areas received over 300 mm of rain from 3 September 2023 to 7 September 2023, overwhelming the existing drainage infrastructure [31]. The Pineios River had a dramatic rise in water levels, with its peak flow reaching approximately 1950 $m^3/s$, leading to severe flooding across the Thessalian Plain. The flooding affected over 720 $km^2$ of the Thessalian Plain, including major towns and villages, leading to extensive damage to homes, infrastructure, and agricultural lands [32]. The rapid accumulation of water led to flash floods, particularly in low-lying areas of the plain. The Pineios River, already characterized by its meandering nature and dynamic fluvial processes, saw an unprecedented rise in water levels. The geomorphological features of the river, including its wide floodplain and alluvial deposits, contributed to the rapid spread of floodwaters [33].

The flood severely damaged the local economy, particularly in the agricultural sector. Thousands of hectares of crops, including cotton, corn, and wheat, were destroyed. Livestock losses were also substantial, with many animals drowning or perishing due to a lack of access to dry land and food [34]. The floodwaters inundated homes, businesses, and public infrastructure, leading to an estimated economic loss of several hundred million euros [34]. Environmentally, the floodwaters carried pollutants and sediments, contaminating water supplies and leading to soil erosion. The ecosystem disruption also affected local wildlife, with many species losing their habitats due to the extensive flooding. The sediment transport and deposition processes inherent to the river's geomorphology played a significant role in redistributing nutrients and contaminants across the floodplain.

An estimated 28,000 people were affected in the impacted areas [35]. The Greek government, along with international aid organizations, provided immediate relief in the form of food, water, medical supplies, and temporary shelter. Long-term recovery efforts focused on rebuilding infrastructure, restoring agricultural productivity, and improving flood defenses.

### 2.2. Fieldwork and Equipment

The DJI Matrice 300 (M300) is a versatile and robust aerial platform for various applications. Equipped with advanced sensors and flight capabilities, up to 55 min of flight time, and various safety features, the M300 has high standards for unmanned aerial vehicle (UAV) performance. In this study, we coupled the drone with the Zenmuse L1 LiDAR sensor, and the M300 became a powerful tool for high-precision aerial surveying and mapping. The Zenmuse L1 integrates a Livox LiDAR module, a high-accuracy IMU, and a camera with a 1-inch CMOS sensor, providing real-time 3D data with centimeter-level accuracy. L1 also features an integrated RGB camera with a 20-megapixel resolution and a 24 mm focal length lens.

This combination allows for efficient data collection in complex environments, making it ideal for topographic mapping, forestry management, and infrastructure inspection applications. Integrating the L1 sensor with the M300 UAV ensures seamless data capture and processing, enhancing the accuracy and efficiency of geospatial data acquisition.

### 2.3. Data Acquisition

This paper focuses on the embankments surrounding the tributaries of the Pineios River in the Thessalian Plain region, specifically between the cities of Karditsa and Trikala in mainland Greece. In the summer of 2023, this area experienced significant damage as flooding the region's rivers destroyed urban infrastructure and crops. The widespread impact of the flood across the entire Thessalian Plain posed significant challenges for the authorities in managing the event. The survey area encompassed four distinct regions within the Thessalian Plain (Figure 2). The initial mapping of the study area necessitated the development of 7 different flight plans.

The data acquisition campaigns were conducted through a series of flights from 24 to 27 February 2024, covering approximately 36 km along linear infrastructures such as river channels and tributaries. To ensure efficient data acquisition, the flights were organized into four corridors. The distribution of the data acquisition flights is as follows: (i) 24 January 2024: 3 flights (Corridor A1); (ii) 25 January 2024: 8 flights (Corridors A1 and A2); (iii) 26 January 2024: 6 flights (Corridor A3); and (iv) 27 January 2024: 3 flights (Corridor A4). The flights were conducted between 08:00 and 13:00 every day. From 24 January to 26 January, the weather conditions remained consistent, characterized by relative cloudiness, low humidity, wind speeds of 1–2 Bf., and temperatures ranging from 3 to 6 degrees °C. The details of these flights are outlined in Table 1 below. Throughout all flight plans, the selected flight altitude was consistently maintained at 80 m above the takeoff point. The flight altitude was chosen to achieve a necessary point cloud density greater than 350 pts/m$^2$.

**Table 1.** Data acquisition details.

| Area | Date | Total Flights | Height of Flight | Point Density (pts/m$^2$) | GSD (cm/pix) | Front Overlap | Length |
|---|---|---|---|---|---|---|---|
| 1 | 24 January 2024 | 7 | 80 m | 393 pts/m$^2$ | 2.18 cm | 80% | 12.199 m |
| 2 | 25 January 2024 | 4 | 80 m | 393 pts/m$^2$ | 2.18 cm | 80% | 7.080 m |
| 3 | 25 January 2024<br>26 January 2024 | 5 | 80 m | 393 pts/m$^2$ | 2.18 cm | 80% | 8.377 m |
| 4 | 26 January 2024<br>27 January 2024 | 4 | 80 m | 393 pts/m$^2$ | 2.18 cm | 80% | 8.055 m |

In Figure 3, a linear flight plan is presented. This study analyzes and presents the results of the flood analysis conducted in this area. All the flood impact geovisualizations pertain to this specific area.

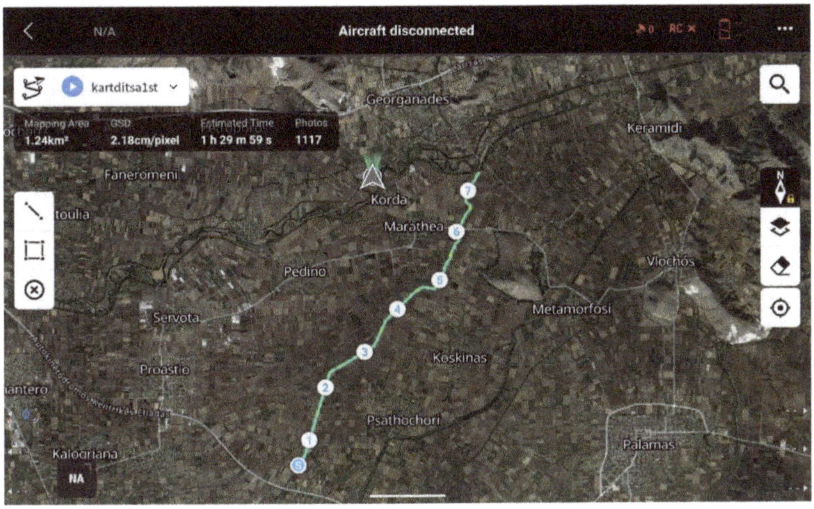


**Figure 3.** Linear flight plan for data acquisition of the study area.

Data acquisition using unmanned aerial vehicles (UAVs) and airborne LiDAR technology presents a highly efficient method for capturing detailed geographic information, especially in areas where access is physically challenging. The embankments surrounding the tributaries of the Pineios River in the Thessalian Plain region are long linear structures. They have heights ranging from approximately 2 to 5 m and lengths spanning from 7 kilometers to 12.2 kilometers across the entire study area. Challenges in this method often include navigating through difficult terrains such as dense forests, mountainous areas, or environmentally sensitive zones where ground access is restricted or impossible. Furthermore, weather conditions, regulations, and sensor calibration can affect UAV operations and data accuracy.

### *2.4. Processing*

#### 2.4.1. Point Cloud Classification

Data processing classified the points collected with airborne LiDAR into two classes: (i) ground and (ii) no ground. This process, leveraging the LiDAR Sensor L1 and point cloud classification techniques within ArcGIS Pro v3.0.0, involves several key steps to classify LiDAR point clouds accurately. Upon acquiring LiDAR point cloud data, the initial preprocessing phase focuses on enhancing data quality. Noise reduction techniques are applied to filter out erroneous points, ensuring a cleaner dataset. Additionally, normalization processes standardize the data, facilitating consistent analysis and interpretation.

Feature extraction is a critical stage where relevant attributes are derived from the point cloud. These features, including elevation, intensity, and point density, provide valuable information for distinguishing ground surfaces from vegetation. By capturing the geometric and spatial characteristics of points, feature extraction lays the foundation for accurate classification.

The subsequent step involves creating a labeled dataset for training the deep learning model. Manual annotation of a point cloud subset assigns labels such as 'ground' and 'vegetation' to points, enabling supervised learning (Figure 4). This annotated dataset serves as the basis for training a convolutional neural network (CNN) within ArcGIS Pro.

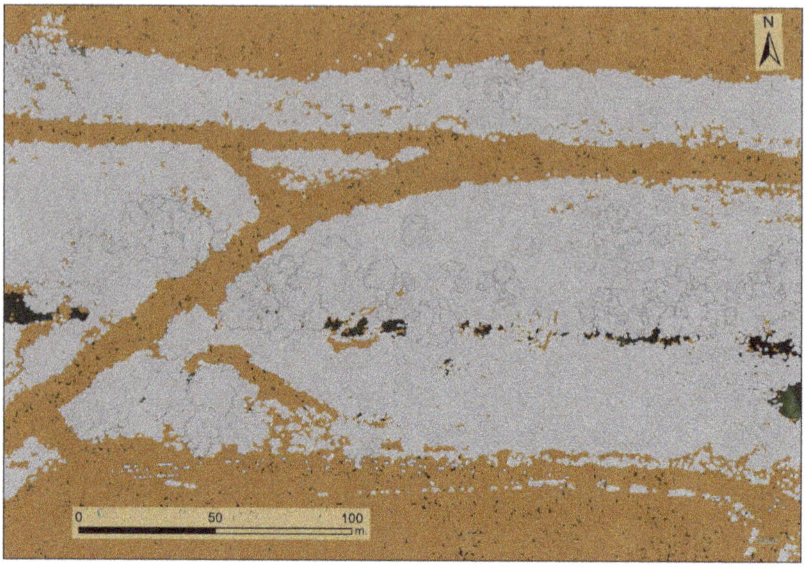


**Figure 4.** Examples of 'Ground' and 'No ground' classes from the manually annotated subset of the point cloud are depicted in gray and brown colors, respectively.

CNN architecture is designed to learn and classify points effectively based on their extracted features. The model learns to differentiate between ground and vegetation points through iterative training using the labeled dataset. Data augmentation and regularization are employed to enhance model performance and prevent overfitting. Integration of the trained CNN into ArcGIS Pro facilitates seamless application to the entire point cloud dataset. The software processes each point, classifying it as either ground or vegetation based on learned patterns. Post-processing steps involve validation and refinement to ensure the accuracy and reliability of the classification results. Using deep learning through ArcGIS Pro, the process of point cloud classification into ground and vegetation from LiDAR Sensor L1 data is a systematic and rigorous methodology. Leveraging advanced computational techniques and geospatial tools, this approach enables precise analysis and interpretation of LiDAR data for various environmental and urban applications.

#### 2.4.2. Image-Based 3D Modeling

According to the methodology, the next stage involved data processing to generate a 3D point cloud, 3D models, and high-resolution orthomosaics of river canals. Photogrammetric and image-based 3D modeling methods were employed to produce cartographic outcomes. The image-based 3D modeling processing steps were consistent across all datasets for each recording date.

Initially, an expert photo interpreter conducted the Very High-Resolution Image (VHRI) quality control visually and subsequently used the Image Quality Index (IQI) algorithm. Images exhibiting blurriness, shaking, overexposure, or containing portions of the horizon were excluded through visual inspections. VHRIs with IQI index values outside the 0.5–1 range were rejected in the subsequent processing steps.

The VHRIs deemed suitable for photogrammetric processing were imported into Agisoft Metashape v1.7.4 software, where image alignment was performed [36]. The alignment process employed Structure-from-Motion (SfM) algorithms, including Scale Invariant Feature Transform (SIFT) and Random Sample Consensus (RANSAC). This process resulted in a sparse point cloud, which was then densified using the Multi-View Stereo (MVS) algorithm. The resulting dense 3D point cloud served as the foundation for creating a 3D mesh. Spatial interpolation connected the points into a Triangulated Irregular Network (TIN), forming a single 3D mesh. This mesh was then textured with photorealistic textures, producing a textured 3D model.

Subsequently, a Digital Surface Model (DSM) was generated to describe the area's elevation, followed by orthorectification of the image pixels to create the orthomosaic. This processing was applied to the data collected across eight recording dates for each of the three different scales. The outcomes of the image-based 3D modeling processing utilized for developing the virtual reality (VR) application included 3D models and orthophoto maps.

### 2.4.3. Flood Analysis

Flood impact analysis is a critical process in mitigating and managing flood risks. Using ArcGIS Pro, this analysis can be conducted effectively with high-resolution digital terrain models (DTMs). This text outlines the process of flood impact analysis using a DTM with a spatial resolution of 10 cm, where the highest height is 133 m and the lowest is 121 m.

First, the high-resolution DTM is imported into ArcGIS Pro. The 10 cm spatial resolution allows for detailed surface modeling, which is crucial for accurately mapping flood extents and depths. The height range of 121 m to 133 m will be used to identify areas at different flood risks. The next step is preprocessing the DTM. This includes filtering out noise and ensuring the data accurately represent the terrain surface. Hydrological tools within ArcGIS Pro, such as "Fill" and "Flow Direction", are then utilized to prepare the DTM for flood simulation. The "Fill" tool corrects small imperfections, such as depressions or pits, that could impact flow accumulation modeling. After preprocessing, flood scenario parameters are defined. These parameters include rainfall intensity, duration, and potential water levels. Using the "Raster Calculator", potential flood extents are modeled based on these parameters. This tool calculates water depth by subtracting the DTM from the expected water surface elevation during a flood event (Figure 5).

**Figure 5.** Three-dimensional visualization of the study area with 3 m WSF flood.

The results are then analyzed to identify critical areas that are most vulnerable to flooding. This includes evaluating the spatial distribution of flood depths and assessing the impact on the flooded area's anthropogenic infrastructure, livestock, and ecological capital. The high-resolution data ensure that even small-scale features are accounted for, providing a comprehensive understanding of flood risks.

## 3. Results

### 3.1. 2D and 3D Results

The 2D and 3D derivatives generated from the study area using the proposed methodology are as follows: (i) 3D dense point cloud, (ii) textured 3D model, (iii) orthomosaic, (iv) digital terrain model (DTM), (v) Digital Surface Model (DSM), and (vi) contour lines (.shp files, 1 m interval). All 2D and 3D cartographic products were in the same coordinate system, WGS84.

For every point cloud created, ground and no ground classification was performed. Furthermore, every 3D point cloud includes (i) intensity values, (ii) return values, (iii) reflectance, (iv) height, (v) RGB, and (iv) classification. The merged 3D point cloud was exported in two reference systems, WGS 84 and UTM 34N, while all the individual 3D point clouds were exported in the UTM 34N reference system. Table 2 presents the spatial derivative results.

**Table 2.** Specifications of the spatial derivatives generated from airborne LiDAR data over the study area.

| Type | Resolution/Density | Coordinate System | Format |
|---|---|---|---|
| 3D point cloud | 525 pts/m$^2$ | WGS 84 (4326) and UTM 34N | .las |
| DTM | 20 cm/pix | WGS 84 (4326) | .tiff |
| DSM | 20 cm/pix | WGS 84 (4326) | .tiff |
| Orthomosaic | 5 cm/pix | WGS 84 (4326) | .tiff |

Figure 6a,b present the 3D point cloud representation (.las) of riverbank areas along the Pineios River. Using the Zenmuse L1 sensor, the initial step involved cleaning the point cloud to remove extraneous terrain and outliers while preserving the targeted area. The LiDAR sensor used to acquire point clouds has a vertical resolution of approximately 3 cm. The DSM for the study area, derived from high-resolution image processing, spans the entire mapped area at a 20 cm spatial resolution. The lowest recorded elevation, 122 m, lies at the riverbank base where the water level resides, while the highest elevation is 153 m. Figure 6c,d display typical examples from the DSM at the same 20 cm spatial resolution. The DTM, generated from analyzing the classified 3D dense point cloud, covers the entire mapped Area 3 with a spatial resolution of 20 cm. At 121.75 m, the lowest elevation in the area lies at the base of the embankments, while the highest elevation recorded is 137.08 m. Figure 6e,f provide illustrative instances from the DTM, displaying the diversity of elevations captured in the dataset.

The orthomosaic created provides a detailed view of the entirety of Area 2, measuring a total length of 7080 m and a width of 100 m. The raster file processes a spatial resolution of 5 cm/pixel, comprises 120,711 $\times$ 48,073 pixels, and covers an area of 1.01 km$^2$. Figure 6g,h present distinctive examples from various sections of the studied area, demonstrating parts of the orthomosaic and its detail. The contour lines were created with a 1 m interval from the DTM, and they underwent additional refinement by a smoothing process applied at 3 m intervals. The implementation of thorough quality control is necessary for the elimination of inaccuracies in the lines. These contours describe elevations that span from 122 m to 133 m throughout the whole area. Figure 6i,j represent an illustrative image of the contour lines designed for the studied area.

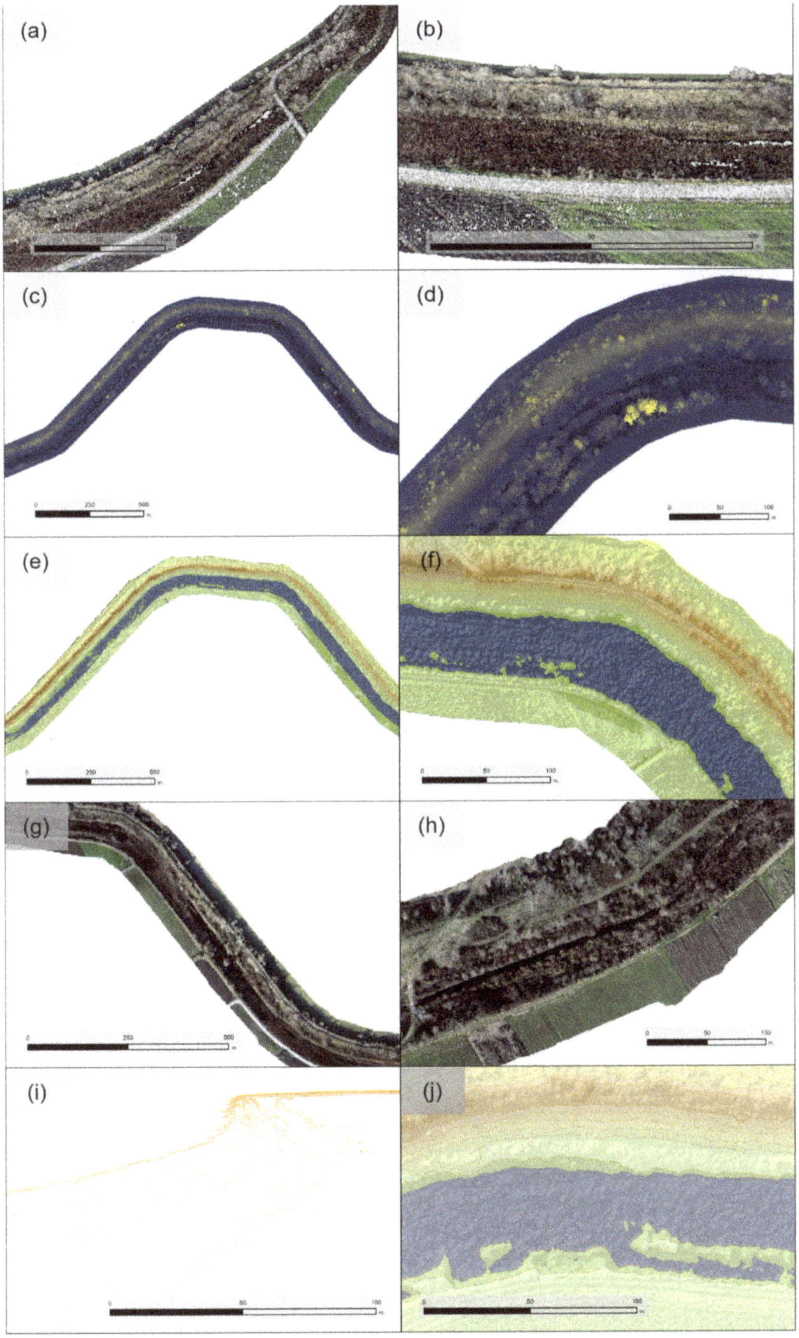


**Figure 6.** Visualizations of the 2D and 3D results of the study area, (**a**) a 3D point cloud of the wider area, (**b**) a 3D point cloud of a representative part, (**c**) a DSM of the wider area, (**d**) a DSM of a representative part, (**e**) a DTM of the wider area, (**f**) an orthomosaic of a representative part, (**g**) an orthomosaic of the wider area, (**h**) a DSM of a representative part, (**i**) contour lines of the wider area, and (**j**) contour lines overlayed in a DTM of a representative part of the study area.

### 3.2. Flood Impact Analysis Results

The results of the flood impact analysis in the Thessalian Plain underscore a gradual yet significant increase in flooding severity as water levels rise along the river canal. This detailed analysis reveals that even a minor increase of 1 m in water level results in noticeable changes in flood dynamics, signaling the onset of significant impacts on the area (Figure 7a). During this initial stage, low-lying areas near the riverbanks begin to experience flooding, although this flooding occurs on a limited scale, affecting approximately 12 hectares of land. This early flooding indicates the potential for more severe impacts if water levels continue to rise, emphasizing the importance of close monitoring and early intervention.

**Figure 7.** Two-dimensional visualization of the flood scenarios at (**a**) 1 m, (**b**) 2 m, (**c**) 3 m, (**d**) 4 m, (**e**) 5 m, and (**f**) 6 m of flood height.

As the water level rises to 2 m, the impact of flooding becomes more pronounced. At this stage, the flood-affected area expands to about 35 hectares. Certain parts of the area begin to be covered by floodwaters more extensively (Figure 7b). Notably, roads crossing the river are among the first infrastructure elements to be affected, leading to disruptions in vital transportation links. With each subsequent meter increase in water level, the extent of the flooding becomes more severe. When the water level reaches 3 m, floodwaters encroach further inland, affecting a broader landscape area, now covering 42 hectares (Figure 7c). The area flooded increased by 13 ha compared to the 2 m scenario, now covering 35 ha of land. A total of 42 ha is flooded, and the impact at this stage is particularly significant. Small hills along the riverbanks become submerged. An important observation from this scenario is the asymmetric expansion of floodwaters. The lower elevation on the eastern side of the river results in a more extensive inundation compared to the higher elevation on the western side. This differential impact highlights the critical role of topography in shaping flood dynamics, with lower elevation areas being more vulnerable to flooding. This topographical influence is particularly notable in urban areas on the eastern side of the river, where substantial rises in the water surface level can pose significant risks to infrastructure and human lives. From the results of this scenario, it is evident that the flooded area has expanded to the road crossing the river, creating a barrier to transportation and isolating residents. The urban area's vulnerability underscores the need for targeted flood protection measures, such as levees and floodwalls, and the importance of integrating flood risk assessments into urban planning and development processes.

As water levels rise to 4 m, the flood-affected area expands, covering 44 hectares (Figure 7d). The flood extent at this stage remains relatively stable compared to the previous level. However, further rises to 5 and 6 m result in an even greater flood coverage. At 5 m (Figure 7e), the covered area remains at 44 hectares, but at 6 m, the floodwaters expand further to cover 56 hectares (Figure 7f). These advanced stages of flooding pose substantial risks to communities and infrastructure, particularly in the eastern, low-lying regions. The increased flood extent towards the east exacerbates the vulnerability of these areas, underscoring the importance of comprehensive flood management strategies. Table 3 further illustrates the impact of rising water levels on the covered area.

**Table 3.** Flooding areas cover the six scenarios realized.

| Scenario Number | Water Surface Level (m) | Cover Area (ha) | Overflow Point of Interest (POI) | Difference in Cover Area (ha) |
|---|---|---|---|---|
| Scenario 1 | 1 m. | 12 ha | - | - |
| Scenario 2 | 2 m. | 35 ha | Eastern embankment | 23 ha |
| Scenario 3 | 3 m. | 42 ha | Crossroads | 13 ha |
| Scenario 4 | 4 m. | 44 ha | Provincial roads | 2 ha |
| Scenario 5 | 5 m. | 44 ha | Provincial roads | 0 ha |
| Scenario 6 | 6 m. | 56 ha | Western embankment | 12 ha |

Table 3 depicts the progressive increase in flood-affected areas corresponding to each meter rise in water levels. Notably, the transition from 1 m to 2 m sees a significant jump in the affected area, emphasizing the non-linear nature of the flood and its escalating impact on the region.

## 4. Geovisualization

### *4.1. 3D Mesh vs. 3D Point Cloud*

For the 3D geovisualization of the results, both 2D and 3D outcomes were utilized. The 3D cartographic derivatives emerged from processing 3D point clouds and 3D mesh models. The cloud-to-mesh distance method was used in CloudCompare software v 2.7.0 to compare 3D point clouds and 3D mesh models. This method involves calculating the distance between each point in one point cloud and the nearest point in the other point cloud. It effectively quantifies the difference between the representations, highlighting areas where the mesh deviates from the true geometry captured by the point cloud. More specifically, the method was applied for distances of 1 m between the surfaces of the 3D model and the points of the 3D point cloud. Figure 8 presents the results of the C2M distance method. Large distances (1 m.) are shown in red, which are observed in the upper part of the trees. This is because the vegetation, especially leaves and tree branches, does not form detailed geometries in a mesh.

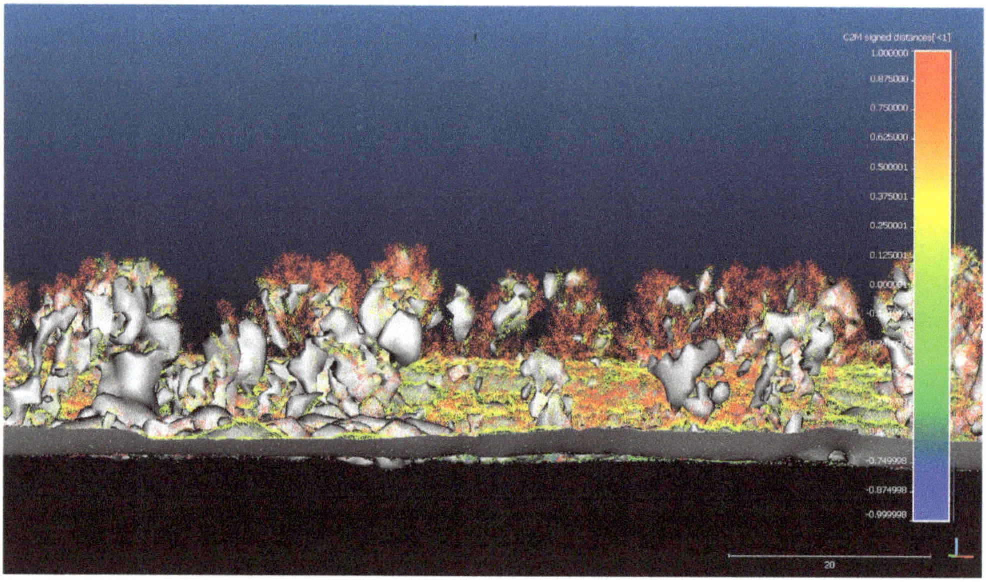


**Figure 8.** Cloud-to-mesh distance comparison. Distances greater than 1 m are depicted in red color.

A 3D mesh model consists of vertices, edges, and faces, forming a continuous surface representing the terrain and vegetation. Although 3D mesh models have been mainly used for VR geovisualization in recent years, they have some limitations [37]. Creating a mesh model from point cloud data can introduce artifacts, especially in areas with dense vegetation. This simplification leads to a loss of geometric accuracy, as the mesh surfaces do not accurately represent the true geometry of the area. This makes the 3D mesh model unsuitable for accurately visualizing the vegetation structure in the Thessalian Plain.

In contrast, a 3D point cloud captures the terrain as a collection of numerous discrete points with precise X, Y, and Z coordinates. This method offers unparalleled geometric accuracy, which is crucial for representing detailed vegetation structures and complex terrain features. Additionally, each point in a 3D point cloud can carry valuable additional information such as the number of returns, intensity value, and class. This information enhances the analysis and visualization capabilities, providing a richer and more detailed dataset. Point clouds allow for a more realistic and accurate visualization, which is particularly important for VR geovisualizations.

Given the goal of achieving the most detailed and accurate VR geovisualization of the river canal in the Thessalian Plain, the decision to continue with the 3D point cloud, supported by the cloud-to-mesh distance analysis performed in CloudCompare, was motivated by its superior geometric precision and the richness of additional data it provides. The point cloud's high level of detail and supplementary information ensures an immersive and realistic experience, which is crucial for comprehensive VR geovisualization tasks.

### 4.2. 3D Mapping

The geovisualization of the results consists of two primary stages: (a) 3D web mapping and (b) the development of the VR application. During the first stage of 3D web mapping, the visual variables of the results in the 3D space were investigated. Specifically, the cartographic results used for creating the geovisualizations included (i) classified 3D point clouds, (ii) DTM, (iii) WSE (water surface elevation), (iv) contour lines, (v) roads, (vi) overflow points, (vii) water level points, and (viii) hazardous areas. Initially, all results were shared on the web through the ArcGIS Online application. Subsequently, the results were transferred to a 3D web scene where their symbolism could be customized. The features used for the development of the geovisualizations include different types of data: (a) vector, (b) raster, and (c) point clouds. This requires a different approach to visualizing them, as the visual variables for these data types are not common.

For the 3D point cloud, the size of the points in relation to the cartographic scale and the thematic information of the points were studied. In a 3D scene environment, the size of the point affects the extent to which the recorded information is referenced. Therefore, when the size of the points increases, there is a risk of overlap. The point sizes examined ranged from 1 to 6 pixels each (Figure 9). The size selected was 3 pixels per point to avoid information loss and overlap (Figure 9c). The thematic information visualized at the 3D point cloud level included RGB value, elevation, and the classification of the points. Color was chosen as the visual variable to depict these thematic layers successfully. As shown in Figure 9, points with higher elevations are displayed in bright red, while points with lower elevations are shown in blue. Two different approaches were used to visualize the classification of points. The first involved removing points classified as vegetation, leaving only ground class points, allowing the user to observe the area without riparian vegetation. The second approach provided the ability to observe points belonging to both classes, differentiating them by color.

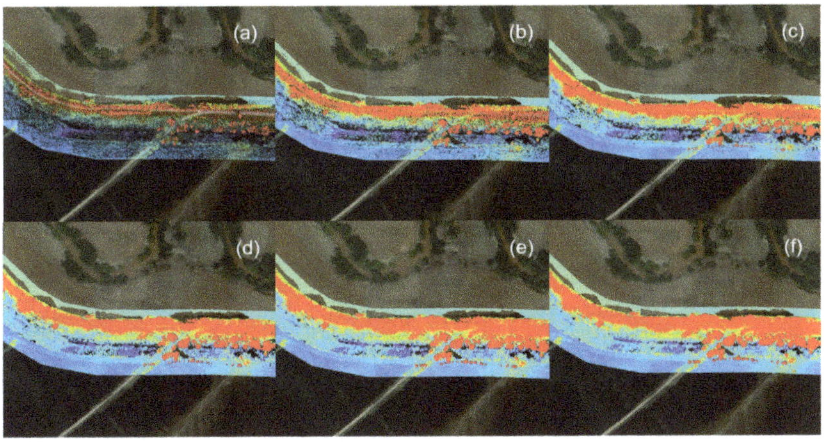


**Figure 9.** Point cloud size variation visualization from 1 to 6 pixels: (**a**) 1 point equal to 1 pixel, (**b**) 2 points equal to 2 pixels, (**c**) 3 points equal to 3 pixels, (**d**) 4 points equal to 4 pixels, (**e**) 5 points equal to 5 pixels, and (**f**) 6 points equal to 6 pixels. To prevent information loss and overlap, using a size of 3 pixels per point is recommended.

The 3D visualization of raster data was primarily accomplished by utilizing their elevation information for extrusion. Additionally, for the DTM, a color gradient was applied to the elevation at 1 m. intervals. At the same time, a specially designed transparent texture was assigned to WSE for river water surfaces, giving a waived animation property to enhance photorealism (Figure 10).

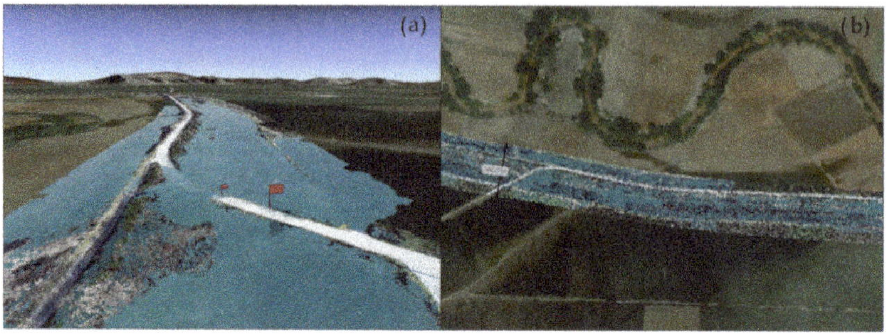


**Figure 10.** VR geovisualization of a flooded road crossing the river. A fly-mode (or drone-mode) approach: (**a**) a user at 50 m above the ground and (**b**) a user at 500 m above the ground. The red flag indicates the affected part, signifying a danger in a 2 m flood scenario.

Although initially created in 2D form, vector data subsequently acquired 3D properties through the 3D scene. Specifically, linear elements such as contour lines were adapted to the DTM, and elevation information was obtained based on their value. The 3D visualization facilitated easy distinction of the rise in water surface levels relative to the area's relief, providing quantitative elevation information. Furthermore, the diameter thickness of the contour lines was chosen to be 10 cm to ensure they were discernible in the 3D space.

### 4.3. VR Web Application

The data processing was conducted using ESRIs ArcGIS Pro software. The processed results were then uploaded to ArcGIS Online, where the 3D web scene was used for the development of the VR application. The thematic information layers, visualized in 2D and 3D dimensions, were also uploaded to the same environment. Finally, the application was adapted to a VR environment using ESRI's online plugin VR 360, which enables the distribution of the application over the internet.

The study area covered in this work spans a geographical extent of 8 km, necessitating its visualization at two different viewing levels. For the development of VR geovisualizations, two scenarios were created: a fly-mode (or drone-mode) scenario and a pedestrian-mode scenario. Each scenario serves a different purpose and addresses different needs.

Fly-mode scenario: The first viewing level caters to the need for an overall observation and identification of high-risk overflow points, allowing users to observe the entire mapped area (Figure 10). In this approach, the user is placed in two flight positions, (i) at 50 m and (ii) at 500 m above ground, denoting a 1:50 and 1:500 cartographic scale, respectively. The scale is calculated based on the camera parameters used in this study. Furthermore, when observing a flooded area or flood scenarios, viewing the affected area at a specific height is crucial to avoid physical obstacles and barriers. This approach provides greater freedom when navigating the scene, allowing for a comprehensive understanding of the flood results.

Additionally, to facilitate the identification of high-risk overflow points, indicator markers (red flags) were added to affected areas or parts of the study area that were vulnerable to the flood. These are anthropogenic infrastructures, livestock, and ecological capital near the riverbanks, such as roads crossing the river, bridges, settlements in proximity, and agricultural or livestock facilities. This drone-level view is crucial for a comprehensive understanding of the region and strategic planning.

Pedestrian-mode scenario: The second viewing level (pedestrian mode) places the user at street level (1:1 scale) and allows them to explore the area up close, observing the condition of the levees. The viewing positions were selected based on their high-risk status due to river overflow. These points were numbered sequentially from east to west. Users of the application can move from one viewing point to the next or select their desired point from the bookmark bar integrated into the application using the VR headset controllers (Figure 11).

**Figure 11.** Three-dimensional visualization of the study area in the VR space using pedestrian mode: (**a**) ground classified point cloud in RGB. A POI is depicted with the red flag: (**b**) ground and no ground classified point cloud in green and brown colors.

Users can change the thematic information they wish to observe at specific viewing points, such as the classification of point cloud points. Additionally, users can choose to observe the river's water level rise incrementally by 1 m and see how this increase affects the area. The application was developed using ArcGIS Online and shared on the web to ensure easy accessibility. More specifically, the application can be used from any mobile device, while VR viewing is supported by Valve Index, HTC Vive, and MetaQuest VR headsets.

## 5. Discussion

VR geovisualizations using 3D point clouds obtained from airborne LiDAR data have offered substantial insights into the condition of river channels and tributaries within the Thessalian Plain following severe flooding in the summer of 2023. This study primarily aimed to utilize advanced data acquisition and visualization geospatial technologies to efficiently illustrate, enhance understanding, and facilitate the management of the impacts of various flooding scenarios in the study area.

One of the main challenges was acquiring high-resolution data in a flood-affected area. Accessing certain regions was difficult due to water damage and debris. However, drone-based LiDAR technology helped overcome some of these obstacles, enabling precise data acquisition from the air. The data processing phase presented additional challenges, especially in classifying point clouds and generating detailed thematic information. The methodology included several stages, beginning with flight plans for comprehensive 3D mapping using airborne LiDAR. The subsequent data processing focused on reducing noise and normalizing the data to ensure high-quality datasets. Classifying point clouds into ground and non-ground points using deep learning techniques was critical. This improved the accuracy of terrain models and facilitated the extraction of essential features such as elevation and point density, which are crucial for flood risk assessment and management. The authors used pre-existing deep learning tools to classify the LiDAR point cloud within ArcGIS Pro. Using airborne LiDAR and deep learning techniques in ArcGIS Pro offers significant advantages over traditional methods in point cloud classification accuracy and efficiency. Conventional methods for point cloud classification often involve manual techniques and basic algorithms such as thresholding, clustering (e.g., k-means), and rule-based methods [38–40]. These methods are being increasingly supplemented or replaced by advanced techniques like deep learning models in software such as ArcGIS Pro, which provides high precision by accurately differentiating ground and non-ground points and analyzing features like elevation, intensity, and point density [41].

The use of spatial data and their VR geovisualization implementation can vary. Rydvanskiy and Hedley [42] use the ArcGIS City Engine but then transfer the data to Unity, resulting in a loss of geographic position and coordinates in VR space. This has been a recurring issue with VR visualizations of geographic data. Additionally, the authors have already tested [20,21] approaches that use spatial data embedded in the VR space, but the final application does not have geographic coordinates.

Other studies [18,43,44] use scaled 3D geovisualizations in the virtual space by depicting 3D models on a virtual table. In the proposed approach, the users are embedded into the virtual space to scales of 1:1 and 1:50, allowing for their immersion into the digital twin of the study area to be enhanced by thematic information of flooding scenarios.

In VR, it is common to adjust the complexity or the level of detail (LoD) of a 3D model based on its distance from the viewer or its importance in the scene. This is crucial because it helps optimize performance by reducing the number of polygons rendered for objects far away or less important while maintaining high detail for objects closer or more critical to the user's experience. When DTM or DSM needs to be presented, highly detailed earth models are simplified, ensuring that the user experience remains immersive without overloading the system. A generalized approach to the 3D models is used [45–47]. By using a LiDAR point cloud integrated into the virtual space to maintain a higher LoD in geovisualization at any scale, this approach ensures a more realistic representation of spatial information without losing details from the Earth's surface 3D visualization.

Developing the VR application significantly enhanced data usability and accessibility by transforming complex geospatial datasets (LiDAR point clouds) into interactive, immersive visualizations. Users can intuitively explore and manipulate 3D point clouds and models, gaining a better understanding of spatial relationships and flood extents in various scenarios. High-resolution DTMs provide detailed and accurate flood risk analysis. The 3D virtual space makes advanced geospatial tools accessible to non-experts, and collaborative features allow multiple users to work together, enhancing the spatial knowledge of the flood phenomenon and thus facilitating planning and decision-making processes. This capability is particularly useful for planning restoration processes and designing mitigation measures, providing a detailed and realistic representation of current conditions.

However, implementing the VR application had limitations. High computational power and specialized VR equipment were needed, restricting its widespread use. Additionally, the accuracy of VR geovisualizations depends heavily on the quality of the initial LiDAR data and the effectiveness of processing algorithms. Inaccuracies in these stages can lead to misrepresentations in the virtual environment. By adopting specific strategies, the authors aim to address potential inaccuracies and enhance the reliability of VR geovisualizations for flood modeling and other geospatial applications. More specifically, to reduce potential inaccuracies in the VR geovisualizations due to the quality of initial LiDAR data or processing algorithms, the authors proposed the following strategies: (1) Data Validation and Calibration: Implementing rigorous validation and calibration processes to ensure the accuracy of the initial LiDAR data before its integration into the VR environment. This includes cross-referencing LiDAR data with other high-precision in-situ measurements. (2) Error Analysis and Correction: Conducting detailed error analysis to identify and correct inaccuracies in the generated DTMs. This may include manual corrections in areas where automated processes fail to deliver the desired accuracy. Concerning VR geovisualizations, the authors propose using user opinions and an iterative improvement approach. This involves incorporating feedback from users interacting with the VR geovisualizations to identify and address any inaccuracies. An iterative development approach allows for continuous improvements based on real-world usage and expert input.

Integrating drone LiDAR data with VR geovisualizations offers numerous benefits for flood impact analysis and management but also presents several challenges. Despite the challenges, drone LiDAR data and thematic information in a VR virtual space can create a valuable tool for managing natural disasters and planning future flood events. This can involve depicting flood scenarios and their impact on infrastructure, livestock, and natural resources.

Compared to traditional 2D maps and static 3D models, VR provides a more engaging and effective tool for flood risk management and urban planning. The combination of drone LiDAR and VR geovisualizations offers significant advantages for flood management and risk assessment. It enables precise and detailed mapping through advanced LiDAR technology, which is crucial for accurate flood modeling and risk assessment. This detailed mapping improves the identification of flood-prone areas and infrastructure planning to minimize flood impacts. VR integration allows for immersive visualizations, enhancing stakeholder engagement and decision-making. Stakeholders can interact with 3D flood scenarios in real time, improving their understanding and response strategies. This integrated approach supports the development of more effective flood mitigation and emergency response plans.

## 6. Conclusions

This study presents a significant advancement in flood risk management and disaster mitigation by integrating airborne LiDAR data with VR geovisualizations, focusing on the Thessalian Plain in Greece, which suffered extensive damage due to severe flooding in the summer of 2023. The methodological approach involved high-resolution 3D mapping of river channels and tributaries, utilizing airborne LiDAR technology for precise data collection and VR for immersive geospatial visualization. Airborne LiDAR facilitated the collection of detailed 3D point clouds, enabling an accurate representation of the terrain and vegetation, which proved invaluable in creating digital elevation and terrain models (DEM-DTM) and other 3D derivatives necessary for flood modeling and risk assessment. By integrating these datasets into a VR platform, this study provided an innovative visualization approach for stakeholders to interact with and understand complex flood scenarios and their affection more intuitively.

Key findings from this study include the efficiency and precision of airborne LiDAR in data acquisition, even in challenging and flood-affected areas, and the effectiveness of advanced classification techniques, including deep learning algorithms, in ensuring accurate differentiation between ground and non-ground points. The VR application offered an immersive experience, allowing users to visualize flood-impacted areas dynamically, observe potential future flood scenarios, and assess the impact on infrastructure and communities, which is crucial for planning restoration processes and designing mitigation measures.

The methodology involved meticulous LiDAR flight planning with the DJI Matrice 300 drone equipped with the Zenmuse L1 LiDAR sensor, achieving a high point density exceeding 300 points per square meter. Data processing included noise reduction, normalization, and classification into ground and non-ground points using deep learning techniques. This resulted in high-quality 3D models and orthomosaics used for flood analysis and VR development. The VR application allowed users to explore the affected areas in both fly and pedestrian modes, observing the impact of incremental water level rises and accessing thematic information interactively.

Integrating airborne LiDAR data with VR geovisualizations represents a promising approach to flood risk management, providing detailed, accurate, and interactive visualizations and tools for disaster preparedness and response. Future work should focus on improving data collection methods in challenging terrains, enhancing the accuracy of point cloud classification, and adding more thematic information into VR, creating a new cartographic approach. Furthermore, the methodology, as described in this study, faces several challenges and limitations in practical flood management and planning applications. This technology demands specialized knowledge and training for stakeholders, who must continually update their skills to utilize VR visualizations effectively. There is potential to enhance and standardize the visualized data in a VR environment, particularly the data spatial analysis and accuracy based on a cartographic scale, which necessitates further research. Developing a data acquisition protocol will aim to standardize UAV LiDAR data used on the cartographic approach of VR geovisualizations.

This study contributes to cartography and geospatial analysis by demonstrating the practical application of advanced technologies in managing natural disasters and planning future flood events. The findings underscore the importance of high-resolution data and innovative visualization tools in enhancing our understanding of environmental challenges, ultimately contributing to more resilient communities and better-informed decision-making processes in the face of natural disasters.

The proposed methodology is versatile and can be applied to various urban applications beyond flood management, aiding in urban planning, environmental monitoring, disaster management, infrastructure inspection, and agricultural management. This approach's high-resolution 3D mapping and advanced analytical capabilities make it a powerful tool for multiple applications.

**Author Contributions:** Conceptualization, E.E.P. and A.P.; methodology, E.E.P. and A.P.; software, E.E.P.; writing—original draft preparation, E.E.P. and A.P.; writing—review and editing, E.E.P. and A.P. All authors have read and agreed to the published version of the manuscript.

**Funding:** This research was funded by ORIENTATE project grant number EX200163.

**Data Availability Statement:** The authors thank HVA International BV for generously providing the license to use the Lidar data for this publication.

**Acknowledgments:** The authors acknowledge the ORIENTATE project (DrOne sensors and immeRsive tEchNologies towards the Thematic MApping of SpaTiotemporal PhEnomena) and CUT internal funding.

**Conflicts of Interest:** The authors declare no conflicts of interest.

## References

1. White, J.C.; Stepper, C.; Tompalski, P.; Coops, N.C.; Wulder, M.A. Comparing ALS and Image-Based Point Cloud Metrics and Modelled Forest Inventory Attributes in a Complex Coastal Forest Environment. *Forests* **2015**, *6*, 3704–3732. [CrossRef]
2. Romanoni, A.; Fiorenti, D.; Matteucci, M. Mesh-Based 3D Textured Urban Mapping. *arXiv* **2017**, arXiv:1708.05543.
3. Javanmardi, M.; Javanmardi, E.; Gu, Y.; Kamijo, S. Towards High-Definition 3D Urban Mapping: Road Feature-Based Registration of Mobile Mapping Systems and Aerial Imagery. *Remote Sens.* **2017**, *9*, 975. [CrossRef]
4. Mohammadzadeh, A.; Valadan Zoej, M.J. A State of Art on Airborne Lidar Application in Hydrology and Oceanography: A Comprehensive Overview. *Int. Arch. Photogramm. Remote Sens. Spat. Inf. Sci.* **2008**, *37*, 315–317.
5. Wedajo, G.K. LiDAR DEM Data for Flood Mapping and Assessment; Opportunities and Challenges: A Review. *J. Remote Sens. GIS* **2017**, *06*, 2015–2018. [CrossRef]
6. Vassilaki, D.I.; Stamos, A.A. TanDEM-X DEM: Comparative Performance Review Employing LIDAR Data and DSMs. *ISPRS J. Photogramm. Remote Sens.* **2020**, *160*, 33–50. [CrossRef]
7. Štular, B.; Lozić, E.; Eichert, S. Airborne LiDAR-Derived Digital Elevation Model for Archaeology. *Remote Sens.* **2021**, *13*, 1855. [CrossRef]
8. Chen, Z.; Li, J.; Yang, B. A Strip Adjustment Method of Uav-Borne Lidar Point Cloud Based on Dem Features for Mountainous Area. *Sensors* **2021**, *21*, 2782. [CrossRef] [PubMed]
9. Yoshida, K.; Pan, S.; Taniguchi, J.; Nishiyama, S.; Kojima, T.; Islam, M.T. Airborne LiDAR-Assisted Deep Learning Methodology for Riparian Land Cover Classification Using Aerial Photographs and Its Application for Flood Modelling. *J. Hydroinform.* **2022**, *24*, 179–201. [CrossRef]
10. Münzinger, M.; Prechtel, N.; Behnisch, M. Mapping the Urban Forest in Detail: From LiDAR Point Clouds to 3D Tree Models. *Urban For. Urban Green.* **2022**, *74*, 127637. [CrossRef]

11. He, Y.; Xu, G.; Kaufmann, H.; Wang, J.; Ma, H.; Liu, T. Integration of InSAR and LiDAR Technologies for a Detailed Urban Subsidence and Hazard Assessment in Shenzhen, China. *Remote Sens.* **2021**, *13*, 2366. [CrossRef]
12. Abdelaziz, N.; El-Rabbany, A. Deep Learning-Aided Inertial/Visual/LiDAR Integration for GNSS-Challenging Environments. *Sensors* **2023**, *23*, 6019. [CrossRef]
13. Ilci, V.; Toth, C. High Definition 3D Map Creation Using GNSS/IMU/LiDAR Sensor Integration to Support Autonomous Vehicle Navigation. *Sensors* **2020**, *20*, 899. [CrossRef]
14. Breen, M.J.; Kebede, A.S.; König, C.S. Assessing Coupled Human-Flood Interactions Using LiDAR Geostatistics and Neighbourhood Analyses. *Geomat. Nat. Hazards Risk* **2024**, *15*, 2361812. [CrossRef]
15. Podhorányi, M.; Unucka, J.; Bobál', P.; Říhová, V. Effects of LIDAR DEM Resolution in Hydrodynamic Modelling: Model Sensitivity for Cross-Sections. *Int. J. Digit. Earth* **2013**, *6*, 3–27. [CrossRef]
16. Muhadi, N.A.; Abdullah, A.F.; Bejo, S.K.; Mahadi, M.R.; Mijic, A. The Use of LiDAR-Derived DEM in Flood Applications: A Review. *Remote Sens.* **2020**, *12*, 2308. [CrossRef]
17. Wu, Y.; Peng, F.; Peng, Y.; Kong, X.; Liang, H.; Li, Q. Dynamic 3D Simulation of Flood Risk Based on the Integration of Spatio-Temporal GIS and Hydrodynamic Models. *ISPRS Int. J. Geo-Inf.* **2019**, *8*, 520. [CrossRef]
18. Kalacska, M.; Arroyo-Mora, J.P.; Lucanus, O. Comparing Uas Lidar and Structure-from-Motion Photogrammetry for Peatland Mapping and Virtual Reality (Vr) Visualization. *Drones* **2021**, *5*, 36. [CrossRef]
19. Yang, Y.; Jenny, B.; Dwyer, T.; Marriott, K.; Chen, H.; Cordeil, M. Maps and Globes in Virtual Reality. *Comput. Graph. Forum* **2018**, *37*, 427–438. [CrossRef]
20. Lütjens, M.; Kersten, T.; Dorschel, B.; Tschirschwitz, F. Virtual Reality in Cartography: Immersive 3D Visualization of the Arctic Clyde Inlet (Canada) Using Digital Elevation Models and Bathymetric Data. *Multimodal Technol. Interact.* **2019**, *3*, 9. [CrossRef]
21. Papadopoulou, E.-E.; Papakonstantinou, A.; Kapogianni, N.-A.; Zouros, N.; Soulakellis, N. VR Multiscale Geovisualization Based on UAS Multitemporal Data: The Case of Geological Monuments. *Remote Sens.* **2022**, *14*, 4259. [CrossRef]
22. Virtanen, J.P.; Julin, A.; Handolin, H.; Rantanen, T.; Maksimainen, M.; Hyyppä, J.; Hyyppä, H. Interactive Geo-Information in Virtual Reality—Observations and Future Challenges. In Proceedings of the 3rd BIM/GIS Integration Workshop and 15th 3D GeoInfo Conference, London, UK, 7–11 September 2020; Volume 44, pp. 159–165. [CrossRef]
23. Havenith, H.-B. 3D Landslide Models in VR. In *Understanding and Reducing Landslide Disaster Risk: Volume 4 Testing, Modeling and Risk Assessment*; Tiwari, B., Sassa, K., Bobrowsky, P.T., Takara, K., Eds.; Springer International Publishing: Cham, Switzerland, 2021; pp. 195–204. ISBN: 978-3-030-60706-7.
24. Hruby, F.; Sánchez, L.F.Á.; Ressl, R.; Escobar-Briones, E.G. An Empirical Study on Spatial Presence in Immersive Geo-Environments. *PFG J. Photogramm. Remote Sens. Geoinf. Sci.* **2020**, *88*, 155–163. [CrossRef]
25. Froehlich, M.; Azhar, S. Investigating Virtual Reality Headset Applications in Construction. In Proceedings of the 52nd ASC Annual International Conference, Provo, UT, USA, 13–16 April 2016; pp. 13–16.
26. Du, J.; Zhu, Q.; Shi, Y.; Wang, Q.; Lin, Y.; Zhao, D. Cognition Digital Twins for Personalized Information Systems of Smart Cities: Proof of Concept. *J. Manag. Eng.* **2020**, *36*, 04019052. [CrossRef]
27. Zhu, Y.; Li, N. Virtual and Augmented Reality Technologies for Emergency Management in the Built Environments: A State-of-the-Art Review. *J. Saf. Sci. Resil.* **2021**, *2*, 1–10. [CrossRef]
28. Minucci, G.; Molinari, D.; Gemini, G.; Pezzoli, S. Enhancing Flood Risk Maps by a Participatory and Collaborative Design Process. *Int. J. Disaster Risk Reduct.* **2020**, *50*, 101747. [CrossRef]
29. Sanders, B.F.; Schubert, J.E.; Goodrich, K.A.; Houston, D.; Feldman, D.L.; Basolo, V.; Luke, A.; Boudreau, D.; Karlin, B.; Cheung, W.; et al. Collaborative Modeling With Fine-Resolution Data Enhances Flood Awareness, Minimizes Differences in Flood Perception, and Produces Actionable Flood Maps. *Earth's Future* **2020**, *8*, e2019EF001391. [CrossRef]
30. Sermet, Y.; Demir, I. GeospatialVR: A Web-Based Virtual Reality Framework for Collaborative Environmental Simulations. *Comput. Geosci.* **2022**, *159*, 105010. [CrossRef]
31. He, K.; Yang, Q.; Shen, X.; Dimitriou, E.; Mentzafou, A.; Papadaki, C.; Stoumboudi, M.; Anagnostou, E.N. Brief Communication: Storm Daniel Flood Impact in Greece 2023: Mapping Crop and Livestock Exposure from SAR. *Nat. Hazards Earth Syst. Sci* **2024**, *24*, 23752382. [CrossRef]
32. Dimitriou, E.; Efstratiadis, A.; Zotou, I.; Papadopoulos, A.; Iliopoulou, T.; Sakki, G.K.; Mazi, K.; Rozos, E.; Koukouvinos, A.; Koussis, A.D.; et al. Post-Analysis of Daniel Extreme Flood Event in Thessaly, Central Greece: Practical Lessons and the Value of State-of-the-Art Water-Monitoring Networks. *Water* **2024**, *16*, 980. [CrossRef]
33. Elhag, M.; Yilmaz, N. Insights of Remote Sensing Data to Surmount Rainfall/Runoff Data Limitations of the Downstream Catchment of Pineios River, Greece. *Environ. Earth Sci.* **2021**, *80*, 35. [CrossRef]
34. Adamopoulos, I.; Frantzana, A.; Syrou, N. Climate Crises Associated with Epidemiological, Environmental, and Ecosystem Effects of a Storm: Flooding, Landslides, and Damage to Urban and Rural Areas (Extreme Weather Events of Storm Daniel in Thessaly, Greece). *Med. Sci. Forum* **2024**, *25*, 7. [CrossRef]
35. HVA. First Report Regarding Post-Disaster Remediation of 2023 Thessaly Flooding. 2023; pp. 1–33. Available online: https://www.government.gov.gr/wp-content/uploads/2023/11/HVA-Fact-Finding-Mission-Report-on-Thessaly-Post-Disaster-Remediation.pdf (accessed on 20 June 2024).
36. Metashape, A. *AgiSoft Metashape Professional Edition, Version 1.7.4*; Agisoft LLC: St. Petersburg, Russia, 2021.

37. Hruby, F.; Ressl, R.; de la Borbolla del Valle, G. Geovisualization with Immersive Virtual Environments in Theory and Practice. *Int. J. Digit. Earth* **2019**, *12*, 123–136. [CrossRef]
38. Diab, A.; Kashef, R.; Shaker, A. Deep Learning for LiDAR Point Cloud Classification in Remote Sensing. *Sensors* **2022**, *22*, 7868. [CrossRef] [PubMed]
39. Yastikli, N.; Cetin, Z. Classification of Raw LiDAR Point Cloud Using Point-Based Methods with Spatial Features for 3D Building Reconstruction. *Arab. J. Geosci.* **2021**, *14*, 146. [CrossRef]
40. Zhang, Z.; Zhang, L.; Tong, X.; Guo, B.; Zhang, L.; Xing, X. Discriminative-Dictionary-Learning-Based Multilevel Point-Cluster Features for ALS Point-Cloud Classification. *IEEE Trans. Geosci. Remote Sens.* **2016**, *54*, 7309–7322. [CrossRef]
41. Esri Inc. *ArcGIS Pro, version 3.0.0*; Software; Desktop; Esri Inc.: Redlands, CA, USA, 2022.
42. Rydvanskiy, R.; Hedley, N. Mixed Reality Flood Visualizations: Reflections on Development and Usability of Current Systems. *ISPRS Int. J. Geo-Inf.* **2021**, *10*, 82. [CrossRef]
43. Lochhead, I.; Hedley, N. Designing Virtual Spaces for Immersive Visual Analytics. *KN—J. Cartogr. Geogr. Inf.* **2021**, *71*, 223–240. [CrossRef]
44. Dong, W.; Yang, T.; Liao, H.; Meng, L. How Does Map Use Differ in Virtual Reality and Desktop-Based Environments? *Int. J. Digit. Earth* **2020**, *13*, 1484–1503. [CrossRef]
45. Haynes, P.; Hehl-Lange, S.; Lange, E. Mobile Augmented Reality for Flood Visualisation. *Environ. Model. Softw.* **2018**, *109*, 380–389. [CrossRef]
46. Keil, J.; Edler, D.; Schmitt, T.; Dickmann, F. Creating Immersive Virtual Environments Based on Open Geospatial Data and Game Engines. *KN—J. Cartogr. Geogr. Inf.* **2021**, *71*, 53–65. [CrossRef]
47. Ma, Y.; Wright, J.; Gopal, S.; Phillips, N. Seeing the Invisible: From Imagined to Virtual Urban Landscapes. *Cities* **2020**, *98*, 102559. [CrossRef]

forests

Article

# Spatio-Temporal Change and Drivers of the Vegetation Trends in Central Asia

**Moyan Li [1], Junqiang Yao [2,3,*] and Jianghua Zheng [4,*]**

[1] College of Geography and Tourism, Xinjiang Laboratory of Lake Environment and Resources in Arid Zone, Xinjiang Normal University, Urumqi 830054, China; limy525@xjnu.edu.cn
[2] Key Laboratory of Tree-Ring Physical and Chemical Research of CMA, Institute of Desert Meteorology, China Meteorological Administration, Urumqi 830002, China
[3] Field Scientific Experiment Base of Akdala Atmospheric Background of CMA, Akdala 836499, China
[4] Key Laboratory of Oasis Ecology of Ministry of Education, Institute of Arid Ecology and Environment, College of Geographical Sciences, Xinjiang University, Urumqi 830046, China
* Correspondence: yaojq@idm.cn (J.Y.); zheng.jianghua@xju.edu.cn (J.Z.)

**Abstract:** The impact of changing climate on vegetation in dryland is a prominent focus of global research. As a typical arid region in the world, Central Asia is an ideal area for studying the associations between climate and arid-area vegetation. Utilizing data from the European Centre for Medium-Range Weather Forecasts fifth-generation reanalysis (ECMWF ERA-5) and normalized difference vegetation index (NDVI) datasets, this study investigates the spatio-temporal variation characteristics of the NDVI in Central Asia. It quantitatively assesses the contribution rates of climatic factors to vegetation changes and elucidates the impact of an increased vapor pressure deficit (VPD) on vegetation changes in Central Asia. The results indicate that the growing seasons' NDVI exhibited a substantial increase in Central Asia during 1982–2015. Specifically, there was a pronounced "greening" process (0.012/10 yr, $p < 0.05$) from 1982 to 1998. However, an insignificant "browning" trend was observed after 1998. Spatially, the vegetation NDVI in the growing seasons exhibited a pattern of "greening in the east and browning in the west" of Central Asia. During spring, the dominant theme was the "greening" of vegetation NDVI, although there was noticeable "browning" observed in southwest region of Central Asia. During summer, the "browning" of vegetation NDVI further expanded eastward and impacted the entire western Central Asia in autumn. According to the estimated results computed via the partial differential equation method, the "browning" trend of vegetation NDVI during the growing seasons was guided by increased VPD and decreased rainfall in western Central Asia. Specifically, the increased VPD contributed 52.3% to the observed vegetation NDVI. Atmospheric drought depicted by the increase in VPD significantly lowers the "greening" trend of vegetation NDVI in arid regions, which further aggravates the "browning" trend of vegetation NDVI.

**Keywords:** vegetation change; normalized difference vegetation index (NDVI); vapor pressure deficit (VPD); central Asia; "greening" trend; "browning" trend

## 1. Introduction

Central Asia is the largest non-zonal arid zone on earth. It has a unique mountain-basin topography comprising desert, oasis, snow, and ice, with over 80% of temperate deserts distributed in Central Asia. The ecosystem, primarily dominated by desert vegetation is fragile and sensitive to climate variations. Hence, this is a key and vulnerable global climate change region and is representative of the arid regions in the world [1–7]. The temperature in Central Asia demonstrated a substantial rise in 1998 and has maintained a fluctuating high-temperature trend ever since [6,8,9]. Because of the high-temperature fluctuations, the Palmer Drought Severity Index (PDSI) in Central Asia has demonstrated a significant decreasing trend since 2000, and approximately 65% of the Central Asian region has experienced increased aridity [8]. Since the onset of the 21st century, ecological and environmental issues have been extremely prominent, including those in the Aral Sea and Lake Balkhash [10].

Consequently, the risks of water resources and crisis have been continuously aggravated, which has severely restricted economic and social development in Central Asia [11]. Additionally, Central Asia is also the key upstream area affecting the weather and climate of China; hence, the ecological and environmental upheaval in this region can severely impact the weather, climate, and environmental health of China [4].

Vegetation is a key link connecting the atmosphere, soil, hydrosphere, and biosphere, with the spatio-temporal change in growing vegetation playing an important role in the regional ecological patterns [12–18]. The response of vegetation in arid areas to the changing climate has garnered substantial attention in the global research arena, but the changes in vegetation and its influence mechanisms in arid areas are still controversial [16,19–22]. Central Asia has fragile ecosystems, with the desert vegetation ecosystem being specifically vulnerable to the impact of climate change. Scientists have found substantial spatial differences in the vegetation changes in Central Asia since the 1980s, among which the sparse vegetation in the western desert and around the Aral Sea has deteriorated severely [8]. In the context of the driving factors that lead to vegetation change, certain studies speculated that hydrological factors potentially dominate the vegetation growth patterns in Central Asia, with air temperature negatively correlated to vegetation change [23]. Hydrothermal change has affected vegetation growth in Central Asia, and climate warming has caused desert vegetation decline and ecosystem degradation [8]. Precipitation is the major climate factor affecting vegetation change in northern Central Asia, while temperature is the primary factor controlling seasonal vegetation changes in mountainous regions and the Aral Sea Basin. Additionally, drought has led to sparse vegetation degradation in deserts [8].

The potential impact of climatic variability on vegetation changes has garnered substantial attention from researchers. Previous studies concentrated more on the effect of high temperatures and drought on vegetation change [24,25], as well as on the impact of decreased precipitation [26,27] and increased $CO_2$ concentration [28,29]. Previous studies have indicated that the rise of $CO_2$ concentration in the atmosphere is a major driving force behind global terrestrial vegetation change and is also a crucial element influencing vegetation growth in arid regions [20]. On the other hand, climate warming is the primary driver of vegetation growth over the Tibetan Plateau and the high latitudes of northern China [29]. Previous studies have established that an increase in $CO_2$ concentration and climate warming are the primary factors regulating vegetation change [29]. However, the vital function of atmospheric water stress caused by global warming has been overlooked. In recent years, scientists have begun to pay attention to the influence of vapor pressure deficit (VPD) and have found that a long-term high value of VPD can result in the large-scale death of trees in forest ecosystems, with the death rate of tropical trees increasing substantially with the aggravation of atmospheric water stress [30–33]. Additionally, VPD also exhibits a positive correlation with wildfire [34,35] and contributes to reduced crop yield [36,37]. Some studies have demonstrated that the reduction in vegetation productivity caused by an increase in VPD since the end of the 1990s has counteracted the "fertilization effect" attributed to the increased $CO_2$ concentration on vegetation growth [38]. In contrast, other studies have concluded that, in comparison to the stress effect induced by a heightened atmospheric moisture deficit on vegetation growth, the physiological impact of $CO_2$ enrichment in the atmosphere exerts a more pronounced effect on vegetation water stress [20].

All of the studies mentioned earlier have focused on climate, ecology, and vegetation changes in Central Asia [1,3,6,39–42]. However, regarding the spatio-temporal distribution of vegetation in Central Asia, there exists a lack of comprehensive understanding, and the study domains are fragmented. For instance, the impact of climate factors, including temperature, rainfall, and VPD, on vegetation growth in Central Asia has been studied [15,17,21], but a limited number of studies have quantified these elements, and the contribution of VPD change in the atmosphere to vegetation dynamic in Central Asia is still uncertain. To address this gap, our study focuses on the spatio-temporal variation trend and the characteristics of the NDVI in Central Asia. Additionally, we quantitatively estimate the contribution rates of climatic elements to vegetation dynamic in Central Asia using the partial derivative equation method and structural equation model (SEM). The research findings will hold scientific significance and practical value, contributing to a deeper understanding of the connection with climate and vegetation in arid regions.

## 2. Materials and Methods

### 2.1. Study Area

Central Asia (CA) is located in the range of 45°–95° E and 34°–54° N in the hinterland of Eurasia, with a total area of approximately 5.66 million km2, comprising five Central Asian countries and the Xinjiang region of China [7]. Central Asia has a mountain-basin landform with a significant elevation difference. The major mountain regions include the Tianshan Mountains, Pamir Plateau, and Altai and Kunlun Mountains, which collectively constitute the "water tower" in Central Asia [3]. Additionally, deserts are extensively distributed in Central Asia, serving as an important source of global sand-dust weather.

### 2.2. Data

#### 2.2.1. The GIMMS-NDVI Data

The NDVI can reveal the surface vegetation growth and is the best index to monitor vegetation change. We employed NASA's Global Inventory Modeling and Mapping Studies NDVI third-generation dataset NDVI3g (GIMMS NDVI3g) from NASA's Goddard Space Flight Center. The GIMMS NDVI3g data are the longest time series of NDVI data, spanning from 1981 to 2015, with a spatial resolution of 1/12° (about 8 km) and a temporal resolution of 15 days [12,15]. GIMMS NDVI 3g data possess several advantages, including a long time series, extensive coverage, and robust characterization capabilities. Generally, an NDVI value below 0.1 is considered the threshold to distinguish whether the NDVI exists or not. Based on this criterion, April to October of each year is regarded as the growing season for Central Asian vegetation.

#### 2.2.2. The ERA5 Dataset

The ERA5 dataset consists of the latest reanalysis data from the European Centre for Medium-Range Weather Forecasting (ECMWF). It combines the largest collection of historical global observations with advanced satellite remote sensing data and uses the latest data assimilation systems and numerical model simulations to obtain a precise atmospheric status [43]. In this study, we choose the ECMWF ERA-5 monthly temperature, relative humidity, precipitation, wind speed, and sunshine hours data, with the temporal coverage spanning from 1982 to 2015.

VPD is an important indicator of the increased atmospheric moisture demand and intensified drought due to rising temperatures [44,45]. The monthly VPD data were calculated based on the monthly temperature and relative humidity of the ERA5 datasets during 1982–2015. The calculation formula for VPD is as follows [33]:

$$\mathrm{VPD} = e_s - e_a \tag{1}$$

where $e_a$ is the actual water vapor pressure (kPa) and $e_s$ is the saturation water vapor pressure (kPa).

The calculation formula for $e_s$ is

$$e_s = 0.611\exp[(17.27 \times Ta)/(Ta + 237.3)] \tag{2}$$

where $Ta$ is the mean temperature (°C).

The calculation formula for $e_a$ is

$$e_a = RH \times \mathrm{es}/100 \tag{3}$$

where $RH$ is relative humidity (%).

*2.3. Methods*

In this study, we utilized the nonparametric Sen method to evaluate the changing climate and vegetation trends and used the Mann–Kendall test (MK) to examine the significance of the changing trends. Additionally, the Pearson correlation analysis was used to reveal the correlation between the NDVI and meteorological elements, expressed by correlation coefficient (CC). Partial correlation analysis (PCA) was employed to study the impact of diverse meteorological elements on NDVI changes, which was expressed as a partial correlation coefficient (PCC) [46].

Since there are difference between the NDVI and climatic elements, it is essential to normalize all variables when revealing the effects of climatic elements on vegetation changes. We use the data normalization method proposed by Li et al. [18]. Based on the partial derivative equation combined with the partial correlation analysis method proposed by Roderick et al. [47] and Naeem et al. [48], Li et al. [18] further implemented the elastic coefficient method and constructed a partial derivative equation method, which can quantitatively estimate the contribution rates of diverse meteorological elements to vegetation changes. In addition, the SEM is a statistical method based on the covariance matrix of variables used to reveal the relationship between variables. In this study, we also employ the SEM method to investigate the effect of climate factors on vegetation changes in Central Asia [49,50].

## 3. Results

*3.1. NDVI Distribution in Central Asia*

Figure 1 depicts the vegetation growth conditions during the growing season, spring, summer, and autumn in Central Asia. The vegetation NDVI in Central Asia during the growing season exhibited a spatial pattern of "high in the north and low in the south, high in the west and low in the east, and high in the mountains and low in the basin". The vegetation coverage area (NDVI $\geq$ 0.1) accounted for 85% of the total acreage in Central Asia, of which the high vegetation coverage area (NDVI $\geq$ 0.5) that accounted for 32% was predominantly distributed in the Tianshan Mountains and the surrounding oasis areas, as well as in the area north of 50° N. The vegetation types were predominantly coniferous forest, scrub, and grassland. The area located south of 50° N was predominantly characterized by grassland and desert steppe, with 16% of the region being occupied by desert and wilderness.

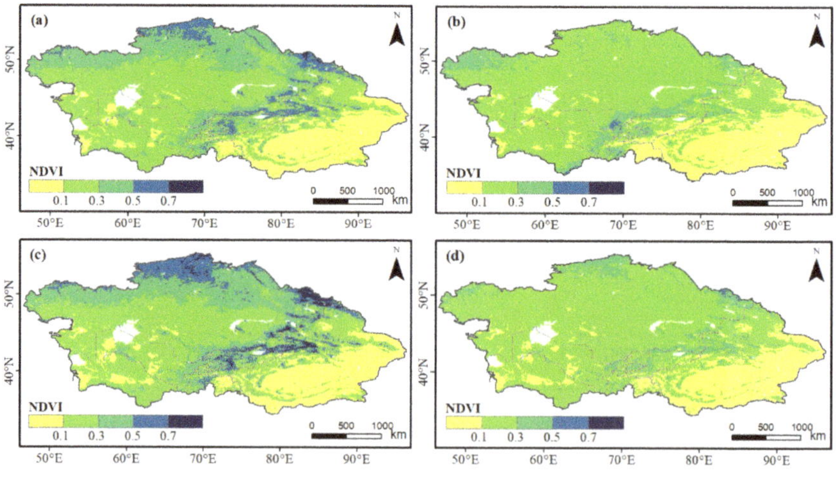


**Figure 1.** Spatial distribution of vegetation NDVI in Central Asia during 1982–2015 (**a**) growing season, (**b**) spring, (**c**) summer, and (**d**) autumn.

In terms of seasonal distribution, the vegetation NDVI exhibited the highest value during summer and was slightly higher during spring in comparison to autumn. During spring and autumn, most regions had NDVI values ranging between 0.1 and 0.3, while the mountainous areas and oasis had NDVI values above 0.3. However, during summer, the vegetation NDVI differed significantly in terms of spatial distribution. Specifically, in the Tianshan Mountains, Ili Valley, Irtysh River Basin, and in areas north of 50° N, the NDVI value exceeded 0.5.

## 3.2. *Temporal Variation of NDVI in Central Asia*

### 3.2.1. NDVI Variations in Vegetation Growing Seasons

The growing season vegetation NDVI showed a slight increasing trend ($p > 0.05$) in Central Asia during 1982–2015, characterized by substantial stage changes. According to the evaluation of the cumulative anomaly curve, the year 1998 marked the turning point of the growing season NDVI (Figure 2). Therefore, this study focused on NDVI changes in Central Asia before and after 1998.

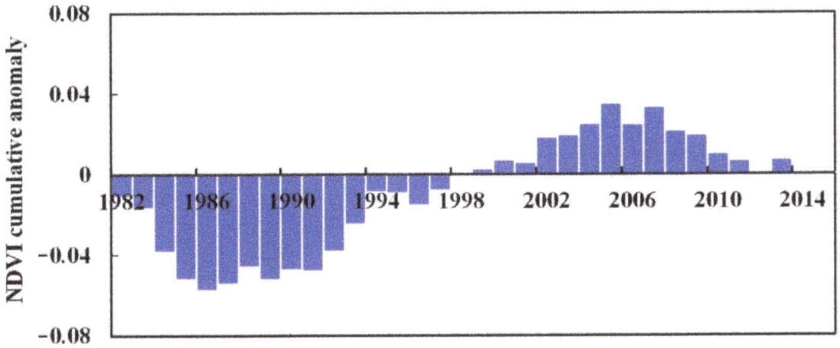


**Figure 2.** The cumulative anomaly curve of vegetation NDVI in Central Asia during the growing seasons of 1982–2015.

During 1982–1998, the vegetation NDVI presented an increasing trend at a rate of 0.012/10 yr ($p < 0.05$), while it exhibited a decreasing trend at a rate of −0.006/10 yr ($p > 0.05$) during 1998–2015. This indicated that the vegetation exhibited a significant "greening" trend during the 1982–1998 growing seasons and an insignificant "browning" after 1998 (Figure 3). Notably, the NDVI declined considerably from 1998 to 2010, and the changing trend was −0.01/10 yr ($p < 0.05$), demonstrating a significant "browning" trend, which gradually increased after 2010.

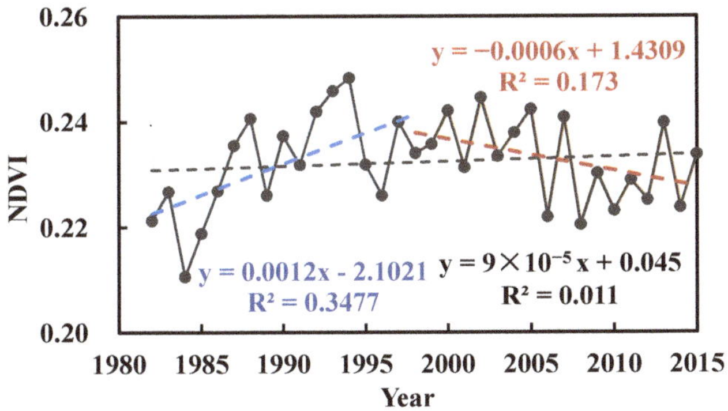


**Figure 3.** Change trends of vegetation NDVI in Central Asia during the growing seasons of 1982–2015 (gray, blue, and red dashed lines represent the variation trends in 1982–2015, 1982–1998, and 1998–2015, respectively).

The NDVI demonstrated an increasing trend from March to June in 1982–2015 but a decreasing or insignificant trend in the remaining months (Figure 4). During 1982–1998, the monthly vegetation NDVI variations exhibited an upward trend in all months except December, indicating the "greening" trend in Central Asia in these years. The greening trend was the highest in May during 1982–1998 (0.021/10 yr, $p < 0.05$). The vegetation "greening" from June to October was approximately 0.01/10 yr ($p < 0.05$), but the variation trend of vegetation from January to March was not significant, which was predominantly associated with the snow cover in the high latitudes and mountainous areas of Central Asia. However, from 1998 to 2015, except for a slight increase in the NDVI in May, the NDVI in other months was mostly decreasing, demonstrating a consistent trend of the "browning" of vegetation. In the growing seasons, the vegetation "browning" trend was the most prominent in July, which was −0.012/10 yr ($p < 0.05$). This was predominantly influenced by a significant increase in temperature, with no significant increase in rainfall in July. Consequently, the effects of drought and water stress on vegetation growth was increased, which led to a substantial stagnation of vegetation growth.

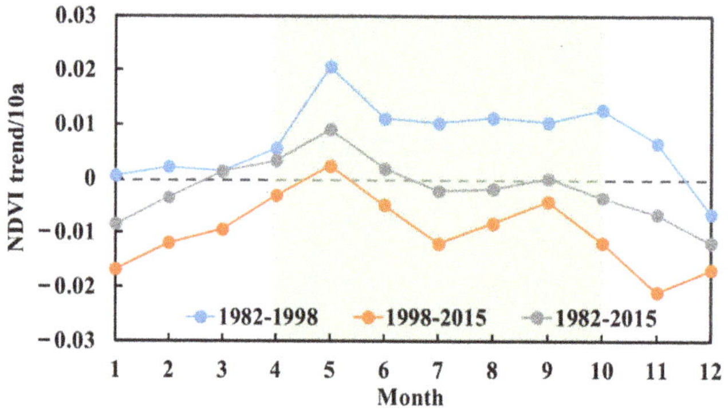


**Figure 4.** Variation trend of monthly NDVI in Central Asia from 1982 to 2015 (shaded area is the growing season).

3.2.2. Seasonal Variation of NDVI

During the spring season of 1982–2015, the vegetation NDVI demonstrated a slight "greening" trend at a rate of 0.005/10 yr ($p > 0.05$). From the point of stages, the NDVI exhibited an increasing trend during 1982–1998, with a rate of 0.009/10 yr ($p > 0.05$), but a decreasing trend from 1998 to 2015 (−0.007/10 yr, $p > 0.05$) (Figure 5a).

During the summer season of 1982–2015, NDVI variation demonstrated an insignificant declining trend (Figure 5b). Between 1982 and 1998, the NDVI showed a consistent increasing trend with a rate of 0.01/10 yr ($p < 0.05$), while the 1998–2015 NDVI exhibited a decrease at the rate of −0.008/10 yr, revealing a significant "browning" trend of vegetation in comparison to the spring season. Particularly, during 1998–2012, the NDVI reduced significantly, exhibiting a changing trend of −0.016/10 yr ($p < 0.05$), representing a significant "browning" trend. Post 2012, the NDVI gradually increased, from 0.24 in 2012 to 0.27 in 2015, indicating a new trend of vegetation change in recent years.

Between 1982 and 2015, the vegetation NDVI during the autumn season typically exhibited a downward trend at the rate of −0.003/10 yr ($p > 0.05$). Between 1982 and 1998, the autumn NDVI showed a significant increasing trend with a rate of 0.01/10 yr ($p < 0.05$), which was similar to that in summer and slightly higher than that in spring. However, during 1998–2015, the NDVI showed a significant downward trend during autumn with a rate of −0.012/10 yr ($p < 0.05$), indicating the evident "browning" trend. Hence, the "browning" trend of vegetation was substantially more significant during autumn than during spring and summer (Figure 5c).

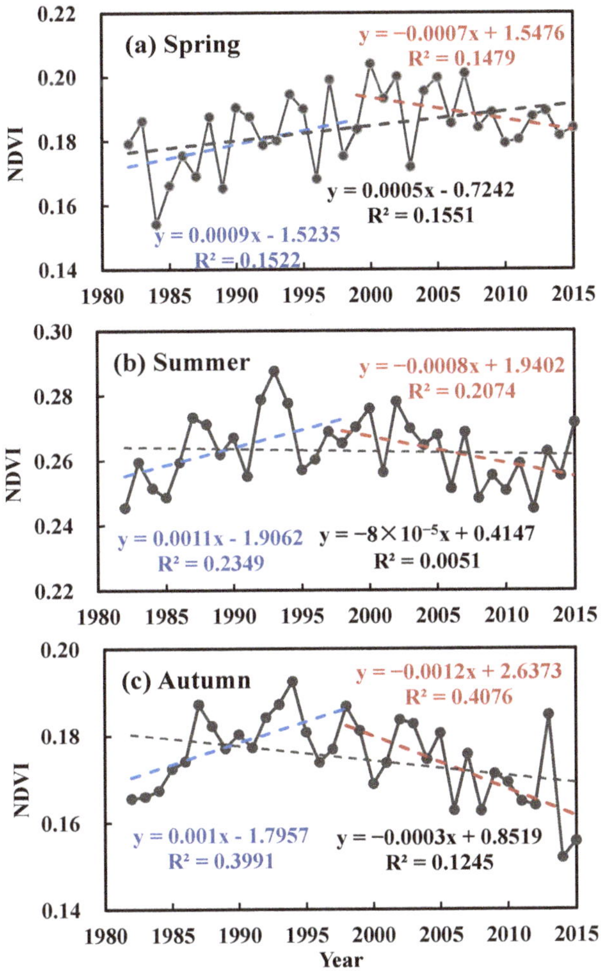


**Figure 5.** Seasonal variation of NDVI in Central Asia from 1982 to 2015 (gray, blue, and red dashed lines represent the variation trend lines during 1982–2015, 1982–1998, and 1998–2015, respectively).

### *3.3. Spatial Distribution of Vegetation NDVI Changes*

#### 3.3.1. During 1982–2015

During 1982–2015, the growing seasons NDVI in Central Asia exhibited a "greening" pattern in the east and a "browning" pattern in the west, with 70° E as the boundary. In most parts to the west located at 70° E, the vegetation demonstrates a dominant "browning" trend, especially in the region located south of 45° N (Figure 6a,b). The predominant regions where the "browning" trend in vegetation was observed are located in northwest Central Asia, including the Turgay Plateau, the Caspian Sea, and the Aral Sea.

The spring NDVI predominantly exhibited a "greening" trend during 1982–2015 (Figure 6c,d). The acreage of vegetation browning was relatively small, predominantly distributed in the surroundings of the Caspian Sea and Aral Sea, the east of Xinjiang, and the lower reaches of the Tarim River. During summer, the regions exhibiting a "browning" trend in the NDVI expanded eastward and westward in Central Asia in comparison to the "browning" trend observed during spring (Figure 6e,f). The significant "browning" areas were predominantly distributed in the lower reaches of the Caspian and Aral Sea, the areas located northwest of Kazakhstan, and the Turgay Plateau. On the other hand, the "greening" areas of vegetation in Central Asia were predominantly concentrated in the hilly areas of Kazakhstan and the Xinjiang, China.

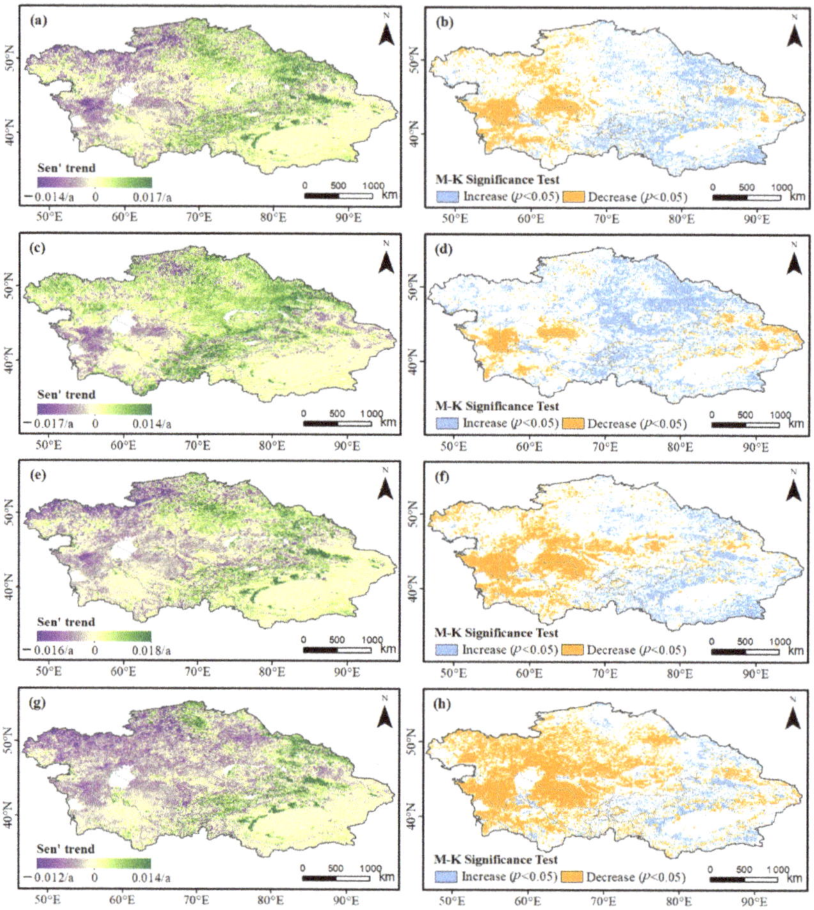


**Figure 6.** Variation trends of NDVI in Central Asia during 1982–2015 (**a**,**c**,**e**,**g**) and the significance test (**b**,**d**,**f**,**h**). (**a**,**b**) growing season, (**c**,**d**) spring, (**e**,**f**) summer, and (**g**,**h**) autumn.

In comparison to the spring and summer vegetation NDVI trend, the autumn "browning" area in Central Asia further expanded, and vegetation "browning" predominated throughout western Central Asia, with patches of "greening" found only in the northern Turgai low-lying land and the western Tianshan Mountains. Conversely, in Xinjiang, located in the east of Central Asia, "greening" was the predominant trend, except for vegetation "browning" observed in the eastern region of Junggar Basin (Figure 6g,h).

In conclusion, the spatial distribution pattern of the growing seasons NDVI was "greening in the east and browning in the west" in Central Asia during 1982–2015. Seasonally, "greening" was the predominant variation trend in Central Asia during spring, excluding the "browning" trend observed in the southwestern region. However, during summer, the vegetation "browning" trend developed eastward and covered the entirety of western Central Asia during autumn. The vegetation in Xinjiang, located in eastern Central Asia, predominantly exhibited the "greening" trend during summer, but during spring and autumn, the vegetation "browning" trend gradually expanded.

### 3.3.2. During 1982–1998

The growing seasons NDVI changing trend was predominantly manifested by vegetation "greening" in Central Asia during 1982–1998, and the significant "greening" regions included the Tianshan Mountains, Altai Mountains, and Ili Valley of Central Asia, as well as parts of northern Central Asia. Simultaneously, patches of "browning" areas were observed exclusively in Turgay Plateau and the lower reaches of the Tarim River (Figure 7a,b).

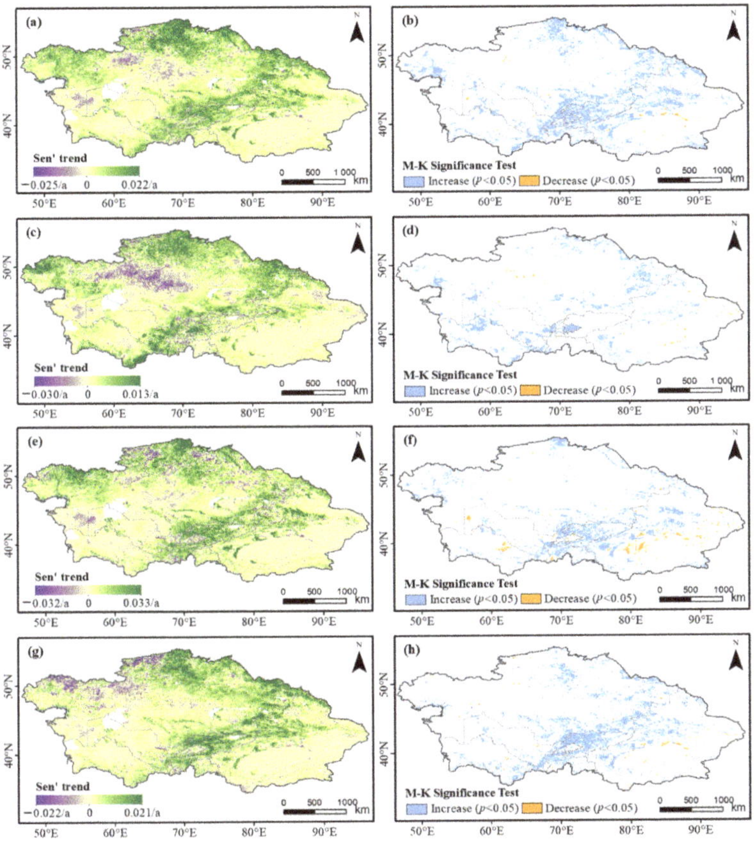


**Figure 7.** Variation trends of vegetation NDVI in Central Asia during 1982–1998 (**a**,**c**,**e**,**g**) and the significance test (**b**,**d**,**f**,**h**). (**a**,**b**) Growing season, (**c**,**d**) spring, (**e**,**f**) summer, and (**g**,**h**) autumn.

From a seasonal perspective, the variation of the NDVI during the spring season between 1982 and 1998 was dominated by vegetation "greening". The significant "greening" regions were present in patches, predominantly found in mountain valleys and plain oases; however, the Turgay Plateau exclusively demonstrated an insignificant "browning" trend (Figure 7c,d). In comparison to the spring season, the acreage of vegetation "greening" during summer was more enlarged and the "greening" trend was enhanced. The areas of significant vegetation "greening" were mainly located on the north and south slopes of the Tianshan Mountains, Ili River Valley, and Syr Darya River Valley, as well as Amu Darya River Valley and oases located on both sides, with the variation rate of the NDVI reaching 0.033/10 yr ($p < 0.05$) (Figure 7e,f). The acreage and "browning" trend further intensified in the Tarim River Basin, Xinjiang. The vegetation "greening" areas during autumn were significantly consistent with those in summer, but the vegetation "browning" trend was enhanced in the surrounding area of the Turgay Plateau, located in northwest Central Asia.

This trend weakened in southern Xinjiang, where the vegetation "browning" state in the lower reaches of Tarim River still existed (Figure 7g,h).

In summary, the growing season NDVI in Central Asia between 1982 and 1998 presented a general "greening" trend. Seasonally, spring was dominated by the "greening" trend of the NDVI, while the "browning" trend was predominantly observed in the Turgay low-lying area. During summer and autumn, the vegetation "greening" trend expanded further, but the vegetation "browning" trend persisted in the lower reaches of the Tarim River.

### 3.3.3. During 1998–2015

During 1998–2015, NDVI variation in growing seasons in Central Asia was dominated by vegetation "browning", especially in the area west of 70° E. The significant "browning" trend predominantly occurred in the area surrounding the Caspian Sea, Aral Sea Basin, and Ili River Basin, whereas the "greening" trend was observed predominantly in the hilly area of eastern Kazakhstan, the northern slope of Tianshan Mountains, and southern Xinjiang (Figure 8a,b).

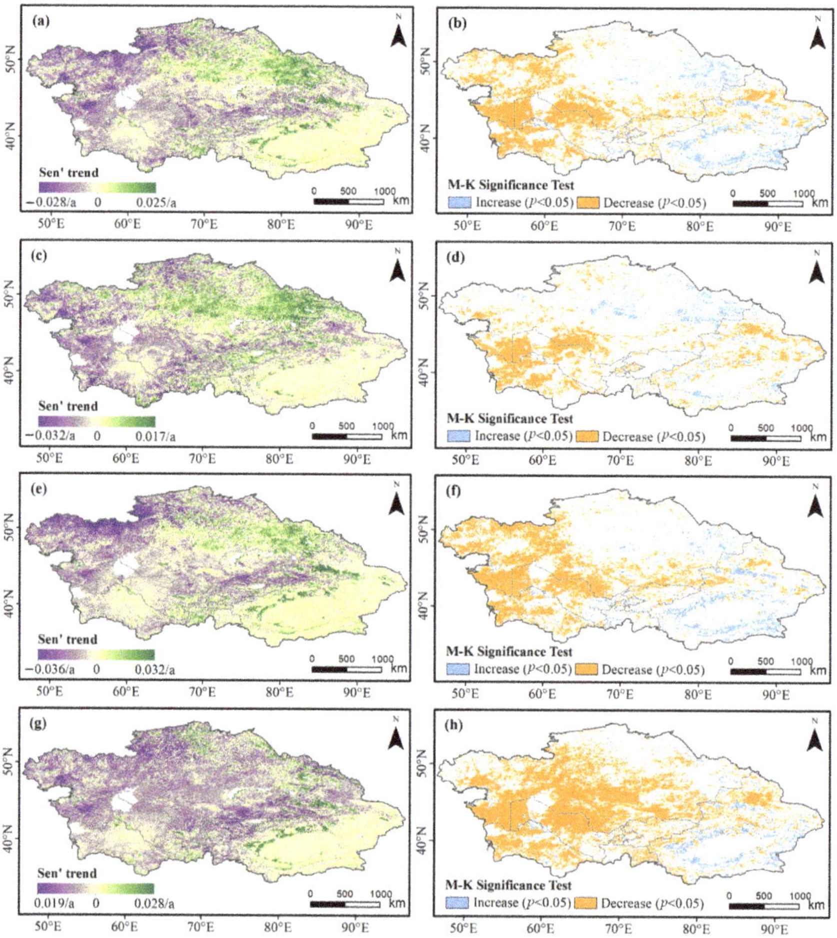


**Figure 8.** Variation trends of vegetation NDVI in Central Asia during 1998–2015 (**a**,**c**,**e**,**g**) and the significance test (**b**,**d**,**f**,**h**). (**a**,**b**) Growing season, (**c**,**d**) spring, (**e**,**f**) summer, and (**g**,**h**) autumn.

In the context of seasons, the spring NDVI in Central Asia from 1998 to 2015 was dominated by vegetation "browning", with significant changes in southern Central Asia and northern Xinjiang. Simultaneously, patchy "browning" areas were observed in the Turgay Plateau in northern Central Asia, the Tianshan Mountains, and the mountain areas of southern Xinjiang. However, the vegetation "greening" areas were predominant in central and eastern Kazakhstan, the north slope of the Tianshan Mountains, and southern Xinjiang, with a variation trend of approximately 0.017/10 yr ($p < 0.05$) (Figure 8c,d). In comparison to the spring season, the area where the vegetation NDVI tended to be significantly "browning" during summer in Central Asia and extended further to the north during spring. The vegetation in the Turgay Plateau showed a significant "browning" trend at a rate exceeding −0.035/10 yr ($p < 0.05$). This was followed by the significant "browning" trend observed in the Ili River Valley to Balkhash Lake Basin. Contrastingly, the summer NDVI exhibited a substantial "greening" trend in comparison to spring in Xinjiang (Figure 8e,f). During autumn, the vegetation "browning" trend significantly expanded in western Central Asia and northern Xinjiang, while the vegetation "greening" trend was predominant in the areas surrounding the Tarim Basin (Figure 8g,h).

As seen from the earlier analysis, the "browning" of the vegetation NDVI in Central Asia during 1998–2015 was the predominant trend, exhibiting significant variations in western Central Asia and northern Xinjiang. During summer and autumn, the "browning" trend enlarged further, with the summer "browning" trend being the most prominent. The sustained vegetation "greening" was observed primarily in southern Xinjiang.

### 3.3.4. Variation Trend of Vegetation NDVI in Western Central Asia

In this study, our analysis demonstrated that there were significant regional differences in vegetation changes in Central Asia, with relatively consistent trends in the west (mainly in the five Central Asian countries) and large differences in trends in eastern Central Asia (Xinjiang). The following section focuses on the variation trend and stage characteristics of the NDVI in western Central Asia.

There was no significant change trend in the growing seasons' vegetation NDVI during 1982–2015 in western Central Asia, but the NDVI showed stage characteristics. When the year 1998 was visualized as the boundary year, the vegetation NDVI during 1982–1998 exhibited a positive "greening" trend at the rate of 0.013/10 yr ($p < 0.05$). Contrastingly, the NDVI during 1998–2015 demonstrated a negative "browning" trend at the rate of −0.009/10 yr ($p > 0.05$) (Figure 9).

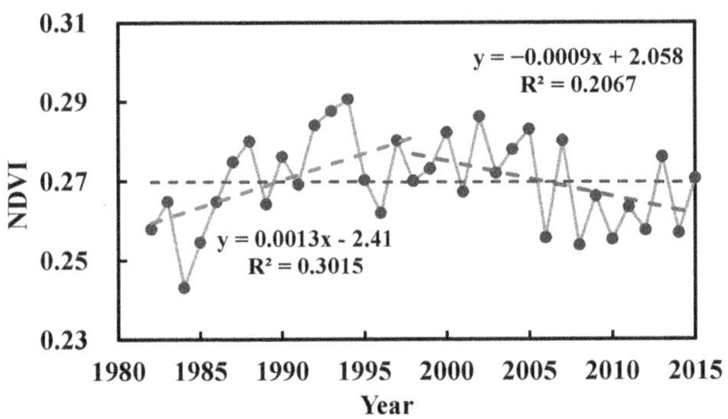


**Figure 9.** Variation trend of the growing seasons' vegetation NDVI in western Central Asia during 1982–2015 (gray, blue, and red dashed lines represent the variation trend lines during 1982–2015, 1982–1998, and 1998–2015, respectively).

When the analysis was conducted from the perspective of seasons, the vegetation NDVI exhibited an overall slight increasing trend (0.006/10 yr, $p > 0.05$) during the spring season between 1982 and 2015, and the vegetation change was consistent. However, a slight turning of the NDVI occurred in the late 1990s, changing from an insignificant "greening" trend to an insignificant "browning" trend (Figure 10a). During summer, the vegetation NDVI typically exhibited a slight decreasing trend, with a rate of −0.003/10 yr ($p > 0.05$) from 1982 to 2015. During this period, the NDVI shifted from the upward trend observed during 1982–1998 (0.011/10 yr, $p > 0.05$) to a significant downward trend observed between 1998 and 2015 (−0.015/10 yr, $p < 0.05$) (Figure 10b). Notably, vegetation has been exhibiting an increasing "greening" trend since 2012. During the autumn seasons between 1982 and 2015, an indistinctive decline in the NDVI was typically observed with a trend rate of −0.005/10 yr ($p > 0.05$). This comprised a significant increasing trend of the NDVI (0.011/10 yr, $p < 0.05$), observed during 1982–1998, and a significant decreasing trend (−0.016/10 yr, $p < 0.05$) observed during 1998–2015 (Figure 10c). In contrast to the NDVI variations observed during spring and summer, all the changes in the trend of the NDVI before and after autumn were significant.

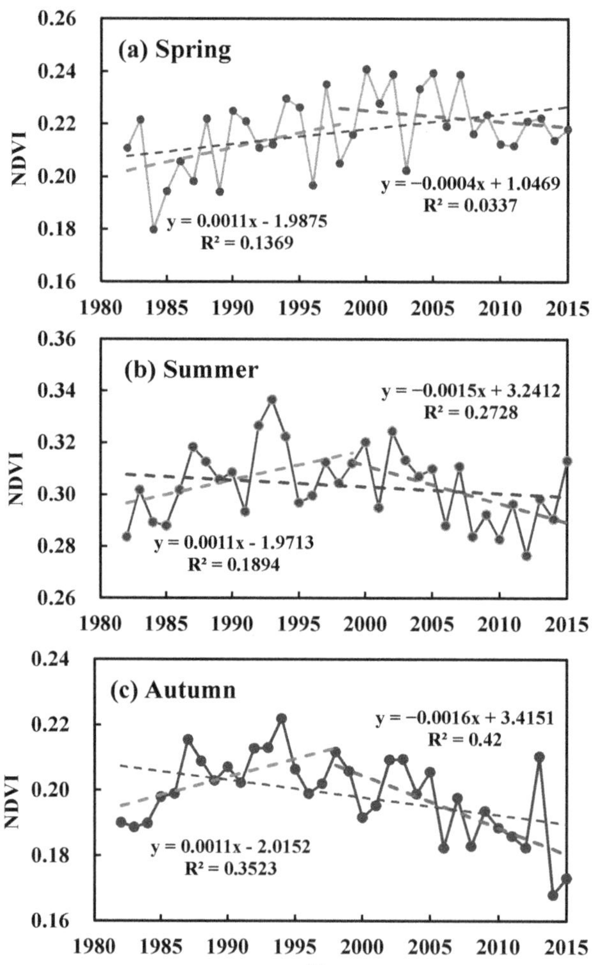


**Figure 10.** Seasonal variation trends of vegetation NDVI in western Central Asia during 1982–2015 (gray, blue, and red dashed lines represent the variation trend lines during 1982–2015, 1982–1998, and 1998–2015, respectively).

## 3.4. Impact Factors of Vegetation Changes in Western Central Asia

### 3.4.1. Correlations between Climate Factors and NDVI

To assess the effect of climate factors on the NDVI changes in western Central Asia, the correlations between climate factors and the NDVI in the growing seasons and the spring, summer, and autumn seasons of 1982–2015, 1982–1998, and 1998–2015 were computed separately. The results illustrated that, from 1982 to 2015, the vegetation NDVI was significantly correlated to VPD (vapor pressure deficit) and precipitation in the growing seasons. Specifically, the growing seasons NDVI showed a significant negative correlation with VPD (CC = −0.54, $p < 0.05$) and a significant positive correlation with precipitation (CC = 0.55, $p < 0.05$). In the context of wind speed and radiation (sunshine hours), the correlations were insignificantly negative. In temporal stages, the vegetation NDVI in the growing seasons exhibited stronger correlations with VPD and precipitation in 1998–2015 than in 1982–1998 (Figure 11), which indicated that the vegetation NDVI was affected significantly by VPD and precipitation in the growing seasons during 1998–2015.

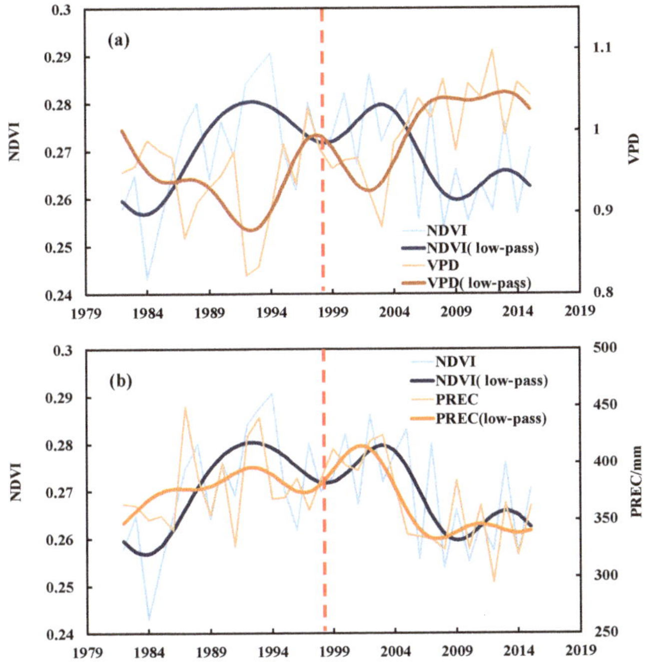


**Figure 11.** Correlation changes of the vegetation NDVI with (**a**) VPD and (**b**) precipitation in western Central Asia during the growing seasons between 1982–2015 (solid line represents a low-pass filtering curve, and red dashed line represents the year 1998).

During spring and autumn, the correlations between the vegetation NDVI and climate factors were not significant, but the variation of the NDVI during summer was significantly correlated to climatic factors. Specifically, the summer NDVI was positively correlated to precipitation during 1982–2015 (CC = 0.73, $p < 0.01$) but negatively correlated to both VPD and radiation (CC = −0.64, $p < 0.05$). The NDVI also exhibited a negative correlation with wind speed, but the correlation was relatively low (CC = =0.51, $p < 0.05$).

To reduce the influence of the mutual relations among climate factors on NDVI variation, the partial correlations between climate factors and the vegetation NDVI during the growing seasons and summers in different periods were computed (Figure 12). The partial correlation analysis suggested that the vegetation NDVI during the growing seasons from 1982 to 1998 was significantly positively correlated to precipitation and radiation and significantly negatively correlated to VPD and wind speed. During the growing seasons from 1998 to 2015, the vegetation NDVI exhibited a significant negative correlation with VPD and a significant positive correlation with precipitation. Moreover, the partial correlation coefficient (PCC) increased in comparison to the PCC in 1982–1998. During the spring seasons of 1998–2015, the vegetation NDVI demonstrated significant positive correlations with precipitation and radiation, while the partial correlations between the NDVI and climate factors during summer were not significantly related. The autumn NDVI during 1982–1998 exhibited a significant negative correlation exclusively with VPD.

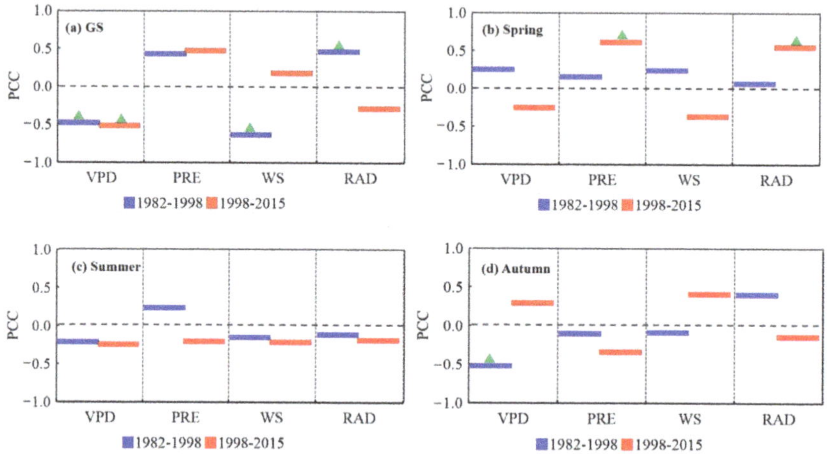


**Figure 12.** Partial correlation coefficients between NDVI and climate factors during the (**a**) growing seasons, (**b**) springs, (**c**) summers, and (**d**) autumns of 1982–1998 and 1998–2015 in western Central Asia (PRE: precipitation; WS: wind speed; RAD: sunshine hours; the triangle represents $p < 0.05$).

### 3.4.2. Quantitatively Evaluating NDVI Changes in Western Central Asia

Based on the constructed partial differential attribution analysis equation, the contributions of major climate factors (VPD, precipitation, wind speed, and radiation) to the NDVI changes during the growing seasons and summers of 1982–1998 and 1998–2015 in western Central Asia were estimated.

#### During the Growing Seasons

From 1982 to 1998, the combination of the different factors, including a decrease in VPD, an increase in precipitation, and a decrease in wind speed and radiation caused vegetation "greening" during the growing seasons in western Central Asia, with the residual term contributing the most to the "greening" of the vegetation NDVI. However, during 1998–2015, an increase in VPD played a prominent role in the growing seasons' "browning" in western Central Asia, which was followed by a decrease in precipitation and an increase in wind speed and radiation. Relative to the period of 1982–1998, the variation trend of climate elements during 1998–2015 exhibited a turning point, and the NDVI also experienced a turning point from vegetation "greening" to "browning". Specifically, VPD decreased by 0.04 kPa/yr during the 1982–1998 growing seasons, resulting in an increase of 0.02/yr in the NDVI, while VPD increased by 0.06 kPa/yr during the 1998–2015 growing seasons, causing a decrease of −0.03/yr in the NDVI. In the context of precipitation, the rainfall amount in the 1982–1998 growing seasons increased by 0.05 (mm/d)/yr, leading to an elevation of 0.03/yr in the NDVI, whereas the rainfall in the 1998–2015 growing seasons decreased by 0.08 (mm/d)/yr, causing a decline of 0.01/yr in the NDVI (Figure 13a).

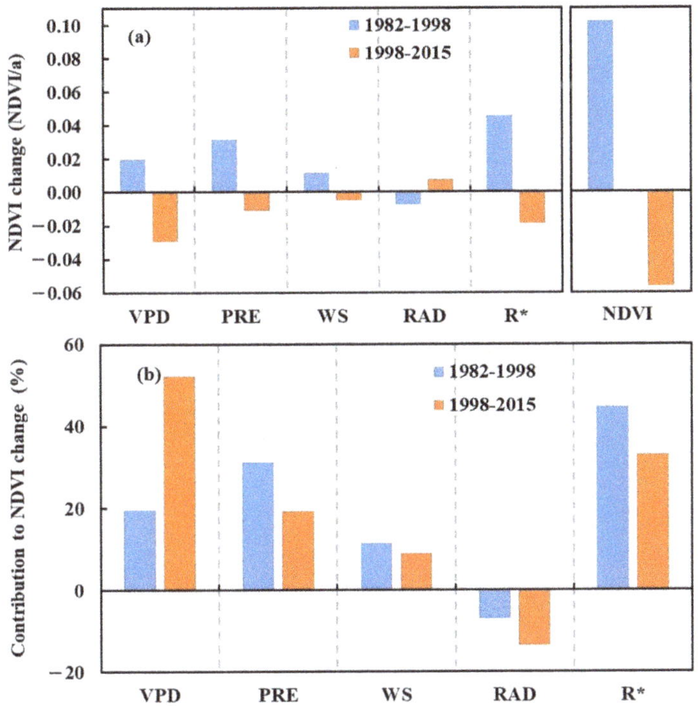


**Figure 13.** (**a**) Contribution and (**b**) relative contribution of climate factors to vegetation NDVI during the growing seasons in western Central Asia (PRE: precipitation; WS: wind speed; RAD: sunshine hours; R*: residual).

Figure 13b depicts the relative contribution of climate factors to the change of the vegetation NDVI during the growing seasons of 1982–1998 and 1998–2015 in western Central Asia. The figure depicts that from 1982 to 1998, the contribution rates of the increased precipitation and decreased VPD to vegetation "greening" were 31.3 and 20%, respectively. However, the highest contributing factor was the residual term, which made more than 40% of the contribution, depicting the significant contribution of non-climatic factors to vegetation "greening". During 1998–2015, the increased VPD contributed 52.3% to vegetation "browning" during the growing seasons, which was followed by decreased precipitation with a contribution rate of 19.3%. Additionally, the contribution rate of residual term reached 33%.

During Summer

Both decreased VPD and increased precipitation were equal primary contributors to the "greening" trend observed during summer in the vegetation NDVI during 1982–1998. Similarly, increased VPD and decreased precipitation played a leading role in the "browning" during 1998–2015. In summary, during 1982–1998, summer VPD reduced by 0.07 kPa/yr, leading to a 0.03 /yr growth in the NDVI, and the precipitation increased by 0.11 (mm/d)/yr, which caused a rise of 0.03/yr in the corresponding NDVI. Comparably, during 1998–2015, an increase of 0.08 kPa/yr in VPD corresponded to a decline of 0.02/yr in the NDVI, while a decrease of 0.10 (mm/d)/yr in precipitation corresponded to a reduction of 0.01/yr in the NDVI (Figure 14a).

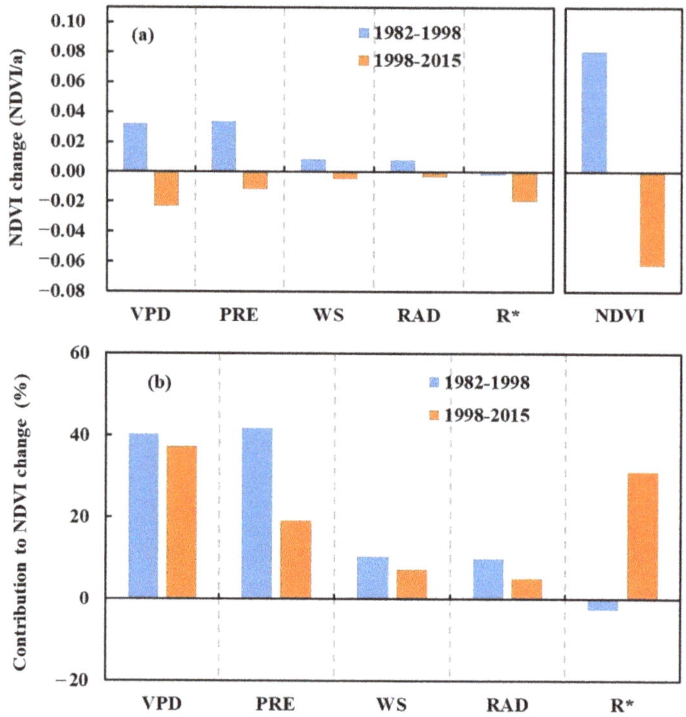


**Figure 14.** (**a**) Contribution and (**b**) relative contribution of climate factors to vegetation NDVI changes in summertime in western Central Asia (PRE: precipitation; WS: wind speed; RAD: sunshine hours; R*: residual).

During 1982–1998, the "greening" of western Central Asia in summer was predominantly controlled by the precipitation and VPD changes, and the contribution rates of the increased precipitation and reduced VPD were 41.9 and 40.3%, respectively. Additionally, wind speed and radiation contributed 10.5 and 9.9%, respectively, towards the "greening" of the vegetation NDVI. Between 1998 and 2015, the vegetation NDVI in western Central Asia changed from "greening" to "browning" during the summer seasons. Attribution analysis indicated that the "browning" was dominated by the increase in VPD and the decrease in precipitation, of which the contribution rate of the increased VPD to the vegetation "browning" in western Central Asia was 37.3%, and that of the reduced precipitation was 19.2%. Furthermore, the contribution rate of the residual term also reached 31.2%, which indicates the influence of human activities and other factors on the vegetation "browning" in western Central Asia (Figure 14b).

3.4.3. Drivers of Vegetation Changes Based on SEM

The SEM is a model for evaluating the causal relationships between variables. We constructed an SEM to evaluate and quantify the climate factors on the vegetation NDVI changes during the growing seasons and summer of 1982–1998 and 1998–2015 in western Central Asia. We assessed the model fit and determined that the established SEM model was capable of reflecting the complex relationship between climate variables and vegetation change. During the growing seasons, the effects of VPD and RAD on vegetation NDVI change are negatively correlated, while those of PRE are positively correlated. PRE is the most influential latent factor during 1982–1998, while the most influential latent factor in 1998–2015 is VPD (Figure 15). Additionally, the WS can indirectly affect vegetation NDVI change through the direct variables. Additionally, the vegetation NDVI structural equation model during the summertime is similar to those in the model for the growing seasons (Figure 16).

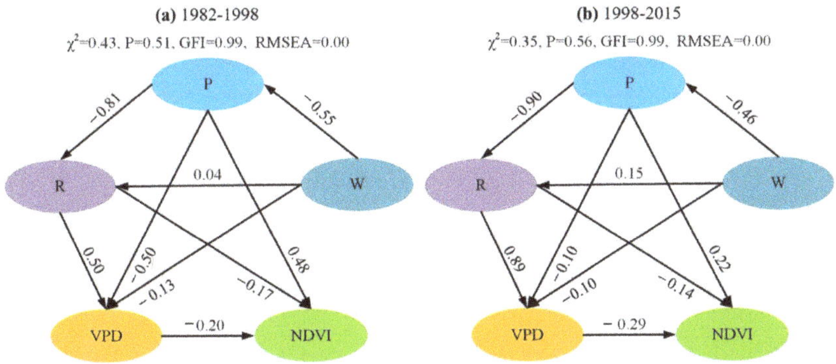


**Figure 15.** The vegetation NDVI structural equation model during the growing seasons in western Central Asia.

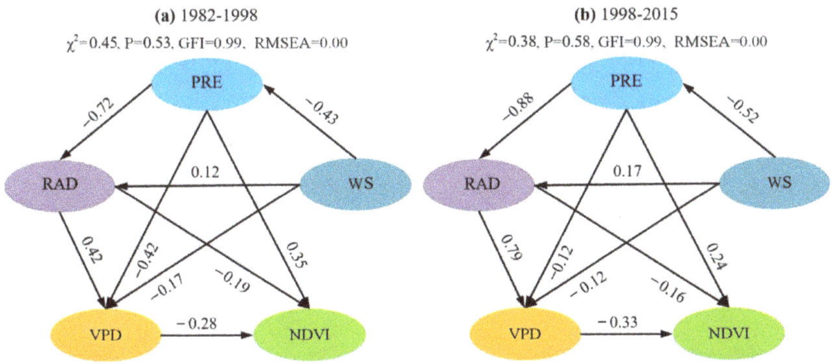


**Figure 16.** The vegetation NDVI structural equation model during the summertime in western Central Asia.

## 4. Discussion and Conclusions

Based on the ECMWF ERA-5 and GIMMS NDVI3g datasets, we analyzed the spatial distribution of the NDVI in Central Asia during 1982–2015 and elucidated the spatial and temporal patterns of vegetation change trends during the growing seasons, springs, summers, and autumns in that period. Additionally, the spatial distribution and difference of the change trends in the NDVI in Central Asia before and after 1998 have been revealed, and the contribution rates of climate factors to the changes in the vegetation NDVI in Central Asia have been quantitatively estimated. The major conclusions are as follows:

(1) The vegetation NDVI during the growing seasons of 1982–2015 in Central Asia demonstrated an insignificant upward trend with apparent stage features, with the vegetation NDVI exhibiting a significant "greening" trend (0.012/10 yr, $p < 0.05$) during the growing seasons from 1982 to 1998. However, this "greening" trend was replaced by an insignificant "browning" trend after 1998. Spatially, the variation trends of the NDVI in Central Asia presented a pattern of "greening in the east and browning in the west". Seasonally, the NDVI exhibited a "greening" trend predominantly during spring, except for the "browning" trend observed in the southwest region of Central Asia. During summer, the "browning" of the vegetation NDVI further increased eastward and expanded to the western Central Asia during autumn. Conversely, the vegetation

NDVI in Xinjiang, China, primarily exhibited a "greening" trend during summer, but a gradually expanding "browning" trend was observed during spring and autumn.

(2) The growing seasons' vegetation NDVI exhibited a general "greening" trend in Central Asia during 1982–1998. Seasonally, spring was dominated by the "greening" of vegetation NDVI, and the "greening" range expanded further during summer and autumn. However, during 1998–2015, "browning" turned out to be the primary variation trend of the NDVI, especially in western Central Asia and northern Xinjiang. The "browning" scopes of the vegetation NDVI were substantially higher during both summer and autumn, and the "browning" trend was the highest during summer. The persistent vegetation "greening" predominantly occurred in the Tarim River Basin of Xinjiang.

(3) Between 1998 and 2015, the increase in VPD and the reduction in precipitation dominated the "browning" trend of the vegetation NDVI during the growing seasons in western Central Asia. The increased VPD contributed 52.3% to the NDVI changes. The increase in VPD reflected the drying condition of the atmosphere, which effectively reduced the "greening" trend in Central Asia, thus leading to the "browning" trend of the vegetation NDVI.

In this study, we quantified the contribution of the increased VPD to the "browning" of vegetation NDVI in Central Asia. Previous studies have demonstrated that increased temperature and decreased relative humidity together lead to an increase in VPD, aggravating atmospheric drought [33]. Water stress caused by atmospheric drought can effectively reduce the "greening" trend of vegetation, decelerating the "greening" trend or resulting in the "browning" trend of vegetation [38]. In this study, we have highlighted the significance of VPD as a limiting factor of the vegetation "greening" trend in arid areas. Therefore, we suggest that the effects of VPD on vegetation growth should be considered when evaluating arid ecosystem function under the global warming background.

Additionally, we focused on evaluating the influence of climate factors, especially the changes in VPD, on the vegetation NDVI in arid areas, without considering the effect of soil moisture on the changes in vegetation. Studies have demonstrated that both the atmospheric dryness caused by increased VPD and soil drought caused by soil moisture deficiency exert negative effects on vegetation growth [32,51].

However, due to the covariation of atmospheric water vapor and soil moisture deficit, it is difficult to distinguish the influence of atmospheric VPD and soil moisture change on vegetation growth [32,51]. This is an important factor that needs to be explored further in future studies. Furthermore, in current Earth system models, negative ecological effects caused by an increase in VPD are underestimated, and this is a common problem [38]. Therefore, it is essential to systematically evaluate whether Earth system models can accurately reflect the impact of the increased VPD on global ecosystems in the future.

**Author Contributions:** Conceptualization, M.L. and J.Y.; methodology, M.L.; software, M.L.; data curation, M.L. and J.Z.; writing—original draft preparation, M.L.; writing—review and editing, J.Y. and J.Z.; visualization, M.L. and J.Y.; project administration, J.Y. All authors have read and agreed to the published version of the manuscript.

**Funding:** This research was funded by the "Tianchi Talents" (Xinjiang) Plan Project for their support to Moyan Li. This research was supported by the Shanghai Cooperation Organization (SCO) Science and Technology Partnership and International S&T Cooperation Program (2023E01022), and the Science and Technology Youth Top-notch Talent Support Program (Tianshan Talents) of Xinjiang (2022TSYCCX0005).

**Data Availability Statement:** The data that support the findings of this study are available by request to the corresponding author.

**Acknowledgments:** The authors would like to thank the data support provided by various data source websites in the article.

**Conflicts of Interest:** The authors declare no conflicts of interest.

## References


1. Chen, F.H.; Huang, W.; Jin, L.Y.; Chen, J.H.; Wang, J.S. Spatiotemporal precipitation variations in the arid Central Asia in the context of global warming. *Sci. China Earth Sci. Chin. J.* **2011**, *41*, 1647–1657. [CrossRef]
2. Chen, Y.; Li, Z.; Fan, Y.; Wang, H.; Deng, H. Progress and prospects of climate change impacts on hydrology in the arid region of northwest China. *Environ. Res.* **2015**, *139*, 11–19. [CrossRef] [PubMed]
3. Chen, Y.; Li, W.; Deng, H.; Fang, G.; Li, Z. Changes in Central Asia's Water Tower: Past, Present and Future. *Sci. Rep.* **2016**, *6*, 35458. [CrossRef] [PubMed]
4. Chen, F.; Chen, J.; Huang, W.; Chen, S.; Huang, X.; Jin, L.; Feng, S. Westerlies Asia and monsoonal Asia: Spatiotemporal differences in climate change and possible mechanisms on decadal to sub-orbital timescales. *Earth Sci. Rev.* **2019**, *192*, 337–354. [CrossRef]
5. IPCC. Climate change 2013: The physical science basis. In *Contribution of Working Group I to the Fifth Assessment Report of the Intergovernmental Panel on Climate Change*; Stocker, T.F., Qin, D., Plattner, G.K., Tignor, M., Allen, S.K., Boschung, J., Nauels, A., Xia, Y., Bex, V., Midgley, P.M., Eds.; Cambridge University Press: Cambridge, UK; New York, NY, USA, 2013.
6. Hu, Z.; Zhang, C.; Hu, Q.S.; Tian, H. Temperature changes in Central Asia from 1979 to 2011 based on multiple datasets. *J. Clim.* **2014**, *27*, 1143–1167. [CrossRef]
7. Yao, J.; Chen, Y.; Chen, J.; Zhao, Y.; Tuoliewubieke, D.; Li, J.; Yang, L.M.; Mao, W. Intensification of extreme precipitation in arid Central Asia. *J. Hydrol.* **2021**, *598*, 125760. [CrossRef]
8. Li, Z.; Chen, Y.N.; Fang, G.H.; Li, Y.P. Multivariate assessment and attribution of droughts in Central Asia. *Sci. Rep.* **2017**, *7*, 1316. [CrossRef]
9. Peng, D.; Zhou, T.; Zhang, L.; Zhang, W.; Chen, X. Observationally constrained projection of the reduced intensification of extreme climate events in Central Asia from 0.5 °C less global warming. *Clim. Dyn.* **2019**, *54*, 543–560. [CrossRef]
10. Yang, S.; Sun, L.; He, J.; Li, C.; Yu, Y. Evolution of the Aral Sea: Crisis and present situation. *Arid. Land Geogr.* **2024**, *47*, 181–191.
11. Wang, X.; Chen, Y.; Li, Z.; Fang, G.; Wang, Y. Development and utilization of water resources and assessment of water security in Central Asia. *Agric. Water Manag.* **2020**, *240*, 106297. [CrossRef]
12. Piao, S.; Fang, J.; Zhou, L.; Guo, Q.; Henderson, M.; Ji, W.; Li, Y.; Tao, S. Interannual variations of monthly and seasonal normalized difference vegetation index (NDVI) in China from 1982 to 1999. *J. Geophys. Res. Atmos.* **2003**, *108*, 4401. [CrossRef]
13. Zhao, W.; Hu, Z.; Guo, Q.; Wu, G.; Chen, R.; Li, S. Contributions of climatic factors to inter-annual variability of vegetation index in northern China grasslands. *J. Clim.* **2020**, *33*, 175–183. [CrossRef]
14. Liu, D.; Wang, T.; Yang, T.; Yan, Z.; Liu, Y.; Zhao, Y.; Piao, S. Deciphering impacts of climate extremes on Tibetan grasslands in the last fifteen years. *Sci. Bull.* **2019**, *64*, 446. [CrossRef] [PubMed]
15. Hao, H.; Chen, Y.; Xu, J.; Li, Z.; Li, Y.; Kayumba, P.M. Water Deficit May Cause Vegetation Browning in Central Asia. *Remote Sens.* **2022**, *14*, 2574. [CrossRef]
16. Yao, J.; Chen, Y.; Guan, X.; Zhao, Y.; Chen, J.; Mao, W. Recent climate and hydrological changes in a mountain-basin system in Xinjiang, China. *Earth-Sci. Rev.* **2022**, *226*, 103957. [CrossRef]
17. Yu, T.; Jiapaer, G.; Bao, A.; Zheng, G.; Zhang, J.; Li, X.; Yuan, Y.; Huang, X.; Umuhoza, J. Disentangling the relative effects of soil moisture and vapor pressure deficit on photosynthesis in dryland Central Asia. *Ecol. Indic.* **2022**, *137*, 108698. [CrossRef]
18. Li, M.; Guan, J.; Zheng, J. Climate Drivers Contribute in Vegetation Greening Stalls of Arid Xinjiang, China: An Atmospheric Water Drying Effect. *Water* **2022**, *14*, 2019. [CrossRef]
19. Fu, B.; Stafford-Smith, M.; Wang, Y.; Wu, B.; Yu, X.; Lv, N.; Ojima, D.S.; Lv, Y.; Fu, C.; Liu, Y.; et al. The Global-DEP conceptual framework—Research on dryland ecosystems to promote sustainability. *Curr. Opin. Environ. Sustain.* **2021**, *48*, 17–28. [CrossRef]
20. Lian, X.; Piao, S.; Chen, A.; Huntingford, C.; Fu, B.; Li, L.Z.; Roderick, M.L. Multifaceted characteristics of dryland aridity changes in a warming world. *Nat. Rev. Earth Environ.* **2021**, *2*, 232–250. [CrossRef]
21. Zhang, W.; Li, Y.; Wu, X.; Chen, Y.; Chen, A.; Schwalm, C.R.; Kimball, J.S. Divergent response of vegetation growth to soil water availability in dry and wet periods over Central Asia. *J. Geophys. Res. Biogeosci.* **2021**, *126*, e2020JG005912. [CrossRef]
22. Zhang, H.; Hu, Z.; Zhang, Z.; Li, Y.; Song, S.; Chen, X. How does vegetation change under the warm–wet tendency across Xinjiang, China? *Int. J. Appl. Earth Obs. Geoinf.* **2024**, *127*, 103664. [CrossRef]
23. Luo, M.; Sa, C.; Meng, F.; Duan, Y.; Liu, T.; Bao, Y. Assessing extreme climatic changes on a monthly scale and their implications for vegetation in Central Asia. *J. Clean. Prod.* **2020**, *271*, 122396. [CrossRef]
24. Lindner, M.; Maroschek, M.; Netherer, S.; Kremer, A.; Barbati, A.; Garcia-Gonzalo, J.; Seidl, R.; Delzon, S.; Corona, P.; Kolström, M.; et al. Climate change impacts, adaptive capacity, and vulnerability of European forest ecosystems. *For. Ecol. Manag.* **2010**, *259*, 698–709. [CrossRef]
25. Way, D.A.; Oren, R. Differential responses to changes in growth temperature between trees from different functional groups and biomes: A review and synthesis of data. *Tree Physiol.* **2010**, *30*, 669–688. [CrossRef] [PubMed]
26. Allen, C.D.; Macalady, A.K.; Chenchouni, H.; Bachelet, D.; McDowell, N.; Vennetier, M.; Kitzberger, T.; Rigling, A.; Breshears, D.D.; Hogg, E.H.; et al. A global overview of drought and heat-induced tree mortality reveals emerging climate change risks for forests. *For. Ecol. Manag.* **2010**, *259*, 660–684. [CrossRef]
27. Knapp, A.K.; Avolio, M.L.; Beier, C.; Carroll, C.J.W.; Collins, S.L.; Dukes, J.S.; Fraser, L.H.; Griffin-Nolan, R.J.; Hoover, D.L.; Jentsch, A.; et al. Pushing precipitation to the extremes in distributed experiments: Recommendations for simulating wet and dry years. *Glob. Chang. Biol.* **2017**, *23*, 1774–1782. [CrossRef]

28. Ainsworth, E.A.; Long, S.P. What have we learned from 15 years of free-air $CO_2$ enrichment (FACE)? A meta-analytic review of the responses of photosynthesis, canopy properties and plant production to rising $CO_2$. *New Phytol.* **2005**, *165*, 351–372. [CrossRef]
29. Zhu, Z.; Piao, S.; Myneni, R.B.; Huang, M.; Zeng, Z.; Canadell, J.G.; Ciais, P.; Sitch, S.; Friedlingstein, P.; Arneth, A.; et al. Greening of the Earth and its drivers. *Nat. Clim. Chang.* **2016**, *6*, 791–795. [CrossRef]
30. Breshears, D.D.; Adams, H.D.; Eamus, D.; McDowell, N.G.; Law, D.J.; Will, R.E.; Zou, C.B. The critical amplifying role of increasing atmospheric moisture demand on tree mortality and associated regional die-off. *Front. Plant Sci.* **2013**, *4*, 266. [CrossRef]
31. Williams, A.P.; Allen, C.D.; Macalady, A.K.; Griffin, D.; Woodhouse, C.A.; Meko, D.M.; McDowell, N.G. Temperature as a potent driver of regional forest drought stress and tree mortality. *Nat. Clim. Chang.* **2013**, *3*, 292–297. [CrossRef]
32. Zhou, S.; Williams, A.P.; Berg, A.M.; Cook, B.I.; Zhang, Y.; Hagemann, S.; Gentine, P. Land–Atmosphere feedbacks exacerbate concurrent soil drought and atmospheric aridity. *Proc. Natl. Acad. Sci. USA* **2019**, *116*, 18848–18853. [CrossRef] [PubMed]
33. Li, M.; Yao, J.Q.; Guan, J.Y.; Zheng, J.H. Observed changes in vapor pressure deficit suggest a systematic drying of the atmosphere in Xinjiang of China. *Atmos Res.* **2020**, *248*, 105199. [CrossRef]
34. Williams, A.P.; Seager, R.; Berkelhammer, M.; Macalady, A.K.; Crimmins, M.A.; Swetnam, T.W.; Rahn, T. Causes and implications of extreme atmospheric moisture demand during the record-breaking 2011 wildfire season in the southwestern United States. *J. Appl. Meteorol. Climatol.* **2014**, *53*, 2671–2684. [CrossRef]
35. Seager, R.; Hooks, A.; Williams, A.P.; Cook, B.; Nakamura, J.; Henderson, N. Climatology, variability, and trends in the U.S. vapor pressure deficit, an important fire-related meteorological quantity. *J. Appl. Meteorol. Climatol.* **2015**, *54*, 1121–1141. [CrossRef]
36. Asseng, S.; Ewert, F.; Martre, P.; Rotter, R.P.; Lobell, D.B.; Cammarano, D.; Kimball, B.A.; Ottman, M.J.; Wall, G.W.; White, J.W.; et al. Rising temperatures reduce global wheat production. *Nat. Clim. Chang.* **2015**, *5*, 143. [CrossRef]
37. Zhao, C.; Liu, B.; Piao, S.; Wang, X.; Lobell, D.B.; Huang, Y.; Huang, M.T.; Yao, Y.T.; Bassu, S.; Ciais, P.; et al. Temperature increase reduces global yields of major crops in four independent estimates. *Proc. Natl. Acad. Sci. USA.* **2017**, *114*, 9326–9331. [CrossRef] [PubMed]
38. Yuan, W.; Zheng, Y.; Piao, S.; Ciais, P.; Lombardozzi, D.; Wang, Y.; Ryu, Y.; Chen, G.; Dong, W.; Hu, Z.; et al. Increased atmospheric vapor pressure deficit reduces global vegetation growth. *Sci. Adv.* **2019**, *5*, eaax1396. [CrossRef]
39. Yao, J.; Chen, Y.; Zhao, Y.; Guan, X.; Mao, W.; Yang, L. Climatic and associated atmospheric water cycle changes over the Xinjiang, China. *J. Hydrol.* **2020**, *585*, 124823. [CrossRef]
40. Li, Z.; Chen, Y.; Li, W.; Deng, H.; Fang, G. Potential impacts of climate change on vegetation dynamics in Central Asia. *J. Geophys. Res. Atmos.* **2015**, *120*, 12345–112356. [CrossRef]
41. Jiang, C.; Ryu, Y.; Fang, H.; Myneni, R.; Claverie, M.; Zhu, Z. Inconsistencies of interannual variability and trends in long-term satellite leaf area index products. *Glob. Chang. Biol.* **2017**, *23*, 4133–4146. [CrossRef]
42. Chen, P.; Yao, J.; Mao, W.; Ma, L.; Chen, J.; Dilinuer, T.; Li, S. Increased Extreme Precipitation in May over Southwestern Xinjiang in Relation to Eurasian Snow Cover in Recent Years. *J. Clim.* **2024**, *37*, 2695–2711. [CrossRef]
43. Hersbach, H.; Bell, B.; Berrisford, P.; Hirahara, S.; Horányi, A.; Muñoz-Sabater, J.; Nicolas, J.; Peubey, C.; Radu, R.; Schepers, D.; et al. The ERA5 global reanalysis. *Q. J. R. Meteorol. Soc.* **2020**, *146*, 1999–2049. [CrossRef]
44. Sherwood, S.; Fu, Q. A drier future? *Science* **2014**, *343*, 737–739. [CrossRef] [PubMed]
45. Ficklin, D.L.; Novick, K.A. Historic and projected changes in vapor pressure deficit suggest a continental-scale drying of the United States atmosphere. *J. Geophys. Res. Atmos.* **2017**, *122*, 2061–2079. [CrossRef]
46. Yadolah, D. *The Concise Encyclopedia of Statistics*; Springer: New York, NY, USA, 2008.
47. Roderick, M.L.; Rotstayn, L.D.; Farquhar, G.D.; Hobbins, M.T. On the attribution of changing pan evaporation. *Geophys. Res. Lett.* **2007**, *34*, L17403. [CrossRef]
48. Naeem, S.; Zhang, Y.; Zhang, X.; Tian, J.; Abbas, S.; Luo, L.; Meresa, H.K. Both climate and socioeconomic drivers contribute in vegetation greening of the loess plateau. *Sci. Bull.* **2021**, *66*, 1160–1163. [CrossRef]
49. Hao, J.; Xu, G.; Luo, L.; Zhang, Z.; Yang, H.; Li, H. Quantifying the relative contribution of natural and human factors to vegetation coverage variation in coastal wetlands in China. *Catena* **2020**, *188*, 104429. [CrossRef]
50. Shao, W.; Wang, Q.; Guan, Q.; Luo, H.; Ma, Y.; Zhang, J. Distribution of soil available nutrients and their response to environmental factors based on path analysis model in arid and semi-arid area of northwest China. *Sci. Total Environ.* **2022**, *827*, 154254. [CrossRef]
51. Zhou, S.; Williams, A.P.; Lintner, B.R.; Berg, A.M.; Zhang, Y.; Keenan, T.F.; Cook, B.I.; Hagemann, S.; Seneviratne, S.I.; Gentine, P. Soil moisture–atmosphere feedbacks mitigate declining water availability in drylands. *Nat. Clim. Chang.* **2021**, *11*, 38–44. [CrossRef]

*forests*

MDPI

*Article*

# Spatial Pattern of Forest Age in China Estimated by the Fusion of Multiscale Information

**Yixin Xu [1,2], Tao Zhou [1,2,*], Jingyu Zeng [1,2], Hui Luo [1,2], Yajie Zhang [1,2], Xia Liu [1,2], Qiaoyu Lin [1,2] and Jingzhou Zhang [1,2]**

1 State Key Laboratory of Earth Surface Processes and Resource Ecology, Beijing Normal University, Beijing 100875, China; 202021051010@mail.bnu.edu.cn (Y.X.); zengjingyufzu@163.com (J.Z.); huiluo@mail.bnu.edu.cn (H.L.); 202131051011@mail.bnu.edu.cn (Y.Z.); 201931051009@mail.bnu.edu.cn (X.L.); 202121051004@mail.bnu.edu.cn (Q.L.); 202031051012@mail.bnu.edu.cn (J.Z.)

2 Key Laboratory of Environmental Change and Natural Disaster of Ministry of Education, Faculty of Geographical Science, Beijing Normal University, Beijing 100875, China

* Correspondence: tzhou@bnu.edu.cn; Tel.: +86-10-5880-7238

**Abstract:** Forest age is one of most important biological factors that determines the magnitude of vegetation carbon sequestration. A spatially explicit forest age dataset is crucial for forest carbon dynamics modeling at the regional scale. However, owing to the high spatial heterogeneity in forest age, accurate high-resolution forest age data are still lacking, which causes uncertainty in carbon sink potential prediction. In this study, we obtained a 1 km resolution forest map based on the fusion of multiscale age information, i.e., the ninth (2014–2018) forest inventory statistics of China, with high accuracy at the province scale, and a field-observed dataset covering 6779 sites, with high accuracy at the site scale. Specifically, we first constructed a random forest (RF) model based on field-observed data. Utilizing this model, we then generated a spatially explicit forest age map with a 1 km resolution (random forest age map, RF map) using remotely sensed data such as tree height, elevation, meteorology, and forest distribution. This was then used as the basis for downscaling the provincial-scale forest inventory statistics of the forest ages and retrieving constrained maps of forest age (forest inventory constrained age maps, FIC map), which exhibit high statistical accuracy at both the province scale and site scale. The main results included the following: (1) RF can be used to estimate the site-scale forest age accurately ($R^2$ = 0.89) and has the potential to predict the spatial pattern of forest age. However, (2) owing to the impacts of sampling error (e.g., field-observed sites are usually located in areas exhibiting relatively favorable environmental conditions) and the spatial mismatch among different datasets, the regional-scale forest age predicted by the RF model could be overestimated by 71.6%. (3) The results of the downscaling of the inventory statistics indicate that the average age of forests in China is 35.1 years (standard deviation of 21.9 years), with high spatial heterogeneity. Specifically, forests are older in mountainous and hilly areas, such as northeast, southwest, and northwest China, than in southern China. The spatially explicit dataset of the forest age retrieved in this study encompasses synthesized multiscale forest age information and is valuable for the research community in assessing the carbon sink potential and modeling carbon dynamics.

**Keywords:** China; forest age; spatial pattern; forest inventory; random forest

## 1. Introduction

Forests are an important component of terrestrial ecosystems, storing 33%–46% of terrestrial carbon [1,2], and forest carbon exchange plays an important role in the global carbon balance. During the 1990–2009 period, the global forest cumulative carbon sink reached approximately 116 Pg C, which could offset 90% of the cumulative fossil emissions (126 Pg C) [3]. Obviously, forest carbon sinks are an effective way to offset fossil fuel emissions and could greatly mitigate climate change caused by elevated $CO_2$ concentrations [4]. The State of the World's Forests 2022 report published by the Food and Agriculture Organization

(FAO) forecasts that maintaining forest areas between 2020 and 2050 could achieve the target of reducing emissions by 3.6 billion tons of $CO_2$ equivalent annually, contributing a 14% reduction toward keeping global warming below 1.5 °C. However, there is very high uncertainty in the estimation of forest carbon stocks; therefore, how to accurately predict and model forest carbon sinks has become the main issue of widespread concern in the academic community.

Forest carbon sinks are influenced by a variety of biotic and abiotic factors, such as forest origin, forest type, forest age, geography, climate, and soil [5]. As a very important internal biological factor, forest age affects aboveground biomass, primary productivity, litter decomposition, etc., and plays a significant role in determining carbon stocks and sinks in forest ecosystems [6–9]. In general, aboveground biomass and primary productivity increase with forest age, reaching a maximum when the canopy is closed after adulthood, and thereafter decrease, resulting in an S-shaped growth curve [10,11]. The balance between the photosynthesis and respiration of trees also fundamentally changes with age [12], causing a notable variation in the forest biomass carbon density among different stand ages, and it has been shown that the average biomass carbon density of old-growth forests can reach 3.7 times that of young-growth forests [13]. In mature forest scenario-based research, it has also been determined that the forest carbon density first increases but then decreases with increasing forest age [14,15]. Model parameterization based on mature forests commonly leads to the underestimation of forest carbon sequestration, especially in regions with many plantations and young forests. Moreover, owing to the lack of spatial distribution data on forest age, this variable has not been given enough attention in most previous studies, which has resulted in increased uncertainty in model estimation and difficulty in accurately predicting the future carbon sequestration potential of forests [7,15–17].

Although the importance of stand age for forest carbon sinks and trends has been widely recognized, accurately reflecting the spatial distribution pattern of forest age is an enormous challenge [18–21]. Stand age generally refers to the average age of a forest stand and is also often considered the time experienced since the last strong disturbance. Thus, stand age is affected not only by natural factors such as site conditions, climate, fire, and pests but also by human factors such as logging and thinning [10,22]. Traditional stand age determination is focused mainly on individual stands, and the direct determination method is mainly used. Although this method is considered the most accurate forest age determination method, obtaining the forest age at large spatial scales is not only difficult but also time-consuming [23]. Although satellite remote sensing can be employed to dynamically monitor forest changes, such as the forest tree height estimated from LiDAR remote sensing [11,16,24], the NDVI retrieved from optical remote sensing [25], land cover change [26], and the phenology [23], which can be used to estimate forest age, the spatial distributions of different forest age datasets greatly vary, and uncertainty is high because of differences in data quality levels or sampling errors across observation sites. Although forest subcompartment data from the Forest Management Inventory can provide forest age information, the corresponding update frequency is low, and data acquisition is difficult. The commonly used forest age inventory statistics are often based on the summary results of administrative regions with low spatial resolutions (e.g., province-scale data), making it difficult to accurately describe their spatial pattern [27]. How to combine forest inventory statistics with remotely sensed information to reliably estimate forest age is currently a hot topic in the academic community [25]. With the wide application of machine learning methods, combining these methods with remotely sensed information and forest inventory statistics is a feasible development direction [11,28], which could help to obtain forest age datasets that are consistent with forest inventory statistics in terms of quantity and exhibit a reasonable spatial distribution pattern.

At present, forest age estimation in China is mainly based on the relationship between tree height and forest age [11,16,29]. In most forest ecosystems, there is a good correlation between forest age and forest structure, and among them, the vertical structure information of forest tree height is usually considered the most closely related variable to forest age [30,31]. It has been found that there exists a suitable correlation between forest age and tree height, which suggests that tree height is a very important explanatory variable in statistical models for estimating forest age [32]. In previous studies, forest origin was not considered when estimating forest age or only the age of plantations was estimated without obtaining a forest age distribution map covering all of China. China has been committed to afforestation since the 1970s, and the area of planted forests is the largest in the world. There are large differences in forest age between planted and natural forests. If the origin of forests is not considered, the age of forests will be greatly over- or underestimated, which will lead to the under- or overestimation, respectively, of carbon sinks in China. In addition, China's forests also have characteristics such as a low forest age and low carbon density, which implies that there is a significant potential for increasing carbon sinks in the future [19]. Therefore, to accurately map forest age in China and effectively evaluate forest carbon sink potential to support the realization of China's carbon peaking and carbon neutrality targets, it is necessary to consider both tree height and stand characteristics when estimating forest age in China by fusing forest inventory and remotely sensed information with a machine learning method.

In this study, we aim to develop a forest age estimation method that integrates forest inventory data with multivariate data. This method integrates the quantitative accuracy of forest inventory age statistical information at the provincial scale with the spatial continuity advantage of spatially continuous remote sensing information. Through downscaling methods, we obtain spatially distributed products of China's forest age that are reasonable in terms of both quantitative and spatial distribution patterns. This included (1) the random forest (RF) model being used as a basic tool to obtain forest age products with an initial resolution of 1 km in China. This product exhibits favorable spatial continuity and can reflect the spatial change in stand age in China but possible statistical discrepancies due to sampling errors caused by the nonuniformity of observation sites. (2) This was then used as the basis for downscaling the provincial forest ages estimated from the ninth (2014–2018) forest inventory statistics, and optimized maps of forest age that exhibited satisfactory spatial continuity and high accuracy at both the provincial and national scales were retrieved. (3) In addition, we compared the optimized forest age map with other forest age datasets (i.e., the MPI-BGC dataset).

## 2. Data and Methods

### *2.1. Data*

(1) Field observation data

A total of 6779 publicly available field data, including forest type and plot conditions, were collected, with ages ranging from 1 to 360 years (Figure 1). The recorded information for each sample plot included six variables: longitude, latitude, average annual temperature, annual precipitation, forest age, and tree height and altitude [33,34]. These data were collected from stands that were typical of the area and did not experience disturbances, such as pruning, thinning, fire, and other factors. There are missing values in the collected station data of the mean annual precipitation (MAP) and mean annual temperature (MAT), and some values are extracted from remotely sensed data; however, the datasets used are different. To ensure data uniformity, the MAP and MAT data were uniformly extracted across all stations from 1970 to 2014 in China on the basis of the MAT and MAP at the 1 km scale [35].

(2) National forest inventory data

The forest resource inventory data (FID) were obtained from the China Forest Resources Report 2014–2018 published by the China Forestry Publishing House [36]. The forest inventory is a national survey of forest resources conducted every five years in China since 1973. The FID includes forest type, forest area, stock volume, etc. However, there is no specific age of stand, and it is only divided into five grades: young forest, middle-aged forest, near-mature forest, mature forest, and overmature forest (Table S1). According to the tree species (group) classification table in the Technical Regulations for Continuous Forest Inventory, the dominant tree species in each province in the arbor forest statistical table of the forest inventory data were divided into coniferous forests, broad-leaved forests, and coniferous and broad-leaved mixed forests. Then, the average age of the three major forest types (Table S2) and the average age in each province were calculated by the method of Dai Ming [25], and the method is described in the Supplementary Material.

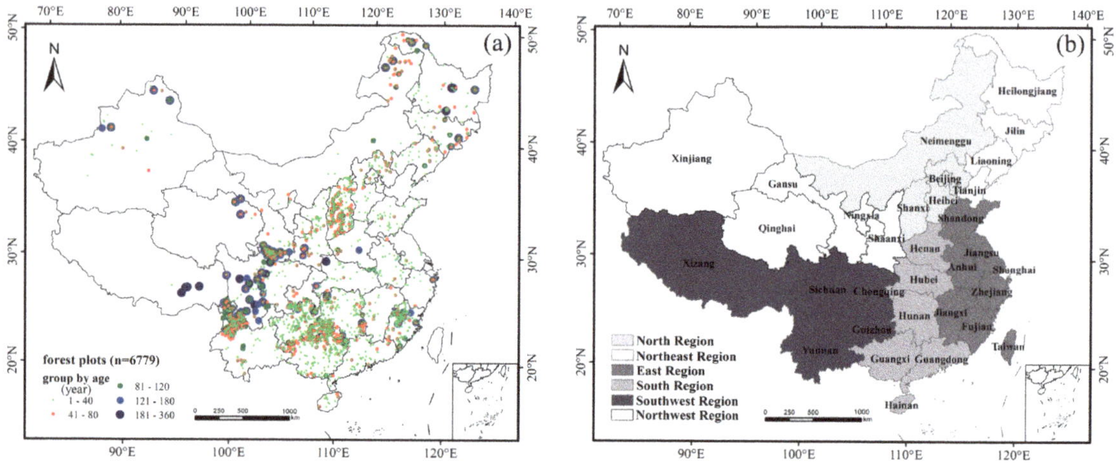


**Figure 1.** Distribution of the sample plots and forest regions in China. (**a**) Geographical distribution of the sample plots; (**b**) commonly used geographical and administrative divisions in China (due to data limitations, forest age estimation in this paper does not include Shanghai, Hong Kong, Macao, or Taiwan).

(3) Remotely sensed tree height data

The remotely sensed tree height data used in this paper originated from the forest height map of China with a 30 m resolution generated by Liu et al. by fusing the Global Ecosystem Dynamics Investigation (GEDI), Ice, Cloud, and land Elevation Satellite-2 (ICESat-2) Advanced Topographic Laser Altimeter System (ATLAS) and Sentinel-2 images, and the data were interpolated by a neural network (NNGI) [37]. The application of these data greatly reduces the saturation effect when using optical images, and regression models are used to determine forest canopy height; when interpolation is used, this generates more continuous and complete tree height spatial data, which increases the accuracy of the data [31,37].

(4) Forest distribution data

The forest distribution data of China were obtained from the 8th forest inventory data of China from 2009 to 2013. ArcGIS 10.2 and ENVI 5.3 software were used to process the forest distribution map of the 8th forest inventory data, including digitalization, spatial registration, reprojection, and resampling steps. Finally, a forest type map of China with a spatial resolution of 1 km was obtained. In these data, China's forests were divided into forest stands, economic forests, and bamboo forests. On the basis of their origin, forest stands were divided into planted and natural forests, including three forest types: coniferous forests, mixed coniferous and broad-leaved forests, and broad-leaved forests. The forest types covered in this study included only stands, excluding economic forests and bamboo forests. In these data, coniferous forests account for 35.4%, while broad-leaved forests account for 59.5%. According to the results of the 9th National Forest Inventory, the total area of arbor forest in China is 180 million hectares, with the area proportions of the three forest types being 34.4%, 57.7%, and 7.9%, respectively. When the two sets of data are compared, the difference in the results is not significant.

(5) Meteorological data and elevation data

Climate factors are important in the process of forest growth and determine the type and growth rate of forests on a regional scale [38,39]. The meteorological data originated from the monthly climate dataset of China from 1901 to 2017 and had a spatial resolution of 1 km [40,41]. According to monthly data processing, the MAT and MAP from 1970 to 2014 were obtained. This dataset was generated by the delta spatial downscaling method using the meteorological dataset with a spatial resolution of 0.5° released by the CRU and the climate dataset released by WorldClim. The reliability of the data was also verified on the basis of data from independent meteorological observation stations across China, and the accuracy of the data was suitably guaranteed [35]. Influenced by the monsoon climate, China exhibits significant spatial heterogeneity in temperature and precipitation (Figure S1). The elevation data were retrieved from the Chinese 1 km resolution digital elevation model (DEM) data provided by Big Earth Data for the Three Poles, which can reflect local topographic features [42]. Table 1 provides detailed information about the data.

**Table 1.** Data sources used in this study.

| Types of Data | Data Names | Resolution/Scale | Data Sources |
|---|---|---|---|
| Forest statistical data | Forest inventory data (2014–2018) | | *China Forest Resources Report 2014–2018* |
| Forest height data | NNGI-Forest Canopy Height (2019) | 30 m | [37] |
| Forest distribution data | forest type and origin map (2009–2013) | 1 km | |
| Meteorological data | Annual average temperature (1970–2014) | 0.0083333°(~1 km) | https://poles.tpdc.ac.cn/zh-hans/ (last accessed on 8 July 2024) |
| | Annual average precipitation (1970–2014) | 0.0083333°(~1 km) | |
| Elevation data | Digital elevation model of China | 1 km | |
| Field observation data | Field observation data (1978–2008) | >100 m$^2$ | [33,34] |
| Administrative division | Chinese administrative division (2020) | | |

### 2.2. *Methods*

#### 2.2.1. Observation Site-Based Random Forest Model

The RF algorithm is a nonlinear statistical ensemble method that is classified as a machine learning method [43]. In ecological studies, collinearity exists among many variables, and multiple factors may act on a single variable. In such cases, RF is an excellent choice, as it is insensitive to multicollinearity and can effectively predict the effects of up to several thousand explanatory variables without being prone to overfitting the data [44–46]. RF often performs well even in comparisons of multiple machine learning models [31].

In most forest ecosystems, forest age is highly correlated with forest structure. Among them, tree height, which contains information on the vertical structure of forests, is commonly considered the most closely related variable to stand age [30]. It has been widely shown that forest age and tree height are highly correlated, with correlation coefficients up to 0.752 [32]. Therefore, tree height is essential for estimating forest age.

Moreover, forests are the products of environmental co-action. To estimate forest age, we considered the correlations between forest age and six variables: tree height, forest type, forest origin, MAT, MAE, and elevation. Using the randomForest package in R 4.1.2, we constructed a random forest regression model that incorporates these variables to predict forest age (age ~ H, forest type, forest origin, MAT, MAP, and elevation). To evaluate the model's performance in predicting forest age, two key metrics were employed: root mean square error (*RMSE*) and mean absolute error (*MAE*).

$$RMSE = \sqrt{\frac{1}{m}\sum_{i}^{m}(y_i - \hat{y}_i)^2} \quad (1)$$

$$MAE = \frac{1}{m}\sum_{i=1}^{m}|y_i - \hat{y}_i| \quad (2)$$

where *RMSE* is the root mean square error, *MAE* is the mean absolute error, $m$ is the number of data points, $y_i$ is the true value, and $\hat{y}_i$ is the predicted value.

2.2.2. Downscaling

On the basis of RF and processed remote sensing data maps, we retrieved an initial spatially explicit map of forest age (RF age map) at a 1 km resolution. Because of the spatial scale difference between the tree height and environmental data observed at the station and the tree height and environmental data retrieved via remote sensing, there is a possible deviation in the absolute magnitude of forest age at the grid scale, but the relative age between the different grid points is comparable. In order to ensure more accurate data in terms of quantity, we used the RF age map as a basis and applied Formula (3) for downscaling the FID at the provincial scale to obtain a 1 km resolution forest age product, which is not only consistent with the forest inventory statistics (mean value) but also exhibits consistency in the spatial distribution of the relative magnitude of forest age derived from remote sensing inversion. Figure 2 shows a detailed framework for the forest age estimation proposed in this study.

$$age_{x(ptr)} = mean_{pt} + \left(\frac{Age_{x(ptr)} - Age_{min(ptr)}}{Age_{max(ptr)} - Age_{min(ptr)}} \times 2 - 1\right) \times rag \quad (3)$$

where $x$ denotes a pixel; $mean_{pt}$ denotes the average age of forest type $t$ in province $p$ calculated from the FID; $Age_{x(ptr)}$ denotes forest age in a given pixel predicted by the random forest model of forest type $t$ in province $p$; $Age_{min(ptr)}$ and $Age_{max(ptr)}$ denote the minimum and maximum ages, respectively, predicted by the random forest model for forest type $t$ in province $p$; and $rag$ is half of the age range of each forest age group ($rag = 0.5 \times$ (maximum forest age-minimum forest age). When the average forest age of the age group is older than 10 years ($mean_{pt} > 10$), a constant value of 10 is adopted according to the *Technical Regulations for Continuous Forest Inventories*. When the average forest age of the age group is younger than 10 years ($0 < mean_{pt} < 10$), rag is set to a constant value of $mean_{pt}$-1 (assuming that the minimum forest age of young forests is 1).

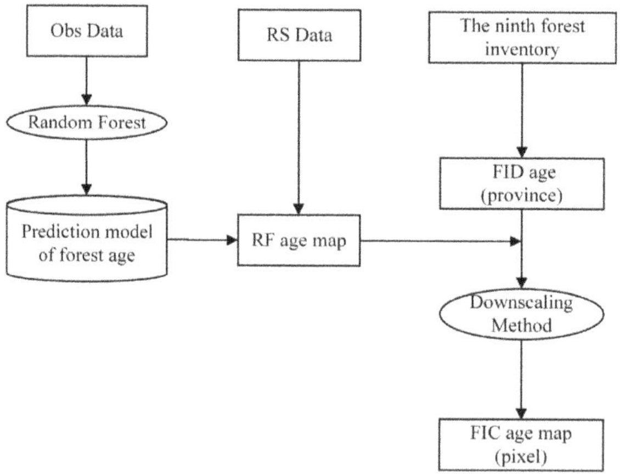


**Figure 2.** Technology roadmap for retrieving forest age by fusing multiscale age information. Obs Data denotes field observation data. RS Data denotes remotely sensed data. RF age map denotes the initial spatially explicit forest age map estimated by the random forest model based on the field observation age dataset, and FIC age map denotes the downscaled forest age map through the forest inventory statistics.

## 3. Results

### 3.1. Random Forest Prediction of Forest Age in China

The random forest regression model exhibited good performance in modeling forest age, with $R^2$ values of 0.92 and 0.78 for the training and testing sets, respectively. Overall, the model explained 85% of the total variation in forest age, with an MAE value less than 13.0 (Figure 3). These results indicated that RF could be used to estimate forest age in China favorably at the site scale.

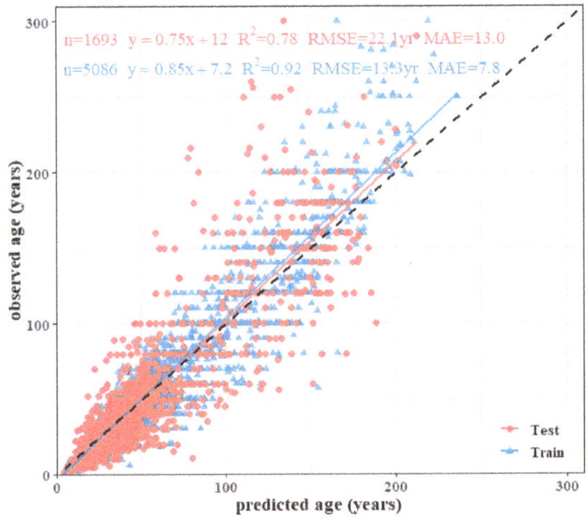


**Figure 3.** Model performance evaluation. The blue points denote the training set data, totaling 5086 points, the red points indicate the test set data, totaling 1693 points, and the black dashed line is the 1:1 line.

On the basis of RF, we generated a spatially explicit forest age map for China (Figure 4). The map shows that forest age clearly exhibits spatial heterogeneity, with older forest age in the northeastern and southwestern regions and younger forest ages in the southeast hilly area. The forests in mountainous areas are generally older, which agrees with the distribution pattern of forest age in China. In the China forest age map predicted by RF, forests aged between 20 and 80 years old dominate in China, accounting for 78% of the total forest area. Forests 0–20 years and over 120 years in age each account for less than 4% of the total forests (Figure 5).

Although this method has good prediction accuracy at the site scale, the forest age obtained via RF is older than that based on the FID at the provincial scale (Figure 6). In comparison, the degree of overestimation of forest age in Guangxi is the most, reaching 156.45%, while Shaanxi Province exhibits the best prediction effect, where the estimated forest age is merely 11.78% older than the actual. On a national scale, the RF forest age is older than that based on the FID, and the predicted frequency of young-age forests (0–40 years old) is much younger than that on the basis of the FID, whereas the frequency of old-age forests (>40 years old) is older than that based on the FID (Figure 5). The possible reasons for these results could be that the sample site survey was conducted with a preference for well-grown stands and the influence of extremely high values at the various sites, which could lead to the overestimation of stand age. Therefore, the RF prediction results cannot be adopted as final results, which were used in this study as the basis for downscaling the forest age of provincial FID to obtain reliable forest age data for China at the regional scale (i.e., at the provincial and national scales).

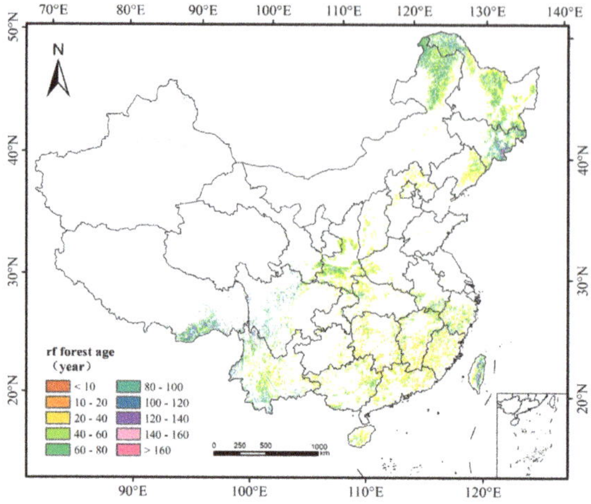


**Figure 4.** Initial spatial distribution of forest age estimated by the RF model.

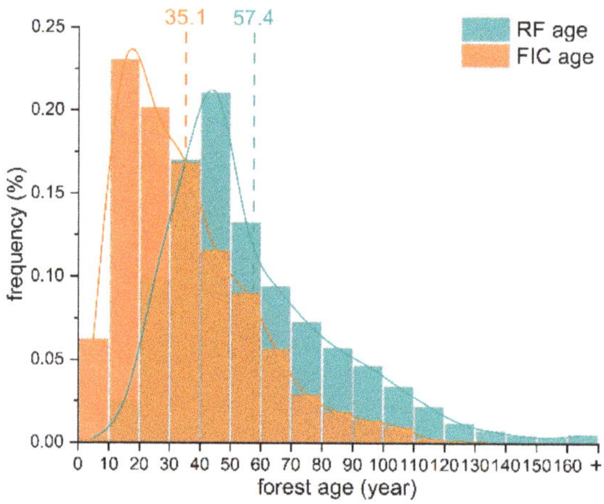


**Figure 5.** Frequency of RF forest ages and FIC forest ages.

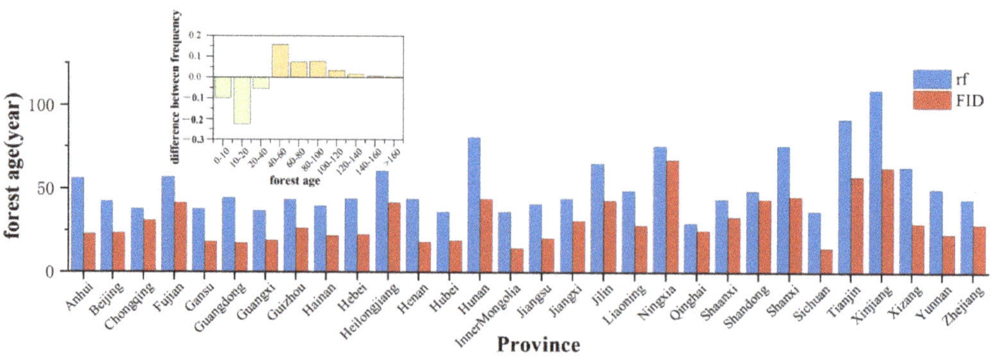


**Figure 6.** Deviations in the predicted forest age between the RF model and FID at the provincial scale. The top left small plot shows the age frequency difference between the FID and RF results.

### 3.2. Forest Age Maps after Forest Inventory Downscaling

We downscaled the provincial FID based on the RF age map to obtain an FIC age map (Figure 7). The overall forest age of Chinese forests decreased from 57.4 (standard deviation of 28.5 years) to 35.1 years (standard deviation of 21.9 years) (Figure 5). The average ages of China's coniferous forests, broad-leaved forests, and mixed coniferous and broad-leaved forests were 38.6 years, 32.8 years, and 37.6 years, respectively. The decrease in forest age over 60 years and the increase in forest age ranging from 0 to 60 years led to an increase in the proportion of middle-aged and young forests and a decrease in the proportion of old forests. Forests younger than 60 years account for 86.9% of the total national forests, among which forests aged 20–40 years account for the highest proportion, at 37.0%. These forests are distributed mainly in eastern and southern-central China, which is a densely populated region where forests are mostly planted, exhibiting a short growth period. Forests older than 60 years are distributed mainly in Heilongjiang, Jilin, Liaoning, Inner Mongolia, Shaanxi, Sichuan, and Tibet. These areas encompass the main mountain ranges in China, which are far from human settlements and are less disturbed. Trees in these areas grow and die naturally, and they encompass abundant age classes and are relatively old.

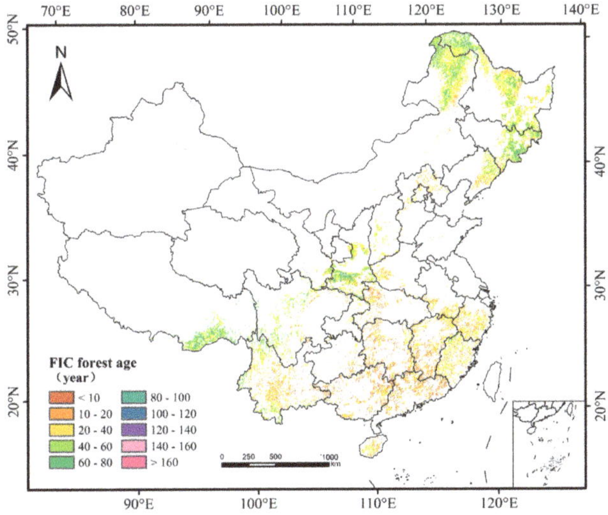


**Figure 7.** Spatial distribution map of FIC forest age. Because forest inventory is a time-consuming survey, the timing of forest inventory is not consistent across provinces (Table S3). To be precise, the applicable time for each province in this map is consistent with the survey time. We chose the middle year as the representative time, so this map applies to 2016.

At the provincial scale, the average forest age in the six provinces of Tianjin, Guangxi, Guangdong, Hunan, Hubei, and Guizhou is less than 20 years, whereas in three provinces of Qinghai, Tibet, and Xinjiang, the average forest age is more than 50 years. Among these provinces, Xinjiang Province has the oldest forest, reaching 94.3 years, whereas Tianjin Province has the youngest forest, at 17.3 years. Moreover, 50% of the provinces had an average forest age between 20 and 50 years.

The forest ages obtained from RF are generally older than those of the FID, and the FIC forest ages are similar to the FID results. However, the results for two provinces, namely Qinghai Province and Xinjiang Uygur Autonomous Region, are relatively poor, with ages that are 12 years older and 31 years younger, respectively, than the FID estimates, but the forest area in these two provinces is small, and the error can be accepted (Figure 8).

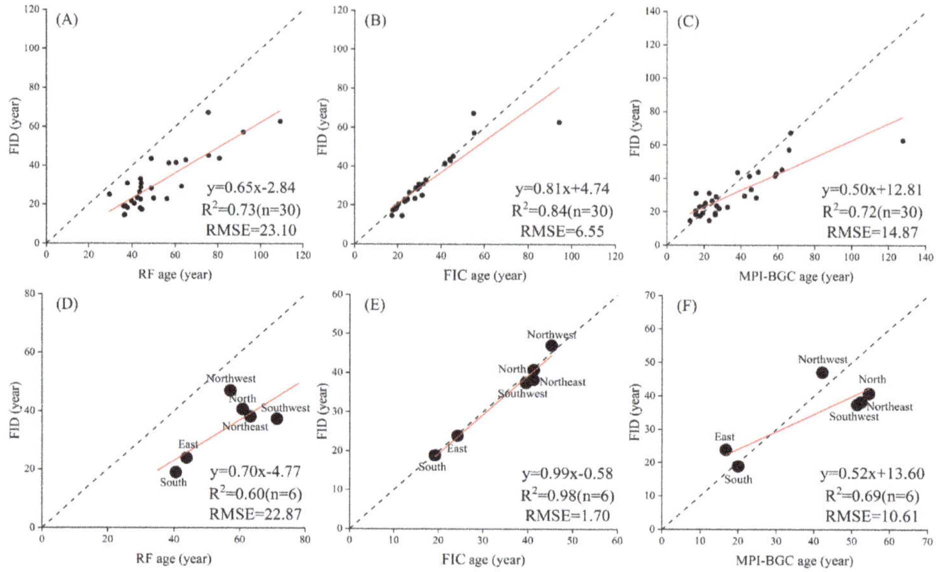


**Figure 8.** Differences between the real forest age dataset and the three forest age datasets at different spatial scales. (**A–C**) show the provincial-scale results, and (**D–F**) show the regional-scale results; the red solid line is the regression line, and the black dashed line is the 1:1 line.

## 4. Discussion

### 4.1. Differences between the Different Forest Age Datasets

In this study, a reliable 1 km resolution forest age dataset was obtained on the basis of a logical framework of data fusion and downscaling, and we compared the obtained dataset with the RF age dataset and the MPI-BGC dataset [47] at the provincial and regional scales. It was found that the FIC forest age dataset yielded better results than the other two datasets at the provincial scale, with reductions in prediction errors of 71.65% and 55.95%, respectively. It is noteworthy that although RF could estimate forest age well at the site scale, it suffers the potential problem of high systematic deviation at the large regional scale (e.g., at the provincial and national scales). This is because it was easily influenced by the sampling point data, so it produced regional estimation deviation [48]. The MPI-BGC forest age dataset, which is a global-scale forest age dataset, provided an acceptable forest age prediction effect in China. Specifically, for 21 out of 30 provinces, the prediction errors are less than 10 years compared to FID. However, a significant prediction error is also found in the Xinjiang Uygur Autonomous Region, where the predicted age is approximately double the actual age.

At the regional scale, the FIC forest age dataset was also the best, with $R^2$ and RMSE values of 0.98 and 1.7 years, respectively. Further analysis reveals that the forest ages in the northern and southwestern parts were overestimated by 3.1 and 2.1 years, respectively, compared to the FID estimates. Conversely, the forest age in the northwestern part was 1.7 years younger than the FID age, whereas the forest ages in the other three regions differed by no more than 0.5 years from the FID age.

In contrast, the RF age dataset performed the worst among the three datasets because it only integrated the remote sensing and ground station data without calibration based on the FID. The MPI-BGC dataset fused with the FID performed well in the China forest age estimation, but it does not consider tree height, which is highly correlated with forest age, so the estimation effect was also inferior to that of our FIC forest age dataset.

We also compared the RF forest age dataset and the MPI-BGC forest age dataset with the FIC forest age dataset in space (Figure 9). It was found that the forest ages predicted by RF were younger in Shanxi and Jiangsu Provinces than the FIC ages were, and the RF prediction values for the other 28 provinces were mostly older, especially in regions with older forests, such as the Changbai Mountain region in Jilin Province and the Hengduan Mountains region in southwest China.

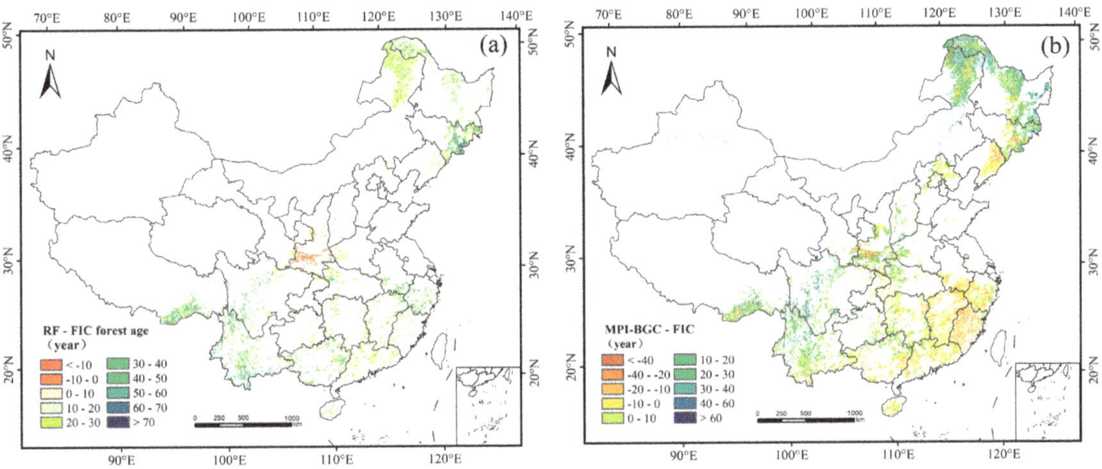


**Figure 9.** Spatial differences between the RF, MPI-BGC, and FIC forest ages. (**a**) shows spatial differences in RF and FIC forest ages, and (**b**) shows spatial differences in MPI-BGC and FIC forest ages.

A comparison between our optimized forest age (i.e., FIC age) dataset and the MPI-BGC forest age dataset revealed that the use of the latter mainly caused overestimation. The areas with forest age overestimation were distributed mainly in the northeastern and southwestern regions, whereas the areas with forest age underestimation were distributed mainly in the southeastern region. According to the percentage deviation of forest age, the age based on the MPI-BGC dataset was notably older for young forests, and the age in some pixels was even overestimated by more than a factor of 10 (Figure S3). This suggests that the forest carbon sink potential in China could be underestimated because the older the forest is, the lower its carbon sequestration capacity is [9,49]. Moreover, the FIC map was obtained by downscaling the FID, which suitably agreed with the FID forest age, while the MPI-BGC forest age deviated from the FID age, with nine provinces of the observed provincial age difference of more than 10 years (Figure S4). This discrepancy may be related not only to the variables selected for prediction but also to the FIC forest age dataset used, which provides a high spatial accuracy by fusing tree height data and the FID at the provincial scale.

Compared with existing forest age products, it is observed that many of them rely solely on remote sensing and field observation data, without incorporating forest inventory statistics as constraints, which leads to the inaccurate estimation of forest age at the national scale [11,16]. Therefore, it is imperative to employ FID as constraints for forest age estimation.

### 4.2. Analysis of the RF Parameters

Six variables were selected for estimating forest stand age in RF. The importance of the six variables decreased in the following order: tree height, MAT, forest type, forest origin, elevation, and MAP. Consistent with the findings of Zhang et al. and Chen et.al. [11,50], our results show that tree height is a key factor of the forest age model, and the importance and degree of influence on the estimation results were much greater than those of the other variables (Figure S2).

This is because canopy height is not only related to the growth period of trees but also the result of a combination of temperature, precipitation, and other environmental variables [31,51]. Canopy height already contains some environmental information. Many researchers and scholars have estimated forest age by extracting the time of disturbance (e.g., fire, logging) occurrence [50,52,53], but in fact, the change in canopy height also reveals a partial degree of information about such disturbances [54]. Therefore, it is effective to choose canopy height as a variable to predict forest age. Additionally, different forest attributes (e.g., forest type) also determine the age of forests, indicating the need to consider not only environmental variables but also forest attributes when estimating stand age.

### 4.3. Uncertainty and Prospects

Sufficient and representative site data are needed to estimate the distribution pattern of forest age by fusing remotely sensed data and site data via the RF model. Although most site data collected in this study pertain to forests younger than 200 years, there are also sites with forest ages greater than 200 years, constituting 1.26% of the total. However, the research scale of this study is 1 km, which is much larger than that of the 10 m × 10 m and 20 m × 20 m rectangular sample plots commonly used in investigations. With respect to forests that are constantly updated with successional characteristics, the overall age of forest stands is typically not very old. Therefore, sites without old forest ages do not greatly influence the spatial distribution trend in forest age. Moreover, the collected site data cover all age groups, which can suitably reflect the relationships between forest age and tree height and other related variables.

The FID reflects growth in Chinese forests through the investigation of fixed sample plots, but unfortunately, it is regrettable that only aggregated data of provincial administrative regions and forest age are classified into five classes, namely young, middle-aged, near-mature, mature, and overmature forests, with no specific forest age. The median age of each age class was used as the forest age of the considered period. However, this may be subject to uncertainty because the forest age of a given period is not necessarily uniformly distributed. With respect to age estimation for overmature forests, we only added an age class value to the minimum age, which may lead to the underestimation of the age of older forests. Fortunately, the area of old-growth forests in China constitutes a relatively small proportion, with overmature forests accounting for merely 6.4% of the total [36].

There is uncertainty in the forest area, especially the area of planted forests [55]. Since the 1970s, China has been committed to afforestation activities, and the area of planted forests has continuously increased. According to the ninth National Forest Resources Inventory of China, the areas of planted and natural forests are 57 million hectares and 123 million hectares, accounting for 31.8% and 68.2%, respectively, of the total forest area of China. However, the forest base map used only contains a plantation area of 25 million hectares, accounting for 11.7% of the total forest area (Figure S5). The age of planted forests in China is 13.9 years, while the age of natural forests is 37.5 years. Notably, there is a large difference between these two types of forests. Treating planted forests as natural forests could lead to forest age overestimation; for example, the overestimation of the age of planted forests was very obvious with the MPI-BGC forest age dataset, with 60.0% of planted forests overestimated by more than 50%, which is approximately twice as high as that of the natural forests (34.0%). The age of planted forests is also influenced by human activities, and the difference in age between managed and unmanaged planted forests can range from 5 to 10 years [29]. This fact further complicates the accurate assessment of the ecological characteristics of planted forests. Many studies have estimated carbon sinks in Chinese forests based on the relationship between stand age and biomass [19,21], and the overestimation of stand age in planted forests will undoubtedly challenge the accuracy of these estimates, mainly leading to the underestimation of carbon sinks. Therefore, when using the forest age map, it should be used in conjunction with suitable datasets to avoid inaccurate results due to information mismatch [16].

Chinese planted forests are currently dominated by young and middle-aged forests [36], which provide a larger carbon sink potential than older forests. However, most of the existing models do not consider the potential impact of forest age on carbon sinks [17]. In terms of model development, the parameterization of process models is usually based on mature forests [14], while few models have focused sufficiently on the role of forest age and succession stage when estimating forest carbon dynamics, making it difficult for these models to accurately reproduce dynamic trajectories of changes in carbon cycle components, such as biomass, or other age-related changes in forest age [56]. Therefore, the spatial distribution map of forest age based on RF and FID could provide an important basis for the accurate prediction of carbon sinks in China. In the future, more detailed forest vegetation type maps and additional factors affecting forest age should be considered to obtain more detailed and accurate forest age distribution maps.

## 5. Conclusions

In this study, we retrieved a high-accuracy and 1 km resolution forest map of China by fusing multiscale age information. RF can effectively reveal the intrinsic relationships between variables such as forest age and tree height in China ($R^2$ = 0.78,RMSE = 13.0). However, there is a certain degree of overestimation, primarily due to the significant spatial variability in forest age and the tendency to select well-growing stands during plot sampling. Therefore, large-scale statistical corrections are necessary when simulating spatial patterns. According to the FIC map, it is highly consistent with FID at the provincial scale and national scale. The average age of forests in China is 35.1 years, with high spatial heterogeneity. Forest age is generally older in the northeastern and southwestern mountain regions than in the southern and eastern regions. Overall, the forest age distribution map is important for estimating the current status and potential of forest carbon sinks in China.

**Supplementary Materials:** The following supporting information can be downloaded at: https://www.mdpi.com/article/10.3390/f15081290/s1, Figure S1: Temperature and precipitation in China, 1970–2014; Figure S2: Importance of random forest variables; Figure S3: Percentage deviation between MPI-BGC age and FIC age; Figure S4: Differences between the different forest age datasets at the provincial scale; Figure S5: Proportion of artificial and natural forests in different provinces; Table S1: Criteria for Surveying Dominant Tree Species (Groups) by Age Class; Table S2: The average age of different forest age classes of dominant tree species; Table S3: Year of Forest Inventory in Each Province.

**Author Contributions:** Conceptualization, T.Z.; Methodology, Y.X. and J.Z. (Jingyu Zeng); Software, Y.Z. and Q.L.; Validation, X.L.; Formal analysis, Y.X., Q.L. and J.Z. (Jingzhou Zhang); Investigation, Y.Z. and J.Z. (Jingzhou Zhang); Resources, J.Z. (Jingyu Zeng), H.L. and X.L.; Data curation, Y.X. and H.L.; Writing—original draft, Y.X.; Writing—review & editing, T.Z. and J.Z. (Jingyu Zeng); Visualization, Y.X.; Supervision, T.Z.; Project administration, T.Z.; Funding acquisition, T.Z. All authors have read and agreed to the published version of the manuscript.

**Funding:** This study was supported by the National Natural Science Foundation of China (42277206) and the second Tibetan Plateau Scientific Expedition and Research Program (2019QZKK0405).

**Data Availability Statement:** The data underlying this article will be shared on reasonable request to the corresponding author.

**Acknowledgments:** We thank all the data providers for the field observations and explicit environmental dataset.

**Conflicts of Interest:** The authors declare no conflict of interest.

## References

1. Bonan, G.B. Forests and Climate Change: Forcings, Feedbacks, and the Climate Benefits of Forests. *Science* **2008**, *320*, 1444–1449. [CrossRef] [PubMed]
2. Liu, W.; Wang, X.; Lu, F.; Ouyang, Z. Regional and global estimates of carbon stocks and carbon sequestration capacity in forest ecosystems: A review. *Chin. J. Appl. Ecol.* **2015**, *26*, 2881–2890. [CrossRef]

3. Pan, Y.; Birdsey, R.A.; Fang, J.; Houghton, R.; Kauppi, P.E.; Kurz, W.A.; Phillips, O.L.; Shvidenko, A.; Lewis, S.L.; Canadell, J.G.; et al. A Large and Persistent Carbon Sink in the World's Forests. *Science* **2011**, *333*, 988–993. [CrossRef] [PubMed]
4. Yu, G.; Chen, Z.; Piao, S.; Peng, C.; Ciais, P.; Wang, Q.; Li, X.; Zhu, X. High Carbon Dioxide Uptake by Subtropical Forest Ecosystems in the East Asian Monsoon Region. *Proc. Natl. Acad. Sci. USA* **2014**, *111*, 4910–4915. [CrossRef] [PubMed]
5. Xu, L.; Shi, Y.; Fang, H.; Zhou, G.; Xu, X.; Zhou, Y.; Tao, J.; Ji, B.; Xu, J.; Li, C.; et al. Vegetation Carbon Stocks Driven by Canopy Density and Forest Age in Subtropical Forest Ecosystems. *Sci. Total Environ.* **2018**, *631–632*, 619–626. [CrossRef] [PubMed]
6. Pan, Y.; Luo, T.; Birdsey, R.; Hom, J.; Melillo, J. New Estimates of Carbon Storage and Sequestration in China's Forests: Effects of Age-Class and Method On Inventory-Based Carbon Estimation. *Clim. Chang.* **2004**, *67*, 211–236. [CrossRef]
7. Liu, Y.; Yu, G.; Wang, Q.; Zhang, Y. How Temperature, Precipitation and Stand Age Control the Biomass Carbon Density of Global Mature Forests: Controls on Biomass Carbon Density of Global Mature Forests. *Glob. Ecol. Biogeogr.* **2014**, *23*, 323–333. [CrossRef]
8. Tang, X.; Li, H.; Ma, M.; Yao, L.; Peichl, M.; Arain, A.; Xu, X.; Goulden, M. How Do Disturbances and Climate Effects on Carbon and Water Fluxes Differ between Multi-Aged and Even-Aged Coniferous Forests? *Sci. Total Environ.* **2017**, *599–600*, 1583–1597. [CrossRef]
9. Chen, Z.; Yu, G.; Wang, Q. Effects of Climate and Forest Age on the Ecosystem Carbon Exchange of Afforestation. *J. For. Res.* **2020**, *31*, 365–374. [CrossRef]
10. Pan, Y.; Chen, J.M.; Birdsey, R.; McCullough, K.; He, L.; Deng, F. Age Structure and Disturbance Legacy of North American Forests. *Biogeosciences* **2011**, *8*, 715–732. [CrossRef]
11. Zhang, Y.; Yao, Y.; Wang, X.; Liu, Y.; Piao, S. Mapping Spatial Distribution of Forest Age in China. *Earth Space Sci.* **2017**, *4*, 108–116. [CrossRef]
12. Köhl, M.; Neupane, P.R.; Lotfiomran, N. The impact of tree age on biomass growth and carbon accumulation capacity: A retrospective analysis using tree ring data of three tropical tree species grown in natural forests of Suriname. *PLoS ONE* **2017**, *12*, e0181187. [CrossRef] [PubMed]
13. Guo, Z.; Hu, H.; Li, P.; Li, N.; Fang, J. Spatio-Temporal Changes in Biomass Carbon Sinks in China's Forests from 1977 to 2008. *Sci. China Life Sci.* **2013**, *56*, 661–671. [CrossRef] [PubMed]
14. Liu, Y.; Yu, G.; Wang, Q.; Zhang, Y.; Xu, Z. Carbon Carry Capacity and Carbon Sequestration Potential in China Based on an Integrated Analysis of Mature Forest Biomass. *Sci. China Life Sci.* **2014**, *57*, 1218–1229. [CrossRef] [PubMed]
15. Shang, R.; Chen, J.M.; Xu, M.; Lin, X.; Li, P.; Yu, G.; He, N.; Xu, L.; Gong, P.; Liu, L.; et al. China's Current Forest Age Structure Will Lead to Weakened Carbon Sinks in the near Future. *Innovation* **2023**, *4*, 100515. [CrossRef]
16. Zhang, C.; Ju, W.; Chen, J.M.; Li, D.; Wang, X.; Fan, W.; Li, M.; Zan, M. Mapping Forest Stand Age in China Using Remotely Sensed Forest Height and Observation Data. *J. Geophys. Res. Biogeosci.* **2014**, *119*, 1163–1179. [CrossRef]
17. Yao, Y.; Li, Z.; Wang, T.; Chen, A.; Wang, X.; Du, M.; Jia, G.; Li, Y.; Li, H.; Luo, W.; et al. A New Estimation of China's Net Ecosystem Productivity Based on Eddy Covariance Measurements and a Model Tree Ensemble Approach. *Agric. For. Meteorol.* **2018**, *253*, 84–93. [CrossRef]
18. Alexandrov, G.A. Carbon Stock Growth in a Forest Stand: The Power of Age. *Carbon Balance Manag.* **2007**, *2*, 4. [CrossRef]
19. Xu, B.; Guo, Z.; Piao, S.; Fang, J. Biomass Carbon Stocks in China's Forests between 2000 and 2050: A Prediction Based on Forest Biomass-Age Relationships. *Sci. China Life Sci.* **2010**, *53*, 776–783. [CrossRef]
20. He, N.; Wen, D.; Zhu, J.; Tang, X.; Xu, L.; Zhang, L.; Hu, H.; Huang, M.; Yu, G. Vegetation Carbon Sequestration in Chinese Forests from 2010 to 2050. *Glob. Chang. Biol.* **2017**, *23*, 1575–1584. [CrossRef]
21. Zhang, C.; Ju, W.; Chen, J.; Fang, M.; Wu, M.; Chang, X.; Wang, T.; Wang, X. Sustained Biomass Carbon Sequestration by China's Forests from 2010 to 2050. *Forests* **2018**, *9*, 689. [CrossRef]
22. Pregitzer, K.S.; Euskirchen, E.S. Carbon Cycling and Storage in World Forests: Biome Patterns Related to Forest Age. *Glob. Change Biol.* **2004**, *10*, 2052–2077. [CrossRef]
23. Li, D.; Ju, W.; Fan, W.; Gu, Z. Estimating the Age of Deciduous Forests in Northeast China with Enhanced Thematic Mapper Plus Data Acquired in Different Phenological Seasons. *JARS* **2014**, *8*, 083670. [CrossRef]
24. Spracklen, B.; Spracklen, D.V. Synergistic Use of Sentinel-1 and Sentinel-2 to Map Natural Forest and Acacia Plantation and Stand Ages in North-Central Vietnam. *Remote Sens.* **2021**, *13*, 185. [CrossRef]
25. Dai, M.; Zhou, T.; Yang, L.; Jia, G. Spatial pattern of forest ages in China retrieved from national-level inventory and remote sensing imageries. *Geogr. Res.* **2011**, *30*, 172–184. [CrossRef]
26. Xiao, Y.; Wang, Q.; Tong, X.; Atkinson, P.M. Thirty-Meter Map of Young Forest Age in China. *Earth Syst. Sci. Data* **2023**, *15*, 3365–3386. [CrossRef]
27. Piao, S.; He, Y.; Wang, X.; Chen, F. Estimation of China's Terrestrial Ecosystem Carbon Sink: Methods, Progress and Prospects. *Sci. China Earth Sci.* **2022**, *65*, 641–651. [CrossRef]
28. Xu, W.; Jin, X.; Liu, J.; Yang, X.; Ren, J.; Zhou, Y. Analysis of Spatio-Temporal Changes in Forest Biomass in China. *J. For. Res.* **2022**, *33*, 261–278. [CrossRef]
29. Yu, Z.; Zhao, H.; Liu, S.; Zhou, G.; Fang, J.; Yu, G.; Tang, X.; Wang, W.; Yan, J.; Wang, G.; et al. Mapping Forest Type and Age in China's Plantations. *Sci. Total Environ.* **2020**, *744*, 140790. [CrossRef]
30. Racine, E.B.; Coops, N.C.; St-Onge, B.; Bégin, J. Estimating Forest Stand Age from LiDAR-Derived Predictors and Nearest Neighbor Imputation. *For. Sci.* **2014**, *60*, 128–136. [CrossRef]

31. Lin, X.; Shang, R.; Chen, J.M.; Zhao, G.; Zhang, X.; Huang, Y.; Yu, G.; He, N.; Xu, L.; Jiao, W. High-Resolution Forest Age Mapping Based on Forest Height Maps Derived from GEDI and ICESat-2 Space-Borne Lidar Data. *Agric. For. Meteorol.* **2023**, *339*, 109592. [CrossRef]
32. Yang, X.; Liu, Y.; Wu, Z.; Yu, Y.; Li, F.; Fan, W. Forest Age Mapping Based on Multiple-Resource Remote Sensing Data. *Environ. Monit Assess* **2020**, *192*, 734. [CrossRef] [PubMed]
33. Guo, Q.; Ren, H. Productivity as Related to Diversity and Age in Planted versus Natural Forests. *Glob. Ecol. Biogeogr.* **2014**, *23*, 1461–1471. [CrossRef]
34. Luo, Y.; Zhang, X.; Wang, X.; Lu, F. Biomass and Its Allocation of Chinese Forest Ecosystems. *Ecology* **2014**, *95*, 2026. [CrossRef]
35. Peng, S.; Ding, Y.; Liu, W.; Li, Z. 1 Km Monthly Temperature and Precipitation Dataset for China from 1901 to 2017. *Earth Syst. Sci. Data* **2019**, *11*, 1931–1946. [CrossRef]
36. Zhang, J. *National Forestry and Grassland Administration China Forest Resources Report (2014–2018)*; China Forestry Publishing House: Beijing, China, 2019; ISBN 978-7-5038-9982-9.
37. Liu, X.; Su, Y.; Hu, T.; Yang, Q.; Liu, B.; Deng, Y.; Tang, H.; Tang, Z.; Fang, J.; Guo, Q. Neural Network Guided Interpolation for Mapping Canopy Height of China's Forests by Integrating GEDI and ICESat-2 Data. *Remote Sens. Environ.* **2022**, *269*, 112844. [CrossRef]
38. Thom, D.; Rammer, W.; Seidl, R. The Impact of Future Forest Dynamics on Climate: Interactive Effects of Changing Vegetation and Disturbance Regimes. *Ecol. Monogr.* **2017**, *87*, 665–684. [CrossRef] [PubMed]
39. Seidl, R.; Thom, D.; Kautz, M.; Martin-Benito, D.; Peltoniemi, M.; Vacchiano, G.; Wild, J.; Ascoli, D.; Petr, M.; Honkaniemi, J.; et al. Forest Disturbances under Climate Change. *Natrue Clim. Chang.* **2017**, *7*, 395–402. [CrossRef]
40. Peng, S. 1-Km Monthly Precipitation Dataset for China (1901–2021) [Dataset]. A Big Earth Data Platform for Three Poles. 2020. Available online: https://poles.tpdc.ac.cn/en/data/faae7605-a0f2-4d18-b28f-5cee413766a2/ (accessed on 8 July 2024).
41. Peng, S. *1-Km Monthly Mean Temperature Dataset for China (1901–2022)*; National Tibetan Plateau Data Center: Beijing, China, 2019. [CrossRef]
42. Tang, G. Digital Elevation Model of China (1KM). National Tibetan Plateau/Third Pole Environment Data Center. 2019. Available online: https://data.tpdc.ac.cn/zh-hans/data/12e91073-0181-44bf-8308-c50e5bd9a734/ (accessed on 8 July 2024).
43. Breiman, L. Random Forests. *Mach. Learn.* **2001**, *45*, 5–32. [CrossRef]
44. Zeng, J.; Zhou, T.; Wang, Q.; Xu, Y.; Lin, Q.; Zhang, Y.; Wu, X.; Zhang, J.; Liu, X. Spatial Patterns of China's Carbon Sinks Estimated from the Fusion of Remote Sensing and Field-Observed Net Primary Productivity and Heterotrophic Respiration. *Ecol. Inform.* **2023**, *76*, 102152. [CrossRef]
45. Wang, Y.; Peng, Y.; Hu, X.; Zhang, P. Fine-Resolution Forest Height Estimation by Integrating ICESat-2 and Landsat 8 OLI Data with a Spatial Downscaling Method for Aboveground Biomass Quantification. *Forests* **2023**, *14*, 1414. [CrossRef]
46. Xu, H.; He, B.; Guo, L.; Yan, X.; Zeng, Y.; Yuan, W.; Zhong, Z.; Tang, R.; Yang, Y.; Liu, H.; et al. Global Forest Plantations Mapping and Biomass Carbon Estimation. *J. Geophys. Res. Biogeosciences* **2024**, *129*, e2023JG007441. [CrossRef]
47. Besnard, S.; Koirala, S.; Santoro, M.; Weber, U.; Nelson, J.; Gütter, J.; Herault, B.; Kassi, J.; N'Guessan, A.; Neigh, C.; et al. Mapping Global Forest Age from Forest Inventories, Biomass and Climate Data. *Earth Syst. Sci. Data* **2021**, *13*, 4881–4896. [CrossRef]
48. Couronné, R.; Probst, P.; Boulesteix, A.-L. Random Forest versus Logistic Regression: A Large-Scale Benchmark Experiment. *BMC Bioinform.* **2018**, *19*, 270. [CrossRef] [PubMed]
49. Peichl, M.; Arain, M.A.; Brodeur, J.J. Age Effects on Carbon Fluxes in Temperate Pine Forests. *Agric. For. Meteorol.* **2010**, *150*, 1090–1101. [CrossRef]
50. Cheng, K.; Chen, Y.; Xiang, T.; Yang, H.; Liu, W.; Ren, Y.; Guan, H.; Hu, T.; Ma, Q.; Guo, Q. A 2020 Forest Age Map for China with 30 m Resolution. *Earth Syst. Sci. Data* **2024**, *16*, 803–819. [CrossRef]
51. Schumacher, J.; Hauglin, M.; Astrup, R.; Breidenbach, J. Mapping Forest Age Using National Forest Inventory, Airborne Laser Scanning, and Sentinel-2 Data. *For. Ecosyst.* **2020**, *7*, 60. [CrossRef]
52. Li, P.; Li, H.; Si, B.; Zhou, T.; Zhang, C.; Li, M. Mapping Planted Forest Age Using LandTrendr Algorithm and Landsat 5–8 on the Loess Plateau, China. *Agric. For. Meteorol.* **2024**, *344*, 109795. [CrossRef]
53. Zhang, S.; Xu, H.; Liu, A.; Qi, S.; Hu, B.; Huang, M.; Luo, J. Mapping of Secondary Forest Age in China Using Stacked Generalization and Landsat Time Series. *Sci. Data* **2024**, *11*, 302. [CrossRef]
54. Senf, C.; Mori, A.S.; Müller, J.; Seidl, R. The Response of Canopy Height Diversity to Natural Disturbances in Two Temperate Forest Landscapes. *Landsc. Ecol.* **2020**, *35*, 2101–2112. [CrossRef]
55. Wang, W.; Tang, X.; Huang, M.; Zhou, G.; Ying, G.; Wang, J.; Wen, D. *Forest Ecosystem Carbon Stocks in China—Dynamics and Mechanisms*; Science Press: Beijing, China, 2018; ISBN 978-7-5088-5393-2.
56. Anderson-Teixeira, K.J.; Miller, A.D.; Mohan, J.E.; Hudiburg, T.W.; Duval, B.D.; DeLucia, E.H. Altered Dynamics of Forest Recovery under a Changing Climate. *Glob. Chang. Biol.* **2013**, *19*, 2001–2021. [CrossRef] [PubMed]

*land*

MDPI

*Article*

# Construction of Ecological Security Patterns and Evaluation of Ecological Network Stability under Multi-Scenario Simulation: A Case Study in Desert–Oasis Area of the Yellow River Basin, China

**Wenhao Cheng [1,2], Caihong Ma [2,*], Tongsheng Li [1] and Yuanyuan Liu [2]**

1 College of Urban and Environment Sciences, Northwest University, Xi'an 710127, China; chengwh@nxu.edu.cn (W.C.); leetang@nwu.edu.cn (T.L.)

2 College of Geography and Planning, Ningxia University, No.539 Helanshan West Road, Yinchuan 750021, China; 12022130985@stu.nxu.edu.cn

* Correspondence: mchyanni@nxu.edu.cn

**Abstract:** Land use change has a significant impact on the sustainability of ecosystems, and ecological security patterns (ESPs) can improve environmental quality through spatial planning. This study explored a multi-scenario ESP framework by integrating future land use simulation (FLUS) and minimum cumulative resistance (MCR) for urban agglomeration along the Yellow River Basin (YRB) in Ningxia. The research involved simulating land use change in 2035 under four development scenarios, identifying ecological security networks, and evaluating network stability for each scenario. The study revealed that the ecological sources under different development scenarios, including a natural development scenario (NDS), an economic development scenario (EDS), a food security scenario (FSS), and an ecological protection scenario (EPS), were 834.82 km$^2$, 715.46 km$^2$, 785.56 km$^2$, and 1091.43 km$^2$, respectively. The overall connectivity values ($O_G$) for these scenarios were 0.351, 0.466, 0.334, and 0.520, respectively. It was found that under an EPS, the ESPs had the largest area of ecological sources and the most stable ecological network structure, which can effectively protect natural habitats. This study provides a valuable method for identifying ESPs that can respond to diversity and the uncertainty of future development. It can assist decision-makers in enhancing the ecological quality of the study area while considering various development scenarios.

**Keywords:** ecological protection; scenario analysis; minimum cumulative resistance; ecological stability evaluation; Yellow River Basin

## 1. Introduction

With rapid increases in economic development and urbanization, land use has undergone unprecedented change, leading to a significant disruption of the ecosystem [1]. This has resulted in landscape fragmentation, urban heat islands, and biodiversity loss, all of which directly or indirectly affect the stability and sustainability of ecological patterns [2,3]. Therefore, the construction of ecological security pattern (ESP) and the evaluation of ecological quality have been developed to address regional ecological issues and provide valuable support for regional management and planning [4].

The establishment of ESP, derived from landscape ecological planning and combined with land use patterns for the construction and optimization of ecological networks, serves as an efficient spatial pathway for balancing ecological conservation and socioeconomic development [5]. Currently, ESP research has evolved into a general paradigm of "identifying the source areas, constructing the resistance surfaces, and extracting the corridors" [6]. There are two major approaches to identify ecological sources: structure-oriented methods, which considers the structural connectivity between different patches [7], and function-oriented methods, which are based on the significance of the indicators; patches with high ecological significance are identified as different from the surrounding landscape matrix [8]. Compared to function-oriented methods, structure-oriented methods used for establishing ESPs have been demonstrated to enhance network connectivity despite changes in scale [9]. The MSPA approach is widely utilized, as it automatically classifies the pixel data of focal feature classes into a new structural connectivity feature class [10]. When assessing ecological resistance surfaces, a comprehensive resistance surface map is created using the analytic hierarchy process (AHP) and Delphi method, which have supplemented expert-assigned resistance values [11]. Ecological corridors are key channels for ecological flow and diffusion, and the minimum cumulative resistance (MCR) model has been extensively applied because it considers landscape heterogeneity and horizontal connections between landscape units [12]. However, the construction of ESP based on the current perspective often fails to reflect spatiotemporal variations in the impacts of natural and artificial factors on the landscape. Models such as the future land use simulation (FLUS) can utilize self-adaptive inertia and competition mechanisms to analyze the complexities and indeterminacies of interactions and competition among different land types, offering better support for simulating future land use change across various periods and scenarios [13]. Gradually, an increasing number of studies have simulated future ESPs by exploring potential ecological sources and corridors. It subsequently offers recommendations for the restoration and optimization of ecological networks.

The evaluation of ESP quality is a prerequisite for enhancing the structural stability of the landscape. Based on graph theory, the metrics used to evaluate ESPs for structure-oriented methods can be classified into centrality metrics and connectivity metrics. Centrality metrics describe the centrality degree and comprehensive influence of a local area in the overall network, such as the $K_p$ index, $L_p$ index, and $M_p$ index [14]. Connectivity metrics reflect the ecological process and ecological connection degree of the landscape, such as the *ɋ* index, *ß* index, ɤ index, and $G$ index [15]. Shen and Wu developed the evaluation system to refer to the aspects of centrality and connectivity. It contained the indexes of average node degree ($k$), average shortest path ($I$), global efficiency ($E_G$), and clustering coefficient ($cc$). Moreover, an overall value ($O_G$) was introduced to comprehensively quantify the structure of the ESPs, representing an equally weighted result of the assessment of $k$, $I$, $E_G$, and $cc$ [9].

The Yellow River Basin (YRB) is the cradle of Chinese civilization and has played a crucial role in maintaining ecological and socioeconomic sustainability [16]. The urban agglomeration along the YRB in Ningxia is a unique desert–oasis area, and holds strategic importance for ensuring stability in the northwestern Chinese frontier. However, due to fragile natural conditions and extensive development, human–land conflict relationships have emerged as a major challenge [17,18]. Therefore, it is necessary to undertake the construction and optimization of the ESPs of urban agglomeration along the YRB in Ningxia as a urgent task in the future [19]. Given the complexity and uncertainty of future development, the land use change and ESP construction could potentially impact a range of various future scenarios. Hence, we attempted to solve this knowledge gap by addressing two key scientific inquiries: how can we predict future land use change under various scenarios, and how can we construct and evaluate ESPs? With a focus on these questions, this study developed a multi-scenario ESP framework to accommodate future land use through the integration of FLUS and MCR models. The major objectives of this study were as follows: (1) to simulate land use change under various scenarios in 2035, the year when China is projected to reach peak urbanization, using the FLUS model; (2) to develop the ESPs by identifying ecological sources through MSPA and landscape connectivity analysis (LCA), as well as extracting ecological corridors using MCR and gravity models; and (3) to assess the stability of the ESPs from comprehensive perspectives.

## 2. Materials and Methods

### 2.1. Study Area and Data

#### 2.1.1. Study Area

The Ningxia Hui Autonomous Region is located in the upper reach of the YRB in northwestern China, ranging from 104°17′–107°39′ E longitude and 35°14′–39°23′ N latitude. It is characterized as a temperate continental climate and classified as an arid or semiarid region. Based on the guiding document, *Development Plan of the Urban Belt along the Yellow River Economic Zone*, the urban agglomeration along the YRB in Ningxia includes 13 counties (districts) (Figure 1), with a total area of 22,700 km$^2$, a resident population of 5.05 million and a GDP of 44.95 billion US dollars in 2020. Although the study area only represents 34.2% of the total area in Ningxia, it accounts for 70.1% of the population and 88.2% of the GDP. The region is traversed by the Yellow River from southwest to northeast, creating several wetland parks with diverse biodiversity [20,21], despite being bordered by the Tengger Desert, Ulan Buh Desert, and Mu Us Sandy Land. It is delineated by Qingtong Canyon into the Yinchuan Plain in the north and the Weining Plain in the south. The primary land use types are grassland and cropland, which make up 45.8% and 24.4%, respectively. Other land use types, such as unused land, construction land, forestland, and water body, account for 10.6%, 8.4%, 5.6%, and 5.1%, respectively. The forestland and water body are classified as the ecological land use types, providing the foods and habitats for different kinds of mammals, birds, and fishes. It promotes biological diversity by species diffusion. Cropland is primarily utilized to produce grain and meat for human beings. Construction land, including urban and rural transportation land types, serves as the main location for human activities. Grassland is used for grazing or remains unused due to low vegetation cover in arid and semi-arid regions.

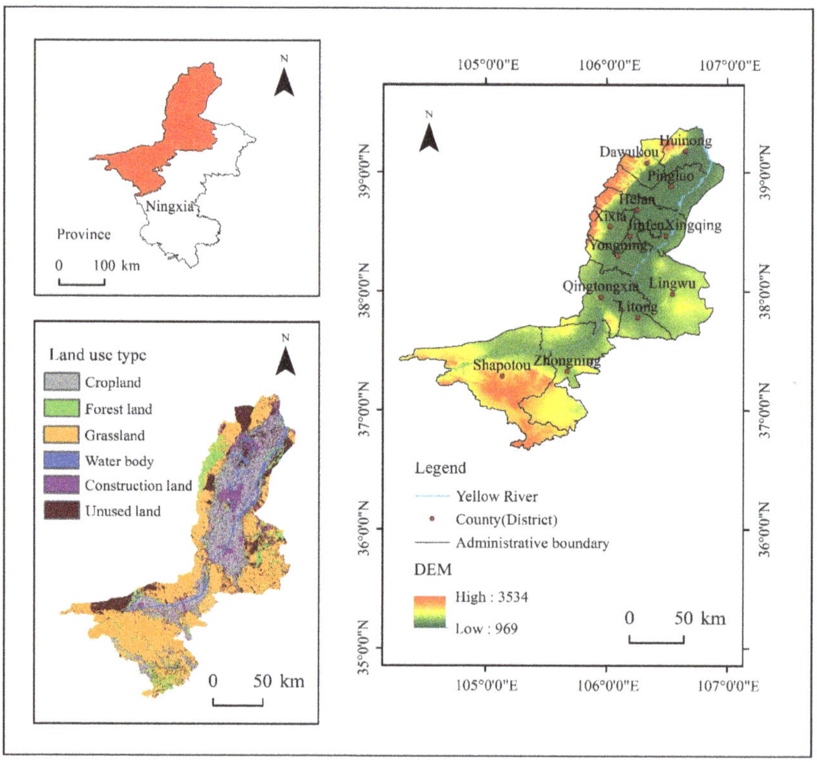


**Figure 1.** Study area.

2.1.2. Data Sources and Processing

The main data and their formats, sources and other key information were shown in Table 1. In this study, the land use types were divided into six categories: cropland, forestland, grassland, water body, construction land and unused land. The grids used here were resampled to 90 m × 90 m for further calculation.

**Table 1.** Main data used in the study.

| Data | Format | Source |
|---|---|---|
| Land use type | Vector data in 2020 | Department of Ningxia Nature Resource |
| Digital elevation model (DEM) | Grid at 30 m in 2020 | http://www.gscloud.cn/ |
| Slope | Grid at 90 m in 2020 | http://www.gscloud.cn/ |
| Soil erosion | Grid at 1000 m in 2020 | http://www.resdc.cn |
| NDVI | Grid at 1000 m in 2020 | http://www.resdc.cn |
| Road and railway | Line in 2020 | http://openstreetmap.org/ |
| River | Line in 2020 | http://openstreetmap.org/ |
| Population density | Grid at 1000 m in 2019 | http://www.worldpop.org/ |
| GDP density | Grid at 1000 m in 2019 | http://www.resdc.cn |
| Night light data | Grid at 1000 m in 2019 | http://www.resdc.cn |
| Precipitation | Point in 2011–2020 | http://www.cma.cn/ |
| Administrative boundary | Polygons in 2020 | http://www.hydrosheds.org/ |

*2.2. Methods*

The flowchart of this study is shown in Figure 2, which consists of three major steps: multi-scenario land use simulation (see Section 2.2.1), ESP construction (see Section 2.2.2), and ESP network connectivity evaluation (see Section 2.2.3).

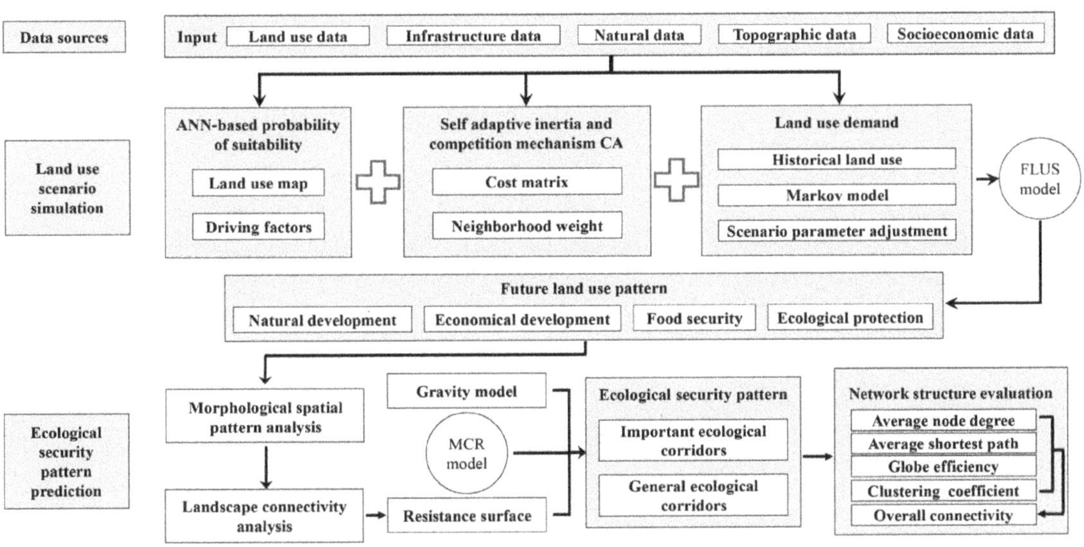


**Figure 2.** Framework of the integration of future ecological security patterns (ESPs) for land use simulation.

2.2.1. Land Use Simulation under Multiple Scenarios

The FLUS model, based on the traditional cellular automata principle (CA-Markov), can effectively simulate future land use change. It contains three modules: (1) an artificial neural network (ANN) algorithm to obtain the transition probability of each type; (2) self-adaptive internal and competition mechanisms to analyze the mutual competition and interaction between land use types according to a transfer matrix; and (3) a Markov model to predict the demand of future land use types [13,22].

Transition Probability of Land Use Change

The transition probability of land use change is calculated in the output layer using the ANN algorithm combined with driving factors. The equations are as follows [22,23]:

$$p(p,k,t) = \sum_j w_j k \times sigmoid\left(net_j(p,t)\right)$$

$$sigmoid\left(net_j(p,t)\right) = \frac{1}{1+e^{-net_j(p,t)}} \qquad net_j(p,t) = \sum_i w_{ij} \times \left(x_j(p,t)\right)$$

where $p(p,k,t)$ means the transition probability of land use type $k$ on grid $p$ at time $t$; $w_j$ $k$ means the weight between layers of hidden and output; sigmoid means the function; $net_j(p,t)$ means the receipt signal of neuron $j$; $w_{ij}$ means the weight between layers of hidden and input; and $x_i(_{p,t})$ is the variable of input neuron $i$.

The total probability from one type converted to another is computed as follows [24]:

$$TP^t_{p,k} = P^t_{p,k} \times \alpha^t_{p,k} \times Inertia^t_k \times (1 - SC_{c\to k})$$

where $TP^t{}_{p,k}$ means the total probability from one land use type to another land use type $k$ on grid $p$ at time $t$; $a^t{}_{p,k}$ means neighborhood effect; $Inertia^t{}_k$ means the inertia coefficient from one land use type to type $k$; $SC_{c\to k}$ means the cost from $c$ to $k$.

Considering the existing research results and the specific situation in the study area [19,22], seven factors, namely the DEM, slope, NDVI, nightlight index, distance from roads, GDP and population density, were selected to calculate the probability of land suitability.

FLUS Models' Parameter Setting

The parameter setting for the self-adaptive internal and competition mechanisms of CA contains the land use transfer matrix, and neighborhood effect.

The land use transfer matrix is used to express the mutual conversion of land use types [25]. A value of 1 indicates that a land type could be converted, and a value of 0 indicates the opposite. The land use transfer matrices were determined according to the actual conditions of the urban agglomeration along the YRB in Ningxia (Table 2).

**Table 2.** Land use transfer matrices of the four scenarios.

| | **NDS** | | | | | **EDS** | | | | | **FSS** | | | | | **EPS** | | | | |
|---|---|---|---|---|---|---|---|---|---|---|---|---|---|---|---|---|---|---|---|---|
| | **a** | **b** | **c** | **d** | **e** | **a** | **b** | **c** | **d** | **e** | **a** | **b** | **c** | **d** | **e** | **a** | **b** | **c** | **d** | **e** |
| **a** | 1 | 1 | 0 | 1 | 1 | 1 | 1 | 0 | 1 | 1 | 1 | 0 | 0 | 0 | 0 | 1 | 1 | 0 | 1 | 1 |
| **b** | 1 | 1 | 0 | 1 | 1 | 1 | 1 | 0 | 1 | 1 | 1 | 1 | 0 | 1 | 1 | 0 | 1 | 0 | 1 | 0 |
| **c** | 1 | 1 | 1 | 1 | 1 | 1 | 1 | 1 | 1 | 1 | 1 | 1 | 1 | 1 | 1 | 0 | 1 | 1 | 1 | 0 |
| **d** | 1 | 1 | 0 | 1 | 1 | 1 | 1 | 0 | 1 | 1 | 1 | 1 | 0 | 1 | 1 | 0 | 1 | 0 | 1 | 0 |
| **e** | 1 | 1 | 0 | 1 | 1 | 0 | 0 | 0 | 0 | 1 | 1 | 0 | 0 | 0 | 1 | 0 | 1 | 0 | 1 | 1 |

Note: a–e represent cropland, forestland, grassland, water body and construction land, respectively. Under the four scenarios, the transfer directions between unused land and other land use types were the same: unused land could be converted to other land use types, but other land use types could not be converted to unused land.

The neighborhood effect not only directly reflects the expansion intensity of land use types but also reflects the expansion ability of each land use type driven by external factors [26]. Within the range of 0–1, the higher the value is, the greater the capacity for expansion. The neighborhood weights of each land use type were designed based on related literature and expert knowledge (Table 3).

**Table 3.** Neighborhood effect of each land use type.

| Type | Cropland | Forestland | Grassland | Water Body | Construction Land | Unused Land |
|---|---|---|---|---|---|---|
| Coefficient | 0.5 | 0.5 | 0.2 | 0.3 | 0.8 | 0.1 |

Forecasting Land Use Demand under Multiple Scenarios

The Markov model can be utilized for trend prediction by analyzing the initial probability of various land use changes and the transition probability between different land use changes. The quantity of land use in the future is computed as follows [27]:

$$S_{(t+1)} = P_{ij} \times S_{(t)}$$

where the $S_{(t)}$ and $S_{(t+1)}$ are the land use transfer matrices at time $t$ and $t + 1$; $P_{ij}$ is the transition from type $i$ to $j$.

Referring to relevant studies, we established four development scenarios in this study [22], which integrated the current natural and social conditions of the study area with the spatial planning and environmental conservation policy of the YRB: (1) the natural development scenario (NDS), based on the tendency of historical land use and predicting future land use; (2) the economic development scenario (EDS), which prioritized economic development and accelerated urban–rural integration, resulting in a 40% increase in conversion to construction land and a 10% reduction in other land types; (3) the food security scenario (FSS), which strictly protected cropland, resulting in a 30% increase in the likelihood of conversion to this land use type and a 10% reduction in conversion to other land use types; and (4) the ecological protection scenario (EPS), which prioritized ecological security, cropland, and construction land in accordance with the NDS, leading to a 40% increase in the likelihood of conversion to forestland and water body and a 10% reduction in conversion to other land use types. Forestland was considered as the restricted area in the first three scenarios, while forestland, water body, and natural reserves were designated as the restricted areas in the EPS.

Model Validation

The kappa coefficient is an important indicator to test the accuracy of land use prediction and is used to assess the consistency between simulated and actual land use data. A kappa value exceeding 0.8 indicates a high level of consistency [28]. The simulation consistency was validated against real land use data for 2015–2020, resulting in a kappa coefficient of 86.52% (an overall accuracy of 90.47%).

#### 2.2.2. ESP Construction

Ecological Source Identification

The morphological spatial pattern analysis (MSPA) method is a mapping algorithm based on the principle of mathematical morphology that uses erosion, expansion, open operation, and closed operation to determine the spatial pattern of raster images [29]. For this study, two values were generated, with forestland and water body used as foreground data and other land use types utilized as background data. The binary raster data were imported into Guidos Toolbox 3.0 software using the eight-neighborhood analysis method [30]. The edge width was set to the default value of "1", the transition was set to the default value of "off", and intext was set to the default value of "on". Finally, the land use types were classified into 7 basic spatial elements: core, islet, perforation, edge, loop, bridge, and branch. The core consists of large, more complete habitat patches, that were selected as alternative ecological patches.

Landscape connectivity analysis (LCA) illustrates the continuity of a landscape's spatial structure and can calculate how easy or difficult it is for organisms to move between ecological patches in a special field [31,32]. Based on the MSPA results, Conefor 2.6 software was used to further select habitat patches with good ecological functions, good development trends, internal homogeneity, and the ability to spread in all directions. Finally, the core areas larger than 2 km$^2$ were selected to quantitatively measure the patch connectivity index (*dIIC*) and patch importance index (*dPC*), allowing for the ranking of the patches in terms of importance. The higher the *dPC* index is, the greater the importance of the patch, and numerous core patches that can be used as ecological sources were extracted accordingly. The equations are as follows [7,19]:

$$dIIC = \frac{\sum_{i=1}^{n}\sum_{j=1}^{n}\frac{a_i a_j}{1+nl_{ij}}}{A_L^2} \qquad dPC = \frac{PC - PC_{remove}}{PC} * 100\%$$

where *dIIC* denotes the connectivity probability, $n$ denotes the number of total patches, $a_i$ and $a_j$ denote the respective areas of patches $i$ and $j$, $l_{ij}$ denotes the maximum diffusion probability between patch $i$ and patch $j$, and $A_L$ denotes the total area of the study region. $PC_{remove}$ is the connectivity probability of the remaining patches after removing a patch, and *dPC* indicates the degree of importance of the patch. Finally, the patches in the *dPC* index which they were greater than 0.2 were selected as the ecological sources.

Resistance Surface Establishment

The key to accurately characterizing ESPs is taking reasonable approach to estimating its ecological resistance value. This value should not only reflect the horizontal resistance of ecological processes but also represent the challenges faced by a species when moving through the landscape [33]. Based on the selection of previous resistance factors [19,34], and according to the condition of study area and the data that can be obtained easily, we selected seven resistance indicators, including land use, DEM, slope, soil erosion, NDVI, distance from roads, and distance from rivers. The ecological resistance factors were reclassified into five levels, assigned as 10, 30, 50, 70, and 90, and the weights of the seven factors were determined using AHP (Table 4). Finally, the resistant surfaces were reclassified into five levels: lower, low, medium, high and higher; the lower the resistance level was, the weaker the impact of external disturbance on the ecosystem [35].

**Table 4.** Resistance values and weights of the ecological resistance factors.

| Evaluation | Resistance Value | | | | | Weight |
|---|---|---|---|---|---|---|
| | 10 | 30 | 50 | 70 | 90 | |
| Land use | Forestland | Grassland, Cropland | Water body | Unused land | Construction land | 0.302 |
| DEM (m) | <1234 | 1235–1478 | 1479–1767 | 1768–2204 | >2205 | 0.120 |
| Slope (°) | <2.89 | 2.90–7.76 | 7.77–15.47 | 15.48–26.17 | 26.17–61.78 | 0.165 |
| Soil erosion | Normal | Mild | Moderate | High | Extremely | 0.176 |
| NDVI | 0.70–0.99 | 0.51–0.69 | 0.35–0.50 | 0.29–0.34 | 0.00–0.28 | 0.145 |
| Distance from roads (m) | >2051 | 1051–2050 | 551–1050 | 51–550 | 100 (itself) | 0.046 |
| Distance from rivers (m) | >7001 | 5001–7000 | 3001–5000 | 1001–3000 | 2000 (itself) | 0.042 |

Ecological Corridor Extraction

The MCR model is widely applied to extract ecological corridors, as it can compute the cost of species migration from a source to a target. The equation is as follows [36]:

$$MCR = fmin\sum_{i=1}^{m}\sum_{j=1}^{n} D_{ij}R_i$$

where *MCR* denotes the minimum cumulative resistance; *f* indicates the positive correlation function between the minimum cumulative resistance and the ecological process; $D_{ij}$

indicates the spatial distance traversed by source $i$ to unit $j$; and $R_i$ indicates the resistance coefficient of the unit to species diffusion. The corridors were obtained by the GIS distance model for our study.

The gravity model is utilized to categorize ecological corridors, with the significance of these corridors being measured by the ecological gravitational force between different patches. The greater the value of the ecological gravitational force between two patches, the higher the importance of the ecological corridors. The equation is as follows [37]:

$$G_{ab} = \frac{N_a N_b}{D_{ab}^2} = \left(\frac{1}{p_a} \times lnS_a\right)\left(\frac{1}{p_b} \times lnS_b\right) / \left(\frac{L_{ab}}{L_{max}}\right)^2$$

where $G_{ab}$ is the mutual force between patches $a$ and $b$; $D_{ab}$ is the normalized value of the resistance of the ecological corridors between two ecological sources; and $N_a(N_b)$, $p_a(p_b)$, and $S_a(S_b)$ are the weight, resistance value and area of ecological source $a(b)$, respectively. $L_{ab}$ is the cumulative resistance value of ecological sources $a$ and $b$, and $L_{max}$ is the maximum resistance value. In this study, corridors with mutual forces > 10,000 between patches were considered important corridors, and those with mutual forces ranging from 5000–10,000 between patches were considered general corridors.

### 2.2.3. ESPs Stability Evaluation

The Ucinet and Pajek tools are applied to abstract ESPs into complex networks, and to explore the structural characteristics of the establishment of various ESPs in terms of metric change. The metrics are composed of nodes and lines representing landscape entities (ecological sources and corridors). The metrics system was summarized in Table 5 via the integration of previous works by Shen and Wu [9].

**Table 5.** Metrics for quantifying the stability of ESPs.

| Metric | Equation | Description | Type | References |
|---|---|---|---|---|
| Average node degree ($k$) | $k = \frac{1}{N} \sum_{i \neq j \in G} a_{ij}$ | $K$ describes the average node degree of nodes in network $G$, $N$ is the number of total nodes, $a_{ij}$ is the number of nodes $i$ directly connected to nodes $j$; a higher $k$ value indicates a better the convenience between nodes. | Centrality | [38,39] |
| Average shortest path ($I$) | $I = \frac{2}{N(N-1)} \sum_{i \neq j \in G} d_{ij}$ | $I$ denotes the migration cost among different nodes, which is negatively correlated with connectivity, and $d_{ij}$ is the distance of the shortest path connecting the nodes. | Connectivity | [40] |
| Global efficiency ($E_G$) | $E_G = \frac{1}{N(N-1)} \sum_{i \neq j \in G} \frac{1}{d_{ij}}$ | $E_G$ refers to the efficiency of movement in the whole network, which can avoid the divergence of the $I$ index; the higher the $E_G$ index is, the less energy is consumed for moving between nodes. | Centrality | [41,42] |

**Table 5.** *Cont.*

| Metric | Equation | Description | Type | References |
|---|---|---|---|---|
| Clustering coefficient ($cc$) | $cc = \frac{1}{N}\sum_{i=1}^{N} \frac{2E_i}{k_i(k_{i-1})}$ | $cc$ measures the agglomeration degree of nodes in network $G$, $E_i$ is the actual number of connecting lines between the neighboring nodes of node $i$, and $k_i$ is the degree of node $i$; a higher $cc$ value indicates a higher agglomeration of network $G$. | Connectivity | [34,43] |
| Overall connectivity ($O_G$) | $O_G = \left(k' + \frac{1}{I'} + E_G' + cc'\right)/4$ | The $O_G$ denotes the overall connectivity of network $G$, which is the equally weighted result of the assessment of $k$, $I$, $E_G$ and $cc$; $k'$, $I'$, $E_G'$, $cc'$ are the normalized values of the four metrics; the higher $O_G$ value is, the better the comprehensive stability. | Comprehensiveness | [9] |

## 3. Results

### 3.1. Land Use Simulation under Multiple Scenarios

Figure 3 depicts the transition probability matrix of land in the study area from 2015–2020. The primary change in land use involved a conversion of 11.39% of cropland and 24.01% of grassland into construction land due to rapid urbanization. It also showed that 37.34% of cropland and 40.84% of construction land were converted into forestland or water body for ecological protection and human welfare, and 40.46% of unused land was converted into cropland to meet the requirement for a minimum cropland area exceeding 120,000,000 hectares in China.

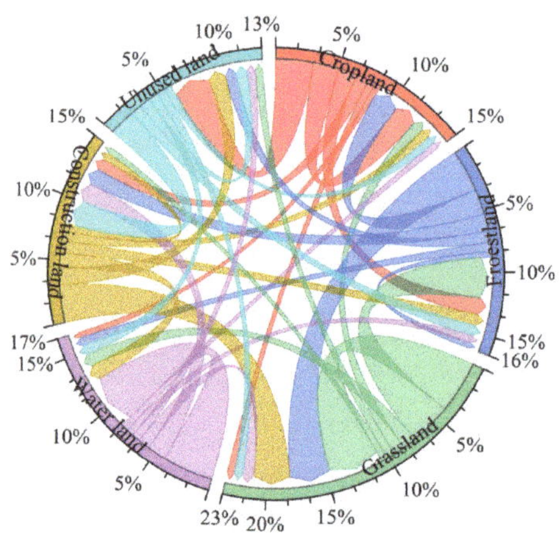


**Figure 3.** Land use transfer probability matrix for 2015–2020.

Table 6 and Figure 4 display the quantities and spatial distributions of land use types under the multiple scenarios simulation in 2035. Grassland and cropland remained the predominant land use types, subjected to the average annual precipitation of nearly 200 mm and strong evaporation. Meanwhile, the presence of the Yellow River throughout the region ensured sufficient water resources for agricultural production. The spatial distributions of forestland and water body were similar across the four scenarios. Forestland was concentrated in the nature reserves and forest farms. These regions were located at the altitudes 1200–1800 m, which have suitable factors for macrophanerophytes, such as sunlight, temperature, precipitation, and soil fertility. Water body was distributed in the mainstream and its tributaries of the Yellow River, as well as in lakes and wetlands in Yinchuan Plain. The terrain of Weining Plain resulted in a narrower main stream of the Yellow River compared to that in Yinchuan Plain, which in turn caused fewer oxbow lakes. The differences among the various scenarios were the areas and scales of the patches. These differences can be attributed to the distinct development aims and strategies, resulting in varying proportions of each land use type across the scenarios. For instance, urban expansion in the EDS led to larger areas of construction land at the expense of cropland, while the FSS prioritized cropland protection. Furthermore, there was a decrease in ecological security land use types under NDS, EDS, and FSS scenarios but an increase under EPS from 2020 to 2035.

**Table 6.** Area ($km^2$) of various land use types under multiple years/scenarios.

| Year/Land Use Type | 2015 | 2020 | NDS in 2035 | EDS in 2035 | FSS in 2035 | EPS in 2035 |
|---|---|---|---|---|---|---|
| Cropland | 5372 | 5532 | 5861 | 5276 | 7620 | 5861 |
| Forestland | 1298 | 1270 | 1211 | 1090 | 1090 | 1696 |
| Grassland | 10,512 | 10,378 | 10,000 | 9000 | 9000 | 9000 |
| Water body | 1203 | 1157 | 1147 | 1037 | 1037 | 1605 |
| Construction land | 1758 | 1901 | 2220 | 3108 | 1994 | 2220 |
| Unused land | 2496 | 2403 | 2198 | 1979 | 1978 | 1978 |

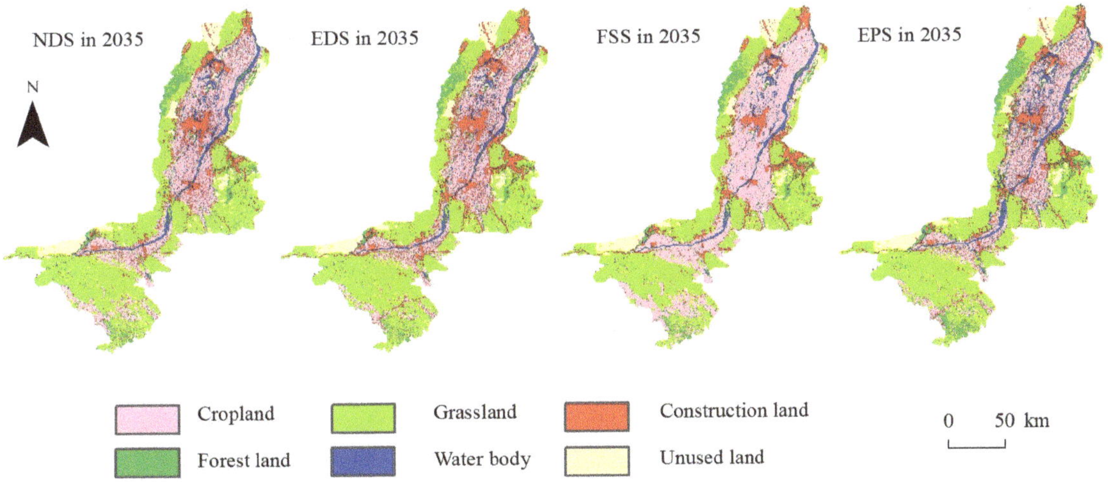


**Figure 4.** Distribution of land use types under the simulation of multiple scenarios.

## 3.2. ESP Identification

### 3.2.1. Ecological Sources

The spatial patterns of the ecological patches under multiple scenarios were displayed in Figure 5. The numbers of ecological sources in the study area were 20, 12, 17, and 15 under the NDS, EDS, FSS, and EPS respectively. The corresponding areas for each scenario were 834.82 $km^2$, 715.46 $km^2$, 785.56 $km^2$, and 1091.43 $km^2$ respectively. The ecological resources in the Yinchuan Plain encompassed the Shizuishan Wetland Park, Yinchuan Wetland Park, and Wuzhong Wetland Park, all situated along the main stream of the Yellow River. Other water bod, such as Sha Lake Province Reserve, and Zhenshu Lake Wetland in northern Yinchuan Plain also contributed to ecological sources. Additionally, the forestland, including Suyukou Forest Park, Gunzhongkou Forest Park within the Helan Mountain

Nature Reserve, the wind sheltering and sand fixing forest belt on the edge of the MuUs Sandy Land and the core areas of Baijitan Nature Reserve were identified as ecological sources. The ecological sources in the Weining Plain included the core of the Shapotou Forest Farm, and Zhongwei Wetland Park in the main stream of the Yellow River. The Qingtong Canyon Bird Island Nature Reserve was located on the junction between the two plains.

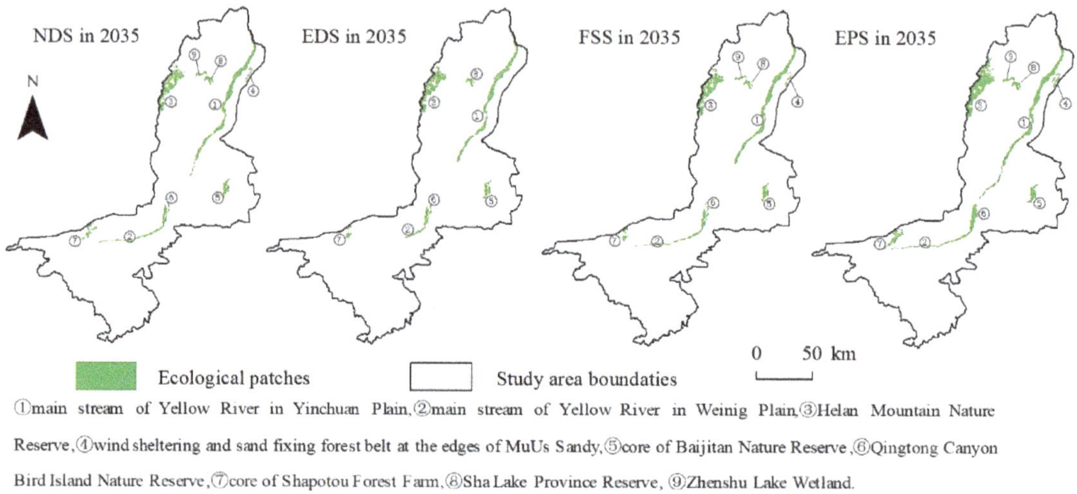


**Figure 5.** Ecological sources under multiple scenarios.

Compared to the other scenarios, ecological sources under the EDS exhibited the lowest numbers and areas. Additionally, certain ecological patches that fell below specified thresholds in terms of size or landscape connectivity were not classified as ecological sources. Under the EPS, the ecological sources were prominently identified. Wetlands along the main stream of the Yellow River were recognized as whole ecological sources in the Yinchuan Plain and Weining Plain, and all nature reserves were acknowledged as complete ecological sources. Under the NDS and FSS, efforts were made to identify ecological sources as completely as possible; however, there was fragmentation in terms of patch characteristics, with all ecological sources identified under the EPS being divided into several patches under NDS and FSS.

### 3.2.2. Ecological Resistance

According to the calculated proportions of the ecological resistance zones (Figure 6), the resistance levels decreased from the periphery zones to the center zones. Zones with the lowest and highest levels of resistance were nearly absent, while zones with medium resistance level had the greatest proportion, accounting for 61.86% of the total area, and were consistent with grassland, water body, and construction land among the land use types. The resistance would be dispersed among many ecological patches. Zones with low levels of resistance accounted for 29.57% of the total area. The resistance was located in low-elevation areas and consistent with cropland. Zones with high resistance level accounted for 8.54% of the total area. The resistance was primarily found in high-elevation areas and regions with little vegetation, including the Tengger Desert, MuUs Sandy Land, core of Baijitan, and Ningdong Energy Chemical Industry Base.

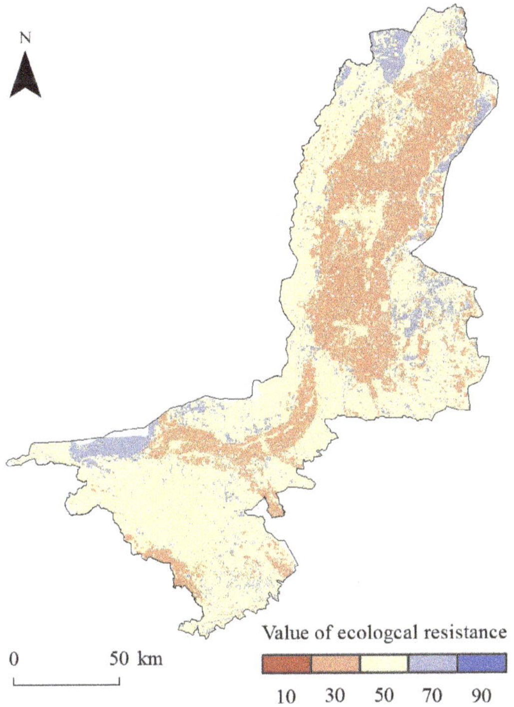


**Figure 6.** Surfaces indicating ecological resistance.

3.2.3. Ecological Corridors

The MCR and gravity models identified a total of 22, 14, 22, and 14 important ecological corridors with lengths of 950.01 km, 721.09 km, 1007.74 km, and 619.20 km for the NDS, EDS, FSS, and EPS respectively. In addition, there were also a total of 8, 5, 7, and 5 general ecological corridors extracted with lengths of 210.65 km, 111.61 km, 246.67 km, and 272.95 km respectively (Figure 7). The ESPs exhibited similar characteristics across the four scenarios, forming an integrated network in the Yinchuan Plain and being dispersed throughout the Weining Plain. The ecological corridors under the EPS were distributed evenly and efficiently. There were important ecological corridors connecting the Qingtong Canyon Bird Island Nature Reserve to the Helan Mountain Nature Reserve and the main stream of the Yellow River in the Yinchuan Plain. Furthermore, these ecological corridors were able to connect ecological sources in the Weining Plain. On the contrary, there was only a general ecological corridor connecting the Qingtong Canyon Bird Island Nature Reserve to the Yinchuan Wetland Park, and no ecological corridor to connect the ecological sources in the Weining Plain under the EDS. The ecological corridors under the NDS and FSS exhibited greater numbers and lengths compared to those under the EPS and EDS scenarios. This was particularly evident in the northern Yinchuan Plain, where ecological sources were dispersed throughout the region and resistance levels were relatively low.

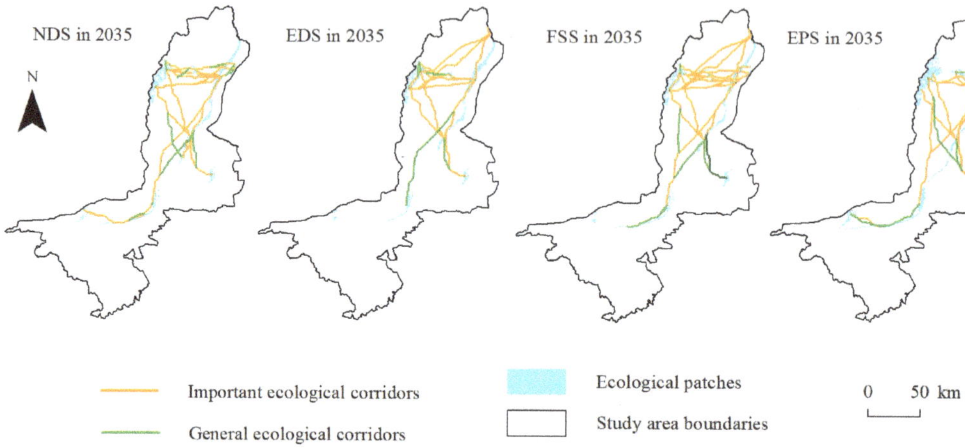


**Figure 7.** Ecological corridors in 2035 under multiple scenarios.

### 3.3. Evaluation of Network Stability of ESPs

Table 7 showed the network stability of the ESPs of the urban agglomeration along the YRB in Ningxia under multiple scenarios. Based on the ecological networks obtained from the multiple scenarios, we attempted to upgrade the general corridors into important corridors for ecological security and then calculated the metrics, depicted in Table 5. The results showed that single metrics exhibited deviations in the assessment of network stability, due to variations in evaluation factors and varying values of single metrics across different scenarios. The NDS had the highest average shortest path ($I$) and the lowest global efficiency ($E_G$) values, indicating that the higher the number of corridors, the stronger the connectivity, and the weaker the centrality. Conversely, the EDS demonstrated the opposite pattern to the NDS, with the lowest path ($I$) and global efficiency ($E_G$) values. The EPS obtained the highest overall connectivity ($O_G$) value; however, it exhibited lower values in single metrics compared to other scenarios, demonstrating that the comprehensive method could avoid this deficiency through the trade-offs between metrics. In addition, the numbers and lengths of ecological corridors in EPS were less than those in NDS and FSS, also indicating the importance of considering both ecological network stability and construction costs in reality.

**Table 7.** Network evaluation under multiple scenarios.

| | NDS | EDS | FSS | EPS |
|---|---|---|---|---|
| $k$ | 2.000 | 1.900 | 2.071 | 1.580 |
| $I$ | 11.054 | 18.616 | 13.785 | 14.153 |
| $E_G$ | 3.663 | 13.800 | 4.579 | 8.417 |
| $cc$ | 4.000 | 4.444 | 3.487 | 8.293 |
| $O_G$ | 0.351 | 0.466 | 0.334 | 0.520 |

## 4. Discussion

### 4.1. Necessity of Simulating Future ESPs under Multiple Scenarios

The role of land use in ESPs has attracted increasing attention, indicating that the distribution of key elements of ESPs affects dynamic alterations in ecosystem structure and function. The ESP construction methods are generally divided into current and future perspectives, with the former relying on a base year and the latter relying on a future year. Compared with studies from the current perspective, the construction of ESPs from the future perspective not only maintains landscape connectivity and provides ecological value services, but also responds to future environmental change [44]. To address the challenge of future development, it is necessary to predict various scenarios of land use with different objectives. Previous studies on ESPs in typical regions of China, such as the Pearl River Delta, Yellow River Basin, and Qinghai–Tibet Plateau, had shown that ESPs under different scenarios presented distinct spatial characteristics. Generally, the stability of ESPs decreased in the following order: EPS > NDS (BS, baseline scenario) > FSS (CLP, cultivated land protection scenario) > EDS (US, urban development scenario). However, there were significant variations in the identification results of ESPs among study areas due to environmental diversity. For example, the spatial distributions of ecological sources and corridors under the EPS were even spread throughout the Pearl River Delta, whereas in other scenarios, these components were only located in the periphery of the central region [45]. A comparison of ESPs under various scenarios revealed that the EPS predicted the best spatial structure of ecological patches and corridors for promoting material circulation and biological migration in the YRB [46]. However, effective environmental management and protection measures were still necessary to ensure a high quality and the sustainability of the entire basin [47]. The dual protection scenarios combining EPS and FSS were the best way forward in the Qinghai–Tibet Plateau [48], as ecosystem function and structure played important roles as ecological security barriers in water conservation, climate stability, and carbon balance in China, and even in Asia [49,50]. In our study, a model of "One Belt, Five Districts and Multiple Points" was developed based on the spatial distribution of ecological sources, which was similar to previous research on the same study area from a current perspective [19,51,52]. Meanwhile, it showed that coupling land use change with the construction of ESPs under multiple scenarios could balance ecological protection, urban expansion, and economic development in different areas. In the urban agglomeration along the YRB in Ningxia, the main stream of the Yellow River served as the ecological security guarantee in the area and was the focus for landscape integration and wetland biodiversity. It facilitated species migration and energy exchange in the Yinchuan Plain. Helan Mountain was designated as an ecological woodland protection zone, contributing to improved water and soil conservation through increased vegetation cover. Shapotou, MuUs Sandy Land, and Baijitan were located at the edge of deserts, leading to enhanced psammon biodiversity. Unfortunately, we identified 884.62 $km^2$ of ecological sources in 2018 [53], which was larger than the area of ecological sources under the NDS, EDS, and FSS but smaller than that under the EPS in 2035 in this study. Thus, it was imperative to take measure to prevent environmental deterioration due to the changes in natural and artificial factors.

In addition, based on the predictions of future land use change under multiple scenarios, we simulated carbon storage [54], habitat quality [55], urban heat island intensity [56], ecological risk [57], ecosystem services [58], and climate change [59]. In this way, if we complete the quantitative simulations of future scenarios, it could facilitate planning strategies and achieve sustainable development [60].

### 4.2. Proposed Development Scenario for the Future of Urban Agglomeration along the YRB in Ningxia

Based on Pepper's world hypotheses, scholars have developed a research framework for urban and regional studies, incorporating the root metaphors of correspondence, causal adjustment, organic whole, and spontaneous (history) [61]. This framework has facilitated the establishment of connections among concepts, models, and reality in our study.

Correspondence refers to the research paradigm and structure pattern of ESPs. Causal adjustment means accounting for the characteristics of the types, quantity, and spatial distribution of future land use, as well as other elements influenced by it. Organic whole refers to the formation and development of the YRB, and the role of urban agglomeration along the YRB in Ningxia. Spontaneous (history) involves implementing relevant planning schemes, policies, measures, and building engineering projects for urban agglomeration along the YRB in Ningxia. The recent paradigm of ESP construction has evolved to identify ecological sources, establish resistance surfaces, and extract corridors in theory, based on the suitability of land use. This study integrated future land use into ESP construction. It not only reflected the continuous effects of natural and anthropogenic factors on landscapes, but also provided potential ecological pathways for improving self-regulation and ecosystem resilience. In 2021, the China Central Government issued a document—*Outline of the Yellow River Basin's Ecological Protection and High-quality Development Plan*. It emphasized steadily implementing a strategy for formulating a green way of production and living. The development design must be limited by the carrying capacity of the YRB. It should be calculated according to the simulation of land use change [62], a calculation of the environmental resource carrying capacity [63], an assessment of landscape ecological risk [64], an evaluation of vegetation restoration [65], an identification of ecological networks [46], and the optimization of urban growth boundaries [66]. The urban agglomeration along the YRB in Ningxia was located in the overlap region of Upper YRB and the Jiziwan Metropolitan Area of the YRB, which were characterized by ecological fragility and natural resource enrichment. Hence, it was essential to address the conflicts arising from ecological preservation and human activities [47], upholding the principle of the utilization of integrated water and mineral resources, agriculture and industry development within permissible limits, as well as the protection of mountains, rivers, forests, farmlands, lakes, and grasslands to prevent ecological degradation and promote ecological restoration [67]. Furthermore, it was crucial to trade off various ESPs to promote practical sustainable development [19,51].

### 4.3. Limitations and Future Research

The proposed framework in this study could be applied as a decision-making tool for ecological conservation by integrating land use for identifying ecological sources and extracting ecological corridors. It also has the potential to inform adjustments to development policies and management measures. However, it was notable that the study had certain limitations.

First, *dIIC* and *dPC* were utilized for a landscape connectivity analysis when identifying ecological patches. However, there is currently no standardized system for assigning values in this context [7]. In our study, we established a distance threshold of 1000 m and a connectivity probability of 0.2 based on the existing literature. Nevertheless, it was crucial to further scrutinize the reasonableness of these values.

Second, the MCR and gravity models only describe ecological corridors as line segments; in actuality, the corridors were band-like units with specific widths that vary depending on the species [68].

Third, there was a lack of a comprehensive system for assessing ecological sources. We employed landscape metrics such as percentage of landscape (PLAND), largest patch index (LPI), landscape shape index (LSI), aggregation index (AI), and landscape shape index (LSI) for evaluation. These metrics were limited to calculating the landscape structure of ecological sources from the functional perspective. However, comprehensive comparisons of the results obtained under different scenarios were not feasible.

Furthermore, the national spatial planning in China delineates three distinct types of land use—agricultural zones, ecological preservation zones, and urban development zones—through the implementation of three lines: a red line for protecting permanent basic cropland, a red line for ecological conservation, and a boundary line for urban development. It was necessary to further explore methods that can regulate the balance of these three delineated zones from a land use perspective [69].

## 5. Conclusions

The FLUS model, landscape ecological theory, and graph-theoretic method were employed in this study to construct ESPs under four (2035) scenarios—NDS, EDS, FSS and EPS—of urban agglomeration along the YRB in Ningxia. Subsequently, the ecological network stability were evaluated comprehensively. The primary conclusions were as follows:

The predominant land use types remained grassland and cropland, given that the study area was located in an arid or semiarid region. The forestland and water body, designated as ecological protective zones, decreased in NDS, EDS, and FSS but increased in EPS. This indicated that it was imperative to intensify ecological restoration and optimization efforts to address the threat of land use change. The spatial characteristics of the ecological sources and ecological corridors varied greatly among the different scenarios. Notably, the EPS exhibited the greatest extent of ecological sources (1091.43 km$^2$) and the highest overall connectivity (0.520), while it did not achieve the optimal value of single metrics and the longest ecological corridors. The results also indicated that both the ecological network stability and the construction cost should be considered. A model of "One Belt, Five Districts and Multiple Points" was developed based on the spatial distribution of ecological sources in the study. The Yellow River was the ecological security guarantee in the area, which served as the most important passageway for species migration and energy exchange. Helan Mountain was designated as an ecological woodland protection zone, while Shapotou, MuUs Sandy Land, and Baijitan were utilized to enhanced psammonphyte biodiversity.

Compared to the current perspective, this study predicted the development diversity of land use change from a future perspective. It not only responded to the various ESPs under multiple scenarios, but also balanced the contractions among the three zones delineated by the three lines for spatial planning in the future.

**Author Contributions:** W.C.: Conceptualization, Formal analysis, Writing—original draft. C.M.: Methodology, Supervision, Writing—review and Funding acquisition. T.L.: Methodology, Writing—review and editing. Y.L.: Software, Resources. All authors have read and agreed to the published version of the manuscript.

**Funding:** This research was funded by [National Natural Science Foundation of China (42061037)]. This research was also funded by [the Ningxia Science Research Project Foundation of China (2021BEG03019)]. This research was also funded by [National Science Research Project Foundation of China (2023YFF1304705)].

**Data Availability Statement:** Data will be made available on request.

**Acknowledgments:** We thank the reviewers for providing constructive comment on this manuscript. We also thank the editors for their carefully work.

**Conflicts of Interest:** The authors declare that they have no known competing financial interest or personal relationships that could have appeared to influence the work reported in this paper.

## References

1. He, Y.; Liang, Y.; Liu, L.; Yin, Z.; Huang, J. Loss of green landscapes due to urban expansion in China. *Resour. Conserv. Recycl.* **2023**, *199*, 107228. [CrossRef]
2. Rahaman, Z.A.; Kafy, A.A.; Saha, M.; Rahim, A.A.; Almulhim, A.I.; Rahaman, S.N.; Fattah, A.; Rahman, M.T.; S, K.; Faisal, A.-A.; et al. Assessing the impacts of vegetation cover loss on surface temperature, urban heat island and carbon emission in Penang city, Malaysia. *Build. Environ.* **2022**, *222*, 109335. [CrossRef]
3. Rachman, F.; Huang, J.; Xue, X.; Marfai, M.A. Insights from 30 Years of Land Use/Land Cover Transitions in Jakarta, Indonesia, via Intensity Analysis. *Land* **2024**, *13*, 545. [CrossRef]
4. Dong, J.; Jiang, H.; Gu, T.; Liu, Y.; Peng, J. Sustainable landscape pattern: A landscape approach to serving spatial planning. *Landsc. Ecol.* **2021**, *37*, 31–42. [CrossRef]
5. Fan, X.; Cheng, Y.; Tan, F.; Zhao, T. Construction and Optimization of the Ecological Security Pattern in Liyang, China. *Land* **2022**, *11*, 1641. [CrossRef]
6. Li, X.; Chen, Y. Projecting the future impacts of China's cropland balance policy on ecosystem services under the shared socioeconomic pathways. *J. Clean. Prod.* **2020**, *250*, 119489. [CrossRef]
7. Qian, W.; Zhao, Y.; Li, X. Construction of ecological security pattern in coastal urban areas: A case study in Qingdao, China. *Ecol. Indic.* **2023**, *154*, 110754. [CrossRef]
8. Cunha, N.S.; Magalhães, M.R. Methodology for mapping the national ecological network to mainland Portugal: A planning tool towards a green infrastructure. *Ecol. Indic.* **2019**, *104*, 802–818. [CrossRef]
9. Shen, Z.; Wu, W.; Tian, S.; Wang, J. A multi-scale analysis framework of different methods used in establishing ecological networks. *Landsc. Urban Plan.* **2022**, *228*, 104579. [CrossRef]

10. Deng, Z.; Quan, B.; Zhang, H.; Xie, H.; Zhou, Z. Scenario Simulation of Land Use and Cover under Safeguarding Ecological Security: A Case Study of Chang-Zhu-Tan Metropolitan Area, China. *Forests* **2023**, *14*, 2131. [CrossRef]
11. Liu, X.; Wei, M.; Li, Z.; Zeng, J. Multi-scenario simulation of urban growth boundaries with an ESP-FLUS model: A case study of the Min Delta region, China. *Ecol. Indic.* **2022**, *135*, 108538. [CrossRef]
12. Dai, L.; Liu, Y.; Luo, X. Integrating the MCR and DOI models to construct an ecological security network for the urban agglomeration around Poyang Lake, China. *Sci Total Environ.* **2021**, *754*, 141868. [CrossRef]
13. Liang, X.; Liu, X.; Li, X.; Chen, Y.; Tian, H.; Yao, Y. Delineating multi-scenario urban growth boundaries with a CA-based FLUS model and morphological method. *Landsc. Urban Plan.* **2018**, *177*, 47–63. [CrossRef]
14. Liu, J.; Zhang, L.; Zhang, Q.; Zhang, G.; Teng, J. Predicting the surface urban heat island intensity of future urban green space development using a multi-scenario simulation. *Sustain. Cities Soc.* **2021**, *66*, 102698. [CrossRef]
15. Mao, H.; Zhou, L.; Ying, R.; Pan, D. Time Preferences and green agricultural technology adoption: Field evidence from rice farmers in China. *Land Use Policy* **2021**, *109*, 105627. [CrossRef]
16. Fang, C.; Liu, H.; Wang, S. The coupling curve between urbanization and the eco-environment: China's urban agglomeration as a case study. *Ecol. Indic.* **2021**, *130*, 108107. [CrossRef]
17. Albers, L.T.; Schyns, J.F.; Booij, M.J.; Zhuo, L. Blue water footprint caps per sub-catchment to mitigate water scarcity in a large river basin: The case of the Yellow River in China. *J. Hydrol.* **2021**, *603*, 126992. [CrossRef]
18. Jia, G.; Hu, W.; Zhang, B.; Li, G.; Shen, S.; Gao, Z.; Li, Y. Assessing impacts of the Ecological Retreat project on water conservation in the Yellow River Basin. *Sci Total Environ.* **2022**, *828*, 154483. [CrossRef]
19. Guan, H.; Bai, Y.; Tang, y.; Zhang, C.; Zou, J. Spatial identification and optimization of ecological network in desert-oasis area of Yellow River Basin, China. *Ecol. Indic.* **2023**, *147*, 109999. [CrossRef]
20. Singh, D.K.; Xu, M.; Singh, N.; Lei, F. Perspectives on emerging pressures and their integrated impact on large river systems: An insight from the Yellow River basin. *J Environ. Manag.* **2021**, *298*, 113423. [CrossRef]
21. Lyu, R.; Zhao, W.; Tian, X.; Zhang, J. Non-linearity impacts of landscape pattern on ecosystem services and their trade-offs: A case study in the City Belt along the Yellow River in Ningxia, China. *Ecol. Indic.* **2022**, *136*, 108608. [CrossRef]
22. Li, L.; Huang, X.; Wu, D.; Yang, H. Construction of ecological security pattern adapting to future land use change in Pearl River Delta, China. *Appl. Geogr.* **2023**, *154*, 102946. [CrossRef]
23. Liang, X.; Liu, X.; Li, D.; Zhao, H.; Chen, G. Urban growth simulation by incorporating planning policies into a CA-based future land-use simulation model. *Int. J. Geogr. Inf. Sci.* **2018**, *32*, 2294–2316. [CrossRef]
24. Lin, W.; Sun, Y.; Nijhuis, S.; Wang, Z. Scenario-based flood risk assessment for urbanizing deltas using future land-use simulation (FLUS): Guangzhou Metropolitan Area as a case study. *Sci Total Environ.* **2020**, *739*, 139899. [CrossRef]
25. Mondal, M.S.; Sharma, N.; Garg, P.K.; Kappas, M. Statistical independence test and validation of CA Markov land use land cover (LULC) prediction results. *Egypt. J. Remote Sens. Space Sci.* **2016**, *19*, 259–272. [CrossRef]
26. Lin, J.; He, P.; Yang, L.; He, X.; Lu, S.; Liu, D. Predicting future urban waterlogging-prone areas by coupling the maximum entropy and FLUS model. *Sustain. Cities Soc.* **2022**, *80*, 103812. [CrossRef]
27. Jokar Arsanjani, J.; Helbich, M.; Kainz, W.; Darvishi Boloorani, A. Integration of logistic regression, Markov chain and cellular automata models to simulate urban expansion. *Int. J. Appl. Earth Obs. Geoinf.* **2013**, *21*, 265–275. [CrossRef]
28. Liang, X.; Guan, Q.; Clarke, K.C.; Liu, S.; Wang, B.; Yao, Y. Understanding the drivers of sustainable land expansion using a patch-generating land use simulation (PLUS) model: A case study in Wuhan, China. *Comput. Environ. Urban Syst.* **2021**, *85*, 101569. [CrossRef]
29. Vogt, P.; Riitters, K. GuidosToolbox: Universal digital image object analysis. *Eur. J. Remote Sens.* **2017**, *50*, 352–361. [CrossRef]
30. Rincón, V.; Velázquez, J.; Gutiérrez, J.; Hernando, A.; Khoroshev, A.; Gómez, I.; Herráez, F.; Sánchez, B.; Pablo Luque, J.; García-abril, A.; et al. Proposal of new Natura 2000 network boundaries in Spain based on the value of importance for biodiversity and connectivity analysis for its improvement. *Ecol. Indic.* **2021**, *129*, 108024. [CrossRef]
31. Saura, S.; Torné, J. Conefor Sensinode 2.2: A software package for quantifying the importance of habitat patches for landscape connectivity. *Environ. Model. Softw.* **2009**, *24*, 135–139. [CrossRef]
32. Qian, W.; Li, X. A cold island connectivity and network perspective to mitigate the urban heat island effect. *Sustain. Cities Soc.* **2023**, *94*, 104525. [CrossRef]
33. Wang, T.; Li, H.; Huang, Y. The complex ecological network's resilience of the Wuhan metropolitan area. *Ecol. Indic.* **2021**, *130*, 108101. [CrossRef]
34. Huang, X.; Ye, Y.; Zhao, X.; Guo, X.; Ding, H. Identification and stability analysis of critical ecological land: Case study of a hilly county in southern China. *Ecol. Indic.* **2022**, *141*, 109091. [CrossRef]
35. Tang, Y.; Gao, C.; Wu, X. Urban Ecological Corridor Network Construction: An Integration of the Least Cost Path Model and the InVEST Model. *ISPRS Int. J. Geo-Inf.* **2020**, *9*, 33. [CrossRef]
36. Santos, J.S.; Leite, C.C.C.; Viana, J.C.C.; dos Santos, A.R.; Fernandes, M.M.; de Souza Abreu, V.; do Nascimento, T.P.; dos Santos, L.S.; de Moura Fernandes, M.R.; da Silva, G.F.; et al. Delimitation of ecological corridors in the Brazilian Atlantic Forest. *Ecol. Indic.* **2018**, *88*, 414–424. [CrossRef]
37. Nie, W.; Shi, Y.; Siaw, M.J.; Yang, F.; Wu, R.; Wu, X.; Zheng, X.; Bao, Z. Constructing and optimizing ecological network at county and town Scale: The case of Anji County, China. *Ecol. Indic.* **2021**, *132*, 108294. [CrossRef]

38. Rayfield, B.; Fortin, M.J.; Fall, A. Connectivity for conservation: A framework to classify network measures. *Ecology* **2011**, *92*, 847–858. [CrossRef]
39. Liccari, F.; Boscutti, F.; Bacaro, G.; Sigura, M. Connectivity, landscape structure, and plant diversity across agricultural landscapes: Novel insight into effective ecological network planning. *J Environ. Manag.* **2022**, *317*, 115358. [CrossRef]
40. Upadhyay, S.; Roy, A.; Ramprakash, M.; Idiculla, J.; Kumar, A.S.; Bhattacharya, S. A network theoretic study of ecological connectivity in Western Himalayas. *Ecol. Model.* **2017**, *359*, 246–257. [CrossRef]
41. Luo, Y.; Wu, J.; Wang, X.; Wang, Z.; Zhao, Y. Can policy maintain habitat connectivity under landscape fragmentation? A case study of Shenzhen, China. *Sci Total Environ.* **2020**, *715*, 136829. [CrossRef]
42. Kim, B.; Park, J. Random ecological networks that depend on ephemeral wetland complexes. *Ecol. Eng.* **2020**, *156*, 105972. [CrossRef]
43. De Montis, A.; Caschili, S.; Mulas, M.; Modica, G.; Ganciu, A.; Bardi, A.; Ledda, A.; Dessena, L.; Laudari, L.; Fichera, C.R. Urban–rural ecological networks for landscape planning. *Land Use Policy* **2016**, *50*, 312–327. [CrossRef]
44. Zhang, M.; Ao, Y.; Liu, M.; Zhao, Y.; Lin, K.; Cheng, T. Ecological security assessment based on ecosystem service value and ecological footprint in the Pearl River Delta urban agglomeration, China. *Ecol. Indic.* **2022**, *144*, 109528. [CrossRef]
45. Jiang, H.; Peng, J.; Dong, J.; Zhang, Z.; Xu, Z.; Meersmans, J. Linking ecological background and demand to identify ecological security patterns across the Guangdong-Hong Kong-Macao Greater Bay Area in China. *Landsc. Ecol.* **2021**, *36*, 2135–2150. [CrossRef]
46. Men, D.; Pan, J. Ecological network identification and connectivity robustness evaluation in the Yellow River Basin under a multi-scenario simulation. *Ecol. Model.* **2023**, *482*, 110384. [CrossRef]
47. Wei, L.; Zhou, L.; Sun, D.; Yuan, B.; Hu, F. Evaluating the impact of urban expansion on the habitat quality and constructing ecological security patterns: A case study of Jiziwan in the Yellow River Basin, China. *Ecol. Indic.* **2022**, *145*, 109544. [CrossRef]
48. Ma, R.; Zhou, W.; Ren, J.; Huang, Y.; Wang, H. Multi-scenario simulation and optimization control of ecological security based on GeoSOS-FLUS model in ecological fragile area in northeast Qinghai-Tibet Plateau, China. *Ecol. Indic.* **2023**, *151*, 110324. [CrossRef]
49. Yao, T. Tackling on environmental changes in Tibetan Plateau with focus on water, ecosystem and adaptation. *Sci. Bull.* **2019**, *64*, 417. [CrossRef]
50. Yu, Q.; Lu, H.; Yao, T.; Xue, Y.; Feng, W. Enhancing sustainability of vegetation ecosystems through ecological engineering: A case study in the Qinghai-Tibet Plateau. *J. Environ. Manag.* **2023**, *325*, 116576. [CrossRef]
51. Tang, H.; Peng, J.; Jiang, H.; Lin, Y.; Dong, J.; Liu, M.; Meersmans, J. Spatial analysis enables priority selection in conservation practices for landscapes that need ecological security. *J Environ. Manag.* **2023**, *345*, 118888. [CrossRef] [PubMed]
52. Jiang, H.; Peng, J.; Xu, D.; Tang, H. Constructing ecological security patterns with differentiated management intensity based on multifunctional landscape identification and multi-criteria decision-making. *Glob. Ecol. Conserv.* **2024**, *50*, e02862. [CrossRef]
53. Cheng, W.; Li, T.; Ma, C. Construction of an ecological security pattern based on MCR model and DO index:a case study of the ecological and economic belt along the Yellow River in Ningxia. *J. Northwest Univ. China* **2022**, *52*, 433–443.
54. Carpio, A.; Ponce-Lopez, R.; Lozano-García, D.F. Urban form, land use, and cover change and their impact on carbon emissions in the Monterrey Metropolitan area, Mexico. *Urban Clim.* **2021**, *39*, 100947. [CrossRef]
55. Yu, B.; Zang, Y.; Wu, C.; Zhao, Z. Spatiotemporal dynamics of wetlands and their future multi-scenario simulation in the Yellow River Delta, China. *J Environ. Manag.* **2024**, *353*, 120193. [CrossRef] [PubMed]
56. Lin, J.; Qiu, S.; Tan, X.; Zhuang, Y. Measuring the relationship between morphological spatial pattern of green space and urban heat island using machine learning methods. *Build. Environ.* **2023**, *228*, 109910. [CrossRef]
57. Gao, L.; Tao, F.; Liu, R.; Wang, Z.; Leng, H.; Zhou, T. Multi-scenario simulation and ecological risk analysis of land use based on the PLUS model: A case study of Nanjing. *Sustain. Cities Soc.* **2022**, *85*, 104055. [CrossRef]
58. Sun, Q.; Qi, W.; Yu, X. Impacts of land use change on ecosystem services in the intensive agricultural area of North China based on Multi-scenario analysis. *Alex. Eng. J.* **2021**, *60*, 1703–1716. [CrossRef]
59. Hanberry, B.B.; Abrams, M.D.; Nowacki, G.J. Potential Interactions between Climate Change and Land Use for Forest Issues in the Eastern United States. *Land* **2024**, *13*, 398. [CrossRef]
60. Pan, W.; Liu, Y.; Wu, Z.; Liu, Z.; Han, W.; Yu, Z. Simulation of changes in land use distribution and biodiversity under different development scenarios in Qinghai Province. *Biodivers. Sci.* **2022**, *30*, 21425. [CrossRef]
61. Yang, X. (Ed.) *Urban Remote Sensing Monitoring, Synthesis and Modeling in the Urban Environment*; Wiley & Sons, Ltd.: Hoboken, NJ, USA, 2011.
62. Lou, Y.; Yang, D.; Zhang, P.; Zhang, Y.; Song, M.; Huang, Y.; Jing, W. Multi-Scenario Simulation of Land Use Changes with Ecosystem Service Value in the Yellow River Basin. *Land* **2022**, *11*, 992. [CrossRef]
63. Chen, G.; Wang, S. Evaluation of Urban Resource Environmental Carrying Capacity and Land Spatial Development Suitability in a Semiarid Area of the Yellow River Basin. *Sustainability* **2023**, *15*, 12411. [CrossRef]
64. Du, L.; Dong, C.; Kang, X.; Qian, X.; Gu, L. Spatiotemporal evolution of land cover changes and landscape ecological risk assessment in the Yellow River Basin, 2015–2020. *J Environ. Manag.* **2023**, *332*, 117149. [CrossRef]
65. Wang, G.; Yue, D.; Niu, T.; Yu, Q. Regulated Ecosystem Services Trade-Offs: Synergy Research and Driver Identification in the Vegetation Restoration Area of the Middle Stream of the Yellow River. *Remote Sens.* **2022**, *14*, 718. [CrossRef]
66. Chen, Y.; Zhu, M.; Lu, J.; Zhou, Q.; Ma, W. Evaluation of ecological city and analysis of obstacle factors under the background of high-quality development: Taking cities in the Yellow River Basin as examples. *Ecol. Indic.* **2020**, *118*, 106771. [CrossRef]

67. Shen, W.; Li, Y.; Qin, Y. Research on the influencing factors and multi-scale regulatory pathway of ecosystem health: A case study in the Middle Reaches of the Yellow River, China. *J. Clean. Prod.* **2023**, *406*, 137038. [CrossRef]
68. Huang, L.; Wang, J.; Cheng, H. Spatiotemporal changes in ecological network resilience in the Shandong Peninsula urban agglomeration. *J. Clean. Prod.* **2022**, *339*, 130681. [CrossRef]
69. Mu, H.; Li, X.; Ma, H.; Du, X.; Huang, J.; Su, W.; Yu, Z.; Xu, C.; Liu, H.; Yin, D.; et al. Evaluation of the policy-driven ecological network in the Three-North Shelterbelt region of China. *Landsc. Urban Plan.* **2022**, *218*, 104305. [CrossRef]

land

MDPI

Article

# Spatio-Temporal Land-Use/Cover Change Dynamics Using Spatiotemporal Data Fusion Model and Google Earth Engine in Jilin Province, China

Zhuxin Liu [1,†], Yang Han [1,*,†], Ruifei Zhu [2], Chunmei Qu [2], Peng Zhang [2], Yaping Xu [3], Jiani Zhang [1], Lijuan Zhuang [1], Feiyu Wang [1] and Fang Huang [1,*]

1 Laboratory of Geographical Processes and Ecological Security in Changbai Mountains, Ministry of Education, School of Geographical Sciences, Northeast Normal University, Changchun 130024, China
2 Chang Guang Satellite Technology Company Ltd., Changchun 130000, China
3 Department of Environmental and Geosciences, Sam Houston State University, Huntsville, TX 77340, USA
* Correspondence: hany025@nenu.edu.cn (Y.H.); huangf837@nenu.edu.cn (F.H.)
† These authors contributed equally to this work.

**Abstract:** Jilin Province is located in the northeast of China, and has fragile ecosystems, and a vulnerable environment. Large-scale, long time series, high-precision land-use/cover change (LU/CC) data are important for spatial planning and environmental protection in areas with high surface heterogeneity. In this paper, based on the high temporal and spatial fusion data of Landsat and MODIS and the Google Earth Engine (GEE), long time series LU/CC mapping and spatio-temporal analysis for the period 2000–2023 were realized using the random forest remote sensing image classification method, which integrates remote sensing indices. The prediction results using the OL-STARFM method were very close to the real images and better contained the spatial image information, allowing its application to the subsequent classification. The average overall accuracy and kappa coefficient of the random forest classification products obtained using the fused remote sensing index were 95.11% and 0.9394, respectively. During the study period, the area of cultivated land and unused land decreased as a whole. The area of grassland, forest, and water fluctuated, while building land increased to 13,442.27 km$^2$ in 2023. In terms of land transfer, cultivated land was the most important source of transfers, and the total area share decreased from 42.98% to 38.39%. Cultivated land was mainly transferred to grassland, forest land, and building land, with transfer areas of 7682.48 km$^2$, 8374.11 km$^2$, and 7244.52 km$^2$, respectively. Grassland was the largest source of land transfer into cultivated land, and the land transfer among other feature types was relatively small, at less than 3300 km$^2$. This study provides data support for the scientific management of land resources in Jilin Province, and the resulting LU/CC dataset is of great significance for regional sustainable development.

**Keywords:** spatio-temporal data fusion; OL-STARFM; LU/CC; GEE; random forest; spatio-temporal characteristics

## 1. Introduction

Land-use/cover change involves a process whereby humans analyze the natural characteristics of land, utilizing biological, technological, and legal policies, and subsequently modify socio-economic activities [1,2]. Land-use/cover change can directly affect the diversity of biological species, contribute to climate change and ecosystem transformation [3], and increase human–environment interactions [4]. With the development of science and technology and the rapid growth of the global population, human activities have caused distinct changes in global land cover [5,6]. Conducting land-use analysis is vital for understanding the intricate relationship between humans and nature, offering insights into current land-use patterns, and guiding the implementation of appropriate interventions.

The study of human land-use and the resulting land-cover change allows for the assessment of current human land-use policies and the prediction of their consequences, in order to achieve sustainable land use [7,8].

LUCC is a critical subject in the realms of global climate change and sustainable development, emerging as a key factor in human–land relationship studies. In recent years, numerous researchers have applied remote sensing technology to investigate land-use changes. Recently, the development of domestic and foreign earth observation technology has entered a new stage, and a wide range of multi-temporal, multi-dimensional, and multi-scale remote sensing data have become available [9]. The traditional techniques for extracting information on land-use/cover changes are based on remote sensing data provided by a single sensor, such as MODIS, Landsat, etc. [10]. Due to the limitations of the physical properties of the remote sensing satellites themselves, the spatial and temporal resolutions of single remote sensing data cannot both be prioritized. Although high-spatial-resolution remote sensing images have obvious advantages in terms of the richness of the information expression of surface details, their spatio-temporal data are missing to a large extent due to the long revisit cycle of the data itself, the lower temporal resolution, and the greater influence of clouds, rain, and snow, etc. [11]. This makes it difficult to achieve continuous monitoring of the same region at the temporal or spatial scales, and, thus, leads to the inability to meet the requirements of continuous dynamic tracking and monitoring [12]. For high-temporal-resolution remote sensing data, due to its low spatial resolution in areas with fragmented or heterogeneous surface landscapes, one image element range may contain many different land-cover types, thus seriously affecting its remote sensing classification accuracy [13]. Therefore, the challenge of combining the advantages of different remote sensing satellite sensors to obtain remote sensing data with a high temporal and spatial resolution and to ensure the spatial and temporal continuity of 30 m remote sensing images is a significant problem.

Spatio-temporal fusion technology refers to the generation of high-spatio-temporal-resolution data through fusion algorithms based on high temporal–low spatial- and high spatial–low temporal-resolution remote sensing data [14,15]. Without changing the existing observation conditions, the method of fusion of remote sensing data with different spatial and temporal resolutions can improve the spatial and temporal resolution of remote sensing images [16–18]. In recent years, research on high-temporal- and high-spatial-resolution data fusion methods has mainly focused on constructing fusion models. The initial fusion models primarily utilized a linear hybrid algorithm, but the heterogeneity and complexity of the actual ground cover resulted in the algorithm no longer being applicable [19,20]. On this basis, ZhuKov et al. [21] and Lorenzo et al. [22] proposed an improved model by quantitatively analyzing the multiple differences that exist between the target and neighboring image elements. Gao et al. [16] proposed the STARFM fusion algorithm, which combines the advantages of Landsat's high spatial resolution and MODIS's high temporal resolution to generate high-spatial- and high-temporal-resolution synthetic data, and then verified the correlation between the fused data and Landsat data. The validation results show that the STARFM fusion algorithm can estimate the reflectance data with a high spatial and temporal resolution in the region. Since then, many researchers have improved the STARFM fusion algorithm. Hilker et al. [23] proposed the STAARCH algorithm based on the STARFM algorithm to complete the mapping of forest disturbances. Zhu et al. [18] developed the STARFM model and proposed the ESTARFM model, which is more suitable for monitoring vegetation dynamics in complex mountainous areas. Zhu et al. [24] also proposed the FSDAF model, which is a flexible spatio-temporal fusion model. It can detect rapid surface changes, i.e., it is able to capture reflectance changes caused by the conversion of land-cover types, and, thus, can effectively predict changes in land-cover types. Guo et al. [25] proposed a flexible object-level (OL) spatio-temporal adaptive reflectance fusion model (STARFM) processing strategy, which can better preserve the structural information of the image, and is 102.89 times faster than the original algorithm with high computational efficiency, and analyzed and verified the effectiveness of the OL processing strategy.

Relatively few studies have been conducted on the application of multi-source spatial and temporal fusion techniques to long time series land-cover and its related products. For example, Hao et al. [26] used the ESTARFM fusion method to extract the lake water body area information of Siling Co Lake in Tibet during the period 2000–2014, in order to monitor the spatial and temporal changes of the lake with spatio-temporal fusion data. Liu Han and Gong Peng [27] proposed spatial and temporal fusion and reconstruction techniques based on virtual constellations to build a seamless 30 m day-by-day data cube SDC land-cover and land-use dynamic mapping of China in the 21st century.

With the development of geographic information, big data analytics, and other technological tools, machine learning methods are expanding in the field of land-cover change research, showing their ability to integrate and process training data [28]. Algorithms such as support vector machine (SVM), classification and regression tree (CART), and random forest (RF) have been widely used for land-cover classification [29–31]. However, there is still room for optimization of machine learning algorithms in terms of their performance in processing different satellite datasets and the improvement in the algorithmic classification accuracy [32,33]. Some commonly used remote sensing indices have good spatial resolution for vegetation, water, buildings, etc. Using these indices can address the limitation of relying solely on original image data, and can improve the classification accuracy of the remotely sensed images and the accuracy of information extraction [34,35].

Ensuring the continuous production of LU/CC products with high spatiotemporal resolution requires continuous monitoring on a large scale and with high accuracy, and traditional geographic information and remote sensing methods are challenged by their high computational complexity and processing costs [36]. The emergence and expansion of the GEE satellite remote sensing cloud storage and cloud computing platform integrates global remote sensing images from multiple sources and scales, and realizes functions such as full-waveband, high-intensity parallel processing, time series analyses, and mapping [37,38], enabling efficient and continuous remote sensing monitoring. Land-use dynamics changes have been extensively studied on the GEE platform, which has a very high computational efficiency [39].

In this study, in the area of Jilin Province, based on Landsat and MODIS fused long time series high-temporal- and high-spatial-resolution satellite images and the GEE platform, the random forest classification algorithm was used to fuse remote sensing indices to map the LU/CC dataset from 2000 to 2023, and to clarify the temporal trend and spatial characteristics of each land-use type in Jilin Province. This study provides technical support for the accurate acquisition of LU/CC information, and the resulting LU/CC land-use change data in Jilin Province are of great significance for regional sustainable development.

## 2. Study Area and Data Source

### 2.1. Study Area

Jilin Province is located in northeastern China, between longitude 121°38′–131°19′ E and latitude 40°50′–46°19′ N, in the geometric center of northeast Asia, which is composed of Japan, Russia, South Korea, North Korea, Mongolia, and northeastern China. It is adjacent to North Korea and Russia in the east, Heilongjiang Province in the northeast, and Inner Mongolia Autonomous Region and Liaoning Province in the southwest, with a total border length of 1438.7 km. Jilin Province has significant differences in geomorphology, with the terrain tilting from southeast to northwest, bounded by the Big Black Mountain in the center, mountains and low hills in the east, and plains in the northwest. This gives it the characteristics of the three major plates, and it is rich in arable land resources, with a vast area of forests. It is a key forestry province of the country, and the Changbai Mountainous Area in the northeast is an important ecological barrier in China. Since the 21st century, intense human activities have led to drastic changes in Jilin Province, which have had a large impact on resources and the environment. Jilin Province has eight prefectural-level cities, namely, Changchun, Jilin, Siping, Liaoyuan, Tonghua, Baicheng, Baishan, and Songyuan, and the Yanbian Korean Autonomous Prefecture, with Changchun as the provincial capital, covering a total area of 187.4 thousand square kilometers, as shown in Figure 1.

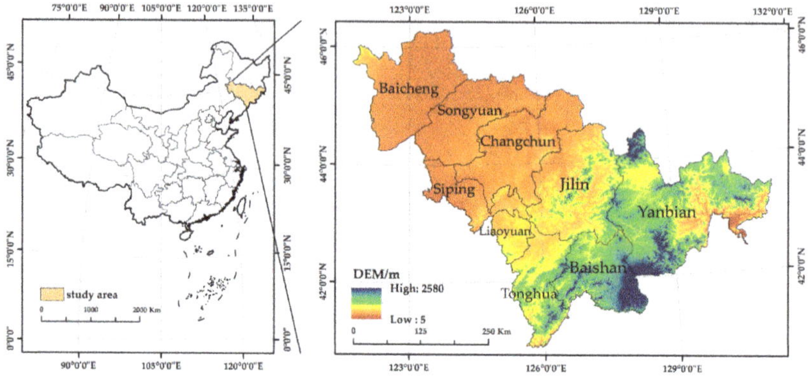


**Figure 1.** Location map of Jilin Province.

## 2.2. Data Sources

The GEE platform is a global-scale geospatial analysis tool that not only integrates massive geographic and remote sensing data resources, but also possesses powerful cloud computing capabilities, which provides an effective way to study geoscience-related issues [40,41].

### 2.2.1. Landsat Data

The satellite remote sensing data used in this paper were the Landsat Surface Reflectance SR dataset for the years 2000–2023. The 2005 and 2010 images are from Landsat 5 Thematic Mapper TM Imagery, the image from 2000 is from Landsat 7 Enhanced Thematic Mapper Plus (ETM+), and imagery from 2015, 2020, and 2023 is from Landsat 8 Operational Land Imager (OLI), as shown in Table 1. The data products from the TM and ETM+ sensors were atmospherically corrected using the Landsat Ecosystem Disturbance Adaptive Processing System (LEDAPS). The OLI sensor data products were atmospherically corrected using the Landsat 8 Surface Reflectance Code (LaSRC), which integrates USGS internal algorithms [42,43].

**Table 1.** Landsat satellite remote sensing data used in the study.

| Data Set | Satellite | Sensor | Resolution | Year | Selected Bands |
|---|---|---|---|---|---|
| LANDSAT/LT05/C01/T1_SR | Landsat 5 | TM | 30 m | 2005, 2010 | Blue, green, red, NIR, SWIR1, SWIR2 |
| LANDSAT/LE07/C01/T1_SR | Landsat 7 | ETM+ | 30 m | 2000 | Blue, green, red, NIR, SWIR1, SWIR2 |
| LANDSAT/LC08/C01/T1_SR | Landsat 8 | OLI | 30 m | 2015, 2020, 2023 | Blue, green, red, NIR, SWIR1, SWIR2 |

### 2.2.2. MODIS Data

MODIS, on board the Terra and Aqua satellites, is the most widely used and important optical sensor in the U.S. Earth Observation Satellite System (EOS) series; these were launched by the National Aeronautics and Space Administration (NASA) in December 1999 and May 2000, respectively [44]. As a continuation of NOAA/AVHRR, MODIS data have been greatly improved in terms of performance and quality. Its scanning width is 2330 km, and the maximum spatial resolution can reach 250 m. It can transit scan up to four times a day, so it has a better time resolution. There are 36 spectral bands, realizing the full spectral coverage of visible to red light bands, which greatly improves the spectral resolution, and strengthens the monitoring and identifying of information regarding the change of land surface cover [45]. MODIS data are rapidly updated, easily accessible, and globally free of charge, which greatly improves the data quality and timeliness of remote sensing monitoring [46,47].

The data selected for this study are MOD09A1 products, which are the standard products of the terrestrial L2 level, and the acquired time range is from 2000 to 2023, with a spatial resolution of 500 m and a temporal resolution of 8 days. The data contain seven major bands, the latter six of which were mainly selected to participate in the subsequent spatial–temporal fusion calculation, and data were acquired against Landsat data at intervals of no more than 3 days. Since the image data were from different sensors, it was necessary to adjust the order of the image bands to match (Table 2).

**Table 2.** The range of different sensor bands.

| Landsat 5 (TM) | | Landsat7 (ETM) | | Landsat 8 (OLI) | | MOD09A1 | |
|---|---|---|---|---|---|---|---|
| Band Name | Bandwidth/μm | Band Name | Bandwidth/μm | Band Name | Bandwidth/μm | Band Name | Bandwidth/μm |
| SR_B1 (blue) | 0.45–0.52 | SR_B1 (blue) | 0.45–0.52 | B2 (blue) | 0.452–0.512 | sur_refl_b03 | 0.459–0.479 |
| SR_B2 (green) | 0.52–0.60 | SR_B2 (green) | 0.52–0.60 | B3 (green) | 0.533–0.590 | sur_refl_b04 | 0.545–0.565 |
| SR_B3 (red) | 0.63–0.69 | SR_B3 (red) | 0.63–0.69 | B4 (red) | 0.636–0.673 | sur_refl_b01 | 0.620–0.670 |
| SR_B4 (nir) | 0.77–0.90 | SR_B4 (nir) | 0.77–0.90 | B5 (nir) | 0.851–0.879 | sur_refl_b02 | 0.841–0.876 |
| SR_B5 (SWIR1) | 1.55–1.75 | SR_B5 (SWIR1) | 1.55–1.75 | B6 (SWIR1) | 1.566–1.651 | sur_refl_b06 | 1.628–1.652 |
| SR_B7 (SWIR2) | 2.08–2.35 | SR_B7 (SWIR2) | 2.08–2.35 | B7 (SWIR2) | 2.107–2.294 | sur_refl_b07 | 2.105–2.155 |

## 3. Methods

In this paper, oriented to regional characteristics, the OL-STARFM fusion model was selected to generate high temporal and spatial fusion data of Landsat and MODIS based on the platform of the Google Earth Engine (GEE). The long time series LU/CC mapping and spatial and temporal pattern analysis from 2000 to 2023 were realized by using the random forest remote sensing image classification method that fuses remote sensing indices.

### 3.1. Data Selection and Pre-Processing

After selecting the satellite remote sensing data of the corresponding time phase in the GEE platform, shadow removal and cloud mask operations were carried out using cloud mask [48]. After that, the images of the selected time were synthesized using median filtering, i.e., assigning a median value to each pixel of the entire image stack, so that all remote sensing images of the study year were synthesized into one, and remote sensing images that clearly and completely displayed the surface coverage information of the study area were obtained [49,50].

### 3.2. OL-STARFM High-Spatial-and-Temporal-Resolution Data Fusion Model

STARFM combines spectrally similar neighborhood pixel variation information for spatio-temporal fusion, which can be expressed as:

$$\hat{F}_P(x,y,B) = \sum_{b=1}^{M}\sum_{i=1}^{N} W_b(x_i,y_i,B) \times [F_b(x_i,y_i,B) + C_P(x_i,y_i,B) - C_B(x_i,y_i,B)] \quad (1)$$

The coarse image was formed by resampling to a fine resolution by nearest neighbor interpolation. The principle of the OL processing version of STARFM (OL-STARFM) [25] after segmentation to process the auxiliary fine image can be expressed as:

$$F_p'(s,B) = F_b(s,B) + M[C_P(s,B) - C_b(s,B)] \quad (2)$$

where M[·] refers to taking the median of all pixel values in the object. OL-STARFM uses the median value rather than the expected value to eliminate the impact of poor-quality pixels for the whole segmented object. Especially for fusing high-quality observations, a residual compensation (RC) step is added afterwards to avoid generating uncertainties and to enhance the initial fusion results:

$$\hat{F}_P = \hat{F}_P' + R \quad (3)$$

The residual R represents the differences between the temporal changes of coarse images and fine images. The bicubic interpolation was adopted to remove blocky artifacts in the residual. The pixel-level computation performed in (3) does not incorporate adjacent similar information and, therefore, takes less time.

### 3.3. LU/CC Information Extraction Based on Random Forests with Fused Remote Sensing Indices

According to the actual situation of land use in the study area, land-use types were divided into six categories: cultivated land, grassland, forest, water, building land, and unused land, as shown in Table 3.

**Table 3.** Characteristics of land-use category and remote sensing image representation.

| Category | Definition | Features of Remote Sensing Images (Standard False Color Composition) |
| --- | --- | --- |
| Cultivated land | Farmland that can be cultivated normally in normal years |  |
| Grassland | Land dominated by herbaceous vegetation | 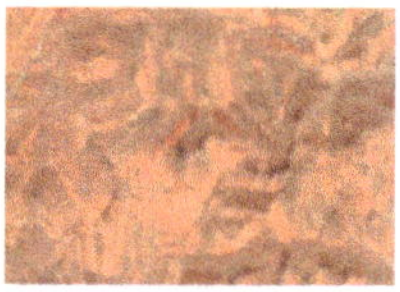 |
| Forest | Trees, shrubs, and other woodlands |  |
| Water | Natural land waters, land for water conservancy facilities and aquaculture ponds | 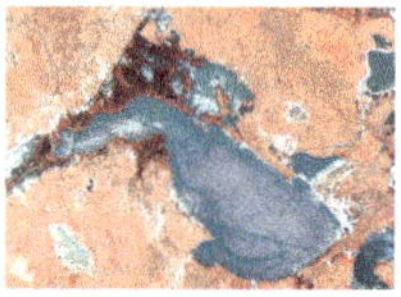 |
| Building | Urban and rural residential areas and industrial and mining transportation land | 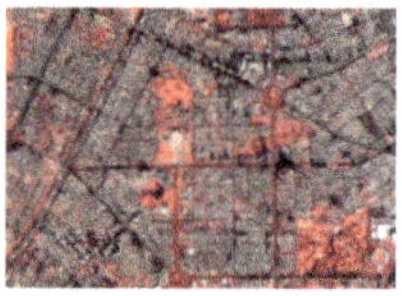 |
| Unused land | The surface is soil or rock, the land is largely unvegetated | 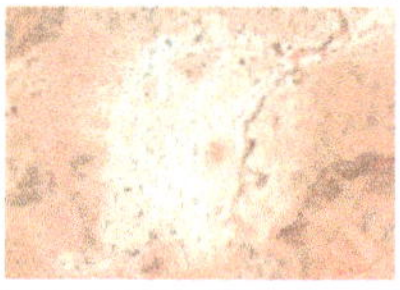 |

#### 3.3.1. Integration of Remote Sensing Indices

In order to improve the accuracy of classification, in addition to the surface reflectance data, this study used various types of remote sensing index information. In this study, in order to better learn the characteristics of each category, the normalized difference vegetation index (NDVI), the normalized difference building index (NDBI), and the modified normalized difference water index (MNDWI) were imported into the fused raw images to enhance the distinguishability of each feature category and improve the classification accuracy. The formula for calculating each index is as follows:

$$\text{NDVI} = \frac{NIR - RED}{NIR + RED} \tag{4}$$

$$\text{NDBI} = \frac{\text{SWIR} - \text{NIR}}{\text{SWIR} + \text{NIR}} \tag{5}$$

$$\text{MNDWI} = \frac{\text{GREEN} - \text{SWIR}}{\text{GREEN} + \text{SWIR}} \tag{6}$$

where NIR, RED, GREEN, and SWIR denote the surface reflectance in the near-infrared, red, green, and short-wave infrared bands of Landsat images, respectively.

#### 3.3.2. Classifier Construction Based on the Random Forest Algorithm

The random forest (RF) algorithm is a nonparametric regression method [51]. In contrast to other regression methods, the RF algorithm does not overfit, does not require variable selection, and includes built-in cross-validation methods that do not require a separate test dataset to evaluate performance [51]. In addition, the RF algorithm makes no distributional assumptions about predictor or response variables and can handle situations where the number of predictor variables greatly exceeds the number of observations. In this study, the number of decision trees was set to 100, the highest accuracy decision tree is selected for classification, and default values were selected for the remaining parameters such as the number of split variables, maximum leaf nodes, and randomized seeds. For sample points, 80% was used for classifier training and 20% for accuracy verification.

#### 3.3.3. Evaluation of the Accuracy of Classification Results

Accuracy evaluation is an essential step in remote sensing information extraction and target recognition, which not only evaluates the accuracy of the results, but also serves as a reference for evaluating the performance of the method and optimizing the parameters [52,53]. Overall accuracy (OA) refers to the distribution of the number of correctly classified image elements along the diagonal of the confusion matrix, which directly reflects the proportion of correctly classified elements, and the kappa coefficient is used to test whether the model prediction results and the actual classification results are consistent. The formulae for the OA and the kappa coefficient are shown below:

$$\text{OA} = \left(\frac{\text{P}_\text{C}}{\text{P}_\text{N}}\right) \cdot 100 \tag{7}$$

$$\text{Kappa} = \frac{\text{N}\sum_{i=1}^{r} x_{ii} - \sum_{i=1}^{r}(x_{i+} \times x_{+i})}{\text{N}^2 - \sum_{i=1}^{r}(x_{i+} \times x_{+i})} \tag{8}$$

where $\text{P}_\text{c}$ is the number of correctly categorized pixels, $\text{P}_\text{N}$ is the total number of pixels. r is the number of rows and columns in the error matrix, $x_{ii}$ is the number of observations in row i and column i, $x_{i+}$ is the marginal total number of rows i, $x_{+i}$ is the marginal total number of columns i, and N is the total number of observations.

## 4. Results

### *4.1. Accuracy Test of the OL-STARFM Spatio-Temporal Fusion Data*

In order to know whether the fusion-generated high-temporal-and-spatial-resolution surface reflectance data can accurately and reliably reflect the real surface cover changes and participate in the subsequent land-cover classification, we need to verify the accuracy of the prediction results. In this study, the Landsat_SR data image taken on 7 October 2020 and the MODIS_SR surface reflectance data images on taken 7 October 2020 and 23 October 2020 were selected as input for into the OL-STARFM model, and the predicted results were verified using the Landsat_SR data image taken on 23 October 2020.

In order to carry out the visual inspection accurately, we selected areas of the same size at typical land types such as cultivated land, forest land, water area, and urban area for detailed observation (Table 4). The visual inspection showed that the prediction results are closer to the real image, and also highlighted the spatial detail distribution characterized by the 30 m spatial resolution of the Landsat data.

**Table 4.** Comparison of different ground conditions between real and fusion images.

| Category | Real Image (True Color Composition) | Fusion Image (True Color Composition) |
|---|---|---|
| Cultivated land |  |  |
| Grassland |  |  |
| Forest | 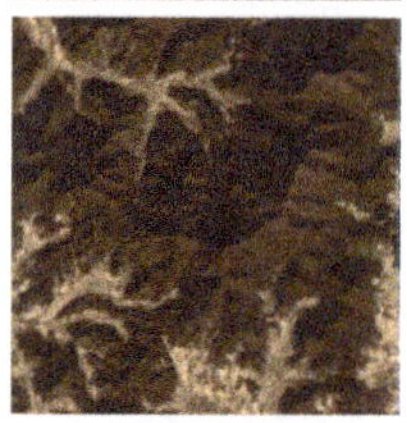 |  |
| Water | 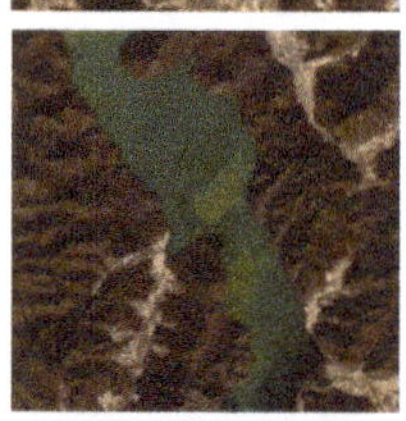 | 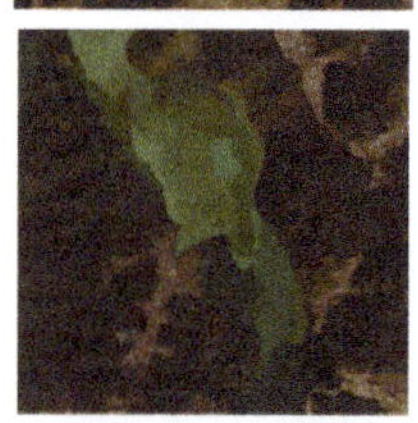 |
| Building |  |  |
| Unused land | 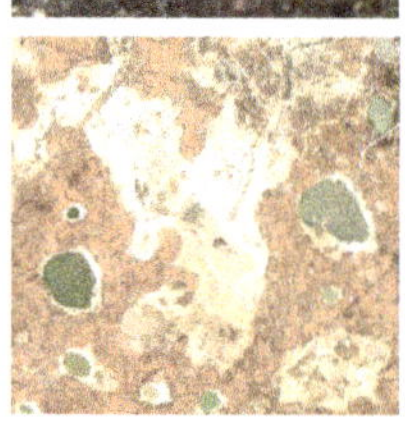 | 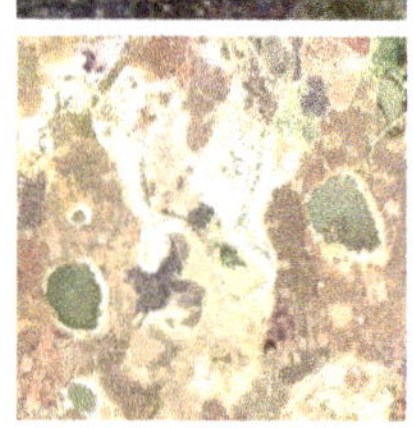 |

The results from visual observation alone did not accurately represent the precision of the fusion results, so we needed to quantitatively analyze the predicted results. The correlation coefficients between the predicted data bands and Landsat_SR data bands were calculated by randomly selecting 1000 points in the study area with different corresponding bands, and the correlation coefficients between the different bands were positive, with Pearson correlation coefficients of more than 0.85, representing a high degree of correlation. After that, the Pearson correlation test was also performed on the calculated and fused remote sensing indices, and all three indices showed a positive correlation. The Pearson correlation coefficient of the NDVI was 0.8959, the Pearson correlation coefficient of the MNDWI was 0.8918, and the Pearson correlation coefficient of the NDBI was 0.6857, which were all relatively good correlations. The above results show that the fusion results better contain the high temporal–low spatial- and high spatial–low temporal-resolution image information, and could be applied to the subsequent land-cover classification study.

Figures 2 and 3 show the fitting results between different bands and remote sensing indices in the real and predicted images.

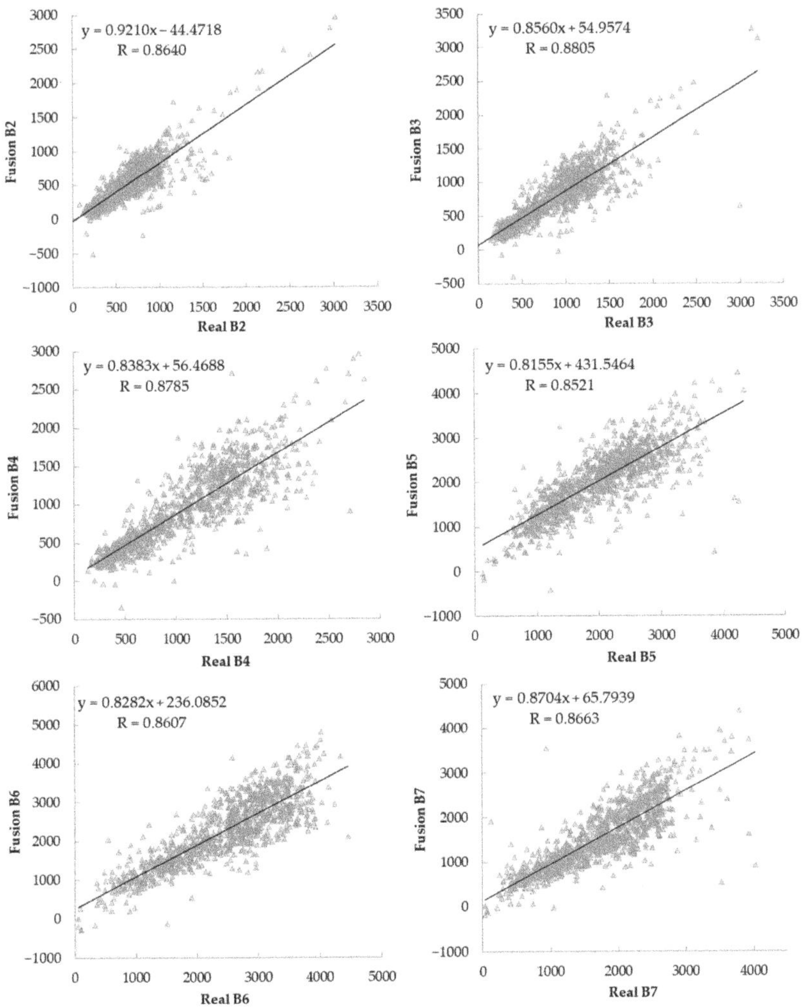


**Figure 2.** Correlation of different bands in the real and fusion images.

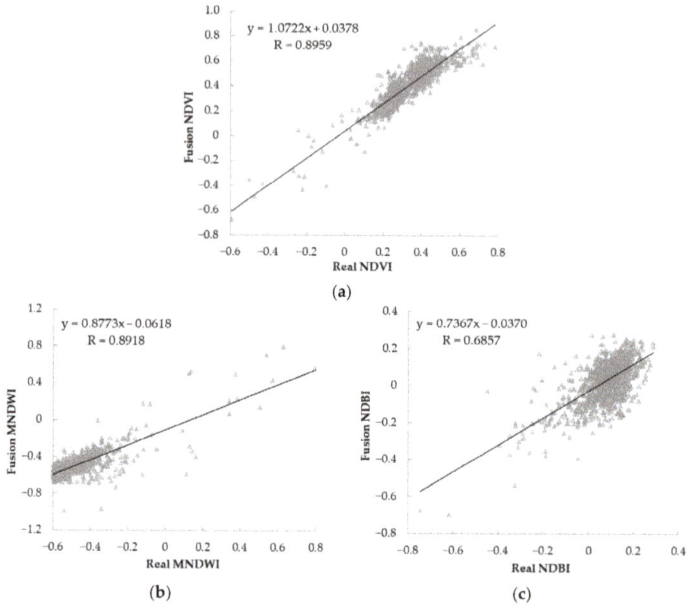


**Figure 3.** Correlation between different remote sensing indices of the real and fusion images (**a**) correlation between the NDVI of the real and fusion images; (**b**) correlation between the MNDWI of the real and fusion images; (**c**) correlation between the NDBI of the real and fusion images.

### *4.2. Results of Remote Sensing Extraction of Land-Use/Cover Change Information in Jilin Province*

In this study, we developed a random forest remote sensing image classification method based on the fusion of Landsat and Modis data with long time series high-spatial- and high-temporal-resolution and fusion of remote sensing indices, and extracted LU/CC information with high spatial and temporal resolution of multiple time-phases in the study area, with the support of the GEE platform, and finally synthesized the annual average data, as shown in Figure 4.

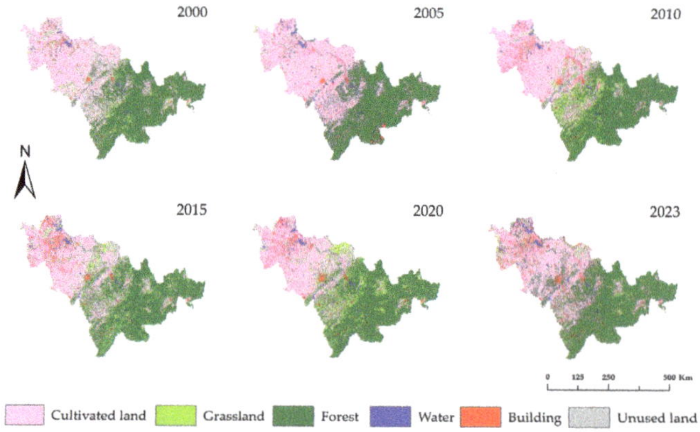


**Figure 4.** Classification of LU/CC in Jilin Province, 2000–2023.

The images, after fusing the remote sensing indices, show enhanced distinguishability of each feature class and improved classification accuracy. Table 5 demonstrates the significant improvement in the accuracy of the classification results of the OA and kappa coefficient evaluation of the confusion matrix before and after fusion of the remote sensing indices, when the training samples were consistent with all other parameters.

**Table 5.** Comparison of accuracy before and after fusion of the remote sensing indices; 1 denotes cultivated land, 2 denotes grassland., 3 denotes forest, 4 denotes water, 5 denotes building land, and 6 denotes unused land, using the 23 October 2020 fusion data as an example.

| **Before Fusion Remote Sensing Index** | | | | | | |
|---|---|---|---|---|---|---|
| | 1 | 2 | 3 | 4 | 5 | 6 |
| 1 | 17 | 0 | 0 | 0 | 1 | 0 |
| 2 | 4 | 10 | 0 | 0 | 0 | 0 |
| 3 | 0 | 1 | 22 | 0 | 1 | 0 |
| 4 | 0 | 0 | 0 | 14 | 0 | 0 |
| 5 | 0 | 0 | 2 | 0 | 17 | 2 |
| 6 | 0 | 0 | 0 | 0 | 0 | 17 |
| Overall Accuracy = 89.81% | | | | | | |
| Kappa = 0.8767 | | | | | | |

| **After Fusion Remote Sensing Index** | | | | | | |
|---|---|---|---|---|---|---|
| | 1 | 2 | 3 | 4 | 5 | 6 |
| 1 | 20 | 2 | 0 | 0 | 0 | 0 |
| 2 | 2 | 13 | 0 | 0 | 0 | 0 |
| 3 | 0 | 0 | 19 | 0 | 0 | 0 |
| 4 | 0 | 0 | 0 | 20 | 0 | 0 |
| 5 | 0 | 1 | 1 | 0 | 17 | 1 |
| 6 | 0 | 0 | 0 | 0 | 0 | 20 |
| Overall Accuracy = 93.97% | | | | | | |
| Kappa = 0.9274 | | | | | | |

The OA and kappa coefficients were calculated based on the confusion matrix to evaluate the accuracy of the classification results, and the confusion matrix was normalized. The results are shown in Table 6 and Figure 5. As can be seen from the table, the OA and kappa coefficients from 2000 to 2023 are greater than 92.96% and 0.9134, respectively, and the average values are 95.11% and 0.9394, respectively, indicating that the classification results are more accurate and could provide more accurate data support for the subsequent analysis of the evolution of spatio-temporal patterns.

**Table 6.** Classification accuracy.

| Year | 2000 | 2005 | 2010 | 2015 | 2020 | 2023 | Mean |
|---|---|---|---|---|---|---|---|
| Overall accuracy/% | 93.09 | 94.55 | 92.96 | 96.48 | 97.24 | 96.31 | 95.11 |
| Kappa | 0.9148 | 0.9323 | 0.9134 | 0.9563 | 0.9658 | 0.9537 | 0.9394 |

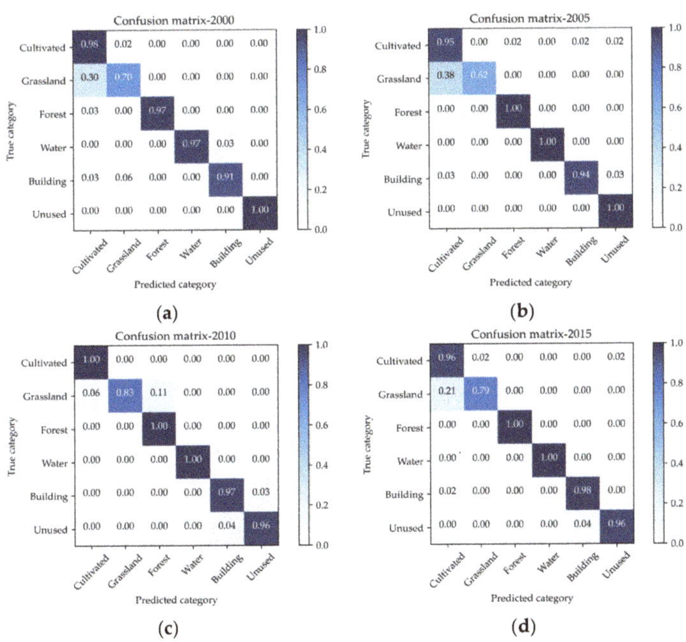


**Figure 5.** *Cont.*

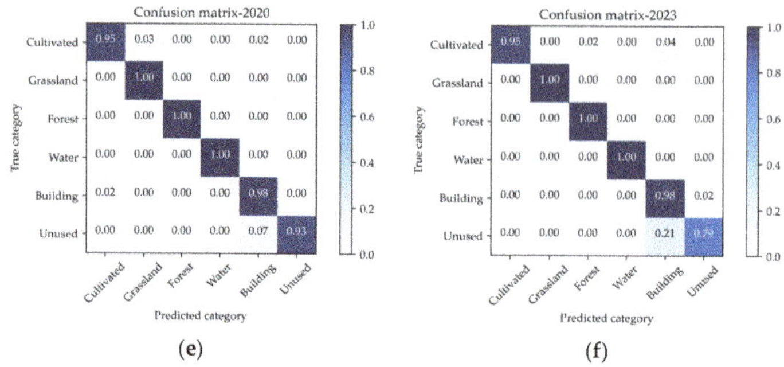


**Figure 5.** Period-by-period normalized confusion matrix; (**a**) 2000 normalized confusion matrix; (**b**) 2005 normalized confusion matrix; (**c**) 2010 normalized confusion matrix; (**d**) 2015 normalized confusion matrix; (**e**) 2020 normalized confusion matrix; (**f**) 2023 normalized confusion matrix.

## *4.3. Evolution of the Spatial and Temporal Patterns of Land-Use/Cover Change in Jilin Province*

### 4.3.1. Changes in Time Trend

In this study, we counted indicators such as the area, the amount of area change, and the rate of area change of cultivated land, grassland, forest, water, building land, and unused land, and the results are shown in Table 7.

**Table 7.** Land-use type area, area change, and rate of area change by period.

| Year | Index | Cultivated Land | Grassland | Forest | Water | Building | Unused Land |
|---|---|---|---|---|---|---|---|
| 2000 | Area (km$^2$) | 80,538.54 | 17,443.88 | 76,389.85 | 2718.03 | 4718 | 5591.7 |
| 2005 | | 91,559.24 | 4176.38 | 77,037.34 | 3009.07 | 6947.44 | 4670.53 |
| 2010 | | 78,762.75 | 16,415.19 | 78,576.78 | 2526.83 | 8811.64 | 2306.81 |
| 2015 | | 72,362.71 | 22,538.86 | 76,247.99 | 3346.58 | 9253.92 | 3649.94 |
| 2020 | | 67,083.93 | 22,832.24 | 80,925.95 | 3410.7 | 10,841.82 | 2305.34 |
| 2023 | | 71,951.81 | 11,773.05 | 82,513.3 | 4158.24 | 13,442.27 | 3561.3 |
| 2000–2005 | Area change /km$^2$ | 11,020.7 | −13,267.5 | 647.49 | 291.04 | 2229.45 | −921.17 |
| 2005–2010 | | −12,796.5 | 12,238.8 | 1539.45 | −482.23 | 1864.19 | −2363.73 |
| 2010–2015 | | −6400.04 | 6123.67 | −2328.79 | 819.74 | 442.29 | 1343.13 |
| 2015–2020 | | −5278.78 | 293.38 | 4677.95 | 64.13 | 1587.9 | −1344.6 |
| 2020–2023 | | 4867.89 | −11,059.2 | 1587.36 | 747.54 | 2600.45 | 1255.97 |
| 2000–2005 | Rate of area change/% | 13.68 | −76.06 | 0.85 | 10.71 | 47.25 | −16.47 |
| 2005–2010 | | −13.98 | 293.05 | 2 | −16.03 | 26.83 | −50.61 |
| 2010–2015 | | −8.13 | 37.30 | −2.96 | 32.44 | 5.02 | 58.22 |
| 2015–2020 | | −7.29 | 1.30 | 6.14 | 1.92 | 17.16 | −36.84 |
| 2020–2023 | | 7.26 | −48.44 | 1.96 | 21.92 | 23.99 | 54.48 |

(1) Cultivated land. The area and the amount of change in the area of cultivated land in different study years were counted, as shown in Figure 6. Through Figure 6 and Table 7, it can be seen that the area of cultivated land increased from 80,538.54 km$^2$ in 2000 to 91,559.24 km$^2$ in 2005, and then decreased continuously to 67,083.93 km$^2$ in 2020. It then increased slightly to 71,951.81 km$^2$ in 2023. Overall, cultivated land decreased by an average of 357.78 km$^2$ per year, with rates of change of 13.68%, −13.98%, −8.13%, −7.29%, and 7.26% in each phase.

(2) Grassland. The area and the amount of change in the area of grassland in different study years were counted, as shown in Figure 7. As can be seen from Figure 7 and Table 7, during the study period, contrary to cultivated land, the area of grassland as a whole showed a trend of decreasing, then increasing, and then decreasing, decreasing from 17,443.88 km$^2$ in 2000 to 4176.38 km$^2$ in 2005, then increasing to 22,832.24 km$^2$ in 2020, and decreasing to 11,773.05 km$^2$ in 2023. Overall, grassland decreased by an

average of 236.28 km$^2$ per year, with rates of change of −76.06%, 293.05%, 37.30%, 1.30%, and −48.44% in each phase.

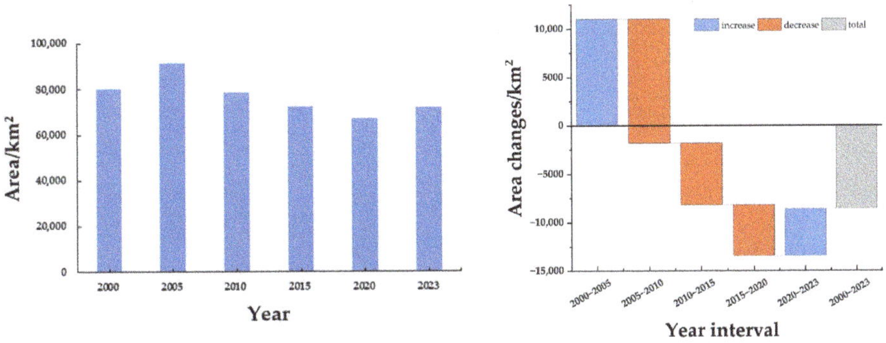


**Figure 6.** Changes in the area of cultivated land in Jilin Province, 2000–2023.

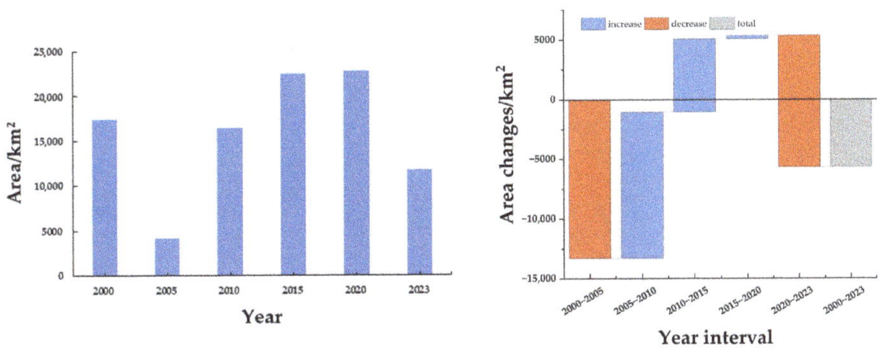


**Figure 7.** Changes in the area of grassland in Jilin Province, 2000–2023.

(3) Forest. The area and the amount of change in the area of forest land in different study years were counted, as shown in Figure 8. As can be seen from Figure 8 and Table 7, the area of forest increased steadily, with an average annual growth of 255.14 km$^2$, and rates of change of 0.85%, 2%, −2.96%, 6.14%, and 1.96% in each phase.

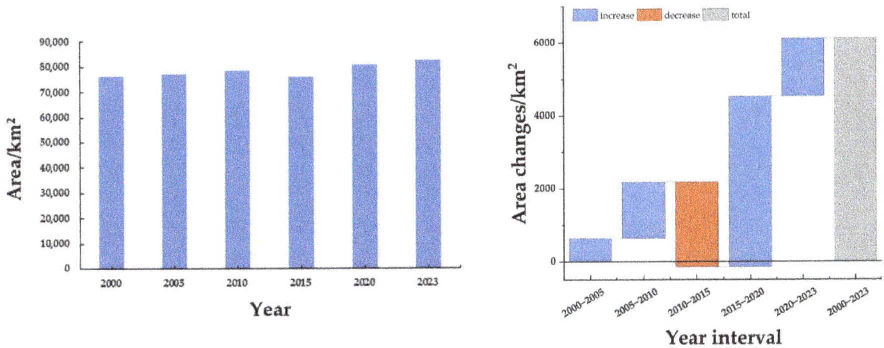


**Figure 8.** Changes in the area of forest in Jilin Province, 2000–2023.

(4) Water. The area and the amount of change in the area of the water body in different study years were counted, as shown in Figure 9. From Figure 9 and Table 7, it can

be seen that the overall trend of water area was increasing, with an average annual increase of 60.01 km$^2$, but there was a fluctuating trend in the area of the water in individual study years, with a rate of change of 10.71%, −16.03%, 32.44%, 1.92%, and 21.92% for the five phases.

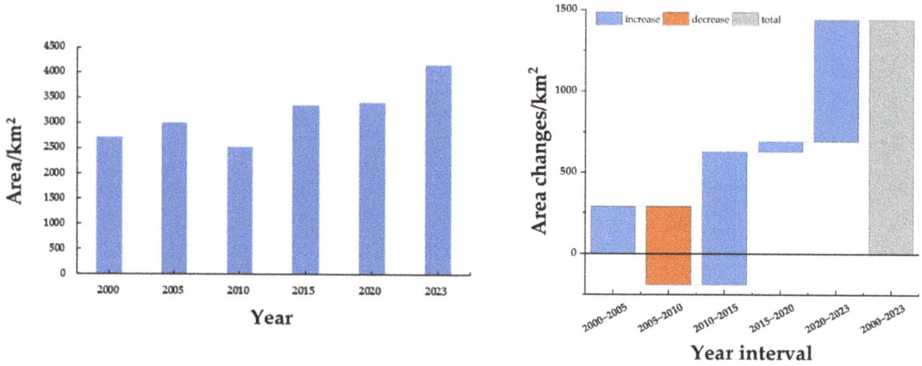


**Figure 9.** Changes in the area of water in Jilin Province, 2000–2023.

(5) Building land. The area of construction land and the amount of area change in different study years were counted, as shown in Figure 10. It can be seen from Figure 10 and Table 7 that during the study period, the area of building land continuously increased from 4718 km$^2$ in 2000 to 13,442.27 km$^2$ in 2023, with an average annual increase of 363.51 km$^2$. The area of building land in the five change phases increased by 2229.45 km$^2$, 1864.19 km$^2$, 442.29 km$^2$, 1587.9 km$^2$, and 2600.45 km$^2$, corresponding to a rate of change of 47.25%, 26.83%, 5.02%, 17.16%, and 23.99%, respectively.

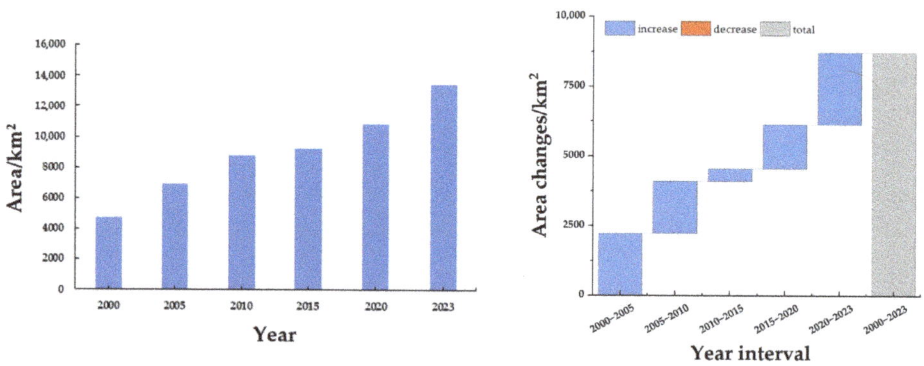


**Figure 10.** Changes in the area of building land in Jilin Province, 2000–2023.

(6) Unused land. The area of bare land and the amount of change in the area in different study years were counted, as shown in Figure 11. As can be seen from Figure 11 and Table 7, the unused land area showed an overall decreasing trend during the study period, with an average annual decrease of 84.6 km$^2$, but there were repeated increases and decreases in 2010–2023, with rates of change of −16.47%, −50.61%, 58.22%, −36.84%, and 54.48% in each phase.

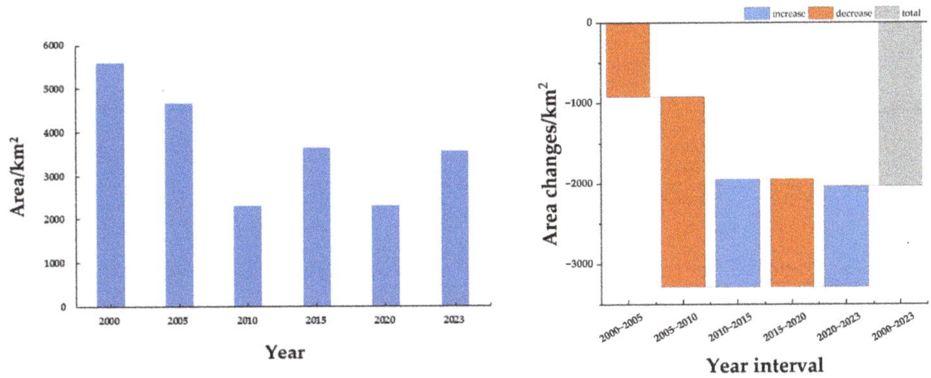


**Figure 11.** Changes in the area of unused land in Jilin Province, 2000–2023.

### 4.3.2. Changes in Spatial Characteristics

The land-use transfer matrix can visualize the type and amount of land use that has changed, which has significance in clarifying the characteristics and laws of transformation between land-use types in each period. This study extracted the land-use change map and land-use transfer matrix for the first and last study years and the interval study years during the period 2000–2023, as shown in Figure 12 and Table 8.

**Table 8.** Land-use transfer matrix for different phases.

| Type of transfer-out | Type of transfer-in | | | | | | | | | | | |
|---|---|---|---|---|---|---|---|---|---|---|---|---|
| | Land transfer matrix (2000–2005) | | | | | | Land transfer matrix (2005–2010) | | | | | |
| | Cultivated | Grassland | Forest | Water | Building | Unused | Cultivated | Grassland | Forest | Water | Building | Unused |
| Cultivated | 68,926.42 | 3260.86 | 4182.32 | 310.6 | 2624.38 | 395.86 | 68,244.91 | 13,182.89 | 5392.66 | 88.72 | 3725.98 | 92.32 |
| Grassland | 13,368.64 | 654.9 | 2553.62 | 128.25 | 665.89 | 33.66 | 3445.67 | 468.72 | 117.58 | 5.34 | 64.67 | 1.33 |
| Forest | 5582.88 | 113.83 | 71,152.05 | 72.69 | 523.05 | 69.79 | 3026.1 | 2555.99 | 72,257.95 | 100.93 | 201.58 | 3.78 |
| Water | 236.08 | 9.21 | 179.73 | 2043.33 | 143.44 | 67.28 | 259.6 | 12.48 | 97.12 | 2207.91 | 291.96 | 95.57 |
| Building | 1512.71 | 39.03 | 76.44 | 310.81 | 2193.76 | 513.48 | 2721.17 | 177.56 | 583.96 | 102.91 | 3152.45 | 155.21 |
| Unused | 1102.4 | 25.22 | 1.99 | 97.82 | 742.42 | 3485.16 | 1061.3 | 18.12 | 130.23 | 22.76 | 1373.27 | 1959.3 |
| | Land transfer matrix (2010–2015) | | | | | | Land transfer matrix (2015–2020) | | | | | |
| Cultivated | 57,420.24 | 10,969.68 | 4057.02 | 535.46 | 4573.56 | 1202.78 | 53,214.63 | 12,413.53 | 1877.7 | 147.49 | 3715.31 | 236.69 |
| Grassland | 7826.63 | 5076.89 | 3241.43 | 43.51 | 211.54 | 15.75 | 6828.34 | 8478.7 | 6022.84 | 67.24 | 1110.28 | 6.81 |
| Forest | 2753.04 | 5665.19 | 69,850.89 | 190.83 | 104.77 | 14.79 | 1359.45 | 1596.81 | 74,129.29 | 19.28 | 205.76 | 0.92 |
| Water | 81.22 | 53.95 | 40.83 | 2152.89 | 112.55 | 87.12 | 284.7 | 40.19 | 47.46 | 2792.33 | 125.8 | 12.18 |
| Building | 3365.56 | 741.96 | 120.92 | 317.86 | 3438.63 | 824.99 | 3439.04 | 268.88 | 19.94 | 213.38 | 4589.41 | 569.17 |
| Unused | 158.65 | 6.54 | 0.43 | 62.1 | 658.77 | 1421.02 | 1041.09 | 14.14 | 2.69 | 120.96 | 961.57 | 1426 |
| | Land transfer matrix (2020–2023) | | | | | | Land transfer matrix (2000–2023) | | | | | |
| Cultivated | 48,756.27 | 5294.46 | 4465.19 | 608.86 | 6015.31 | 1027.17 | 54,730.78 | 7682.48 | 8374.11 | 857.29 | 7244.52 | 811.25 |
| Grassland | 14,038 | 4180.71 | 2902.48 | 71.8 | 1597.71 | 21.55 | 9942.33 | 1979.99 | 3261.63 | 359.51 | 1788.16 | 73.33 |
| Forest | 5050.48 | 1161.81 | 75,706.55 | 18.62 | 158.86 | 3.6 | 4265.32 | 1190.8 | 71,663.49 | 76.78 | 303.06 | 14.86 |
| Water | 22.76 | 75.56 | 78.87 | 2892.59 | 114.3 | 176.59 | 143.65 | 104.74 | 95.3 | 2021.27 | 141.08 | 173.01 |
| Building | 3160.69 | 953.43 | 435 | 415.28 | 4730.36 | 1013.37 | 1004.21 | 447.4 | 173.7 | 495.34 | 2021.19 | 504.4 |
| Unused | 250.91 | 22.95 | 4 | 80.31 | 657.77 | 1235.84 | 1194.46 | 283.16 | 24.72 | 275.88 | 1775.43 | 1901.37 |

As can be seen from Figure 12, the land-use pattern in Jilin Province underwent significant changes between 2000 and 2023. Cultivated land was the most important source of land transfer-out, decreasing from 80,538.54 $km^2$ in 2000 to 71,951.81 $km^2$ in 2023, with the proportion of the total area decreasing from 42.98% to 38.39%. During the 24-year period, cultivated land was mainly transferred to grassland, forest, and building land, with areas of transferring out of 7682.48 $km^2$, 8374.11 $km^2$, and 7244.52 $km^2$, respectively. Cultivated land was transferred out to other feature types and also received the transfer-in of other features, among which grassland and forest were the most significant, with an area of 9942.33 $km^2$ and 4265.32 $km^2$. The land transfer among the other feature types was relatively small, at less than 3300 $km^2$.

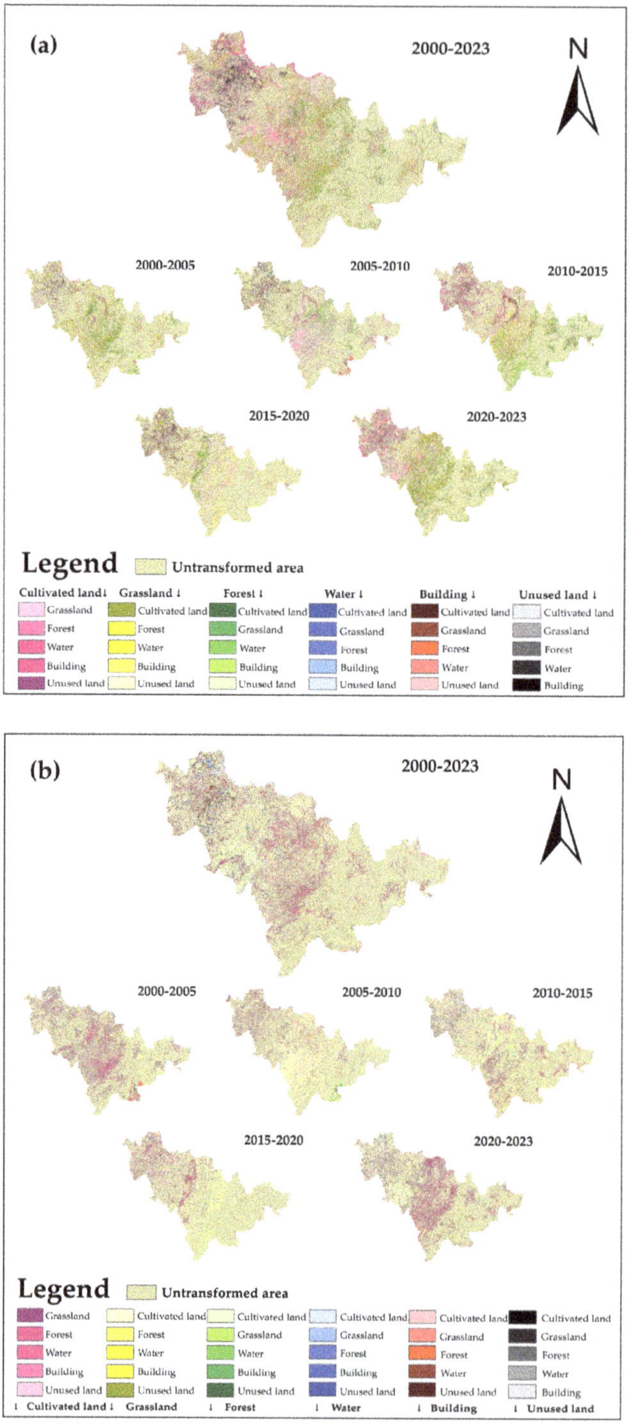


**Figure 12.** Map of land-use transfers out and in at different phases. (**a**) map of land-use transfers out; (**b**) map of land-use transfers in. The "↓" indicates: "transfer to".

In order to more clearly describe the flow and diversity of LU/CC changes, the land-use transfer was categorized into three types, namely, transfer-in greater than transfer-out (grasslands, water, and building land), transfer-out is greater than transfer-in (cultivated land, and unused land), and transfer-in transfer-out change class (forest). The conversion relationship between the LU/CC types in different periods was quantitatively expressed using chordal diagrams, as shown in Figure 13.

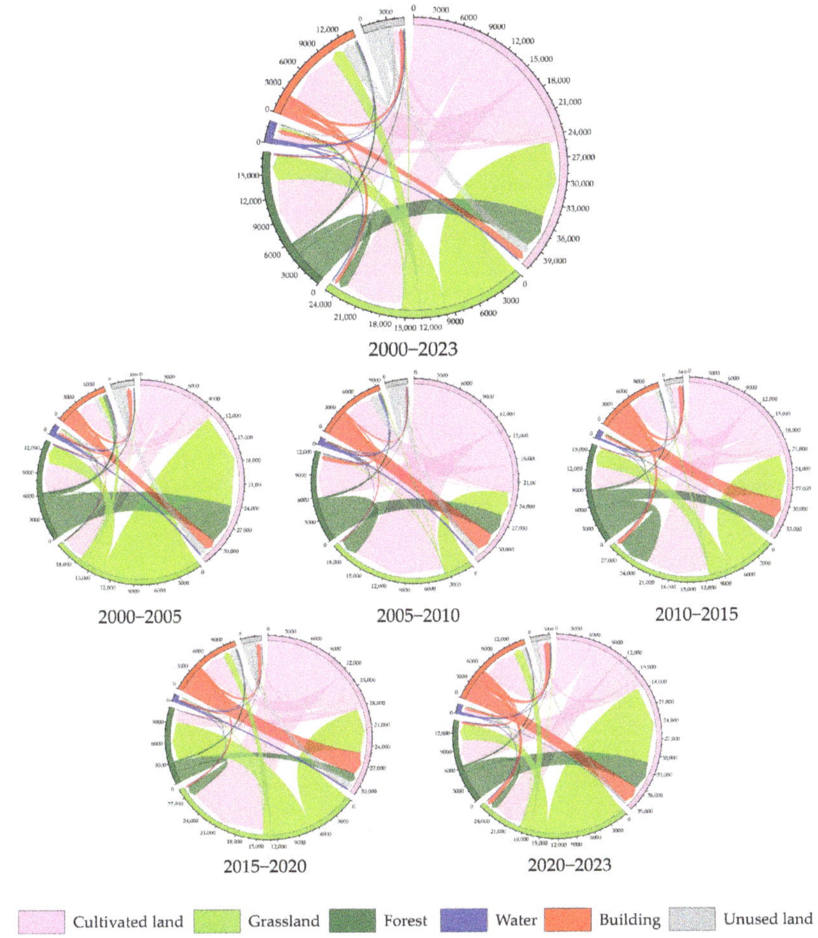


**Figure 13.** Land transfer chord diagrams at different phases.

As can be seen from Figures 12 and 13, during the study period, construction land in Jilin Province increased rapidly with the expansion of urban agglomeration and constantly encroached on the surrounding cultivated land, and the land transfer between cultivated land and grassland, cultivated land and forest land, and cultivated land and building land was the most obvious.

(1) Cultivated land. The total area of cultivated land decreased, mainly in the western part of Jilin Province, where agriculture predominates, and in the areas of Changchun City, Jilin City, and Yanbian Prefecture, which corresponded with the expansion of construction land and tourism development. During the study period, a total of 24,969.65 $km^2$ was transferred, of which 33.54% was converted to forest, 30.77% was converted to grassland, and 29.01% was converted to building land. In terms of

transfer-in, water, building land, and unused land have always been the stable transfer-in types of cultivated land, and the largest areas transformed into cultivated land were 284.7 km$^2$ (2015–2020), 3439.04 km$^2$ (2015–2020), and 1102.4 km$^2$ (2000–2005). The area of grassland and forest transferred to cultivated land was unstable and fluctuated.

(2) Grassland. Grassland makes up a relatively small proportion of Jilin Province, and is mainly distributed at in the eastern forest edge area of Jilin City and a small part of Changchun City. With the implementation of the policy of returning cultivated land to forest and grassland and the reconstruction of the ecological environment, the area of grassland increased between 2005 and 2020, but declined in 2020–2023. In the whole study period, under the influence of the policy of returning farmland to forest and grassland, the main transfer of grassland area occurred between cultivated land and grassland. In 24a, both the maximum area transferred in and the maximum area transferred out of grassland were cultivated land, at 7682.48 km$^2$ and 9942.33 km$^2$. In the interval study, the area of cultivated land transferred to grassland was the largest, with a maximum interval change of 13,182.89 km$^2$ (2005–2010), which was strongly influenced by institutional factors. The largest type of grassland transferred was also cultivated land, with an area of 13,368.64 km$^2$ in 2000–2005.

(3) Forest. The forest land is mainly concentrated in the eastern part of the study area and in the green spaces of a few cities in the central and western regions, such as Changchun, Jilin, and Liaoyuan. The spatial pattern is relatively stable, and it is a land type with a low frequency of change. The main types of transfer-in and transfer-out are cultivated land. In terms of transfer-in, the main transfer-in source of forest is cultivated land, and the largest transfer-in area was 5392.66 km$^2$ (2005–2010). The largest transfer source was grassland from 2015 to 2020, at 6172.14 km$^2$. In terms of transfer-out, forest was mainly diverted to grassland and cultivated land, with maximum diversion areas of 5665.19 km$^2$ (2010–2015) and 5582.88 km$^2$ (2000–2005), respectively.

(4) Water. The changes in the water body area in the study area are relatively stable, and mainly distributed in lakes in the western platform plain area of Jilin Province and the five river systems flowing through the province: Songhua River, Liao River, Yalu River, Tumen River, and Suifen River. The main reason for the changes in water was the mutual transformation between water and other land features such as cultivated land. In terms of transfer-in, the main sources were cultivated land and building land, and the largest transfer-in areas were 608.86 km$^2$ (2020–2023) and 415.28 km$^2$ (2020–2023). In terms of transfer-out, building land and cultivated land were the main transfer-out areas, and the largest transfer-out areas were 291.96 km$^2$ (2005–2010) and 284.7 km$^2$ (2015–2020), respectively.

(5) Building land. Building land showed a trend of continuous expansion, and the development tended to start from a city cluster and spread around. The expansion area was mainly concentrated in central and western Jilin Province, with the urban expansion of Changchun City and the western platform plain cities being the most obvious. The area of building land increased year by year. In terms of transfer-in, the main sources of building land transfer-in were cultivated land, grassland, and unused land, and the largest transfer areas were 6015.315 km$^2$ (2020–2023), 1597.71 km$^2$ (2020–2023), and 1373.27 km$^2$ (2005–2010). In terms of transfer-out, the transfer-out area of building land was mainly cultivated land and unused land, and the maximum transfer-out area was 3439.04 km$^2$ (2015–2020) and 1013.37 km$^2$ (2020–2023).

(6) Unused land. The unused land area decreased, occasionally fluctuated, and the spatial distribution was relatively small and mainly distributed in the west. Saline–alkali land was more likely to change. In terms of transfer-out, unused land was mainly transferred to building land and cultivated land, and the largest transfer areas were 1373.27 km$^2$ (2005–2010) and 1102.4 km$^2$ (2000–2005). In terms of transfer-in, the main sources of transfer were also cultivated land and building land, with the largest transfer areas of 1202.78 km$^2$ (2010–2015) and 1013.37 km$^2$ (2020–2023).

## 5. Discussion

### *5.1. Attitudes towards Land-Use Change Dynamics*

Table 9 shows the one-way dynamic attitudes of different land-use types in Jilin Province in each time period. Grassland had higher values of unidirectional motivation in 2000–2010, especially in 2005–2010, when the unidirectional motivation was as high as 58.61%; the next largest change was in unused land, which began to produce large recurring changes after 2005. The highest single-movement attitudes for building land were 9.45% in the period 2000–2005, followed by 8% in the period 2020–2023. Changes in cultivated land and forests were relatively small in all periods; the single dynamic attitude of water underwent repeated changes. The land-use type with the largest single-movement attitude over the 24a was building land at 8.04%, and the smallest was forest land at just 0.35%.

**Table 9.** Single-movement attitudes to land-use change at different phases.

| | Cultivated | Grassland | Forest | Water | Building | Unused |
|---|---|---|---|---|---|---|
| 2000–2005 | 2.74% | −15.21% | 0.17% | 2.14% | 9.45% | −3.29% |
| 2005–2010 | −2.80% | 58.61% | 0.40% | −3.21% | 5.37% | −10.12% |
| 2010–2015 | −1.63% | 7.46% | −0.59% | 6.49% | 1.00% | 11.64% |
| 2015–2020 | −1.46% | 0.26% | 1.23% | 0.38% | 3.43% | −7.37% |
| 2020–2023 | 2.42% | −16.15% | 0.65% | 7.31% | 8.00% | 18.16% |
| 2000–2023 | −0.46% | −1.41% | 0.35% | 2.30% | 8.04% | −1.58% |

The calculated integrated land-use dynamics for the five time periods 2000–2005, 2005–2010, 2010–2015, 2015–2020, and 2020–2023 were 1.51%, 1.67%, 0.93%, 0.71%, and 1.97%, in that order (as shown in Table 10). Although the individual land-use types showed different rates of change in different time periods, at the regional level, the combined dynamics of the five time periods showed an increasing, then decreasing, and then increasing trend. During the first 20 years, the integrated motivation was greatest during the period 2000–2005, indicating that land use in the region changed strongly and gradually intensified during the first period, while it gradually decreased during the latter period. Therefore, the rate of overall land-use change in the region has gradually slowed down since then, which suggests that the intensity of the combined effect of natural and anthropogenic factors on the land-use system in the region has been weakening.

**Table 10.** Integrated dynamic attitudes to land-use change at different phases.

| | 2000–2005 | 2005–2010 | 2010–2015 | 2015–2020 | 2020–2023 | 2000–2023 |
|---|---|---|---|---|---|---|
| Integration of dynamic attitudes | 1.51% | 1.67% | 0.93% | 0.71% | 1.97% | 0.38% |

### *5.2. Frequency Analysis of Land-Use Change*

By calculating the number of changes in land-use types, land-use data for five time-phases were frequency-calculated and classified into four categories: no-change areas, low-change areas with 1–2 changes, middle-change areas with 3–4 changes, and high-change areas with 5 changes. Further frequency analysis of land-use changes was carried out.

Figure 14 shows that the frequency of land-use change in Jilin Province was dominated by no-change areas, accounting for 50.84% of the province's land area, mainly distributed in the forests in the eastern part of Jilin Province, and Songyuan, which is dominated by cultivated land. Cultivated land and forests constitute two of the largest land-use types in Jilin Province, and their land-use types are stable. The low-change area account for 29.82% of the province's area, dominating the total change area, and the main land-use changes were 1–2 type changes. These were mainly distributed in urban areas and the edge of cultivated land and forests, as well as part of the unused land in Baicheng City in western Jilin Province, which is mainly due to the expansion of agriculture and the process of urbanization leading to change to cultivated land and building land, and at the same time the improvement in unused land. The proportion of middle-change areas is relatively small, accounting for 17.54%, and the middle-change areas were mainly located in the northeast of Baicheng, the central part of Changchun, and the northwest of Yanbian. The areas of high change were the smallest, at just 1.8%. There were fewer areas where land-use changes occurred many times in 24a, and all land-use types were basically stable. Constant change leads to unstable land-use types, which may lead to damage to the land. The results show that frequent land type planning is being carried out in areas of high change, and that more attention should be paid to land in areas of high frequency change in order to carry out accurate planning assessments.

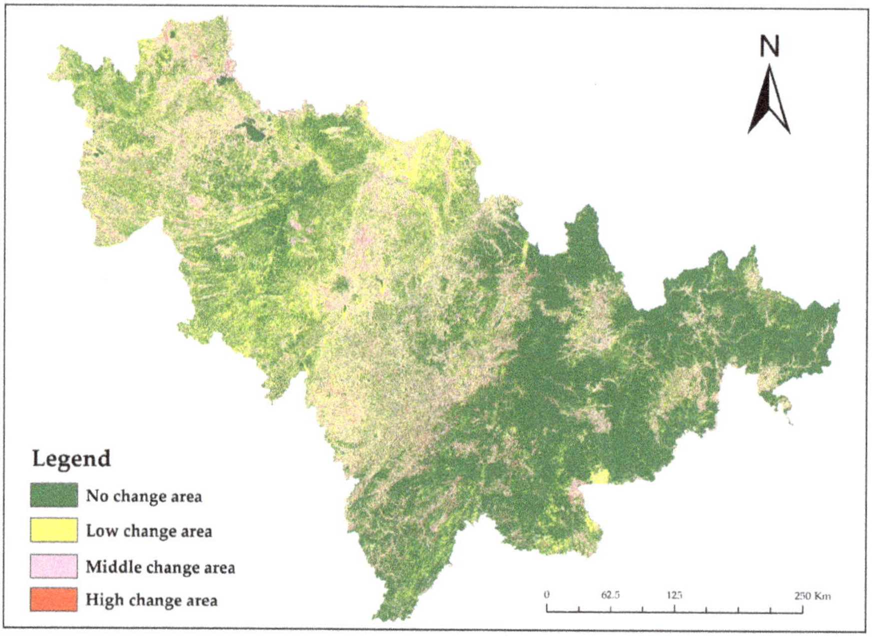


**Figure 14.** Frequency map of land-use changes.

The statistic and comparison of different frequency areas across various cities (autonomous prefecture) in Jilin Province were illustrated in Figure 15. The no-change area is the largest in YB, covering 32,382.73 km$^2$. The low-change area is most widespread in BC and CC, while the middle-change area is largest in BC. The high-change area is most extensive in Jilin, indicating more frequent land-use planning activities. Additionally, it is observed that in three cities (autonomous prefecture) in the eastern part of Jilin Province (BS, TH, YB), the unchanged area is larger than the changed area. In contrast, the remaining cities have a larger changed area than the unchanged area.

By analyzing the frequency of land-use changes, we can better understand the causes and impacts of those changes and understand their patterns and trends, so as to provide a scientific basis for future land-use planning and policymaking, and predict future land use. The study of land-use change, especially its environmental and social impacts, needs to be strengthened in future research.

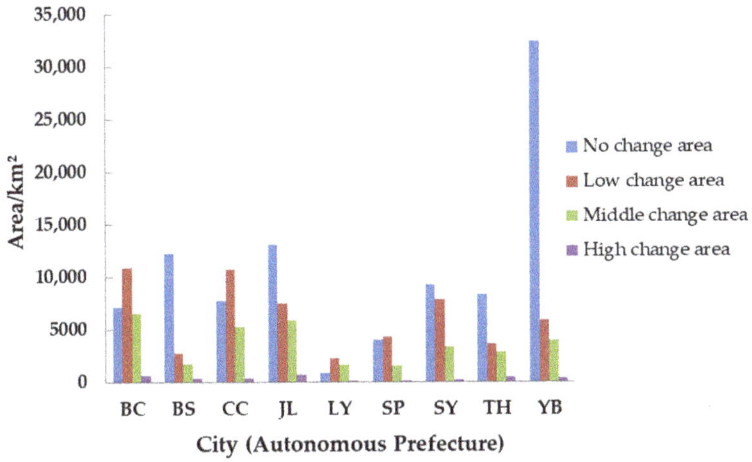


**Figure 15.** Comparison of frequency change area in different cities (autonomous prefecture) in Jilin Province. Baicheng City (BC); Baishan City (BS); Changchun City (CC); Jilin City (JL); Liaoyuan City (LY); Siping City (SP); Songyuan City (SY); Tonghua City (TH); Yanbian Korean Autonomous Prefecture (YB).

## 6. Conclusions

Based on long time series and high spatio-temporal resolution satellite images and the GEE platform fusion of Landsat and MODIS, this study developed an exponential random forest classification algorithm integrating remote sensing, carried out land-use classification research from 2000 to 2023, and analyzed the time trend and spatial characteristics of each land-use type. It provides basic information for the comprehensive management and sustainable development of Jilin Province. The main conclusions are as follows:

(1) The prediction results using the OL-STARFM method were close to the real images, and also highlighted more spatially detailed features. The different bands obtained by spatio-temporal data fusion showed a positive correlation, and the Pearson correlation coefficient reached more than 0.85.

(2) The land-use type maps with 30 m resolution for six time phases in Jilin Province from 2000 to 2023 were drawn by using the classification method of random forest remote sensing images based on the remote sensing index. The average overall accuracy and kappa coefficient of the products were 95.11% and 0.9394, respectively.

(3) During the study period, the cultivated land area increased from 2000 to in 2005, then decreased continuously until 2020, and then increased slightly to 71,951.81 km$^2$ in 2023, with an average annual decrease of 357.78 km$^2$.The overall grassland area showed a trend of decreasing, then increasing, then decreasing, for an average annual decrease of 236.28 km$^2$. The area of forest increased steadily, with an average annual increase of 255.14 km$^2$. The overall water area showed an upward trend, with an average annual increase of 38.96 km$^2$, but it showed fluctuations in individual study years. The area of building land continuously increased, with an average annual growth of 363.51 km$^2$. The overall unused land area showed a downward trend during the study period, with an average annual decrease of 84.6 km$^2$, but there were fluctuating changes in 2010–2023.

(4) During the study period, the pattern of land use in Jilin Province changed significantly. Cultivated land was the most important source of transfer-out, and the proportion of the total area decreased from 42.98% to 38.39%. Grassland, forest, and building land were mainly transferred out, and the transfer-out areas were 7682.48 km$^2$, 8374.11 km$^2$, and 7244.52 km$^2$, respectively. Grassland was the largest source of transfer-in, at

9942.33 km$^2$. The land transfer among other feature types was relatively small, at less than 3300 km$^2$ each.

**Author Contributions:** Conceptualization, Z.L. and Y.H.; methodology, Z.L., F.W. and Y.H.; validation, Y.X., J.Z. and L.Z.; formal analysis, Z.L.; investigation, J.Z.; resources, F.H.; data curation, Z.L. and R.Z.; writing—original draft preparation, Z.L. and Y.H.; writing—review and editing, Y.X. and F.W.; visualization, Z.L., P.Z., C.Q., J.Z. and L.Z.; supervision, Y.H. and Y.X.; project administration, Y.H.; funding acquisition, Y.H. and F.H. All authors have read and agreed to the published version of the manuscript.

**Funding:** This work was supported by the Project of Jilin Province Science and Technology Development Plan under grant 20210101101JC.

**Data Availability Statement:** The raw data supporting the conclusions of this article will be made available by the authors on request.

**Acknowledgments:** Thanks to the anonymous reviewers for their pertinent suggestions!

**Conflicts of Interest:** Authors Ruifei Zhu, Chunmei Qu and Peng Zhang were employed by the company Chang Guang Satellite Technology Company Ltd. The remaining authors declare that the research was conducted in the absence of any commercial or financial relationships that could be construed as a potential conflict of interest.

## References

1. Rashidi, P.; Patil, S.D.; Schipper, A.M.; Alkemade, R.; Rosa, I. Downscaling Global Land-Use Scenario Data to the National Level: A Case Study for Belgium. *Land* **2023**, *12*, 1740. [CrossRef]
2. Jing, Q.Q.; He, J.J.; Li, Y.R.; Yang, X.Y.; Peng, Y.; Wang, H.; Yu, F.; Wu, J.; Gong, S.L.; Che, H.Z.; et al. Analysis of the spatiotemporal changes in global land cover from 2001 to 2020. *Sci. Total Environ.* **2024**, *908*, 168354. [CrossRef] [PubMed]
3. Zhang, Z.; Wei, Y.; Li, X.; Wan, D.; Shi, Z. Study on Tianjin Land-Cover Dynamic Changes, Driving Factor Analysis, and Forecasting. *Land* **2024**, *13*, 726. [CrossRef]
4. Zhang, M.M.; Zhang, C.; Kafy, A.A.; Tan, S.K. Simulating the Relationship between Land Use/Cover Change and Urban Thermal Environment Using Machine Learning Algorithms in Wuhan City, China. *Land* **2022**, *11*, 14. [CrossRef]
5. Mendoza-Ponce, A.; Corona-Nunez, R.; Kraxner, F.; Leduc, S.; Patrizio, P. Identifying effects of land use cover changes and climate change on terrestrial ecosystems and carbon stocks in Mexico. *Glob. Environ. Chang.—Hum. Policy Dimens.* **2018**, *53*, 12–23. [CrossRef]
6. Zhao, Y.; An, R.; Xiong, N.; Ou, D.; Jiang, C. Spatio-Temporal Land-Use/Land-Cover Change Dynamics in Coastal Plains in Hangzhou Bay Area, China from 2009 to 2020 Using Google Earth Engine. *Land* **2021**, *10*, 1149. [CrossRef]
7. Chen, C.; Chen, H.; Chen, D.; Zhang, Z.; Zhang, X.; Zhuang, Y.; Chu, Y.; Chen, J.; Zheng, H. Coastline extraction and spatial—temporal variations using remote sensing technology in Zhoushan Islands. *Remote Sens. Land Resour.* **2021**, *33*, 141–152.
8. Du, Z.R.; Yu, L.; Chen, X.; Gao, B.B.; Yang, J.Y.; Fu, H.H.; Gong, P. Land use/cover and land degradation across the Eurasian steppe: Dynamics, patterns and driving factors. *Sci. Total Environ.* **2024**, *909*, 168593. [CrossRef] [PubMed]
9. Cruz, J.D.; Blanco, C.J.C.; de Oliveira, J.F. Modeling of land use and land cover change dynamics for future projection of the Amazon number curve. *Sci. Total Environ.* **2022**, *811*, 15234. [CrossRef]
10. Wang, C.Y.; Yang, K.; Yang, W.; Qiang, H.Y.; Xue, H.Y.; Lu, B.B.; Zhou, P. R-MFNet: Analysis of Urban Carbon Stock Change against the Background of Land-Use Change Based on a Residual Multi-Module Fusion Network. *Remote Sens.* **2023**, *15*, 2823. [CrossRef]
11. Li, X.; Huang, C.L.; Che, T.; Jin, R.; Wang, S.G.; Wang, J.M.; Gao, F.; Zhang, S.W.; Qiu, C.J.; Wang, C.H. Development of a Chinese land data assimilation system: Its progress and prospects. *Prog. Nat. Sci.—Mater. Int.* **2007**, *17*, 881–892.
12. You, Q.X.; Deng, W.X.; Tang, X.; Liu, Y.; Lei, P.; Chen, J.J.; You, H.T. Monitoring of mangrove dynamic change in Beibu Gulf of Guangxi based on reconstructed time series images. *Sci. Total Environ.* **2024**, *917*, 170395. [CrossRef]
13. Mu, S.; Li, B.; Yao, J.; Yang, G.; Wan, R.; Xu, X. Monitoring the spatio-temporal dynamics of the wetland vegetation in Poyang Lake by Landsat and MODIS observations. *Sci. Total Environ.* **2020**, *725*, 138096. [CrossRef] [PubMed]
14. Jing, W.; Lou, T.; Wang, Z.; Zou, W.; Xu, Z.; Mohaisen, L.; Li, C.; Wang, J. A Rigorously-Incremental Spatiotemporal Data Fusion Method for Fusing Remote Sensing Images. *IEEE J. Sel. Top. Appl. Earth Obs. Remote Sens.* **2023**, *16*, 6723–6738. [CrossRef]
15. Cheng, Q.; Zhang, Z.; Liang, D.; Ye, F. Fine Spatial and Temporal Ice/Snow Surface Temperature Generation: Evaluation Spatiotemporal Fusion Methods in Greenland Ice Sheet. *IEEE J. Sel. Top. Appl. Earth Obs. Remote Sens.* **2023**, *16*, 10216–10229. [CrossRef]
16. Gao, F.; Masek, J.; Schwaller, M.; Hall, F. On the blending of the Landsat and MODIS surface reflectance: Predicting daily Landsat surface reflectance. *IEEE Trans. Geosci. Remote Sens.* **2006**, *44*, 2207–2218. [CrossRef]

17. Wu, M.Q.; Wang, J.; Niu, Z.; Zhao, Y.Q.; Wang, C.Y. A model for spatial and temporal data fusion. *J. Infrared Millim. Waves* **2012**, *31*, 80–84. [CrossRef]
18. Zhu, X.L.; Chen, J.; Gao, F.; Chen, X.H.; Masek, J.G. An enhanced spatial and temporal adaptive reflectance fusion model for complex heterogeneous regions. *Remote Sens. Environ.* **2010**, *114*, 2610–2623. [CrossRef]
19. Lin, L.; Shen, Y.; Wu, J.; Nan, F. CAFE: A Cross-Attention Based Adaptive Weighting Fusion Network for MODIS and Landsat Spatiotemporal Fusion. *IEEE Geosci. Remote Sens. Lett.* **2023**, *20*, 1–5. [CrossRef]
20. Liu, Q.; Meng, X.; Li, X.; Shao, F. Detail Injection-Based Spatio-Temporal Fusion for Remote Sensing Images with Land Cover Changes. *IEEE Trans. Geosci. Remote Sens.* **2023**, *61*, 1–14. [CrossRef]
21. Zhukov, B.; Oertel, D.; Lanzl, F.; Reinhäckel, G. Unmixing-based multisensor multiresolution image fusion. *IEEE Trans. Geosci. Remote Sens.* **1999**, *37*, 1212–1226. [CrossRef]
22. Busetto, L.; Meroni, M.; Colombo, R. Combining medium and coarse spatial resolution satellite data to improve the estimation of sub-pixel NDVI time series. *Remote Sens. Environ.* **2008**, *112*, 118–131. [CrossRef]
23. Hilker, T.; Wulder, M.A.; Coops, N.C.; Linke, J.; McDermid, G.; Masek, J.G.; Gao, F.; White, J.C. A new data fusion model for high spatial- and temporal-resolution mapping of forest disturbance based on Landsat and MODIS. *Remote Sens. Environ.* **2009**, *113*, 1613–1627. [CrossRef]
24. Zhu, X.L.; Helmer, E.H.; Gao, F.; Liu, D.S.; Chen, J.; Lefsky, M.A. A flexible spatiotemporal method for fusing satellite images with different resolutions. *Remote Sens. Environ.* **2016**, *172*, 165–177. [CrossRef]
25. Guo, D.; Shi, W.; Zhang, H.; Hao, M. A Flexible Object-Level Processing Strategy to Enhance the Weight Function-Based Spatiotemporal Fusion Method. *IEEE Trans. Geosci. Remote Sens.* **2022**, *60*, 165–177. [CrossRef]
26. Guo, D.; Shi, W.; Zhang, H.; Hao, M. Temporal and spatial variation analysis of the area of Siling Co lake in Tibet based on ESTARFM (1976–2014). *J. Geo-Inf. Sci.* **2016**, *18*, 833–846.
27. Liu, H.; Gong, P. 21st century daily seamless data cube reconstruction and seasonal to annual land cover and land use dynamics mapping-iMap(China)1.0. *J. Remote Sens.* **2021**, *25*, 126–147. [CrossRef]
28. Wang, J.; Bretz, M.; Dewan, M.A.A.; Delavar, M.A. Machine learning in modelling land-use and land cover-change (LULCC): Current status, challenges and prospects. *Sci. Total Environ.* **2022**, *822*, 153559. [CrossRef]
29. Da Silva, G.M.; Arai, E.; Hoffmann, T.B.; Duarte, V.; Martini, P.R.; Dutra, A.C.; Mataveli, G.; Cassol, H.L.G.; Shimabukuro, Y.E. Land Use and Land Cover Classification in Sao Paulo, Brazil, Using Landsat-8 Oli Images and Derived Spectral Indices. In Proceedings of the IEEE International Geoscience and Remote Sensing Symposium (IGARSS), Pasadena, CA, USA, 16–21 July 2023; pp. 2961–2964.
30. Dastour, H.; Hassan, Q.K. A Comparison of Deep Transfer Learning Methods for Land Use and Land Cover Classification. *Sustainability* **2023**, *15*, 7854. [CrossRef]
31. Xi, X.Y.; Liu, Z.M.; Sun, L.; Xie, S.; Wang, Z.H. High-Confidence Sample Generation Technology and Application for Global Land-Cover Classification. *IEEE J. Sel. Top. Appl. Earth Obs. Remote Sens.* **2023**, *16*, 3248–3263. [CrossRef]
32. Peng, J.; Sun, W.; Wei, T.; Fan, W. A modified correlation alignment algorithm for the domain adaptation of GF-5 hyperspectral image. *J. Remote Sens.* **2020**, *24*, 417–426. [CrossRef]
33. Sun, W.; Zhang, D.; Yang, G.; Li, W. Band selection for hyperspectral imagery based on weighted probabilistic archetypal analysis. *J. Remote Sens.* **2018**, *22*, 110–118. [CrossRef]
34. Zhou, D.; Zhang, L.; Hao, L.; Sun, G.; Xiao, J.; Li, X. Large discrepancies among remote sensing indices for characterizing vegetation growth dynamics in Nepal. *Agric. For. Meteorol.* **2023**, *339*, 109546. [CrossRef]
35. Belay, H.; Melesse, A.M.; Tegegne, G. Scenario-Based Land Use and Land Cover Change Detection and Prediction Using the Cellular Automata-Markov Model in the Gumara Watershed, Upper Blue Nile Basin, Ethiopia. *Land* **2024**, *13*, 396. [CrossRef]
36. Ghorbanian, A.; Kakooei, M.; Amani, M.; Mahdavi, S.; Mohammadzadeh, A.; Hasanlou, M. Improved land cover map of Iran using Sentinel imagery within Google Earth Engine and a novel automatic workflow for land cover classification using migrated training samples. *ISPRS J. Photogramm. Remote Sens.* **2020**, *167*, 276–288. [CrossRef]
37. Rodrigues, S.W.P.; Souza-Filho, P.W.M. Use of Multi-Sensor Data to Identify and Map Tropical Coastal Wetlands in the Amazon of Northern Brazil. *Wetlands* **2011**, *31*, 11–23. [CrossRef]
38. Akar, ö.; Güngö, o. Classification of multi-spectral images using Random Forest algorithm. *J. Geod. Geoinf.* **2012**, *1*, 105–112. [CrossRef]
39. Ji, Y.W.; Zhang, L.; Gu, X.C.; Zhang, L. Study on the Spatial and Temporal Evolution of Regional Green Space Morphology Outside Built-Up Areas based on the Google Earth Engine and Biophysical Component Modeling. *Land* **2023**, *12*, 2184. [CrossRef]
40. Ning, X.; Chang, W.; Wang, H.; Zhang, H.; Zhu, Q. Extraction of marsh wetland in Heilongjiang Basin based on GEE and multi-source remote sensing data. *J. Remote Sens.* **2022**, *26*, 386–396. [CrossRef]
41. Shen, W.; Zhang, J.; Wang, K.; Zhang, Z. Identifying the spatio-temporal dynamics of regional ecological risk based on Google Earth Engine: A case study from Loess Plateau, China. *Sci. Total Environ.* **2023**, *873*, 162346. [CrossRef]
42. Yin, Z.; Li, J.; Liu, Y.; Xie, Y.; Zhang, F.; Wang, S.; Sun, X.; Zhang, B. Water clarity changes in Lake Taihu over 36 years based on Landsat TM and OLI observations. *Int. J. Appl. Earth Obs. Geoinf.* **2021**, *102*, 10457. [CrossRef]
43. Sivrikaya, F.; Gunlu, A.; Kucuk, O.; Urker, O. Forest fire risk mapping with Landsat 8 OLI images: Evaluation of the potential use of vegetation indices. *Ecol. Inform.* **2024**, *79*, 102461. [CrossRef]

44. Nguyen, L.H.; Joshi, D.R.; Clay, D.E.; Henebry, G.M. Characterizing land cover/land use from multiple years of Landsat and MODIS time series: A novel approach using land surface phenology modeling and random forest classifier. *Remote Sens. Environ.* **2020**, *238*, 111017. [CrossRef]
45. Roman, M.O.; Justice, C.; Paynter, I.; Boucher, P.B.; Devadiga, S.; Endsley, A.; Erb, A.; Friedl, M.; Gao, H.; Giglio, L.; et al. Continuity between NASA MODIS Collection 6.1 and VIIRS Collection 2 land products. *Remote Sens. Environ.* **2024**, *302*, 113963. [CrossRef]
46. Gurugnanam, S.S.C.B. Spatiotemporal detection of land use/land cover changes and land surface temperature using Landsat and MODIS data across the coastal Kanyakumari district, India. *Geod. Geodyn.* **2023**, *14*, 172–181.
47. Li, J.; Zhang, B.; Huang, X. A hierarchical category structure based convolutional recurrent neural network (HCS-ConvRNN) for Land-Cover classification using dense MODIS Time-Series data. *Int. J. Appl. Earth Obs. Geoinf.* **2022**, *108*, 102744. [CrossRef]
48. Mateo-Garcia, G.; Gomez-Chova, L.; Amoros-Lopez, J.; Munoz-Mari, J.; Camps-Valls, G. Multitemporal Cloud Masking in the Google Earth Engine. *Remote Sens.* **2018**, *10*, 1079. [CrossRef]
49. Wang, C.; Jia, M.; Chen, N.; Wang, W. Long-Term Surface Water Dynamics Analysis Based on Landsat Imagery and the Google Earth Engine Platform: A Case Study in the Middle Yangtze River Basin. *Remote Sens.* **2018**, *10*, 1635. [CrossRef]
50. Zurqani, H.A.; Post, C.J.; Mikhailova, E.A.; Schlautman, M.A.; Sharp, J.L. Geospatial analysis of land use change in the Savannah River Basin using Google Earth Engine. *Int. J. Appl. Earth Obs. Geoinf.* **2018**, *69*, 175–185. [CrossRef]
51. Breiman, L. Random forests. *Mach. Learn.* **2001**, *45*, 5–32. [CrossRef]
52. Chen, C.; Fu, J.; Sui, X.; Lu, X.; Tan, A. Construction and application of knowledge decision tree after a disaster for water body information extraction from remote sensing images. *J. Remote Sens.* **2018**, *22*, 792–801. [CrossRef]
53. Sun, W.; Yang, G.; Chen, C.; Chang, M.; Huang, K.; Meng, X.; Liu, L. Development status and literature analysis of China's earth observation remote sensing satellites. *J. Remote Sens.* **2020**, *24*, 479–510. [CrossRef]

land

MDPI

Article

# Analysis of Spatial and Temporal Pattern Evolution and Decoupling Relationships of Land Use Functions Based on Ecological Protection and High-Quality Development: A Case Study of the Yellow River Basin, China

**Hanwen Du [1,2,*], Zhanqi Wang [1,2], Haiyang Li [1,2] and Chen Zhang [1,2]**

1 Department of Land Resource Management, School of Public Administration, China University of Geosciences, Wuhan 430074, China
2 Key Laboratory of Law Evaluation, Ministry of Land and Resources of China, 388 Lumo Road, Hongshan District, Wuhan 430074, China
* Correspondence: hwdu@cug.edu.cn

**Abstract:** With rapid industrialization and urbanization, the contradiction between the human exploitation of land production and living functions and natural ecosystem service functions has intensified. The issues of how to coordinate the exploitation and conservation functions of land and guide the rational distribution of human activities have become important for global sustainable development, especially considering the realization that multifunctional land use is an effective way to relieve land pressure and improve land use efficiency, that land multifunction has significant spatio-temporal heterogeneity, and that there is a mutual promotion and stress relationship between multifunctional land use. However, few existing studies have discussed the decoupling relationship among land use functions. In this study, a system of 10 sub-functions and 25 indicators was established based on the production function (PDF), living function (LVF), and ecological function (ELF) for 59 cities in the Yellow River Basin (YRB). There are both subjective and objective procedures employed to determine the weights, while an exploratory spatial data analysis is used to analyze the time-based and territorial changes in various functions of land use in the study area from 2000 to 2020. The decoupling relationship between the three functions is detected utilizing the theoretical foundation of the decoupling analysis. The results show that land use is multifunctional, LUFs develop unevenly, and their spatial distribution varies substantially. The results of the decoupling analysis demonstrate that the predominant types of correlations among the land use ELF and PDF and LVF over the research period are strong decoupling and strong negative decoupling correlations, with the former being a dilemma and the latter being a sustainable type of development.

**Keywords:** LUFs; ecological–production–living functions; decoupling relationship; Yellow River Basin

## 1. Introduction

Over the past decades, China's economic output-oriented land space development model has exacerbated the contradictions among cities, agriculture, and ecology [1,2], generating a series of derivative problems, such as the unbalanced development of land space, overconsumption of resources, serious deterioration of the ecological environment, and uncoordinated development [3,4]. With the proposal of unified territorial spatial planning in China, optimizing and coordinating the spatial pattern of agriculture, towns, and ecology have become the core content of the sustainable development of territorial space [5]. The essence of homeland space is the interactive functional space determined by LUFs [6]. As the basis of land resource allocation and homeland spatial planning [7,8], revealing the evolution and interaction of LUFs is of great significance for optimizing and controlling homeland space.

Land use runs through the whole development process of human society [9]. Resources from the land are not only used to produce components of production such as water [10], food, as well as other necessities for human survival [11], but also are important carriers for various economic and social activities of human beings, and also have ecological environment elements such as soil, hydrology, climate, and vegetation, which deter-mine the multiple functional attributes of land use, i.e., [12–15], multifunctional land use.

Originating from agricultural multifunctionality, LUFs encompass the economic, social, and environmental dimensions of regional sustainable development, which refers to the ability to provide private and public goods or services through diversified land use [16,17], and typically include economic, social, and environmental functions [18]. In recent years, research on LUFs has been more extensive, focusing on the conceptual and theoretical framework of LUFs [17], the construction of the indicator system [19], spatial identification [8], spatio-temporal changes [16,17], and other aspects, as well as the spatio-temporal evolution of LUFs [20,21], influencing factors [22,23], trade-off analysis, and sustainable land management [24,25]. Existing studies mainly focus on a single province or city as the basic research scope, with relatively few studies at the watershed scale; in the determination of LUFs, the differences in functions between regions are not taken into account; and the judgment of the coordination of LUFs mostly relies on the coupled coordination degree model, which is too homogeneous in its approach. We argue that the scale of the land use framework is similar to that of ecosystem services [26], with stakeholders at different scales focusing on different priorities. At small scales, more attention is paid to the economic and social functions of land use, such as economic output and employment support [27]; at large scales, more attention is paid to major regional functions, such as the ecological barrier function of the Yunnan–Guizhou Plateau and the food production function of the Jianghan Plain [28]. Therefore, in a large region such as the Yellow River Basin (YRB), we should use a higher-level administrative unit as the evaluation unit to study the impacts of land use change.

Existing studies have mainly focused on the assessment of land use functions and their multifunctional characteristics [8,19], or the trade-off relationship between different types of land use functions and their influence mechanisms [24,25]. The mechanism of inter-functionality of land use has not yet been finalized., and in addition to the relationship between trade-offs and synergies, there may also be a "decoupling" relationship between the multifunctionality of land use in the Yellow River Basin in view of the large geographic and developmental differences between different regions in the upper, middle, and lower reaches of the Basin. As a measure of the coupling degree between human activities and resources and the environment, decoupling theory has attracted much attention worldwide, and is now mainly used to measure the degree of decoupling between energy, environment, water resources, etc., and economic development [29–34]. There are not many studies on the decoupling relationship between the multifunctionality of land use; the decoupling analysis method can be used to analyze the response relationship between two or more elements, which can effectively reveal the degree of influence between the production, living, and ecological functions of land use, and according to the different state of decoupling, we can distinguish the development process and evolution trend of regional land use. Therefore, we want to identify the decoupling relationship between the complex and diverse land use functions in the YRB and highlight the interaction mechanism of land use functions and regional heterogeneity.

In light of this, the primary goals of this work are as follows:

(1) To develop a research framework on the assessment of LUMF and the relationship between different functions, taking into account temporal and regional effects [35,36].
(2) To construct a comprehensive evaluation index system with a large scale and multiple indicators to evaluate LUMF [37].
(3) To examine spatial disparities and the emergence of multifunctional land use across time.
(4) To analyze the decoupling relationship between the PDF, LVF, and ELF in a long time series [38].

(5) Analysing the decoupling of ELF from PDF and LVF over long time series to decipher the evolution of the decoupling status of land use functions in each region.
(6) To propose governance proposals and policy recommendations based on the results of land use function research [18].

## 2. Methods

### *2.1. Research Framework*

Land usage in China is facing problems of disorderly development, structural imbalance, functional degradation, and ecological environment destruction in the midst of a major urbanization process. Through making wise use of the land, different controls on agricultural, urban, and ecological spaces are carried out so as to maximize the land use function.

Figure 1 displays the theoretical structure used in this study. Different units in the study area have different natural resource endowments, socio-economic conditions, and other factor inputs, which are differentiated into agricultural production, non-agricultural production, livelihood security, and ecological land use services. The various functions are interdependent and also influence each other. The PDF provides agricultural, economic, and transportation products and services for the LVF; the LVF provides motivation for the PDF [25], but excessive demand can impose restrictions on the PDF; the ELF provides natural and physical conditions for the PDF and LVF, but the development of the LVF and PDF may lead to resource depletion and damage to the ecological structure [39]. A comprehensive evaluation using multi-source data from multiple years is combined with an exploratory spatial analysis to summarize the multifunctional dynamic changes and spatial differentiation of regional land use. Scholars often use decoupling models to look into the relationship among resource usage and economic expansion; this study innovatively explores the decoupling link between LUMF on the basis of spatial differentiation [40]. Therefore, this paper reveals the horizontal and vertical comparisons of regional LUMF, makes a judgment on the development trends of different functions, points out the development priorities of each part of the study area, and makes suggestions for national policies.

### *2.2. The Multi-Dimension Evaluation*

On the one hand, the method of extreme difference standardisation is used to process each index for the evaluation of land functions in a dimensionless manner. Secondly, based on multiplicative synthesis, the multi-expert decision making hierarchical analysis method, independence weight coefficient method, and entropy value method are adopted for combined weight calculation, and the combination of subjective assignment methods and objective assignment methods can obtain a better evaluation effect. The independence weight coefficient method and the entropy value method represent two ideas of the objective assignment method, respectively. The former determines the indicator weights based on each indicator's intricate correlation coefficients with other indicators [41], and the larger the complex correlation coefficient, the smaller the independence of the indicator and the smaller the weight value. The latter determines the index weight based on the information entropy of the index; the greater the information entropy, the greater the information of the index, the greater the influence on the comprehensive evaluation, and the greater the weight. Finally, each function is evaluated by using (1).

$$f_j = \sum_{j=1}^{n} W_{ij} P_{ij} \tag{1}$$

The specific steps of the entropy weighting method (EWM) to determine the index weights are as follows.

(1) Standardization of data

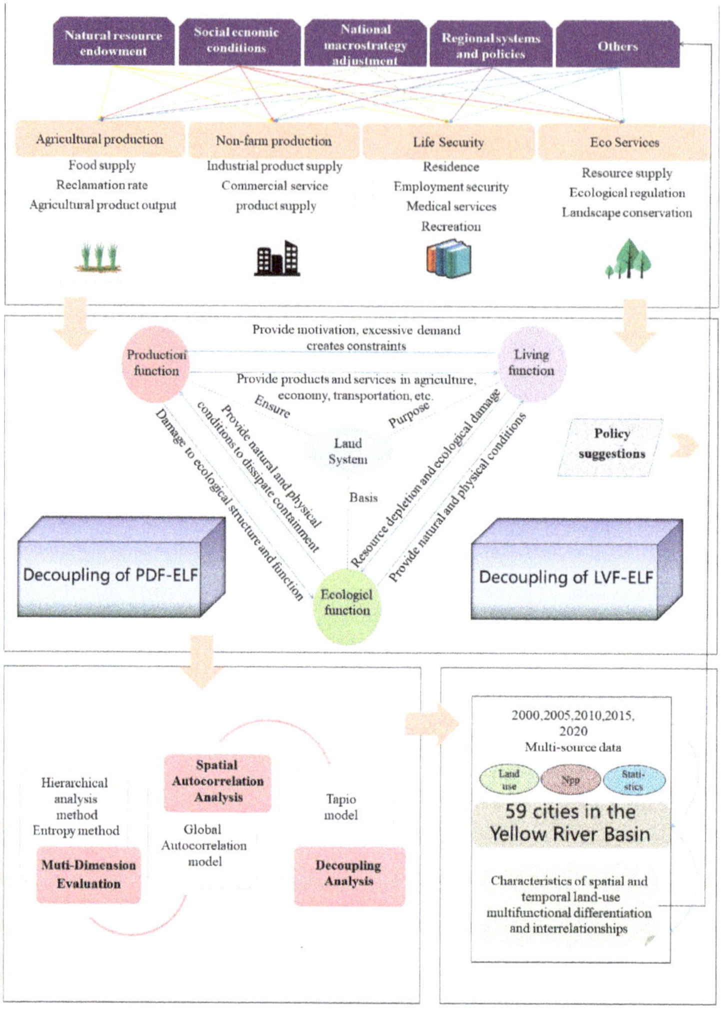


**Figure 1.** Research framework.

The 25 indicators were normalized using the extreme value method and calculated as follows to eliminate the effect of scale [41]:

$$X_{ij} = \frac{X_{ij} - MinX_{ij}}{MaxX_{ij} - MinX_{ij}} \tag{2}$$

where $X_{ij}$ is the indicator's starting value, $i = 1 \ldots m$, $j = 1 \ldots n$; the numbers $m$ and $n$ represent the number of research units and indicators; the standardized value of $x_{ij}$ is $X_{ij}$; and Max $X_{ij}$ and Min $X_{ij}$ are the maximum value and minimum value of the $j$ indicator of city $i$, respectively. In this article, all indicators are positive.

(2) $Z_{ij}$ composite standardized value calculation

$$Z_{ij} = \frac{X_{ij}}{\sum_{j=1}^{n} X_{ij}} \tag{3}$$

(3) Entropy number calculation for the $j$th indicator

$$H_j = -(\ln m)^{-1}\sum_{i=1}^{m} Z_{ij}\ln Z_{ij} \quad (4)$$

The jth index's information entropy value is $H_j$. Although a logarithm is not used for the original data in this study, if $Z_{ij} = 0$, a minimal value (0.00000001) is given to $Z_{ij}$ in order to make $lnZ_{ij}$ relevant.

(4) Calculation of the indicator $j$'s weight

$$W_j = (1 - H_j)/\left(k - \sum_{j=1}^{k} H_j\right) \quad (5)$$

The number of indices used to evaluate a single function is $k$, and $W_j$ is the weight value of the $j$th indices. Table 1 displays the weight values.

**Table 1.** PDF/LVF-ELF Decoupling Index Comparison Table.

| Decoupling Categories | %ΔELF | %ΔPDF/LVF | DI |
|---|---|---|---|
| Strong decoupling | <0 | >0 | DI < 0 |
| Weak decoupling | >0 | >0 | 0 ≤ DI < 0.8 |
| Recessive decoupling | <0 | <0 | DI > 1.2 |
| Growth connection | >0 | >0 | 0.8 ≤ DI ≤ 1.2 |
| Declining connection | <0 | <0 | 0.8 ≤ DI ≤ 1.2 |
| Expansive negative decoupling | >0 | >0 | DI > 1.2 |
| Weak negative decoupling | <0 | <0 | 0 ≤ DI < 0.8 |
| Strong negative decoupling | >0 | <0 | DI < 0 |

### 2.3. Spatial Autocorrelation Analysis

The exploratory spatial data analysis methods used in this paper include the global Moran index and the cold and hot spot analysis methods [42]. The global Moran index can visually reflect the type of multifunctional spatial correlation of land use in the study area and its spatial distribution characteristics, and better determine whether its distribution belongs to the cluster type, discrete type, or random type.

$$Global\ Moran'I = \frac{\sum_i^n \sum_{j\neq i}^n W_{ij}(X_i - X)(X_j - X)}{S^2\sum_i^n \sum_{j\neq i}^n W_{ij}} \quad (6)$$

In Equation (6), $n$ is the number of study units, indicating the mean value of study unit attributes; $X_i$ and $X_j$ are the feature values of research modules $i$ and $j$, respectively; $S^2$ is the range of score values; $W_{ij}$, the element of the matrix of geographical weights, is the link relationship between the $i$th and $j$th points of spatial objects, which usually have contiguity weight and distance weight, and this paper will be based on the criterion of contiguity weight.

Based on this, this study uses Getis-Ord Gi* on the ArcGIS platform to analyze the spatial clustering of each secondary indicator that affects LUMF by performing a cold and hotspot analysis to better understand where the spatial clustering of high or low value items lies.. Hot spots are areas in the study area with high values of land use functions and vice versa. The specific formula for Getis-Ord Gi* is given below [43].

$$G_i^* = \frac{\sum_{j=1}^{n} W_{ij}X_j - \overline{X}\sum_{j=1}^{n} W_{ij}}{S\sqrt{\frac{\left[n\sum_{j=1}^{n} w_{ij}^2 - \left(\sum_{j=1}^{n} w_{ij}\right)^2\right]}{n-1}}} \quad (7)$$

where $n$ is equivalent to the total amount of characteristics, $X_j$ is a value for an attribute for feature $j$, and $W_{ij}$ is the distance and weighting between item $i$ and $j$:

$$\overline{X} = \frac{\sum_{j=1}^{n} X_j}{n} \tag{8}$$

$$S = \sqrt{\frac{\sum_{j=1}^{n} x_j}{n} - (\overline{X})^2} \tag{9}$$

The Gi* statistics returned for each feature in the dataset is a z-score [40]. The resulting z-scores show where characteristics with high or low values geographically cluster. The clustering of high values is more pronounced for statistically significant positive z-scores with bigger z-scores (hot spots). For statistically significant negative z-scores, the smaller the z-score is, the more intense the clustering of low values (cold spots).

*2.4. Tapio Decoupling Model*

In the fields of agriculture, energy, building expansion, transportation, and carbon emissions, the decoupling analysis method has been widely applied; relatively few research projects have used the decoupling method to assess the link between various land use functions. Additionally, this work uses a decoupling analysis to examine the connections between the three. In this research, the link between the three functions is investigated using the decoupling analysis method [44]. Currently, there are various methods used to study the decoupling state, such as the integrated analysis of change method and the IPAT model method [45]. Among them, the Tapio elasticity analysis method is widely used in the fields of economic growth and resource environments. In this paper, we use the Tapio elasticity analysis method to measure the elasticity coefficients between production ecology and living ecology in the study area from the perspective of the elasticity concept. The decoupling conceptual foundation is separated into eight logical options based on the extension and elaboration of decoupling theory by Tapio [46] and OECD [47]. They include strong decoupling, strong negative decoupling, weak decoupling, expansion connection, expansion-negative decoupling, decline decoupling, decline connection, and weak negative decoupling types. The significance of each specific type is given in reference and is calculated as follows:

$$DI_{t2-t1} = \frac{\%\Delta ELF}{\%\Delta PDF} = \frac{(ELF_{t2} - ELF_{t1})/ELF_{t1}}{(PDF_{t2} - PDF_{t1})PDF_{t1}} \tag{10}$$

$$DI_{t2-t1} = \frac{\%\Delta ELF}{\%\Delta LVF} = \frac{(ELF_{t2} - ELF_{t1})/ELF_{t1}}{(LVF_{t2} - LVF_{t1})LVF_{t1}} \tag{11}$$

where $DI_{t2-t1}$ is an indicator of the decoupling between years $t1$ and $t2$; $ELF_{t1}$ and $ELF_{t2}$ represent the ecosystem service intensity index in year $t1$ and $t2$, respectively; $PDF_{t1}$, $PDF_{t2}$, and $LVF_{t1}$, $LVF_{t2}$ represent the change rates of PDF and life function in year $t1$ and $t2$, respectively.

## 3. Research Area and Data

*3.1. Research Area*

In this study, 59 cities in eight provinces along the YRB (32°35′~43°22′ N, 100°52′~119°19′ N) (Figures 2 and 3) are studied, encompassing 952,600 sq. kilometers, or 9.89% of China's total land area; 213 million people are living there as of 2020. The research area consists of five main urban agglomerations that have a strong agricultural and pastoral sector foundation, accounting for one-third of China's output value; these agglomerations have very rich energy resource reserves, and are also the source of several regional and historical cultures [47,48]. The area is the primary conduit for regional economic growth poles and the layout of population and productivity in the YRB [49].

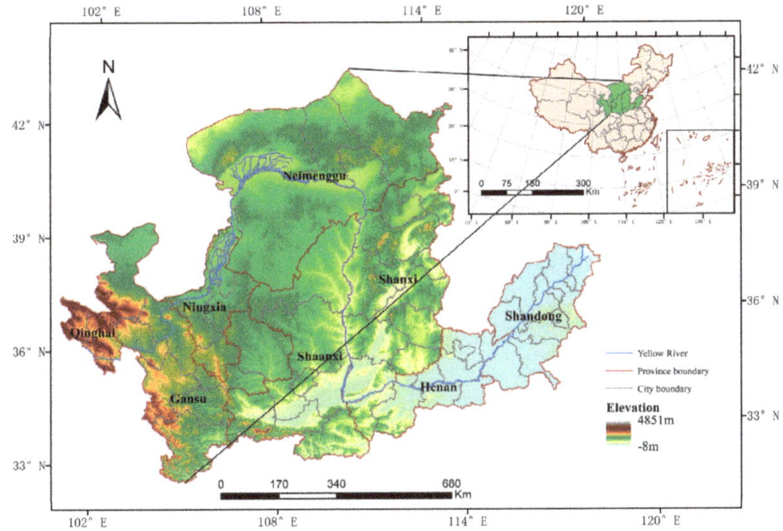


**Figure 2.** Study area.

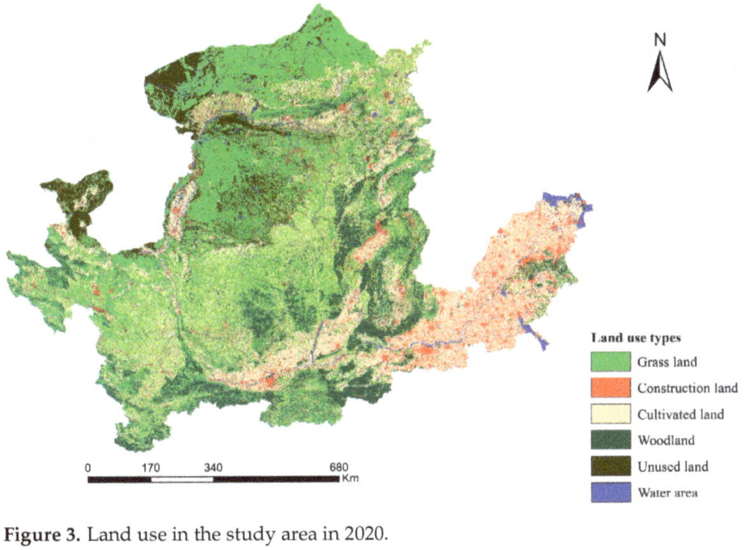


**Figure 3.** Land use in the study area in 2020.

*3.2. Data Sources*

Five research years, ranging from 2000 to 2020, were chosen for this investigation. The data types and sources used in this paper are as follows: (1) land use data and data on the administrative boundaries of Chinese municipalities in 2015 are from CAS (http://www.resdc.cn/) (accessed on 4 January 2023). Land use data are for the years 2000, 2005, 2010, 2015, and 2020, with a spatial resolution of 30 m; (2) NPP data from the National Oceanic and Atmospheric Administration (https://www.noaa.gov/) (accessed on 4 January 2020), with a spatial resolution of 1 km; (3) socio-economic data from statistical yearbooks published by provinces and municipalities and statistics on the nation's economic and social progress; and (4) nature reserve boundary data from the China Nature Reserve Specimen Resource Sharing Platform (http://cnpapc.zrbhq.cn/) (accessed on 10 January 2020).

### 3.3. Building of Indicator System

Based on the principles of comprehensiveness, representativeness, scientificity, and accessibility, the multifunctional land evaluation index system of the YRB was constructed with reference to the existing research results (Table 2), taking into account the land use characteristics of the study area [50–54].

**Table 2.** Evaluation index system of LUFs. (+ indicates that the indicator has a positive effect on the system).

| Target Layer | Factor Layers and Weights | Indicator Layer | Description | Properties | Index Weight |
|---|---|---|---|---|---|
| PDFs (0.40) | Agricultural production function (0.437) | Average land grain production | Total grain output/cultivated land area (kg·ha) | + | 0.0797 |
| | | Food production per capita | Total grain output/total population (kg) | + | 0.1892 |
| | | Reclamation rate | Cultivated land area/total land area (%) | + | 0.1637 |
| | | Meat production per capita | Meat output per capita (kg) | + | 0.2261 |
| | | Total output value of agriculture, forestry, animal husbandry, and fishery | Total output value of agriculture-forestry-stockbreeding-fishery (million) | + | 0.3413 |
| | Economic development function (0.563) | Economic Density | Regional GDP/total area (billion yuan·km$^{-2}$) | + | 0.6742 |
| | | The proportion of secondary and tertiary industries | Output value of secondary and tertiary industries/GDP (%) | + | 0.0324 |
| | | GDP per capita | Regional GDP/total population (million) | + | 0.2934 |
| LVFs (0.40) | Traffic carrying function (0.129) | Road network density | Road mileage/total area of the region | + | 1.0000 |
| | Residential home function (0.346) | Population density | Total population/total area of the region (person·km$^{-2}$) | + | 0.4897 |
| | | Construction land ratio | Construction land area/total land area (%) | + | 0.5103 |
| | Employment support function (0.282) | Population density in the workforce | Total employed population/total area of the region (person·km$^{-2}$) | + | 0.9074 |
| | | Tertiary sector as a percentage of employed population | Tertiary sector employed population/total employed population (%) | + | 0.0926 |
| | Social security function (0.151) | Urban-rural income balance index | Rural per capita income/Urban per capita income (%) | + | 0.5344 |
| | | Per capita income of rural residents | Per capita income of rural residents (yuan) | + | 0.4578 |
| | | Number of beds in health facilities for 10,000 people | (Number of beds in health facilities/total population)/(beds/million people) | + | 0.0078 |
| | Cultural and leisure function (0.092) | Tourist arrivals | Regional annual tourist arrivals (million people) | + | 0.5874 |
| | | Greening coverage of built-up areas | Urban built-up area covered by greenery/built-up area (%) | + | 0.0446 |
| | | Number of books per capita in public libraries | Number of books in public libraries/total population (volume) | + | 0.3680 |

Table 2. *Cont.*

| Target Layer | Factor Layers and Weights | Indicator Layer | Description | Properties | Index Weight |
|---|---|---|---|---|---|
| ELFs (0.30) | Resource supply function (0.251) | Water resources per capita | Total water resources/total population ($m^3$) | + | 1.0000 |
| | Ecological regulation function (0.423) | Value of Ecosystem Services | Referring to the improved method of Xie et al. [4], paddy field, dry land, garden land, wooded land, shrubland, grassland, and water were counted at 3.89, 4.01, 26.96, 22.95, 15.22, 19.69, and 125.61 each (Yuan·ha) | + | 0.2585 |
| | | npp | The average content of carbon in atmospheric $CO_2$ fixed by vegetation on a regional scale | + | 0.2892 |
| | | Percentage of ecological space | (Water area + forest area + garden area + grassland area)/village land area (%) | + | 0.4523 |
| | Landscape conservation function (0.326) | Nature reserve area share | Nature reserve area/land area (%) | + | 0.7334 |
| | | Forest cover | Woodland area/land area (%) | + | 0.2666 |

## 4. Results

### 4.1. Changes in Land Use Functions through Time and Space

Firstly, the study area was divided into three spatial levels: upper, center, and lower, and the total value of LUFs was calculated. From 2000 to 2020, the total value of the PDF–LVF-ELF showed a lower Yellow River > middle Yellow River > upper Yellow River (Figure 4). Among them, the overall total value of LUFs in the upper and center cities of the Yellow River does not change much over time, and the total value of LUFs in the Yellow River's lower cities decreases over time.

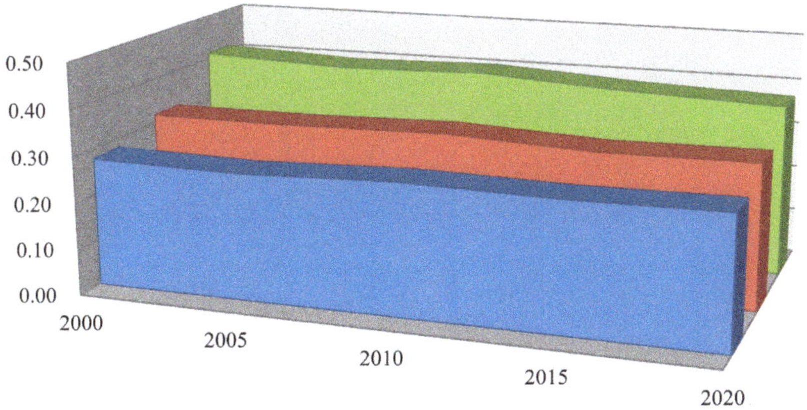


**Figure 4.** Changes in LUFs in various parts of the YRB. (Green for downstream, red for midstream, blue for upstream).

According to the comprehensive evaluation results, the temporal and geographical variations in the evaluation results for functions of production, living, and ecology in the study area were mapped using GIS technology (Figure 5), and it was found that the spatial distribution of the change values for the PDF, LVF, and ELF showed non-equilibrium type characteristics. Regarding the change in the PDF, it shows obvious spatial aggregation characteristics during 2000–2020, with numerous cities within the Yellow River's upstream and middle reaches increasing their PDF, among which Yulin in Shaanxi Province and

Guyuan in Ningxia Hui Autonomous Region have the highest growth rates of 0.78 and 0.74, respectively, while a series of cities in the downstream of the Yellow River, Henan Province, and Shandong Province show a decreasing trend in the value of PDF from 2000 to 2020. The value of the LVF in the study area during 2000–2020 is unbalanced, among which Erdos, Zhongwei, Guyuan, and Baiyin have the fastest growth rates, while Xining, Qingyang, Xinzhou, and Datong in the upper and middle reaches of the Yellow River have a decreasing trend. In the period 2000–2020, the overall ELF of the study area changed slowly, among which Liaocheng City, Dezhou City, and Weinan City in the Yellow River's lower and central sections increased more, while Dingxi City and Ordos City decreased slightly.

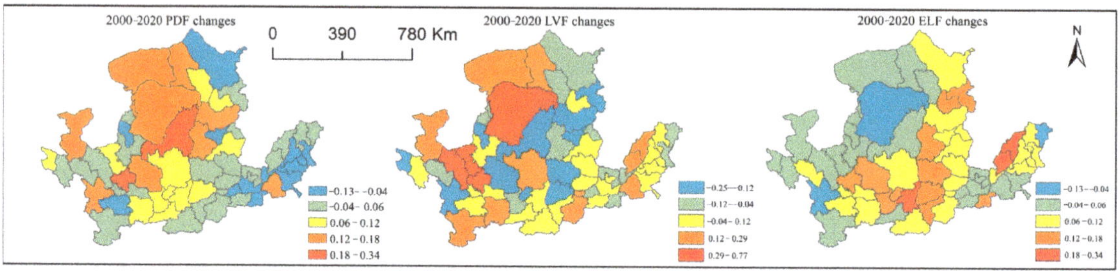


**Figure 5.** Spatial and temporal rates of change of LUFs.

### 4.2. Spatial Aggregation Analysis of LUFs

Spatial Aggregation Analysis of Land Use Functions

The distribution of the land's productivity, habitation, and ecological functions was calculated using the global Moran's indicator and a cold/hot spot analysis (Figure 6). The calculation results of the global Moran's I index confirm that the three have obvious positive spatial correlation [50], i.e., a strong interaction of adjacent units and agglomerative spatial phenomena. Specifically analyzing the spatial cold and hot spots of each secondary index, the hot spots of the agricultural production function are located in most cities in Shandong Province, and the cold spots are located in Taiyuan City and the surrounding cities; there are two hot spots of the economic development function, namely, the city cluster around Jinan City in Shandong Province and the city cluster around Zhengzhou City in Henan Province, and the cold spots are the areas centered on Baiyin City, Guyuan City, and Pingliang City in the west; the traffic carrying the hot spots of ecological regulation function are located in Tianshui City, Longnan City, and the nearby cities, as well as four cities in the heart of the province of Shanxi, and the cold spots are located in most cities in the Yellow River's lower reaches in the southeastern part of the study area; the research area's south contains the hotspots for landscape conservation function, and the cold spots are five cities on the border between the provinces of Henan and Shandong.

### 4.3. Production Ecology and Living Ecology Decoupling Analysis

In order to fully reveal the relationship between the ELF, PDF, and LVF, the scatter plots between the ELF, PDF, and LVF were first analyzed (Figure 7). From the scatter plots, it can be seen that there is a negative linear relationship between the ELF, PDF, and LVF, but the slope of the linear relationship between the two is small when the values of the PDF and LVF are low, and the slope gradually increases as the values rise. This shows that when the land use PDF and LVF are low, there is a negative but not obvious effect on the ELF, and as the land use PDF and LVF rise to a certain threshold, their negative effect on the ELF is reflected more obviously.

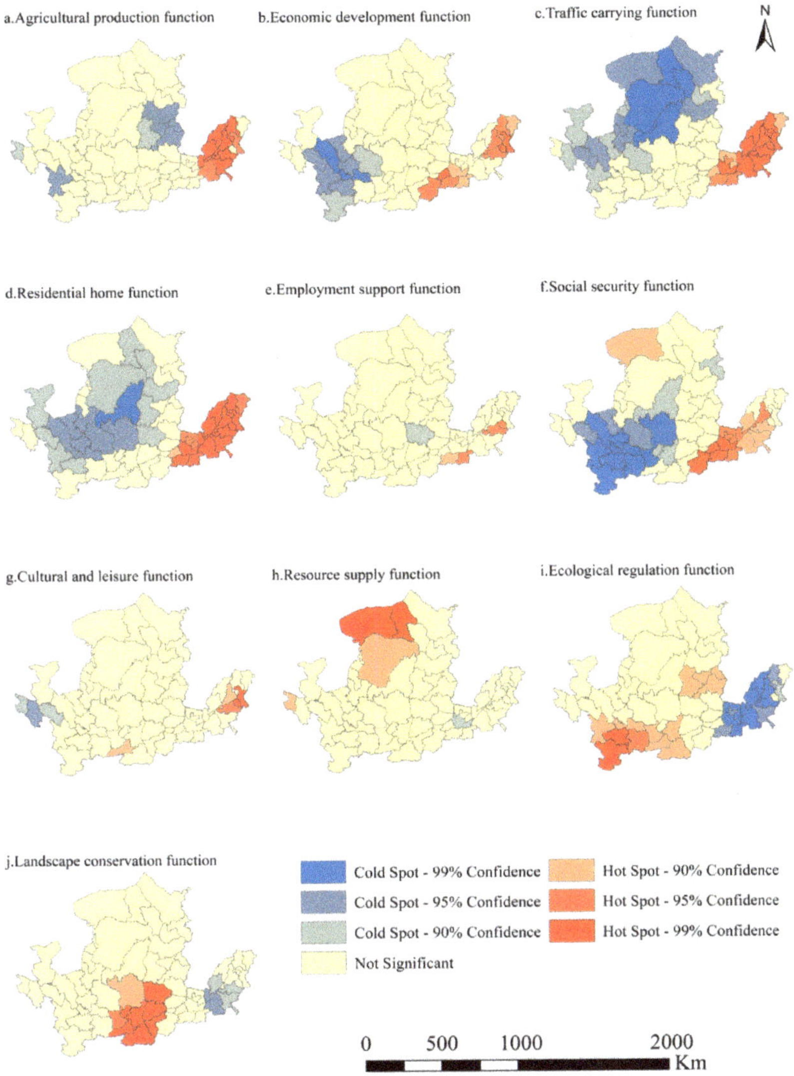


**Figure 6.** Spatial aggregation of secondary functions.

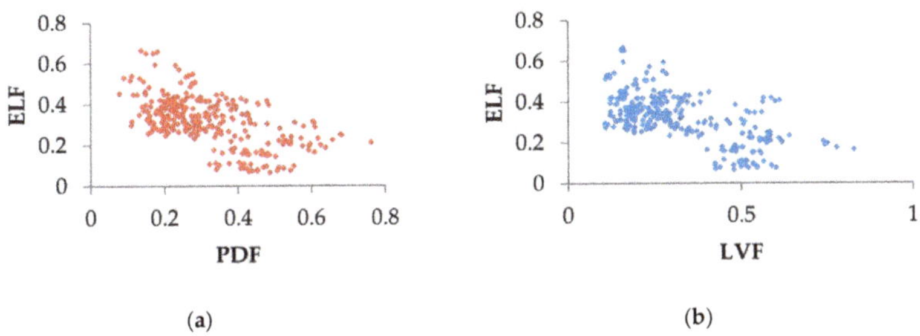


**Figure 7.** Scatterplot of land use functions in the YRB from 2000 to 2020 (**a**) PDF-ELF scatter plot; (**b**) LVF-ELF scatter plot.

In order to further explain the relationship between the ELF, PDF and LVF of land use in different regions of the area of investigation, the decoupling relationship between the ELF, PDF, and LVF from 2000 to 2020 was analyzed (Figure 8), and the statistical results of its spatial distribution characteristics and the proportion of different logical types of municipalities were obtained (Table 3). Overall, the PDF-ELF and LVF-ELF, which are in strong decoupling and strong negative decoupling relationships, are higher, reaching more than 50% in all years. The strong decoupling type represents the enhancement of the PDF and LVF leading to the reduction in the ELF, and this development model is in a dilemma, while the strong negative decoupling type represents the reduction in the PDF and LVF, leading to the improvement in the ELF, and this type is a sustainable development type. After 2015, the proportion of this type of expansion-negative decoupling increases more, and expansion-negative decoupling represents that the increase in the PDF or LVF leads to the increase in the ELF, and the increase in the ELF is higher than the increase in the PDF or LVF, which is the most desirable development model.

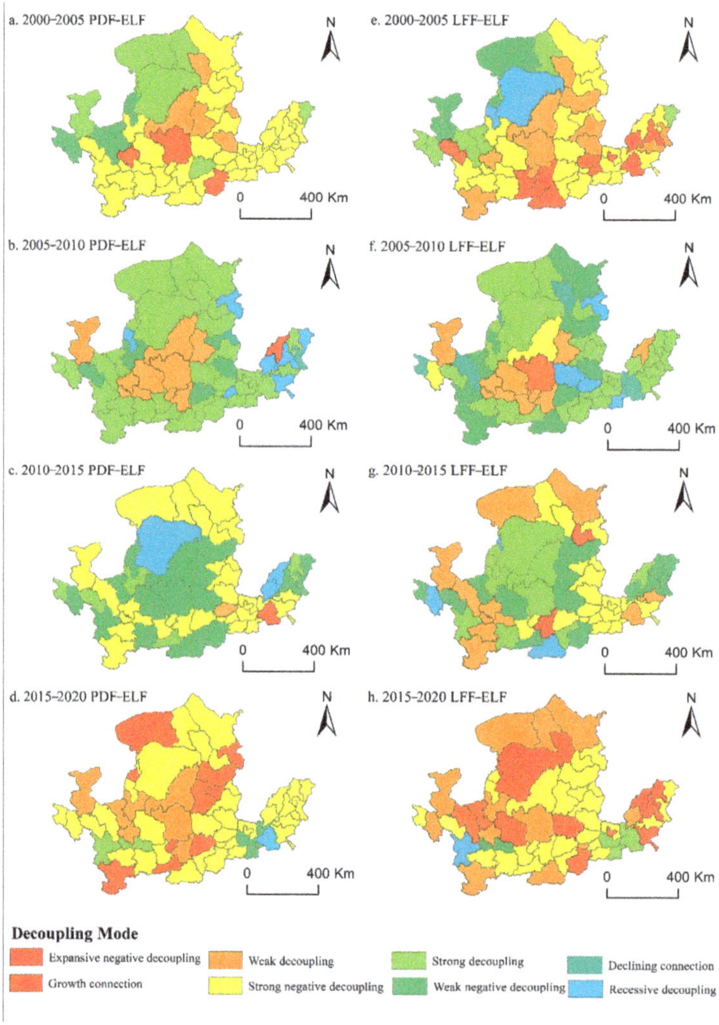


**Figure 8.** Decoupling of LUFs.

Table 3. Logical possibilities statistics.

| Decoupling Categories | 2000–2005 (PDF-ELF) | 2005–2010 (PDF-ELF) | 2010–2015 (PDF-ELF) | 2015–2020 (PDF-ELF) | 2000–2005 (LVF-ELF) | 2005–2010 (LVF-ELF) | 2010–2015 (LVF-ELF) | 2015–2020 (LVF-ELF) |
|---|---|---|---|---|---|---|---|---|
| Expansive negative decoupling | 5.08 | 1.69 | 1.69 | 10.17 | 16.95 | 0.00 | 3.39 | 15.25 |
| Growth connection | 0.00 | 0.00 | 0.00 | 6.78 | 1.69 | 1.69 | 0.00 | 6.78 |
| Weak decoupling | 8.47 | 15.25 | 1.69 | 11.86 | 18.64 | 15.25 | 16.95 | 20.34 |
| Strong negative decoupling | 64.41 | 0.00 | 50.85 | 61.02 | 42.37 | 3.39 | 33.90 | 47.46 |
| Strong decoupling | 11.86 | 59.32 | 8.47 | 3.39 | 11.86 | 47.46 | 16.95 | 6.78 |
| Weak negative decoupling | 10.17 | 8.47 | 30.51 | 5.08 | 5.08 | 15.25 | 23.73 | 1.69 |
| Declining connection | 0.00 | 3.39 | 1.69 | 0.00 | 0.00 | 10.17 | 0.00 | 0.00 |
| Recessive decoupling | 0.00 | 11.86 | 5.08 | 1.69 | 3.39 | 6.78 | 5.08 | 1.69 |

At the same time, some cities have this type of decoupling. Decline decoupling is also an important type of decoupling logic, and the decrease in the PDF or LVF in this type of municipal unit leads to the decrease in the ELF, and the decrease in the ELF is greater than the decrease in the PDF or LVF, so this type of decoupling is unsustainable.

## 5. Discussion

### 5.1. Land Use Function and Its Decoupling Relationship

The uncoordinated development of land use change will bring challenges to build a balanced pattern of agriculture, city, and ecology [55]. In this study, 10 sub-functions and 25 indicator systems were established based on the production function (PDF), living function (LVF), and ecological function (ELF) of 59 cities in the Yellow River Basin. The spatial and temporal changes in various functions of land use in the study area from 2000 to 2020 were analyzed by using the subjective–objective combination of the empowerment method and exploratory spatial data analysis method. The theoretical basis of decoupling analysis was utilized to detect the decoupling relationship among the three functions.

According to the findings of the comprehensive assessment, in the study region, the total value of LUFs in general shows downstream > midstream > upstream. Meanwhile, our study concludes that the land use level of the ELF in the YRB cities has improved as a whole from 2000 to 2020; many towns downstream of the YRB have different degrees of decline in land PDF; Xining, Qingyang, Xinzhou, Datong, and other cities upstream of the YRB have an obvious decline in LVF. These indications are consistent with the research of other scholars [56].

This study explored the decoupling relationship between land use functions, and in general, the PDF-ELF and LVF-ELF were in the state of strong decoupling and strong negative decoupling relationships, respectively, and the proportion was high, both exceeding 50% of the cities, which indicated that the process of industrial and agricultural production and ecological protection in the whole Yellow River Basin were still in a dilemma, which was due to the fact that industrial relocation and urban expansion have led to the loss and fragmentation of agricultural land and natural landscapes, making it difficult to maintain ecological functions [57]. However, the development process of the land to give residents living functions and the health of the ecosystem are in a sustainable and benign state, and the provision of ecological services and comprehensive remediation projects on agricultural and forest lands have led to a certain degree of ecosystem preservation [58]. The expansion-negative decoupling state appeared considerably more frequently in both groups of subjects after 2015, indicating that an increase in the PDF or the LVF led to an increase in the ELF and the increase in the ELF is higher than the increase in the PDF or LVF, which is the most favorable development pattern for high-quality and sustainable development. In recent years, in the Yellow River Basin as a whole, in the active promotion of vegetation restoration projects and returning farmland to forest and grassland projects, soil erosion has been weakened and the vegetation cover in the watershed has increased, and at the same time, through comprehensive management and ecological restoration and management, the economic benefits have been maximally retained [59].

### 5.2. Spatial Guidance of Land Use Functions and Policy Recommendations

The study claims that the YRB 's ecological conservation and management approach has contributed in some way in recent years, through the setting up of major ecological protection and restoration projects, which has led to an overall steady rise in the ELF during the study period, with only three cities showing a significant decline. The Yellow River's central reaches have obvious advantages in landscape conservation and resource supply functions, but the overall level of ecological regulation function is not high and should be improved through soil and water conservation and desertification control. PDF development in the study area is uneven, with higher levels of agricultural production in Shandong Province than in other regions, while non-agricultural economic development revolves around two urban clusters around the cities of Zhengzhou and Jinan, with an overall rise in other parts. However, PDF is still inadequate for developing the following high quality [60]. The inadequate development of people's livelihood is a major weakness in the development of the YRB, with more gaps in public services and infrastructure in all provinces and regions. Medical and health facilities are insufficient, the disparity between urban and rural populations' earnings is lower than the national average, and the LVF level of most cities has increased during the study period; upstream of the YRB, the LVF level has improved more significantly, which is a positive trend. In 2021, the Communist Party of China Central Committee and the State Council published *the Yellow River Basin Outline of the Plan for Ecological Protection and High-Quality Development*, and in 2022, four departments, including the Ministry of Ecology and Environment and the Ministry of Natural Resources, jointly issued *the Plan for Ecological and Environmental Protection of the Yellow River Basin*. The above plans and national strategies are more conducive to the coordinated development of LUFs to boost the overall level of LUFs on the basis of further enhancing ecological protection along the Yellow River, to increase the effectiveness of land use, to compensate for the weaknesses of LUFs, and to raise the overall level of LUFs [61].

According to the spatial distribution results of the LUF, the upper reaches of the YRB have a superior ecological environment and should take eco-tourism, high-value agriculture, etc., as the main development direction, rationally develop mountain and water resources, and build a tourism development model combining ecology, culture, and industry; the middle reaches of the YRB should deepen the optimization of the spatial layout to form an interdependent and synergistic urban belt to connect the exchanges between the upper and lower reaches; and the downstream areas of the YRB should strengthen the development of the coastal economy, build an efficient industrial chain of ports, logistics, and industrial support, and promote the transformation of old and new kinetic energy. At the same time, the downstream area should also pay attention to the protection of the ecological environment and strengthen the construction of sponge cities.

### 5.3. Limitations and Future Research Directions

In this research, more indicators have been selected for the evaluation of LUFs at a large scale and the special geographical conditions of the YRB, but these indicators still cannot cover all LUFs, and more indicators still need to be explored in the future due to the availability of data [62].

In the future, we will use spatial data such as high-resolution images, POI data, and more quantitative models to improve the accuracy of the evaluation results of land use functions and try to explore the interaction between functions from multiple dimensions and scales in order to deepen the study of land use functions and provide a basis for decision making in land spatial planning and management.

## 6. Conclusions

The multifunctionality of land has been widely recognized. However, the spatial and temporal evolution of LUFs and their decoupling from each other is often overlooked by urban planners and policy makers.

This study carefully examines the geographical and temporal differentiation properties of various land use functions in each section of the YRB, using the YRB as a bigger research scope. Based on the outline of the plan for ecological protection and high-quality development in the YRB, as well as the geographical environment and development degree of the YRB, the evaluation models and indicators of land use functions in each city were constructed from the three dimensions of production, living, and ecology, which can better reflect the complex interaction between multiple functions of land use. Further, the decoupling relationship between different functions of land use has been studied in depth. Compared with previous studies, this study not only considered a single function but also integrated different functions of multiple indicators. In terms of scale, the municipal administrative unit and a larger span of the study area were chosen so that this study can be used as an example of a multifunctional study of land use on a large scale.

In terms of research results, the LUFs values of cities upstream of the YRB and midstream of the YRB have changed less, and the LUFs values of cities downstream of the YRB have decreased, while the ELFs in the study area have significantly increased after 2010, and the decoupling relationship between the ELF, PDF, and LVF has increasingly developed to the good side, thanks to the systematic management and policy guidance of ecological problems in the YRB.

The empirical results of the YRB case considered in this study provide relevant insights into how the LUF decoupling analysis can contribute to supporting spatial planning and land use management, thereby better establishing the coordinated development of urban–rural areas, especially in developing countries such as China, where land availability is limited.

The development status and degree of coupling and coordination of the "three living spaces" in the upper, middle, and lower reaches of the YRB are different, and the government should formulate a clear low-carbon development plan based on the current development status, resource endowment, and industrial demand of the localities, actively guide the transformation and upgrading of the industrial structure of the cities, and strengthen the inputs in the fields of energy, transportation, and pollution control.

**Author Contributions:** Conceptualization, H.D.; Methodology, H.D.; Software, H.D. and H.L.; Validation, H.D. and H.L.; Formal analysis, H.D.; Investigation, H.D.; Data curation, H.D.; Writing—original draft, H.D.; Writing—review & editing, H.D. and Z.W.; Visualization, H.D. and H.L.; Supervision, H.D. and C.Z.; Project administration, Z.W. and C.Z.; Funding acquisition, Z.W. All authors have read and agreed to the published version of the manuscript.

**Funding:** This research was funded by [the Key Project from the National Social Science Foundation of China] grant number [23AZD058].

**Data Availability Statement:** The original contributions presented in the study are included in the article, further inquiries can be directed to the corresponding author.

**Conflicts of Interest:** The authors declare no conflicts of interest.

## References

1. Liu, Y.S.; Fang, F.; Li, Y.H. Key issues of land use in China and implications for policy making. *Land Use Policy* **2014**, *40*, 6–12. [CrossRef]
2. Jiang, S.; Meng, J.J.; Zhu, L.K. Spatial and temporal analyses of potential land use conflict under the constraints of water resources in the middle reaches of the Heihe River. *Land Use Policy* **2020**, *97*, 104773. [CrossRef]
3. Cao, S.X.; Lv, Y.; Zheng, H.R.; Wang, X. Challenges facing China's unbalanced urbanization strategy. *Land Use Policy* **2014**, *39*, 412–415. [CrossRef]
4. Zhang, X.; Chen, N.C.; Sheng, H.; Ip, C.; Yang, L.; Chen, Y.; Sang, Z.; Tadesse, T.; Lim, T.P.Y.; Rajabifard, A.; et al. Urban drought challenge to 2030 sustainable development goals. *Sci. Total Environ.* **2019**, *693*, 133536. [CrossRef]
5. Liu, Y.S.; Zhou, Y. Territory spatial planning and national governance system in China. *Land Use Policy* **2021**, *102*, 105288. [CrossRef]
6. Lin, G.; Jiang, D.; Fu, J.Y.; Zhao, Y. A Review on the Overall Optimization of Production-Living-Ecological Space: Theoretical Basis and Conceptual Framework. *Land* **2022**, *11*, 345. [CrossRef]

7. Slee, B. Social indicators of multifunctional rural land use: The case of forestry in the UK. *Agric. Ecosyst. Environ.* **2007**, *120*, 31–40. [CrossRef]
8. Zou, L.L.; Liu, Y.S.; Yang, J.X.; Yang, S.F.; Wang, Y.S.; Zhi, C.; Hu, X.D. Quantitative identification and spatial analysis of land use ecological-production-living functions in rural areas on China's southeast coast. *Habitat Int.* **2020**, *100*, 102182. [CrossRef]
9. Cai, F.; Vliet, J.v.; Verburg, P.H.; Pu, L. Land use change and farmer behavior in reclaimed land in the middle Jiangsu coast, China. *Ocean Coast. Manag.* **2017**, *137*, 107–117. [CrossRef]
10. Jiang, Y.; Long, H.; Ives, C.D.; Deng, W.; Chen, K.; Zhang, Y. Modes and practices of rural vitalisation promoted by land consolidation in a rapidly urbanising China: A perspective of multifunctionality. *Habitat Int.* **2022**, *121*, 102514. [CrossRef]
11. Lyu, R.; Zhang, J.; Xu, M.; Li, J. Impacts of urbanization on ecosystem services and their temporal relations: A case study in Northern Ningxia, China. *Land Use Policy* **2018**, *77*, 163–173. [CrossRef]
12. Liu, T.; Liu, H.; Qi, Y. Construction land expansion and cultivated land protection in urbanizing China: Insights from national land surveys, 1996–2006. *Habitat Int.* **2015**, *46*, 13–22. [CrossRef]
13. Ruimin, Y.; Zhanqi, W.; Ji, C.; Yunxiao, G.; Feng, X. The Evolution and Response of Space Utilization Efficiency and Carbon Emissions: A Comparative Analysis of Spaces and Regions. *Land* **2022**, *11*, 438. [CrossRef]
14. Groot, R.d. Function-analysis and valuation as a tool to assess land use conflicts in planning for sustainable, multi-functional landscapes. *Landsc. Urban Plan.* **2005**, *75*, 175–186. [CrossRef]
15. Long, H.; Qu, Y. Land use transitions and land management: A mutual feedback perspective. *Land Use Policy* **2018**, *74*, 111–120. [CrossRef]
16. Liu, C.; Xu, Y.Q.; Lu, X.H.; Han, J. Trade-offs and driving forces of land use functions in ecologically fragile areas of northern Hebei Province: Spatiotemporal analysis. *Land Use Policy* **2021**, *104*, 105387. [CrossRef]
17. Paracchini, M.L.; Pacini, C.; Jones, M.L.M.; Pérez-Soba, M. An aggregation framework to link indicators associated with multifunctional land use to the stakeholder evaluation of policy options. *Ecol. Indic.* **2011**, *11*, 71–80. [CrossRef]
18. Zhang, Y.N.; Long, H.L.; Tu, S.S.; Ge, D.Z.; Ma, L.; Wang, L.Z. Spatial identification of land use functions and their trade-offs/synergies in China: Implications for sustainable land management. *Ecol. Indic.* **2019**, *107*, 105550. [CrossRef]
19. Fan, Y.T.; Jin, X.B.; Gan, L.; Jessup, L.H.; Pijanowski, B.C.; Yang, X.H.; Xiang, X.M.; Zhou, Y.K. Spatial identification and dynamic analysis of land use functions reveals distinct zones of multiple functions in eastern China. *Sci. Total Environ.* **2018**, *642*, 33–44. [CrossRef]
20. Zhou, D.; Xu, J.C.; Lin, Z.L. Conflict or coordination? Assessing land use multi-functionalization using production-living-ecology analysis. *Sci. Total Environ.* **2017**, *577*, 136–147. [CrossRef]
21. Liu, C.; Xu, Y.Q.; Huang, A.; Li, Y.X.; Wang, H.; Lu, L.H.; Sun, P.L.; Zheng, W. Spatial identification of land use multifunctionality at grid scale in farming-pastoral area: A case study of Zhangjiakou City, China. *Habitat Int.* **2018**, *76*, 48–61. [CrossRef]
22. Fan, Y.T.; Gan, L.; Hong, C.Q.; Jessup, L.H.; Jin, X.B.; Pijanowski, B.C.; Sun, Y.; Lv, L.G. Spatial identification and determinants of trade-offs among multiple land use functions in Jiangsu Province, China. *Sci. Total Environ.* **2021**, *772*, 145022. [CrossRef] [PubMed]
23. Loiselle, A.; Proulx, R.; Larocque, M.; Pellerin, S. Synergies and trade-offs among ecosystems functions and services for three types of lake-edge wetlands. *Ecol. Indic.* **2023**, *154*, 110547. [CrossRef]
24. Xue, Z.; Zhen, L.; Miah, M.G.; Shoyama, K. Impact assessment of land use functions on the sustainable regional development of representative Asian countries—A comparative study in Bangladesh, China and Japan. *Sci. Total Environ.* **2019**, *694*, 133689. [CrossRef] [PubMed]
25. Zou, L.L.; Liu, Y.S.; Wang, J.Y.; Yang, Y.Y. An analysis of land use conflict potentials based on ecological-production-living function in the southeast coastal area of China. *Ecol. Indic.* **2021**, *122*, 107297. [CrossRef]
26. Hein, L.; van Koppen, K.; de Groot, R.S.; van Ierland, E.C. Spatial scales, stakeholders and the valuation of ecosystem services. *Ecol. Econ.* **2006**, *57*, 209–228. [CrossRef]
27. Meng, J.J.; Cheng, H.R.; Li, F.; Han, Z.Y.; Wei, C.J.; Wu, Y.D.; You, N.W.; Zhu, L.K. Spatial-temporal trade-offs of land multi-functionality and function zoning at finer township scale in the middle reaches of the Heihe River. *Land Use Policy* **2022**, *115*, 106019. [CrossRef]
28. Yang, X.; Zhou, X.H.; Shang, G.Y.; Zhang, A.L. An evaluation on farmland ecological service in Jianghan Plain, China-from farmers' heterogeneous preference perspective. *Ecol. Indic.* **2022**, *136*, 108665. [CrossRef]
29. Jiang, Y.F.; Tian, S.; Xu, Z.C.; Gao, L.; Xiao, L.J.; Chen, S.C.; Xu, K.; Chang, J.F.; Luo, Z.K.; Shi, Z. Decoupling environmental impact from economic growth to achieve Sustainable Development Goals in China. *J. Environ. Manag.* **2022**, *312*, 114978. [CrossRef]
30. Wang, F.; Zhang, Z.X. Decoupling economic growth from energy consumption in top five energy consumer economies: A technological and urbanization perspective. *J. Clean. Prod.* **2022**, *357*, 131890. [CrossRef]
31. Zhang, Y.Z.; Chen, X.P. Spatial and nonlinear effects of new-type urbanization and technological innovation on industrial carbon dioxide emission in the Yangtze River Delta. Environ. *Sci. Pollut. Res.* **2023**, *30*, 29243–29257. [CrossRef] [PubMed]
32. Yang, Y.; Jia, J.S.; Devlin, A.T.; Zhou, Y.M.; Xie, D.M.; Ju, M. Decoupling and Decomposition Analysis of Residential Energy Consumption from Economic Growth during 2000–2017: A Comparative Study of Urban and Rural Guangdong, China. *Energies* **2020**, *13*, 4461. [CrossRef]
33. Song, X.; Jia, J.S.; Hu, W.H.; Ju, M. Provincial Contributions Analysis of the Slowdown in the Growth of China's Industrial $CO_2$ Emissions in the "New Normal". *Pol. J. Environ. Stud.* **2021**, *30*, 2737–2753. [CrossRef] [PubMed]

34. Wu, Y.; Zhu, Q.W.; Zhu, B.Z. Decoupling analysis of world economic growth and $CO_2$ emissions: A study comparing developed and developing countries. *J. Clean. Prod.* **2018**, *190*, 94–103. [CrossRef]
35. Bennett, E.M.; Peterson, G.D.; Gordon, L.J. Understanding relationships among multiple ecosystem services. *Ecol. Lett.* **2009**, *12*, 1394–1404. [CrossRef] [PubMed]
36. Aguilera, M.A.; Tapia, J.; Gallardo, C.; Núñez, P.; Varas-Belemmi, K. Loss of coastal ecosystem spatial connectivity and services by urbanization: Natural-to-urban integration for bay management. *J. Environ. Manag.* **2020**, *276*, 111297. [CrossRef] [PubMed]
37. Liu, C.; Xu, Y.; Wang, Y.; Cheng, L.; Lu, X.; Yang, Q. Analyzing the Value and Evolution of Land Use Functions from "Demand–Function–Value" Perspective: A Framework and Case Study from Zhangjiakou City, China. *Land* **2021**, *11*, 53. [CrossRef]
38. Hualou, L. *Land Use Transitions and Rural Restructuring in China*; Springer: Singapore, 2020; pp. 415–458, ISBN 978-981-15-4923-6. [CrossRef]
39. Huang, S.; Wang, Y.; Liu, R.; Jiang, Y.; Qie, L.; Pu, L. Identification of Land Use Function Bundles and Their Spatiotemporal Trade-Offs/Synergies: A Case Study in Jiangsu Coast, China. *Land* **2022**, *11*, 286. [CrossRef]
40. Zhu, C.; Dong, B.; Li, S.; Lin, Y.; Shahtahmassebi, A.; You, S.; Zhang, J.; Gan, M.; Yang, L.; Wang, K. Identifying the trade-offs and synergies among land use functions and their influencing factors from a geospatial perspective: A case study in Hangzhou, China. *J. Clean. Prod.* **2021**, *314*, 128026. [CrossRef]
41. Luo, J.; Zhang, X.; Shi, P. Land Use Multi-Functionality and Zoning Governance Strategy of Densely Populated Areas in the Upper Reaches of the Yellow River: A Case Study of the Lanzhou–Xining Region, China. *Land* **2022**, *11*, 897. [CrossRef]
42. Chen, Y.; Zhu, M. Spatiotemporal Evolution and Driving Mechanism of "Production-Living-Ecology" Functions in China: A Case of Both Sides of Hu Line. *Int. J. Environ. Res. Public Health* **2022**, *19*, 3488. [CrossRef]
43. Xu, F.; Shao, Y.; Xu, B.; Li, H.; Xie, X.; Xu, Y.; Pu, L. Evaluation and Zoning of Cultivated Land Quality Based on a Space–Function–Environment. *Land* **2023**, *12*, 174. [CrossRef]
44. Chen, W.; Yan, S. The decoupling relationship between $CO_2$ emissions and economic growth in the Chinese mining industry under the context of carbon neutrality. *J. Clean. Prod.* **2022**, *379*, 134692. [CrossRef]
45. Qian, Y.; Zhao, J.; Lyu, Y.; Liu, Y.; Tian, J.; Chen, L. Uncovering the roadmap of decoupling economic growth and $CO_2$ emissions targeting energy-resource-emission-intensive industrial parks located nearby large river: Practices and implications from China. *J. Clean. Prod.* **2023**, *393*, 136306. [CrossRef]
46. Tapio, P. Towards a theory of decoupling: Degrees of decoupling in the EU and the case of road traffic in Finland between 1970 and 2001. *Transp. Policy* **2005**, *12*, 137–151. [CrossRef]
47. Organisation for Economic Co-operation and Development. OECD Environment Ministers adopt an environmental strategy for the next decade. Paris, 16 May 2001. *Sci. Total Environ.* **2001**, *275*, 137–138. [CrossRef] [PubMed]
48. Lu, X.; Qu, Y.; Sun, P.; Yu, W.; Peng, W. Green Transition of Cultivated Land Use in the Yellow River Basin: A Perspective of Green Utilization Efficiency Evaluation. *Land* **2020**, *9*, 475. [CrossRef]
49. Ren, J.; Li, Y.; Zhang, J.; Zhao, L.; Guo, J.; Du, M.; Hao, C.; Hu, R. Analysis of coordinated development of "society-ecology-policy" and spatio-temporal variation of people's livelihoods and well-being in the Yellow River basin, China. *Ecol. Indic.* **2023**, *148*, 1327–1335. [CrossRef]
50. Li, Q.; Zhou, Y.; Xu, T.; Wang, L.; Zuo, Q.; Liu, J.; Su, X.; He, N.; Wu, Z. Trade-offs/Synergies in Land-use Function Changes in Central China from 2000 to 2015. *Chin. Geogr. Sci.* **2021**, *31*, 711–726. [CrossRef]
51. Li, S.; Zhu, C.; Lin, Y.; Dong, B.; Chen, B.; Si, B.; Li, Y.; Deng, X.; Gan, M.; Zhang, J.; et al. Conflicts between agricultural and ecological functions and their driving mechanisms in agroforestry ecotone areas from the perspective of land use functions. *J. Clean. Prod.* **2021**, *317*, 128453. [CrossRef]
52. Liu, Y.S.; Liu, Y.; Chen, Y.F. Territorial multi-functionality evaluation and decision-making mechanism at county scale in China. *Acta Geogr. Sin.* **2011**, *66*, 1379–1389. (In Chinese) [CrossRef]
53. Zhang, J.; Li, S.; Lin, N.; Lin, Y.; Yuan, S.; Zhang, L.; Zhu, J.; Wang, K.; Gan, M.; Zhu, C. Spatial identification and trade-off analysis of land use functions improve spatial zoning management in rapid urbanized areas, China. *Land Use Policy* **2022**, *116*, 106058. [CrossRef]
54. Xie, G.; Zhang, C.; Zhen, L.; Zhang, L. Dynamic changes in the value of China's ecosystem services. *Ecosyst. Serv.* **2017**, *26*, 146–154. [CrossRef]
55. Zhang, J.J.; Fu, M.C.; Zhang, Z.Y.; Tao, J.; Fu, W. A trade-off approach of optimal land allocation between socio-economic development and ecological stability. *Ecol. Model.* **2014**, *272*, 175–187. [CrossRef]
56. Liu, B.; Pan, L.B.; Qi, Y.; Guan, X.; Li, J.S. Land Use and Land Cover Change in the Yellow River Basin from 1980 to 2015 and Its Impact on the Ecosystem Services. *Land* **2021**, *10*, 1080. [CrossRef]
57. Zhu, C.M.; Zhang, X.L.; Zhou, M.M.; He, S.; Gan, M.Y.; Yang, L.X.; Wang, K. Impacts of urbanization and landscape pattern on habitat quality using OLS and GWR models in Hangzhou, China. *Ecol. Indic.* **2020**, *117*, 106654. [CrossRef]
58. Qiu, L.F.; Pan, Y.; Zhu, J.X.; Amable, G.S.; Xu, B.G. Integrated analysis of urbanization-triggered land use change trajectory and implications for ecological land management: A case study in Fuyang, China. *Sci. Total Environ.* **2019**, *660*, 209–217. [CrossRef]
59. Li, W.M.; Cai, Z.Y. Spatiotemporal differences and influencing factors of high-quality utilization of land resources in the Yellow River Basin of China. *Environ. Sci. Pollut. Res.* **2022**, *29*, 89438–89448. [CrossRef] [PubMed]
60. Yang, R.; Lin, Y. Rural spatial transformation and governance from the perspective of land development rights: A case study of Fenghe village in Guangzhou. *Growth Change* **2021**, *53*, 1102–1121. [CrossRef]

61. Hartmann, T.; Jehling, M. From diversity to justice—Unraveling pluralistic rationalities in urban design. *Cities* **2019**, *91*, 58–63. [CrossRef]
62. Nian, P.; Cai, Y.; Ma, S.; Zhuang, L.; Zhang, L. Review of spatial comprehensive zoning in China. *China Land Sci.* **2014**, *28*, 20–25. [CrossRef]

MDPI AG
Grosspeteranlage 5
4052 Basel
Switzerland
Tel.: +41 61 683 77 34

MDPI Books Editorial Office
E-mail: books@mdpi.com
www.mdpi.com/books

www.ingramcontent.com/pod-product-compliance
Lightning Source LLC
Chambersburg PA
CBHW041934260926
51850CB00018B/163
* 9 7 8 3 7 2 5 8 4 7 0 9 9 *